T0344861

Optimal MODIFIED CONTINUOUS Galerkin CFD

Optimal MODIFIED CONTINUOUS Galerkin CFD

A. J. Baker
Professor Emeritus
The University of Tennessee, USA

WILEY

This edition first published 2014

Registered office
John Wiley & Sons Ltd, The Atrium, Southern Gate, Chichester, West Sussex, PO19 8SQ, United Kingdom

For details of our global editorial offices, for customer services and for information about how to apply for permission to reuse the copyright material in this book please see our website at www.wiley.com.

Library of Congress Cataloging-in-Publication Data

Baker, A. J., 1936-
Optimal modified continuous Galerkin CFD / A. J. Baker.
pages cm
Includes bibliographical references and index.
ISBN 978-1-119-94049-4 (hardback)
1. Fluid mechanics. 2. Finite element method. 3. Galerkin methods. I. Title.
TA357.B273 2014
518′.63–dc23

2013049543

A catalogue record for this book is available from the British Library.

ISBN 9781119940494

Set in 10/12pt TimesLTStd-Roman by Thomson Digital, Noida, India.
Printed and bound in Malaysia by Vivar Printing Sdn Bhd

1 2014

Yogi Berra is quoted,
"If you come to a fork in the road take it."
With Mary Ellen's agreement,
following this guidance
I found myself at the
dawning of weak form CFD

Contents

Preface

Fluid dynamics, with heat/mass transport, is the engineering sciences discipline wherein explicit *nonlinearity* fundamentally challenges analytical theorization. Prior to digital computer emergence, hence *computational fluid dynamics* (CFD), the subject of this text, the regularly revised monograph *Boundary Layer Theory*, Schlichting (1951, 1955, 1960, 1968, 1979) archived Navier–Stokes (NS) knowledge analytical progress. Updates focused on advances in characterizing *turbulence*, the continuum phenomenon permeating *genuine* fluid dynamics. The classic companion for NS simplified to the *hyperbolic* form, which omits viscous-turbulent phenomena while admitting *non-smooth* solutions, is Courant et al. (1928).

The analytical subject of CFD is rigorously addressed herein via what has matured as optimal modified continuous Galerkin weak form theory. The predecessor burst onto the CFD scene in the early 1970s disguised as the weighted-residuals *finite element* (FE) alternative to *finite difference* (FD) CFD. Weighted-residuals obvious connections to variational calculus prompted mathematical formalization, whence emerged continuum *weak form theory*. It is this theory, discretely implemented, herein validated precisely pertinent to nonlinear(!) NS, and *time averaged* and *space filtered* alternatives, elliptic boundary value (EBV) partial differential equation (PDE) systems.

Pioneering weighted-residuals CFD solutions proved reasonable compared with expectation and comparative data. Reasonable was soon replaced with *rigor*, first via laminar *and* turbulent boundary layer (BL) *a posteriori* data which *validated* linear weak form theory-predicted *optimal* performance within the discrete peer group, Soliman and Baker (1981a,b). Thus matured NS weak form theorization in *continuum* form, whence discrete implementation became a post-theory decision. As thoroughly detailed herein, the FE trial space basis choice is validated *optimal* in classic and weak form theory-identified norms. Further, this decision uniquely retains calculus and vector field theory supporting *computable form* generation precision.

Text focus is derivation and thorough quantitative assessment of *optimal* modified continuous Galerkin CFD algorithms for incompressible laminar-thermal NS plus the manipulations for turbulent and transitional flow prediction. Optimality accrues to *continuum* alteration of classic text NS PDE statements via rigorously derived nonlinear differential terms. Referenced as *m*odified PDE (*m*PDE) theory, wide ranging *a posteriori* data quantitatively validate the theory-generated dispersive/anti-dispersive operands annihilate significant order discrete approximation error in space *and* time, leading to monotone solution prediction without an artificial diffusion operator.

Weak formulations in the computational engineering sciences, especially fluid dynamics, have a storied history of international contributions. Your author's early 1970s participation culminated in leaving the Bell Aerospace principal research scientist position in 1975 to initiate the University of Tennessee (UT) Engineering Science graduate program focusing in weak form CFD. UT CFD Laboratory, formed in 1982, fostered collaboration among aerospace research technical colleagues, graduate students, commercial industry and the UT Joint Institute for Computational Science (JICS), upon its founding in 1993.

As successor to the 1983 text *Finite Element Computational Fluid Mechanics*, this book organizes the ensuing three decades of research generating theory advances leading to *rigorous*, efficient, optimal performance Galerkin CFD algorithm identification. The book is organized into 10 chapters, Chapter 1 introducing the subject content in perspective with an historical overview. Since postgraduate level mathematics are involved, Chapter 2 provides pertinent subject content overview to assist the reader in gaining the appropriate analytical dexterity. Chapters 3 and 4 document weak interaction aerodynamics, the union of potential flow NS with Reynolds-ordered BL theory, laminar and time averaged turbulent, with extension to parabolic NS (PNS) with PNS-ordered full Reynolds stress tensor algebraic closure. Linearity of the potential EBV enables a thoroughly formal derivation of continuum weak form theory via bilinear forms. Content concludes with optimal algorithm identification with an isentropic (weak) shock validation. An Appendix extends the theory to a Reynolds-ordered turbulent hypersonic shock layer aerothermodynamics formulation (PRaNS).

Chapter 5 presents a thorough derivation of *m*PDE theory generating the weak form optimal performance modified Galerkin algorithm, in time for linear through cubic trial space bases, and in space for optimally efficient linear basis. Theory assertion of optimality within the discrete peer group is quantitatively verified/validated. Chapter 6 validates the algorithm for laminar-thermal NS PDE system arranged to well-posed using vector field theory. Chapter 7 complements content with algorithm validation for the classic state variable laminar-thermal NS system, rendered well-posed via pressure projection theory with a genuine pressure weak formulation pertinent to multiply-connected domains. Content derives/validates a Galerkin theory for radiosity theory replacing Stephan–Boltzmann, also an ALE algorithm for thermo-solid-fluid interaction with melting and solidification.

Chapter 8 directly extends Chapter 7 content to time averaged NS (RaNS) for single Reynolds stress tensor closure models, standard deviatoric and full Reynolds stress model (RSM). Chapter 9 addresses space filtered NS (LES) with focus the Reynolds stress *quadruple* formally generated by filtering. Manipulations rendering RaNS and LES EBV statements identical lead to closure summary via subgrid stress (SGS) tensor modeling. The alternative completely *model-free* closure (*ar*LES) for the full tensor quadruple is derived via union of rational LES (RLES) and *m*PDE theories. Thus is generated an $O(1, \delta^2, \delta^3)$ member state variable for gaussian filter uniform measure δ *a priori* defining unresolved scale threshold. Extended to bounded domains, *ar*LES EBV system including boundary convolution error (BCE) integrals is rendered well-posed via derivation of non-homogeneous Dirichlet BCs for the complete state variable. The *ar*LES theory is validated applicable $\forall$ Re, generates δ-ordered resolved-unresolved scale diagnostic *a posteriori* data, and confirms model-free prediction of laminar-turbulent wall attached resolved scale velocity transition.

Chapter 10 collates text content under the US National Academy of Sciences (NAS) large scale computing identification "Verification, Validation, Uncertainly Quantification"

(VVUQ). Observed in context is replacement of legacy CFD algorithm numerical diffusion formulations with proven *m*PDE operand superior performance. More fundamental is the ∀Re model-free *ar*LES theory specific responses to NAS-cited requirements:

- error *quantification*
- *a posteriori* error *estimation*
- error *bounding*
- spectral content accuracy *extremization*
- phase selective *dispersion* error annihilation
- *monotone* solution generation
- error *extremization* optimal mesh quantification
- mesh resolution *inadequacy* measure
- efficient optimal *radiosity* theory with error *bound*

which in summary address in completeness VVUQ.

Your author must acknowledge that the content of this text is the result of collaborative activities conducted over three decades under the umbrella of the UT CFD Lab, especially that resulting from PhD research. Content herein is originally published in the dissertations of Doctors Soliman (1978), Kim (1987), Noronha (1988), Iannelli (1991), Freels (1992), Williams (1993), Roy (1994), Wong (1995), Zhang (1995), Chaffin (1997), Kolesnikov (2000), Barton (2000), Chambers (2000), Grubert (2006), Sahu (2006) and Sekachev (2013), the last one completed in the third year of my retirement. During 1977–2006 the UT CFD Lab research code enabling weak form theorization transition to *a posteriori* data generation was the brainchild of Mr Joe Orzechowski, the maturation of a CFD technical association initiated in 1971 at Bell Aerospace. The unsteady fully 3-D *a posteriori* data validating *ar*LES theory was generated using the open source, massively parallel PICMSS (Parallel Interoperable Computational Mechanics Systems Simulator) platform, a CFD Lab collaborative development led by Dr Kwai Wong, Research Scientist at JICS.

Teams get the job done – this text is proof positive.

A. J. Baker
Knoxville, TN
November 2013

About the Author

A. J. Baker, PhD, PE, left commercial aerospace research to join The University of Tennessee College of Engineering in 1975, with the goal to initiate a graduate academic research program in the exciting new field of CFD. Now Professor Emeritus and Director Emeritus of the University's CFD Laboratory (http://cfdlab.utk.edu), his professional career started in 1958 as a mechanical engineer with Union Carbide Corp. He departed after five years to enter graduate school full time to "learn what a computer was and could do." A summer 1967 digital analyst internship with Bell Aerospace Company led to the 1968 technical report "A Numerical Solution Technique for a Class of Two-Dimensional Problems in Fluid Dynamics Formulated via Discrete Elements," a pioneering expose in the fledgling field of finite-element (FE) CFD. Finishing his (plasma physics topic) dissertation in 1970, he joined Bell Aerospace as Principal Research Scientist to pursue fulltime FE CFD theorization. NASA Langley Research Center stints led to summer appointments at their Institute for Computer Applications in Science and Engineering (ICASE), which in turn led to a 1974–1975 visiting professorship at Old Dominion University. He transitioned directly to UT and in the process founded Computational Mechanics Consultants, Inc., with two Bell Aerospace colleagues, with the mission to convert FE CFD theory academic research progress into computing practice.

Notations

a	expansion coefficient; speed of sound; characteristics coefficient
A	plane area; 1-D FE matrix prefix; coefficient
AD	approximate deconvolution
ADBC	approximate deconvolution boundary condition algorithm
AF	approximate factorization algorithm
ALE	arbitrary-lagrangian-eulerian algorithm
[A]	factored global matrix, RLES theory auxiliary problem matrix operator
*ar*LES	essentially analytic LES closure theory
b	coefficient; boundary condition subscript
{b}	global data matrix
B	2-D FE matrix prefix
B(•)	bilinear form
B	body force
BC	boundary condition
BCE	boundary commutation error integral
BHE	borehole heat exchanger
BiSec	bisected borehole heat exchanger
BL	boundary layer
c	coefficient; specific heat
c	phase velocity vector
C	3-D matrix prefix; coefficient; chord; Courant number $\equiv \mathrm{U}\Delta t/\Delta x$
C_{ij}	cross stress tensor
$\mathrm{C_p}$	aerodynamic pressure coefficient, $= p/\rho u^2/2$
$\mathrm{C_S}$	Smagorinsky constant, its generalization
CFD	computational fluid dynamics
CFL	Courant number
C_f	skin friction coefficient
CF/2	boundary layer skin friction coefficient
CNFD	Crank–Nicolson finite difference
CS	control surface
CV	control volume
d(•)	ordinary derivative, differential element
d	coefficient; FE matrix basis degree label, RSM distance; characteristics coefficient
D	binary diffusion coefficient; diagonal matrix

D	dimensionality, non-D diffusion coefficient $\equiv \Delta t/\mathrm{Pa}h^2$
$D(\bullet)$	differential definition
$\mathrm{D}(\bullet)$	substantial derivative
$\mathrm{D}^{\mathrm{m}}(\bullet)$	modified substantial derivative
DES	detached eddy simulation
DE	conservation of energy PDE
DG	discontinuous Galerkin weak form theory
DM	conservation of mass PDE
D**P**	conservation of momentum PDE
DY	conservation of species mass fraction PDE
$\mathbf{D}(\mathbf{u}, P)$	NS full stress tensor, $\equiv -\nabla P + (2/\mathrm{Re})\nabla \cdot \mathbf{S}(\mathbf{u})$
diag[•]	diagonal matrix
[DIFF]	laplacian diffusion matrix
DNS	direct numerical simulation
e	specific internal energy; element-dependent (subscript)
$e(\cdot)$	error
e^N	continuum approximation error
e^h	discrete approximation error
e_{ijk}	alternating tensor
e_{KL}	curl alternator on $n=2$
EBV	elliptic boundary value
Ec	Eckert number
eta_{ji}	coordinate transformation data
E	thermal energy; energy semi-norm (subscript)
f_j	flux vector
f_n	normal flux
$f(\bullet)$	function of the argument
$f(\mathrm{vf}, \epsilon)$	radiation *view factor*
$F(\bullet)$	Fourier transform
{F}	weak form terminal algebraic statement
$F(k \rightarrow i)$	Lambert's cosine law viewfactor
FD	finite difference
FE	finite element
FV	finite volume
$\mathbf{f}$	efflux vector on $\partial\Omega$
$\mathbf{F}$	applied force
g	gravity magnitude; amplification factor; spatial filter function; characteristics enthalpy ratio
$\mathbf{g}$	gravity
Gr	Grashoff number $\equiv \mathrm{g}\beta\Delta\mathrm{TL}^3/\nu^2$
$G_{k\rightarrow i}$	Gebhart viewfactor
GHP	ground source heat pump
GLS	Galerkin least squares algorithm
GWS	Galerkin weak statement
h	mesh measure; discrete (superscript), heat transfer coefficient
H	boundary layer shape factor

H	Gauss quadrature weight; Hilbert space
H.O.T.	truncated Taylor series higher order terms
i	summation index; mesh node
$\hat{\mathbf{i}}$	unit vector parallel to x
I	discrete matrix summation index, identity matrix
iff	if and only if
I-EBV	initial-elliptic boundary value
j	summation index, mesh node
$\hat{\mathbf{j}}$	unit vector parallel to y
J	discrete matrix summation index
[J]	coordinate transformation jacobian
[JAC]	matrix statement jacobian
k	thermal conductivity; basis degree; index; diffusion coefficient
k_{ij}	element of the [DIFF] matrix
$\overline{k}$	average value of conductivity
$\hat{\mathbf{k}}$	unit vector parallel to z
K	discrete matrix summation index
ℓ	element length; summation index
$\ell(\cdot)$	differential operator on $\partial\Omega$
L	reference length scale
L	discrete matrix summation index
$\mathcal{L}(\cdot)$	differential operator on Ω
L_{ij}	Leonard stress tensor
LES	large eddy simulation, convolved Navier–Stokes PDEs
m	non-D wavenumber $\equiv \kappa h$, integer
$[m]$	mass matrix
m_i	point mass; discrete matrix summation index
M	particle system mass; domain matrix prefix; elements in Ω^h
M_i	molecular mass
[M]	*m*PDE theory altered mass matrix
Ma	Mach number
*m*GWS	optimal modified Galerkin weak form
*m*PDE	modified partial differential equation
*m*ODE	modified ordinary differential equation
MLT	mixing length theory
n	index; normal subscript; dimension of domain Ω; integer
n-D	n-dimensional, $1 \leq n \leq 3$
$\hat{\mathbf{n}}$	outward pointing unit vector normal to $\partial\Omega$
N	Neumann BC matrix prefix
N	summation termination; approximation (superscript)
NC	natural coordinate basis
NWM	near wall modeling LES BCs
NWR	near wall resolution LES algorithm
NS	Navier–Stokes
$\{N_k\}$	finite element basis of degree k
$O(\bullet)$	order of argument $(\bullet)$

p	pressure
P	kinematic pressure
P	Gauss quadrature order
P	linear momentum
Pa	placeholder for non-D groups Re, Pr, Gr, Ec
$\{P\}$	intermediate computed matrix
PDE	partial differential equation
Pe	Peclet number = RePr
PNS	parabolic Navier–Stokes
PRaNS	hypersonic parabolic Reynolds-averaged Navier–Stokes
Pr	Prandtl number $\equiv \rho_0 \nu c_p / k$
pr	non-uniform mesh progression ratio
q	generalized dependent variable
q	heat flux vector
Q	discretized dependent variable; heat added
$\{Q\}$	nodal coefficient column matrix
r	reference state subscript; radius
R	perfect gas constant, temperature degrees Rankine
R	radiosity
$\mathcal{R}$	universal gas constant
R_{ij}	Reynolds subfilter scale tensor
RaNS	Reynolds-averaged Navier–Stokes
Re	Reynolds number $\equiv$ UL/ν
Re^t	turbulent Reynolds number $\equiv \nu^t / \nu$
Re^+	compressible turbulent BL similarity coordinate $= \rho u_\tau y / \mu$
$\mathfrak{R}^n$	euclidean space of dimension n
RSM	Reynolds stress model
$\{\mathrm{RES}\}$	weak form terminal matrix statement residual
s	source term on Ω; heat added
s	unit vector tangent to $\partial\Omega$
S	entropy
$\overline{\mathbf{S}}$	filtered Stokes tensor dyadic
S_e	matrix assembly operator
$\mathrm{S}_{i,j,k}$	stencil assembly operator
$\{\mathrm{S}\}$	computational matrix
Sc	Schmidt number $\equiv$ D/ν
SFS_{ij}	subfilter scale tensor
SGS_{ij}	subgrid scale tensor
St	Stanton number $\equiv \tau$U/L
SUPG	Streamline upwind Petrov Galerkin
sym	symmetric
t	time; turbulent (superscript)
T	temperature
$\overline{T}(z)$	BHE conduit temperature distribution
TE	Taylor series truncation error
TG	Taylor Galerkin algorithm

TP	tensor product basis
$\mathbf{T}$	surface traction vector
T^N	continuum approximate temperature solution
TWS	Taylor weak statement
$\mathbf{u}$	displacement vector; velocity vector
u	velocity x component; speed
$\overline{u_i' u_j'}$	time averaged NS Reynolds stress tensor
$\tilde{u}_j$	Favre time averaged velocity
U	reference velocity scale
UQ	uncertainty quantification
U	discretized speed nodal value
v	velocity y component
$\mathbf{v}_g(\boldsymbol{\kappa})$	group velocity, $\equiv \nabla_\kappa \omega$
v_j	LES theory scalar state variable closure vector
V	volume
VBV	verification, benchmarking, validation
VVUQ	verification, validation, uncertainty quantification
$\mathbf{V}$	velocity
VLES	very large eddy simulation
w	weight function; fin thickness; velocity z component
W	weight; work done by system
WF	weak form
WR	weighted residuals
WS	weak statement
x, x_i	cartesian coordinate, coordinate system $1 \leq i \leq n$
$\overline{x}$	transformed local coordinate
X	discrete cartesian coordinate
y	displacement; cartesian coordinate
y^+	incompressible turbulent BL similarity coordinate $= u_\tau y/\nu$
Y	mass fraction; discrete cartesian coordinate
z	cartesian coordinate
Z	thickness ratio; discrete cartesian coordinate
∇	gradient differential operator
∇^2	laplacian operator
$\mathrm{d}(\cdot)/\mathrm{d}x$	ordinary derivative
$\partial(\cdot)/\partial x$	partial derivative
$(\cdot)$	scalar (number)
$\{\cdot\}$	column matrix
$\{\cdot\}^T$	row matrix
$[\cdot]$	square matrix
$\mathrm{diag}[\cdot]$	diagonal square matrix
$\Vert\cdot\Vert$	norm
$\cup$	union (non-overlapping sum)
$\cap$	intersection
$\det\,[\cdot]$	matrix determinant
$\forall$	denotes "for all"

Symbol	Description
$\in$	inclusion
$\subset$	belongs to
$*$	complex conjugate multiplication
$\otimes$	matrix tensor product
α	coefficient, thermal diffusivity ratio
β	absolute temperature; coefficient
γ	specific heat ratio, coefficient, gaussian filter shape factor
δ	boundary layer thickness, coefficient, spatial filter measure, bow shock standoff distance
δ^*	boundary layer displacement thickness
δ_{ij}	Kronecker delta
Δ	discrete increment
ϵ	isotropic dissipation function. emissivity
ϵ_{ij}	cartesian alternator
ϕ	velocity potential function
$\phi(\cdot)$	trial space function
Φ	potential function
$\Phi_\beta(\mathbf{x})$	test space
$\Psi_\alpha(\mathbf{x})$	trial space
$\mathbf{\Psi}$	vector streamfunction
ψ	streamfunction scalar component
η	transform space, wave vector angle
η_i	tensor product coordinate system
κ	thermal diffusivity, Karman constant = 0.435
κ^T	turbulent thermal diffusivity
$\boldsymbol{\kappa}$	wavenumber vector
$\boldsymbol{\kappa}_{\alpha\beta}$	element of a square matrix
λ	Lagrange multiplier, wavelength, Lame' parameter
μ	absolute viscosity
ν	kinematic viscosity
ν^t	kinematic eddy viscosity
π	pi (3.1415926...)
θ	TS implicitness factor, BL momentum thickness
Θ	potential temperature $\equiv (T - \mathrm{T}_{\min})/(\mathrm{T}_{\max} - \mathrm{T}_{\min})$
ρ	density
σ	Stefan-Boltzmann coefficient = 5.67 E-08 $\mathrm{w/m^2K^4}$
$d\sigma$	differential element on $\partial\Omega$
τ	time scale
τ_{ij}	Reynolds stress tensor
τ_{ij}^{D}	deviatoric Reynolds stress tensor
ω	frequency, Van Driest damping function, vorticity scalar
$\mathbf{\Omega}$	vorticity vector
Ω	domain of differential equation
Ω_e	finite element domain
Ω^h	discretization of Ω
$\partial\Omega$	boundary of Ω
ζ_α	natural coordinate system

1

Introduction

1.1 About This Book

This text is the successor to *Finite Element Computational Fluid Mechanics* published in 1983. It thoroughly organizes and documents the subsequent three decades of progress in *weak form* theory derivation of *optimal performance* CFD algorithms for the infamous Navier–Stokes (NS) *nonlinear* partial differential equation (PDE) systems. The text content addresses the complete range of NS and *filtered* NS (for addressing turbulence) PDE systems in the incompressible fluid-thermal sciences. Appendix B extends subject NS content to a weak form algorithm addressing hypersonic shock layer aerothermodynamics.

As perspective color and dynamic computer graphics are support *imperatives* for CFD *a posteriori* data assimilation, and hence interpretation, www.wiley.com/go/baker/GalerkinCFD renders available the full color graphics content absent herein. The website also contains detailed academic course lecture content at advanced graduate levels in support of outreach and theory exposure/implementation.

Weak form theory is *the* mathematically *elegant* process for generating *approximate solutions* to *nonlinear* NS PDE systems. Theoretical formalities are always conducted in the *continuum*, and *only* after such musings are completed are space and time discretization decisions made. This final step is a *matter of choice*, with a *finite element* (FE) spatial semi-discretization retaining use of calculus and vector field theory throughout conversion to terminal *computable form*. This choice enables implementing weak form theory precision into an *optimal* performance compute engine, eliminating any need for *heurism*.

The text tenor assumes that the reader remembers some calculus and is adequately versed in fluid mechanics and heat and mass transport at a post-baccalaureate level. It further assumes that this individual is neither comfortable with nor adept at formal mathematical manipulations. Therefore, text fluid mechanics subject exposure sequentially enables *just-in-time* exposure to essential mathematical concepts and methodology, in progressively addressing more detailed NS PDE systems and closure formulations.

Potential flow enables elementary weak form theory exposure, with subsequent theorization modifications becoming progressively more involved in addressing NS pathological nonlinearity. The exposure process is sequentially supported by *a posteriori* data from

Optimal MODIFIED CONTINUOUS Galerkin CFD, First Edition. A. J. Baker.

Companion Website: www.wiley.com/go/baker/GalerkinCFD

precisely designed computational experiments, enabling *quantitative validation* of theory predictions of *accuracy*, *convergence*, *stability* and *error estimation/distribution* which, in concert, lead to confirmation of *optimal mesh* solution existence.

Text content firmly quantifies the practice preference for an FE semi-discrete spatial implementation. The apparent simplicity of finite volume (FV) and finite difference (FD) discretizations engendered the FV/FD commercial CFD code legacy practice. However, as documented herein, FV/FD spatial discretizations constitute non-Galerkin weak form decisions leading to nonlinear *schemes* via *heuristic* arguments. This is totally obviated in converting FE algorithms to computable syntax using calculus and vector field theory. This aspect hopefully further prompts the reader's interest in acquiring knowledge of these elegant *practice* aspects, such that assimilating FE constructs proves to be worth the effort.

The progression within each chapter, hence throughout the text, sequentially addresses more detailed fluid/thermal NS PDE systems, each chapter building on prior material. The elegant uniformity of weak form theory facilitates this approach with mathematical formalities *never* requiring an *ad hoc* scheme decision. In his reflections on teaching the finite element method Bruce Irons is quoted, "Most people, mathematicians apart, abhor *abstraction*." Booker T. Washington concurred, "An ounce of application is worth a ton of abstraction." These precepts guide the development and exposition strategies in this text, with abstraction *never* taking precedence over developing a firm engineering-based theoretical exposure.

Summarizing, *modified continuous* Galerkin weak formulations for fluid/thermal sciences CFD generate practical computational algorithms fully validated as *optimal* in performance as predicted by a rich theory. Conception and practice goals always lead to the theoretical exposition, to convince the reader that its comprehension is a worthwhile goal, paying the requisite dividend.

1.2 The Navier–Stokes Conservation Principles System

Computational fluid-thermal system simulation involves seeking a *solution* to the *nonlinear* PDE systems generated from the basic conservation observations in engineering mechanics. In the lagrangian (point mass) perspective, these principles state

$$\text{conservation of mass:}\ \ \mathrm{d}M = 0, M = \Sigma m_i \tag{1.1}$$

$$\text{Newton's second law:}\ \ \mathrm{d}\mathbf{P} = \Sigma\mathbf{F}, \mathbf{P} = M\mathbf{V} \tag{1.2}$$

$$\text{thermodynamics, first law:}\ \ \mathrm{d}E = \mathrm{d}Q - \mathrm{d}W \tag{1.3}$$

$$\text{thermodynamic process:}\ \ \mathrm{d}S \geq 0 \tag{1.4}$$

In (1.1) m_i denotes a point mass, M is total mass of a particle system, $\mathbf{V}$ is velocity of that system and $\mathbf{F}$ denotes applied (external) forces. Equations (1.3–1.4) are statements of the first and second law of thermodynamics where E is system total energy, Q is heat added, W is work done by the system and S is entropy.

Practical CFD applications almost never involve addressing the conservation principles in lagrangian form. Instead, the transition to the *continuum* (eulerian) description is made, wherein one assumes that there exist so many mass points per characteristic volume V that

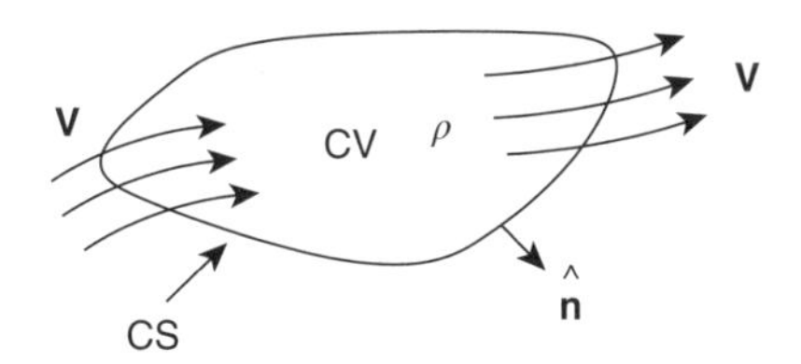

Figure 1.1 Control volume for Reynolds transport theorem

a *density* function ρ can be defined

$$\rho(\boldsymbol{x}, t) = \lim_{V \to 0} \frac{1}{V} \sum_i m_i \tag{1.5}$$

One then identifies a control volume CV, with bounding control surface CS, Figure 1.1, and transforms the conservation principles from lagrangian to eulerian viewpoint via *Reynolds transport theorem*

$$d() \Rightarrow D() \equiv \frac{\partial}{\partial t} \int_{cv} (\cdot)d\tau + \oint_{cs} (\cdot)\mathbf{V} \cdot \hat{\mathbf{n}}\, d\sigma \tag{1.6}$$

Thus is produced a precise mathematical statement of the conservation principles for *continuum* descriptions as a system of integro-differential equations

$$DM = \frac{\partial}{\partial t} \int_{cv} \rho d\tau + \oint_{cs} \rho \mathbf{V} \cdot \hat{\mathbf{n}}\, d\sigma = 0 \tag{1.7}$$

$$D\mathbf{P} \Rightarrow \frac{\partial}{\partial t} \int_{cv} \rho \mathbf{V} d\tau + \oint_{cs} \mathbf{V} \rho \mathbf{V} \cdot \hat{\mathbf{n}}\, d\sigma = \int_{cv} \rho \mathbf{B} d\tau + \int_{cs} \mathbf{T} d\sigma \tag{1.8}$$

$$DE \Rightarrow \frac{\partial}{\partial t} \int_{cv} \rho e d\tau + \oint_{cs} (e + p/\rho)\rho\, \mathbf{V} \cdot \hat{\mathbf{n}}\, d\sigma = \int_{cv} s d\tau + \int_{cs} (W - \mathbf{q} \cdot \hat{\mathbf{n}})d\sigma \tag{1.9}$$

Note the eulerian "filling in" of the right hand sides for D**P** and DE with $\sum \mathbf{F} \Rightarrow$ body forces $\mathbf{B}$ + surface tractions $\mathbf{T}$, and $dQ - dW \Rightarrow$ heat added s, bounding surface heat efflux $\mathbf{q} \cdot \hat{\mathbf{n}}$ and work done W.

From (1.7–1.9), one easily develops the PDE statements of direct use for CFD formulations by assuming that the control volume CV is stationary, followed by invoking the *divergence theorem* for the identified surface integrals. For example for (1.7)

$$\oint_{cs} \rho \mathbf{V} \cdot \hat{\mathbf{n}}\, d\sigma = \int_{cv} \nabla \cdot \rho\, \mathbf{V} d\tau \tag{1.10}$$

where ∇ is the *gradient* (vector) differential operator.

Via the divergence theorem the integro-differential equation system (1.7–1.9) is uniformly re-expressed as integrals vanishing identically on the CV. Such expressions can hold in general if and only if (*iff*) the *integrand* vanishes identically, whereupon D*M*, **DP** and D*E* morph to the *nonlinear* PDE system

$$\text{D}M\text{:}\quad \frac{\partial \rho}{\partial t} + \nabla \cdot \rho \mathbf{u} = 0 \tag{1.11}$$

$$\textbf{DP:}\quad \frac{\partial \rho \mathbf{u}}{\partial t} + \nabla \cdot \rho \mathbf{u}\mathbf{u} = \rho \mathbf{g} + \nabla \mathbf{T} \tag{1.12}$$

$$\text{D}E\text{:}\quad \frac{\partial \rho e}{\partial t} + \nabla \cdot (\rho e + p)\mathbf{u} = s - \nabla \cdot \mathbf{q} \tag{1.13}$$

Herein the velocity vector label **V** in the preceding equations is replaced with the more conventional symbol **u**.

It remains to simplify (1.11–1.13) for constant density ρ_0 and to identify *constitutive closure* for traction vector **T** and heat flux vector **q**. For laminar flow **T** contains pressure and a fluid viscosity hypothesis involving the Stokes strain rate tensor. For constant density ρ_0 and multiplied through by ∇ the resultant vector statement is

$$\nabla \mathbf{T} = -\nabla p + \nabla \bullet \mu \nabla \mathbf{u} \tag{1.14}$$

where p is pressure and μ is fluid absolute viscosity. The Fourier conduction hypothesis for heat flux vector **q** is

$$\nabla \cdot \mathbf{q} = -\nabla \bullet k \nabla T \tag{1.15}$$

where k is fluid thermal conductivity and T is temperature.

Substituting these closure *models* and enforcing that density and specific heat are assumed constant converts (1.11–1.13) to the very familiar textbook appearance of Navier–Stokes. Herein the *homogeneous* form preference leads to the subject incompressible NS PDE system

$$\text{D}M\text{:}\quad \nabla \cdot \mathbf{u} = 0 \tag{1.16}$$

$$\textbf{DP:}\quad \frac{\partial \mathbf{u}}{\partial t} + \nabla \cdot \mathbf{u}\mathbf{u} + \rho_0 \nabla p - \nabla \cdot \nu \nabla \mathbf{u} + (\rho/\rho_0)\mathbf{g} = 0 \tag{1.17}$$

$$\text{D}E\text{:}\quad \frac{\partial T}{\partial t} + \nabla \cdot \mathbf{u}T - \nabla \cdot \kappa \nabla T - s/\rho_0 c_p = 0 \tag{1.18}$$

In (1.17), $\nu = \mu/\rho_0$ is fluid kinematic viscosity with density assumed as the constant ρ_0 except for thermally induced impact in the gravity body force term in (1.17). Finally, in (1.18) $\kappa = k/\rho c_p$ is fluid *thermal diffusivity*.

Thermo-fluid system performance is thus characterized by a balance between unsteadiness and convective and diffusive processes. This identification is precisely established by non-dimensionalizing (1.16–1.18). The reference time, length and velocity scales are τ, L, and U, respectively, with the (potential) temperature scale definition $\Theta = (T - \mathrm{T_{min}})/(\mathrm{T_{max}} - \mathrm{T_{min}})$. Then implementing the Boussinesq buoyancy model for the gravity body force, the *non*-D incompressible NS PDE system for thermal-laminar flow with mass transport is

$$\text{D}M\text{:} \quad \nabla \cdot \mathbf{u} = 0 \tag{1.19}$$

$$\text{D}\mathbf{P}\text{:} \quad \text{St}\frac{\partial \mathbf{u}}{\partial t} + \nabla \cdot \mathbf{uu} + \nabla P - \frac{1}{\text{Re}}\nabla \cdot \nabla \mathbf{u} + \frac{\text{Gr}}{\text{Re}^2}\Theta\,\widehat{\mathbf{g}} = 0 \tag{1.20}$$

$$\text{D}E\text{:} \quad \text{St}\frac{\partial \Theta}{\partial t} + \nabla \cdot \mathbf{u}\Theta - \frac{1}{\text{RePr}}\nabla \cdot \nabla \Theta - s_\Theta = 0 \tag{1.21}$$

$$\text{D}Y\text{:} \quad \text{St}\frac{\partial Y}{\partial t} + \nabla \cdot \mathbf{u}Y - \frac{1}{\text{ReSc}}\nabla \cdot \nabla Y - s_Y = 0 \tag{1.22}$$

The unknowns in PDE system (1.19–1.22) are the non-D velocity vector $\mathbf{u}$, kinematic pressure P, temperature Θ and mass fraction Y. No special notation emphasizes that these are non-D, which will always be the case. These variables as a group are hereinafter referenced as the NS PDE system *state variable* symbolized as the column matrix $\{q(\mathbf{x},t)\} = \{\mathbf{u}, P, \Theta, Y\}^T$.

The definitions for Stanton, Reynolds, Grashoff, Prandtl and Schmidt numbers are conventional as $\text{St} \equiv \tau\text{U/L}$, $\text{Re} \equiv \text{UL}/\nu$, $\text{Gr} \equiv \text{g}\beta\Delta\text{TL}^3/\nu^2$, $\text{Pr} \equiv \rho_0 \nu c_p/k$ and $\text{Sc} \equiv \text{D}/\nu$, where D is the binary diffusion coefficient. Additionally $\Delta\text{T} \equiv (\text{T}_{\text{max}} - \text{T}_{\text{min}})$, $\beta \equiv 1/\text{T}_{\text{abs}}$ and $P \equiv p/\rho_0$. St is defined unity, $\tau \equiv \text{L/U}$, except when addressing flowfields exhibiting harmonic oscillation, and the Peclet number $\text{Pe} \equiv \text{RePr}$ is the common replacement in D*E*.

1.3 Navier–Stokes PDE System Manipulations

The NS PDE system (1.19–1.22) is universally accepted as an accurate descriptor of fluid-thermal phenomena for all Reynolds numbers Re. However, it is also universally recognized that for $\text{Re} \geq O(\sim\text{E+04})$, where $O(\bullet)$ signifies *order*, the resultant NS flowfields will be characterized as *turbulent*.

The CFD simulation procedure that addresses the expressed NS PDE system for all Re is called *direct numerical simulation* (DNS). Even with massive computer resources the DNS approach to solution of practical NS problem statements is contraindicated, cf. Dubois *et al.* (1999). The DNS approach is not addressed herein, although these algorithms do enjoy an identical weak form theoretical basis.

Instead, generating computational simulation algorithms for NS PDE statements for practical Re requires manipulations of (1.19–1.22). In the CFD community, the operation of *time averaging* generates the Reynolds-averaged NS (RaNS) PDE system. The alternative is *spatial filtering* via convolution with a filter function that produces the large eddy simulation (LES) NS PDE system. A union of the two has been termed *very large eddy simulation* (VLES), or a RaNS-LES hybrid termed *detached eddy simulation* (DES).

In each instance the mathematical manipulations introduce *a priori* unknown variables into the resultant PDE system *state variable* due to the *nonlinear* convection terms in (1.20–1.22). Time averaging resolves $\{q(\mathbf{x},t)\}=\{\mathbf{u}, p, \Theta, Y\}^T$ into the time-independent steady component and the fluctuation (in time) thereabout. In tensor index notation the resolution statement for velocity vector $\mathbf{u}$ is

$$\mathbf{u}(\mathbf{x}, t) \Rightarrow u_i(x_k, t) \equiv \overline{u}_i(x_k) + u_i'(x_k, t) \tag{1.23}$$

Time averaging the convection term in (1.20) produces

$$\overline{u_j u_i} = \overline{u}_j \overline{u}_i + \overline{u'_j u'_i} \tag{1.24}$$

that is, the tensor product of time averaged velocity plus the time average of the tensor product of velocity fluctuations about the steady average.

The second term in (1.24) is called the RaNS *Reynolds stress tensor*, a model for which must be constructed to close the RaNS PDE system state variable. Similar operations on DE and DY produce mean convection term products plus the fluctuating product *Reynolds vectors*

$$\overline{u_j \Theta} = \overline{u}_j \overline{\Theta} + \overline{u'_j \Theta'} \quad \text{and} \quad \overline{u_j Y} = \overline{u}_j \overline{Y} + \overline{u'_j Y'} \tag{1.25}$$

which must also be modeled to achieve closure.

The specifics of RaNS closure model development are derived in Chapter 4, then further detailed in Chapters 8 and Appendix B. It is sufficient here to note RaNS closure models emulate the force-flux form of the Stokes and Fourier fluid-dependent constitutive closures (1.14–1.15). The correlation coefficient becomes a *turbulent eddy viscosity,* υ^t*, extended* to turbulent heat/mass flux vector closure via turbulent Pr and Sc number assumptions.

The non-D turbulent eddy viscosity defines the *turbulent Reynolds number*

$$\mathrm{Re}^t \equiv \nu^t / \nu \tag{1.26}$$

and the non-D RaNS PDE system alternative to laminar NS PDEs (1.19–1.22), assuming unit Stanton number is

$$\mathrm{D}M: \quad \nabla \cdot \overline{\mathbf{u}} = 0 \tag{1.27}$$

$$\mathbf{DP:} \quad \frac{\partial \overline{\mathbf{u}}}{\partial t} + \nabla \cdot \overline{\mathbf{u}}\overline{\mathbf{u}} + \nabla \overline{P} - \frac{1}{\mathrm{Re}} \nabla \cdot \left[(1 + \mathrm{Re}^t) \nabla \overline{\mathbf{u}} \right] + \frac{\mathrm{Gr}}{\mathrm{Re}^2} \overline{\Theta} = 0 \tag{1.28}$$

$$\mathrm{D}E: \quad \frac{\partial \overline{\Theta}}{\partial t} + \nabla \cdot \overline{\mathbf{u}}\overline{\Theta} - \frac{1}{\mathrm{Re}} \nabla \cdot \left[\left(\frac{1}{\mathrm{Pr}} + \frac{\mathrm{Re}^t}{\mathrm{Pr}^t} \right) \nabla \overline{\Theta} \right] - \overline{s}_\Theta = 0 \tag{1.29}$$

$$\mathrm{D}Y: \quad \frac{\partial \overline{Y}}{\partial t} + \nabla \cdot \overline{\mathbf{u}}\overline{Y} - \frac{1}{\mathrm{Re}} \nabla \cdot \left[\left(\frac{1}{\mathrm{Sc}} + \frac{\mathrm{Re}^t}{\mathrm{Sc}^t} \right) \nabla \overline{Y} \right] - \overline{s}_Y = 0 \tag{1.30}$$

The time averaging alternative of spatial filtering employs the mathematical operation of *convolution*. In one dimension for velocity scalar component u the space filtered velocity definition is

$$\overline{u}(x, t) \equiv \int_{-\infty}^{\infty} g(y)\, u(x - y, t) dy \tag{1.31}$$

where $g(y)$ denotes the filter function. The filtered velocity remains time dependent, and in n dimensions the CFD literature notation for spatial filtering

$$\overline{u}_i(x_j, t) \equiv g_\delta {}^* u_i(x_j, t) \tag{1.32}$$

is symbolized by * with δ denoting the *measure* (diameter) of the filter function g.

Spatial filtering the NS PDE system (1.19–1.22) produces the LES PDE system addressed in Chapter 9. Following convolution, Fourier transformation and deconvolution spatial filtering of the convection term in (1.20) generates the *a priori* unknown *stress tensor quadruple*

$$\overline{u_j u_i} = \overline{\overline{u}_j \overline{u}_i} + \overline{u'_j \overline{u}_i} + \overline{\overline{u}_j u'_i} + \overline{u'_j u'_i} \tag{1.33}$$

Generating closure models for the LES PDE system typically involves consequential approximations in (1.33). For example, adding and subtracting the filtered velocity tensor product in (1.33) produces the *triple decomposition* approximation

$$\overline{u_j u_i} = \overline{u}_j \overline{u}_i + \underbrace{\left(\overline{\overline{u}_j \overline{u}_i} - \overline{u}_j \overline{u}_i\right)}_{L_{ij}} + \underbrace{\left(\overline{\overline{u}_j u'_i} + \overline{u'_j \overline{u}_i}\right)}_{C_{ij}} + \underbrace{\overline{u'_j u'_i}}_{R_{ij}} \tag{1.34}$$

for L_{ij} termed the *Leonard stress*, C_{ij} the *cross stress* and R_{ij} the *Reynolds subfilter scale* tensors.

LES theory states that the latter accounts for energetic dissipation at the *unresolved scale threshold*, the spatial scale defined by filter measure δ. This is the LES theory equivalent of viscous dissipation in the unfiltered NS system at molecular scale. The legacy published *subgrid scale* (SGS) tensor models again are of force-flux mathematical form, (1.27–1.30), with many variations, Piomelli (1999). As with time averaging, spatial filtering the NS DE and DY PDEs generates the companion unknown filtered thermal and mass *vector quadruples*.

An NS PDE system manipulation particularly pertinent to aerodynamics CFD applications alters the steady form of (1.19–1.22) assuming the velocity vector field is *unidirectional*. The end results are the famous *boundary layer* (BL) PDE system, and the n-dimensional generalization *parabolic Navier–Stokes* (PNS) PDE system. Both systems possess initial-value character in the direction of dominant flow. Importantly, both have largely supported development and validation of Reynolds stress tensor/vector closure models for the RaNS PDE system. The compressible turbulent PNS (PRaNS) PDE system is applicable to hypersonic external shock layer aerothermodynamics, detailed in Appendix B.

1.4 Weak Form Overview

The incompressible NS PDEs, also the model closed BL, PNS, RaNS and LES PDE systems, are elliptic boundary value (EBV) with selective initial-value character. Weak form theory is a *thoroughly formal process* for constructing approximate solutions to I-EBV/EBV PDE systems. The mathematical *hooker* in these PDEs is that each contains the *differential constraint* of DM, (1.19), which requires that the velocity field be *divergence-free*. This is *the* fundamental theoretical issue in identifying PDE systems with boundary conditions (BCs) that are well-posed, fully detailed in subsequent chapters.

The fundamental axiom of weak form theory is that one *indeed* seeks to construct an *approximate solution*. By very definition, a PDE solution is a *function* of space (and time perhaps) distributed continuously and smoothly, or possessing a finite number of finite discontinuities, over the PDE domain and on its boundaries. Thereby, the *prime requirement* for a weak form CFD algorithm is to clearly identify *the* candidate approximate solution.

This is *completely distinct* from historical FD/FV CFD approaches which generate a union of *stencils* via Taylor series approximations to PDE derivatives rather than stating the sought-for solution.

So, the starting point for a weak form construction is to identify a set of functions, called the *trial space*, endowed with properties appropriate for supporting an approximate solution. With certified existence of a trial space, the following questions come to mind:

- How good (accurate) an approximate solution can be supported by the selected trial space?
- How does the trial space supporting a finite element (FE) approximation differ from that for a finite difference (FD) scheme or a finite volume (FV) integral construction?
- Can these trial spaces be identical, and, if not, what are the distinguishing issues?
- Bottom line: is the *error* in the approximation related to a specific trial space selection?

Weak form theory possesses an elegant formalism for defining approximate solution error, developed in thoroughness in the next few chapters. The premier realization is that *approximation error* is a function(!) distributed over the PDE domain and its boundaries, as are the approximate *and* the exact NS solutions, with the latter of course never known. The key precept of weak form theory is to establish an integral *constraint on error* which requires definition of another set of functions termed the *test space*, against which the approximation error can be "tested." In English:

weak form theory formalizes the ingredients of
approximate solution construction in terms
of a trial space, a test space and the rendering
of the resultant approximation error an extremum
in an appropriate integral measure called a norm

Weak form theory practice is no more complicated than implementing this statement. Thereby, one must immediately enquire whether there exists an *optimal* test space, such that the approximation *error is the smallest* possible for *any* trial space selection. Again weak form theory provides the answer:

Weak form theory predicts the approximation
error is extremized, in practice minimized, when
the trial and test spaces contain the identical
members, which is termed the Galerkin criterion

Weak form theoretical musings always occur in the *continuum* and fully utilize calculus and vector field theory for generality with precision. Of pertinence, the weak form continuum construction holds for the PDE + BCs analytical solution as well as *any* approximation! Once the theory statement is formed, the remaining decision is trial space selection, hence identical test space, for error extremization. The key trial space requirement is for members to possess differentiability sufficient to enable integrals of their PDE derivatives to exist. This is really no problem, so the completion issue is forming the integrals that the weak form generates.

The continuum trial space contains functions spanning the global extent of the PDE domain. Finding suitable functions that one can integrate is nigh impossible (*the* challenge in DNS), hence the solution is to *discretize* the PDE domain and its boundaries, and hence identify much smaller *subsets* of the global trial (and test) space. Underlying is interpolation theory with net result a *discrete approximation trial space basis*. Trial space bases possess support *only* in the *generic* discretization cell, the *union* (non-overlapping sum) of which constitutes the computational mesh. This process admits full geometric generality, accurate evaluation of weak form integrals and a rigorous path to progressively more accurate formulations. Of ultimate importance is that it supports *analytical* formation of nonlinear algebraic matrix statements *amenable to computing*.

Summarizing, weak form theory involves *clear* organization of the sequence of decisions required for *approximate solution error extremization*, prior to definition of any specific discrete trial space basis. Thus, any given discrete solution methodology, specifically FE, FD or FV, becomes clearly identifiable among its peers by the sequence of decisions exposed in the weak formulation process. This text develops the subject in thoroughness, across a broad spectrum of fluid-thermal NS and manipulated NS PDE systems for which an *optimal performance* CFD algorithm is sought.

1.5 A Brief History of Finite Element CFD

Finite difference methods in CFD were first reported in the late 1920s, Courant *et al.* (1928), with fundamental theoretical developments emerging from the Courant Institute following World War II, Lax (1954), Lax and Wendroff (1960). Thereafter, many contributions to CFD emerged from the Los Alamos Scientific Laboratory (LASL), Amsden and Harlow (1970), Harlow (1971), coincident with the Imperial College team's development of the "SIMPLE" algorithm, Gosman *et al.* (1969). The NASA Ames Research Center (ARC) picked up the lead on compressible aerodynamics CFD, MacCormack (1969), Beam and Warming (1976), with timing coincident with CFD maturing to an international research topic with hundreds of contributors.

The progenitor of weak form theory is the *finite element method* developed in practice in the late 1950s by aeronautical engineers to analyze aircraft structural components. Exploratory musings preceded this; for example Hrenikoff (1941) developed an elasticity solution for torsion problems based on triangles, Courant (1943) developed a variational formulation for problems in vibrations. Turner *et al.* (1956) first derived the *stiffness matrix* for truss and beam analysis, and Clough (1960) coined the term *finite element*. Argyris (1963) published the first monograph detailing a precise mathematical foundation for the engineer's newly emerging finite element analysis capability.

The finite element method's first application to the non-structural problem of unsteady heat conduction required convolution, Zienkiewicz and Cheung (1965). A formal addressing of the wider problem class in nonlinear mechanics followed, Oden (1972). As finite element structural theory and methodology matured, the *mechanistic* engineering precepts became replaced by a rich mathematical basis founded in the variational calculus and Rayleigh–Ritz methods, Rayleigh (1877), Ritz (1909). This classic theory base for finite element structural analyses grew rapidly, including many fundamental contributions, Babuska and Aziz (1972), Ciarlet and Raviart (1972), Aubin (1972), Lions and Magenes (1972), Strang and Fix (1973) and Oden and Reddy (1976).

The direct extension of classic variational mechanics to CFD algorithm construction for fluid/thermal descriptions is not possible. The impediment is the conservation principle eulerian reference frame, which renders momentum conservation **DP** explicitly nonlinear. For this reason, at least, pioneering CFD procedures employed replacement of derivatives by divided differences, that is, *finite differences*, Richtmyer and Morton (1967), Roache (1972).

The FD successor *finite volume* CFD developments involved direct integration of each PDE over cells of a domain discretization, followed by the divergence theorem which exposed cell face fluxes. These were evaluated via finite difference quotients coupled with an approximate enforcement for D*M*, Patankar (1980). The *particle-in-cell* method, Evans and Harlow (1957), also employed a cell flux concept and used pseudo-lagrangian particle distributions for approximate satisfaction of the D*M* constraint.

Most FD/FV quotient-based CFD algorithms were discovered recoverable as specific criteria selections within the *discrete weighted residuals* (WR) framework, with Finlayson (1972) the pioneering exposition. In WR theory, approximation error was constrained by requiring local integrals containing "weights" to vanish. Within the class, the *collocation* method weights were the Dirac delta, which exactly reproduced classic FD quotients. A *finite volume* algorithm was retrieved for a constant weight. Generalizing the weights to functions, and defining them identical to the discrete trial space basis reproduced the *Galerkin* method, named after the (non-discrete, non-CFD) procedure of B.G. Galerkin (1915). Finally, defining the weights to be the PDE operator itself reproduced the *least squares* method.

Pioneering FE CFD algorithms employed various WR criteria. Oden (1969, 1972) was among the first to derive the basic theoretical analog for the NS PDE system. Using a Galerkin WR FE formulation, Baker (1971, 1973, 1974) reported two-dimensional compressible and incompressible flow simulations with recirculation regions. Olson (1972) detailed a pseudo-variational (hence Galerkin) FE algorithm for the streamfunction biharmonic PDE equivalent of the two-dimensional NS PDE system.

Lynn (1974) used the least squares criterion for an FE laminar boundary layer flow algorithm that retained a symmetric matrix structure. Popinski and Baker (1974) published an FE Galerkin WR algorithm for laminar boundary layer flow and reported accuracy and convergence in an appropriate norm, including direct comparisons with the Crank–Nicolson (1947) FD algorithm. Chung and Chiou (1976) published an FE Galerkin WR algorithm for compressible laminar boundary layer development behind an impinging shock. Connor and Brebbia (1974) authored *Finite Element Techniques for Fluid Flow*, the first monograph for a restricted problem class. Then followed *The Finite Element Method for Engineers*, Huebner (1975), with limited FE Galerkin WR content, and *Finite Element Simulation in Surface and Subsurface Hydrology*, Pinder and Gray (1977), with content restricted to the title category.

From the mid 1970s through the 1980s burgeoning interest in weak form FE CFD algorithm research and applications sparked annual international conferences. This resulted in the Wiley conference monograph series *Finite Elements in Fluids*, with Gallagher, Oden and Zienkiewicz principal editors, Gallagher *et al.* (1975–1988). Publication of FE discrete CFD algorithms for NS systems moved to textbooks. *Finite Element Computational Fluid Mechanics*, Baker (1983), was the first topical treatise to specifically include asymptotic convergence theory validation for RaNS systems. (Therein *weak form* never appeared, owing to the author's opinion that engineers didn't do weak things!)

Fletcher (1984) authored *Computational Galerkin Methods* containing a handful of NS solutions. With a sharp mathematical focus, but containing applications restricted to very small Reynolds number, Gunzburger (1989) published *Finite Element Methods for Viscous Incompressible Flows*. Similarly, Pironneau (1989) published *Finite Element Methods for Fluids*, an exceptionally formal mathematical treatise which included laminar flow benchmark problem validation data. Coincident with PC capability maturation, Baker and Pepper (1991) authored *Finite Elements* 1–2–3 which included a 5¼ inch floppy disc containing Fortran code.

It's quite alarming looking into these early texts and viewing solution graphics computed on *incredibly coarse* meshes! Much larger scale PC computing emerged in the 1990s, as the FE CFD textbook litany moved forward. First prize for size (over 1000 pages!) goes to Gresho and Sani (1998) for *Incompressible Flow and the Finite Element Method*, which contained incredible mathematical detail for laminar isothermal incompressible flows only! Reddy and Gartling (2001) published *The Finite Element Method in Heat Transfer and Fluid Dynamics*, with applications to low-Re non-Newtonian fluids.

Chung (2002) followed with *Computational Fluid Dynamics*, also scaling in at over 1000 pages, covering a broad spectrum in FD/FV CFD theory and practice but with a very limited FE content. Donea and Huerta (2003) followed with *Finite Element Methods for Flow Problems*, with validations restricted to two-dimensional laminar NS benchmarks. Recently, *Finite Elements ⇔ Computational Engineering Sciences*, Baker (2012), details introductory coverage of weak form theory broadly applied to heat transfer, structural mechanics, vibrations, fluid mechanics and heat/mass transport. Text content assimilation includes still/dynamic color graphics and academic course organization support at www.wiley.com/go/baker/finite. Hands-on PC computing is Matlab- and COMSOL-enabled with focus generation of *a posteriori* data validating asymptotic convergence theory, error estimation, the impact of *non-smooth* data and design optimization.

1.6 A Brief Summary

So, does the CFD community need another textbook on weak form FE-implemented algorithms? I believe the answer is yes, as the broad duplication of historical, small Re constructions reported can now be replaced by a single comprehensive weak form theory text with precise attention to performance estimation and *validation*. This current knowledge base is applicable across the complete spectrum of NS, BL, PNS, RaNS, LES and PRaNS PDE systems.

Weak form theory, of necessity *linear*, predicts that the solution associated with the stationary coordinate of the Galerkin criterion is *optimal*. Its failure to address dispersive error instability and phase accuracy compromising issues associated with pioneering Galerkin NS constructions is now fully resolved. Via weak form manipulations based on limiting independent coordinate-derived Taylor series, differential term additions to the *classic text* NS PDE statements generate *modified* (*m*PDE) systems. The analysis framework ultimately leads to a precise theory recovering essentially *all* prior FD/FV/FE constructions, but most importantly theoretical prediction of an *optimal* construction.

Hence, CFD algorithms for all NS, BL, PNS, RaNS, LES and PRaNS *m*PDE systems herein are based on this theoretically sound *optimal modified continuous Galerkin weak* *form* CFD theory. The key required *confirmation* of this assertion, as detailed herein, is that linear weak form theory *is* the *accurate predictor* of nonlinear NS PDE/*m*PDE CFD algorithm performance. Had this not occurred, this text would not have been written!

To summarize, weak form theory enables removal of essentially all elements of *mystery* surrounding algorithm constructions in NS fluid-thermal CFD. Comparison to legacy and/or current discrete CFD formulations is direct, simply by selecting the non-Galerkin, non-augmented PDE weak form criteria appropriate to reproducing that algorithm. This text is the culmination of the author's decades-held premise that the (multiple) hundreds of published CFD algorithms do not produce a corresponding number of *linearly independent* constructions! That the cogent approach to the subject can shed light uniformly across the spectrum was my academic research objective. This text is the result, hopefully written in a manner making the exposé enjoyable and rewarding. *Bon voyage*!

References

Amsden, A. and Harlow, F. (1970). "The SMAC Method: A Numerical Technique for Calculating Incompressible Fluid Flow," Univ. Calif. LASL Technical Report LA-4370.

Argyris, J.H. (1963). *Recent Advances in Matrix Methods of Structural Analysis*, Pergamon Press, Elmsford, NY.

Aubin, J.P. (1972). *Approximation of Elliptic Boundary Value Problems*, Wiley-Interscience, New York.

Babuska, I. and Aziz, A.K. (1972). "Lectures on the Mathematical Foundations of the Finite Element Method," in A.K. Aziz (ed.), *Mathematical Foundations of the Finite Element Method with Applications to Partial Differential Eqns*, Academic, NY, pp. 1–345.

Baker, A.J. (1971). "A Finite Element Computational Theory for the Mechanics and Thermodynamics of a Viscous Compressible Multi-Species Fluid," Bell Aerospace Research Rpt. 9500–920200.

Baker, A.J. (1973). "Finite Element Solution Algorithm for Viscous Incompressible Fluid Dynamics," *J. Numerical Methods Engineering*, V6, pp. 89–101.

Baker, A.J. (1974). "A Finite Element Algorithm for the Navier–Stokes Equations," NASA Technical Report CR-2391.

Baker, A.J. (1983). *Finite Element Computational Fluid Mechanics*, Hemisphere/McGraw-Hill, New York.

Baker, A.J. and Pepper, D.W. (1991). *Finite Elements* 1–2–3, McGraw-Hill, New York.

Baker, A.J. (2012). *Finite Elements ⇔ Computational Engineering Sciences*, John Wiley and Sons, London.

Beam, R. and Warming. R. (1976). "An Implicit Finite-Difference Algorithm for Hyperbolic Systems in Conservation Law Form," *J. Computational Physics* V22, pp. 87–110.

Chung, T.J. and Chiou, J.N. (1976). "Analysis of Unsteady Compressible Boundary Layer Flow Via Finite Elements," *Computers & Fluids*, V4, pp. 1–12.

Chung, T.J. (2002). *Computational Fluid Dynamics*, Cambridge University Press, UK.

Ciarlet, P.G. and Raviart, P.A. (1972). "General Lagrange and Hermite Interpolation in R^n with Applications to the Finite Element Method," *Arch. Rat. Mech. Anal.*, V46, pp. 177–199.

Clough, R.W. (1960). "The Finite Element Method in Plane Stress Analysis," *Proc. 2nd Conf. Electronic Computation*, Amer. Soc. Civil Engineers, Pittsburgh, PA, pp. 345–378.

Connor, J.J. and Brebbia, C.A. (1974). *Finite Element Techniques for Fluid Flow*, Newnes-Butterworths, UK.

Courant, R., Friedrichs, K. and Lewy, H. (1928). "Uber die Partiellen Differenzengleichungen der Mathematischen Physik," *Mathematische Annalen*, V100, pp. 32–74.

Courant, R. (1943). "Variational Methods for Solution of Problems of Equilibrium and Vibrations," *Bull. Amer. Math. Soc.*, V49, pp. 1–23.

Crank, J. and Nicolson, P. (1947). "A Practical Method for Numerical Evaluation of Solutions of Partial Differential Equations of Heat-Conduction Type," *Proc. Cambridge Philosophical Soc.*, V43, p. 50.

Donea, J. and Huerta, A. (2003). *Finite Element Methods for Flow Problems*, Wiley, West Sussex.

Dubois, T., Jauberteau, F. and Temam, R. (1999). *Dynamic Multilevel Methods and the Numerical Simulation of Turbulence*, Cambridge University Press, London.

Evans, M.E. and Harlow, F.H. (1957). "The Particle-in-Cell Method for Hydrodynamic Calculations," Univ. Calif., LASL Technical Report. LA-2139.

Finlayson, B.A. (1972). *The Method of Weighted Residuals and Variational Principles*, Academic, New York.
Fletcher, C.A.J. (1984). *Computational Galerkin Methods*, Springer-Verlag, Berlin.
Gallagher, R.H, Oden, J.T. and Zienkiewicz, O.C. (1975 –1988). Principal Editors. *Finite Elements in Fluids*, Volumes 1–7, Wiley Interscience, London.
Galerkin, B.G. (1915). "Series Occurring in Some Problems of Elastic Stability of Rods and Plates," *Eng. Bull.*, V19, pp. 897–908.
Gosman, A., Pun, W., Runchal, A., Spalding, D. and Wolfshtein, M. (1969). *Heat and Mass Transfer in Recirculating Flows*, Academic, London.
Gresho, P.M. and Sani, R.L. (1998). *Incompressible Flow and the Finite Element Method; Advection-Diffusion and Isothermal Laminar Flow*, Wiley, England.
Gunzburger, M.D. (1989). *Finite Element Methods for Viscous Incompressible Flows*, Academic, New York.
Harlow, F. (1971). "Fluid Dynamics – a LASL Monograph," Univ. Calif., LASL Technical Report LA-4700.
Hrenikoff, A. (1941). "Solution of Problems in Elasticity by the Frame Work Method." *J. Appl. Mechanics, ASME Trans.*, V8, pp. 169–175.
Huebner, K.H. (1975). *The Finite Element Method for Engineers*, Wiley, New York.
Lions, J.L. and Magenes, E. (1972). "Non-Homogeneous Boundary-Value Problems and Applications, Vol. I," (Trans., 1963 French edition by P. Kenneth), Springer-Verlag, Germany.
Lax, P. (1954). Weak Solutions of Nonlinear Hyperbolic Equations and their Numerical Computation," *Comm. Pure & Appl. Math*, V7, p. 159.
Lax, P. and Wendroff, B. (1960). Systems of Conservation Laws," *Comm. Pure & Applied Math*, V13, pp. 217–237.
Lynn, P.O. (1974). "Least Squares Finite Element Analysis of Laminar Boundary Layers," *J. Numerical Methods Engineer*, V8, p. 865.
MacCormack, R. (1969). "The Effect of Viscosity in Hypervelocity Impact Cratering," AIAA Technical Paper 69–354.
Oden, J.T. (1969). "Finite Element Applications in Fluid Dynamics," ASCE, *J. Engr. Mechanics Div.*, V95, EM3, pp. 821–826.
Oden, J.T. (1972). *Finite Elements of Nonlinear Continua*, McGraw-Hill, New York.
Oden, J.T. and Reddy, J.N. (1976). *Introduction to the Mathematical Theory of Finite Elements*, Wiley, New York.
Olson, M.E. (1972). "Formulation of a Variational Principle-Finite Element Method for Viscous Flows," *Proc. Variational Methods Engineering*, Southampton University, pp. 5.27–5.38.
Patankar, S.V. (1980). *Numerical Heat Transfer and Fluid Flow*, Hemisphere, Washington, DC.
Pinder, G.F. and Gray, W.G. (1977). *Finite Element Simulation in Surface and Subsurface Hydrology*, Academic, New York.
Piomelli, U. (1999). "Large Eddy Simulation: Achievements and Challenges," *Prog. Aerospace Sciences*, V35, pp. 335–362.
Pironneau, O. (1989). *Finite Element Methods for Fluids*, Wiley, Paris.
Popinski, Z. and Baker, A.J. (1974). "An Implicit Finite Element Algorithm for the Boundary Layer Equations," *J. Computational Physics*, V21, p. 55.
Rayleigh, J.W.S. (1877). *Theory of Sound*, 1st Edition Revised (1945), Dover, New York.
Reddy, J.N. and Gartling, D.K. (2001). *The Finite Element Method in Heat Transfer and Fluid Dynamics*, CRC Press, Boca Raton.
Richtmyer, R.D. and Morton, K.W. (1967). *Difference Methods of Initial-Value Problems*, 2nd Edition, Interscience, New York.
Ritz, W. (1909). "Uber Eine Neue Methode zur Losung Gewisser Variations-Probleme der Mathematischen Physik," *J. Reine Angew. Math.*, V135, p. 1.
Roache, P.J. (1972). *Computational Fluid Mechanics*, Hermosa Press, Albuquerque, NM.
Strang, G. and Fix, G.J. (1973). *An Analysis of the Finite Element Method*, Prentice Hall, Englewood Cliffs, NJ.
Turner, M., Clough, R., Martin, H. and Topp, L. (1956). "Stiffness and Deflection Analysis of Complex Structures," *J. Aeronautical Sciences*, V23(9), pp. 805–823.
Zienkiewicz, O.C. and Cheung, Y.K. (1965). "Finite Elements in the Solution of Field Problems," *The Engineer*, pp. 507–510.

2

Concepts, terminology, methodology

2.1 Overview

As Chapter 1 implies, *weak form theory* constitutes a *thoroughly formal process* for constructing approximate solutions to initial-boundary value partial differential equation (PDE) systems in engineering, specifically in the fluid-thermal sciences. The historical discrete finite element CFD implementations, developed as weighted residuals extensions on the classical mechanics variational concept, are now verified to constitute but a specific set of choices in a *continuum* (non-discrete) weak form, subsequently implemented using finite element trial space bases.

Similarly, essentially *all*(!) discrete CFD solution algorithms independently developed over the past 70 years for NS PDE systems, are also constituted of (no more than) a specific set of choices in a weak form. The hundreds of methods published over the history of CFD algorithm developments cannot be, and in fact *are not, linearly independent*. Instead, each exhibits fundamental decisions which become transparent when viewed under the *weak form* theoretical umbrella.

Experience confirms that students and practitioners of CFD are well-versed in the engineering aspects, but are less comfortable with the formal mathematical processes supporting analysis. This observation understandably stems from the engineer's relative lack of use of the calculus "learned" in the freshman-sophomore course sequence. Conversely, mathematicians writing on this subject delight in abstraction, hence communicate in a specialized language in which the CFD practitioner is not literate, conversant or even interested!

This chapter is designed to bridge the communication gap between the engineer, or student of CFD, and the mathematical elegance underlying weak form theory. It provides a thorough but brief presentation of fundamental mathematical concepts, notions and notation of use in weak form developments. The operations of Taylor series, a stationary point, interpolation and separation of variables are certainly a review of undergraduate material. Conversely, functionals, function spaces, orthogonality, completeness, quadratic forms, extremization, error estimation, norms, stability, and so on, are concepts lying well outside the typical engineer's cognizance.

Optimal MODIFIED CONTINUOUS Galerkin CFD, First Edition. A. J. Baker.

Companion Website: www.wiley.com/go/baker/GalerkinCFD

Serious thought was given to relegating this material to an appendix. However, decades of classroom teaching experience confirm that its practical importance contraindicates such a move. Readers with recallable calculus will experience a refresher. Those whose calculus is rusty, and others less conversant, will benefit greatly with the level of comfort generated with notions and notation exposition. Alternatively, one could proceed directly to the next chapter, after perusing the following two sections, then return here on an as-needed basis.

2.2 Steady D*E* Weak Form Completion

The dialogue to this point contains the oft repeated word *solution*. What is thus implied; specifically what are the *attributes of a solution* to a PDE? Quite simply, a solution is a function – not an array of numbers – distributed over the domain of definition of the differential equation and its boundary that exhibits precise satisfaction everywhere, as constrained by all information given beforehand. This given information, formally termed *data*, includes the domain span, the properties of materials in the domain, an applied source and values that must be matched by the solution on the domain boundary.

Herein, the Galerkin weak form is constructed for D*E*, (1.18), simplified to steady conduction. The temperature distribution due to conduction throughout the *solution domain*, the region Ω lying on ($\subset$) n-dimensional euclidean space (denoted $\Omega \subset \Re^n$), subject to a distributed source and appropriate heat transfer boundary conditions (BCs), is sought. This PDE + BCs definition is

$$\text{D}E\text{:}\ \mathcal{L}(T) = -\nabla \cdot \kappa \nabla T - s/\rho_0 c_p = 0, \quad \text{on}\ \Omega \subset \Re^n \tag{2.1}$$

$$\text{BCs:}\ \ell(T) = k\nabla T \bullet \widehat{\mathbf{n}} + h(T - T_r) - f_n = 0, \quad \text{on}\ \partial\Omega_R \subset \Re^{n-1} \tag{2.2}$$

$$T(\mathbf{x}_b) = T_b, \quad \text{on}\ \partial\Omega_D \subset \Re^{n-1} \tag{2.3}$$

This D*E* statement introduces the PDE + BCs notation used in this text. Specifically, $\mathcal{L}(\cdot)$ denotes a *differential equation* expressed in *homogeneous* form on the *state variable*, the argument in the parentheses $(\cdot)$, while $\ell(\cdot)$ signifies the *boundary conditions* constraining the state variable *normal* derivative on that portion $\partial\Omega_R$ (subscript R for *R*obin) of the boundary $\partial\Omega$ of the domain Ω of $\mathcal{L}(\cdot)$. In particular, (2.2) is the flux BC common to heat transfer analyses, where h is the convection coefficient, T_r is the convective heat exchange medium temperature and f_n is a given applied heat flux. Finally, fixing the temperature on select boundary segments $\partial\Omega_D$ of the domain boundary $\partial\Omega$ is permissible, where subscript D denotes a *D*irichlet BC, (2.3). The *data* for this D*E* statement include Ω, $\partial\Omega_R$, $\partial\Omega_D$, κ, s, ρ_0, c_p, k, h, T_r, f_n and T_b and the *state variable* is $T(\mathbf{x})$.

Recalling the Chapter 1 commentary, the first step in generating an approximate solution is to select the *trial space* upon which it will be projected. The set of functions constituting the trial space is assigned the symbol $\Psi_\alpha(\mathbf{x})$, $1 \le \alpha \le N$, the members and attributes of which will become identified later. The associated approximation $T^N(\mathbf{x})$ to the state variable $T(\mathbf{x})$ is then clearly expressed as

$$\begin{aligned} T^N(\mathbf{x}) &\equiv \sum_{\alpha=1}^{N} \Psi_\alpha(\mathbf{x}) Q_\alpha \\ &= Q_1\Psi_1(\mathbf{x}) + Q_2\Psi_2(\mathbf{x}) + \cdots + Q_N\Psi_N(\mathbf{x}) \end{aligned} \tag{2.4}$$

With the approximate solution $T^N(\mathbf{x})$ unequivocally identified, the fundamental question is, "How can one constrain the error in $T^N(\mathbf{x})$ to be as small as possible?" This requires a clear identification of *error*, a function(!) distributed on the domain and its boundary, $\Omega \cup \partial\Omega$, as are the exact and approximate solutions. Clearly

$$e^N(\mathbf{x}) \equiv T(\mathbf{x}) - T^N(\mathbf{x}) \tag{2.5}$$

is *the* precise statement of error associated with approximation $T^N(\mathbf{x})$.

Following the weak form protocol the error (2.5) is "measurable" using *test functions*, each of which the theory requires to be *absolutely arbitrary*! Mathematically, the measurement process consists of substituting (2.4) into DE, (2.1), then requiring that the integral of this expression over the domain Ω, when multiplied by *any* test function $w(\mathbf{x})$, vanishes identically.

Therefore, the *weak form* (WF) statement for testing error in the approximation T^N is

$$\begin{aligned} \mathrm{WF} &\equiv \int_\Omega w(\mathbf{x})\mathcal{L}(e^N)\mathrm{d}\tau \equiv 0, \quad \text{for any } w(\mathbf{x}) \\ &= \int_\Omega w(\mathbf{x})\mathcal{L}(T - T^N)\mathrm{d}\tau = \int_\Omega w(\mathbf{x})\mathcal{L}(T^N)\mathrm{d}\tau \equiv 0 \end{aligned} \tag{2.6}$$

The final form results since $\mathcal{L}(T) = 0$, hence also is the integral of the product of $\mathcal{L}(T)$ with any function $w(\mathbf{x})$. The integral over the domain Ω has associated a differential element, such as $\mathrm{d}\tau = \mathrm{d}x\mathrm{d}y\mathrm{d}x$ for rectangular cartesian coordinates.

The immediate question is, "How can one evaluate (2.6) for *all* test functions $w(\mathbf{x})$?" The answer is, "It's not possible!" Hence, an operation is needed to render (2.6) a "computable form." Since $w(\mathbf{x})$ is a *known* function (you choose it), then interpolation theory (detailed later) can be used to approximate $w(\mathbf{x})$ to any level of accuracy desired. A set of *interpolation polynomials* $\Phi_\beta(\mathbf{x})$ is appropriate with interpolation symbolized by superscript I

$$w(\mathbf{x}) \approx w^I(\mathbf{x}) \equiv \sum_{\beta=1}^{I} \Phi_\beta(\mathbf{x}) W_\beta \tag{2.7}$$

Of significance, the associated interpolation coefficient set W_β, $1 \le \beta \le I$, is *a priori* known constant *data* (you choose it!). Substituting (2.7) into (2.6), and extracting the *constant data* W_β from within the integrand, alters (2.6) to the form

$$\mathrm{WF} = \sum_\beta W_\beta \int_\Omega \Phi_\beta(\mathbf{x})\mathcal{L}(T^N)\mathrm{d}\tau \equiv 0, \quad \text{for } 1 \le \beta \le I \tag{2.8}$$

For any test function $w(\mathbf{x})$, hence definition of the coefficient set W_β, $1 \le \beta \le I$, the weak form WF (2.8) is a number. From (2.7), the requirement for test function *arbitrariness* is now carried entirely by the definition of these W_β. The weak form theory states that the WF *sought* is the *extremum*! Recalling elementary calculus, an extremum corresponds to a *stationary point*, identified as the coordinate (in W_β space) where the first derivative of (2.8) vanishes.

Since the WF is a linear function the W_β, forming the first derivative is elementary, which produces the *weak statement* (WS) corresponding to the weak form extremum. Thereby,

$$
\begin{aligned}
\mathrm{WS} &\equiv \partial \mathrm{WF}/\partial W_\beta \\
&= \partial\left[\sum_\beta W_\beta \int_\Omega \Phi_\beta(\mathbf{x})\mathcal{L}(T^N)\mathrm{d}\tau\right]\Big/\partial W_\beta \\
&= \int_\Omega \Phi_\beta(\mathbf{x})\mathcal{L}(T^N)\mathrm{d}\tau = \{0\}, \quad \text{for } 1 \le \beta \le N
\end{aligned}
\tag{2.9}
$$

There exists an absolutely *key distinction* between WS and WF. Specifically, (2.8) is a single (scalar) number while there are N integrals defined in (2.9), as β is no longer a summation index! Hence, (2.9) is equated to the (column) matrix of zeros $\{0\}$, rather than the scalar 0.

Finally, weak form theory proves that the global extremum for the error constraint (2.9) accrues to selection of the *Galerkin criterion*. This occurs when the test space set $\Phi_\beta(\mathbf{x})$ is identical with the approximation trial space $\Psi_\alpha(\mathbf{x})$, $1 \le \alpha \le N$. This alteration to (2.8) leads to the *Galerkin weak form* (GWF)

$$
\mathrm{GWF} = \sum_\beta W_\beta \int_\Omega \Psi_\beta(\mathbf{x})\mathcal{L}(T^N)\mathrm{d}\tau \equiv 0, \quad \text{for } 1 \le \beta \le N \tag{2.10}
$$

and its extremum is the *Galerkin weak statement* (GWS)

$$
\begin{aligned}
\mathrm{GWS} &= \int_\Omega \Psi_\beta(\mathbf{x})\mathcal{L}(T^N)\mathrm{d}\tau = \{0\} \\
&= \int_\Omega \Psi_\beta(\mathrm{x})\left[-\nabla\cdot\kappa\nabla T^N - s/\rho_0 c_p\right]\mathrm{d}\tau \\
&= \{0\}, \quad \text{for } 1 \le \beta \le N
\end{aligned}
\tag{2.11}
$$

with second line the result of direct substitution of (2.1).

The terminal operation leading to the *computable* GWS accrues to the fact that trial space members $\Psi_\alpha(\mathbf{x})$ must be twice differentiable, due to the thermal diffusivity (κ) term in (2.11), while the same functions $\Psi_\beta(\mathbf{x})$ populating the test space are not differentiated at all. This is remedied by implementing the Green–Gauss form of the Divergence Theorem, Kreyszig (1967), which corresponds to multi-dimensional integration by parts. This symmetrizes the differentiability requirement in (2.11) and directly inserts the state variable efflux over the boundary of Ω

$$
\begin{aligned}
\mathrm{GWS} &= \int_\Omega \Psi_\beta(\mathbf{x})\left[-\nabla\cdot\kappa\nabla T^N - s/\rho_0 c_p\right]\mathrm{d}\tau \\
&= \int_\Omega \nabla\Psi_\beta \bullet \kappa\nabla T^N\mathrm{d}\tau - \int_\Omega \Psi_\beta(s/\rho_0 c_p)\mathrm{d}\tau \\
&\quad - \int_{\partial\Omega} \Psi_\beta\kappa\nabla T^N \bullet \widehat{\mathbf{n}}\mathrm{d}\sigma = \{0\}, \quad \text{for } 1 \le \beta \le N
\end{aligned}
\tag{2.12}
$$

In (2.12), $\mathrm{d}\tau$ remains the differential element on Ω while $\mathrm{d}\sigma$ corresponds for integrals on the boundary $\partial\Omega$ of Ω. The Robin BC portion of the integral is immediately substituted on

$\partial\Omega_R$ of $\partial\Omega$ via (2.2). This yields the *final form* of the *continuum Galerkin weak statement* for the n-D steady D*E* PDE + BCs system (2.1–2.3)

$$\begin{aligned}
\mathrm{GWS}^N &= \int_\Omega \Psi_\beta(\mathbf{x})\left[-\nabla\cdot\kappa\nabla T^N - s/\rho_0 c_p\right]\mathrm{d}\tau \\
&= \int_\Omega \nabla\Psi_\beta \bullet \kappa\nabla T^N \mathrm{d}\tau - \int_\Omega \Psi_\beta(s/\rho_0 c_p)\mathrm{d}\tau \\
&+ \int_{\partial\Omega_R\cap\partial\Omega} \Psi_\beta \frac{k}{\kappa}\left[h(T^N - T_r) + f_n\right]\mathrm{d}\sigma \\
&- \int_{\partial\Omega_D\cap\partial\Omega} \Psi_\beta \kappa\nabla T^N \bullet \widehat{\mathbf{n}}\mathrm{d}\sigma = \{0\}, \quad \text{for } 1 \le \beta \le N
\end{aligned} \tag{2.13}$$

Superscript N has now been added to the GWS (2.12) to notationally *enforce* its specific connection to the sought approximate solution $T^N(\mathbf{x})$.

2.3 Steady D*E* GWS^N Discrete FE Implementation

The D*E* Galerkin weak statement (2.13) is now pushed through discrete implementation to generation of an approximate solution for an elementary, operationally informative 1-D example. Specifically, find the temperature distribution due to heat conduction through a slab of thermal diffusivity κ, spanning $\mathrm{a} \le x \le \mathrm{b}$, subject to an internal heat source s, a specified heat flux f_n on the left boundary and a fixed temperature $T = T_b$ on the right boundary. For simplicity, assume that the product $\rho_0 c_p$ is unity, hence $\kappa = k$. This D*E* problem statement is

$$\mathrm{D}E\text{: } \mathcal{L}(T) = -\frac{\mathrm{d}}{\mathrm{d}x}\left(k\frac{\mathrm{d}T}{\mathrm{d}x}\right) - s = 0, \text{ on a} < x < b \tag{2.14}$$

$$\text{BCs: } \ell(T) = -k\frac{\mathrm{d}T}{\mathrm{d}x} - f_n = 0, \text{ at } x = \mathrm{a} \tag{2.15}$$

$$T = T_b, \text{ at } x = b \tag{2.16}$$

Figure 2.1 illustrates the problem: the *data* are *k*, *q*, T_b, *s* and $\mathrm{L} \equiv \mathrm{b} - \mathrm{a}$ and the *state variable* is *T*(*x*).

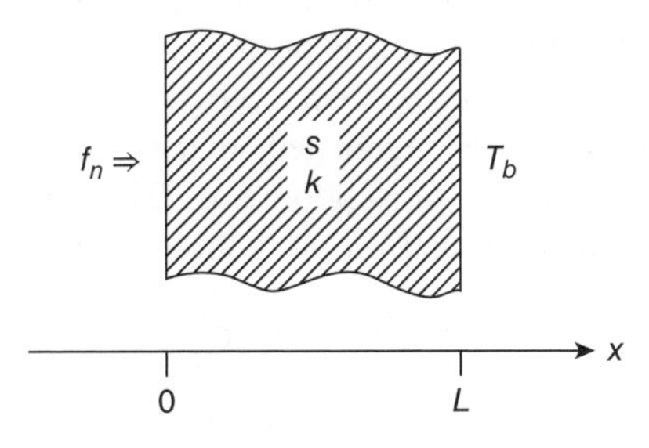

Figure 2.1 D*E* example problem definition

For the given data uniformly constant, the suggested exercise confirms that the exact solution to (2.14–2.16) is

$$T(x, data) = \frac{sL^2}{2k}\left[1 - \left(\frac{x}{L}\right)^2\right] + \frac{f_n L}{k}\left(1 - \frac{x}{L}\right) + T_b \tag{2.17}$$

That (2.4) is indeed a *viable* representation of a (*any*!) solution is readily illustrated by observing that (2.4) is a statement of the exact solution for the trial space and coefficient set definitions

$$\begin{aligned} Q_1 &= \frac{sL^2}{2k}, \quad \Psi_1(x) \equiv 1 - \left(\frac{x}{L}\right)^2 \\ Q_2 &= \frac{f_n L}{k}, \quad \Psi_2(x) \equiv 1 - \left(\frac{x}{L}\right) \\ Q_3 &= T_b, \quad \Psi_3(x) \equiv 1 \end{aligned} \tag{2.18}$$

Note that trial space members are functions *only* of the spatial coordinate x normalized by domain span L, while the expansion coefficients contain all given *data*!

The GWS^N specific to (2.14–2.15) is

$$\text{GWS}^N = \int_\Omega \frac{d\Psi_\beta}{dx} k \frac{dT^N}{dx} dx - \int_\Omega \Psi_\beta s dx - \Psi_\beta k \frac{dT^N}{dx}\Big|_a^b = 0, \quad \text{for } 1 \le \beta \le N \tag{2.19}$$

Substituting the approximation (2.1) and inserting the defined flux BC (2.15) yields the terminal *continuum* GWS^N

$$\begin{aligned} \text{GWS}^N = \sum_{\alpha=1}^{N} \left(Q_\alpha \int_\Omega \frac{d\Psi_\beta}{dx} k \frac{d\Psi_\alpha}{dx} dx \right) - \int_\Omega \Psi_\beta s dx \\ - \Psi_N k \frac{dT^N}{dx}\Big|^b - \Psi_1 f_n\big|_a = 0 \end{aligned} \tag{2.20}$$

All that remains is to identify a suitable trial space $\Psi_\alpha(x)$ for (2.20) then evaluate the defined integrals, noting that $\Psi_\beta(x)$ is the same variable set. Since the Ψ_α span the global solution domain Ω, finding such a function set with genuine versatility is a challenge, especially for CFD. The alternative option is to organize a spatially *discrete implementation* which requires identification of a *trial space basis*.

Thereby, the solution domain Ω is segmented, that is discretized, into the *union* (non-overlapping sum, symbolized as $\cup$) of subdomains. Throughout this text superscript h denotes spatially discrete. Domain discretization is the operation

$$\Omega \Rightarrow \Omega^h \equiv \cup_e \Omega_e \tag{2.21}$$

which creates a *computational mesh* consisting of the union of subdomains Ω_e. For a discretization finite element (FE) implemented, each Ω_e is called a *finite element*. The solution notation for the associated *FE discrete weak form approximation* is

$$T^N(x) \equiv \sum_{\alpha=1}^{N} \Psi_\alpha(x)Q_\alpha \Rightarrow T^h(x) = \cup_e T_e(x) \tag{2.22}$$

An FE discrete implementation hinges on identification of the functional form for $T_e(x)$, (2.22). In this text, the Ω_e contribution to the approximate solution $T^h(x)$ is the matrix scalar product

$$T_e(x) \equiv \{N_k(\bullet)\}^T \{Q\}_e \tag{2.23}$$

wherein $\{\bullet\}$ denotes a column matrix and $\{\bullet\}^T$ the transpose, a row matrix; hence (2.23) is indeed a *scalar*. The matrix function $\{N_k(\bullet)\}$ is a *finite element trial space basis* with argument $(\bullet)$ typically involving k^{th} degree polynomials written on local coordinates normalized.

The *linear* $k=1$ FE basis $\{N_1(x)\}$ spanning one dimension is readily established from the global trial space $\Psi_\alpha(x)$ by considering a uniform $M=2$ element discretization of the domain Ω of span $a \le x \le b$. Figure 2.2 illustrates this mesh with the generated three geometric *nodes* labeled X1, X2, and X3. Thereby, the continuum approximation (2.4) must possess three trial functions $\Psi_\alpha(x)$. i.e.,

$$T^N(x) \equiv \sum_{\alpha=1}^{N=3} \Psi_\alpha(x)Q_\alpha = \Psi_1(x)Q_1 + \Psi_2(x)Q_2 + \Psi_3(x)Q_3 \tag{2.24}$$

assuming the approximate solution coefficients Q_α are co-located with the geometric nodes, the typical choice.

The key constraint on trial space members $\Psi_\alpha(x)$ is that the expansion coefficients Q_α in (2.24) must possess units of temperature. Specifically, $T^N(x=a)$ must equate to Q_1; correspondingly $\Psi_1(x=a)$ must be unity, while $\Psi_2(a)$ and $\Psi_3(a)$ must each vanish, and so on. Ultimate simplicity accrues to assuming piecewise linear dependence for each $\Psi_\alpha(x)$. The middle graphs in Figure 2.3 illustrate the resultant distributions for each $\Psi_\alpha(x)$, $1 \le \alpha \le 3$.

In subscript notation for familiarity, $XJ \Rightarrow X_j$, the linear (lagrange) monomial associated with $\Psi_\alpha(x)$ centered at the generic node X_j is

$$\Psi(x) = \begin{cases} \dfrac{x - X_{j-l}}{X_j - X_{j-l}}, & \text{for } X_{j-l} \le x \le X_j \\ \dfrac{X_{j+l} - x}{X_{j+l} - X_j}, & \text{for } X_j \le x \le X_{j+l} \\ 0, \quad \text{for } x < X_{j-l} \text{ and/or } x > X_{j+l} \end{cases} \tag{2.25}$$

Figure 2.2 Nodes of $M=2$ element discretization Ω^h of domain Ω

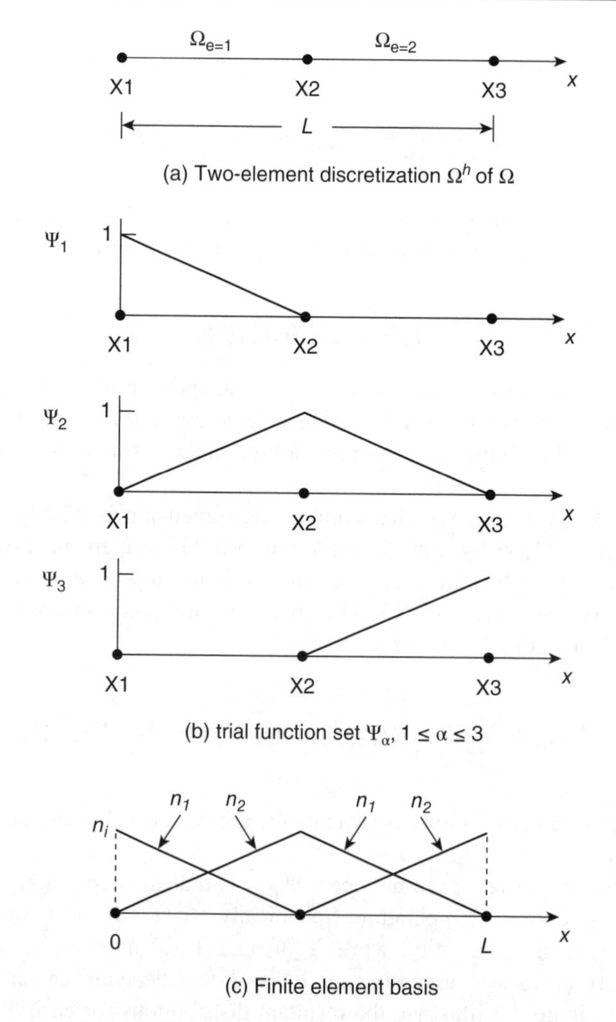

Figure 2.3 Construction of the FE linear trial space basis $\{N_1\}$ from the piecewise continuous global span trial space $\Psi_\alpha(x)$, $1 \le \alpha \le 3$

The last line in (2.25) emphasizes that *each* $\Psi_\alpha(x)$ is of global span, that is, it is defined on the entirety of $a \le x \le b$!

The $k = 1$ FE trial space basis associated with global span trial space $\Psi_\alpha(x)$ is identified by superposing the *non-zero* components of each Ψ_α onto every finite element Ω_e of the mesh. The bottom graph in Figure 2.3 is the result and by direct observation the FE *linear trial space basis* is

$$\{N_1(x)\} \equiv \begin{Bmatrix} n_1 \\ n_2 \end{Bmatrix} = \frac{1}{(\mathrm{XR} - \mathrm{XL})_e} \begin{Bmatrix} \mathrm{XR} - x \\ x - \mathrm{XL} \end{Bmatrix}_e \tag{2.26}$$

where XR and XL denote the *right* and *left* nodal coordinates of any Ω_e.

Observation also confirms the $\{N_1\}$ basis spanning *every* Ω_e is absolutely the same function, and hence is independent of a specific Ω_e! Functional notation is clarified by imposing an affine coordinate translation $\bar{x}$ with origin at the left node of *each* Ω_e, that is, $\bar{x} = x - \mathrm{XL}_e$. Then for l_e the span of Ω_e, (2.26) takes the *element-generic* normalized form

$$\{N_1(\bar{x})\} \equiv \begin{Bmatrix} n_1 \\ n_2 \end{Bmatrix} = \frac{1}{l_e}\begin{Bmatrix} 1 - \bar{x}/l_e \\ \bar{x}/l_e \end{Bmatrix} \tag{2.27}$$

Since (2.27) defines a function strictly local to Ω_e, not global, and since $\{N_1(\bar{x}/l_e)\}$ contains *all* pertinent $\Psi_\alpha(x)$ information on Ω_e, the global integrals in GWS^N, (2.20), can be replaced by a summation of local integrals over the M elements of the mesh supporting GWS^h. Specifically

$$\begin{aligned} \mathrm{GWS}^N &\equiv \int_\Omega \Psi_\beta(x)\mathcal{L}(T^N)\mathrm{d}x \Rightarrow \mathrm{GWS}^h \\ &\equiv \sum_e^{\mathrm{M}} \left[\int_{\Omega_e} \{N_1(\bar{x}/l_e)\}\mathcal{L}(T_e)\mathrm{d}x \right] = \{0\} \end{aligned} \tag{2.28}$$

with T_e correspondingly replacing T^N as the differential equation argument. Note also that integrand insertion of the column matrix $\{N_1\}$ equates notationally with the column matrix $\{0\}$, that is, (2.28) is indeed a matrix statement as is its progenitor (2.13).

Recalling globally formed GWS^N, (2.19), then substituting T_e for T^N therein produces the GWS^h in analytical form

$$\begin{aligned} \mathrm{GWS}^h = \sum_e^{\mathrm{M}} &\left[\int_{\Omega_e} \frac{\mathrm{d}\{N_1\}}{\mathrm{d}x} k \frac{\mathrm{d}\{N_1\}^T}{\mathrm{d}x} \mathrm{d}x\{Q\}_e - \int_{\Omega_e} \{N_1\} s \mathrm{d}x \right] \\ &- k\frac{\mathrm{d}T_e}{\mathrm{d}x}\{N_1\}\Big|^{\mathrm{b}} - \{N_1\} f_n\big|_{\mathrm{a}} = \{0\} \end{aligned} \tag{2.29}$$

It is immediately apparent in (2.29) that forming *all* integrals in GWS^h is an operation done *once* on the *generic* element since $\{N_1\}$ is *universal* to every Ω_e. The lead integral in (2.29) requires the spatial derivative of $\{N_1(\bar{x}/l_e)\}$, *exactly* generated from its definition (2.27) via the calculus chain rule

$$\frac{\mathrm{d}\{N_1(\bar{x}/l_e)\}}{\mathrm{d}x} = \frac{\mathrm{d}\{N_1(\bar{x}/l_e)\}}{\mathrm{d}\bar{x}}\frac{\mathrm{d}\bar{x}}{\mathrm{d}x} = \frac{1}{l_e}\begin{Bmatrix} -1 \\ 1 \end{Bmatrix} \tag{2.30}$$

Therefore the first integral in (2.29) for *any* (every!) finite element domain Ω_e computes to

$$\begin{aligned} \int_{\Omega_e} \frac{\mathrm{d}\{N_1\}}{\mathrm{d}x} k \frac{\mathrm{d}\{N_1\}^T}{\mathrm{d}x}\{Q\}_e \mathrm{d}x &= \int_{\Omega_e} \frac{1}{l_e}\begin{Bmatrix} -1 \\ 1 \end{Bmatrix}\frac{1}{l_e}\{-1 \quad 1\} k \mathrm{d}x\{Q\}_e \\ &= \frac{1}{l_e^2}\begin{bmatrix} 1 & -1 \\ -1 & 1 \end{bmatrix}\int_{\Omega_e} k\mathrm{d}x\{Q\}_e = \frac{\bar{k}_e}{l_e}\begin{bmatrix} 1 & -1 \\ -1 & 1 \end{bmatrix}\{Q\}_e \end{aligned} \tag{2.31}$$

where $\bar{k}_e$, the average thermal conductivity on Ω_e, multiplied by l_e is the *exact* result for the integral defined in the third expression in (2.31).

The second term in GWS^h, (2.29), requires an integration of $\{N_1(\bar{x}/l_e)\}$ over Ω_e. The calculus exercise confirms that the result is

$$\int_{\Omega_e}\{N_1\}s\mathrm{d}x = s\int_{\Omega_e}\begin{Bmatrix} 1-\bar{x}/l_e \\ \bar{x}/l_e \end{Bmatrix}\mathrm{d}x = \frac{sl_e}{2}\begin{Bmatrix} 1 \\ 1 \end{Bmatrix} \tag{2.32}$$

The two remaining terms in (2.29) require only evaluation at the end points of Ω_e. These term evaluations become

$$\begin{aligned} -k\frac{\mathrm{d}T_e}{\mathrm{d}x}\{N_1\}\bigg|^{\mathrm{b}} &= -k\left|\frac{\mathrm{d}T_e}{\mathrm{d}x}\begin{Bmatrix} 1-(\bar{x}=l_e)/l_e \\ (\bar{x}=l_e)/l_e \end{Bmatrix}_{e=M}\right. = -k\frac{\mathrm{d}T_e}{\mathrm{d}x}\begin{Bmatrix} 0 \\ 1 \end{Bmatrix}_{e=M} \\ -\{N_1\}f_n|_{\mathrm{a}} &= -f_n\begin{Bmatrix} 1-(\bar{x}=0)/l_e \\ (\bar{x}=0)/l_e \end{Bmatrix}_{e=1} = -f_n\begin{Bmatrix} 1 \\ 0 \end{Bmatrix}_{e=1} \end{aligned} \tag{2.33}$$

since the elements of $\{N_1(\bar{x}/l_e)\}$ take on (0,1) values at nodes of Ω_e. Hence, (2.33) organizes boundary data imposition *only* at the right node of $\Omega_{e=\mathrm{M}}$ and the left node of $\Omega_{e=1}$.

The GWS^h integral statement for linear FE trial space basis $\{N_1\}$ implementation is complete. Hence (2.29) morphs to the sought *computable statement*, valid for *any* discretization with M elements

$$\begin{aligned} \mathrm{GWS}^h &\equiv \sum_e^{\mathrm{M}}\{\mathrm{WS}\}_e = \{0\} \\ &= \sum_e^{\mathrm{M}}\left(\frac{\bar{k}_e}{l_e}\begin{bmatrix} 1 & -1 \\ -1 & 1 \end{bmatrix}\{Q\}_e - \frac{sl_e}{2}\begin{Bmatrix} 1 \\ 1 \end{Bmatrix} - k\frac{\mathrm{d}T_e}{\mathrm{d}x}\begin{Bmatrix} 0 \\ \delta_{e\mathrm{M}} \end{Bmatrix} - f_n\begin{Bmatrix} \delta_{e1} \\ 0 \end{Bmatrix}\right) \end{aligned} \tag{2.34}$$

Terms in (2.34) with subscript e contain element-dependent data. Each *Kronecker delta* δ_{e1}, $\delta_{e\mathrm{M}}$ is a "switch" that takes the value unity *only* when $e=1$ or $e=\mathrm{M}$, respectively. These clearly define the matrix location where given and/or unknown flux BCs are inserted.

For the $\mathrm{M}=2$ uniform mesh example the *measure* (length) of each Ω_e is $l_e=\mathrm{L}/2$. Assuming $\bar{k}_e = k$ is a constant, the $\mathrm{M}=2$ element mesh contributions to GWS^h (2.34) are

$$\begin{aligned} \text{for } e=1: \{\mathrm{WS}\}_1 &= \frac{k}{l_1}\begin{bmatrix} 1 & -1 \\ -1 & 1 \end{bmatrix}\{Q\}_{e=1} - \frac{sl_1}{2}\begin{Bmatrix} 1 \\ 1 \end{Bmatrix} - f_n\begin{Bmatrix} \delta_{11} \\ 0 \end{Bmatrix} \\ &= \frac{k}{\mathrm{L}/2}\begin{bmatrix} 1 & -1 \\ -1 & 1 \end{bmatrix}\begin{Bmatrix} Q_1 \\ Q_2 \end{Bmatrix} - \frac{s\mathrm{L}/2}{2}\begin{Bmatrix} 1 \\ 1 \end{Bmatrix} - \begin{Bmatrix} f_n \\ 0 \end{Bmatrix} \end{aligned} \tag{2.35}$$

$$\begin{aligned} \text{for } e=2: \{\mathrm{WS}\}_2 &= \frac{k}{l_2}\begin{bmatrix} 1 & -1 \\ -1 & 1 \end{bmatrix}\{Q\}_2 - \frac{sl_2}{2}\begin{Bmatrix} 1 \\ 1 \end{Bmatrix} - k\frac{\mathrm{d}T}{\mathrm{d}x}\begin{Bmatrix} 0 \\ \delta_{22} \end{Bmatrix} \\ &= \frac{2k}{\mathrm{L}}\begin{bmatrix} 1 & -1 \\ -1 & 1 \end{bmatrix}\begin{Bmatrix} Q_2 \\ Q_3 \end{Bmatrix} - \frac{s\mathrm{L}}{4}\begin{Bmatrix} 1 \\ 1 \end{Bmatrix} + \begin{Bmatrix} 0 \\ F_3 \end{Bmatrix} \end{aligned} \tag{2.36}$$

In (2.36), F_3 is the label assigned to the *unknown* heat flux $-k\mathrm{d}T/\mathrm{d}x$ at the boundary $x = \mathrm{b}$ where the fixed temperature BC T_b is applied.

The summation of matrix statements (2.35) and (2.36) form the (global) matrix statement GWS^h, (2.34). Adding subscripts to denote the order of these matrices

$$\mathrm{GWS}^h \equiv \sum_{e=1}^{\mathrm{M}=2} \{\mathrm{WS}\}_e \Rightarrow [\mathrm{Matrix}]_{3\times3} \begin{Bmatrix} Q_1 \\ Q_2 \\ Q_3 \end{Bmatrix}_{3\times1} - \{\mathrm{b}\}_{3\times1} = \{0\}_{3\times1} \tag{2.37}$$

how does one sum the 2×2 matrices in (2.35–2.36) into the 3×3 square [Matrix] in (2.37), also the 2×1 column matrices into the 3×1 column matrix $\{\mathrm{b}\}$? The answer is to simply insert row-column pairs of zeros into each element matrix statement corresponding to locations in $\{Q\}_e$ missing in the global matrix $\{Q\} = \{Q_1, Q_2, Q_3\}^T$. Thereby

$$[\mathrm{Matrix}] = \sum_{e=1}^{2} [\mathrm{Matrix}]_e = \frac{2k}{\mathrm{L}} \left(\begin{bmatrix} 1 & -1 & 0 \\ -1 & 1 & 0 \\ 0 & 0 & 0 \end{bmatrix} + \begin{bmatrix} 0 & 0 & 0 \\ 0 & 1 & -1 \\ 0 & -1 & 1 \end{bmatrix} \right)$$

$$= \frac{2k}{\mathrm{L}} \begin{bmatrix} 1 & -1 & 0 \\ -1 & 2 & -1 \\ 0 & -1 & 1 \end{bmatrix} \tag{2.38}$$

$$\{\mathrm{b}\} = \sum_{e=1}^{2} \{\mathrm{b}\}_e = \frac{s\mathrm{L}}{4} \left(\begin{Bmatrix} 1 \\ 1 \\ 0 \end{Bmatrix} + \begin{Bmatrix} 0 \\ 1 \\ 1 \end{Bmatrix} \right) + \begin{Bmatrix} f_n \\ 0 \\ 0 \end{Bmatrix} + \begin{Bmatrix} 0 \\ 0 \\ -F_3 \end{Bmatrix} = \frac{s\mathrm{L}}{4} \begin{Bmatrix} 1 \\ 2 \\ 1 \end{Bmatrix} + \begin{Bmatrix} f_n \\ 0 \\ -F_3 \end{Bmatrix} \tag{2.39}$$

Completing the substitutions, dividing through by $2k/\mathrm{L}$ and moving the data matrix $\{\mathrm{b}\}$ across the equal sign produces the *assembled* global GWS^h matrix statement

$$\mathrm{GWS}^h \Rightarrow \begin{bmatrix} 1 & -1 & 0 \\ -1 & 2 & -1 \\ 0 & -1 & 1 \end{bmatrix} \begin{Bmatrix} Q_1 \\ Q_2 \\ Q_3 \end{Bmatrix} = \frac{sL^2}{8k} \begin{Bmatrix} 1 \\ 2 \\ 1 \end{Bmatrix} + \frac{L}{2k} \begin{Bmatrix} f_n \\ 0 \\ -F_3 \end{Bmatrix} \tag{2.40}$$

It remains to enforce the Dirichlet BC $Q_3 \equiv T_b$, and weak form theory has identified that the *unknown* in the last matrix row of (2.40) is *not* Q_3 but the heat flux F_3! Moving F_3 to the matrix of unknowns and substituting $Q_3 \equiv T_b$ produces the *final* computable matrix statement

$$\mathrm{GWS}^h \Rightarrow \begin{bmatrix} 1 & -1 & 0 \\ -1 & 2 & 0 \\ 0 & -1 & L/2k \end{bmatrix} \begin{Bmatrix} Q_1 \\ Q_2 \\ F_3 \end{Bmatrix} = \frac{sL^2}{8k} \begin{Bmatrix} 1 \\ 2 \\ 1 \end{Bmatrix} + \frac{L}{2k} \begin{Bmatrix} f_n \\ 0 \\ 0 \end{Bmatrix} + \begin{Bmatrix} 0 \\ T_b \\ -T_b \end{Bmatrix} \tag{2.41}$$

Enforcing Dirichlet BC data *decouples* the solution for unknown temperatures $\{Q1, Q2\}^T$ from F_3. The reduced matrix statement is

$$\begin{bmatrix} 1 & -1 \\ -1 & 2 \end{bmatrix} \begin{Bmatrix} Q_1 \\ Q_2 \end{Bmatrix} = \begin{Bmatrix} \dfrac{\mathrm{L}}{2k} \left(\dfrac{s\mathrm{L}}{4} + f_n \right) \\ \dfrac{s\mathrm{L}^2}{4k} + T_b \end{Bmatrix} \tag{2.42}$$

and solution via Cramer's rule yields

$$\begin{aligned} Q_1 &= \frac{sL^2}{2k} + \frac{f_n L}{k} + T_b \\ Q_2 &= \frac{3sL^2}{8k} + \frac{f_n L}{2k} + T_b \end{aligned} \tag{2.43}$$

which provides the data needed to solve the third row equation in (2.41), which yields $F_3 = sL + f_n$.

This elementary M = 2 FE $\{N_1\}$ basis implemented GWS^h example illustrates several key weak form attributes. The discrete solution T^h coefficient matrix $\{Q\} = \{Q_1, Q_2, Q_3\}^T$ contains all the *data* defining the D*E* problem statement, recall the sentence following (2.18). The discrete T^h approximate solution is a *distribution*, specifically a *function* of x

$$\begin{aligned} T^N(x) &\equiv \sum_{\alpha=1}^{N=3} \Psi_\alpha(x) Q_\alpha = \Psi_1(x)Q_1 + \Psi_2(x)Q_2 + \Psi_3(x)Q_3 \\ &\equiv T^h(x) = \cup_e^M \left(\{N_1(\bar{x})\}^T \begin{Bmatrix} Q_1 \\ Q_2 \end{Bmatrix}_{e=1}, \{N_1(\bar{x})\}^T \begin{Bmatrix} Q_2 \\ Q_3 \end{Bmatrix}_{e=2} \right) \\ &\equiv \cup_e \{N_1(\bar{x})\}^T \left(\begin{Bmatrix} sL^2/2k + f_n L/k + T_b \\ 3sL^2/8k + f_n L/2k + T_b \end{Bmatrix}_{e=1}, \begin{Bmatrix} 3sL^2/8k + f_n L/2k + T_b \\ T_b \end{Bmatrix}_{e=2} \right) \end{aligned} \tag{2.44}$$

as is the analytical solution, (2.17), and its $T^N(x)$ interpretation (2.18). However, this discrete FE *approximate* solution T^h cannot be exact (i.e., identical with either), due to absence of a quadratic member in the FE trial space basis. Finally, for "real problems" the values for the data are inserted into (2.44) yielding *numbers* for $\{Q\} = \{Q_1, Q_2, Q_3\}^T$ replacing the data symbols.

The difference between the FE $\{N_1\}$ basis discrete approximate solution T^h and the analytical solution $T(x)$ is the *approximation error* $e^h(x)$, replacing superscript N by h in (2.5). This error, a *distribution*, is graphed in Figure 2.4 for the source s positive and negative. The GWS^h theory M = 2 $\{N_1\}$ basis solution generates *nodally exact* temperatures (a suggested

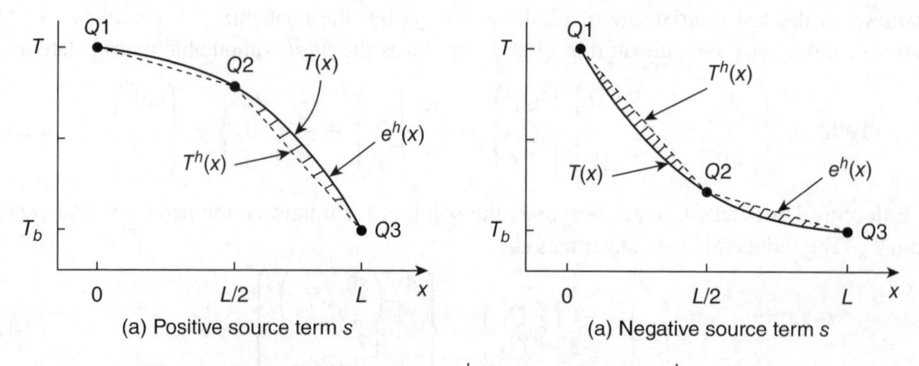

Figure 2.4 Distribution of error $e^h(x)$ associated with $T^h(x)$, M = 2

verification), but observe that $e^h(x)$ will persist non-zero for T^h generated on *any* mesh M using the linear FE trial space basis!

That this T^h solution is nodally exact validates the assertion that weak form theory generates the *optimal* solution possible for the $\{N_1\}$ trial space basis and M = 2 mesh! How could any piecewise continuous linear approximation be closer to $T(x)$? In fact, *all* GWS^h solutions generated for any M in (2.34) will produce exact nodal temperatures with the associated error $e^h(x)$ monotonically decreasing in a predictable manner (later) with increasing M.

Finally, fixing a boundary temperature T_b always(!) generates the *a priori* unknown heat flux at this location via the Green–Gauss divergence theorem; this was precisely recognized in forming GWS^N and fully illustrated in the M = 2 GWS^h example. As a further measure of *optimality*, F_3 computed from the M = 2 GWS^h matrix statement (2.41) is in *exact agreement* with the analytical solution with confirmation a suggested exercise. Conversely, computing this flux via a Taylor series using Q_2 and Q_3 yields $F_3 = 3s\text{L}/4 + f_n$ which is comparatively inaccurate due to the truncation error associated with this low-order TS.

In summary, this example has clearly illustrated each step in the process WF ⇒ WS ⇒ $\text{GWS}^N \Rightarrow \text{GWS}^h$, hence clear interpretation of transition from the global span trial space $\Psi_\alpha(x)$ to the FE *trial space basis* $\{N_1(\bar{x}/l_e)\}$. While very elementary, it nonetheless provides solid confirmation that weak form discrete approximate solutions $T^h(x)$ are the *absolute best* achievable using $\{N_1\}$ on any mesh M for these *data*. It is equally clear that $T^h(x)$ possesses approximation error $e^h(x)$ for any mesh M when generated using the $\{N_1\}$ basis. These observations hopefully whet the reader's appetite to develop an understanding of the elegant theory underlying weak form methodology.

2.4 PDE Solutions, Classical Concepts

With the weak form elementary example completed, the underlying theoretical robustness is further illustrated by comparison with the classical PDE solution process, a sophomore calculus topic. Alter (1.18) to unsteady 1-D conduction with constant properties, fixed temperature BCs and an initial condition (IC). This D*E* PDE + BCs + IC system for state variable $T(x, t)$ is

$$\text{D}E\text{: } \mathcal{L}(T) = \frac{\partial T}{\partial t} - \frac{\partial}{\partial x}\left(\kappa\frac{\partial T}{\partial x}\right) = 0, \quad \text{on a} < x < \text{b for } t > t_0 \tag{2.45}$$

$$\text{BCs: } T(x = \text{a}, t) = 0 = T(x = b, t) \tag{2.46}$$

$$\text{IC: } T(x, t = t_0) = f(x) \tag{2.47}$$

For this problem, the statement *data* are a, b, κ, T (x = a), T(x = b) and $f(x)$. The classical *separation of variables* (SOV) process, cf. Kreyszig (1967), starts by assuming that time and space are separable, which admits the candidate solution statement

$$T(x, t) = F(x)G(t) \tag{2.48}$$

Substituting (2.48) into (2.45), then dividing through by (2.48), since $T(x, t)$ cannot be zero everywhere, yields

$$\frac{1}{\kappa G}\frac{\text{d}G}{\text{d}t} = \frac{1}{F}\frac{\text{d}^2F}{\text{d}^2x} \equiv \pm\beta^2 \tag{2.49}$$

as non-trivial functions of the independent variables x and t can be equal *iff* they are equal to a constant, denoted $\pm\beta^2$ in (2.49). Thus (2.49) is an ordinary differential equation (ODE) pair. Since $G(t)$ must decay in time, the negative sign for β^2 is required. A simple exercise then verifies that the ODE solutions for (2.49) transition (2.48) to

$$T(x,t) = (\mathrm{A}\cos(\beta x) + \mathrm{B}\sin(\beta x))\exp(-\kappa\beta^2 t) \tag{2.50}$$

where A and B are the (combined) constants of integration.

As observed, algebraic manipulations are simplified by the affine transformation $x \Rightarrow \bar{x}$ with origin at $x = \mathrm{a}$. For $\mathrm{L} \equiv \mathrm{b} - \mathrm{a}$ and imposing the first of the Dirichlet BCs, (2.46), confirms that A must vanish for all time t. Since B cannot also be zero, the second BC requires $\sin(\beta \mathrm{b}) = 0$ for all t. This can be enforced by observing

$$\beta \Rightarrow \beta_n \equiv n\pi/\mathrm{L}, \quad \text{for all integers } n = 0, 1, 2, \ldots, N, \ldots \tag{2.51}$$

meets the vanishing requirement for any n. A coefficient β_n is called an *eigenvalue*, or in English a *principal* value.

That the separation constant in (2.49) is now parameterized by the integer n significantly enlarges the class of solutions represented by (2.48). In fact, Dirichlet BC enforcement has produced what is termed the *characteristic solution*

$$T_n(x,t) = \mathrm{B}\sin(n\pi x/\mathrm{L}))\exp(-\kappa(n\pi x/\mathrm{L})^2 t) \tag{2.52}$$

Then, since (2.45) is linear and homogeneous, the fundamental theorem, Kreyszig (1967), proves that the general solution replacing (2.48) is a linear combination of *all* characteristic solutions

$$T(x,t) \equiv \sum_{n=1}^{\infty} \mathrm{B}_n \sin(n\pi x/\mathrm{L}))\exp(-\kappa(n\pi x/\mathrm{L})^2 t) \equiv \sum_{n=1}^{\infty} \mathrm{B}_n \Psi_n(x,t) \tag{2.53}$$

This constitutes a definition of the SOV solution *trial space* $\Psi_n(x,t)$, parameterized by n rather than α, (2.5). Note that all trial space members are of global span. The coefficient set B_n is determined by enforcing the IC, (2.47). Hence for all n and assuming $t_0 = 0$ for simplicity, (2.53) must satisfy

$$T(x, t=0) \equiv f(x) = \sum_{n=1}^{\infty} \mathrm{B}_n \sin(n\pi x/\mathrm{L}) \tag{2.54}$$

Thus is generated a *single* equation for determination of an infinite number of coefficients B_n! This is readily resolved, however, because the *trial space* $\sin(n\pi x/\mathrm{L})$ contains members that are solutions to a Sturm–Liouville equation (next section). Hence they possess the very desirable mathematical properties of *orthogonality* and *completeness*.

Determination of B_n for all n employs orthogonality, statement generated by substituting (2.54) into a GWS to extremize the "distance" across the equality therein using $\sin(m\pi x/\mathrm{L})$

as the space of test functions. From (2.13) the GWS is

$$\begin{aligned}\mathrm{GWS} &\equiv \int_{\Omega} \Psi_m(x)\left[f(x) - \sum_{n=1}^{\infty} \mathrm{B}_n \sin(n\pi x/\mathrm{L})\right] \mathrm{d}x = 0, \quad 1 \leq m < \infty \\ &= \int_{\mathrm{L}} \left[\sin(m\pi x/\mathrm{L}) f(x) - \sum_{n=1}^{\infty} \mathrm{B}_n \sin(m\pi x/\mathrm{L}) \sin(n\pi x/\mathrm{L})\right] \mathrm{d}x\end{aligned} \tag{2.55}$$

Then, since $\sin(m\pi x/\mathrm{L})$ and $\sin(n\pi x/\mathrm{L})$ are orthogonal on L the integral of their product reduces to $\delta_{nm}(\mathrm{L}/2)$. Recalling that the Kronecker delta δ_{nm} equals unity *only* for $n = m$, the solution to (2.55) for any n is

$$\mathrm{B}_n = \frac{2}{\mathrm{L}} \int_{\mathrm{L}} f(x) \sin(n\pi x/\mathrm{L}) \mathrm{d}x, \quad \text{for } 1 \leq n < \infty \tag{2.56}$$

Thereby, (2.53) with (2.56) is the *analytical solution* to the D*E* statement (2.45–2.47). There exists a practical flaw however, as the series in (2.53) is infinite, an unachievable operation! Thereby, an *approximation* to this exact solution must be sought, generated by truncating the series at some practical N

$$T(x,t) \approx T^N(x,t) \equiv \sum_{n=1}^{N} B_n \sin(n\pi x/\mathrm{L}))\exp(-\kappa(n\pi x/\mathrm{L})^2 t) \tag{2.57}$$

The *completeness* property of the trial space $\Psi_n(x,t) = \sin(n\pi x/\mathrm{L})$ admits series truncation at a specified N with an accuracy guarantee, since the convergence of the infinite series in (2.53) is *monotone*.

You will shortly observe that weak form theory for constructing approximate solutions to CFD PDE systems exactly employs these *analytical* mathematical foundations and operations. The key distinction in generating a discrete approximate solution is that determining the equivalent of coefficient set B_n, (2.53), that is, the approximation *degrees-of-freedom* (DOF) array $\{Q\}$, involves a fully coupled matrix solution process.

At the risk of belaboring the issue, but for completeness, consider the SOV process for the n-D steady D*E* PDE + BCs statement

$$\mathrm{D}E\colon \mathcal{L}(T) = -\nabla \cdot \kappa \nabla T = 0, \quad \text{on}\Omega \subset \Re^n \tag{2.58}$$

$$\mathrm{BCs}\colon T(\mathbf{x}_b) = 0, \quad \text{on } \partial\Omega_{x,y} \subset \Re^{n=1} \tag{2.59}$$

$$T(x,y,z=c) = f(x,y), \quad \text{on } \partial\Omega_z \subset \Re^{n=1} \tag{2.60}$$

The solution domain Ω is the interior of a rectangular hexahedron with boundary $\partial\Omega$, the coordinate planes $0 = x = \mathrm{a}$, $0 = y = \mathrm{b}$ and $0 = z = \mathrm{c}$. The temperature is set to zero on all planes in (2.59) except the upper z plane where it varies as $f(x,y)$, (2.60).

As always, the SOV process starts with the assumption

$$T(x,y,z) = F(x)G(y)H(z) \tag{2.61}$$

Proceeding through the algebraic exercise similar to (2.49–2.51), and applying all homogeneous BCs (2.59), produces the *characteristic* solution

$$T_{n,m}(x, y, z) = \sin(n\pi x/\mathrm{a})\sin(m\pi y/\mathrm{b})\sinh(\beta_{n,m}z) \tag{2.62}$$

parameterized for all integers m, $n > 0$ with *eigenvalues*

$$\beta_{n,m} \equiv \pi\sqrt{(n/\mathrm{a})^2 + (m/\mathrm{b})^2} \tag{2.63}$$

Thus from the fundamental theorem the general solution is

$$T(x, y, z) \equiv \sum_{n=1}^{\infty}\sum_{m=1}^{\infty} \mathrm{B}_{n,m}T_{n,m} \tag{2.64}$$

Determination of the expansion coefficient set $\mathrm{B}_{n,m}$ accrues to equating (2.64) to the non-homogeneous BC (2.60), then writing a GWS using $\Phi_{p,q}(x,y) = \sin(p\pi x/\mathrm{a})\sin(q\pi y/\mathrm{b})$ as the space of test functions. Analogous to (2.55) orthogonality of the Galerkin trial-test space selection leads to the coefficient solution

$$\mathrm{B}_{n,m} = \frac{4}{\mathrm{ab}(\sinh(\beta_{n,m}\mathrm{c}))}\int_x\int_y f(x, y)\sin(n\pi x/\mathrm{a})\sin(m\pi y/\mathrm{b})\mathrm{d}x\mathrm{d}y \tag{2.65}$$
$$\text{for } 1 \le n, m < \infty$$

The doubly infinite sum in (2.64–2.65) is impractical. Completeness of the trial space members $\sin(n\pi x/\mathrm{a})\sin(m\pi y/\mathrm{b})$ admits truncation of (2.64) at limits N, M with guaranteed accuracy.

2.5 The Sturm–Liouville Equation, Orthogonality, Completeness

The analytical SOV solution process relies totally on the mathematical properties of the solution trial space, as amply illustrated. This theoretical elegance warrants further exposition to illustrate weak form GWS theory connections available to support CFD fluid/thermal problem classes examined in this text.

The second-order ODEs generated in these SOV illustrations are but special forms of the *Sturm–Liouville* (SL) ODE system

$$\mathcal{L}(u) = \frac{\mathrm{d}}{\mathrm{d}x}\left(p(x)\frac{\mathrm{d}u}{\mathrm{d}x}\right) + (q(x) + \lambda r(x))u = 0, \text{on } \Omega \tag{2.66}$$

$$l(u) = a_{i1}\frac{\mathrm{d}u}{\mathrm{d}x} + a_{i2}u = 0, \quad \text{on } \partial\Omega_i, i = 1, 2 \tag{2.67}$$

In (2.66), the state variable is $u(x)$, the data $p(x) > 0$ and $r(x) \ge 0$ are assumed everywhere continuous, λ is a known parameter and the a_{ij} coefficients in (2.67) are also given data. The characteristic solutions of SL systems are well known, such as sines, cosines, Bessel functions, Legendre polynomials and spherical harmonics, Kreyszig (1967).

The fundamental solutions to SL ODE systems are called *eigenfunctions*, in English "principal functions," which universally exhibit the two unique and highly useful properties of *orthogonality* and *completeness*, as illustrated. Weak form theory embraces these attributes in organizing the solution approximation, to ensure error minimization and guaranteed convergence to an accurate solution because the trial space is either mathematically enriched or increased in size.

A brief formalizing of these solution properties is thereby appropriate. A differential operator $\mathcal{L}(\cdot)$ is defined *self-adjoint*, provided that, for any pair of functions $u(x)$ and $w(x)$ that are solutions to $\mathcal{L}(\cdot)$, the following identity holds

$$\int_{x_1}^{x_2} [w\mathcal{L}(u) - u\mathcal{L}(w)]\mathrm{d}x = 0 \tag{2.68}$$

The SL operator $\mathcal{L}(\cdot)$, (2.66), is indeed self-adjoint, as direct substitution into (2.68) yields

$$\begin{aligned}\int_{x_1}^{x_2} [w\mathcal{L}(u) - u\mathcal{L}(w)]\mathrm{d}x &= \int_{x_1}^{x_2} \left[w\frac{\mathrm{d}}{\mathrm{d}x}\left(p(x)\frac{\mathrm{d}u}{\mathrm{d}x}\right) - u\frac{\mathrm{d}}{\mathrm{d}x}\left(p(x)\frac{\mathrm{d}w}{\mathrm{d}x}\right)\right]\mathrm{d}x \\ &+ \int_{x_1}^{x_2} [w(q(x) + \lambda r(x))u - u(q(x) + \lambda r(x))w]\,\mathrm{d}x\end{aligned} \tag{2.69}$$

which equates to zero via: (1) the last two terms canceling identically, and (2) integrating by parts the first two terms, then substituting the BCs (2.67), produces the remaining cancellation.

As the consequence of being self-adjoint, the integer-parameterized SL system *eigenfunctions* $u_n(x)$ exist for all $n = 0, 1, \ldots, N, \ldots$, and for an infinite set of monotonically increasing *numbers* λ_n, $\lambda_0 < \lambda_1 < \lambda_2 < \ldots < \lambda_N < \ldots$, called *eigenvalues*. Switching notation from β to λ, per mathematical convention, for the illustrated SOV-generated solutions, $\lambda_n = n\pi/L$ and $\lambda_{n,m} \equiv \pi\sqrt{(n/\mathrm{a})^2 + (m/\mathrm{b})^2}$ are the eigenvalues while $u_n = \sin(\lambda_n x)$ and $u_m = \sin(\lambda_m y)$ are the corresponding eigenfunctions, both for all n, m.

As fundamental solutions of an SL ODE system, eigenfunctions exhibit both *orthogonality* and *completeness*. Since any solution $u(x)$ and/or $w(x)$ to (2.66–2.67) is the sum of the characteristic solutions u_n and u_m, then substituting this representation into (2.68) yields

$$(\lambda_n - \lambda_m)\int_{x_1}^{x_2} r(x)u_n(x)u_m(x)\mathrm{d}x = 0 \tag{2.70}$$

Only for $n = m$ does the left side of (2.70) vanish identically. Otherwise, since λ_n and λ_m can never be equal, the integral itself must vanish for all $n \neq m$. Recalling the Kronecker delta definition, (2.70) states

$$\int_{x_1}^{x_2} r(x)u_n(x)u_m(x)\mathrm{d}x = a_n\delta_{nm} \tag{2.71}$$

with the constant a_n, the value taken by the integral in (2.70) for $n = m$.

The formal mathematical statement of *orthogonality* is (2.71). Specifically, the functions $u_n(x)$ and $u_m(x)$ are defined as *orthogonal on the interval* $x_1 \leq x \leq x_2$, if the integral of their product, with respect to the weight $r(x)$ appearing in the Sturm–Liouville equation, (2.66), vanishes for all $n \neq m$. It is this property that admitted analytical determination of the SOV examples expansion coefficient sets B_n and B_{nm}.

The eigenfunction role appears in *weak form* theory as the trial and test spaces fundamental to WF, (2.5–2.8), complemented by interpolation theory, (2.9), to handle the requirement of test function arbitrariness. Forming the *extremum* converts WF, a number, into the *weak statement* WS, (2.11), which is a set of constraints equal in number to the functions residing in the test space interpolation. Then defining the test and trial space members as identical produces the *Galerkin weak statement* GWS, (2.13), which, referring to (2.70), requires the *error* in the approximate solution $T^N(\mathbf{x})$, (2.1), to be *orthogonal* to its trial space on the span of Ω. This duality is hopefully patently apparent!

The second useful property of the eigenfunctions of an SL equation is *completeness.* A function set $u_n(x)$ is defined as *complete on* $x_1 \leq x \leq x_2$, if for every constant >0, there exists a number $N>0$ such that the difference (*error*) between an arbitrary square-integrable function $f(x)$, and a series expansion in the eigenfunctions $u_n(x)$, with expansion coefficients C_n, can be made arbitrarily small.

The formal statement is

$$\int_{x_1}^{x_2} \left[f(x) - \sum_{n}^{N} C_n u_n(x) \right]^2 \mathrm{d}x < \delta \tag{2.72}$$

for some $N>0$. Via (2.71), (2.72) is *absolutely guaranteed* by defining evaluation of the expansion coefficient set C_n, $1 \leq n \leq N$, as

$$C_n \equiv \frac{\int_{x_1}^{x_2} f(x) r(x) u_n(x) \mathrm{d}x}{\int_{x_1}^{x_2} u_n(x) u_n(x) \mathrm{d}x} = \frac{1}{a_n} \int_{x_1}^{x_2} f(x) r(x) u_n(x) \mathrm{d}x \tag{2.73}$$

Note that (2.73) is the end result of the GWS operation determining the coefficient sets B_n, (2.56), and B_{nm}, (2.65), in the SOV exposition.

The SOV process produces an *infinite series* to evaluate, for determination of the analytical solution, which can never be completed in practice. However, with the underlying *eigenfunction* support, and the property of *completeness*, (2.72–2.73), an analytical solution *approximation* of known accuracy can be generated by truncating the series at some $n = N$. Importantly, one is guaranteed that any level of accuracy demanded is achievable, by selecting N sufficiently large, provided that the trial space $u_n(x)$ is *complete*.

The Galerkin weak statement thus assimilates both *orthogonality* and *completeness* principles. Orthogonality is intrinsic to the GWS error constraint, (2.13) with (2.5), for *any* trial space $\Psi_\alpha(\mathbf{x})$, $1 \leq \alpha \leq N$. Completeness underlies the available asymptotic error estimates, which predict the rate at which the approximation approaches the exact solution as a function of trial space complexity, and the number N of functions contained therein, (2.5).

In CFD practice, the GWS implementation employs a domain discretization, hence uses very elementary polynomial FE trial space basis functions $\{N_k(\mathbf{x})\}$. Orthogonality is not affected, and the theory predicts optimal performance for FE trial space bases complete to polynomial degree k. Further, asymptotic convergence under mesh refinement is dependent on basis completeness degree k *and* mesh size M, which replaces the series limit N for given accuracy.

This brief exposition of Sturm–Liouville and its connection to the classical separation of variables theory hopefully provides a basis to appreciate the elegance *and* practical utility of the mathematical underpinnings of weak form theory. Applications to CFD fluid/thermal problem statements in this text directly benefit from assimilation of this robust mathematical foundation.

2.6 Classical Variational Calculus

For some of those interested in weak form theory for CFD, hence its discrete FE implementation, a cause for uneasiness is the practice of *extremization*. A concise summary of the weak form process is:

> *weak form methodology returns the calculus, vector field theory, and the classical concepts of extremization, orthogonality and completeness, to formulation of* CFD *approximate solution algorithms amenable to digital computing*

Formation of the extremum is intrinsic to the variational calculus as well as weak form practice. Section 2.2 clearly stated that the solution to the D*E* PDE + BCs statement (2.1–2.3) is a *function*, not an array of numbers, distributed over the domain of definition and its boundary. The illustrations to this point confirm that a D*E* solution is of the form $T(x,y,z) = T(\mathbf{x})$. Recalling the basic calculus *delta process*, the test to determine whether a function is stationary involves writing a Taylor series (TS) at some coordinate $\mathbf{x}$, then testing behavior of the function in the near vicinity of $\mathbf{x}$ measured by distance $\Delta\mathbf{x}$. The TS statement is

$$T(\mathbf{x} + \Delta\mathbf{x}) = T(\mathbf{x}) + \nabla T(\mathbf{x}) \bullet \Delta x + O(\Delta\mathbf{x}^2) \tag{2.74}$$

Viewing (2.74), it is obvious that an *extremum*, also called a *stationary point*, of a function exists *if and only if* its first derivative vanishes at $\mathbf{x}$. An extremum may correspond to a local or global minimum or maximum, or a point of inflection, as determined by the sign of the $O(\Delta x^2)$ term in (2.74). Irrespective, existence of an extremum requires

$$\begin{aligned}\nabla T(\mathbf{x}) &\equiv \frac{\partial T(\mathbf{x})}{\partial x_i}\widehat{\mathbf{e}}_i = \mathbf{0} \\ &= \frac{\partial T(\mathbf{x})}{\partial x}\widehat{\mathbf{i}} + \frac{\partial T(\mathbf{x})}{\partial y}\widehat{\mathbf{j}} + \frac{\partial T(\mathbf{x})}{\partial z}\widehat{\mathbf{k}}\end{aligned} \tag{2.75}$$

In (2.74–2.75), boldface denotes a vector, superscript "hat" identifies it as a unit vector, and the last line expands the vector gradient (∇) operation in rectangular cartesian coordinates for illustration.

Consider the analytical 1-D D*E* solution (2.17). Inserting it into (2.75) produces

$$\begin{aligned}\nabla T(x, data) &= \widehat{\mathbf{i}}\,\frac{\partial}{\partial x}\left[\frac{s\mathrm{L}^2}{2k}\left[1 - \left(\frac{x}{\mathrm{L}}\right)^2\right] + \frac{f_n\mathrm{L}}{k}\left(1 - \frac{x}{\mathrm{L}}\right) + T_b\right] \\ &= -\widehat{\mathbf{i}}\left[\frac{sx}{k} + \frac{f_n}{k}\right] \neq \mathbf{0}\end{aligned} \tag{2.76}$$

which cannot vanish unless both the boundary flux f_n and source s are zero. That this function (solution) does not possess a *stationary point* (extremum) is readily confirmed by its graph, Figure 2.4.

In contrast to the extremum of a function, the *variational calculus* seeks determination of the extremum of a function of functions, called a *functional*. For 1-D exposition, as well as pertinence to (2.16–2.18), seek the *functional* $f(\bullet)$ that renders the integral

$$I = \int_{x_1}^{x_2} f(x, T(x), T_x(x))\mathrm{d}x \tag{2.77}$$

an *extremum*, which is a number. Note that I depends one-to-one on $f(\bullet)$, a *functional* of the functions $T(x)$ *and* the first spatial derivative $T_x(x) \equiv \mathrm{d}T/\mathrm{d}x$, as well as the independent variable x.

The search process is formalized by definition of a one-parameter family of test functions $\phi(x)$ that lie arbitrarily close to the sought function $T(x)$, Weinstock (1952), Huebner and Thornton (1982). For a parameter ε, the entire class of eligible *test functions* is

$$\phi(x) \equiv T(x) + \varepsilon\eta(x) \tag{2.78}$$

In (2.78), $\eta(x)$ is absolutely arbitrary, except that $\eta(x_1)$ and $\eta(x_2)$ must vanish if BC (2.18) is given *data* for both $T(x_1)$ and $T(x_2)$. Specifically, $\phi(x)$ cannot differ from $T(x)$ at any location fixed by a Dirichlet BC.

The parameter ε controls the "closeness" of $\phi(x)$ to $T(x)$ for any function $\eta(x)$. Hence, for all $\eta(x)$ there exists an ε such that the (distributed) distance between $\phi(x)$ and $T(x)$ can be made arbitrarily small. Mathematically, and in conceptual agreement with the eigenfunction completeness property, (2.73), this implies

$$|\phi(x) - T(x)| < \delta \tag{2.79}$$

for δ an arbitrarily small constant, and recall $|\bullet|$ denotes magnitude.

With definition (2.78), determining the extremum of I, (2.77), is replaced by seeking the stationary value (extremum) of $I(\varepsilon)$,

$$I(\varepsilon) = \int_{x_1}^{x_2} f(x, \phi(x, \varepsilon), \phi_x(x, \varepsilon))\mathrm{d}x \tag{2.80}$$

with respect to ε for *all* eligible $\eta(x)$. For this definition, the generated integrals are now a function of ε and the corresponding TS is

$$I(\varepsilon) = I(\varepsilon = 0) + \left.\frac{\mathrm{d}I}{\mathrm{d}\varepsilon}\right|_{\varepsilon=0} \Delta\varepsilon + \left.\frac{1}{2}\frac{\mathrm{d}^2 I}{\mathrm{d}\varepsilon^2}\right|_{\varepsilon=0} \Delta\varepsilon^2 + O(\Delta\varepsilon^3) \tag{2.81}$$

The *necessary condition* for $I(\varepsilon)$ to be stationary is the first derivative in (2.81) vanishing at $\varepsilon = 0$, whereupon (2.78) states that $\phi(x)$ is identical with $T(x)$, the sought solution.

Substituting (2.78) into (2.77) and using the chain rule for differentiation

$$\begin{aligned}\frac{\mathrm{d}I}{\mathrm{d}\varepsilon} &= \frac{\mathrm{d}}{\mathrm{d}\varepsilon}\int_{x_1}^{x_2} f(x, \phi(x,\varepsilon), \phi_x(x,\varepsilon))\mathrm{d}x \\ &= \int_{x_1}^{x_2}\left[\frac{\partial f}{\partial \phi}\frac{\partial \phi}{\partial \varepsilon} + \frac{\partial f}{\partial \phi_x}\frac{\partial \phi_x}{\partial \varepsilon}\right]\mathrm{d}x = \int_{x_1}^{x_2}\left[\frac{\partial f}{\partial \phi}\eta(x) + \frac{\partial f}{\partial \phi_x}\frac{\mathrm{d}\eta}{\mathrm{d}x}\right]\mathrm{d}x \\ &= \int_{x_1}^{x_2}\left[\frac{\partial f}{\partial \phi} - \frac{\mathrm{d}}{\mathrm{d}x}\left(\frac{\partial f}{\partial \phi_x}\right)\right]\eta(x)\mathrm{d}x + \eta(x)\frac{\partial f}{\partial \phi_x}\bigg|_{x_1}^{x_2}\end{aligned} \tag{2.82}$$

The last line in (2.82) results from integration by parts, and the generated second term vanishes identically since $\eta(x_1) \equiv 0 \equiv \eta(x_2)$ in meeting the Dirichlet BC requirement. Since $\eta(x)$ remains completely arbitrary, (2.82) can be rendered zero in general only for the *integrand* vanishing identically. Enforcing the constraint $\varepsilon = 0$ then simply amounts to replacing $\phi(x)$ with $T(x)$, whereupon the extremum of I, (2.77), occurs for the *functional* $f(\bullet)$ satisfying the partial differential equation

$$\frac{\partial f}{\partial T} - \frac{\mathrm{d}}{\mathrm{d}x}\left(\frac{\partial f}{\partial T_x}\right) = 0 \tag{2.83}$$

Equation (2.83) is the famous *Euler–Lagrange equation* of the variational calculus. It is the necessary condition for I, (2.77), to possess an extremum. Should $f(\bullet) = f(x)$ only, (2.77) reduces to an elementary integration, hence (2.83) is identically zero and no extremum occurs. For $f(\bullet) = f(x, T_x)$ only, then $\partial f/\partial T = 0$ and (2.83) integrated once yields

$$\frac{\partial f}{\partial T_x} = \mathrm{C}_1 \tag{2.84}$$

for C_1 some constant. For the further restriction $f(\bullet) = f(T_x)$, (2.83) predicts that $f(\bullet)$ is linear in T_x. Integrating once produces the solution $T(x) = \mathrm{C}_2 x + \mathrm{C}_3$ as the function that renders I stationary for $f(\bullet) = f(T_x)$.

The connection of interest is between the D*E* conservation principle (2.16) and the Euler–Lagrange equation (2.83). Factually, (2.16) corresponds exactly to (2.83) for a functional $f(\bullet)$ which is readily determined. Equating (2.83) and (2.16) produces

$$\frac{\partial f}{\partial T} - \frac{\mathrm{d}}{\mathrm{d}x}\left(\frac{\partial f}{\partial T_x}\right) = -s - \frac{\mathrm{d}}{\mathrm{d}x}\left(k\frac{\mathrm{d}T}{\mathrm{d}x}\right) \tag{2.85}$$

exchanging the order of terms in D*E*. Since T and T_x are *independent* functions in the variational calculus, integrating (2.85) by terms leads to

$$\frac{\partial f}{\partial T} = -s \Rightarrow f(T) = -sT + g_1(T_x) \tag{2.86}$$

$$\frac{\partial f}{\partial T_x} = k\frac{\mathrm{d}T}{\mathrm{d}x} \Rightarrow f(T_x) = \frac{1}{2}k\left(\frac{\mathrm{d}T}{\mathrm{d}x}\right)^2 + g_2(T) \tag{2.87}$$

where $g_1(\bullet)$ and $g_2(\bullet)$ are at most functions of the indicated arguments. Since the two right side terms must sum to zero, via (2.83), it must be that $g_1(T_x) = 1/2kTx^2$ and $g_2(T) = -sT$.

Thus, the functional $f(\bullet)$ extremizing the variational principle I, (2.77), that is, satisfies the Euler–Lagrange equation (2.83), leading identically to the DE principle in 1-D is

$$f(T, T_x) = \frac{1}{2}kT_x^2 - sT \tag{2.88}$$

Thereby, knowledge of a conservation principle in differential equation form, along with the Euler–Lagrange equation connection to the extremum, enables the inverse process of finding the functional $f(\bullet)$ that correlates the two expressions. Of primary significance, this duality leads to a cogent understanding of integral *norms* being the measure for weak form error estimation and prediction of optimal mesh solutions.

2.7 Variational Calculus, Weak Form Duality

Extension of variational calculus theory to linear, second-order PDE + BCs conservation statements in n dimensions is direct. Continuing DE, (2.1–2.3), as the example, (2.77) generalized to n dimensions and *all* admissable BCs is

$$I = \int_\Omega f(\mathbf{x}, T(\mathbf{x}), \nabla T(\mathbf{x}))d\tau + \lambda \int_{\partial\Omega_n} g(\mathbf{x}, T(\mathbf{x}))d\sigma \tag{2.89}$$

where $d\tau$ and $d\sigma$ are differential elements on Ω and its boundary $\partial\Omega$, and λ is the *Lagrange multiplier* for implementing *natural* (normal derivative) BC constraints in the extremization. The arbitrary nearby function is $\eta(\mathbf{x})$, and proceeding through the illustrated ε-parametric process confirms that the resultant *Euler–Lagrange equation system* is

$$\frac{\partial f}{\partial T} - \nabla \bullet \left(\frac{\partial f}{\partial(\nabla T)}\right) = 0, \quad \text{on } \Omega \tag{2.90}$$

$$\lambda\frac{\partial g}{\partial T} - \frac{\partial f}{\partial(\nabla T)} \bullet \hat{\mathbf{n}} = 0, \quad \text{on } \partial\Omega_n \tag{2.91}$$

Equation (2.90) holds on Ω, a region in n-dimensional euclidean space $\Re^n$, while (2.91) holds on a *natural* BC surface $\partial\Omega_n$ lying on $\Re^{n-1}$ characterized by outwards-pointing unit normal vector $\hat{\mathbf{n}}$.

For the DE system (2.1–2.2), the functionals $f(\bullet)$ and $g(\bullet)$ that reproduce this PDE + BC statement, via (2.90–2.91), are

$$f(T, \nabla T, data) = \frac{1}{2}\kappa\nabla T \bullet \nabla T - Ts/\rho_0 c_p \tag{2.92}$$

$$g(T, data) = \frac{1}{2}hT^2 - hTT_r - f_n T \tag{2.93}$$

for the definition $\lambda \equiv \kappa/k$. Therefore, the variational principle extremum that reproduces (2.1–2.2) to within the Dirichlet BC (2.3) is

$$
\begin{aligned}
I(T,\nabla T, data) = & \int_{\Omega}\left[\frac{1}{2}\kappa\nabla T \bullet \nabla T - Ts/\rho_0 c_p\right]\mathrm{d}\tau \\
& + \frac{\kappa}{k}\int_{\partial\Omega_n}\left[\frac{1}{2}hT^2 - hTT_r - Tf_n\right]\mathrm{d}\sigma
\end{aligned}
\tag{2.94}
$$

The variational–weak form duality emerges upon approximating the solution to D*E* via (2.94), rather than using GWS^N, (2.15), for the PDE + BCs form (2.1–2.3). Recalling (2.5) as the generic statement of approximation, T^N substituted into (2.94) produces

$$
\begin{aligned}
I \approx I^N \equiv & \int_{\Omega}\left[\frac{1}{2}\kappa\nabla T^N \bullet \nabla T^N - T^N s/\rho_0 c_p\right]\mathrm{d}\tau \\
& + \frac{\kappa}{k}\int_{\partial\Omega_n}\left[\frac{1}{2}hT^{N^2} - hT^N T_r - T^N f_n\right]\mathrm{d}\sigma \\
= & \sum_{\alpha}^{N}\sum_{\beta}^{N} Q_\beta\left[\begin{array}{l}\int_{\Omega}\left[\frac{1}{2}\kappa\nabla\Psi_\beta \bullet \nabla\Psi_\alpha Q_\alpha - \Psi_\beta(s/\rho_0 c_p)\right]\mathrm{d}\tau \\ + \frac{\kappa}{k}\int_{\partial\Omega_n}\left[\frac{1}{2}h\Psi_\beta\Psi_\alpha Q_\alpha - \Psi_\beta(hT_r - f_n)\right]\mathrm{d}\sigma\end{array}\right]
\end{aligned}
\tag{2.95}
$$

possessing dual summation indices. Once the trial space Ψ_α, $1 \le \alpha \le N$, is identified, integrating (conceptually) the square bracket terms leads to an (α,β)-size set of numbers premultiplied by Q_β with select Q_α products with data.

Thereby, I^N is a *function only* of the coefficient set (Q_α, Q_β), in clear distinction to I, (2.94), being a *functional* of $T(\mathbf{x})$ and $\nabla T(\mathbf{x})$. Since *only* the extremum of I corresponds to D*E* in PDE + BCs form, then finding the stationary point of (2.95) in (Q_α, Q_β) space is required. This operation, identical to that used to form WS from WF, (2.11), yields

$$
\begin{aligned}
\frac{\partial I^N}{\partial(Q_\alpha, Q_\beta)} \equiv & \frac{\partial}{\partial(Q_\alpha, Q_\beta)}\sum_{\alpha}^{N-D}\sum_{\beta}^{N-D} Q_\beta\left[\begin{array}{l}\int_{\Omega}\left[\frac{1}{2}\kappa\nabla\Psi_\beta \bullet \nabla\Psi_\alpha Q_\alpha - \Psi_\beta(s/\rho_0 c_p)\mathrm{d}\tau\right] \\ + \frac{\kappa}{k}\int_{\partial\Omega_n}\left[\frac{1}{2}h\Psi_\beta\Psi_\alpha Q_\alpha - \Psi_\beta(hT_r - f_n)\right]\mathrm{d}\sigma\end{array}\right] \\
= & \sum_{\alpha}^{N-D}\left[\begin{array}{l}\int_{\Omega}\left[\kappa\nabla\Psi_\beta \bullet \nabla\Psi_\alpha Q_\alpha - \Psi_\beta(s/\rho_0 c_p)\right]\mathrm{d}\tau \\ + \frac{\kappa}{k}\int_{\partial\Omega_n}\left[h\Psi_\beta\Psi_\alpha Q_\alpha - \Psi_\beta(hT_r - f_n)\right]\mathrm{d}\sigma\end{array}\right]
\end{aligned}
\tag{2.96}
$$

The key constraint on forming (2.96) is that *only* those coefficients in (Q_α, Q_β) space not associated with approximation DOF on Dirichlet BC segments $\partial\Omega_d$, recall (2.15), are eligible for "variation." This constraint is identical with that placed on the *functions* $\eta(x)$ and $\eta(\mathbf{x})$ in the classical development, and is identified in (2.96) by the summation limit $N-D$, where "*D*" symbolizes the members in (Q_α, Q_β) fixed by *D*irichlet BCs. Additionally, note the disappearance of the ½ multipliers on the quadratic (Q_α, Q_β) terms in (2.95), which being symmetric admit term transposition to the form given in (2.96).

The final step is to compare (2.96) with the GWS^N statement (2.15) for DE in PDE + BCs form. Substituting the generic statement of approximation (2.5), (2.95) becomes

$$
\begin{aligned}
\frac{\partial I^N}{\partial(Q_\alpha, Q_\beta)} &= \int_\Omega \nabla\Psi_\beta \bullet \kappa\nabla T^N \mathrm{d}\tau - \int_\Omega \Psi_\beta(s/\rho_0 c_p)\mathrm{d}\tau \\
&\quad + \frac{\kappa}{k}\int_{\partial\Omega_n \cap \partial\Omega} \Psi_\beta\left[h(T^N - T_r) + f_n\right] \bullet \hat{\mathbf{n}}\mathrm{d}\sigma \\
&= \{0\}, \quad \text{for } 1 \le \beta \le N
\end{aligned}
\tag{2.97}
$$

which is *identical* to (2.15) to within: (1) omission of the boundary integral exposing the unknown heat flux present on *all* Dirichlet BC surfaces, and (2) the Lagrange multiplier κ/k appearing external to the surface integral integrand, which implies that these data are constants.

Thereby, the GWS^N construction is formally identical to the extremum of the approximate evaluation of the variational principle, I^N, for the linear restrictions associated with variational theory. Such linearity restrictions *are absent* in the nonlinear GWS^N formulations required for CFD algorithm development!

2.8 Quadratic Forms, Norms, Error Estimation

In summary, for the approximation definition (2.4), the extremized Galerkin weak form GWS^N*exactly reproduces* the variational equivalent of DE in PDE + BCs form, (2.1–2.2), additionally inserting the precise impact of Dirichlet BCs. Thereby, the elegant variational-founded theories on accuracy, convergence and optimality for linear elliptic boundary value problems are undoubtedly extensible to GWS^N, hence GWS^h constructions in the fluid-thermal sciences. Rest assured this text would not be written if this supposition proved flawed, including extension to the much more challenging nonlinear PDE + BCs + IC for laminar/transitional/turbulent CFD problem statements of interest.

In full distinction to the Taylor series (TS) concept of *order-of-accuracy* error estimation, accuracy/convergence statements generated by the theoreticians employ an integral expression called a *norm*, symbolized as $\|\bullet\|$, encompassing the entire solution. One norm directly extracted from a variational integral includes all terms quadratic in the integrands. Continuing the DE example, the *quadratic form* embedded within $I(T, \nabla T, data)$, (2.94), is

$$
E(T, \nabla T) \equiv \frac{1}{2}\int_\Omega [\kappa\nabla T \bullet \nabla T]\mathrm{d}\tau + \frac{\kappa}{2k}\int_{\partial\Omega_n} \left[hT^2\right]\mathrm{d}\sigma \tag{2.98}
$$

which is called the *energy functional* for the DE conservation principle written in variational form.

When evaluated using a solution approximation T^N, (2.98) is termed the *energy norm* with definition

$$
\|T^N, T^N\|_E \equiv \frac{1}{2}\int_\Omega \left[\kappa\nabla T^N \bullet \nabla T^N\right]\mathrm{d}\tau + \frac{\kappa}{2}\int_{\partial\Omega_n} \left[\frac{h}{k}(T^N)^2\right]\mathrm{d}\sigma \tag{2.99}
$$

Strictly speaking (2.99) defines a *semi-norm*, with precise distinction deferred to the next chapter. The standard symbolic notation for a semi-norm is $|\bullet|$, not used in this text to avoid confusion with magnitude. Additionally, the theoretician's energy norm definition extracts the *data*, that is, κ, k and h, from the integrands in (2.99), and replaces the action by bounds on the range of these data. These details and their impact are deferred to Chapter 3.

With these caveats, substitution of an approximation T^N, (2.5), into (2.99) confirms that the energy norm is not a function but the single *number*

$$\left|\left|T^N\right|\right|_E = \frac{1}{2}\sum_{\alpha}^{N}\sum_{\beta}^{N} Q_\beta \begin{bmatrix} \int_\Omega \left[\kappa\nabla\Psi_\beta \bullet \nabla\Psi_\alpha\right] d\tau \\ +\kappa\int_{\partial\Omega_n}\left[\frac{h}{k}\Psi_\beta\Psi_\alpha\right] d\sigma \end{bmatrix} Q_\alpha \tag{2.100}$$

This fact is emphasized in (2.100) by replacing the dual argument $\left|\left|T^N, T^N\right|\right|_E$ in (2.99) with $\left|\left|T^N\right|\right|_E$. The actual evaluation of (2.100) requires that the T^N coefficient set (Q_α, Q_β) is known. This is not germane, however, since weak form implementations always employ a domain discretization for determination of the DOF coefficient set in a genuine CFD application.

The transition from T^N to T^h is fully illustrated in the D*E* example, Section 2.3. Generalizing to n dimensions and a kth degree FE trial space basis, the GWS^h solution $T^h(\mathbf{x})$ to be computed is

$$T^N(\mathbf{x}) \equiv \sum_{\alpha}^{N} \Psi_\alpha(\mathbf{x})Q_\alpha \Rightarrow T^h(\mathbf{x}) \equiv \cup_e^{\mathrm{M}}\{N_k(\boldsymbol{\eta}(\mathbf{x}))\}^T\{Q\}_e \tag{2.101}$$

recall (2.44). The notation $\{N_k(\boldsymbol{\eta}(\mathbf{x}))\}$ in (2.101) emphasizes that an FE trial space basis is *always* expressed in local coordinates $\boldsymbol{\eta}$, generated by a coordinate transformation from the global system spanning Ω.

Evaluating the energy norm with $T^h(\mathbf{x})$ replaces the double summation over N, (2.100), with a single summation on M, the number of finite elements in the discretization Ω^h. Further, all integrals move from global to local evaluation on the generic domain Ω_e. Thereby, for a GWS^h discrete solution $T^h(\mathbf{x})$, the *energy norm* definition is

$$\left|\left|T^h\right|\right|_E \equiv \frac{1}{2}\sum_{e}^{\mathrm{M}}\{Q\}_e^T \begin{bmatrix} \int_{\Omega_e}\left[\kappa\nabla\{N_k\}^T \bullet \nabla\{N_k\}\right] d\tau \\ +\int_{\partial\Omega_e\cap\partial\Omega_n}\left[\frac{\kappa h}{k}\{N_k\}^T\{N_k\}\right] d\sigma \end{bmatrix}\{Q\}_e \tag{2.102}$$

The $T^h(\mathbf{x})$ solution energy norm (2.102) is central to estimation of *error*, hence approximation accuracy. Recall that the *order-of-accuracy* theory underlying FD schemes is based on TS truncation error, the essential form of which is $error^h \approx O(\Delta x^{\mathrm{p}})$. Here, O signifies *order*, Δx is the size (*measure*) of the FD mesh, and p denotes the exponent of the significant retained term. The TS concept underlying this is thoroughly linear, and produces no prediction on how other factors such as data non-smoothness affect accuracy estimation.

In distinction, the *asymptotic error estimate* for a GWS^h solution is expressed as a bound on the error distributed over the *entire solution domain* Ω^h via the error norm $\left|\left|e^h(\mathbf{x})\right|\right|$. Generalizing the definition of approximation error (2.7) to $e^h(\mathbf{x})$ is obvious, hence also is identification of error specified in the energy norm

$$\left|\left|e^h\right|\right|_E \equiv \|T\|_E - \left|\left|T^h\right|\right|_E \tag{2.103}$$

For the $T^h(\mathbf{x})$ solution definition (2.101), the theoreticians prove, as soon detailed, the rate of error reduction hence *convergence* to the exact solution for a GWS^h algorithm solution for D*E* is

$$\left|\left|e^h\right|\right|_E \leq Ch^{2\gamma}\|\text{data}\|_{\mathrm{L2},\Omega,\partial\Omega}, \quad \gamma = \min(k, r-1) \tag{2.104}$$

In (2.104), C is a constant and h is a *measure* of the "size" of the finite elements Ω_e constituting the mesh Ω^h. The exponent γ determines the convergence rate as the minimum of the FE basis *completeness degree* k and the factor $(r-1)$, a measure of *data non-smoothness* impact.

The actual error magnitude is determined by the L2 norm of these same data over the *entirety* of Ω and $\partial\Omega$. Recall that the L2 norm is simply the integral of the square of the argument, a number. Finally, note that (2.104) predicts that the linear FE ($k=1$) basis GWS^h solution converges as h^2 *only* for sufficiently smooth data, in accord with the FD prediction of second-order accuracy.

The asymptotic error estimate (2.104) enables precise *quantification* of the error in T^h via a *regular* (detailed later) mesh refinement process. For example, upon halving the mesh measure h for each refinement produces the approximate solution sequence

$$\left|\left|T^h\right|\right|_E + \left|\left|e^h\right|\right|_E = \|T\|_E = \left|\left|T^{h/2}\right|\right|_E + \left|\left|e^{h/2}\right|\right|_E = \ldots \tag{2.105}$$

Substituting (2.104) for both error norms in (2.105) and rearranging cancels out C and the data L2 norm, both of which are generally unknown. Thus results the finer mesh solution *error estimate*

$$\left|\left|e^{h/2}\right|\right|_E = \frac{\Delta\left|\left|T^{h/2}\right|\right|_E}{2^{2k}-1}, \quad for\ \Delta\left|\left|T^{h/2}\right|\right|_E \equiv \left|\left|T^{h/2}\right|\right|_E - \left|\left|T^h\right|\right|_E \tag{2.106}$$

For (2.106) to be accurate requires that meshes Ω^h and $\Omega^{h/2}$ be sufficiently refined such that GWS^h solutions T^h*and* $T^{h/2}$ both adhere to (2.104), another detail addressed later. In this instance, (2.106) precisely quantifies *error*, hence *accuracy* of the solution $T^{h/2}$.

2.9 Theory Illustrations for Non-Smooth, Nonlinear Data

Baker (2012, Chapter 4) details an elementary D*E* GWS^h example that fully illustrates these developments and predictions. The steady temperature distribution is sought for convection heat transfer in the fin cylinder geometry characteristic of a small air-cooled engine, Figure 2.5. Assuming axial conduction is negligible, the D*E* solution domain Ω is reduced to an axisymmetric 1-D, fin-pitch cylinder section with geometric discontinuity at the cylinder–fin intersection.

There are no Dirichlet BCs, hence the D*E* statement (2.1–2.2) specialized to radial heat conduction with Robin BCs is

$$\mathcal{L}(T) = -\frac{1}{r}\frac{\mathrm{d}}{\mathrm{d}r}\left[rk\frac{\mathrm{d}T}{\mathrm{d}r}\right] = 0, \quad \text{on } \Omega \subset \Re^1 \tag{2.107}$$

$$\ell(T) = k\frac{\mathrm{d}T}{\mathrm{d}n} + h(T - T_r)_{fin+cyl} = 0, \quad \text{on } \partial\Omega_h \subset \Re^1 \tag{2.108}$$

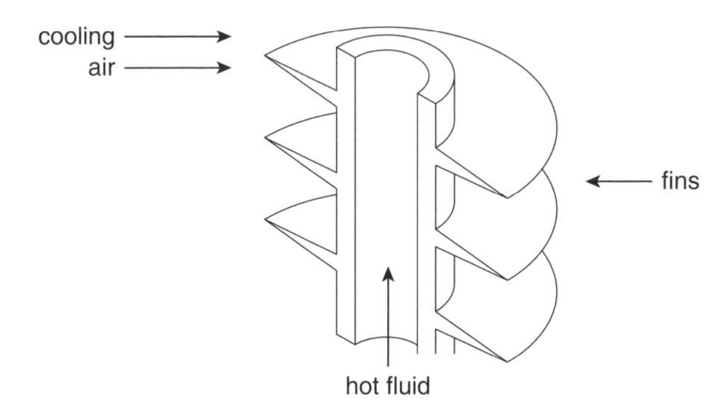

Figure 2.5 Air-cooled gas engine fin cylinder geometry

In (2.107), r is radius with origin at the cylinder axis, while n in (2.108) denotes normal to the cylinder wall as well as both fin surfaces.

To support accurate formulation of the convection heat transfer BCs on the cylinder and both sides of the fin, with unit normal pair $\pm\widehat{\mathbf{k}}$, the GWS^N is formed using a 3-D differential element

$$\mathrm{GWS}^N = \int_{\Omega} \Psi_\beta \mathcal{L}(T^N) r\mathrm{d}r\,\mathrm{d}\theta\,\mathrm{d}z = 0, \forall\beta \tag{2.109}$$

Temperature distributions are sought, predicted for a fin of uniform thickness, then for the *optimal geometry* of fin tapered to a point, Incropera and Dewitt (2004).

These solutions, graphed in Figure 2.6 for uniform $\mathrm{M}=8$ and $\mathrm{M}=128$ linear basis meshings confirm that the solution for uniform thickness fin is visually insensitive to mesh number, while that for the tapered fin does respond to mesh refinement. In both cases note that the computed temperature distribution is noticeably only *piecewise continuous* at the cylinder–fin juncture, the result of the wall section–fin geometry thickness discontinuity.

The mesh refinement study *a posteriori* data enable precise error prediction via (2.106), as all $\mathrm{M}>16$ uniform mesh solutions adhere to (2.104). Figure 2.7 graphs the computed

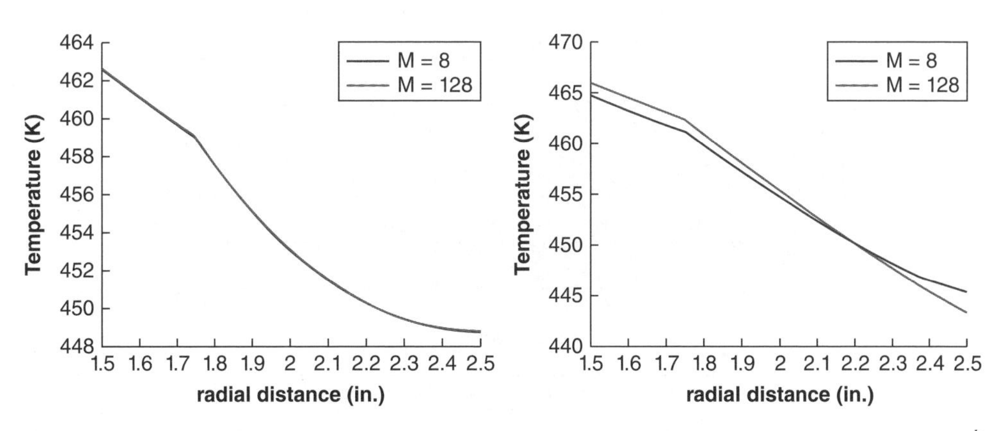

Figure 2.6 Cylinder–fin temperature distributions, uniform mesh refinement *a posteriori* data, GWS^h $k=1$ basis algorithm; left, uniform thickness fin; right, tapered fin

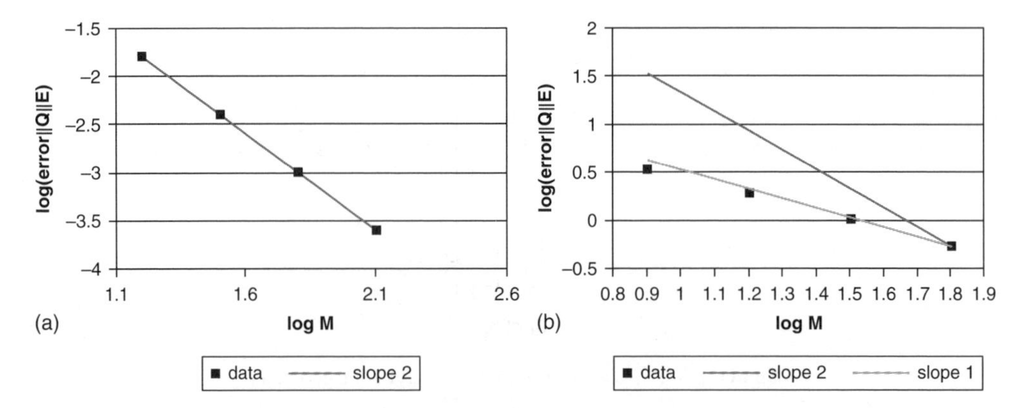

Figure 2.7 Asymptotic convergence, cylinder-fin uniform mesh refinement *a posteriori* data, GWS^h $k=1$ basis algorithm: left, uniform thickness fin; right, tapered fin

asymptotic error estimates which confirm that the uniform thickness fin solution converges at the quadratic rate while that for the tapered fin is about linear. These data *precisely* confirm that FE basis degree $k=1$ domination in γ is replaced by $(r-1)$, due solely to the truly modest geometry alteration from uniform to tapered fin! This occurs despite the TS theory predicting that the FE $k=1$ basis approximate solution algorithm is "second order accurate."

Of course, GWS^h algorithm (2.109) can be implemented using *any* completeness degree k FE basis. Choosing quadratic ($k=2$), the tapered fin asymptotic convergence rate remains about linear, Figure 2.8. Therefore, the theory (2.106) accurately predicts that for *non-smooth data* the $(r-1)$ term will dominate convergence rate for *any* completeness degree k FE basis implementation. This predicts that the compute cost associated with generating $k>1$ basis solutions may not be warranted for genuine nonlinear CFD statements. Conversely, although the convergence rates are suboptimal, the accuracy of a $k>1$ basis solution may be an improvement on a sufficiently refined mesh M, as clearly illustrated in Figure 2.8.

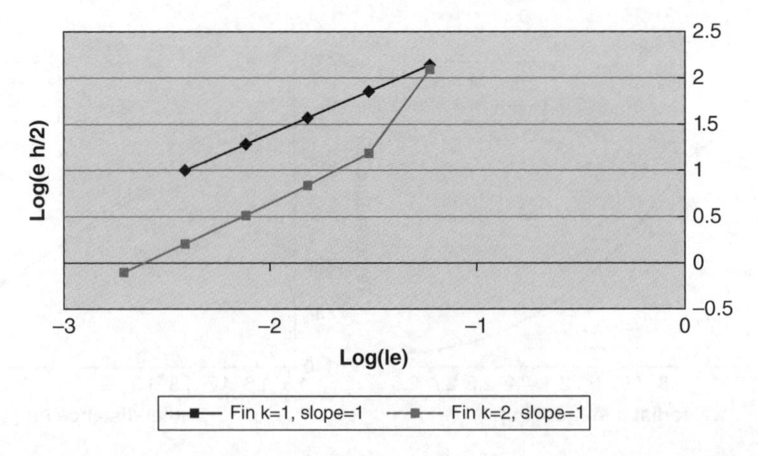

Figure 2.8 Asymptotic convergence, cylinder-tapered fin uniform mesh refinement data, GWS^h $k=1,2$ FE basis algorithms

For genuine multi-dimensional statements the domain boundary $\partial\Omega$ is rarely devoid of geometric irregularities. While TS theory predicts that the FE $k=1$ basis D*E* GWS^h algorithm remains second-order accurate in n-D, the theory (2.106) admits that sub-optimal convergence rates can result. As illustration, Figure 2.9 graphs the steady D*E* FE $k=1$ basis temperature solution distribution on an $n=2$ L-shaped domain. Uniform mesh refinement *a posteriori* data confirm that the convergence rate is ~5/3 rather than quadratic (2).

Bottom line, *non-smooth data* can lead to *suboptimal* convergence rates requiring sufficiently refined meshes M to generate engineering accuracy. Measuring error via (2.104) fully quantifies this issue in clear distinction to TS order-of-accuracy estimates. For D*E* with *nonlinear data*, GWS^h can also generate optimally accurate solutions. As illustration, a DOE-specified legacy 1-D two-phase nonlinear reactor heat transport code validation specification was constant velocity steady D*E* with thermal conductivity linearly dependent on temperature, $k = k(T) = \mathrm{a}T + \mathrm{b}$.

The resultant D*E* FE-implemented GWS^h algorithm is explicitly nonlinear, requiring a matrix iterative solution process, addressed shortly. Surprisingly, all GWS^h FE $k=1$ basis solutions for *any* M ≥ 2 mesh produced *exact* nodal temperatures (the *optimal* solution), this time confirmed by these DOF data being independent of M. Additionally, these GWS^h solutions were predicted to *not converge* under uniform mesh refinement as measured in the *nonlinear* form of energy norm (2.102). Conversely, extracting $\kappa(k(T))$ from the

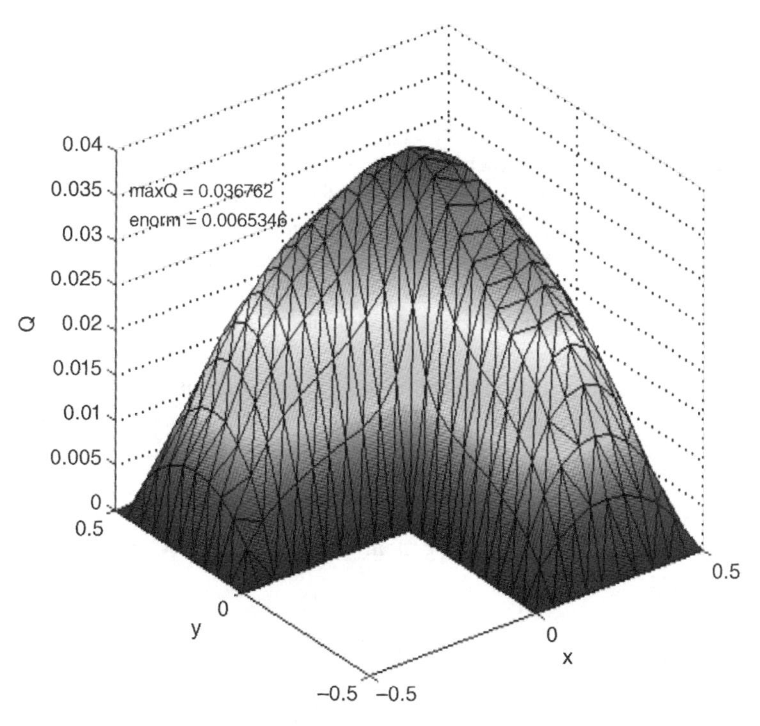

Figure 2.9 Temperature distribution on an L-shape region.

integrands in (2.102) and replacing them by a scalar average multiplier (the theoretician's approach, detailed next chapter) did result in the FE $k=1$ basis solution *a posteriori* data sequence exhibiting quadratic convergence!

To summarize, the impact of *non-smooth* data and explicit *nonlinearity* intrinsic to genuine CFD statements are directly accounted for in weak form theory error estimation, hence GWS^h algorithm accuracy and convergence quantification.

2.10 Matrix Algebra, Notation

The remaining pertinent topics are matrix algebra and equation solving. As illustrated, GWS^h algorithms generate a *solvable* algebraic equation system, in distinction to FD/FV methods expressing the result as a *stencil*. The essence of matrix nomenclature and basic operations are briefly reviewed; detail is available in classic treatises, cf. Kreyszig (1967, Ch.7), Varga (1962).

GWS^h linear statements generate standard matrix expressions while nonlinear statements produce *hypermatrices*, the elements of which are matrices. Conventional notation for a matrix, an $m \times n$ rectangular array of numbers (or functions) that obeys certain rules of manipulation is a capital letter, sometimes enclosed in a square bracket, e.g., [A]. The bracket is employed herein when clarity is required.

The elements of A are denoted a_{ij}, for $1 \leq i \leq m$ and $1 \leq j \leq n$, with lowercase the convention. The *rows* of a matrix are the horizontal arrays, labeled by subscript i, while the *columns* are the vertical arrays keyed to subscript j. Names given to matrix A possessing certain properties include:

$$
\begin{aligned}
\text{real:} &\quad \text{all } a_{ij} \text{ have zero imaginary parts}\\
\text{square:} &\quad m = n\\
\text{symmetric:} &\quad a_{ij} = a_{ji}\\
\text{skew-symmetric:} &\quad a_{ij} = -a_{ji}, \text{hence } a_{ii} = 0\\
\text{triangular (upper, lower):} &\quad a_{ij} = 0 \text{ for } i < j \text{ or } j < i\\
\text{diagonal:} &\quad a_{ij} = 0 \text{ for } i \neq j\\
\text{identity:} &\quad \text{a diagonal matrix with } a_{ii} = 1\\
\text{row matrix:} &\quad a_{ij} = 0 \text{ for } i > 1\\
\text{column matrix:} &\quad a_{ij} = 0 \text{ for } j > 1
\end{aligned}
$$

Matrices obey the familiar rules of arithmetic. The addition of two matrices, denoted by +, defines a new matrix, e.g., $A+B=C$. The elements of the C are formed as

$$c_{ij} = a_{ij} + b_{ij}, \quad \text{for all } i, j \tag{2.110}$$

The product of a matrix with a scalar g defines a new matrix $C=gB$, each element of which is the product

$$c_{ij} = gb_{ij}, \quad \text{for all } i, j \tag{2.111}$$

The difference between two matrices accrues to setting $g=-1$ in (2.111), e.g., $C=A+(-1)B$. The arithmetic operations of addition and subtraction are associative, commutative and distributive, therefore

$$D = (A + B) + C = A + (B + C) = (A + C) + B \tag{2.112}$$

The transpose of matrix A is denoted A^T, or $[A]^T$, with elements

$$a_{ij}^T = a_{ji}, \quad \text{for all } i, j \tag{2.113}$$

The matrix formed by the product of one matrix by another does not use an arithmetic symbol, hence C = AB or [C] = [A][B]. The elements of C are formed by summing products of rows of A with columns of B as

$$c_{ij} = \sum_{k}^{p} a_{ik} b_{kj}, \quad \text{for all } i, j \tag{2.114}$$

where the integer p is the number of columns in A, which must equal the number of rows in B. Subscript notation with brackets adds clarity

$$[C]_{m\times n} = [A]_{m\times p}[B]_{p\times n} \tag{2.115}$$

where subscript $(\cdot) \times (\cdot)$ signifies the number of columns and rows in each matrix. Multiplication is associative and distributive, hence

$$\begin{aligned} (AB)C &= A(BC) \\ A(B + C) &= AB + AC \end{aligned} \tag{2.116}$$

but not commutative

$$AB \neq BA \tag{2.117}$$

A matrix with only one row or one column is denoted by an italic lowercase letter, sometimes enclosed in a bracket $\{\cdot\}$. Herein, the bracket denotes a column matrix, with the companion row matrix using the transpose notation $\{\cdot\}^T$. Matrix products of specific GWS^h utility include

$$\begin{aligned} \text{row} \times \text{square:} &\quad \{a\}_{1\times m}^T [B]_{m\times n} = \{c\}_{1\times n}^T \\ \text{row} \times \text{column:} &\quad \{a\}_{1\times m}^T \{b\}_{m\times 1} = c_{1\times 1} \\ \text{column} \times \text{row:} &\quad \{a\}_{m\times 1}^T \{b\}_{1\times n} = [C]_{m\times n} \\ \text{identity} \times \text{square:} &\quad [I]_{m\times m}[B]_{m\times m} = [B]_{m\times m} \end{aligned} \tag{2.118}$$

Standard operations on a square matrix include forming the inverse and the determinant. The inverse A^{-1} of a square matrix A is the matrix C, with elements c_{ij}, which are equal to the cofactors of A, transposed and divided by the determinant det A. Hence

$$A^{-1} \equiv C = \frac{1}{\det A}[\text{Cofactors of } A]^T \tag{2.119}$$

The scalar $\mathrm{d} \equiv \det \mathrm{A}$ is the *determinant* of an $r \times r$ square matrix A, and is evaluated via expansion by cofactors on any row or column of A, e.g., by columns

$$\mathrm{d} = a_{i1}c_{i1} + a_{i2}c_{i2} + \cdots + a_{ir}c_{ir}, \quad \text{for } 1 \leq i \leq r \tag{2.120}$$

An element c_{ij} of the matrix C is evaluated as the signed minor of the element a_{ij} of A, that is,

$$c_{ij} \equiv \mathrm{d}^{-1}(-1)^{i+j}\det \mathrm{M} \tag{2.121}$$

where the *minor* M is the matrix formed by deleting the ith row and j^{th} column from matrix A. Hence, computing c_{ij} involves r determinants of order $r-1$, which are each evaluated by $r-1$ determinants of order $r-2$, and so on, until one arrives at evaluation of a second-order determinant which is a number.

Quadratic forms play a central role in weak form theory, as introduced. Given the n variables x_i, $1 \leq i \leq n$, hence the column and row matrices $\{x\}$ and $\{x\}^T$, along with the $n \times n$ square symmetric matrix A, a *quadratic form* is the scalar

$$g = g(x) \equiv \frac{1}{2}\{x\}^T_{1\times n}[\mathrm{A}]_{n\times n}\{x\}_{n\times 1} \tag{2.122}$$

In terms of the elements of $\{x\}$ and [A], the *algebraic form* of (2.122) is

$$g(x) \equiv \frac{1}{2}\sum_{i=1}^{n}\sum_{j=1}^{n} a_{ij}x_i x_j \tag{2.123}$$

Generating the extremum of a quadratic form, as with the weak form, requires differentiation by the pertinent array or function. Forming the extremum of (2.122), as assisted by (2.123), for A symmetric yields

$$\begin{aligned}
\frac{\partial g(x)}{\partial x} &\equiv \frac{1}{2}\frac{\partial}{\partial\{x\}}(\{x\}^T_{1\times n}[\mathrm{A}]_{n\times n}\{x\}_{n\times 1}) = \frac{1}{2}\frac{\partial}{\partial x_k}\sum_{i=1}^{n}\sum_{j=1}^{n} a_{ij}\, x_i\, x_j \\
&= \frac{1}{2}\sum_{i=1}^{n} a_{ij}x_i\,\delta_{jk} + \frac{1}{2}\sum_{j=1}^{n} a_{ij}\, x_j\,\delta_{ik} \\
&= \frac{1}{2}\sum_{i=1}^{n} a_{ik}\, x_i + \frac{1}{2}\sum_{j=1}^{n} a_{kj}\, x_j = \frac{1}{2}\sum_{i=1}^{n}(a_{ik} + a_{ki})x_i = \sum_{i=1}^{n} a_{ik}\, x_i \\
&= [\mathrm{A}]_{n\times n}\{x\}_{n\times 1}
\end{aligned} \tag{2.124}$$

2.11 Equation Solving, Linear Algebra

GWS^N theory (2.15), implemented as GWS^h, is an *integral transformation* of a conservation principle PDE + BCs system into a matrix statement, a *solvable form*, recall the DE example, (2.40–2.43), with (2.42) solved via Cramer's rule. Terminology describing properties of matrix algebraic equation systems includes

order (n):	the number of rows or columns of A
submatrix:	a matrix formed by deleting rows and/or columns of A
rank (r):	the order of the largest submatrix of A with non-vanishing determinant
non-singular:	a matrix A for which $r = n$
singular:	a matrix A for which $r < n$

A GWS^h matrix statement in fluid/thermal sciences CFD is explicitly *nonlinear*, hence of the form $[A(x)]\{x\} = \{b\}$. The Cramer's rule solution $\{x\} = [A]^{-1}\{b\}$ is *never* used in practice. Therefore, an absolutely key ingredient of CFD algorithm design and implementation is an *iterative linear algebra* solution procedure, now briefly reviewed.

The GWS^h, always written in homogeneous form (2.13), generates the matrix algebra starting point

$$\mathrm{GWS}^h \Rightarrow f(x) \equiv [A(x)]x - b = 0 \tag{2.125}$$

All linear algebra procedures exchange the Cramer's rule appearance of A^{-1} for a set of simpler operations producing a sequence of *solution approximations* x^p, x^{p-1}, . . . , where p denotes the *iteration index*. The simplest starting point assumes that (2.125) admits extraction of a term linear in x, whereby the GWS^h statement can be expressed as

$$f(x) \equiv [A(x)]x - b \equiv x - g(x) = 0 \tag{2.126}$$

A *linear iteration* is the (2.126) operational sequence

$$x^{p+1} = g(x^p), \quad \text{for } p = 0, 1, 2, 3, \ldots \tag{2.127}$$

starting with $x \equiv x^0$ as the (guessed) initial value to the root (solution) $x = \alpha$; specifically, $f(\alpha) = 0$ *is* the solution to (2.125). The key practice issue is determination that the sequence x^p converges to α for p bounded. *Convergence* of the linear iteration (2.127) is assured, on the interval I of the x axis containing the coordinate $x = \alpha$, provided the initial guess is sufficiently close, i.e., $|x^0 - \alpha| < \varepsilon$, a small number.

The proof is direct. Assuming that $g(x)$ and $g'(x) \equiv \partial g/\partial x$ are continuous on I, then if $|g'(x)| \leq K < 1$ at all coordinates on I, and if x^0 also lies on I, (2.127) will indeed converge to the root $x = \alpha$. Obviously $\alpha = g(\alpha)$, and subtracting this from (2.127) yields

$$x^{p+1} - \alpha = g(x^p) - g(\alpha) = g(\bar{x})(x^p - \alpha) \tag{2.128}$$

via the mean value theorem of calculus with $x^{p+1} \leq \bar{x} \leq x^p$. For $p = 0$, and recalling the triangle inequality

$$|x^1 - \alpha| \leq g'(\bar{x})|x^0 - \alpha| < |x^0 - \alpha| \tag{2.129}$$

since $\bar{x}$ lies in I and $|g(\bar{x})| < 1$ by assumption. Then $|x^1 - \alpha| < \varepsilon$ and x^1 lies in I. By induction then, every x^p lies in I for $|g'(\bar{x})| \leq \mathrm{K} < 1$ for every p, hence x^p converges to α for finite p.

The *rate of convergence* of x^p to α is linear, as easily verified. Substituting the pth iterate error definition $e^p \equiv x^p - \alpha$ into (2.129) states $e^{p+1} = \mathrm{g}'(\bar{x})e^p$, and for the assumption $|\mathrm{g}(\bar{x})| \leq \mathrm{K}$, then

$$|e^{p+1}| \leq \mathrm{K}|e^p|, \quad p = 0, 1, 2, \ldots \tag{2.130}$$

Applying this inequality recursively yields

$$|e^{p+1}| \leq \mathrm{K}|e^p| \leq \mathrm{K}^2|e^{p-1}| \leq \ldots \leq \mathrm{K}^{p+1}|e^0| \tag{2.131}$$

where e^0 is the error in the initial guess. Since $|\mathrm{K}| < 1$ is assumed, then in the limit p approaching a sufficiently large number

$$\lim_{p \to \infty} |e^{p+1}| = 0 \tag{2.132}$$

regardless of the initial guess error e^0, provided the iteration converges, that is, x^0 lies in I.

Then, for this convergent process, from (2.129)

$$\lim_{p \to \infty} \left| \frac{e^{p+1}}{e^p} \right| = \lim_{p \to \infty} |g'(\bar{x})| = g'(\alpha) \tag{2.133}$$

since $x^{p+1} \leq \bar{x} \leq x^p$ and $g'(x)$ is assumed continuous for x in the vicinity *of* $\bar{x}$. Thus, (2.133) clearly states that the error in the $(p+1)^{\text{st}}$ iterate is linearly proportional to the error in the pth iterate with constant of proportionality approximately equal to $g'(\alpha)$. Thereby, (2.133) is proof that (2.127) defines a *linear iteration*.

The class of *stationary iterations* results for $g(x)$ in (2.127), expressible as the matrix product $\mathrm{B}(x^p)x^p$. If the iteration matrix B is independent of x^p, there results the *linear stationary iteration*

$$x^{p+1} = \mathrm{B}x^p + \mathrm{M}b, \quad \text{for}\, p = 0, 1, 2, 3, \ldots \tag{2.134}$$

The family of *point iterative* methods, that is, those requiring no matrix solves for (2.134), centers around the partition of A into

$$\mathrm{A} \equiv \mathrm{L} + \mathrm{D} + \mathrm{U} \tag{2.135}$$

where L, D and U are each square matrices having the same elements as A below the main diagonal (L), on the main diagonal (D) and above it (U). Recalling matrix nomenclature, L and U are lower and upper triangular matrices, while D is diagonal.

The most familiar iteration algorithms belonging to the point linear stationary class are *Picard* (*Jacobi*), *Gauss–Seidel* and *successive over-relaxation* (SOR). The Jacobi form for (2.134) is

$$x^{p+1} = -D^{-1}[L + U]x^p + D^{-1}b \tag{2.136}$$

as obtained for the definitions $B \equiv -D^{-1}[(L+U]$ and $M \equiv D^{-1}$, both of which are very easy to form. Further, no matrix operations of the rank r of A are required, since forming the elements of D^{-1} is the trivial operation $1/a_{ii}$ for $1 \le i \le r$. The detraction to its ultimate simplicity is a ponderously slow rate of convergence, since $|K|$ in (2.130) is near unity.

The *Gauss–Seidel* iteration differs from Jacobi in that each new iterate is used immediately in calculating successive solution estimate entries in x^{p+1}. The Gauss–Seidel form for (2.134) is

$$x^{p+1} = -[L - D]^{-1}Ux^p + [L + D]^{-1}b \tag{2.137}$$

hence $B \equiv -[L+D]^{-1}U$ and $M \equiv [L+D]^{-1}$, which give the appearance of requiring a (non-diagonal) matrix inverse. This is not the case, however, as verified upon casting (2.137) in its *algebraic* form

$$x^{p+1} = -D^{-1}Lx^{p+1} - D^{-1}Ux^p + D^{-1}b \tag{2.138}$$

The key distinction between *Gauss–Seidel* and *Jacobi* iteration is apparent in comparing (2.136) and (2.138), which results in an improved convergence rate. Variations on Gauss–Seidel (GS) iteration are very common in CFD code practice.

The third linear-stationary iteration procedure is named *successive-over-relaxation* (SOR), a variation on Gauss–Seidel involving a relaxation parameter ω. The SOR form of (2.134) is

$$x^{p+1} = -[D + \omega L]^{-1}[(1 - \omega)D - \omega U]x^p + \omega[D + \omega L]^{-1}b \tag{2.139}$$

The useful parameter range is $1 \le \omega \le 2$, and $\omega = 1$ reproduces Gauss–Seidel. Forming $[D + \omega L]^{-1}$ involves operations on a lower triangular matrix only, a computationally easy operation.

The iteration methods belonging to (2.134) are characterized by a linear convergence rate, that is, the error in the $(p+1)^{st}$ iterate (2.134) is a linear multiple of its predecessor. Consequently, it may require several hundred iterations to converge the solution $x^{p+1} \cong \alpha$ to an acceptable tolerance. The alternative is a *Newton* (or Newton–Raphson) iteration algorithm, characterized by a robust *quadratic* convergence rate which accrues at the expense of manipulating matrices of full rank r.

A Newton algorithm is developed starting with a Taylor Series (TS) for (2.125)

$$f(x + \Delta x) \equiv f(x) + \left.\frac{\partial f}{\partial x}\right|_x \Delta x + \frac{1}{2}\left.\frac{\partial^2 f}{\partial x^2}\right|_x \Delta x^2 + \cdots \tag{2.140}$$

For $\Delta x \equiv x^{p+1} - x^p$, (2.140) becomes

$$f(x^{p+1}) = f(x^p) + \left.\frac{\partial f}{\partial x}\right|_{x^p} (x^{p+1} - x^p) + \frac{1}{2}\left.\frac{\partial^2 f}{\partial x^2}\right|_{x^p} (x^{p+1} - x^p)^2 \tag{2.141}$$

Then, for x^{p+1} sufficiently close to the root α, the higher order term can be neglected, and since $f(\alpha)=0$ by definition, the TS final form is

$$x^{p+1} = x^p - \frac{f(x^p)}{f'(x^p)} \tag{2.142}$$

where $f'(x_p)$ denotes the *jacobian*, the derivative matrix $\partial f/\partial x$.

If the sequence of iterates x^{p+1}, $p=0, 1, 2, 3, \ldots$ converges to $x=\alpha$, and if $f''(\alpha)\neq 0$, then clearly $f(\alpha)=0$ and (2.142) defines the solution to (2.125). An estimate of conditions for which the Newton iteration converges is required. Comparing (2.142) to the generic linear iteration statement (2.126) yields

$$g(x) \equiv x - \frac{f(x)}{f'(x)} \tag{2.143}$$

On some interval of the x axis containing α, let I denote the set of coordinates satisfying $|x-\alpha|<\varepsilon$ for some ε. Since $f''(\alpha)\neq 0$, there must exist continuity of $f(x)$ on a subset I_1 of I, such that $f'(x)\neq 0$ for all x on I_1. Then, (2.143) predicts g(x) is continuous on I_1 hence differentiable

$$g'(x) = 1 - \frac{(f')^2 - ff''}{(f')^2} = \frac{f(x)f''(x)}{(f'(x))^2} \tag{2.144}$$

Assuming $f''(x)$ is continuous on I_1, and since $f'(x)\neq 0$, then $g'(x)$ is also continuous on I and $g'(\alpha)=0$. Then, from continuity there must exist some subset of I, call it I_2, on which $|g'(x)|<1$. Then, if I contains coordinates common to both I_1 and I_2, thereupon g(x) and $g'(x)$ are continuous, hence $|g'(x)|<1$. If x^0 is then chosen to lie in I, specifically x^0 is chosen such that

$$|g'(x^0)| = \left|\frac{f(x^0)f''(x^0)}{(f'(x^0))^2}\right| < 1 \tag{2.145}$$

then the *Newton* iteration algorithm (2.142) is *convergent*.

The key feature of a Newton iteration is its *quadratic* rate of convergence, readily proved. Substituting the root $x=\alpha$ into (2.145), and subtracting it from (2.145) written for the pth iteration exactly generates (2.128). Expanding $g(x^p)$ in a TS about $x=\alpha$ yields

$$g(x^p) = g(\alpha) + g'(\alpha)(x^p - \alpha) + \frac{1}{2}g''(\alpha)(x^p - \alpha)^2 + \cdots \tag{2.146}$$

Recalling that $g'(\alpha)=0$, and for the pth iteration error definition $e^p \equiv x^p - \alpha$, (2.128) and (2.146) combine to predict

$$e^{p+1} = g''(\alpha)(e^p)^2 \tag{2.147}$$

which confirms that the error in any Newton iterate is proportional to the square of the error in the previous iterate. Hence, convergence is quadratic for $g''(\bar{x})$ continuous in the vicinity $|\bar{x} - \alpha| < \varepsilon$.

The final topic is *equation solving* via matrix *elimination* processes, universally applied in CFD code practice to implement the GS and SOR point iteration algorithms, (2.138) and (2.139), as well as the Newton iteration algorithm (2.142). Development proceeds by rewriting the GS and SOR algorithms in the matrix form

$$\begin{aligned} [\mathrm{D} + \omega \mathrm{L}]x^{p+1} &\equiv -r^p \\ r^p &\equiv [(1-\omega)\mathrm{D} - \omega \mathrm{U}]x^p - \omega b \end{aligned} \tag{2.148}$$

where r denotes the algorithm *residual*, which progresses towards zero as the iteration converges. The corresponding re-expression of the Newton algorithm (2.142) is

$$\begin{aligned} [\mathrm{JAC}]\delta x^{p+1} &\equiv -f^p \\ f^p &\equiv [\mathrm{A}(x^p)]x^p - b \end{aligned} \tag{2.149}$$

for the definition $\delta x^{p+1} \equiv x^{p+1} - x^p$, and f^p is the pth formation of the GWS^h statement, (2.125), for which the progression to zero is sought.

Considering (2.148) first, denote as d_{ii} and l_{ij} the elements of the diagonal (D) and lower triangle (L) matrix partitions of A respectively. Then, the explicit matrix statement is

$$\begin{bmatrix} d_{11}, & 0, & 0, \ldots\ldots, & 0 \\ \omega l_{21}, & d_{22}, & 0, \ldots\ldots, & 0 \\ \omega l_{31}, & \omega l_{32}, & d_{22}, \ 0, \ldots, & 0 \\ \cdots & & & \\ \omega l_{n1}, & \omega l_{n2}, & \omega l_{n3}, \ldots\ldots, & d_{nn} \end{bmatrix} x^{p+1} = -r^p \tag{2.150}$$

Proceeding by rows in $[\mathrm{D} + \omega \mathrm{L}]$, the equation solving sequence for the unknown array x^{p+1} is patently apparent as

$$\begin{aligned} x_1^{p+1} &= -r_1^p / d_{11} \\ x_2^{p+1} &= -(r_2^p + \omega l_{21} x_1^{p+1}) / d_{22} \\ x_3^{p+1} &= -(r_3^p + \omega l_{31} x_1^{p+1} + \omega l_{32} x_2^{p+1}) / d_{33} \\ &\cdots \\ x_n^{p+1} &= -(r_n^p + \cdots) / d_{nn} \end{aligned} \tag{2.151}$$

Thereby, the inverse contained in the theory statements (2.138–2.139) is trivially replaced.

The equation solving procedure for the Newton algorithm, while not as easy due to [JAC] being a full rank matrix, is nevertheless direct via *Gauss elimination*. This is a sequence of algebraic manipulations that resolves [JAC] into the lower-upper triangular matrix product LU, also termed *LU decomposition*. The Newton algorithm (2.149) becomes

$$\mathrm{LU}\delta x^{p+1} = -f^p \tag{2.152}$$

the resultant solution statement is

$$\delta x^{p+1} = -\mathrm{U}^{-1}\mathrm{L}^{-1}f^m \tag{2.153}$$

accomplished by a forward and back substitution algebraic process.

Forming the LU decomposition involves a progression of row-by-row algebraic manipulations on [JAC]. Denoting the matrix elements as jac_{ij}, the second row in [JAC] is multiplied through by jac_{11}/jac_{21} followed by subtraction of the first row. This replaces jac_{21} by zero and modifies (2.149) to the form

$$\begin{bmatrix} jac_{11}, & jac_{12}, & jac_{13}, \ldots, & jac_{1n} \\ 0, & jac_{22}^m, & jac_{23}^m, \ldots, & jac_{2n}^m \\ jac_{13}, & jac_{23}, & jac_{33}, \ldots, & jac_{3n} \\ \ldots & & & \\ jac_{1n}, & jac_{2n}, & jac_{3n}, \ldots, & jac_{nn} \end{bmatrix} \delta x^{p+1} = -\begin{Bmatrix} f_1^p \\ f_2^m \\ f_3^p \\ \cdot \\ f_n^p \end{Bmatrix} \tag{2.154}$$

where superscript *m* denotes *m*odified entries. The next step subtracts row one from the third row, after multiplying through by jac_{11}/jac_{13}, which replaces jac_{13} with a zero, and so on. This elimination process is executed $N-1$ times, yielding (2.154) with column one modified to all zeros except in row one.

The Gauss elimination process then moves to the second column in modified (2.153), hence all entries except jac_{22}^m become replaced by zeros with the residual matrix $\{f_i^m\}$ further modified. The endpoint of *Gauss elimination* generates

$$\mathrm{U} = \begin{bmatrix} u_{11}, & u_{12}, & u_{13}, \ldots, & u_{1n} \\ 0, & u_{22}, & u_{23}, \ldots, & u_{2n} \\ 0, & 0, & u_{33}, \ldots, & u_{3n} \\ \ldots & & & \\ 0, & 0, & 0, \ldots, & u_{nn} \end{bmatrix} \tag{2.155}$$

as required for (2.152).

The formation of L, hence the (2.153) solution operation $\mathrm{L}^{-1}f^m$, is accomplished coincident with the forward elimination process. Then, implementing the U^{-1} solution step to produce δx^{p+1} is direct via *back substitution*, that is, sequential division by the diagonal entries u_{ii}, starting with u_{nn}, and moving from the bottom to the top row of U, (2.155).

Gauss elimination is a standard CFD code operation. Mechanisms exist to numerically destabilize the elimination process such as a small diagonal entry in [JAC]. Should this occur, [JAC] may be row reordered via *pivoting* to move the small entry off the diagonal. For [JAC] a tri-diagonal matrix, Gauss elimination is called the *Thomas algorithm*. Varga (1962) presents full substance and variations on the linear algebra processes discussed.

2.12 Krylov Sparse Matrix Solver Methodology

Starting in the 1990s, sparse matrix solvers began replacing elimination methods for nonlinear matrix equation system solution. The original construction for symmetric positive definite matrices is the *Preconditioned Conjugate Gradient* (PCG) solver, Golub and van Loan (1989). The ensuing extensions to non-symmetric matrices are key to CFD utility, all of which are *Krylov* subspace methods, Freund *et al.* (1991), based on the non-symmetric Lanczos process. A comprehensive presentation of Krylov subspace iterative solvers is given in Barrett *et al.* (1994).

The original non-symmetric matrix solver is GMRES, the acronym for *Generalized Minimal Residuals*, Saad and Schulz (1985). The convergence performance of Krylov solvers applied to $A(x)x = b$ is enhanced by through multiplication by a *preconditioning* matrix. The optimal preconditioner is A^{-1} which of course is never used! The computationally efficient preconditioner is D^{-1}, the coefficients $1/a_{kk}$ of the diagonal of A, (2.135), which alters convergence rate very modestly. The more effective preconditioner is the *incomplete* LU (ILU) factorization of A, where "incomplete" neglects entries distant from D.

In scalar computing environments, ILU-GMRES has supported generation of optimal weak form CFD algorithm *a posteriori* verification/validation data. In parallel computing environments, Wong (1995), weak form code implementation migrated to the parallel-efficient *BiConjugate Gradient Squared Stabilized* (BiCGSTAB) algorithm, van der Vorst (1992). The BiCGSTAB algorithm pseudo-code for iterative solution of $A(x)x = b$, Dias da Cunha and Hopkins (1992), for Q a preconditioning matrix formed from A is

BiCGSTAB Krylov solver

choose x_0

$r_0 = Q(b - Ax_0)$

$s_0 = r_0$

$p_{.0} = v_0 = 0$

$\rho_0 = \alpha_0 = \omega_0 = 1$

for $k = 1, 2, 3, \ldots$

$\rho_k = s_0^T\, r_{k-1}$

$\beta_0 = (\rho_k \alpha_{k-1})/(\rho_{k-1}\omega_{k-1})$

$p_k = r_{k-1} + \beta_k(p_{k-1} - \omega_{k-1} v_{k-1})$

$v_k = QA\, p_k$

$\xi_k = s_0^T v_k$

$\alpha_k = \rho_k/\xi_k$

$c_k = r_{k-1} - \alpha_k v_k$

if $\|c\| < 10^{-16}; x_k = x_{k-1} + \alpha_k p_k$ *stop*

$t_k = QA\, c_k$

$\omega_k = (t_k^T\, ck)/(t_k^T\, t_k)$

$x_k = x_{k-1} + \alpha_k\, p_k + \omega_k\, c_k$

$r_k = c_k - \omega_k\, t_k$

check stopping criteria

End BiCGSTAB

While the ILU preconditioner is more effective, its parallel implementation lacks efficiency, hence $Q = D^{-1}$ is the preconditioning matrix employed for weak form CFD algorithm validation *a posteriori* data herein.

2.13 Summary

This chapter's goal is the introduction of fundamental notions underlying weak form theory, in the process establishing connections to classical mathematical concepts. The energy conservation principle D*E*, assumed familiar to the majority of readers, supports expositions. The conversion of the weak form for D*E*, via *interpolation* and *extremization* to the basic GWS^N, (2.13), then into computational form GWS^h is thoroughly detailed in the M = 2 example. This illustration exposes, step by step, the ingredients of approximation *trial space*, discrete implementation via an FE *trial space basis*, matrix *assembly* to the computable statement, the role of *data* and finally *approximation error*.

Thereupon, a brief excursion into classic separation of variables introduces infinite series truncation, hence the connection to the weak form. This in turn leads to a brief exposure to the Sturm–Liouville equation, hence functions that exhibit *orthogonality* and *completeness*, principles underlying weak form *optimality* and error estimation. This leads naturally to the *variational calculus*, hence connections between energy principle *extremization* and the GWS^N for D*E*. This exposure paved the way to *quadratic forms*, hence the energy norm as a fundamental measure of *approximation error*, hence *accuracy and convergence*. These founding principles are universally illustrated in D*E* examples, including the impact of *non-smooth data* on asymptotic convergence and *nonlinearity* issues with select norms.

Finally, since the discrete FE implementation GWS^h of GWS^N produces the *computable form*, an algebraic matrix statement, the remaining material addresses basic notions in matrix algebra and linear algebra leading to a brief exposure to both legacy and sparse matrix equation solving procedures. Hopefully, this chapter serves the intended purpose, in generating a sense of familiarity with concepts, terminology and methodology fundamental to *optimal performance* continuous Galerkin weak form CFD algorithms derived, implemented and *validated* herein.

Exercises

2.3.1 Verify the exact solution (2.17) to D*E*, (2.14–2.16).

2.3.2 Confirm that the Ψ_α and Q_α given in (2.18) agree with (2.17).

2.3.3 Verify that the Green–Gauss divergence theorem forming (2.13), reduces to integration-by-parts in forming (2.19).

2.3.4 Starting with (2.25), confirm the functional form of the FE linear trial space basis $\{N_1(\bar{x})\}$, (2.27).

2.3.5 Verify the GWS^h M = 2 matrix operations leading to (2.34).

2.3.6 Confirm the M = 2 $\{WS\}_e$ matrices (2.35–2.36), hence their assembly into the GWS^h final form (2.38–2.39).

2.3.7 Verify the algebraic solution (2.42–2.43).

2.3.8 Verify the M = 2 GWS^h boundary flux solution $F_3 = sL + f_n$, and the FD approximation $F_3 = 3sL/4 + f_n$.

2.3.9 Generate a linear basis uniform M = 3 mesh GWS^h solution; hence confirm that it generates exact nodal temperatures and the exact heat flux at the fixed temperature node.

2.4.1 Assume the opposite sign for the separation constant β in (2.49), hence verify that the given BCs cannot be satisfied by (2.48) in a non-trivial manner.

2.4.2 Consider a rod of length L, initially at uniform temperature T1. At $t=0$, the end-point temperature $T(\mathrm{L},t)$ is raised to T2 and held constant. Find the transient and steady-state temperature distributions for $T(0,t)=\mathrm{T1}$ held fixed. (Hint: solve for the difference between the steady-state solution and the unsteady temperature evolution, which properly simplifies the BCs.)

2.4.3 Via the SOV process, express the general solution $\phi(x,y)$ to the 2-D laplacian PDE $\partial^2\phi/\partial x^2+\partial^2\phi/\partial y^2=0$ on the interior of the domain $0\le x\le \mathrm{a}$ and $0\le y\le \mathrm{b}$. Then determine the particular solution for the BCs $\phi(0,y)=0$, $\partial\phi(x,0)/\partial y=0=\partial\phi(x,\mathrm{b})/\partial y$ and $\phi(\mathrm{a},y)=\cos(3\pi y/\mathrm{b})$.

2.4.4 Find the solution to the Laplace PDE $-\nabla_2\phi=0$ on the interior of a right circular cylinder of length L, centered on and including the z axis of radius $r=\mathrm{a}$, for the following BCs:

$$
\begin{aligned}
&\text{a)}\quad \phi(r,\theta,0)=0=\phi(\mathrm{a},\theta,z),\phi(r,\theta,\mathrm{L})=\mathrm{V}(r), && \textit{easy}\\
&\text{b)}\quad \phi(r,\theta,0)=0=\phi(\mathrm{a},\theta,z),\phi(r,\theta,\mathrm{L})=\mathrm{V}(r,\theta), && \textit{harder}\\
&\text{c)}\quad \phi(r,\theta,0)=0=\phi(r,\theta,\mathrm{L}),\phi(\mathrm{a},\theta,z)=\mathrm{V}(\theta,z), && \textit{hardest}
\end{aligned}
$$

where $\mathrm{V}(\cdot)$ is some specified potential distribution.

2.5.1 Starting with (2.68), complete the verification that (2.66) is a self-adjoint operator.

2.5.2 Verify that definition (2.73) satisfies the requirement expressed in (2.72).

2.6.1 Verify that (2.82) evaluated at $\varepsilon=0$ yields (2.83).

2.6.2 For $f(\bullet)=f(T_x)$ only, verify that $T(x)=\mathrm{C}_2x+\mathrm{C}_3$ is the function extremizing (2.77).

2.6.3 Determine the functional $f(\bullet)$ such that the extremum of (2.77) corresponds to the SOV process Sturm–Liouville equation $\mathrm{d}^2F/\mathrm{d}x^2+\beta^2F=0$, recall (2.49).

2.7.1 Verify that analytical extremization of the DE variational functional (2.94) produces (2.1) on Ω and (2.2) on $\partial\Omega_n$.

2.7.2 Proceed formally through the differentiation process, hence confirm that the extremum of I^N is correctly expressed in (2.96)

2.8.1 Substitute the generic GWS^h solution (2.5) into (2.99), hence convince yourself that the energy norm of the approximate solution (2.100) is a function, not a functional.

2.8.2 From the definition of the GWS^h discrete solution T^h, (2.101), verify that (2.102) correctly expresses formation of the associated energy norm.

2.8.3 Substitute the asymptotic error estimate (2.104) into the uniform mesh refinement statement (2.105), hence verify that (2.106) is an explicit statement of the $T^{h/2}$ solution error.

2.10.1 Given the matrices

$$\mathrm{A}=\begin{bmatrix}2, & 1\\ 1, & 3\end{bmatrix},\quad \mathrm{B}=\begin{bmatrix}-2, & 0\\ 3, & 4\end{bmatrix},\quad \mathrm{C}=\begin{bmatrix}3, & 2, & 0\\ 1, & 0, & 4\end{bmatrix}$$

evaluate the following, if they exist

a) A + B, B + A, b) $(\mathrm{A}+\mathrm{B})^T$, $\mathrm{A}^T+\mathrm{B}^T$, c) AB
d) BC, e) AC^T, f) $\mathrm{AB}^T\mathrm{C}$, g) det A, h) det AB

2.10.2 Given [D], confirm $\mathrm{D}^{-1}\mathrm{D}=[\mathrm{I}]$, $\mathrm{D}=\begin{bmatrix}0, & 0, & 1\\ 0, & 1, & 0\\ 1, & 0, & 0\end{bmatrix}$

2.11.1 Determine the rank r of the matrix D.

$$\mathrm{D}=\begin{bmatrix}6, & 1, & 8, & 3\\ 2, & 3, & 0, & 2\\ 4, & -1, & -8, & -3\end{bmatrix}$$

2.11.2 Given the nonlinear algebraic equation $f(x)=x^2-3x-4=0$, determine the roots α using any of the point linear iteration algorithms, (2.148), and Newton iteration (2.149). Use the initial guesses $x^0=0$ and $x^0=2$. Confirm the convergence rate for each algorithm and interpret results in terms of the convergence proofs, (2.133) and (2.147), hence the associated convergence intervals I.

2.11.3 Solve the following system using Gauss elimination.

$$\begin{bmatrix}1, & -1, & 3, & -3\\ -5, & 2, & -5, & 4\\ -3, & -4, & 7, & -2\\ 2, & 3, & 1, & -11\end{bmatrix}\begin{Bmatrix}w\\ x\\ y\\ z\end{Bmatrix}=\begin{Bmatrix}3\\ -5\\ 7\\ 1\end{Bmatrix}$$

References

Baker, A.J. (2012). *Finite Elements ⇔ Computational Engineering Sciences*, Wiley, England.

Barrett, R., Berry, M., Chan, T., Demmel, J., Donato, J., Dongarra, J., Eijkhout, V., Pozo, R., Romine, C. and van der Vorst, H. (1994). *Templates for the Solution of Linear Systems: Building Blocks for Iterative Methods*, SIAM, Philadelphia PA.

Dias da Cunha, R. and Hopkins, T. (1992). "PIM 1.0: the Parallel Iterative Methods Package for Systems of Linear Equations, User's Guide (Fortran 77 Version)," University of Kent, Canterbury, England.

Freund, R., Golub, G. and Nachtigal, N. (1991). "Iterative Solution of Linear Systems," Tech. Report NA-91-05, Stanford University, Stanford CA.

Golub, G. and vanLoan, C. (1989). *Matrix Computations*, The Johns Hopkins University Press, Baltimore MD.

Huebner, K.H. and Thornton, E.A. (1982). *The Finite Element Method for Engineers*, 2nd Ed., Wiley-Interscience, NY.

Incropera, F.P. and Dewitt, D.P. (2004). *Fundamentals of Heat and Mass Transfer*, Wiley, NY.

Kreyszig, E. (1967). *Advanced Engineering Mathematics*, Wiley, NY.

Saad, Y. and Schulz, M.H. (1985). "GMRES: a generalized minimal residual algorithm for solving non-symmetric linear systems," *SIAM J. Scientific & Statistical Computing*, V 7

van der Vorst, H., (1992). "BiCGSTAB: a Fast and Smoothly Converging Variant of Bi-CG for Solution of Non-symmetric Linear Systems," *SIAM J. Scientific & Statistical Computing*, V.13, pp. 631–644.
Varga, R.S. (1962). *Matrix Iterative Analysis*, Prentice-Hall, NJ.
Weinstock, R. (1952). *The Calculus of Variations*, McGraw-Hill, NY.
Wong, K.L. (1995). "A Parallel Finite Element Algorithm for 3D Incompressible Flow in Velocity-Vorticity Form," PhD dissertation, University of Tennessee.

3

Aerodynamics I: Potential flow, GWS^h theory exposition, transonic flow *m*PDE shock capturing

3.1 Aerodynamics, Weak Interaction

The prediction of aerodynamic flowfields via potential theory was one of the very first applications in CFD. Potential theory, established by the mathematicians in the mid 1800s, describes the irrotational flow of a perfect fluid (inviscid, non-heat conducting) via a linear laplacian PDE. Fluid mechanics texts of past decades included a chapter devoted to the theory, leading to superposition of sources, sinks and doublets with circulation to generate pressure distributions pertinent to aerodynamic-appearing profiles.

About the mid 1960s, potential flow CFD panel methods came into practice, a computer-based implementation of the Green's function integral solution of the linear laplacian. Factually, throughout the 1990s, Boeing employed nonlinear transonic panel code potential flow predictions to optimize forebody–wing juncture geometries for minimum pressure drag of their 7-series aircraft.

The complement to aerodynamic potential theory is boundary layer (BL) theory, a simplification of the Navier–Stokes (NS) PDE system for steady, unidirectional viscous and turbulent flow of a fluid in direct contact with an aerodynamic surface. The coupling of a farfield potential solution with nearfield boundary layer predictions constitutes *weak interaction theory*. Therein, the aerodynamic surface for the potential flow prediction is translated to the distribution of the boundary layer displacement thickness $\delta^*(\mathbf{s})$, an integral parameter computed from the detailed BL solution, with coupling via imposition of the δ^* potential solution pressure distribution onto the boundary layer solution. In the case of wing–body juncture region flow, or for ducted flows, the 3-D generalization of the BL system is termed the parabolic Navier–Stokes (PNS) system.

Development of $\text{GWS}^N \Rightarrow \text{GWS}^h \Rightarrow m\text{GWS}^h$ algorithms for the potential fluid mechanics problem class provides the unique opportunity to thoroughly identify the elegantly formal weak form accuracy/convergence error quantification theory foundations. The next chapter

Optimal MODIFIED CONTINUOUS Galerkin CFD, First Edition. A. J. Baker.

Companion Website: www.wiley.com/go/baker/GalerkinCFD

content on aerodynamic BL-PNS PDE systems for laminar and turbulent flows facilitates completion of weak interaction theory, and *quantitative validation* of pertinence to this weak form linear theory to $\mathrm{GWS}^N \Rightarrow \mathrm{GWS}^h$ algorithm handling of explicit nonlinearity.

3.2 Navier–Stokes Manipulations for Aerodynamics

Computational aerodynamics involves seeking a *solution* to appropriate manipulations of the PDE + BCs + IC systems for the basic conservation principles for mass, momentum and energy.

Recalling Chapter 1, the starting point is

$$\mathrm{D}M: \quad \frac{\partial \rho}{\partial t} + \nabla \cdot \rho \mathbf{u} = 0 \tag{3.1}$$

$$\mathbf{DP}: \quad \frac{\partial \rho \mathbf{u}}{\partial t} + \nabla \cdot \rho \mathbf{u}\mathbf{u} + \nabla p = \rho \mathbf{g} + \nabla \mathbf{T} \tag{3.2}$$

$$\mathrm{D}E: \quad \frac{\partial \rho e}{\partial t} + \nabla \cdot (\rho e + p)\mathbf{u} = s - \nabla \cdot \mathbf{q} \tag{3.3}$$

with *state variable* $\{q(\mathbf{x}, t)\} = \{\rho, \mathbf{u}, p, e\}^T$ containing density, velocity vector, pressure and specific internal energy.

One common assumption in aerodynamics is that the flow is *isentropic*, which implies a perfect fluid without viscosity and thermal conductivity effects and with negligible buoyancy and heat source contributions. The resulting alteration to (3.2) and (3.3)

$$\mathbf{DP}: \frac{\partial \rho \mathbf{u}}{\partial t} + \nabla \cdot \rho \mathbf{u}\mathbf{u} + \nabla p = 0 \tag{3.4}$$

$$\mathrm{D}E: \frac{\partial \rho e}{\partial t} + \nabla \cdot (\rho e + p)\mathbf{u} = 0 \tag{3.5}$$

augmented with (3.1) constitutes the *Euler* PDE system simplification to compressible fluid NS conservation principles.

The assumption of potential flow adds the *irrotational* assumption to isentropic. Mathematically the velocity vector is thereby definable as the gradient of a potential function $\varphi(\mathbf{x})$

$$\mathbf{u} \equiv -\nabla\varphi \tag{3.6}$$

Subtracting (3.1) from (3.4), employing a vector identity for the convection term, then inserting (3.6) and integrating along a stream path in Ω defines the Cauchy–Lagrange integral, the mechanical energy integro-differential equivalent of (3.5)

$$\int_{\Omega} \mathbf{DP} \bullet \mathrm{d}\mathbf{r} = \frac{\partial \varphi}{\partial t} + \frac{1}{2}\nabla\varphi \cdot \nabla\varphi + \int \frac{\mathrm{d}p}{\rho} - \left(\frac{\partial \varphi}{\partial t} + \frac{1}{2}\nabla\varphi \cdot \nabla\varphi\right)_{ref} = 0 \tag{3.7}$$

where $(\cdot)_{ref}$ denotes the reference state.

For a perfect gas, the pressure integral in (3.7) transitions to $(\gamma/(\gamma-1))(T-T_{ref})$. Then choosing as reference state the farfield (subscript ∞) where $(\partial\varphi/\partial t)_{ref}$ vanishes, the NS conservation principles PDE system for isentropic irrotational flow of a perfect gas is

$$\mathrm{D}M\text{:}\ \frac{\partial\rho}{\partial t}+\nabla\cdot\rho\nabla\varphi=0 \tag{3.8}$$

$$\int \mathrm{DP}\text{:}\ \frac{\partial\varphi}{\partial t}+\frac{1}{2}\nabla\varphi\cdot\nabla\varphi+\left(\frac{\gamma T_\infty}{\gamma-1}\right)\left(\frac{\rho}{\rho_\infty}\right)^{\gamma-1}=\frac{\gamma T_{0\infty}}{\gamma-1} \tag{3.9}$$

The state variable for first order in time and space PDE system (3.8–3.9) is $\{q(\mathbf{x},t)\}=\{\rho,\varphi\}^T$. A necessary condition for the flow to be potential is that the farfield is potential. This PDE system is appropriate across the Mach number spectrum to transonic and modest supersonic; however, if a shock is generated the description is valid *only* for a shock sufficiently weak to admit its isentropic approximation.

Simplifications exist that substantially alter the complexity of (3.8–3.9). Assuming steady flow (3.9) reverts to an algebraic equation for density as a function of potential, and (3.8) becomes the elliptic boundary value (EBV) PDE for φ

$$\mathrm{D}M\text{:}\quad -\nabla\cdot\rho\nabla\varphi=0 \tag{3.10}$$

For the further assumption of low subsonic onset, that is, $\mathrm{M}_\infty<0.3$, density variations are negligible and DM transitions to the linear laplacian

$$\mathrm{D}M\text{:}\quad -\nabla^2\varphi=0 \tag{3.11}$$

Finally, for onset flow Mach number Ma near sonic with direction parallel to the x axis, (3.10) can be approximated as

$$\mathrm{D}M\text{:}\quad -\nabla^2\varphi+\mathrm{Ma}_\infty^2\left[1+\mathrm{Ma}_\infty^2\left[(1+\gamma)/\mathrm{U}_\infty\right]\frac{\partial\varphi}{\partial x}\right]\frac{\partial^2\varphi}{\partial x^2}=0 \tag{3.12}$$

which introduces an explicit nonlinearity. The reference Mach number is $\mathrm{Ma}_\infty\equiv(\mathrm{U}_\infty^2/\gamma\mathrm{R}T_\infty)^{1/2}$ for U_∞ onset speed and R the gas constant. Note that (3.12) unaltered is valid only for potential flow devoid of shocks.

The complement to potential flow aerodynamics is *boundary layer* (BL) theory. The fundamental assumption is steady unidirectional flow, Schlichting (1979). Figure 3.1, a

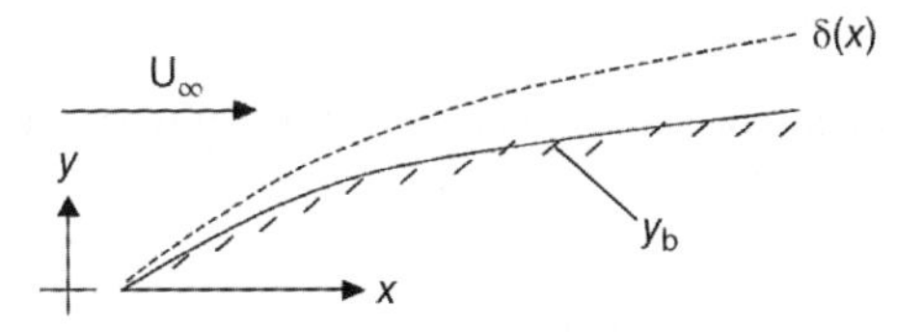

Figure 3.1 Boundary layer flow domain with bounding surfaces

topical section of an airplane wing, graphs the BL domain as the region bounded by the aerodynamic surface and another surface displaced therefrom called the *boundary layer thickness*, $\delta(x)$. For swept wing flow this becomes $\delta(x,z)$, an *a priori* unknown function determined by solution of the BL PDE system. The surface-normal coordinate system is (s,n,z), but since the curvature of s is a second-order effect, Cebeci and Smith (1974), the BL system is usually expressed in the cartesian (x,y,z) system with x the surface coordinate nominally paralleling the dominant flow direction.

Reynolds ordering transforms the initial-boundary value NS PDE system (3.1–3.3) into a *parabolic* PDE system, initial value in x, boundary value on $0 \le y \le \delta$ with z dependence parametric. Deferring details to the next chapter, and restricting to incompressible, the BL PDE system complementary to the potential flow conservation principles PDE systems (3.8–3.12) is

$$\mathrm{D}M\colon \frac{\partial v}{\partial y} + \left(\frac{\partial u}{\partial x} + \frac{\partial w}{\partial z}\right) = 0 \tag{3.13}$$

$$\mathbf{DP} \bullet \widehat{\mathbf{i}} : \mathbf{u} \cdot \nabla u - \frac{1}{\mathrm{Re}}\frac{\partial^2 u}{\partial y^2} + \frac{\partial P^I}{\partial x} + \frac{\mathrm{Gr}}{\mathrm{Re}^2}\Theta\,(\widehat{\mathbf{g}} \bullet \widehat{\mathbf{i}}\,) = 0 \tag{3.14}$$

$$\mathbf{DP} \bullet \widehat{\mathbf{k}} : \mathbf{u} \cdot \nabla w - \frac{1}{\mathrm{Re}}\frac{\partial^2 w}{\partial y^2} + \frac{\partial P^I}{\partial z} = 0 \tag{3.15}$$

$$\mathrm{D}E\colon \mathbf{u} \cdot \nabla\Theta - \frac{1}{\mathrm{RePr}}\frac{\partial^2 \Theta}{\partial y^2} - s_\Theta = 0 \tag{3.16}$$

The *state variable* for the BL PDE system is $\{q(\mathbf{x})\} = \{\mathbf{u},\ \Theta\}^T$ as (kinematic) pressure $P = p/\rho$ now appears *solely* as *data*, denoted by superscript I in (3.14–3.15). This constitutes the *truly significant*(!) mathematical consequence of Reynolds ordering as well as enabling weak interaction theory connection to potential flow.

3.3 Steady Potential Flow GWS

GWS algorithm development starts with the DM steady potential PDE (3.10). Now written in differential operator form

$$\mathcal{L}(\varphi) = -\nabla \bullet \rho\nabla\varphi = 0 \tag{3.17}$$

Recalling the GWS formulation process, Section 2.1, for any potential function approximation

$$\varphi^N(\mathbf{x}) \equiv \sum_{\alpha=1}^{N} \Psi_\alpha(\mathbf{x})Q_\alpha \tag{3.18}$$

the associated *continuum* GWS^N is

$$\begin{aligned}\mathrm{GWS}^N &= \int_\Omega \Psi_\beta(\mathbf{x})\mathcal{L}(\varphi^N)\mathrm{d}\tau = \{0\}\\ &= \int_\Omega \Psi_\beta(\mathbf{x})\left[-\nabla\cdot\rho\nabla\varphi^N\right]\mathrm{d}\tau\\ &= \int_\Omega \nabla\Psi_\beta \bullet \rho\nabla\varphi^N\mathrm{d}\tau - \int_{\partial\Omega}\Psi_\beta\rho\nabla\varphi^N \bullet \widehat{\mathbf{n}}\mathrm{d}\sigma\\ &= \{0\},\quad for\quad 1\le\beta\le N\end{aligned}\tag{3.19}$$

The Green–Gauss divergence theorem application in (3.19) exposes the requirement for a Neumann BC, provided by an elementary operation on the potential definition (3.6)

$$\ell(\varphi) = \rho\nabla\varphi\cdot\hat{\mathbf{n}} - \rho\mathbf{u}\cdot\hat{\mathbf{n}} = 0\tag{3.20}$$

Then since the discrete implementation GWS^h is sought, recall (2.44) and (2.36), (3.18–3.20) transitions to

$$\varphi^N(\mathbf{x}) \equiv \sum_{\alpha=1}^{N}\Psi_\alpha(\mathbf{x})Q_\alpha \Rightarrow \varphi^h(\mathbf{x}) \equiv \cup_e^{\mathrm{M}}\{N_k(\boldsymbol{\nu}(\mathbf{x}))\}^T\{Q\}_e\tag{3.21}$$

$$\begin{aligned}\mathrm{GWS}^h &= \mathrm{S}_e\left[\begin{array}{l}\displaystyle\int_{\Omega_e}\nabla\{N_k(\boldsymbol{\nu}(\mathbf{x}))\}\bullet\rho_e\nabla\{N_k(\boldsymbol{\nu}(\mathbf{x}))\}^T\{Q\}_e\mathrm{d}\tau\\ \displaystyle\quad-\int_{\partial\Omega_N\cap\partial\Omega_e}\{N_k(\boldsymbol{\nu}(\mathbf{x}))\}\rho_e\mathbf{u}\bullet\widehat{\mathbf{n}}\mathrm{d}\sigma\end{array}\right]\\ &\equiv [\mathrm{DIFF}]\{Q\} - \{\mathrm{b}(data)\} = \{0\}\end{aligned}\tag{3.22}$$

for S_e symbolizing *assembly*, the matrix row-wise addition process over $1\le e\le \mathrm{M}$ forming the *global* matrices [DIFF] and {b} defined in the last line of (3.22). Assembly is valid for all $n\le 3$; (2.38–2.40) highlights the $n=1$ operation.

From (3.21) the last requirement is to define a suitable FE trial space basis $\{N_k(\boldsymbol{\nu}(\mathbf{x}))\}$, hence evaluate the integrals defined in (3.22) to generate the computable matrix statement. The geometrical shapes for $1\le n\le 3$ dimensional domains associated with FE trial space bases $\{N_k(\boldsymbol{\nu}(\mathbf{x}))\}$, Figure 3.2, are

$n = 1$: line segments
$n = 2$: triangles and quadrilaterals
$n = 3$: tetrahedra and hexahedra

Each FE domain possesses associated GWS^h approximation coefficient sets, called *degrees of freedom* (DOF), with one at least at each geometric singularity on the element boundary, that is, the vertices. Development of higher (than linear) degree FE trial space bases adds either non-vertex DOF, Zienkiewicz and Taylor (1996), or embedded local enrichments called "*p*-elements" that are "condensed out" prior to the algebraic solution process, Oden (1994).

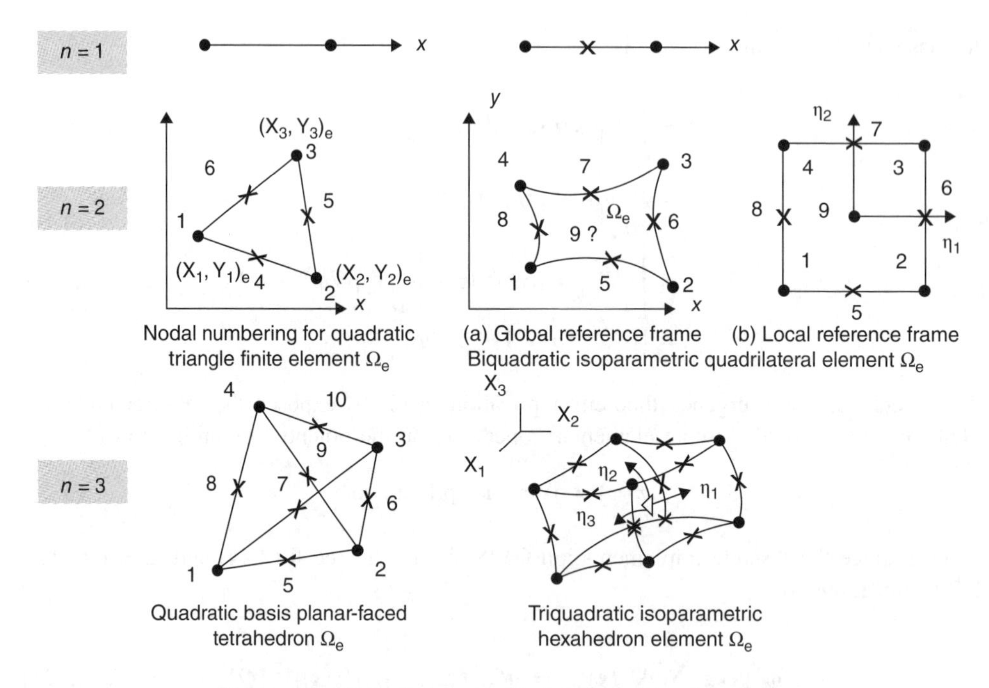

Figure 3.2 FE basis domains Ω_e for $1 \leq n \leq 3$ and $k = 1, 2$

The notation $\{N_k(\boldsymbol{\nu}(\mathbf{x}))\}$ emphasizes that an FE trial space basis is *always* expressed in one of two *local* coordinate systems:

- *natural coordinate* system ζ_α, $1 \leq \alpha \leq n+1$, for lines, triangles and tetrahedra
- *tensor product* system η_i, $1 \leq i \leq n$, for lines, quadrilaterals and hexahedra

The linear ($k = 1$) FE basis always defines these local coordinate systems as well as the *coordinate transformation* from global to the local system, $x_i = x_i(\nu_j) = f(\zeta_\alpha, \eta_j)$. The creation of $k > 1$ FE bases involves relatively elementary algebraic processes, all with direct connection to Lagrange and Hermite interpolation polynomials.

The natural coordinate (NC) FE $k = 1$ bases result directly from geometric observation leading to the definitions

$$\begin{aligned} n = 1&: \{N_1\}^T \equiv \{\zeta_1, \zeta_2\} \\ n = 2&: \{N_1\}^T \equiv \{\zeta_1, \zeta_2, \zeta_3\} \\ n = 3&: \{N_1\}^T \equiv \{\zeta_1, \zeta_2, \zeta_3, \zeta_4\} \end{aligned} \tag{3.23}$$

Each entry ζ_α in matrices $\{N_1\}^T$ is non-dimensional and normalized. The NC coordinate systems are *linearly dependent* for any n, as expressed by the constraint

$$\sum_{\alpha=1}^{n+1} \zeta_\alpha \equiv 1 \tag{3.24}$$

Via (3.23), each ζ_α is zero at *every* node of Ω_e except *that* node with index "α" whereat ζ_α is equal to unity.

Higher degree NC bases $\{N_k(\zeta_\alpha)\}$, $k>1$, employ non-vertex DOF and involve polynomials on the ζ_α defined in the $n=1$ basis. Specifically, the FE $k=2$ basis functions $\{N_2(\zeta_\alpha)\}$ for $1 \le n \le 3$, for the node numberings and DOF graphed in Figure 3.2, are

$$n=1: \left\{ \begin{matrix} \zeta_1(2\zeta_1-1) \\ \zeta_2(2\zeta_2-1) \\ 4\zeta_1\zeta_2 \end{matrix} \right\}, \quad n=2: \left\{ \begin{matrix} \zeta_1(2\zeta_1-1) \\ \zeta_2(2\zeta_2-1) \\ \zeta_3(2\zeta_3-1) \\ 4\zeta_1\zeta_2 \\ 4\zeta_2\zeta_3 \\ 4\zeta_3\zeta_1 \end{matrix} \right\}, \quad n=3: \left\{ \begin{matrix} \zeta_1(2\zeta_1-1) \\ \zeta_2(2\zeta_2-1) \\ \zeta_3(2\zeta_3-1) \\ \zeta_4(2\zeta_4-1) \\ 4\zeta_1\zeta_2 \\ 4\zeta_2\zeta_3 \\ 4\zeta_3\zeta_1 \\ 4\zeta_1\zeta_4 \\ 4\zeta_2\zeta_4 \\ 4\zeta_3\zeta_4 \end{matrix} \right\} \tag{3.25}$$

While triangles and tetrahedra obviously admit meshing of regions with arbitrarily shaped boundaries, the quadrilateral-hexahedron family is also versatile and quite amenable to non-regular boundary closures via boundary-fitted coordinate transformations, Thompson *et al.* (1985). For these element geometries, Figure 3.2, the corresponding FE bases are denoted $\{N_k^+(\eta_j)\}$ in this text. The associated *tensor product* (TP) coordinate systems η_j are *linearly independent*, $1 \le j \le n$, and appear rectangular cartesian on the element unit geometry in η-space.

The FE $k=1$ TP bases for $n=2$, 3 are

$$\{N_1^+(\eta)\} = \frac{1}{4} \left\{ \begin{matrix} (1-\eta_1)(1-\eta_2) \\ (1+\eta_1)(1-\eta_2) \\ (1+\eta_1)(1+\eta_2) \\ (1-\eta_1)(1+\eta_2) \end{matrix} \right\}, \quad \{N_1^+(\eta)\} = \frac{1}{8} \left\{ \begin{matrix} (1-\eta_1)(1-\eta_2)(1-\eta_3) \\ (1+\eta_1)(1-\eta_2)(1-\eta_3) \\ (1+\eta_1)(1+\eta_2)(1-\eta_3) \\ (1-\eta_1)(1+\eta_2)(1-\eta_3) \\ (1-\eta_1)(1-\eta_2)(1+\eta_3) \\ (1+\eta_1)(1-\eta_2)(1+\eta_3) \\ (1+\eta_1)(1+\eta_2)(1+\eta_3) \\ (1-\eta_1)(1+\eta_2)(1+\eta_3) \end{matrix} \right\} \tag{3.26}$$

with progression to FE $k>1$ TP bases similarly straightforward. For example, for the node numbering and DOF designations in Figure 3.2, the $n=2=k$ *serendipity element* TP basis, which omits the centroid DOF in Figure 3.2, is

$$\{N_2^+(\eta)\} = \frac{1}{4}\begin{Bmatrix} (1-\eta_1)(1-\eta_2)(-\eta_1-\eta_2-1) \\ (1+\eta_1)(1-\eta_2)(\eta_1-\eta_2-1) \\ (1+\eta_1)(1+\eta_2)(\eta_1+\eta_2-1) \\ (1-\eta_1)(1+\eta_2)(-\eta_1+\eta_2-1) \\ 2(1-\eta_1^2)(1-\eta_2) \\ 2(1+\eta_1)(1-\eta_2^2) \\ 2(1-\eta_1^2)(1+\eta_2) \\ 2(1-\eta_1)(1-\eta_2^2) \end{Bmatrix} \tag{3.27}$$

3.4 Accuracy, Convergence, Mathematical Preliminaries

This library of FE trial space bases $\{N_k(\zeta_\alpha,\eta_j)\}$ enables the potential flow GWS^h implementation (3.22). Selection is guided by theoretically predicted performance, principally *asymptotic convergence* under *regular* mesh refinement which leads to *quantified error estimation*. The engineer or student will find the following three chapter sections "*mathematically heavy*," The equation of *final* pertinence is (3.60), Section 3.7, should you find this theoretical content (that never(!) appears in the engineering CFD literature) exhausting.

The theoretical characterizing of *approximation error* generated by weak form implementation GWS^h involves bounding by interpolation error. The mathematicians accomplish this elegantly, Oden and Demkowicz (1996), using integral forms expressed as inner products of functions populating a suitable *Sobolev space* symbolized as H^m. Prior to detailing these formalities it is appropriate to develop preliminaries.

Maintaining notation generality, functions $u(\mathbf{x})$ and $v(\mathbf{x})$ with m derivatives that are square integrable on some region Ω of euclidean space, for $\mathbf{x}$ spanning n dimensions, are precisely characterized by the *inner product* in the space of functions $H^m(\Omega)$

$$(u,v)_{H^m} \equiv \int_\Omega u(\mathbf{x})v(\mathbf{x})d\tau + \sum_{\alpha,\beta,\gamma,\delta=1}^{m}\int_\Omega \frac{\partial^\alpha u}{\partial \mathbf{x}^\beta}\frac{\partial^\gamma v}{\partial \mathbf{x}^\delta}d\tau \tag{3.28}$$

Setting $v=u$, the square root of this inner product produces what is termed the H^m *norm* symbolized as

$$\|u\|_{H^m(\Omega)} \equiv (u,u)^{1/2}_{H^m(\Omega)} \tag{3.29}$$

The familiar L2 norm, recall (2.104), is thus the square of the H^0 norm

$$\|u\|_{L2(\Omega)} = \|u\|^2_{H^0(\Omega)} = (u,u)_{H^0(\Omega)} = \int_\Omega u(\mathbf{x})u(\mathbf{x})d\tau \tag{3.30}$$

The smoothness of a function, such as the GWS^h solution $\varphi^h(\mathbf{x})$, plays a central role in error estimation theory. The mathematics community expresses *function smoothness* via *inclusion* (symbol $\in$) in a Sobolev space H^m. Specifically, $u \in H^m$ states that $u(\mathbf{x})$ is sufficiently smooth such that *all* derivative combinations, up to and including m^{th} order, are square integrable. Table 3.1 summarizes this hierarchy with terminology.

In generating an error estimate in H^m, *data* from the PDE + BCs statement must be included since ultimately *data non-smoothness* can be *the* controlling influence on asymptotic convergence rate, recall Section 2.9. The mathematicians' preference is to include *data* as a scalar bound on a norm, the scalar being some representative average. In CFD applications, the origin of such data may be linear, for example the fin taper examined in Section 2.9, or very nonlinear, such as the *distribution* of turbulent eddy viscosity impact on the solution of the time averaged BL PDE + BCs statement.

Anticipating this issue is what prompted identification of the D*E* energy norm, (2.99), also introduction of quadratic forms in Chapter 2. A norm can be defined for any function or set of functions; upon deleting the approximation superscript N the fundamental D*E* principal energy norm is

$$\|T, T\|_E \equiv \frac{1}{2}\int_\Omega [\kappa \nabla T \bullet \nabla T]\mathrm{d}\tau + \frac{\kappa}{2}\int_{\partial\Omega_n}\left[\frac{h}{k}T^2\right]\mathrm{d}\sigma \tag{3.31}$$

This *quadratic form* exhibits essential correspondence with the H^1 norm (3.28–3.29) to within:

- embedding of the data κ within the PDE integrand
- omission of all $m = 1$ cross-derivatives present in (3.28)
- addition of Neumann BC quadratic form with data k, h plus κ
- omission of the H^0 contribution contained in H^1, and finally
- deletion of the square root in the definition (3.29).

The energy norm for the steady potential flow EBV PDE statement (3.17) is

$$\|\varphi, \varphi\|_E \equiv \frac{1}{2}\int_\Omega [\rho\nabla\varphi \cdot \nabla\varphi]\mathrm{d}\tau \tag{3.32}$$

which is devoid of a Neumann BC contribution since (3.20) contains no quadratic term. Density ρ exhibits the apparent role of *data* in (3.32), but in truth is an explicit quadratic

Table 3.1 Properties of functions $u(\mathbf{x}) \in H^m$

m	Symbol	Verbalization	Smoothness requirement
0	$\|u\|_{H^0}$	"L-two"	None, function can be piecewise discontinuous denumerable times
1	$\|u\|_{H^1}$	"H-one"	u sufficiently continuous such that *all* first derivatives lie in L2, that is, are square integrable
2	$\|u\|_{H^2}$	"H-two"	u sufficiently continuous such that *all* first and second derivatives lie in L2
:	:	:	:
m	$\|u\|_{H^m}$	"H-m"	all derivatives to and including order m exist and are square integrable

nonlinearity as readily confirmed by the steady form of ∫**DP**, (3.9)

$$\int \mathbf{DP}: \rho = \rho_\infty \left[\frac{T_{0\infty}}{T_\infty} - \frac{\gamma - 1}{2\gamma T_\infty} \nabla\varphi \cdot \nabla\varphi \right]^{1/(\gamma-1)} \tag{3.33}$$

Recall the Section 2.9 D*E* discussion wherein GWS^h FE $k=1$ basis solutions yielded nodally exact temperatures for any mesh M for conductivity a linear function of *T*. Hence, convergence was not measurable in the energy norm but was confirmed in what is now understood as the Sobolev H^1 norm with *k*(*T*) extracted as an average. In agreement, for potential flow in a venturi geometry with Dirichlet BCs, Baker (2012, Ch. 5), the GWS^h FE $k=1$ basis algorithm (3.22) solution exhibits quadratic convergence in the energy norm (3.32), also linear convergence in the Sobolev H^1 norm, $m=1$ in (3.29), for density assumed constant.

The nonlinearities pervasive in NS PDE statements generate significant distributions of *nonlinear parametric data*. In this text *a posteriori* data error estimation is the focus in CFD algorithm predictions in genuine fluid/thermal sciences applications. Since rigorous theory is invariably limited to linearized analyses, theory validation assessments must and do rely on precisely designed and executed computational experiments.

As thoroughly detailed throughout this text, thus generated *a posteriori* data analysis clearly validates the available elegant (linear) theory accurately predicts GWS^h discrete implementation performance in nonlinear NS applications including precise guidance on generation/confirmation of *optimal mesh solution* existence.

3.5 Accuracy, Galerkin Weak Form Optimality

The requirement is to rigorously prove the Galerkin weak form discrete implementation GWS^h for a linear PDE + BCs statement, (3.19–3.21), (2.1–2.3), generates an approximate solution exhibiting *optimal accuracy* among its peers. Specifically, any other discrete implementation of an alternatively extremized WF, (2.8–2.11), will generate a solution possessing a measure of error not smaller than that associated with GWS^h for the equivalent trial space basis complexity and DOF (in English, on the same mesh M).

The mathematical proof centers on manipulations of a generic *bilinear form* B(*u*,*v*). The mathematicians' preference is non-homogeneous written on arbitrary functions *u*(**x**) and *v*(**x**)

$$\mathrm{B}(u,v) \equiv \int_\Omega \nabla u \cdot k_1 \nabla v \, d\tau + \int_{\partial\Omega_n} k_2 uv \, d\sigma = \mathrm{b}(v) \tag{3.34}$$

The integrals in (3.34) lie on a region Ω of euclidean space including boundary segments $\partial\Omega_n$ upon which a Neumann BC is imposed. The parameters k_1 and k_2 are given data distributions and b(*v*) contains all other data (independent of *u*,*v*) defining the statement.

Note that the Sobolev inner product (3.28) is a bilinear form devoid of data. Conversely, the energy norms $\|T, T\|_E$ and $\|\varphi, \varphi\|_E$, (3.31–3.32), are each symmetric quadratic bilinear forms that contain data. For specific reference, the bilinear form terms for the steady D*E* and DM potential aerodynamics GWS^N algorithms, (2.15), (3.19), are

$$\begin{aligned} \mathrm{B}(u \to T^N, v \to \Psi_\beta) &= \int_\Omega \kappa \nabla T^N \cdot \nabla \Psi_\beta \, \mathrm{d}\tau + \int_{\partial\Omega_n} \frac{\kappa h}{k} T^N \Psi_\beta \, \mathrm{d}\sigma \\ \mathrm{b}(v \to \Psi_\beta) &= \int_\Omega \Psi_\beta (s/\rho_0 c_p) \mathrm{d}\tau + \int_{\partial\Omega_n} \Psi_\beta \frac{\kappa}{k} \left[hT_r - f_n \right] \mathrm{d}\sigma \end{aligned} \tag{3.35}$$

$$\begin{aligned} \mathrm{B}(u \to \varphi^N, v \to \Psi_\beta) &= \int_\Omega \rho \nabla \varphi^N \cdot \nabla \Psi_\beta \, \mathrm{d}\tau \\ \mathrm{b}(v \to \Psi_\beta) &= \int_{\partial\Omega_n} \Psi_\beta \rho \mathbf{u} \bullet \hat{\mathbf{n}} \mathrm{d}\sigma \end{aligned} \tag{3.36}$$

The abstract weak form theory replaces the requirement of finding the solution approximation $T^N(\mathbf{x})$ or $\varphi^N(\mathbf{x})$ to their EBV PDE + BCs elliptic boundary value statements with the following:

$$\text{find } u \in H^1 \text{ such that } \mathrm{B}(u, v) = \mathrm{b}(v) \tag{3.37}$$

for all $v \in H^1$ and $v(\partial\Omega_D) \equiv 0$, where $\partial\Omega_D$ denotes domain boundary segments upon which *D*irichlet data are specified. This data restriction is symbolized as $v \in H_0^1$ and accounts for the fact that fluxes on such boundary segments are *a priori* unknown, recall (2.41).

The process of "converting" (3.37) into computable form – the discretely implemented weak statement WS^h – is to first replace v by its interpolant, recall (2.9). Then u is replaced by its spatially discrete approximation $u^h(\mathbf{x})$, a process also based on interpolation, recall (2.27) and Figure 2.3. These *much smaller* function spaces v^h and u^h are each termed a *finite dimensional subspace*, symbolized as H^h, and *all* such functions in H^h also reside in H^1 for (3.34), hence (3.35–3.36).

Implementing the v interpolation first, and denoting it v^h for clarity, the resultant abstract weak form (3.37) becomes

$$\text{find } u \in H^1 \text{ such that } \mathrm{B}(u, v^h) = \mathrm{b}(v^h), \forall v^h \subset H^h \in H_0^1 \tag{3.38}$$

where $\forall$ denotes "for all." Then restricting u to $u^h \subset H^h \in H^1$ via an interpolation process produces the terminal abstract statement

$$\begin{aligned} &\text{find } u^h \subset H^h \text{ such that } \mathrm{B}(u^h, v^h) = \mathrm{b}(v^h), \\ &\forall v^h \subset H^h \in H_0^1 \text{ and } u^h \subset H^h \in H^1 \end{aligned} \tag{3.39}$$

Equations (3.35–3.39) provide the mathematical concepts and tools necessary to *measure accuracy* with generality, hence quantify the opportunity for *optimal* accuracy. The first step is to subtract (3.39) from (3.38) yielding

$$\mathrm{B}(u - u^h, v^h) = 0, \forall u \in H^1, v^h \subset H^h \in H_0^1, u^h \subset H^h \in H^1 \tag{3.40}$$

Since $u - u^h \equiv e^h$ is the definition of approximation error in a weak form solution to (3.35) and/or (3.36), (3.40) states that this error is *orthogonal* in H^h, the finite dimensional subspace of functions in H^1 from which *all eligible* functions can be extracted. Specifically, (3.40) holds for functions supporting a non-Galerkin WS^h, and *all* test functions $v^h \subset H^h$ still remain absolutely arbitrary.

The fact that e^h is orthogonal in H^h, as formed via WS^h for *any* v^h, naturally leads to the question, "Is there an *optimal* choice for v^h such that the *best approximation* u^h results?" Figure 3.3 illustrates this mathematical query geometrically. The space H^h is represented as a plane which contains *all* eligible functions u^h and v^h graphically interpreted as vectors lying in the plane. Since H^h is not large enough to contain the exact solution to $B(u,v)$, u is represented as an out-of-plane vector. Thereby, u is resolvable into the projection $u^h \subset H^h$ lying in the plane, and the plane-orthogonal complement labeled $u - Pu^h$, where "P" denotes "perpendicular" to the plane H^h.

This illustration indicates that vector addition of u^h with $u - Pu^h$ is "equal to" u for all u^h in H^h. Any candidate for v^h also lies in H^h, and its associated projection $u - P^*v^h$, also sketched in Figure 3.3, has magnitude larger than $u - Pu^h$, since $u - P^*v^h$ is not plane-perpendicular in H^h. Thereby, u^h must be the *optimal* choice for v^h which constitutes the Galerkin criterion for GWS^h identified in (2.11).

The geometric interpretation yields to mathematical precision. The bilinear form (3.38) can be written for any argument pair, hence replace both u and v^h therein with $u - v^h$, which can be interpreted as the error associated with v^h being the interpolant of u. For these arguments, (3.40) becomes

$$B(u - v^h, u - v^h) = 2\|u - v^h\|_E \tag{3.41}$$

which equals precisely twice the energy norm of the associated interpolation error. Adding and subtracting u^h from each argument in (3.41) produces no alteration but admits the manipulation

$$\begin{aligned} 2\|u - v^h\|_E &= B\left(u - u^h - v^h + u^h, u - u^h - v^h + u^h\right) \\ &= B\left(u - u^h, u - u^h, u^h - v^h, u^h - v^h\right) \\ &= B\left(e^h, e^h\right) + B\left(u^h - v^h, u^h - v^h\right) \end{aligned} \tag{3.42}$$

Again, $e^h \equiv u - u^h$ remains the *error* in the weak form solution u^h for *any* v^h. Since the right side terms in (3.42) are symmetric quadratic forms each is uniformly non-negative.

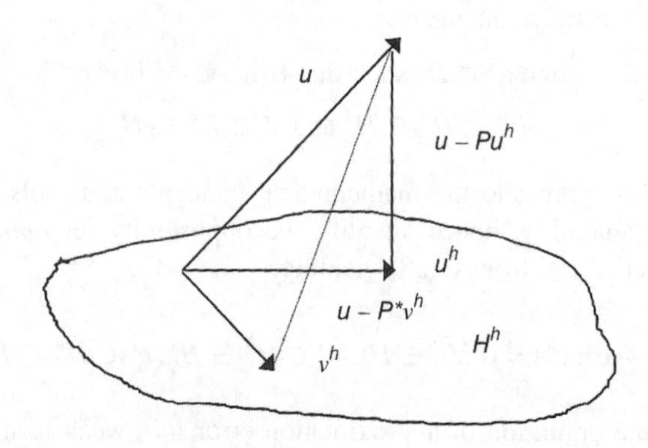

Figure 3.3 Geometric interpretation of best approximation

Hence, the energy norm measuring the distance between u and v^h is a minimum when u^h and v^h are identical, which causes the second term to vanish.

Thus, the error in the approximate solution u^h to $\mathrm{B}(u^h, v^h)$ is the *smallest possible*(!) for *all* v^h when v^h and u^h are *identical*. This abstract weak form theory thus precisely predicts that the *Galerkin criterion* for any (linear) WF is *optimal*, as first asserted in Chapter 2, recall (2.10–2.13). Thereby, the generated Galerkin weak statement solution u^h is the *best possible* among all candidates in H^h, as measured in the energy norm $||u^h||_E$, and/or the Sobolev H^1 norm $\|u^h\|_{H^1}$, since u^h is the *perpendicular projection* of u onto H^h in the abstract sense.

3.6 Accuracy, GWS^h Error Bound

It remains to establish a *bound* on the error $e^h \equiv u - u^h$ associated with the GWS^h solution in terms of the given *data* specifications. From the abstract bilinear form (3.41–3.42), the Galerkin criterion solution error for any function $u^h \subset H^h$ equals exactly twice the energy norm

$$\mathrm{B}\left(u-u^h, u-u^h\right) = \mathrm{B}\left(e^h, e^h\right) = 2\left\|e^h\right\|_E \tag{3.43}$$

Of key importance, recall that $\mathrm{B}(u, v)$, (3.34), for either the DE or DM PDE + BCs statement, contains the *data* $k_1(\mathbf{x}) \subset \Omega$, $k_2(\mathbf{x}) \subset \partial\Omega$, while $\mathrm{b}(v)$ contains all remaining distributed source data on Ω.

For (3.43) to be mathematically *well-posed* requires that certain properties exist, specifically *continuity* and *coerciveness*. In the continuum, where $(u, v) \in H^1$ with associated *measures* $\|u\|_{H^1}$ and $\|v\|_{H_0^1}$, *continuity* requires that the corresponding bilinear form (3.43) be bounded. Following Oden (1994), for the steady potential flow DM bilinear form (3.36), the bounding statement is

$$\mathrm{B}(\rho\nabla\varphi, \nabla\Psi) \leq max\{\rho(\varphi(\mathbf{x}))\}\mathrm{B}(\nabla\varphi, \nabla\Psi) \tag{3.44}$$

which extracts the intrinsic nonlinearity via its extremum. Conversely, for the steady DE principle bilinear form (3.35) the bounding statement must include data in the Robin BC

$$\mathrm{B}(\kappa\nabla T, \nabla\Psi, (h/k)T, \Psi) \leq max\{\kappa, max(h(\mathbf{x}_b)/k(\mathbf{x}))\}\mathrm{B}(\nabla T, \nabla\Psi, T, \Psi) \tag{3.45}$$

Denoting by β the resultant of the maximum search in (3.44) or (3.45), with generalized notion of *continuity* the well-posed EBV PDE + BCs statement requires

$$\mathrm{B}(u, v, \kappa, h, k) \leq \beta\mathrm{B}(u, v) \leq \beta\|u, v\|_{H^1} \leq \beta\|u\|_{H^1}\|v\|_{H^1} \tag{3.46}$$

where the final expression employs the triangle inequality, Oden and Demkowicz (1996).

The principle of *coerciveness* is the counterpart of *continuity*. Again, following Oden (1994), for the steady potential flow DM symmetric bilinear form, coincident with the energy norm (3.32), a data minimum bound must exist of the form

$$\mathrm{B}(\rho\nabla\varphi, \nabla\varphi) \leq min\{\rho(\varphi(\mathbf{x}))\}\mathrm{B}(\nabla\varphi, \nabla\varphi) \tag{3.47}$$

Similarly, for the steady D*E* symmetric bilinear form the bound is

$$\mathrm{B}(\kappa\nabla T, \nabla T, (h/k)T, T) \leq min\{\kappa, min(h(\mathbf{x}_b)/k(\mathbf{x}))\}\mathrm{B}(\nabla T, \nabla T, T, T) \tag{3.48}$$

Generalizing notation and letting α denote the limit of the minimum search process in (3.47) or (3.48), *coerciveness* requires

$$\mathrm{B}(u, u) \geq \alpha\|u\|_{H^1}^2, \quad \forall\, u \in H^1 \tag{3.49}$$

With *continuity* and *coerciveness* thus quantified, the sought bound on e^h is readily established. Combining (3.43) with (3.49) and implementing the argument exchange $u \to e^h$, coerciveness requires

$$\left\|e^h\right\|_E \geq \alpha\left\|e^h\right\|_{H^1}^2 \tag{3.50}$$

The right side of (3.50) admits the useful manipulation

$$\begin{aligned} \left\|e^h\right\|_{H^1}^2 &= \mathrm{B}\left(e^h, e^h\right) = \mathrm{B}\left(u - u^h, u - u^h \pm v^h\right) \\ &= \mathrm{B}\left(e^h, (u - v^h) + (v^h - u^h)\right) \\ &= \mathrm{B}\left(e^h, u - v^h\right) + \mathrm{B}\left(e^h, v^h - u^h\right) \end{aligned} \tag{3.51}$$

and the last line second term vanishes for the Galerkin criterion. Then substituting (3.51) into (3.50) and rearranging yields, for arbitrary v^h

$$\begin{aligned} \left\|e^h\right\|_{H^1}^2 &= \mathrm{B}\left(e^h, e^h\right) \leq \frac{1}{\alpha}\mathrm{B}\left(e^h, u - v^h\right) \\ &\leq \frac{\beta}{\alpha}\left\|e^h\right\|_{H^1}\left\|u - v^h\right\|_{H^1} \end{aligned} \tag{3.52}$$

using the constraint of continuity (3.46).

Making the obvious cancellation the terminal expression for the bound on the error for $u^h \subset H^h \in H^1$ is

$$\left\|e^h\right\|_{H^1} \leq \frac{\beta}{\alpha}\left\|u - v^h\right\|_{H^1}, \quad \forall\, v^h \tag{3.53}$$

Hence, the GWS^h error e^h is bounded above by the material *data* constant β/α times the distance between the exact solution u and *any* function $v^h \subset H^h \in H^1$. The smaller the magnitude of α the larger is the bound, and one anticipates that as $\alpha \to 0$ the problem will become unsolvable or unstable. Finally, since v^h is absolutely arbitrary, selecting it as the *interpolant* u^I of u, recall (2.9), bounds e^h by the error associated with interpolation of the exact solution(!) for reasonable β/α.

3.7 Accuracy, GWS^h Asymptotic Convergence

Having determined that WS^N generates the *optimal* solution for the Galerkin criterion $\Phi_\beta \equiv \Psi_\beta$, also the associated error bound (3.53), it remains to determine the rate at which the GWS^h solution approaches the (unknown) exact solution. The key is establishing a suitable length scale, called a *measure*, of the union of FE domains Ω_e constituting the computational mesh Ω^h. The key resolution is handling the multiple geometric length scales existing in n dimensions, recall Figure 3.2.

For $n = 2$, the geometric scales of the generic triangle domain Ω_e can be reduced to the radius ρ_e of an *inscribed* circle, tangent to each boundary segment, and the radius h_e of the *circumscribed* circle intersecting each vertex node of the triangle, Figure 3.4. For $n = 3$, these become the radius of the smallest inscribed sphere, tangent to at least three of the four bounding planes of Ω_e, and the radius of a circumscribed sphere intersecting at least three of the four vertex nodes while encompassing all. Similarly, two radii are sufficient to quantify the size and aspect ratio of quadrilateral and/or hexahedron Ω_e geometries. Finally, for higher degree FE bases $\{N_k(\zeta_\alpha, \eta_i)\}$, recall (3.25), (3.27), ρ_e and h_e remain the characteristic scales as the additional DOF are always located interior to or on bounding edges (planes) of Ω_e for $n = 2(3)$.

The key to deriving the *asymptotic convergence* rate estimate is bounding the GWS^h approximation error $e^h = T - T^h$ by interpolation error, as elegantly accomplished in the abstract analysis, (3.53). Interpreting the right side of (3.53) as interpolation error, at the level of an individual finite element, Oden and Reddy (1976, Ch. 8) predict

$$\left\| u - v^h \right\|_{H^m,\Omega_e} \leq C_k \frac{h_e^{k+1}}{\rho^m} \|u\|_{r,\Omega_e} \tag{3.54}$$

This states that the interpolation error of the exact solution is bounded in the H^m Sobolev *norm* by the product of a possibly k-dependent (FE basis degree) constant C_k, the element characteristic radii pair raised to the powers $k+1$ and minus m, and the H^r norm of the analytical solution u. The H^r norm constitutes a smoothness constraint, that is, the exact solution u must possess up to order r derivatives, in all combinations, that are square-integrable, recall Table 3.1.

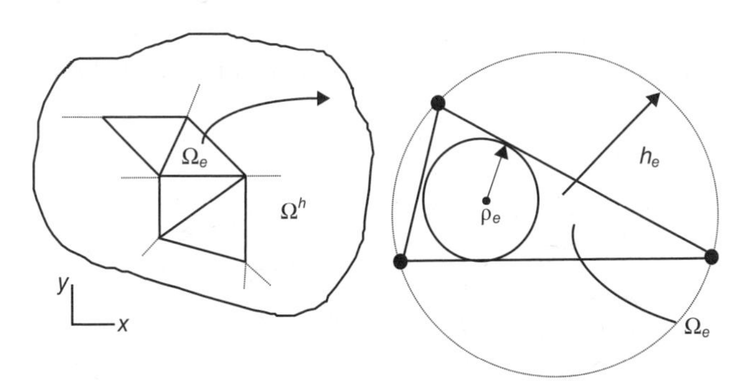

Figure 3.4 Measures of the generic FE domain $\Omega_e \subset \Re^2$

This emphasizes a key feature of the GWS^h procedure. For linear second-order EBV PDE + BCs systems, for example D*E*, (2.1–2.3), D*M*, (3.11), $m = 1$ and the exact solution u must possess two derivatives in all combinations. Conversely, the GWS^h approximate solution u^h need possess only $m = 1$ derivative, as clearly illustrated in Chapter 2 examples. Hence, r is *never smaller* than m in (3.54), and r is further required to grow as order $2k$ in the progression to higher degree FE basis implementations. The precise statement of this *weakened* requirement is

$$u \in H^r, \quad \text{while} \quad u^h \subset H^h \in H^m \tag{3.55}$$

The two *independent* geometric scales ρ_e and h_e appearing in (3.54) can be reduced to a single *measure* by controlling the mesh refinement process, $\Omega^h \Rightarrow \Omega^{h/2} \Rightarrow \Omega^{h/4}$, and so on. Specifically, *regular* mesh refinement requires that the length scale ratio h_e/ρ_e remain a nominal constant for all mesh refinements. The left graph in Figure 3.5 shows a non-regular refinement, wherein h_e increases remarkably (not illustrated) for each refinement while $\rho_e \to 0$. Conversely, the right graph illustrates a *regular* refinement for which the characteristic length ratio remains an essential constant.

Therefore, under *regular* mesh refinement the generic Ω_e length scale ratio is bounded as $h_e/\rho_e \leq \text{constant} \equiv \sigma$. Substituting $\rho_e^{-m} \leq \sigma h_e^{-m}$ into (3.54) produces the revised *local* interpolation error bound

$$\left\| u - v^h \right\|_{H^m, \Omega_e} \leq C_{k\sigma} h_e^{k+1-m} \|u\|_{H^r, \Omega_e} \tag{3.56}$$

with new constant $C_{k\sigma}$ which can depend at most on FE basis degree k. The *global* asymptotic error estimate accrues to summation of (3.56) over all $\Omega_e \subset \Omega^h$. Then, replacing the interpolation v^h by the GWS^h solution u^h, which itself is bounded by interpolation error, (3.53), the *asymptotic convergence rate* for the GWS^h solution approximation error under regular mesh refinement is

$$\left\| e^h \right\|_{H^m, \Omega} \equiv \sum_{e=1}^{M} \left\| u - u^h \right\|_{H^m, \Omega_e} \leq C\, h^{k+1-m} \|u\|_{H^r, \Omega} \tag{3.57}$$

Two key conclusions are thus generated via (3.57). First, the u^h approximation error indeed vanishes asymptotically in a predictable way, with rate dependent on regularized mesh measure h, FE trial space basis degree k and order m of the Sobolev space in

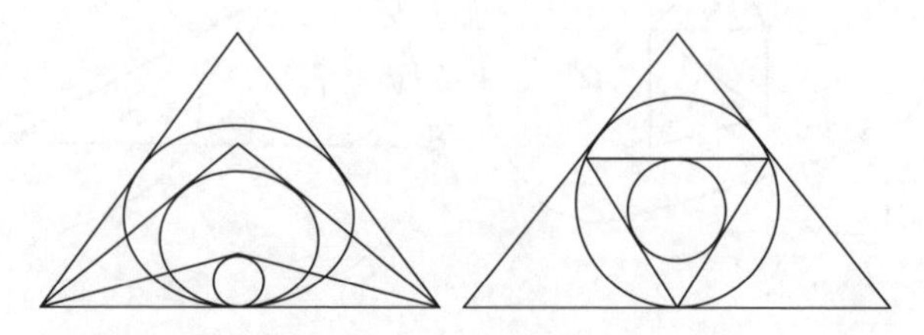

Figure 3.5 Non-regular and regular mesh refinement, $n = 2$

which u^h resides. Second, for this to be valid requires the exact solution u to be *sufficiently smooth* such that it resides in a Sobolev space of order $r > m$. However, the exact solution is never known(!) so this issue must be resolved for (3.57) to be of practical use.

Guidance comes from (3.34), which equates the abstract bilinear form $\mathrm{B}(u,v)$ to $\mathrm{b}(v)$, the placeholder for all remaining *data* driving the problem. For $\mathrm{b}(v)$ containing only algebraic expressions, that is, no spatial derivatives, then the functions in $\mathrm{b}(v)$ all reside in L2, correspondingly H^0, that is, are square integrable. Recalling the D*E* and D*M* definition statements for $\mathrm{b}(v)$, (3.35–3.36), data exist on both Ω and $\partial\Omega$ so the corresponding alteration admitted for (3.57) is

$$\left\|e^h\right\|_{H^m,\Omega} \equiv \sum_{e=1}^{\mathrm{M}} \left\|u - u^h\right\|_{H^m,\Omega_e} \leq \mathrm{C}\, h^{k+1-m}\left[\|\mathrm{data}\|_{H^0,\Omega} + \|\mathrm{data}\|_{H^0,\partial\Omega}\right] \tag{3.58}$$

Problem data are always known, hence the H^0 norms in (3.58) can be evaluated. However, whether the analytical solution u is sufficiently smooth for (3.58) to be valid remains unanswered. This is readily corrected upon noting that the PDE + BCs requirement $u \in H^r$ is replaced by $u \in H^m$ in the weak form, specifically the differentiability requirement has been *weakened*, the source of the name *weak form*!

Therefore, the rate of asymptotic convergence must admit domination by $r - m$ for non-smooth data generating u not possessing the smoothness required by (3.57). This key theory issue is embedded by replacement of the exponent in (3.58) with $\gamma \equiv \min(k+1-m, r-m)$. Thus results the sought asymptotic error estimate

$$\left\|e^h\right\|_{H^m,\Omega} \leq \mathrm{C}\, h^{\gamma}\left[\|\mathrm{data}\|_{H^0,\Omega} + \|\mathrm{data}\|_{H^0,\partial\Omega}\right], \gamma \equiv \min(k+1-m, r-m) \tag{3.59}$$

Theory development is completed by alteration of (3.59) to the CFD-preferred energy norm. The modification is truly elementary: via (3.50) the expression for *asymptotic convergence* in the *energy norm* is

$$\left\|e^h\right\|_E \leq \mathrm{C}\, h^{2\gamma}\left[\|\mathrm{data}\|_{\mathrm{L2},\Omega} + \|\mathrm{data}\|_{\mathrm{L2},\partial\Omega}\right], \gamma \equiv \min(k+1-m, r-m) \tag{3.60}$$

for another constant C and replacing the H^0 norms with L2.

Of note, should problem data be sufficiently smooth such that the dominating term is $(k+1-m)$ then (3.60) predicts asymptotic convergence rates of 2, 4, 6 for $k = 1$, 2, 3 FE trial space basis implementations for $m = 1$ EBV PDEs, that is, the systems generally addressed by CFD. These rates are identical with those predicted via a TS analysis of the associated GWS^h stencils, to be confirmed shortly, hence correlate one-to-one with the FD concept of *order of accuracy* for smooth data (*only*!).

That (3.60) accurately predicts GWS^h algorithm performance for limiting γ was clearly illustrated in the Chapter 2 D*E* examples. Prediction validity will be universally confirmed in progression through the range of addressed nonlinear NS PDE systems.

3.8 GWS^h Natural Coordinate FE Basis Matrices

Implementation of the steady flow D*M* potential flow GWS^h algorithm into computable form employs either the natural coordinate (NC) or tensor product (TP) FE trial space bases, recall Figure 3.2. The vector calculus operation forming the $\nabla \bullet \rho\nabla\varphi$ element matrix, (3.22), employs the chain rule to account for FE bases expressed in local coordinates, (3.23–3.27). Hence, the GWS^h vector matrix operation to be performed on the generic FE domain Ω_e is

$$\begin{aligned}[\mathrm{DIFF}]_e\{Q\}_e &\equiv \int_{\Omega_e} \nabla\{N_k\}\cdot\rho_e\nabla\{N_k\}^T \mathrm{d}\tau\{Q\}_e \\ &= \int_{\Omega_e} \frac{\partial\{N\}}{\partial\nu_j}\left(\frac{\partial\nu_j}{\partial x_i}\right)_e \rho_e \frac{\partial\{N\}^T}{\partial\nu_k}\left(\frac{\partial\nu_k}{\partial x_i}\right)_e \mathrm{det}_e \mathrm{d}\nu\{Q\}_e\end{aligned} \tag{3.61}$$

where det_e is the determinant of the coordinate transformation $\mathbf{x}=\mathbf{x}(\nu)$, with $\mathrm{d}\tau$ and $\mathrm{d}\nu$ the differential elements in global and local coordinates.

Clear exposition results by choosing the linear $(k=1)$ NC basis implementation for $n=2$. Evaluate the linear polynomial $\varphi_e(x,\ y)\equiv \mathrm{a}_1+\mathrm{a}_2x+\mathrm{a}_3y$ at the triangular element vertex nodes, Figure 3.6. Assigning DOF $\{Q\}_e$ entries accordingly generates

$$\begin{aligned}\varphi_e(x=\mathrm{X}_1, y=\mathrm{Y}_1) &\equiv Q_1 = \mathrm{a}_1+\mathrm{a}_2\mathrm{X}_1+\mathrm{a}_3\mathrm{Y}_1\\ \varphi_e(x=\mathrm{X}_2, y=\mathrm{Y}_2) &\equiv Q_2 = \mathrm{a}_1+\mathrm{a}_2\mathrm{X}_2+\mathrm{a}_3\mathrm{Y}_2\\ \varphi_e(x=\mathrm{X}_3, y=\mathrm{Y}_3) &\equiv Q_3 = \mathrm{a}_1+\mathrm{a}_2\mathrm{X}_3+\mathrm{a}_3\mathrm{Y}_3\end{aligned} \tag{3.62}$$

Solving this 3×3 matrix statement via Cramer's rule yields

$$\begin{aligned}\mathrm{a}_1 &= [(\mathrm{X}_2\mathrm{Y}_3-\mathrm{X}_3\mathrm{Y}_2)Q_1+(\mathrm{X}_3\mathrm{Y}_1-\mathrm{X}_1\mathrm{Y}_3)Q_2+(\mathrm{X}_1\mathrm{Y}_2-\mathrm{X}_2\mathrm{Y}_1)Q_3]/2\mathrm{A}_\mathrm{e}\\ \mathrm{a}_2 &= [(\mathrm{Y}_2-\mathrm{Y}_3)Q_1+(\mathrm{Y}_3-\mathrm{Y}_1)Q_2+(\mathrm{Y}_1-\mathrm{Y}_2)Q_3]/2\mathrm{A}_\mathrm{e}\\ \mathrm{a}_3 &= [(\mathrm{X}_3-\mathrm{X}_2)Q_1+(\mathrm{X}_1-\mathrm{X}_3)Q_2+(\mathrm{X}_2-\mathrm{X}_1)Q_3]/2\mathrm{A}_\mathrm{e}\\ \mathrm{det}_e &= 2\mathrm{A}_e = (\mathrm{X}_1\mathrm{Y}_2-\mathrm{X}_2\mathrm{Y}_1)+(\mathrm{X}_3\mathrm{Y}_1-\mathrm{X}_1\mathrm{Y}_3)+(\mathrm{X}_2\mathrm{Y}_3-\mathrm{X}_3\mathrm{Y}_2)\end{aligned} \tag{3.63}$$

The FE basis $\{N_1(\zeta_\alpha)\}$, also the required *inverse* coordinate transformation $\zeta_\alpha=\zeta_\alpha(\mathbf{x})$ results upon extracting $\{Q\}_e$ in (3.63) as a common multiplier

$$\varphi_e(x,y)\equiv\{N_1(\zeta_\alpha)\}^T\{Q\}_e \Rightarrow \{\zeta_1,\zeta_2,\zeta_3\}\{Q\}_e \tag{3.64}$$

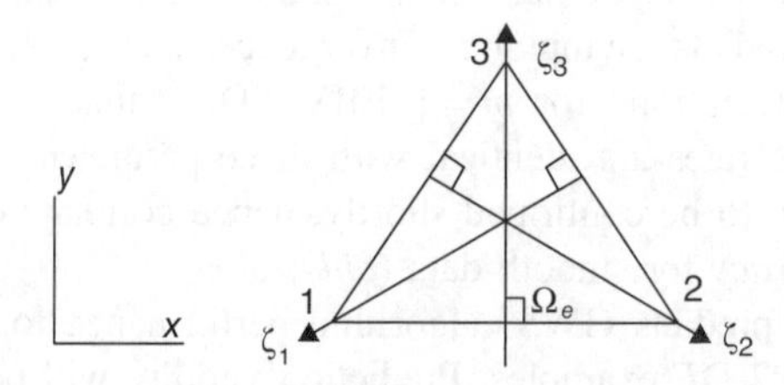

Figure 3.6 Natural coordinate system ζ_α system spanning Ω_e, $n=2$

the completion of which produces

$$
\begin{aligned}
\zeta_1 &= [\{X_2Y_3 - X_3Y_2)_e + (Y_2 - Y_3)_e x + (X_3 - X_2)_e y]/\det_e \\
\zeta_2 &= [\{X_3Y_1 - X_1Y_3)_e + (Y_3 - Y_1)_e x + (X_1 - X_3)_e y]/\det_e \\
\zeta_3 &= [\{X_1Y_2 - X_2Y_1)_e + (Y_1 - Y_2)_e x + (X_2 - X_1)_e y]/\det_e
\end{aligned}
\tag{3.65}
$$

Each coordinate ζ_α is a non-D normalized linear monomial in (x, y) in (3.65). Figure 3.6 graphs the ζ_α system spanning the generic $n = 2\,\Omega_e$. It is visually transparent the ζ_α system satisfies the Boolean (0,1) character at each node of Ω_e as required by trial space basis definition (3.64).

Determining the $n = 3$ natural coordinate system ζ_α replaces (3.62) with a 4×4 matrix statement, hence solution via determinant expansions by cofactors. The fundamental definition is

$$
\varphi_e(x, y, z) = \{N_1(\zeta)\}^T \{Q\}_e = \{\zeta_1, \zeta_2, \zeta_3, \zeta_4\}\{Q\}_e \tag{3.66}
$$

and the determinantal solution statement is

$$
\{N_1(\zeta)\} = \begin{Bmatrix} \zeta_1 \\ \zeta_2 \\ \zeta_3 \\ \zeta_4 \end{Bmatrix} \equiv \frac{1}{6V_e} \begin{Bmatrix} a_1 + b_1 x + c_1 y + d_1 z \\ -(a_2 + b_2 x + c_2 y + d_2 z) \\ a_3 + b_3 x + c_3 y + d_3 z \\ -(a_4 + b_4 x + c_4 y + d_4 z) \end{Bmatrix} \tag{3.67}
$$

$$
\begin{aligned}
a_i &\equiv \det \begin{bmatrix} X_j & Y_j & Z_j \\ X_m & Y_m & Z_m \\ X_p & Y_p & Z_p \end{bmatrix}, \quad b_i \equiv -\det \begin{bmatrix} 1 & Y_j & Z_j \\ 1 & Y_m & Z_m \\ 1 & Y_p & Z_p \end{bmatrix}_e \\
c_i &\equiv \det \begin{bmatrix} X_j & 1 & Z_j \\ X_m & 1 & Z_m \\ X_p & 1 & Z_p \end{bmatrix}_e, \quad d_i \equiv -\det \begin{bmatrix} X_j & Y_j & 1 \\ X_m & Y_m & 1 \\ X_p & Y_p & 1 \end{bmatrix}_e
\end{aligned}
\tag{3.68}
$$

with tetrahedron volume measure

$$
6V_e \equiv \det \begin{bmatrix} 1 & X_1 & Y_1 & Z_1 \\ 1 & X_2 & Y_2 & Z_2 \\ 1 & X_3 & Y_3 & Z_3 \\ 1 & X_4 & Y_4 & Z_4 \end{bmatrix} \tag{3.69}
$$

In (3.67–3.68), the indices (i, j, m, p) permute (1, 2, 3, 4) over the vertex nodes of Ω_e, see Figure 3.7 for node numbering. The ζ_4 coordinate, which lies perpendicular to the plane containing nodes 1, 2 and 3, is graphed and obviously meets the Boolean (0,1) character required by (3.66) at each node of Ω_e, as thus do ζ_1, ζ_2 and ζ_3.

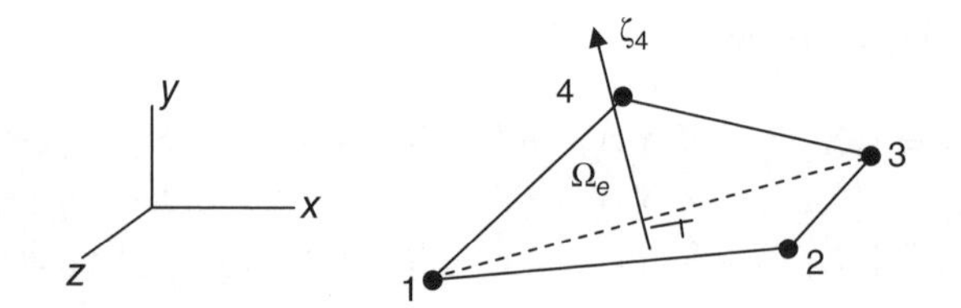

Figure 3.7 Natural coordinate member ζ_4 spanning Ω_e, $n=3$

The consequential observation is the $k=1$ NC basis $\{N_1(\zeta_\alpha)\}$ contains only linear terms which *greatly* simplifies evaluating (3.61)

$$\begin{aligned}[\mathrm{DIFF}]_e\{Q\}_e &= \int_{\Omega_e} \nabla\{N_1\}\cdot\rho_e\nabla\{N_1\}^T d\tau\{Q\}_e \\ &= \nabla\{N_1\}\cdot\nabla\{N_1\}^T\int_{\Omega_e}\rho_e d\tau\{Q\}_e \\ &= \nabla\{N_1\}\cdot\nabla\{N_1\}^T\overline{\rho}_e(\det_e/n!)\{Q\}_e\end{aligned} \tag{3.70}$$

The second line integral exactly generates $\overline{\rho}_e$, the average density on Ω_e, multiplying A_e, V_e for $n=2$, 3, equivalently $(\det_e/n!)$, see (3.63), (3.69).

Via the chain rule, the matrix product in (3.70) expands to

$$\nabla\{N_1\}\cdot\nabla\{N_1\}^T = \left(\frac{\partial\{N_1\}}{\partial\zeta_\alpha}\frac{\partial\{N_1\}^T}{\partial\zeta_\beta}\right)\left(\frac{\partial\zeta_\alpha}{\partial x_i}\frac{\partial\zeta_\beta}{\partial x_i}\right)_e \tag{3.71}$$

The basis derivatives, $1\leq(\alpha,\beta)\leq n+1$, generate elementary matrices

$$\frac{\partial\{N_1\}}{\partial\zeta_\alpha} = \{1,0,0,0,\}^T,\{0,1,0,0,\}^T,\{0,0,1,0,\}^T,\{0,0,0,0,1\}^T \tag{3.72}$$

The matrix product in (3.71) is thus the square matrix

$$\left(\frac{\partial\{N_1\}}{\partial\zeta_\alpha}\frac{\partial\{N_1\}^T}{\partial\zeta_\beta}\right) = \left[\mathrm{M}\delta_{\alpha\beta}\right]_{n+1,n+1} \tag{3.73}$$

the coefficients of which are the Kronecker delta $\delta_{\alpha\beta}$, i.e., unity at column-row location α, β and otherwise zero. Notationally, "[M]" is the prefix used throughout the text to signify "**M**atrix."

Thus, the GWS^h $\nabla\bullet\rho\nabla\varphi$ term in (3.22), when discretely implemented using the $k=1$ NC FE basis, produces the DOF-order element-level matrix product

$$\begin{aligned}[\mathrm{DIFF}]_e\{Q\}_e &= \int_{\Omega_e}\nabla\{N_1\}\cdot\rho_e\nabla\{N_1\}^T d\tau\{Q\}_e \\ &= \overline{\rho}_e(\det_e/n!)\left(\frac{\partial\zeta_\alpha}{\partial x_i}\frac{\partial\zeta_\beta}{\partial x_i}\right)_e\left[\mathrm{M}\delta_{\alpha\beta}\right]\{Q\}_e\end{aligned} \tag{3.74}$$

It remains to evaluate the inverse coordinate transformation, hence scalar products for all $1 \leq (\alpha, \beta) \leq n+1$ and $1 \leq i \leq n$ in (3.74). Formation is an elementary operation on (3.65) or (3.67), for example for $n=2$

$$\left[\frac{\partial\zeta_\alpha}{\partial x_i}\right]_e = \frac{1}{\det_e}\begin{bmatrix} Y_2 - Y_3, & X_3 - X_2 \\ Y_3 - Y_1, & X_1 - X_3 \\ Y_1 - Y_2, & X_2 - X_1 \end{bmatrix}_e \tag{3.75}$$

which clearly illustrates the fundamental role played by the nodal geometry of each FE domain Ω_e forming Ω^h.

Letting $\zeta_{\alpha i}$ denote the data ΔX and ΔY in the coordinate transformation matrix (3.75), the linear NC basis GWS^h algorithm element matrix term for $\nabla \bullet \rho\nabla\varphi$ on $n=2$ is

$$[\mathrm{DIFF}]_e\{Q\}_e = \frac{\bar{\rho}_e}{\det_e n!}\begin{bmatrix} \zeta_{11}^2 + \zeta_{12}^2, & \zeta_{11}\zeta_{21} + \zeta_{12}\zeta_{22}, & \zeta_{11}\zeta_{31} + \zeta_{12}\zeta_{32} \\ \zeta_{21}\zeta_{11} + \zeta_{22}\zeta_{12}, & \zeta_{21}^2 + \zeta_{22}^2, & \zeta_{21}\zeta_{31} + \zeta_{22}\zeta_{32} \\ \zeta_{31}\zeta_{11} + \zeta_{32}\zeta_{12}, & \zeta_{31}\zeta_{21} + \zeta_{32}\zeta_{22}, & \zeta_{31}^2 + \zeta_{32}^2 \end{bmatrix}\{Q\}_e \tag{3.76}$$

The element data set companion to (3.75) for $n=3$ is more detailed, as it involves determinantal expansions, recall (3.68). The essential form is

$$\left[\frac{\partial\zeta_\alpha}{\partial x_i}\right]_e = \frac{1}{\det_e}[f(\Delta X\Delta Y, \Delta X\Delta Z, \Delta Y\Delta Z]_e \tag{3.77}$$

with $\Delta X\Delta Y$, $\Delta Y\Delta Z$, $\Delta X\Delta Z$ the nodal coordinate difference products for the Ω_e geometry analogous to ΔX, ΔY in (3.75). The essential form of the GWS^h element matrix term for $\nabla \bullet \rho\nabla\varphi$ on $n=3$ is

$$[\mathrm{DIFF}]_e\{Q\}_e = \frac{\bar{\rho}_e}{\det_e n!}\begin{bmatrix} \zeta_{11}^2 + \zeta_{12}^2 + \zeta_{13}^2, & \zeta_{11}\zeta_{21} + \zeta_{12}\zeta_{22} + \zeta_{13}\zeta_{23}, \cdots \\ \zeta_{21}\zeta_{11} + \zeta_{22}\zeta_{12} + \zeta_{23}\zeta_{13}, & \zeta_{21}^2 + \zeta_{22}^2 + \zeta_{23}^2, \quad \cdots \\ \zeta_{31}\zeta_{11} + \zeta_{32}\zeta_{12} + \zeta_{33}\zeta_{13}, & \zeta_{31}\zeta_{21} + \zeta_{32}\zeta_{22} + \zeta_{33}\zeta_{23}, \cdots \\ \zeta_{41}\zeta_{11} + \zeta_{42}\zeta_{12} + \zeta_{43}\zeta_{13}, & \zeta_{41}\zeta_{21} + \zeta_{42}\zeta_{22} + \zeta_{43}\zeta_{23}, \cdots \end{bmatrix}\{Q\}_e \tag{3.78}$$

The theory (3.60) predicts a significant improvement in asymptotic convergence rate under regular mesh refinement for a $k>1$ basis implementation, provided the data are sufficiently smooth. This benefit is realizable for subsonic and shock-free (only!) transonic–supersonic flow applications. GWS^h implementation using the quadratic ($k=2$) NC basis is direct for discretizations formed as the union of straight-sided/planar-faced triangles/tetrahedra. The replacement for (3.70) is

$$\begin{aligned}[\mathrm{DIFF}]_e\{Q\}_e &= \int_{\Omega_e} \nabla\{N_2\} \cdot \rho_e \nabla\{N_2\}^T \mathrm{d}\tau\{Q\}_e \\ &= (\det_e/n!)\left(\frac{\partial\zeta_\alpha}{\partial x_i}\frac{\partial\zeta_\beta}{\partial x_i}\right)_e \int_{\Omega_e} \frac{\partial\{N_2\}}{\partial\zeta_\alpha}\rho_e\frac{\partial\{N_2\}^T}{\partial\zeta_\beta}\mathrm{d}\tau\{Q\}_e\end{aligned} \tag{3.79}$$

The inverse coordinate transformation remains identically (3.75) or (3.77), so does the extracted product $\left(\frac{\partial\zeta_\alpha}{\partial x_i}\frac{\partial\zeta_\beta}{\partial x_i}\right)_e$.

Forming integrals of the matrix product of the non-constant first derivatives of the $\{N_2(\zeta_\alpha)\}$ basis, (3.46), multiplying $\rho_e(\mathbf{x})$ are now required. This is greatly simplified by extracting ρ_e from the integrand and replacing it with the element average $\overline{\rho}_e$. This constitutes commission of a data *interpolation error*, which tends to zero faster than approximation error for *smooth data*, Strang and Fix (1974), which is the character of $\rho_e(\mathbf{x})$ for subsonic and shock-free transonic flow.

Expressing the coordinate transformation product $\left(\frac{\partial\zeta_\alpha}{\partial x_i}\frac{\partial\zeta_\beta}{\partial x_i}\right)_e$ in the shorthand $\zeta_{\alpha i}$ form, the GWS^h element matrix term is

$$[\mathrm{DIFF}]_e\{Q\}_e \approx \frac{\overline{\rho}_e}{\det_e n!}(\zeta_{\alpha i}\zeta_{\beta i})_e \int_{\Omega_e} \frac{\partial\{N_2\}}{\partial\zeta_\alpha}\frac{\partial\{N_2\}^T}{\partial\zeta_\beta} d\tau \{Q\}_e \tag{3.80}$$

The defined integrals can now be evaluated once and for all on the generic domain Ω_e since the integrand is devoid of element-dependent (subscript e) data. The closed form solution for *all* such integrals of natural coordinate products for arbitrary n is

$$\int_{\hat{\Omega}_e} \zeta_1^p\zeta_2^q\zeta_3^r\zeta_4^s \, d\tau = \det_e \frac{p!q!r!s!}{(n+p+q+r+s)!} \tag{3.81}$$

Recalling the "[M]" nomenclature for all generated matrices, the $k=2$ basis GWS^h algorithm matrix term for $\nabla \bullet \rho\nabla\varphi$ becomes

$$[\mathrm{DIFF}]_e\{Q\}_e = \frac{\overline{\rho}_e}{\det_e n!}(\zeta_{\alpha i}\zeta_{\beta i})_e[\mathrm{M2}\alpha\beta]\{Q\}_e \tag{3.82}$$

with summation on all repeated indices over ranges $1 \le (\alpha, \beta) \le n+1$ and $1 \le i \le n$. Convenient symmetric matrix sums of [M2αβ] can be generated from (3.82) for the $k=2$ NC basis on $n=2$. The following matrix library constitutes the replacement for $[\mathrm{M}\delta_{\alpha\beta}]$, (3.73), of the $k=1$ basis construction.

$$[\mathrm{M211}] = \frac{1}{3}\begin{bmatrix} 3 & 0 & 0 & 0 & 0 & 0 \\ & 0 & 0 & 0 & 0 & 0 \\ & & 0 & 0 & 0 & 0 \\ & & & 8 & 0 & 4 \\ & & & & 0 & 0 \\ & & & & & 8 \end{bmatrix}, \quad [\mathrm{M222}] = \frac{1}{3}\begin{bmatrix} 0 & 0 & 0 & 0 & 0 & 0 \\ & 3 & 0 & 0 & 0 & 0 \\ & & 0 & 0 & 0 & 0 \\ & & & 8 & 4 & 0 \\ & & & & 8 & 0 \\ & & & & & 0 \end{bmatrix}$$

$$[\mathrm{M233}] = \frac{1}{3}\begin{bmatrix} 0 & 0 & 0 & 0 & 0 & 0 \\ & 0 & 0 & 0 & 0 & 0 \\ & & 3 & 0 & 0 & 0 \\ & & & 0 & 0 & 0 \\ & & & & 8 & 4 \\ & & & & & 8 \end{bmatrix}$$

$$[\text{M212}]+[\text{M221}]=\frac{1}{3}\begin{bmatrix} 0 & -1 & 0 & 4 & 0 & 0 \\ & 0 & 0 & 4 & 0 & 0 \\ & & 0 & 0 & 0 & 0 \\ & & & 8 & 4 & 4 \\ & & & & 0 & 8 \\ & & & & & 0 \end{bmatrix}$$

$$[\text{M213}]+[\text{M231}]=\frac{1}{3}\begin{bmatrix} 0 & 0 & -1 & 0 & 0 & 4 \\ & 0 & 0 & 0 & 0 & 0 \\ & & 0 & 0 & 0 & 4 \\ & & & 0 & 8 & 4 \\ & & & & 0 & 4 \\ & & & & & 8 \end{bmatrix} \tag{3.83}$$

$$[\text{M223}]+[\text{M232}]=\frac{1}{3}\begin{bmatrix} 0 & 0 & 0 & 0 & 0 & 0 \\ & 0 & -1 & 0 & 4 & 0 \\ & & 0 & 0 & 4 & 0 \\ & & & 0 & 4 & 8 \\ & & & & 8 & 4 \\ & & & & & 0 \end{bmatrix}$$

The remaining GWS^h algorithm matrix term is the velocity efflux BC, the surface integral in (3.22), which for PDEs on n-D is integrated on $(n-1)$-D boundary segments $\partial\Omega_e$ of appropriate FE domains Ω_e. The flexible implementation is to interpolate these data onto nodal DOF of $\partial\Omega_e$, that is, $\rho_e\mathbf{u}(\mathbf{x}_b) \Rightarrow \{\rho\mathbf{u}\}_e$, which for arbitrary NC basis degree k produces the *data* matrix

$$\begin{aligned} \{\text{b}\}_e &= \int_{\partial\Omega_N \cap \partial\Omega_e} \{N_k(\zeta_\alpha)\}\rho_e\mathbf{u} \bullet \widehat{\mathbf{n}}\,d\sigma \\ &\approx \int_{\partial\Omega_e} \{N_k(\zeta_\alpha)\}\{N_k(\zeta_\alpha)\}^T\{\rho\mathbf{u}\}_e \bullet \widehat{\mathbf{n}}\,d\sigma \\ &\equiv \text{sgn}(\mathbf{u} \bullet \widehat{\mathbf{n}})\frac{\det_e}{n!}[\text{N200}]\{|\rho\mathbf{u}|\}_e \end{aligned} \tag{3.84}$$

The placeholder is symbolized as $\{\text{b}\}_e$, as in the abstract form (3.36), and $\text{sgn}(\mathbf{u} \bullet \widehat{\mathbf{n}})$ is the sign of the dot product. The matrix symbol prefix "[N]" signifies a **N**eumann BC matrix on $\partial\Omega_e$, the complement to the "[M]" prefix for **M**atrices on Ω_e, while $\det_e$ is pertinent to the $(n-1)$ dimension of $\partial\Omega_e$. Finally, the [N] matrix extension "200" denotes that it is formed by integrating two (2) FE bases, neither of which is differentiated, hence the Boolean index pair (00).

Neumann BC matrices are readily evaluated using (3.81). For the $k=1$ NC basis the $n=1$, 2 BC matrices on $\partial\Omega_e$ for the GWS^h algorithm formed on $n=2$, 3 dimensional domains Ω are

$$[\text{N200}]_{n=1} = \frac{1}{6}\begin{bmatrix} 2 & 1 \\ 1 & 2 \end{bmatrix}, \quad [\text{N200}]_{n=2} = \frac{1}{24}\begin{bmatrix} 2 & 1 & 1 \\ 1 & 2 & 1 \\ 1 & 1 & 2 \end{bmatrix} \tag{3.85}$$

with $\det_e$ in (3.84) equal to l_e or $2\mathrm{A}_e$, respectively the length of the BC line segment $\partial\Omega_e$ of a triangular Ω_e or twice the area of the BC face $\partial\Omega_e$ of the tetrahedron Ω_e. The companion $k=2$ NC basis, $n=1$ matrix is

$$[\text{N200}]_{n=1} = \frac{1}{15}\begin{bmatrix} 4 & 2 & -1 \\ 2 & 16 & 2 \\ -1 & 2 & 4 \end{bmatrix} \tag{3.86}$$

3.9 GWS^h Tensor Product FE Basis Matrices

GWS^h algorithm implementation via the tensor product (TP) basis $\{N_k^+(\boldsymbol{\eta})\}$, (3.26–3.27), requires the chain rule to complete integration of the lead term in (3.22) on the generic FE domain Ω_e

$$\begin{aligned}[\text{DIFF}]_e\{Q\}_e &\equiv \int_{\Omega_e} \nabla\{N_k^+\}\cdot\rho_e\nabla\{N_k^+\}^T d\tau\{Q\}_e \\ &= \int_{\Omega_e} \frac{\partial\{N^+\}}{\partial\eta_j}\left(\frac{\partial\eta_j}{\partial x_i}\right)_e \rho_e \frac{\partial\{N^+\}^T}{\partial\eta_k}\left(\frac{\partial\eta_k}{\partial x_i}\right)_e \det_e d\eta\{Q\}_e\end{aligned} \tag{3.87}$$

As with the NC basis, $\det_e$ is the determinant of the coordinate transformation $\mathbf{x}=\mathbf{x}(\boldsymbol{\eta})$ with $d\tau$ and $d\eta$ the differential elements.

In distinction to the NC basis completion, the TP basis inverse coordinate transformation $\eta_j=\eta_j(x_i)$ is not *a priori* known. The resolution is to interpolate the x_i coordinate system on Ω_e in terms of its nodal coordinate set $\{\mathrm{X}_i\}_e$, $1\le i\le n$, and the TP basis $\{N_k^+(\eta)\}$.

$$x_i = x_i(\eta_j) \equiv \{N_k^+(\eta_j)\}^T\{\mathrm{X}_i\}_e \tag{3.88}$$

Figure 3.8 illustrates the operation whereby the general curvilinear domain Ω_e in physical (x_i) space is transformed to the unit cartesian parallelepiped in transform (η_j) space. (Note: this operation is required for a $k>1$ NC basis implementation should Ω_e be curve-sided/faced.)

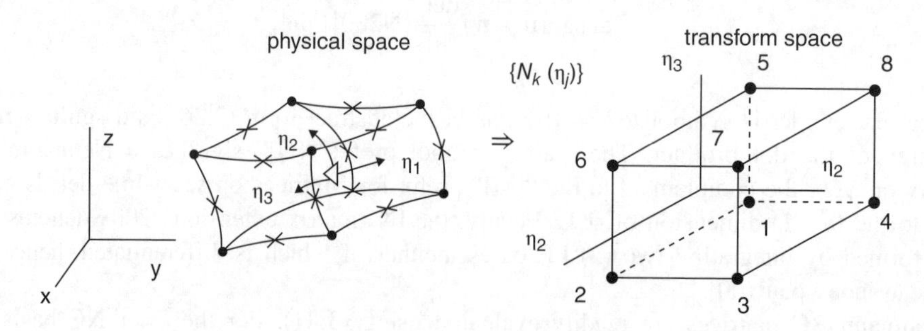

Figure 3.8 Tensor product basis coordinate transformation

With (3.88), the forward coordinate transformation on Ω_e is

$$\left[\frac{\partial x_i}{\partial \eta_j}\right]_e = \frac{\partial}{\partial \eta_j}\{N_k^+(\eta)\}^T\{X_i\}_e \tag{3.89}$$

The inverse transformation required in (3.87) is readily determined from (3.89)

$$\left[\frac{\partial \eta_j}{\partial x_i}\right]_e = \left[\frac{\partial x_i}{\partial \eta_j}\right]_e^{-1} = \frac{1}{\det_e}\begin{bmatrix} \textit{matrix of} \\ \textit{cofactors} \\ \textit{transposed} \end{bmatrix}_e \tag{3.90}$$

The $n=3$ transformation matrix contains cofactors computed via determinants, while the $n=2$ matrix is readily formed

$$\left[\frac{\partial \eta_j}{\partial x_i}\right]_e = \left[\frac{\partial x_i}{\partial \eta_j}\right]_e^{-1} = \frac{1}{\det_e}\begin{bmatrix} \dfrac{\partial x_2}{\partial \eta_2}, & -\dfrac{\partial x_1}{\partial \eta_2} \\ -\dfrac{\partial x_2}{\partial \eta_1}, & \dfrac{\partial x_1}{\partial \eta_1} \end{bmatrix}_e \tag{3.91}$$

Note the elements of the inverse coordinate transformation matrices (3.90–3.91) are η-dependent, in distinction to the NC basis transformation, and so also is $\det_e$. To illustrate, the elements of the $k=1$ TP basis transformation matrix on $n=2$, via (3.88), are

$$\begin{aligned}
\left(\frac{\partial \eta_2}{\partial x_2}\right)_e &= \frac{1}{\det_e}\left(\frac{\partial x_1}{\partial \eta_1}\right)_e = \frac{1}{\det_e}\frac{\partial}{\partial \eta_1}\{N_1^+(\eta)\}^T\{X_1\}_e \\
&= \frac{1}{4\det_e}\left[(1-\eta_2)(X_2-X_1)_e + (1+\eta_2)(X_3-X_4)_e\right] \\
\left(\frac{\partial \eta_1}{\partial x_1}\right)_e &= \frac{1}{4\det_e}\left[(1-\eta_1)(Y_4-Y_1)_e + (1+\eta_1)(Y_3-Y_2)_e\right] \\
\left(\frac{\partial \eta_1}{\partial x_2}\right)_e &= \frac{-1}{4\det_e}\left[(1-\eta_1)(X_4-X_1)_e + (1+\eta_1)(X_3-X_2)_e\right] \\
\left(\frac{\partial \eta_2}{\partial x_1}\right)_e &= \frac{-1}{4\det_e}\left[(1-\eta_2)(Y_2-Y_1)_e + (1+\eta_2)(Y_3-Y_4)_e\right]
\end{aligned} \tag{3.92}$$

The remaining issue for (3.87) is forming TP basis η-derivatives. Elementary expressions result *only* for the bilinear ($k=1$) basis $\{N_1^+(\eta)\}$. From (3.26) for $n=3$

$$\frac{\partial}{\partial \eta_j}\{N_k^+(\eta)\} = \frac{1}{8}\left\{\begin{matrix} -(1-\eta_2)(1-\eta_3) \\ (1-\eta_2)(1-\eta_3) \\ (1+\eta_2)(1-\eta_3) \\ -(1+\eta_2)(1-\eta_3) \\ -(1-\eta_2)(1+\eta_3) \\ (1-\eta_2)(1+\eta_3) \\ (1+\eta_2)(1+\eta_3) \\ -(1+\eta_2)(1+\eta_3) \end{matrix}\right\}_{j=1}, \text{ etc} \tag{3.93}$$

with the other two matrices formed via obvious j index alterations on the $(1 \pm \eta_j)$ term omitted in (3.93). The dual matrix set for $n=2$ is

$$\frac{\partial}{\partial \eta_j}\{N_k^+(\eta)\} = \frac{1}{4}\begin{Bmatrix} -(1-\eta_2) \\ (1-\eta_2) \\ (1+\eta_2) \\ -(1+\eta_2) \end{Bmatrix}_{j=1}, \quad \frac{1}{4}\begin{Bmatrix} -(1-\eta_1) \\ -(1+\eta_1) \\ (1+\eta_1) \\ (1-\eta_1) \end{Bmatrix}_{j=2} \tag{3.94}$$

Thereby, the TP basis derivatives are non-constant on Ω_e for *any* polynomial degree k, in total distinction to the constant $k=1$ NC basis derivatives. This precludes extraction of TP basis derivatives *as well as* the inverse coordinate transformation products from the integrand in (3.87), a truly consequential TP basis complication in forming GWS^h element matrices.

The resolution is using numerical quadrature to form (3.87). Since in transform space every domain Ω_e is rectangular cartesian, Gauss quadrature readily generates integral approximations of *a priori* known accuracy. Recalling $\zeta_{\alpha i}$ as the NC transformation matrix elements, let $\eta_{j\,i}$ be the element placeholder for the TP basis coordinate transformation (3.90). The generic matrix element of $[\mathrm{DIFF}]_e$, (3.87), that must be integrated can then be compactly expressed

$$\begin{aligned}(\mathrm{diff}_{\alpha\beta})_e &\equiv \iiint_{\hat{\Omega}_e} \rho_e(\eta)\frac{\partial N_\alpha}{\partial \eta_j}\frac{\partial N_\beta}{\partial \eta_k}(\eta_{ji}\eta_{ki})_e \det{}_e^{-1} d\eta \\ &\equiv \int_{\hat{\Omega}}\left(\mathrm{integrand}_{j,k,i}^{\alpha,\beta}(\boldsymbol{\eta})\right) d\eta, \quad \textit{for} \quad 1 \le j,k,i \le n\end{aligned} \tag{3.95}$$

Evaluating this matrix element via Gauss quadrature is the summation cascade

$$(\mathrm{diff}_{\alpha\beta})_e \cong \sum_{i=1}^{n}\sum_{j=1}^{n}\sum_{k=1}^{n}\sum_{p=1}^{P}\sum_{q=1}^{Q}\sum_{r=1}^{R} \mathrm{H}_p\mathrm{H}_q\mathrm{H}_r\left(\mathrm{integrand}_{j,k,i}^{\alpha,\beta}(\boldsymbol{\eta} \Rightarrow \eta_{j,k,i}^{p,q,r})\right) \tag{3.96}$$

where p, q, r denote specific η_j coordinate locations in Ω_e. For $n=3$ quadrature summation limits (P, Q, R), Table 3.2 lists the associated $\eta_{j,k,i}$ coordinates and weights $\mathrm{H}_{p,q,r}$. Figure 3.9 graphs these coordinates in plan view, and the geometric patterns extend directly into the third dimension.

The theory underlying Gauss quadrature, Zienkiewicz and Taylor (1996), confirms that a Pth order (P, Q, R in $n=3$) quadrature rule will exactly integrate a polynomial of degree $2\mathrm{P}-1$ on the cartesian TP element $\hat{\Omega}_e$ in transform space. Hence, integers (P, Q, R) can be selected and mixed to meet the integration accuracy requirements dependent on the highest degree polynomial present in (3.96). However, limit decisions require interpretation as the integrand in (3.96) is a *rational* polynominal due to $\det_e^{-1}$, recall the first line in (3.95).

Table 3.2 Gauss numerical quadrature rule data, $n=2, 3$

P, Q, R	*coordinates*, $\eta_{j,k,i}^{p,q,r}$	*weights*, $\mathrm{H}_{p,q,r}$
1	0.0	1.0
2	$\pm 1/\sqrt{3}$	1.0, 1.0
3	0.0, $\pm\sqrt{0.6}$	8/9, 5/9, 5/9

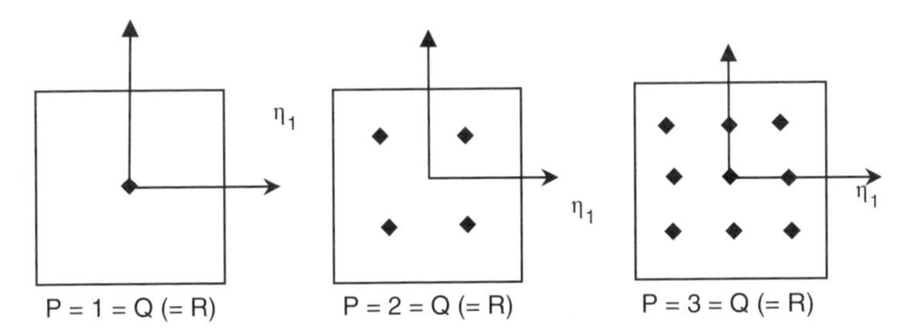

Figure 3.9 Gauss quadrature coordinates in plan view

For the D*M* potential $k=1$ TP basis GWS^h implementation, the (P, Q, R) $\equiv 1$ rule, which approximates the integral as the element centroid average, is inappropriate. The accurate and $k=1$ basis practical choice is (P, Q, R) $= 2$. Completing the quadrature loop for all elements in the $[\mathrm{DIFF}]_e$, matrix for *any* TP basis degree k construction generates the longer summation cascade

$$[\mathrm{DIFF}]_e \cong \sum_{\alpha,\beta}^{k,n} \sum_{i}^{n} \sum_{j}^{n} \sum_{k}^{n} \sum_{p}^{P} \sum_{q}^{Q} \sum_{r}^{R} \mathrm{H}_p \mathrm{H}_q \mathrm{H}_r \left(\mathrm{integrand}_{j,k,i}^{\alpha,\beta} (\boldsymbol{\eta} \Rightarrow \eta_{j,k,i}^{p,q,r}) \right) \tag{3.97}$$

This is a computationally unattractive deeply nested shallow FOR loop. As will become developed, the *modified* Galerkin ($m\mathrm{GWS}^h$) theory predicts no need for CFD implementation to employ other than the $k=1$ NC/TP bases. Thereby, a thoroughly practical completion exists provided the curvilinear body-fitted discretization Ω^h is the union of *essentially* parallelepiped domains Ω_e, the usual case.

For any parallelepiped Ω_e, Figure 3.10, the off-diagonal elements in the transformation matrix (3.90) vanish identically, readily validated via (3.92). Also, the diagonal elements of the transformation matrix (3.90) become η-*in*dependent, for example for $n=2$

$$\begin{aligned} \left(\frac{\partial \eta_1}{\partial x_1}\right)_e &= \frac{1}{4\,\det_e}\left(\frac{\partial x_2}{\partial \eta_2}\right)_e = \frac{1}{4\,\det_e}\left[(1-\eta_1)\Delta \mathrm{Y}_e + (1+\eta_1)\Delta \mathrm{Y}_e\right] = \frac{\Delta \mathrm{Y}_e}{2\,\det_e} \\ \left(\frac{\partial \eta_2}{\partial x_2}\right)_e &= \frac{1}{4\,\det_e}\left(\frac{\partial x_1}{\partial \eta_1}\right)_e = \frac{1}{4\,\det_e}\left[(1-\eta_2)\Delta \mathrm{X}_e + (1+\eta_2)\Delta \mathrm{X}_e\right] = \frac{\Delta \mathrm{X}_e}{2\,\det_e} \end{aligned} \tag{3.98}$$

and thereby $\det_e$ is also η-independent.

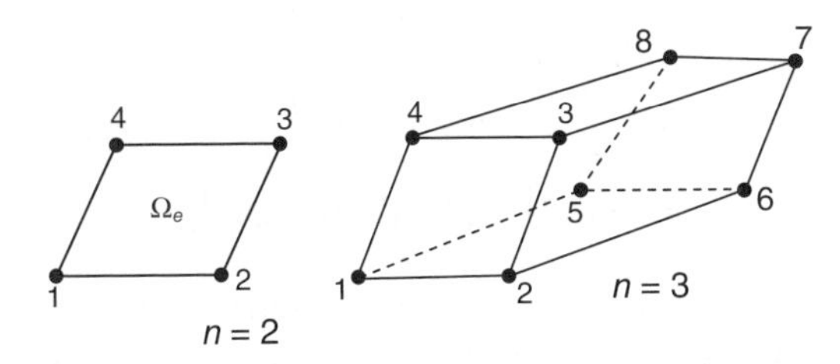

Figure 3.10 Bilinear TP basis parallelepiped Ω_e, $n=2$, 3

For the essential parallelepiped Ω_e constraint, and for the D*M* GWS^h applied to subsonic–transonic flows devoid of shocks, density is adequately approximated by $\bar{\rho}_e$ on Ω_e. Then (3.87) takes the *much* more efficient form

$$\begin{aligned}
[\mathrm{DIFF}]_e\{Q\}_e &= \int_{\hat{\Omega}_e} \frac{\partial\{N_1^+\}}{\partial\eta_j}\left(\frac{\partial\eta_j}{\partial x_i}\right)_e \rho_e \frac{\partial\{N_1^+\}^T}{\partial\eta_k}\left(\frac{\partial\eta_k}{\partial x_i}\right)_e \det_e d\eta\{Q\}_e \\
&\approx \bar{\rho}_e \det_e\left(\frac{\partial\eta_j}{\partial x_i}\frac{\partial\eta_k}{\partial x_i}\right)_e \int_{\hat{\Omega}_e} \frac{\partial\{N_1^+\}}{\partial\eta_j}\frac{\partial\{N_1^+\}^T}{\partial\eta_k} d\eta\{Q\}_e \\
&\equiv \frac{\bar{\rho}_e}{4\det_e}\left(\eta_{ji}\eta_{ki}\right)_e [\mathrm{M2}jk]\{Q\}_e
\end{aligned} \tag{3.99}$$

where η_{ji} and η_{ki} denote the data ΔY_e and ΔX_e, recall the NC basis notation $\zeta_{\alpha i}$ in (3.76). Thereby, the integrand in (3.99) now contains *no* element data which permits the integrals [M2*jk*] to be formed once and for all on the generic element $\hat{\Omega}_e$ in transform space.

Establishing the library of element-*in*dependent matrices [M2*jk*] (3.99) is direct using Gauss quadrature. For the $k=1$ TP basis the highest degree polynomial in η_j is quadratic, hence the (P, Q, R) = 2 rule produces accurate results, as $2P-1=3>2$. The $n=2$ element matrix library is

$$\begin{aligned}
[\mathrm{M211}] &= \frac{1}{3}\begin{bmatrix} 2 & -2 & -1 & 1 \\ -2 & 2 & 1 & -1 \\ -1 & 1 & 2 & -2 \\ 1 & -1 & -2 & 2 \end{bmatrix} \quad [\mathrm{M212}] = \frac{1}{4}\begin{bmatrix} 1 & 1 & -1 & -1 \\ -1 & -1 & 1 & 1 \\ -1 & -1 & 1 & 1 \\ 1 & 1 & -1 & -1 \end{bmatrix} \\
[\mathrm{M221}] &= \frac{1}{4}\begin{bmatrix} 1 & -1 & -1 & 1 \\ 1 & -1 & -1 & 1 \\ -1 & 1 & 1 & -1 \\ -1 & 1 & 1 & -1 \end{bmatrix} \quad [\mathrm{M222}] = \frac{1}{3}\begin{bmatrix} 2 & 1 & -1 & -2 \\ 1 & 2 & -2 & -1 \\ -1 & -2 & 2 & 1 \\ -2 & -1 & 1 & 2 \end{bmatrix}
\end{aligned} \tag{3.100}$$

and the $n=3$ library is formed with equal ease. Note the similarities between (3.100) and the $k=2$ NC basis library (3.83).

The remaining GWS^h element matrix term is the velocity efflux BC, the surface integral in (3.22). The TP basis data matrix definition is the minor alteration on (3.84)

$$\begin{aligned}
\{\mathrm{b}\}_e &= \int_{\partial\Omega_N \cap \partial\Omega_e} \{N_1^+(\eta)\}\rho_e \mathbf{u} \bullet \hat{\mathbf{n}} d\sigma \\
&= \int_{\partial\Omega_e} \{N_1^+(\eta)\}\{N_1^+(\eta)\}^T \{\rho\mathbf{u}\}_e \bullet \hat{\mathbf{n}} d\sigma \\
&\equiv sgn(\mathbf{u} \bullet \hat{\mathbf{n}})\det_e[\mathrm{N200}]\{|\rho\mathbf{u}|\}_e
\end{aligned} \tag{3.101}$$

with the [N200] matrix again formed via Gauss quadrature. For the TP $k=1$ basis, the $n=1$, 2 matrices for $\partial\Omega_e$ segments associated with GWS^h formed on $n=2$, 3 dimensional domains Ω are

$$[\text{N200}]_{n=1} = \frac{1}{3}\begin{bmatrix} 2 & 1 \\ 1 & 2 \end{bmatrix}, \quad [\text{N200}]_{n=2} = \frac{1}{9}\begin{bmatrix} 4 & 2 & 1 & 2 \\ 2 & 4 & 2 & 1 \\ 1 & 2 & 4 & 2 \\ 2 & 1 & 2 & 4 \end{bmatrix} \tag{3.102}$$

In (3.101), $\det_e = l_e/2$ for line segment $\partial\Omega_e$ for Ω_e a quadrilateral, similarly $\det_e = A_e/4$ for the face $\partial\Omega_e$ of a hexahedron Ω_e.

3.10 GWS^h Comparison with Laplacian FD and FV Stencils

For subsonic flow the aerodynamics D*M* principle reduces to the *linear* laplacian PDE, (3.11). It is informative to compare TS second-order accurate FD and WF FV *stencils* on uniform rectangular cartesian (i, j, k) meshes to stencils generated by the GWS^h algorithm for the NC and TP $k=1$ basis matrices $[\text{M}\delta_{jk}]$ and $[\text{M}2jk]$. The well-known $n=1$, 2, 3 finite difference stencils for the linear laplacian on these discretizations are

$$\begin{aligned} -\nabla^2 q &= \frac{-1}{\Delta x^2}\left(Q_{j-1} - 2Q_j + Q_{j+1}\right) + O(\Delta x^2), n = 1 \\ &= \frac{-1}{\Delta x^2}\begin{pmatrix} & Q_{j,k-1} & \\ Q_{j-1,k} & -4Q_{j,k} & Q_{j+1,k} \\ & Q_{j,k+1} & \end{pmatrix} + O(\Delta x^2), n = 2 \\ &= \frac{-1}{\Delta x^2}\left(-6Q_{i,j,k} + Q_{i\pm1,j\pm1,k\pm1}\right) + O(\Delta x^2), n = 3 \end{aligned} \tag{3.103}$$

GWS^h algorithm stencils are generated via *assembly* at a generic node in the mesh, symbolized as "$\text{S}_{i,j,k}$" rather than assembly via S_e into the matrix statement (3.22). Drawing on the Section 2.3 $n=1$ example, (2.38–2.39) are replaced by assembly at the generic node X_j shared by the Ω_e domains to the left and right. The outcome is the middle row-column matrix product in (2.40), in j-index notation

$$\begin{aligned} \text{S}_j\left([\text{DIFF}]_e\{Q\}_e\right) &= \text{S}_j\left(\frac{1}{l_e}\begin{bmatrix} 1 & -1 \\ -1 & 1 \end{bmatrix}\begin{Bmatrix} Q_{j-1} \\ Q_j \end{Bmatrix}, \frac{1}{l_e}\begin{bmatrix} 1 & -1 \\ -1 & 1 \end{bmatrix}\begin{Bmatrix} Q_j \\ Q_{j+1} \end{Bmatrix}\right) \\ &= \frac{1}{\Delta x}\left(-Q_{j-1} + 2Q_j - Q_{j+1}\right) \end{aligned} \tag{3.104}$$

which is identically the second-order accurate $n=1$ laplacian FD stencil, (3.103), multiplied by $\Delta x = l_e$ and -1, the consequence of the integration by parts intrinsic to GWS^h on $n=1$.

Quite tedious algebraic operations are required to node-assemble the $k=1$ NC basis GWS^h stencils for the $n=2$, 3 linear laplacian. For absolutely *arbitrary* orientations of triangle diagonals, or tetrahedron facets, see Figure 3.11, the suggested exercises will verify for $n=2$

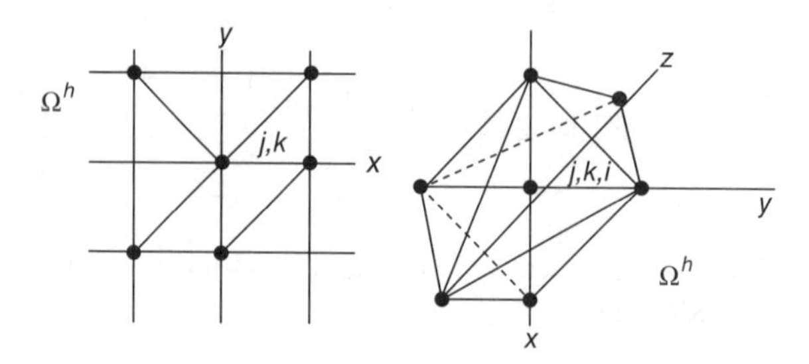

Figure 3.11 Uniform cartesian Ω^h for NC bases GWS^h nodal assembly

$$\mathrm{S}_{j,k}\left([\mathrm{DIFF}]_e\{Q\}_e\right) \Rightarrow \begin{pmatrix} & -Q_{j,k+1} & \\ -Q_{j-1,k} & 4Q_{j,k} & -Q_{j+1,k} \\ & -Q_{j,k-1} & \end{pmatrix} \tag{3.105}$$

and for $n=3$ in the shorthand notation of (3.103)

$$\mathrm{S}_{j,k,i}([\mathrm{DIFF}]_e\{Q\}_e) \Rightarrow \Delta x(6Q_{j,k,i} - Q_{j\pm1,k\pm1,i\pm1}) \tag{3.106}$$

Clearly, the $\{N_1(\zeta_\alpha)\}$ basis implementations of GWS^h *exactly reproduce* uniform rectangular cartesian mesh second-order accurate FD stencils, to within the Green–Gauss divergence theorem-induced sign difference and integration exponent on mesh measure h (Δx). This is a surprise, perhaps, since these FD stencils are constructed on unions of quadrilateral or hexahedron cells. The GWS^h theoretical prediction of quadratic convergence in the energy norm substantiates the prior assertion of FD order-of-accuracy equivalence *for smooth data*.

Difficulty in maintaining FD stencil order-of-accuracy on boundary-fitted curvilinear meshes led the CFD community to develop *finite volume* (FV) integral discrete methodology. An FV construction is readily interpreted as a set of decisions for a weak form WF, recall (2.8), the first of which restricts the *arbitrary* test function $w(\mathbf{x})$ requirement to the set of *all constants*. The resultant interpolation polynomial $\Phi(\mathbf{x})$ is unity, (2.9), yielding the WF extremization WS the scalar statement

$$\mathrm{WS} \equiv \partial\mathrm{WF}/\partial W_\beta = \int_\Omega \mathcal{L}(\varphi^N)\mathrm{d}\tau = 0 \tag{3.107}$$

Derived scalar (3.107) is valueless, cured in FV CFD algorithm practice by *immediately* forming the discretization $\Omega^h = \cup\Omega_v$ then invoking the *divergence theorem*. This transforms (3.107) to the set of surface integrals over *all* discrete cell boundaries defined in the mesh

$$\mathrm{WS} \Rightarrow \mathrm{WS}^h \equiv \mathrm{FV}^h = \int_{\Omega^h} -\nabla^2\varphi^h\mathrm{d}\tau = -\sum_{\partial\Omega_v}^{\Omega^h}\int_{\partial\Omega_v}\nabla\varphi^h \bullet \widehat{\mathbf{n}}\,\mathrm{d}\sigma = 0 \tag{3.108}$$

Implementing FV^h statement (3.108) requires coordinate transformations, as with all $n>1$ GWS^h FE constructions, as illustrated. However, in distinction to weak form use of calculus, (3.108) is *not* endowed *a priori* with an approximation trial space basis to support differentiation. Hence, in the absence of calculus to compute the boundary fluxes in (3.108), implementation relies on Taylor series plus *insight* to establish consistent order accurate discrete representations.

To illustrate the $FV^h \leftrightarrow GWS^h$ stencil comparison requires simplification to an $n=2$ uniform rectangular cartesian mesh. The FV^h construction employs both cell centroid and vertex nodes generating a *staggered mesh*. Figure 3.12 isolates the generic Ω_v union sharing node (j, k), labeled $1-4$ near the centroid nodes, and illustrates the unit normal pair on the two boundary segments $\partial\Omega_v$ shared by finite volumes 1 and 4. Vertex/centroid nodes are symbolized by circles/diamonds.

The FV^h stencil at node (j, k) involves summation of all surface integrals in (3.108) sharing this vertex node. For $n=2$, eight $\partial\Omega_v$ line integrals terminate at node (j, k). In FD notation, with (s,n) denoting local tangent/normal coordinates, the stencil for the FV^h surface integral (3.108) is formed as

$$-\sum_{\partial\Omega_v}^{\Omega^h}\int_{\partial\Omega_v}\nabla\varphi^h \bullet \hat{\mathbf{n}}\,d\sigma\bigg|_{j,k} \Rightarrow \sum_{j,k}\left(-\int_{\partial\Omega_v}\nabla\varphi^h\cdot\hat{\mathbf{n}}\,d\sigma\right)$$
$$\equiv \sum_{\alpha=1}^{8}(\pm\Delta Q_\alpha\Delta s/\Delta n)_{j,k} \tag{3.109}$$

Focusing on the two $\partial\Omega_v$ surfaces common to Ω_v domains 1 and 4, the terms in (3.109) are formed by finite differences of vertex node DOF. To what appear as first-order accurate approximations

$$-\int_{\partial\Omega_{1-4}}(\nabla\varphi^h\cdot\hat{\mathbf{n}})d\sigma \cong -\frac{\Delta x}{2\Delta y}\left[Q_{j,k}+Q_{j-1,k}-(Q_{j,k-1}+Q_{j-1,k-1})\right]$$
$$-\int_{\partial\Omega_{4-1}}(\nabla\varphi^h\cdot\hat{\mathbf{n}})d\sigma \cong -\frac{\Delta x}{2\Delta y}\left[Q_{j-1,k+1}+Q_{j,k+1}-(Q_{j-1,k}+Q_{j,k})\right] \tag{3.110}$$

Thus, efflux integral FD evaluations for all eight domain surfaces communicating with node (j, k) end up containing differences in vertex nodal DOF $Q_{j,k}$ and $Q_{j\pm1,k\pm1}$ multiplied/

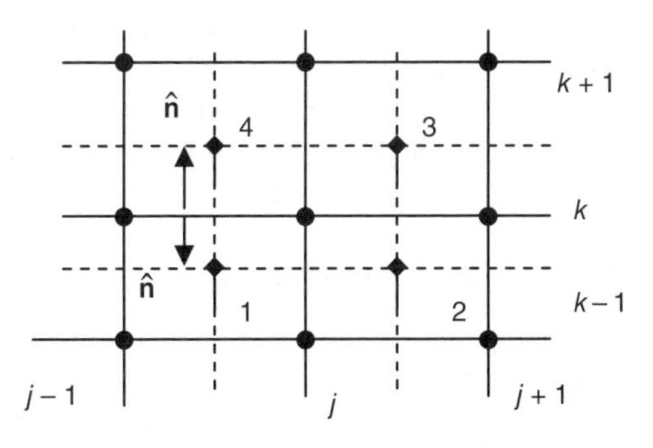

Figure 3.12 Cartesian staggered mesh for FV^h stencil at node (j, k)

divided by Δx and Δy. Summing the eight flux terms, then writing the result as a stencil for restriction a uniform mesh ($\Delta x = \Delta y$) yields

$$\sum_{j,k}\left(-\int_{\partial\Omega_v} \nabla\varphi^h \cdot \hat{\mathbf{n}} d\sigma\right) \approx \frac{1}{2}\begin{pmatrix} -2 & 0 & -2 \\ 0 & 8 & 0 \\ -2 & 0 & -2 \end{pmatrix} Q_{j,k} = \begin{pmatrix} -Q_{j-1,k+1} & 0 & -Q_{j+1,k+1} \\ 0 & 4Q_{j,k} & 0 \\ -Q_{j-1,k-1} & 0 & -Q_{j+1,k-1} \end{pmatrix} \tag{3.111}$$

The key distinguishing feature in (3.111) compared to the FD stencil (3.103) is replacement of the (j, k) row/column DOF with the out-rigger $(j \pm 1, k \pm 1)$ vertex DOF. Again, due to integration over $n = 2$ the mesh measure multiplier has disappeared.

Assembly of the GWS^h laplacian $\{N_1^+(\boldsymbol{\eta})\}$ basis implementation at vertex node (j, k) on this mesh (an exercise) produces

$$\mathrm{S}_{j,k}([\mathrm{DIFF}]_e\{Q\}_e) \Rightarrow \frac{1}{6}\begin{pmatrix} -2 & -2 & -2 \\ -2 & 16 & -2 \\ -2 & -2 & -2 \end{pmatrix} Q_{j,k} \tag{3.112}$$

a stencil connecting *all* DOF $(j \pm 1, k \pm 1)$ to (j, k). Weak form theory predicts that (3.112) is the *optimal* construction. From this viewpoint, since both the FD and FV^h stencils are second-order accurate so is their linear combination, hence

$$\frac{1}{2}(\mathrm{FD} + \mathrm{FV}^h)_{j,k} = \frac{1}{4}\begin{bmatrix} -2 & -2 & -2 \\ -2 & 16 & -2 \\ -2 & -2 & -2 \end{bmatrix} Q_{j,k} \tag{3.113}$$

reproduces the optimal GWS^h stencil (3.112) to within a scalar multiplier.

To summarize, on uniform rectangular cartesian meshes the very familiar FD laplacian stencils are *a priori* known optimally second-order accurate for all n and smooth data. Their exact reproduction via the linear NC basis GWS^h implementation validates the weak form theory assertion of optimal construction. Conversely the most readily formed FV^h algorithm stencil is predicted not optimal, in comparison to that derived by the linear TP basis GWS^h implementation. Interestingly, a linear FD + FV^h combination mimics the optimal TP basis form.

The GWS^h stencil process is broadly utilized in this text to support *optimal modified continuous* Galerkin weak form (mGWS^h) theoretical constructions for nonlinear CFD algorithms. This exercise certainly verifies, for the most elementary non-trivial PDE, the assertion that weak form theory can precisely organize the decisions underlying *independently derived* CFD discrete algorithm constructions.

3.11 Post-Processing Pressure Distributions

The sought output from a D*M* potential flow GWS^h algorithm solution is the pressure distribution to be imposed on the boundary layer-displaced aerodynamic shape, recall Figure 3.1. The conventional approach would employ FD nodal DOF manipulations on the Cauchy–Lagrange integral, (3.7), simplified for steady flow (more familiarly, the

Bernoulli equation)

$$\int \mathbf{DP}: \quad p(\mathbf{x}) = p_\infty(\mathbf{x}) - \frac{1}{2}\rho\mathbf{u}\cdot\mathbf{u} = p_\infty(x) - \frac{1}{2}\rho\nabla\varphi\cdot\nabla\varphi \tag{3.114}$$

The alternative restates (3.114) in homogeneous form leading to the *differential definition* $\mathcal{D}(p) \equiv p - p_\infty + \frac{1}{2}\rho\nabla\varphi\cdot\nabla\varphi = 0$. Forming the *continuum* GWS^N on this restatement of (3.114) renders the approximation error an extremum

$$\begin{aligned}\text{GWS}^N &= \int_\Omega \Psi_\beta(\mathbf{x})\mathcal{D}(p^N)\mathrm{d}\tau = \{0\}\\ &= \int_\Omega \Psi_\beta(\mathbf{x})\left[p^N(\mathbf{x}) - p_\infty(\mathbf{x}) + \frac{1}{2}\rho^N\nabla\varphi^N\cdot\nabla\varphi^N\right]\mathrm{d}\tau\\ &= \{0\}, \quad for \quad 1\le\beta\le N\end{aligned} \tag{3.115}$$

immediately transformable to the FE trial space basis statement

$$\text{GWS}^h = \text{S}_e\left[\begin{array}{c}\int_{\Omega_e}\{N_k\}\{N_k\}^T\{P\}_e\mathrm{d}\tau - \int_{\Omega_e}\{N_k\}P_\infty\mathrm{d}\tau\\ +\int_{\Omega_e}0.5\rho_e(\mathbf{x})\nabla\{N_k\}\{\varphi\}_e \bullet \nabla\{N_k\}^T\{\varphi\}_e\mathrm{d}\tau\end{array}\right] = \{0\} \tag{3.116}$$

Subsonic flow density variations are modest, hence extract ρ_e from the integrand and replace it by the element average. The final GWS^h is

$$\text{GWS}^h = \text{S}_e\left[\begin{array}{c}\int_{\Omega_e}\{N_k\}\{N_k\}^T\mathrm{d}\tau\{P\}_e - P_\infty\int_{\Omega_e}\{N_k\}\mathrm{d}\tau\\ +0.5\bar{\rho}_e\{\varphi\}_e^T\int_{\Omega_e}\nabla\{N_k\} \bullet \nabla\{N_k\}^T\mathrm{d}\tau\{\varphi\}_e\end{array}\right] = \{0\} \tag{3.117}$$

Note the element matrices produced by the integrals in (3.117) are not new, except for the reference pressure term. Selecting the NC linear basis, the matrix statement equivalent for (3.117) is

$$\text{GWS}^h = \text{S}_e\left[\begin{array}{l}\det_e[\text{M200}]\{P\}_e - P_\infty\det_e\{\text{M10}\}\\ +\dfrac{\bar{\rho}_e\det_e}{2n!}\left(\dfrac{\partial\zeta_\alpha}{\partial x_i}\dfrac{\partial\zeta_\beta}{\partial x_i}\right)_e\{\varphi\}_e^T[\text{M}\delta_{\alpha\beta}]\{\varphi\}_e\end{array}\right] = \{0\} \tag{3.118}$$

Alternatively, the TP $k=1$ basis matrix statement for (3.117) is

$$\text{GWS}^h = \text{S}_e\left[\begin{array}{c}\det_e[\text{M200}]\{P\}_e - P_\infty\det_e\{\text{M10}\}\\ +0.5\bar{\rho}_e\det_e\left(\dfrac{\partial\eta_j}{\partial x_i}\dfrac{\partial\eta_k}{\partial x_i}\right)_e\{\varphi\}_e^T[\text{M2}jk]\{\varphi\}_e\end{array}\right] = \{0\} \tag{3.119}$$

For $n=2$ implementations, the NC or TP linear basis [M200] matrix is identically [N200], (3.85) or (3.102). The corresponding $n=3$ NC linear and TP tri-linear basis matrices are

$$[\mathrm{M200}]_{\mathrm{NC}} = \frac{1}{120}\begin{bmatrix} 2 & 1 & 1 & 1 \\ 1 & 2 & 1 & 1 \\ 1 & 1 & 2 & 1 \\ 1 & 1 & 1 & 2 \end{bmatrix}, \quad [\mathrm{M200}]_{\mathrm{TP}} = \frac{1}{27}\begin{bmatrix} 8 & 4 & 2 & 4 & 4 & 2 & 1 & 2 \\ 4 & 8 & 4 & 2 & 2 & 4 & 2 & 1 \\ 2 & 4 & 8 & 4 & 1 & 2 & 4 & 2 \\ 4 & 2 & 4 & 8 & 2 & 1 & 2 & 4 \\ & & & & & \mathrm{etc} & & \end{bmatrix} \tag{3.120}$$

The reference pressure matrix is formed as $\{\mathrm{M10}\} \equiv [\mathrm{M200}]\{\mathrm{one}\}$, for column matrix $\{\mathrm{one}\}$ of length matching the DOF order of [M200].

Baker (2012, Ch. 6.6) documents post-processing pressure via (3.119) for potential flow in a venturi, a duct with variable cross-section area with a minimum (throat). Figure 3.13 graphs uniform mesh refinement *a posteriori* data documenting convergence of centerline potential and pressure distributions. Note that throat pressure is *very* sensitive to small (*nonlinear*!) changes in potential slope. The data confirm that the potential and pressure solutions both exhibit quadratic convergence in the energy norm.

3.12 Transonic Potential Flow, Shock Capturing

For subsonic and shock-free transonic potential flow the GWS^h algorithm k-degree FE basis implementation generates the *optimal* solution which, under *regular mesh refinement*, will exhibit order $2k$ asymptotic convergence in the energy norm, (3.60). The more challenging, also mathematically interesting, problem is transonic potential flow with an embedded shock with caveat a *weak shock* isentropic approximation. A shock is the ultimate example of *solution non-smoothness*, as it corresponds to a step change in the solution over an infinitesimal distance compared to the measure h of *any*(!) mesh.

Derivation of CFD discrete algorithms to "capture" weak shocks for various forms of the potential flow conservation principles DM and $\int$**DP**, (3.8–3.12), dominated the early 1970s. Approaches focused on steady formulations generating the *serious mathematical challenge* of requiring steady DM, (3.10), to switch from elliptic boundary value to hyperbolic in the supersonic flow reach. The additional challenge was maintaining algorithm stability due to supersonic flow abrupt termination in a shock which generates a *dispersive error* mode (detailed in a later chapter).

The inaugural FD algorithm, Murman and Cole (1971), devised a "typed finite difference" scheme for the $n=2$ small perturbation form of steady DM, (3.10), using central differences throughout the subsonic region and an upwind scheme in the supersonic reach. The goal was to emulate the mathematical domains of dependence as DM switched character. Jameson (1975) followed with a "rotated difference scheme" for the steady full potential DM, (3.10), which aligned the FD stencil with flow streamlines.

These typed-FD methods proved cumbersome hence research turned to strictly central difference schemes augmented with numerical *artificial diffusion* operators designed to dissipate discrete solution-generated $O(2h)$ mesh scale oscillations. The community

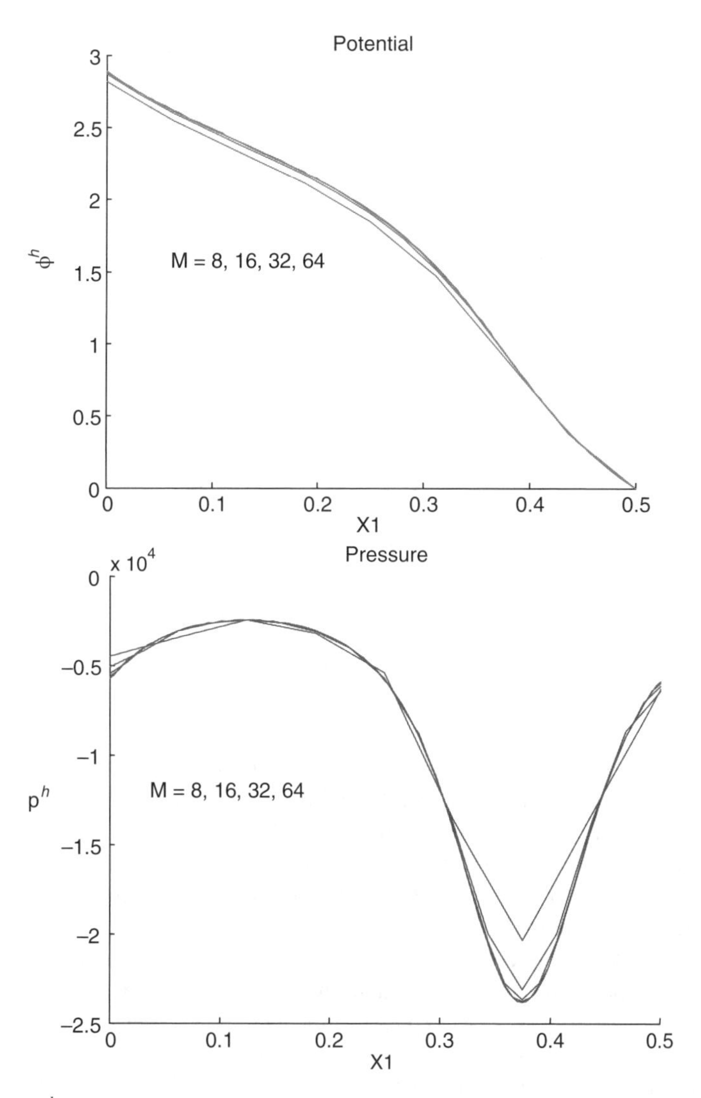

Figure 3.13 GWS^h algorithm, pressure post-processing from potential flow solution, venturi geometry, convergence with uniform axial mesh refinement $8 \leq M \leq 64$: top, centerline potential function; bottom, centerline pressure

transported fledgling NS numerical diffusion concepts to potential flow resulting in an *artificial viscosity* FV algorithm, Jameson & Caughey (1977). Due to computer hardware modesty these algorithms used a segregated iterative solution strategy with steady D*M* first solved for potential using a lagged density, followed by updating density algebraically via steady $\int$D**P**. Maintaining iterative stability also required *upwinding* of the central FD stencil everywhere which constituted an additional artificial diffusion mechanism, Shankar *et al.* (1981).

The 1970s also witnessed derivation of a range of weak form-consistent FE algorithms for $n=2$ transonic potential flows, Chan *et al.* (1975), Glowinski *et al.* (1976), Hafez and Murman (1979). The last observed that an *artificial compressibility* development, Eberle (1977), equated precisely to the Jameson fully conservative artificial diffusion FD operator. The resultant $n=2$ *modifications* to steady (3.8–3.9) are

$$\mathrm{D}M: -\frac{\partial}{\partial x}\left(\tilde{\rho}\frac{\partial \varphi}{\partial x}\right) - \frac{\partial}{\partial y}\left(\tilde{\rho}\frac{\partial \varphi}{\partial y}\right) = 0 \tag{3.121}$$

$$\tilde{\rho} \equiv \rho + \frac{\mu\rho}{a^2}\left[u\frac{\partial u}{\partial x}\Delta x + v\frac{\partial v}{\partial y}\Delta y\right], \quad \mu = max(0, 1 - \mathrm{Ma}^{-2}) \tag{3.122}$$

$$\int \mathbf{DP}: \quad \frac{\rho}{\rho_\infty} = \left[1 - \frac{(\gamma-1)\nabla\varphi\cdot\nabla\varphi}{2a_\infty^2}\right]^{1/(\gamma-1)} \tag{3.123}$$

for a sound speed and μ based on local Mach number Ma in (3.122).

Habashi and Hafez (1981) published a segregated iterative $k=1$ TP basis FE algorithm for (3.121) and (3.123), replacing the FD-based artificial density (3.122) with

$$\tilde{\rho} = \rho - \mu\rho_s\Delta s, \quad \rho_s\Delta s \equiv \left[\frac{u}{q}\frac{\partial \rho}{\partial x}\Delta x + \frac{v}{q}\frac{\partial \rho}{\partial y}\Delta y\right], \quad q = \left(u^2 + v^2\right)^{1/2} \tag{3.124}$$

The refined μ definition involves nodal Mach number on each side of the shock to moderate the excess artificial diffusion associated with (3.122). For a range of Mach number onset flows to an $n=2$ airfoil at zero angle of attack, reported solutions evidencing sharp shock capturing without pre-shock potential overshoots. These results were published as data interpolation without DOF symbols, hence unknown is the number of elements across which the shock was "captured."

The choice of steady conservation principle forms, hence state variable identification in (3.121–3.123), to support discrete "shock-capturing" algorithms creates mathematical complications. Factually unsteady DM (3.8) expresses time evolution of density, yet the steady PDE was solved as a pseudo-laplacian on potential! Unsteady $\int$**DP** expresses the time evolution of potential, which was instead solved algebraically for density. Clearly the steady form PDE state variable selection truly violated conservation principle fundamentals.

Since steady-state iteration is algebraically analogous to time marching, Zhang (1995) reports *a posteriori* data for an implicit time stepping $m\mathrm{GWS}^h + \theta\mathrm{TS}$ algorithm. The continuum potential theory *modified* PDE (mPDE) system annihilates the significant order time truncation error generating a dimensionally consistent replacement for (3.124). The theory starting point is primitive DM (3.1) inserted into a time Taylor series (TS), cf. Donea (1984)

$$\begin{aligned}\rho(t+\Delta t) &= \rho(t) + \Delta t\frac{\partial \rho}{\partial t} + \frac{\Delta t^2}{2}\frac{\partial^2 \rho}{\partial t^2} + O(\Delta t^3)\\ &= \rho(t) - \Delta t\nabla\cdot\rho\mathbf{u} - \frac{\Delta t^2}{2}\frac{\partial}{\partial t}\nabla\cdot\rho\mathbf{u} + O(\Delta t^3)\end{aligned} \tag{3.125}$$

then substituting (3.1). Space and time being independent coordinates, upon interchanging and switching to tensor index notation (3.125) is re-expressed as

$$\rho(t+\Delta t) = \rho(t) - \Delta t\frac{\partial \rho u_j}{\partial x_j} - \frac{\Delta t^2}{2}\frac{\partial}{\partial x_j}\left[\frac{\partial \rho u_j}{\partial t}\right] \rightarrow \left[-u_j u_k \frac{\partial \rho}{\partial x_k}\right] \qquad (3.126)$$

with second bracket term resulting from another insertion of (3.1).

Dividing through by Δt produces

$$\frac{\rho(t+\Delta t)-\rho(t)}{\Delta t} + \frac{\partial \rho u_j}{\partial x_j} - \frac{\Delta t}{2}\frac{\partial}{\partial x_j}\left[u_j u_k \frac{\partial \rho}{\partial x_k}\right] \approx 0 \qquad (3.127)$$

Replacing $\Delta t/2$ by the mesh local time scale $h/|\mathbf{u}|$, absorbing ½ into the Habashi Mach number DOF shock switch $\mu = \max(0, 1 - \{\mathrm{Ma}^{-2}\})$, then taking the limit $\Delta t \Rightarrow 0$ generates the TS *modified* D*M* conservation principle *m*PDE

$$\mathrm{D}^m M\colon \frac{\partial \rho}{\partial t} + \frac{\partial}{\partial x_j}\left[\rho u_j - \mu h \frac{u_j u_k}{|\mathbf{u}|}\frac{\partial \rho}{\partial x_k}\right] = 0, \quad \mu = \max\left(0, 1 - \left\{\mathrm{Ma}^{-2}\right\}\right) \qquad (3.128)$$

Comparing (3.128) to (3.121) or (3.124) clearly confirms that what was historically described as artificial compressibility in a steady transonic D*M* formulation is really a dimensionally homogeneous, nonlinear *semi-positive definite* tensor product differential operator augmentation to the *continuum* unsteady D*M* for state variable member ρ. Closure for this unsteady *modified* continuous Galerkin ($m\mathrm{GWS}^N$) algorithm written for (3.128) involves

$$\int \mathbf{DP}\colon \quad \frac{\partial \varphi}{\partial t} + \frac{1}{2}\frac{\partial \varphi}{\partial x_j}\frac{\partial \varphi}{\partial x_j} + \frac{\gamma}{\gamma - 1}\left[T_\infty\left(\frac{\rho}{\rho_\infty}\right)^{\gamma-1} - T_{0\infty}\right] = 0 \qquad (3.129)$$

an initial-value *nonlinear* PDE operating on state variable member φ.

Note that text title wording appears in this transonic potential flow $m\mathrm{GWS}^N$ formulation designed to admit weak shock capture. Thorough development of the $m\mathrm{GWS}^N$ theory awaits the addressing of complete I-EBV NS PDE systems. As stated earlier, this topic is deferred to Chapter 5 as aerodynamic weak interaction theory completion via GWS^N algorithms for BL/PNS PDE systems do not require stabilization.

Zhang (1995) reports *a posteriori* data for $k=1$ TP FE basis $m\mathrm{GWS}^h$ algorithm implementation of (3.128–3.129) solved fully coupled for state variable $\{q(\mathbf{x}, t)\} = \{\rho, \varphi\}^T$. Integration through time to a steady solution used the backwards ($\theta = 1.0$) Euler implicit algorithm. The reported data are for the $n=2$ benchmark problem of $\mathrm{Ma}_\infty = 0.675$ onset flow to a 15% thick parabolic bump at zero angle of attack. The *body-fitted* solution-adapted TP basis mesh, Figure 3.14 top, is indeed constituted of essentially parallelepiped elements.

Confirmation of $m\mathrm{GWS}^h$ steady solution monotonicity is visualized, Figure 3.14 bottom, by "painting" the TP basis mesh on the Mach number DOF distribution surface

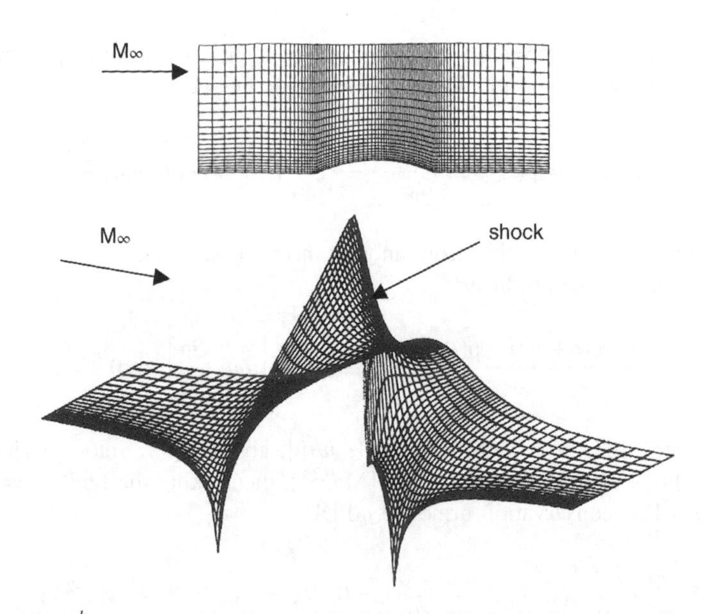

Figure 3.14 mGWS^h unsteady algorithm transonic potential simulation, onset flow $\mathrm{Ma}_\infty = 0.675$: top, $k = 1$ TP basis body fitted mesh; bottom, steady solution Mach number $\{\mathrm{Ma}\}$ DOF distribution perspective

perspective. The shock foot is located at 79% bump chord with wall Mach number DOF directly upstream of the shock of $\{\mathrm{Ma}(x_S)\} = 1.38$. The perspective confirms that the shock is "captured" over the span of three finite elements and exhibits no pre- or post-shock oscillations. Zhang reports that shock $\{\mathrm{Ma}(x_S)\}$ and shock foot location are each within 1% of the benchmark solution.

3.13 Summary

The aerodynamics potential flow DM and ∫**DP** conservation principles constitute the *only* PDE + BCs system in fluid mechanics that can be linear without being trivial. This facilitates rigorous exposition of weak form mathematical theory and the spatially discrete implementation of a Galerkin weak form algorithm GWS^N.

The subsonic–transonic potential flow conservation principle primitive statement is (3.8–3.9)

$$\mathrm{D}M\text{:} \quad \frac{\partial \rho}{\partial t} + \nabla \cdot \rho \nabla \varphi = 0$$

$$\int \mathbf{DP}\text{:} \quad \frac{\partial \varphi}{\partial t} + \frac{1}{2} \nabla \varphi \cdot \nabla \varphi + \frac{\gamma}{\gamma - 1} \frac{T_\infty}{\rho_\infty^{\gamma - 1}} \rho^{\gamma - 1} = \frac{\gamma}{\gamma - 1} T_{0\infty}$$

for which legacy CFD algorithm approaches have simplified to steady flow. In differential operator notation this generates the quasi-linear EBV PDE plus algebraic system (3.17), (3.33) for state variable $\{q(\mathbf{x})\} = \{\varphi, \rho\}^T$

$$\text{D}M: \quad \mathcal{L}(\varphi) = -\nabla \cdot \rho \nabla \varphi = 0$$
$$\int \text{DP}: \quad \rho = \rho_\infty \left[\frac{T_{0\infty}}{T_\infty} - \frac{\gamma - 1}{2\gamma T_\infty} \nabla \varphi \cdot \nabla \varphi \right]^{1/(\gamma-1)}$$

Weak form theory addressing (3.17) starts with defining the approximate solution with trial space, (3.18)

$$\varphi^N(\mathbf{x}) \equiv \sum_{\alpha=1}^{N} \Psi_\alpha(\mathbf{x}) Q_\alpha$$

The Galerkin criterion weak statement GWS^N, the continuum calculus integral statement (3.19), is converted to computable form via domain discretization and definition of GWS^h trial space bases $\{N_k(\boldsymbol{\nu}(\mathbf{x}))\}$ spanning finite element domains Ω_e. Expressed via non-D normalized local coordinate systems, finite elements span line segments, triangles or quadrilaterals and/or tetrahedra or hexahedra for $n = 1, 2, 3$ dimensions.

The precise expression of solution approximation and the resultant GWS^h algorithm for (3.17) is (3.21–3.22)

$$\varphi^N(\mathbf{x}) \equiv \sum_{\alpha=1}^{N} \Psi_\alpha(\mathbf{x}) Q_\alpha \Rightarrow \varphi^h(\mathbf{x}) \equiv \cup_e^{\mathrm{M}} \{N_k(\boldsymbol{\nu}(\mathbf{x}))\}^{\mathrm{T}} \{Q\}_e$$

$$\text{GWS}^h = \text{S}_e \left[\begin{array}{c} \displaystyle\int_{\Omega_e} \nabla \{N_k(\boldsymbol{\nu}(\mathbf{x}))\} \bullet \rho_e \nabla \{N_k(\boldsymbol{\nu}(\mathbf{x}))\}^T \{Q\}_e \mathrm{d}\tau \\ \displaystyle -\int_{\partial\Omega_N \cap \partial\Omega_e} \{N_k(\boldsymbol{\nu}(\mathbf{x}))\} \rho_e \mathbf{u} \bullet \widehat{\mathbf{n}} \mathrm{d}\sigma \end{array} \right] = \{0\}$$

where S_e symbolizes *assembly*, the matrix row-wise addition process forming the global matrix statement from its element-level components.

The fundamental question of why choose a GWS^h construction, as opposed to any other formulation, was rigorously answered via the *linear* theory proof that the GWS^h solution is *optimal* among its discrete implementation peer group. Specifically, no other discrete solution possessing the identical degrees-of-freedom (DOF) will exhibit error, suitably measured, smaller than that produced by the GWS^h algorithm.

By following (perhaps *too*) thoroughly detailed mathematical expositions, it emerges that the *error distribution* $e^h(\mathbf{x})$ associated with shock free GWS^h solutions $\varphi^h(\mathbf{x})$ to (3.18) are rigorously proven to asymptote to zero under *regular mesh refinement* as, (3.60)

$$\left\| e^h \right\|_E \leq \mathrm{C}\, h^{2\gamma} \left[\|\text{data}\|_{\mathrm{L2},\Omega} + \|\text{data}\|_{\mathrm{L2},\partial\Omega} \right], \quad \gamma \equiv \min(k + 1 - m, r - m)$$

where $\|\bullet\|_E$ denotes the *energy norm*, which for (3.17) is (3.32)

$$\|\varphi, \varphi\|_E \equiv \frac{1}{2} \int_\Omega [\rho \nabla \varphi \cdot \nabla \varphi] \mathrm{d}\tau$$

Further, in (3.60) C is a constant, $\|\text{data}\|_{\mathrm{L2}}$ contains the L2 norm of the *data* defining the problem, k is the trial space basis *completeness* degree, $m = 1$ for the laplacian and r is the measure of data *non-smoothness* influence on convergence.

Fully detailed construction of GWS^h algorithm computable forms for $n = 1,2,3$ dimensions, including BCs, are derived. Thereafter, comparisons between second-order accurate FD and FV^h laplacian stencils with GWS^h $k = 1$ NC/TP basis implementation stencils are detailed. The stencil for the $k = 1$ NC basis, not the TP basis, *exactly* reproduces the $O(\Delta x^2)$ rectangular cartesian FD constructions which substantiates weak form theory assertion of *optimal* form creation.

The primary useful output from potential flow analyses are pressure distributions. The GWS^h for *post-processing* pressure from its *differential definition*, (3.130)

$$\mathcal{D}(p) \equiv p(\mathbf{x}) - p_\infty(\mathbf{x}) + \frac{1}{2}\rho\nabla\varphi \cdot \nabla\varphi = 0$$

was derived. Computed *a posteriori* data for a venturi geometry document prediction accuracy sensitivity to the quadratic nonlinearity in (3.130) and quantitatively validate the asymptotic convergence rate predicted by theory (3.60).

The significantly more challenging *and* mathematically interesting issue of transonic potential flow with an embedded *shock* completes the chapter. This enabled identification, without detail, of the $m\mathrm{GWS}^N$ algorithm written on the theory-*modified* *m*PDE conservation principle (3.128) replacement for (3.1)

$$\mathrm{D}^m M\!: \quad \frac{\partial \rho}{\partial t} + \frac{\partial}{\partial x_j}\left[\rho u_j - \mu h \frac{u_j u_k}{|\mathbf{u}|}\frac{\partial \rho}{\partial x_k}\right] = 0, \quad \mu = \max(0, 1 - \{\mathrm{Ma}^{-2}\})$$

with $m\mathrm{GWS}^h + \theta\mathrm{TS}$ algorithm solution a quality approximation to the shock characteristic to a benchmark problem. This appearance of text title wording hopefully prompts reader interest in *modified optimal continuous* Galerkin CFD theory.

Exercises

3.4.1 Expand the $m = 1$ H^m Sobolev norm definition (3.28) for $n = 1, 2, 3$, making sure to generate *all* derivative combinations.

3.4.2 Verify the steady density solution (3.33) starting with (3.9).

3.5.1 Substitute (3.35) into the abstract weak form (3.39), hence confirm that it correctly states the GWS for the steady D*E* principle.

3.5.2 Substitute (3.36) into the abstract weak form (3.39), hence confirm that it correctly states the GWS for the steady aerodynamics potential flow D*M* principle.

3.5.3 Confirm that (3.41) is identically twice the energy norm for both the steady D*E* and D*M* principles, (3.31–3.32).

3.7.1 Confirm that the graphs in Figure 3.5 correspond to non-regular and regular mesh refinement by sketching the progression of length scales ρ_e and h_e therein associated.

3.7.2 For the steady D*E* PDE + BCs system, (3.35), precisely state and evaluate the H^0 and L2 norms presented in (3.59), (3.60).

3.7.3 For the steady DM PDE + BCs system, (3.36), precisely evaluate the H^0 and L2 norms, (3.59) and (3.60).

3.8.1 Using Cramer's rule, solve (3.62), hence verify the solution (3.63). Also validate that $\det_e$ equates to 2 A_e.

3.8.2 For the definition (3.64), confirm the solution (3.65).

3.8.3 Verify that (3.69) is correct and that $\det_e$ equates to 6 V_e.

3.8.4 Using (3.71), generate a few entries in the $k=1$ basis matrix (3.73), hence verify its accuracy.

3.8.5 Verify the inverse coordinate transformation matrix (3.75).

3.8.6 Verify the $k=1$ basis $n=2$ [DIFF] matrix in (3.76).

3.8.7 Fill out the last row in the $k=1$, $n=3$ matrix in (3.78).

3.8.8 Verify the $k=1$ basis Neumann BC matrices (3.85).

3.8.9 Verify the $k=2=n$ Neumann BC matrix (3.86).

3.9.1 Verify the transformation matrix (3.88), then confirm the TP $k=1$ basis $n=2$ data, (3.91).

3.9.2 For the $k=1$ TP basis parallelepiped Ω_e, verify vanishing of the inverse coordinate transformation matrix off-diagonal elements and η-independence of the diagonal elements.

3.9.3 Using (P,Q) = 2 quadrature rule, verify the elements of the [M2jk] matrix (3.100).

3.9.4 Verify the BC matrices (3.102).

3.10.1 Confirm that FD stencils (3.103) correspond to second-order accurate discrete approximations to the laplacian.

3.10.2 For a cartesian uniform mesh bisected into triangles, $n=2$, or tetrahedra, $n=3$, confirm that the linear NC basis GWS^h implementation exactly reproduces the FD stencils in (3.103), independent of bisector or facet orientations.

3.10.3 Confirm the FV^h algorithm stencil (3.111).

3.10.4 Generate the GWS^h bilinear TP basis laplacian stencil (3.112), hence estimate (or form) the $n=3$ construction.

References

Baker, A.J. (2012). *Finite Elements ⇔ Computational Engineering Sciences*, John Wiley and Sons, London.

Cebeci, T. and Smith, A.M.O. (1974). *Analysis of Turbulent Boundary Layers*, Academic, NY.

Chan, S.T.K., Brashears, M.R. and Young, V.Y.C. (1975). "Finite element analysis of transonic flow by the method of weighted residuals," Technical Paper AIAA75:XXX.

Donea, J. (1984). "A Taylor-Galerkin algorithm for hyperbolic conservation laws," *J. Numerical Methods Engineering*, V.20.

Eberle, A. (1977). "Eine Methode Finiter Elements Zur Brechnung der Transsonischen Potential-Stromung um Profile," Technical Report MBB Bericht Nr. UEE 132(0).

Glowinski, R., Periaux, J. and Pironneau, O. (1976). "Transonic flow simulation by the finite element method via optimal control," Proceedings, 2nd *Int. Symposium in Flow Problems*, Santa Margherita, Italy.
Habashi, W.G. and Hafez, M. (1981). "Finite element solution of transonic flow problems," *AIAA Journal*, V.20.
Hafez, M. and Murman, E. (1979). "Artificial compressibility methods for numerical solution of the transonic full potential equation," *AIAA Journal*, V.17.
Jameson, A. (1975). "Transonic potential flow calculations using conservation form," Proc. AIAA 2nd *CFD Conference*, pp. 148–161.
Jameson, A. and Caughey, D. (1977). "A finite volume scheme for transonic potential flow calculations," *AIAA Journal*, V.15.
Murman, E. and Cole, J.D. (1971). "Calculation of plane steady transonic flow," *AIAA Journal*, V.9.
Oden, J.T. and Reddy, J.N. (1976). *An Introduction to the Mathematical Theory of Finite Elements*, Wiley-Interscience, NY.
Oden, J.T. (1994). "Optimal *h-p* finite element methods," *Comp. Methods Applied Mechanics and Engineering*, V.112.
Oden, J.T. and Demkowicz, L.F. (1996). *Applied Functional Analysis*, CRC Press, Boca Raton, FL.
Schlichting, H. (1979). *Boundary Layer Theory*, McGraw-Hill, NY.
Shankar, N.L., Malone, J.B. and Tassa, Y. (1981). "An implicit conservation algorithm for steady and unsteady three-dimensional transonic potential flow," Technical Paper AIAA:81-1016.
Strang, G. and Fix, G.J. (1974). *An Analysis of the Finite Element Method*, Prentice-Hall, NJ.
Thompson, J.F., Warzi, Z.U.A. and Mastin, C.W. (1985). *Numerical Grid Generation, Formulations & Applications*, Elsevier, NY.
Zhang, J. (1995). "A selective Taylor weak statement-based, quadrature-free finite element algorithm for aerodynamic flow," PhD dissertation, University of Tennessee/Knoxville.
Zienkiewicz, O.C. and Taylor, R.L. (1996). *The Finite Element Method*, McGraw-Hill, NY.

4

Aerodynamics II: boundary layers, turbulence closure modeling, parabolic Navier–Stokes

4.1 Aerodynamics, Weak Interaction Reprise

The approximate characterization of aerodynamic flowfields via potential theory was a first CFD practical application. The complement to aerodynamic potential theory is boundary layer (BL) theory, a significant simplification of the Navier–Stokes (NS) PDE system upon assumption of steady unidirectional viscous and turbulent fluid flow in direct contact with an aerodynamic surface.

The coupling of a farfield potential solution with the nearfield boundary layer prediction is termed *weak interaction theory*. Therein, the aerodynamic surface for potential flow prediction is moved to the boundary layer *displacement thickness* (distribution) $\delta^*(x, z)$, an integral parameter computed from the BL solution. Coupling is achieved by imposing the δ^* surface inviscid potential pressure distribution onto the BL solution. For wing–body juncture region flow, completely ducted flow or hypersonic shock layer flow, the 3-D generalization of the BL PDE system is called the *parabolic* Navier–Stokes (PNS) system.

The $\text{GWS}^N \Rightarrow \text{GWS}^h$ algorithms for BL and PNS Reynolds ordered PDE + BCs + IC systems are developed in this chapter. Of true pedagogic impact, the time averaged form of the BL-PNS PDE systems, applicable to *turbulent* aerodynamic flow prediction, have fundamentally supported essentially *all* historical theoretical musings on turbulent flow closure modeling. The associated *very consequential* simplifications of the NS conservation principle system via Reynolds ordering generate well-posed PDE + BCs statements exhibiting initial value character in the direction of dominant flow. Additionally, these systems are not pathologically nonlinear, which obviates the need for numerical *artificial diffusion* stabilization in the absence of singular behavior.

Optimal MODIFIED CONTINUOUS Galerkin CFD, First Edition. A. J. Baker.

Companion Website: www.wiley.com/go/baker/GalerkinCFD

4.2 Navier–Stokes PDE System Reynolds Ordered

Development of $\mathrm{GWS}^N \Rightarrow \mathrm{GWS}^h$ solution methodology for manipulations of NS PDE + BCs + IC basic mass, momentum and energy conservation principles for aerodynamic flow prediction is the objective. The starting point remains (recall Chapter 1)

$$\mathrm{D}M\colon \frac{\partial \rho}{\partial t} + \nabla \cdot \rho \mathbf{u} = 0 \tag{4.1}$$

$$\mathbf{DP\colon} \frac{\partial \rho \mathbf{u}}{\partial t} + \nabla \cdot \rho \mathbf{u}\mathbf{u} = \rho \mathbf{g} + \nabla \mathbf{T} \tag{4.2}$$

$$\mathrm{D}E\colon \frac{\partial \rho e}{\partial t} + \nabla \cdot (\rho e + p)\mathbf{u} = s - \nabla \cdot \mathbf{q} \tag{4.3}$$

for NS state variable $\{q(\mathbf{x},t)\}$ members density ρ, velocity vector $\mathbf{u}$, pressure p and total internal energy e. The chapter focus is the incompressible form of (4.1–4.3). Appendix B addresses the parabolic compressible hypersonic aerodynamics PDE system.

The fundamental BL assumptions are steady, incompressible and *unidirectional* flow, Schlichting (1979). For the added restriction of laminar boundary layer (BL) theory, development starts with the steady non-dimensional (non-D) NS system (1.19–1.22)

$$\mathrm{D}M\colon \nabla \cdot \mathbf{u} = 0 \tag{4.4}$$

$$\mathbf{DP\colon u} \cdot \nabla \mathbf{u} - \frac{1}{\mathrm{Re}} \nabla^2 \mathbf{u} + \nabla P + \frac{\mathrm{Gr}}{\mathrm{Re}^2} \Theta\, \widehat{\mathbf{g}} = 0 \tag{4.5}$$

$$\mathrm{D}E\colon \mathbf{u} \cdot \nabla \Theta - \frac{1}{\mathrm{RePr}} \nabla^2 \Theta - \frac{\mathrm{Ec}}{\mathrm{Re}} s_\Theta = 0 \tag{4.6}$$

The non-D *state variable* is $\{q(\mathbf{x})\} = \{\mathbf{u},\ P,\ \Theta\}^T$ with Reynolds, Prandtl, Grashoff and Eckert non-D parameter definitions remaining $\mathrm{Re} \equiv \mathrm{UL}/\nu$, $\mathrm{Pr} \equiv \rho_0 \nu c_p/k$, $\mathrm{Gr} \equiv \mathrm{g}\beta\Delta \mathrm{TL}^3/\nu^2$ and $\mathrm{Ec} \equiv \mathrm{U}^2/c_p\Delta\mathrm{T}$. Further, $\Delta\mathrm{T} \equiv (T_{\max} - T_{\min})$, $\beta \equiv 1/\mathrm{T}_{\mathrm{abs}}$, the kinematic pressure definition is $P \equiv p/\rho_0$ and in D*E* the Peclet number $\mathrm{Pe} \equiv \mathrm{RePr}$ is common.

The BL system results from (4.4–4.6) via Reynolds ordering arguments enforcing the assumption that the flow is *unidirectional*. Figure 4.1 graphs the BL flow domain, the region bounded by the aerodynamic surface and another surface displaced therefrom called the *boundary layer thickness* denoted $\delta(x, z)$. The location of $\delta(x, z)$ is *a priori* unknown, being determined by the solution of the BL PDE system. The surface-normal coordinate system is (s, n, z); the curvature of s is a second-order effect, Cebeci and Smith (1974),

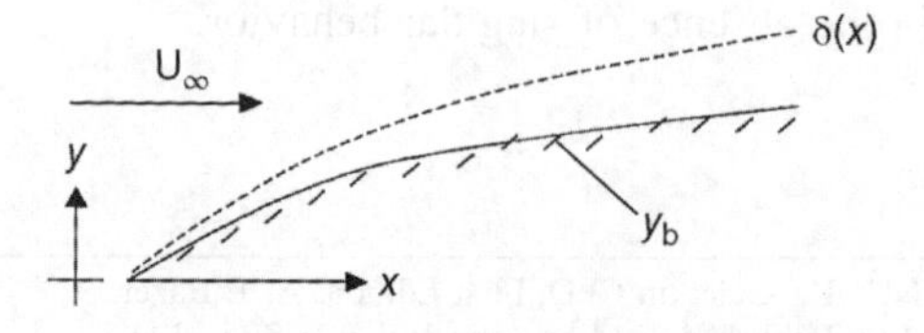

Figure 4.1 Boundary layer flow domain with bounding surfaces

hence the global (x, y, z) system is typically employed with x spanning the onset flow direction.

Reynolds ordering transforms the differentially constrained, elliptic boundary value (EBV) PDE system (4.4–4.6) into a *parabolic* PDE system, initial value in x, boundary value on $0 \le y \le \delta$ with z dependence embedded parametrically. For reference scales U_∞, L and $\delta \ll L$, Reynolds *ordering* (symbolized as "O") of DM, (4.4), yields

$$O(\nabla \cdot \mathbf{u}) = O\left(\frac{u/U_\infty}{x/L}\right) + O\left(\frac{v/U_\infty}{\delta/L}\right) + O\left(\frac{w/U_\infty}{z/L}\right) \tag{4.7}$$

By definition, the numerator and denominator of the lead term are both order unity, hence so must be *all* remaining terms, since DM is a fundamental principle totally independent of the fluid present *and* the flow state. Since by definition $\delta \ll L$, the order of v/U_∞ must thereby be very small. Conversely, z/L is order about unity for a swept wing, hence for x properly aligned the order of w/U_∞ is certainly less than unity but most likely larger than δ/L.

Determination of $O(\delta/L)$ accrues to Reynolds ordering the x component of **DP**, (4.5). In non-D and suppressing the appearance of U_∞ and L, $\mathbf{DP} \bullet \hat{\mathbf{i}}$ thus ordered is

$$\begin{aligned} O(\mathbf{DP} \bullet \hat{\mathbf{i}}) = O\left(\frac{uu}{1}\right) + O\left(\frac{vu}{\delta}\right) + O\left(\frac{wu}{1}\right) + O\left(\frac{P}{1}\right) \\ -\frac{1}{\text{Re}}\left[O\left(\frac{u}{1^2}\right) + O\left(\frac{u}{\delta^2}\right) + O\left(\frac{u}{1^2}\right)\right] + \frac{\text{Gr}}{\text{Re}^2} O\left(\Theta \hat{\mathbf{g}} \bullet \hat{\mathbf{i}}\right) \end{aligned} \tag{4.8}$$

The first two convection terms are each order unity, nothing can be said about the third convection and pressure terms, and via $\hat{\mathbf{g}} \bullet \hat{\mathbf{i}}$ the Gr/Re^2 term order is negligible unless the BL is formed on a nominally vertical surface. The remaining (viscous) term orders cannot exceed unity, hence Re^{-1} must be $O(\delta^2)$, via the middle term, thereby the first and third terms are negligible. The resultant BL PDE for $\mathbf{DP} \bullet \hat{\mathbf{i}}$ is

$$\mathbf{DP} \bullet \hat{\mathbf{i}} : \mathbf{u} \cdot \nabla u + \frac{\partial P}{\partial x} - \frac{1}{\text{Re}} \frac{\partial^2 u}{\partial y^2} + \frac{\text{Gr}}{\text{Re}^2} \Theta(\hat{\mathbf{g}} \bullet \hat{\mathbf{i}}) = 0 \tag{4.9}$$

The second significant result accrues to Reynolds ordering the **DP** PDE for transverse velocity component v. The result is

$$\begin{aligned} O(\mathbf{DP} \bullet \hat{\mathbf{j}}) = O\left(\frac{uv}{1}\right) + O\left(\frac{vv}{\delta}\right) + O\left(\frac{wv}{1}\right) + O\left(\frac{P}{\delta}\right) \\ -\frac{1}{\text{Re}}\left[O\left(\frac{v}{1^2}\right) + O\left(\frac{v}{\delta^2}\right) + O\left(\frac{v}{1^2}\right)\right] + \frac{\text{Gr}}{\text{Re}^2} O\left(\Theta \hat{\mathbf{g}} \bullet \hat{\mathbf{j}}\right) \end{aligned} \tag{4.10}$$

Since v is $O(\delta)$ and Re^{-1} is $O(\delta^2)$, then buoyancy effects and the variation of kinematic pressure P across the boundary layer span must be $O(\delta)$. Thereby ordering predicts that $\mathbf{DP} \bullet \hat{\mathbf{j}}$ is negligible in its entirety! The truly fundamental consequence is that $\partial P/\partial x$ in

(4.9) can be replaced by $\partial P^I/\partial x$. Specifically, pressure is no longer a state variable member being replaced by *imposition* of the farfield potential (*I*nviscid, superscript *I*) flow pressure distribution $P^I(x,z)$ throughout the BL thickness.

Reynolds ordering the remaining NS PDEs converts them to parabolic form. Since $\mathrm{D}\mathbf{P} \bullet \hat{\mathbf{j}}$ is negligible, D*M* becomes the defining PDE for transverse velocity *v*, and the laminar incompressible BL PDE system is

$$\mathrm{D}M: \frac{\partial v}{\partial y} + \left(\frac{\partial u}{\partial x} + \frac{\partial w}{\partial z}\right) = 0 \tag{4.11}$$

$$\mathrm{D}\mathbf{P} \bullet \hat{\mathbf{i}} : \mathbf{u} \cdot \nabla u - \frac{1}{\mathrm{Re}}\frac{\partial^2 u}{\partial y^2} + \frac{\partial P^I}{\partial x} + \frac{\mathrm{Gr}}{\mathrm{Re}^2}\Theta(\hat{\mathbf{g}} \bullet \hat{\mathbf{i}}) = 0 \tag{4.12}$$

$$\mathrm{D}\mathbf{P} \bullet \hat{\mathbf{k}} : \mathbf{u} \cdot \nabla w - \frac{1}{\mathrm{Re}}\frac{\partial^2 w}{\partial y^2} + \frac{\partial P^I}{\partial z} = 0 \tag{4.13}$$

$$\mathrm{D}E: \mathbf{u} \cdot \nabla\Theta - \frac{1}{\mathrm{RePr}}\frac{\partial^2 \Theta}{\partial y^2} - \frac{\mathrm{Ec}}{\mathrm{Re}}\left(\frac{\partial u}{\partial y}\right)^2 = 0 \tag{4.14}$$

In distinction to the NS PDE system (4.4–4.6), the state variable for the laminar-thermal BL PDE system (4.11–4.14) is now $\{q\} = \{\mathbf{u},\Theta\}^T$, as pressure enters only as *data*, *the truly significant* mathematical consequence of Reynolds ordering the NS PDE system to BL form!

4.3 GWS^h, $n = 2$ Laminar-Thermal Boundary Layer

The laminar-thermal BL GWS^h algorithm is detailed for PDE system (4.11–4.14) reduced to $n = 2$. Replacing the D*M*, D**P**·**i** and D*E* designations therein with PDE placeholder notation $\mathcal{L}(\bullet)$, recall Chapter 2, in scalar notation the BL PDE system is

$$\mathcal{L}(u) = u\frac{\partial u}{\partial x} + v\frac{\partial u}{\partial y} - \frac{1}{\mathrm{Re}}\frac{\partial^2 u}{\partial y^2} + \frac{\mathrm{d}P^I}{\mathrm{d}x} + \frac{\mathrm{Gr}}{\mathrm{Re}^2}\Theta(\hat{\mathbf{g}} \bullet \hat{\mathbf{i}}) = 0 \tag{4.15}$$

$$\mathcal{L}(\Theta) = u\frac{\partial \Theta}{\partial x} + v\frac{\partial \Theta}{\partial y} - \frac{1}{\mathrm{RePr}}\frac{\partial^2 \Theta}{\partial y^2} - \frac{\mathrm{Ec}}{\mathrm{Re}}\left(\frac{\partial u}{\partial y}\right)^2 = 0 \tag{4.16}$$

$$\mathcal{L}(v) = \frac{\partial v}{\partial y} + \frac{\partial u}{\partial x} = 0 \tag{4.17}$$

ordered in BL state variable $\{q\} = \{u,\Theta,v\}^T$ member sequence pertinent to each principle. The appearance of (4.17) emphasizes that *y* is an initial value coordinate, confirming it is a first-order ordinary differential equation (ODE).

Deferring consideration of (4.17), the GWS^N for (4.15–4.16) is identically (3.18–3.19) extended for state variable pair $\{q(x, y)\} = \{(u,\Theta\}^T$. For trial space Ψ_α the statements of solution approximation for either member can be

$$q^N(x,y) \equiv \sum_{\alpha=1}^{N} \Psi_\alpha(x,y)Q_\alpha \quad or \quad \sum_{\alpha=1}^{N} \Psi_\alpha(y)Q_\alpha(x) \tag{4.18}$$

which identifies a decision requirement. Specifically, should initial-value coordinate dependence be embedded in the trial space, left, or assigned to the expansion coefficient set, right?

The practical answer is the latter, as the former would require modification of the established library of GWS^h finite element (FE) trial space bases $\{N_k(\boldsymbol{\nu}(\mathbf{x}))\}$. Hence *continuum* GWS^N for (4.15) and/or (4.16) with this decision for (4.18) is

$$\mathrm{GWS}^N = \int_\Omega \Psi_\beta(y)\mathcal{L}(q^N(x,y))\mathrm{d}y \equiv \{0\}, \quad for \quad 1 \le \beta \le N \tag{4.19}$$

For the progression GWS$^N \Rightarrow$ GWS^h, the statement of approximation, (4.18) right, for the state variable *pair* is

$$\{q^N(x,y)\} \equiv \sum_{\alpha=1}^{N} \mathrm{diag}[\Psi_\alpha(y)]\{Q_\alpha(x)\} \Rightarrow \{q^h(x,y)\} \equiv \bigcup\nolimits_e^{\mathrm{M}}\{q(x,y)\}_e \tag{4.20}$$

where diag[•] denotes diagonal matrix and $\{q(x,y)\}_e$ is a discrete solution state variable on the generic FE domain Ω_e in Ω^h. As $\{q\}_e = \{u,\ \Theta,\ v\}^T$ is the BL global matrix statement state variable, the GWS^h trial space basis definition (4.20) extended to the state variable trio is

$$\{q(x,y)\}_e \equiv \{u,\Theta,v\}_e^T = \mathrm{diag}\left[\{N_k(\zeta_\alpha(y))\}^T\right]\begin{Bmatrix} \mathrm{U}(x) \\ \mathrm{T}(x) \\ \mathrm{V}(x) \end{Bmatrix}_e \tag{4.21}$$

Substituting (4.15) into GWS^N, (4.19), and integrating by parts the second derivative term multiplied by Re^{-1}, the GWS^h for axial velocity u^h discrete approximation using (4.21) is

$$\begin{aligned}
\mathrm{GWS}^N\left(\mathcal{L}(u^N)\right) &\equiv \int_\Omega \Psi_\beta(y)\mathcal{L}(u^N)\mathrm{d}y = \{0\}, \forall\beta \\
&\rightarrow S_e\left[\int_{\Omega_e}\{N_k(\zeta_\alpha)\}\mathcal{L}(u_e)\mathrm{d}y\right] \\
&= S_e\left[\begin{array}{l}\displaystyle\int_{\Omega_e}\{N_k(\zeta_\alpha)\}\left(u_e\frac{\partial u_e}{\partial x} + v_e\frac{\partial u_e}{\partial y} + \frac{\partial P^I}{\partial x} + \frac{\mathrm{Gr}}{\mathrm{Re}^2}\Theta_e(\widehat{\mathbf{g}}\bullet\widehat{\mathbf{i}})\right)\mathrm{d}y \\ \displaystyle + \frac{1}{\mathrm{Re}}\int_{\Omega_e}\frac{\mathrm{d}\{N_k(\zeta_\alpha)\}}{\mathrm{d}y}\frac{\partial u_e}{\partial y}\mathrm{d}y - \frac{1}{\mathrm{Re}}\{N_k(\zeta_\alpha)\}\frac{\partial u_e}{\partial y}\bullet\widehat{\mathbf{n}}\Big|_{y_{\mathrm{wall}}}^{\delta}\end{array}\right]
\end{aligned} \tag{4.22}$$

In substituting (4.21) throughout (4.22), the y derivatives are supported by the trial space basis. The x derivative is implemented on the Ω_e DOF matrix $\{\mathrm{U}(x)\}_e$, which being *only* a function of x generates the matrix ordinary derivative $\mathrm{d}\{\mathrm{U}(x)\}_e/\mathrm{d}x$. Now suppressing basis functional notation the element matrix statement for (4.22) is

$$
\begin{aligned}
&\mathrm{GWS}^h\left(\mathcal{L}(u^h)\right) = \{0\}\\
&= S_e\left[\begin{aligned}
&\int_{\Omega_e}\{N_k\}\left(\{N_k\}^T\{\mathrm{U}\}_e\{N_k\}^T\frac{\mathrm{d}\{\mathrm{U}\}_e}{\mathrm{d}x} + \{N_k\}^T\{\mathrm{V}\}_e\frac{\mathrm{d}\{N_k\}^T}{\mathrm{d}y}\{\mathrm{U}\}_e\right)\mathrm{d}y\\
&+\int_{\Omega_e}\{N_k\}\left(\frac{\mathrm{d}P^I}{\mathrm{d}x} + \frac{\mathrm{Gr}}{\mathrm{Re}^2}\{N_k\}^T\{\mathrm{T}\}_e(\widehat{\mathbf{g}}\bullet\widehat{\mathbf{i}})\right)\mathrm{d}y\\
&+\frac{1}{\mathrm{Re}}\int_{\Omega_e}\frac{\mathrm{d}\{N_k\}}{\mathrm{d}y}\frac{\mathrm{d}\{N_k\}^T}{\mathrm{d}y}\{\mathrm{U}\}_e\mathrm{d}y - \frac{1}{\mathrm{Re}}\{N_k\}\frac{\partial u_e}{\partial y}\bullet\widehat{\mathbf{n}}\Big|_{y_{\mathrm{wall}}}^{\delta}
\end{aligned}\right]
\end{aligned}
\tag{4.23}
$$

The concluding step is to extract constant data and parameters from the integrands. Since $\{N_k\}^T\{\mathrm{U}\}_e$ and $\{N_k\}^T\{\mathrm{V}\}_e$ are scalars, they and their transposes can be moved at will, which admits expressing (4.23) in the final form for integration

$$
\begin{aligned}
&\mathrm{GWS}^h\left(\mathcal{L}(u^h)\right) = \{0\}\\
&= \mathrm{S}_e\left[\begin{aligned}
&\{\mathrm{U}\}^T\int_{\Omega_e}\{N_k\}\{N_k\}\{N_k\}^T\mathrm{d}y\frac{\mathrm{d}\{\mathrm{U}\}_e}{\mathrm{d}x}\\
&+\{\mathrm{V}\}^T\int_{\Omega_e}\{N_k\}\{N_k\}\frac{\mathrm{d}\{N_k\}^T}{\mathrm{d}y}\mathrm{d}y\{\mathrm{U}\}_e\\
&+\frac{\mathrm{d}P^I}{\mathrm{d}x}\int_{\Omega_e}\{N_k\}\mathrm{d}y + \frac{\mathrm{Gr}}{\mathrm{Re}^2}\int_{\Omega_e}\{N_k\}\{N_k\}^T\mathrm{d}y\{\mathrm{T}\}_e(\widehat{\mathbf{g}}\bullet\widehat{\mathbf{i}})\\
&+\frac{1}{\mathrm{Re}}\int_{\Omega_e}\frac{\mathrm{d}\{N_k\}}{\mathrm{d}y}\frac{\mathrm{d}\{N_k\}^T}{\mathrm{d}y}\mathrm{d}y\{\mathrm{U}\}_e - \frac{1}{\mathrm{Re}}\{N_k\}\frac{\partial u_e}{\partial y}\bullet\widehat{\mathbf{n}}\Big|_{y_{\mathrm{wall}}}^{\delta}
\end{aligned}\right]
\end{aligned}
\tag{4.24}
$$

The key feature in (4.24) is that each integrand is totally devoid of element data. Hence, all trial space basis product integrals can be evaluated once and for all on generic element Ω_e. Recalling the [M] matrix prefix convention, Section 3.8, the symbolic matrix statement equivalent for the GWS^h algorithm (4.24) using (4.21) is

$$
\mathrm{GWS}^h\left(\mathcal{L}(u^h)\right) = \mathrm{S}_e\left[\begin{aligned}
&\det_e\{\mathrm{U}\}^T[\mathrm{M3000}]\frac{\mathrm{d}\{\mathrm{U}\}_e}{\mathrm{d}x} + \{\mathrm{V}\}_e^T[\mathrm{M3001}]\{\mathrm{U}\}_e\\
&+\frac{\mathrm{d}P^I}{\mathrm{d}x}\det_e\{\mathrm{M10}\} + \frac{\mathrm{Gr}}{\mathrm{Re}^2}\det_e[\mathrm{M200}]\{\mathrm{T}\}_e(\widehat{\mathbf{g}}\bullet\widehat{\mathbf{i}})\\
&+\frac{1}{\det_e\mathrm{Re}}[\mathrm{M211}]\{\mathrm{U}\}_e - \frac{1}{\mathrm{Re}}\{N_k\}\frac{\partial u_e}{\partial y}\bullet\widehat{\mathbf{n}}\Big|_{y_{\mathrm{wall}}}^{\delta}
\end{aligned}\right]
\tag{4.25}
$$

Repeating this process the GWS^h algorithm matrix statement for BL PDE (4.16) is

$$\mathrm{GWS}^h\left(\mathcal{L}(\Theta^h)\right) = \mathrm{S}_e \begin{bmatrix} \det_e\{\mathrm{U}\}^T[\mathrm{M3000}]\dfrac{\mathrm{d}\{\mathrm{T}\}_e}{\mathrm{d}x} + \{\mathrm{V}\}_e^T[\mathrm{M3001}]\{\mathrm{T}\}_e \\ + \dfrac{\mathrm{Ec}}{\det_e\mathrm{Re}}\{\mathrm{U}\}_e^T[\mathrm{M3101}]\{\mathrm{U}\}_e \\ + \dfrac{1}{\det_e\mathrm{RePr}}[\mathrm{M211}]\{\mathrm{T}\}_e - \dfrac{1}{\mathrm{Re}}\{N_k\}\dfrac{\partial\Theta_e}{\partial y}\bullet\widehat{\mathbf{n}}\,\Big|_{y_{\mathrm{wall}}}^{\delta} \end{bmatrix} \tag{4.26}$$

In total distinction to the potential flow GWS^h producing a global algebraic matrix statement (3.22), the BL GWS^h statements (4.25–4.26) are global DOF-dimension ODE systems. Assembling symbolically the global matrix ODEs for DOF $\{Q\} = \{\mathrm{U,T}\}^T$ are

$$\begin{aligned} \mathrm{GWS}^h\left(\mathcal{L}(q^h)\right) &= \{0\} \\ &= [\mathrm{M3000}(\{\mathrm{U}\})]\frac{\mathrm{d}\{Q\}}{\mathrm{d}x} + \{\mathrm{RES\_}\{Q(P^I, \mathrm{Re}, \mathrm{Gr}, \mathrm{Pr}, \mathrm{Ec})\}\} \end{aligned} \tag{4.27}$$

where parentheses enclose functional dependence in $\{\mathrm{RES\_}\{\mathrm{Q}(\bullet)\}\}$ the *residual* column matrix containing all BL GWS^h terms except the first.

The symbolic matrix inverse for (4.27)

$$\frac{\mathrm{d}\{Q\}}{\mathrm{d}x} = -[\mathrm{M3000}(\{\mathrm{U}\})]^{-1}\{\mathrm{RES\_}\{Q(\bullet)\}\} \tag{4.28}$$

defines the derivative required for a Taylor series (TS) to advance information in the x direction. CFD algorithms often select the Euler single step TS family with implicitness parameter θ

$$\{Q(x_{n+1})\} = \{Q(x_n)\} + \Delta x\left[\theta\frac{\mathrm{d}\{Q\}}{\mathrm{d}x}\bigg|_{n+1} + (1-\theta)\frac{\mathrm{d}\{Q\}}{\mathrm{d}x}\bigg|_{n}\right] + O(\Delta x^2, \Delta x^3) \tag{4.29}$$

with TS timing $x_{n+1} = x_n + \Delta x$. The truncation error is $O(\Delta x^2)$ for *explicit* or *backwards* Euler, $\theta = 0, 1$, improving to $O(\Delta x^3)$ for the *trapezoidal rule* definition $\theta = 0.5$.

The BL algorithm decision must be $\theta = 0.5$ as (4.28) contains an *explicitly nonlinear* matrix inverse required accurately evaluated in clearing. Inserting (4.28) into (4.29) with this decision, the $O(\Delta x^3)$ TS algorithm statement is

$$\begin{aligned} \{Q(x_{n+1})\} &= \{Q(x_n)\} \\ &\quad - \frac{\Delta x}{2}[\mathrm{M3000}(\{\mathrm{U}\})]_{n+1/2}^{-1}\left(\{\mathrm{RES\_}\{Q(\bullet)\}|_{n+1} + \{\mathrm{RES\_}\{Q(\bullet)\}|_{n}\right) \end{aligned} \tag{4.30}$$

Then clear the inverse via through multiplication by $[\mathrm{M3000(U)}]_{n+1/2}$ which produces the laminar-thermal BL $\mathrm{GWS}^h + \theta\mathrm{TS}$ algorithm global *nonlinear* algebraic matrix statement

$$\{F\_\{Q\}\} \equiv [M3000(\{U\})]_{n+1/2}\{Q_{n+1} - Q_n\} + \frac{\Delta x}{2}(\{RES\_\{Q(\bullet)\}|_{n+1} + \{RES\_\{Q(\bullet)\}|_n) = \{0\} \tag{4.31}$$

Closure of (4.31) requires a trial space basis form for the D*M* principle ODE (4.17). Since the y coordinate spans the trial space of $GWS^h + \theta TS$, and denoting DOF nodal coordinates by subscript j, the accurate $\theta = 0.5$ TS for (4.17) at axial station x_{n+1} is

$$V(y_{j+1}) = V(y_j) + \frac{\Delta y}{2}\frac{dV}{dy}\bigg|_{j+1/2} = V(y_j) - \frac{\Delta y}{2}\frac{dU}{dx}\bigg|_{j+1/2} + O(\Delta y^3) \tag{4.32}$$

While dU/dx, specifically d{U}/dx, is defined in (4.28) for $\{Q\} \equiv \{U\}$, it is *never formed* in practice due to the inverse. The efficient alternative is the backwards second-order accurate non-uniform panel TS for DOF *ordinary* derivative $d\{U\}/dx \equiv \{U'\} = \{UP\}$ at solution station x_{n+1}

$$\frac{d\{U\}}{dx} \equiv \{UP\}|_{n+1} = a\{U\}|_{n+1} + b\{U\}|_n + c\{U\}|_{n-1} + O(\Delta x^3) \tag{4.33}$$

with coefficients

$$a = \frac{(2\Delta x_{n+1} + \Delta x_n)}{\Delta x_{n+1}(\Delta x_{n+1} + \Delta x_n)}, \quad b = \frac{-(\Delta x_{n+1} + \Delta x_n)}{\Delta x_{n+1}\Delta x_n}, \quad c = \frac{\Delta x_{n+1}}{\Delta x_n(\Delta x_{n+1} + \Delta x_n)}$$

and definitions $\Delta x_{n+1} \equiv x_{n+1} - x_n$, $\Delta x_n \equiv x_n - x_{n-1}$.

4.4 GWS^h + θTS BL Matrix Iteration Algorithm

An iterative algebraic solution procedure is required for the explicitly nonlinear $n=2$ laminar-thermal $GWS^h + \theta TS$ BL algorithm (4.31–4.33). For arbitrary degree k NC basis the starting point is the base Newton algorithm (2.139–2.141)

$$\begin{aligned}
[JAC_\{Q\}]\{\delta Q\}^{p+1} &= -\{F_\{Q\}\}^p_{n+1} \\
\{Q\}^{p+1}_{n+1} &= \{Q\}^p_{n+1} + \{\delta Q\}^{p+1} = \{Q\}_n + \sum_{\alpha=0}^{p}\{\delta Q\}^{\alpha+1} \\
[JAC_\{Q\}] &\equiv \frac{\partial\{F_\{Q\}\}}{\partial\{Q\}}
\end{aligned} \tag{4.34}$$

The matrix {F_{Q}} is defined in (4.31) for DOF {U} and {T}. The jacobian [JAC_{Q}] is the derivative of {F_{Q}} with respect to algorithm DOF $\{Q\} = \{U, T, V\}^T$ and p is the iteration index.

As always for a GWS^h algorithm, the matrices defined in (4.34) are formed at element level then *assembled* into the global statement. The element matrix expressions forming the algorithm algebraic statement {F_{Q}} for (4.25–4.26), from (4.31) are

$$\begin{aligned}\{\{\mathrm{F\_}\{Q\}\}_e &\equiv \det{}_e\{\mathrm{U}_{n+1/2}\}^T[\mathrm{A3000}]\{Q_{n+1}-Q_n\}_e\\ &+\frac{\Delta x}{2}\left(\{\mathrm{RES\_}\{Q(\bullet)\}_e\big|_{n+1}+\{\mathrm{RES\_}\{Q(\bullet)\}_e\big|_n\right)\end{aligned} \tag{4.35}$$

$$\begin{aligned}\{\mathrm{RES\_U}\}_e &= \{\mathrm{V}\}^T[\mathrm{A3001}]\{\mathrm{U}\}_e+\frac{1}{\det_e\mathrm{Re}}[\mathrm{A211}]\{\mathrm{U}\}_e\\ &+\frac{\mathrm{d}P^I}{\mathrm{d}x}\det{}_e\{\mathrm{A10}\}+\frac{\mathrm{Gr}}{\mathrm{Re}^2}\det{}_e[\mathrm{A200}]\{\mathrm{T}\}_e(\widehat{\mathbf{g}}\bullet\widehat{\mathbf{i}})\end{aligned} \tag{4.36}$$

$$\begin{aligned}\{\mathrm{RES\_T}\}_e &= \{\mathrm{V}\}_e^T[\mathrm{A3001}]\{\mathrm{T}\}_e+\frac{1}{\det_e\mathrm{RePr}}[\mathrm{A211}]\{\mathrm{T}\}_e\\ &+\frac{\mathrm{Ec}}{\det_e\mathrm{Re}}\{\mathrm{U}\}_e^T[\mathrm{A3101}]\{\mathrm{U}\}_e\end{aligned} \tag{4.37}$$

wherein the generic [M] matrix prefix has been replaced by [A] to signify formation via the $n=1$ degree k NC basis. (Recall [M] transitions to [B] for $n=2$ matrices and [C] for $n=3$ matrices.) An exercise will verify the rearrangement of the DM TS, (4.32–4.33), into the required element matrix form $\{\mathrm{F\_V}\}_e$ is

$$\{\mathrm{F\_V}\}_e = [\mathrm{A1}]\{\mathrm{V}\}_e+\det{}_e[\mathrm{A2}]\{\mathrm{UP}\}_e \tag{4.38}$$

for definitions $[\mathrm{A1}]\equiv\begin{bmatrix}0 & 0\\ -1 & 1\end{bmatrix}$ and $[\mathrm{A2}]\equiv\frac{1}{2}\begin{bmatrix}0 & 0\\ 1 & 1\end{bmatrix}$.

With (4.35–4.38), deriving the element-level jacobian matrix contributions in (4.34) is via DOF $\{Q\}_e$ *analytical differentiation*. For terms linear in $\{Q\}$ the result is always the unit diagonal matrix

$$\frac{\partial\{Q\}_e}{\partial\{Q\}_e}=\frac{\partial\{Q1,Q2\}_e^T}{\partial\{Q1,Q2\}_e^T}=\begin{bmatrix}\partial\{Q1,Q2\}_e^T/\partial Q1\\ \partial\{Q1,Q2\}_e^T/\partial Q2\end{bmatrix}=\begin{bmatrix}1 & 0\\ 0 & 1\end{bmatrix}\equiv[\mathrm{I}] \tag{4.39}$$

for [I] the identity matrix. For $\{Q\}_e$ the terminal entry in a matrix multiplication string differentiation is trivially completed, for example,

$$\begin{aligned}\partial\{\mathrm{RES\_U}\}_e/\partial\{\mathrm{U}\}_e &= \partial(\{\mathrm{V}\}^T[\mathrm{A3001}]\{\mathrm{U}\}_e)/\partial\{\mathrm{U}\}_e\\ &= \{\mathrm{V}\}^T[\mathrm{A3001}]\partial\{\mathrm{U}\}_e/\partial\{\mathrm{U}\}_e = \{\mathrm{V}\}_e^T[\mathrm{A3001}][\mathrm{I}]\\ &= \{\mathrm{V}\}_e^T[\mathrm{A3001}]\end{aligned} \tag{4.40}$$

For PDE nonlinearities, such as the convection term, GWS^h + θTS algorithms place one term as a matrix string post-multiplier column matrix and the other term a row matrix pre-multiplier to a hypermatrix. Differentiation by DOF appearing as $\{Q\}_e^T$ requires matrix transposition, specifically, for (4.36)

$$\begin{aligned}\partial\{\text{RES\_U}\}_e/\partial\{\text{V}\}_e &= \partial\left(\{\text{V}\}^T[\text{A3001}]\{\text{U}\}_e\right)/\partial\{\text{V}\}_e \\ &= \{\text{U}\}^T[\text{A3100}]\partial\{\text{V}\}_e/\partial\{\text{V}\}_e = \{\text{U}\}^T[\text{A3100}]\end{aligned} \tag{4.41}$$

Denoting the element-level jacobian matrices as $[\text{J\_}QQ]_e$, for example $[\text{J\_UV}]_e \equiv \partial\{\text{F\_U}\}_e/\partial\{\text{V}\}_e$, verification of the laminar-thermal BL $\text{GWS}^h + \theta\text{TS}$ algorithm jacobian element matrix contributions detailed in (4.42) is a suggested exercise. Note specifically the distinct TS valuations for the $\{\text{U}\}_e^T$ nonlinearity in $[\text{J\_UU}]_e$ and $[\text{J\_TT}]_e$, and also the appearance of the $\{\text{UP}\}_e$ TS coefficient a in $[\text{J\_VU}]_e$.

$$\begin{aligned}
[\text{J_UU}]_e &\equiv \det_e\{\text{U}_{n+1}\}^T[\text{A3000}] + \frac{\Delta x}{2}\left[\{\text{V}\}^T[\text{A3001}] + \frac{1}{\det_e \text{Re}}[\text{A211}]\right] \\
[\text{J_UV}]_e &\equiv \frac{\Delta x}{2}\{\text{U}\}^T[\text{A3100}] \\
[\text{J_UT}]_e &\equiv \frac{\Delta x}{2}\frac{\text{Gr}}{\text{Re}^2}\det_e[\text{A200}](\widehat{\mathbf{g}} \bullet \widehat{\mathbf{i}}) \\
[\text{J_VU}]_e &\equiv a \det_e[\text{A2}] \\
[\text{J_VV}]_e &\equiv [\text{A1}] \\
[\text{J_VT}]_e &\equiv [0] \\
[\text{J_TU}]_e &\equiv \det_e\{\text{T}\}_e^T[\text{A3000}] + \frac{\Delta x}{2}\frac{2\text{Ec}}{\det_e \text{Re}}\{\text{U}\}_e^T[\text{A3101}] \\
[\text{J_TV}]_e &\equiv \frac{\Delta x}{2}\{\text{T}\}_e^T[\text{A3100}] \\
[\text{J_TT}]_e &\equiv \det_e\{\text{U}_{n+1/2}\}^T[\text{A3000}] + \frac{\Delta x}{2}\left[\{\text{V}\}^T[\text{A3001}] + \frac{1}{\det_e \text{RePr}}[\text{A211}]\right]
\end{aligned} \tag{4.42}$$

Algorithm completion requires the $n=1$ NC k degree bases [A . . .] matrix library for BL algorithm (4.34–4.42). Table 4.1 lists the $k=1$ basis library while Table 4.2 contains the $k=2$ basis set. Recall that for all $n=1$ NC bases, $\det_e = l_e$, the element length.

Table 4.1 NC basis matrix library, $n=1$, $k=1$

$$[\text{A3000}] = \frac{1}{12}\begin{bmatrix}\begin{Bmatrix}3\\1\end{Bmatrix} & \begin{Bmatrix}1\\1\end{Bmatrix}\\ \begin{Bmatrix}1\\1\end{Bmatrix} & \begin{Bmatrix}1\\3\end{Bmatrix}\end{bmatrix}, [\text{A3001}] = \frac{1}{6}\begin{bmatrix}\begin{Bmatrix}-2\\-1\end{Bmatrix} & \begin{Bmatrix}2\\1\end{Bmatrix}\\ \begin{Bmatrix}-1\\-2\end{Bmatrix} & \begin{Bmatrix}1\\2\end{Bmatrix}\end{bmatrix}$$

$$[\text{A3101}] = \frac{1}{2}\begin{bmatrix}\begin{Bmatrix}1\\-1\end{Bmatrix} & \begin{Bmatrix}-1\\1\end{Bmatrix}\\ \begin{Bmatrix}1\\-1\end{Bmatrix} & \begin{Bmatrix}-1\\1\end{Bmatrix}\end{bmatrix}, [\text{A3100}] = \frac{1}{6}\begin{bmatrix}\begin{Bmatrix}-2\\2\end{Bmatrix} & \begin{Bmatrix}-1\\1\end{Bmatrix}\\ \begin{Bmatrix}-1\\1\end{Bmatrix} & \begin{Bmatrix}-2\\2\end{Bmatrix}\end{bmatrix}$$

$$[\text{A211}] = \begin{bmatrix}1 & -1\\ -1 & 1\end{bmatrix}, [\text{A200}] = \frac{1}{6}\begin{bmatrix}2 & 1\\ 1 & 2\end{bmatrix}, \{\text{A10}\} = \frac{1}{2}\begin{Bmatrix}1\\1\end{Bmatrix}$$

Table 4.2 NC basis matrix library, $n=1$, $k=2$

$$[\mathrm{A3000}]=\frac{1}{420}\begin{bmatrix}\begin{Bmatrix}39\\20\\-3\end{Bmatrix} & \begin{Bmatrix}20\\16\\-8\end{Bmatrix} & \begin{Bmatrix}-3\\-8\\-3\end{Bmatrix}\\ \begin{Bmatrix}20\\16\\-8\end{Bmatrix} & \begin{Bmatrix}16\\192\\16\end{Bmatrix} & \begin{Bmatrix}-8\\16\\20\end{Bmatrix}\\ \begin{Bmatrix}-3\\-8\\-3\end{Bmatrix} & \begin{Bmatrix}-8\\16\\20\end{Bmatrix} & \begin{Bmatrix}-3\\20\\39\end{Bmatrix}\end{bmatrix},[\mathrm{A3001}]=\frac{1}{30}\begin{bmatrix}\begin{Bmatrix}-10\\-6\\1\end{Bmatrix} & \begin{Bmatrix}12\\8\\0\end{Bmatrix} & \begin{Bmatrix}-2\\-2\\-1\end{Bmatrix}\\ \begin{Bmatrix}-6\\-16\\2\end{Bmatrix} & \begin{Bmatrix}8\\0\\8\end{Bmatrix} & \begin{Bmatrix}-2\\16\\6\end{Bmatrix}\\ \begin{Bmatrix}1\\2\\2\end{Bmatrix} & \begin{Bmatrix}0\\-8\\-12\end{Bmatrix} & \begin{Bmatrix}-1\\6\\10\end{Bmatrix}\end{bmatrix}$$

$$[\mathrm{A3101}]=\frac{1}{30}\begin{bmatrix}\begin{Bmatrix}37\\-44\\7\end{Bmatrix} & \begin{Bmatrix}-44\\48\\4\end{Bmatrix} & \begin{Bmatrix}7\\-4\\-3\end{Bmatrix}\\ \begin{Bmatrix}36\\-32\\-4\end{Bmatrix} & \begin{Bmatrix}-32\\64\\-32\end{Bmatrix} & \begin{Bmatrix}-4\\-32\\36\end{Bmatrix}\\ \begin{Bmatrix}-3\\-4\\7\end{Bmatrix} & \begin{Bmatrix}-4\\48\\-44\end{Bmatrix} & \begin{Bmatrix}7\\-44\\37\end{Bmatrix}\end{bmatrix},[\mathrm{A3100}]=\frac{1}{30}\begin{bmatrix}\begin{Bmatrix}-10\\12\\-2\end{Bmatrix} & \begin{Bmatrix}-6\\8\\-2\end{Bmatrix} & \begin{Bmatrix}1\\0\\-1\end{Bmatrix}\\ \begin{Bmatrix}-6\\8\\-2\end{Bmatrix} & \begin{Bmatrix}-16\\0\\16\end{Bmatrix} & \begin{Bmatrix}2\\-8\\6\end{Bmatrix}\\ \begin{Bmatrix}1\\0\\-1\end{Bmatrix} & \begin{Bmatrix}2\\-8\\6\end{Bmatrix} & \begin{Bmatrix}2\\-12\\10\end{Bmatrix}\end{bmatrix}$$

$$[\mathrm{A211}]=\frac{1}{3}\begin{bmatrix}7 & -8 & 1\\-8 & 16 & -8\\1 & -8 & 7\end{bmatrix},[\mathrm{A200}]=\frac{1}{30}\begin{bmatrix}4 & 2 & -1\\2 & 16 & 2\\-1 & 2 & 4\end{bmatrix},\{\mathrm{A10}\}=\frac{1}{6}\begin{Bmatrix}1\\4\\1\end{Bmatrix}$$

4.5 Accuracy, Convergence, Optimal Mesh Solutions

The $n=2$ laminar-thermal BL GWS^h + θTS NC basis algorithm definition is (4.34–4.38) with (4.42). The first requirement is to generalize the weak form linear theory asymptotic error estimate, (3.60), for the initial-value addition to an EBV PDE.

Oden and Reddy (1976, Ch. 8) derive this generalization; the energy norm statement is

$$\begin{aligned}\left\|e^h(x)\right\|_E &\le Ch^{2\gamma}\left[\|\mathrm{data}\|_{\mathrm{L2},\Omega}+\|\mathrm{data}\|_{\mathrm{L2},\partial\Omega}\right]+\mathrm{C}_2\Delta x^{f(\theta)}\left\|q^h(x_0)\right\|^2_{H^m(\Omega)}\\ &\text{for: }\gamma\equiv\min(k+1-m,r-m),\ f(\theta)=(2,3)\end{aligned}\tag{4.43}$$

where C_2 is another constant, exponent $f(\theta)=2$ for $\theta=0,1$, and 3 for $\theta=0.5$. and $q^h(x_0)$ is interpolation of the state variable initial condition (IC) onto the mesh DOF. Recall $m=1$ for a second-order elliptic PDE, to which the BL PDE system (4.11–4.14) is representative.

The principal requirement is to assess the BL GWS^h + θTS NC basis algorithm performance in comparison with theory, (4.43). The *a posteriori* data generation requirements are:

- design compute experiments controlling unknown data in theory statement (4.43)
- via uniform mesh refinement validate the linear theory asymptotic error estimate dependence on basis degree k

- assess impact of *data* on BL algorithm characterization
- via regular solution adapted mesh refinement verify the theory prediction that the $\mathrm{GWS}^h + \theta\mathrm{TS}$ solution is optimal
- assess vertical surface BL flow alterations due to heat transfer

While a BL profile for laminar $u(y, x_0)$ is available, Schlichting (1979), hence its DOF interpolation $\{U(x_0)\}$, this is not true for $\{V(x_0)\}$. Thereby it is *impossible* to define a state variable DOF initial condition (IC) satisfying the BL PDE system. Further, the DOF derivative $\mathrm{d}\{U\}/\mathrm{d}x$ required to evaluate {F_V} is not available at $\mathrm{GWS}^h + \theta\mathrm{TS}$ algorithm integration initiation at x_0.

Selecting u^h the example, to precisely assess accuracy/convergence via uniform mesh refinement *a posteriori* data stems from the *equality*

$$\left\|u^h(x)\right\|_E + \left\|e^h(x)\right\|_E = \|u(x)\|_E = \left\|u^{h/2}(x)\right\|_E + \left\|e^{h/2}(x)\right\|_E = \ldots \tag{4.44}$$

Since C and C_2 are constants in (4.43), assuming that error bound $\left\|q^h(x_0)\right\|^2_{H^1(\Omega)}$ is mesh independent enables pertinent cancellations upon substituting (4.43) into (4.44). Thus is generated a *quantitative* statement of approximation error associated with the finer mesh solution

$$\left\|e^{h/2}(x)\right\|_E = \frac{\Delta\left\|u^{h/2}(x)\right\|_E}{2^{2k}-1}, \quad for \quad \Delta\left\|u^{h/2}(x)\right\|_E \equiv \left\|u^{h/2}(x)\right\|_E - \left\|u^h(x)\right\|_E \tag{4.45}$$

The energy norm definition for BL PDE (4.15) is

$$\|u(x)\|_E \equiv \frac{1}{2\mathrm{Re}}\int_\Omega \left[\frac{\partial u(x,y)}{\partial y}\frac{\partial u(x,y)}{\partial y}\right]\mathrm{d}y \tag{4.46}$$

with evaluation via $\mathrm{GWS}^h + \theta\mathrm{TS}$ solution DOF distribution for arbitrary k degree NC basis

$$\begin{aligned}\left\|u^h(x)\right\|_E &\equiv \frac{1}{2\mathrm{Re}}\sum_{e=1}^{\mathrm{M}}\left[\int_{\Omega_e}\{\mathrm{U}(x)\}_e^T\frac{\mathrm{d}\{N_k\}}{\mathrm{d}y}\frac{\mathrm{d}\{N_k\}^T}{\mathrm{d}y}\{\mathrm{U}(x)\}_e\mathrm{d}y\right]\\ &= \frac{1}{2\mathrm{Re}}\sum_{e=1}^{\mathrm{M}}\left[\det_e^{-1}\{\mathrm{U}(x)\}_e^T[\mathrm{A211}(k)]_e\{\mathrm{U}(x)\}_e\right]\end{aligned} \tag{4.47}$$

The $\|\mathrm{data}\|_{\mathrm{L2},\Omega}$ in (4.43) for isothermal BL flow is solely farfield pressure gradient as $\|\mathrm{data}\|_{\mathrm{L2},\partial\Omega}$ vanishes, since BC data are zero Dirichlet or vanishing Neumann. For $\theta \equiv 0.5$, and *assuming* smooth data, the *magnitude* of error estimate (4.43) is determined by

$$\left\|e^h(x)\right\|_E \le Ch^{2k}\left\|\mathrm{d}P^l/\mathrm{d}x\right\|_{\mathrm{L2},\Omega} + \mathrm{C}_2\Delta x^3\left\|q^h(x_0)\right\|^2_{H^1(\Omega)} \tag{4.48}$$

Thereby compute experiments must be carefully designed to ensure that variation of IC norm $\left\|q^h(x_0)\right\|^2_{H^1(\Omega)}$ in (4.43) is avoided.

Because error estimate (4.44) was unknown at the time of inaugural GWS^h + θTS $k = 1$ basis BL algorithm publication, Popinski and Baker (1974), their *a posteriori* data omitted energy norm characterization. Soliman (1978) rectified this via carefully designed isothermal BL experiments for $1 \leq k \leq 3$ NC basis implementations leading to precise performance quantification via (4.45).

The key requirement for precision is IC DOF $\{U(x_0), V(x_0)\}$ definition such that $\|q^h(x_0)\|^2_{H^1(\Omega)}$ in (4.43), hence (4.48), is a uniform constant for all meshes Ω^h, $\Omega^{h/2}$, ... Figure 4.2 left graphs the defined M-*in*dependent $\{U(x_0)\}$ IC DOF distribution meeting this requirement, used in concert with IC $\{V(x_0)\} = \{0\}$. Via (4.10–4.21), these IC DOF are *not* BL solutions, but DOF $\{Q(x)\} = \{U(x), V(x)\}^T$ do emerge as a BL solution following five Δx integration steps for all M.

The compute experiment specification is Re/L = 0.7 E+06. The BCs are velocity no-slip at the wall and farfield $\{U\}$ vanishing Neumann for domain Ω extending well beyond the boundary layer thickness $\delta(x)$, Figure 4.1. The $k = 1$ uniform mesh refinement *a posteriori* data in $\|u^h(x)\|_E$ for $10 \leq M \leq 80$, Soliman and Baker (1981a), precisely *validate* the linear theory asymptotic error estimate (4.44) predicted slope $2k = 2$, Figure 4.2 right. Additionally, error magnitude is predicted *independent* of the *data* $\|dP^I/dx\|_{L2,\Omega}$ for imposed dP^I/dx inducing thickening, neutral and thinning boundary layer thickness evolution $\delta(x)$, labeled Δp in graph legend.

The *regular* mesh refinement *a posteriori* data quantize impact of wall-normal graded non-uniform solution adapted meshes on solution *accuracy*. Each mesh M is generated maintaining constant the uniform M = 80 mesh wall element span, then increasing element span distribution sequentially by the geometric progression rule $l_{e+1} = \text{pr}\, l_e$ for $1.0 \leq \text{pr}$

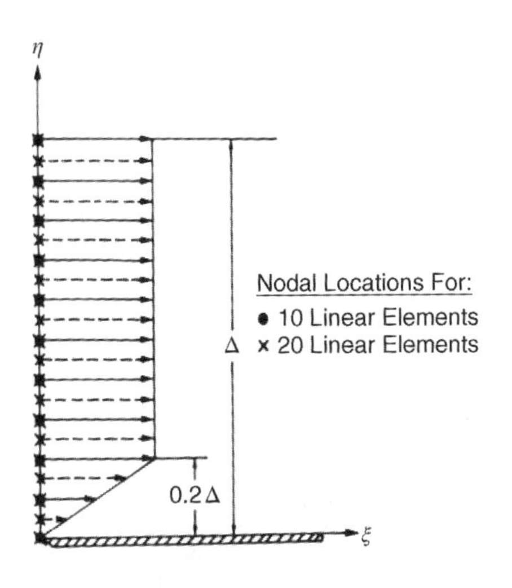

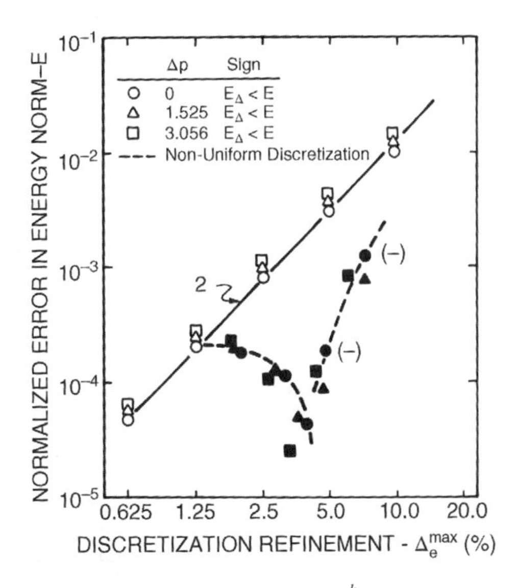

Figure 4.2 Laminar BL simulation: left, IC DOF $\{U(x_0)\}$ for various M; right, GWS^h + θTS algorithm $k = 1$ basis asymptotic convergence in $\|u^h(x)\|_E$ as function of dP^I/dx *data*, Re/L = 0.7 E+06, uniform mesh refinement $10 \leq M \leq 80$, solution adapted regular mesh refinement, $1.0 < \text{pr} < 1.19$, for various M < 80. *Source:* Soliman, M.O and Baker, A.J. 1981a. Reproduced with permission of Elsevier

≤ 1.19. The element number range is $80 \geq \mathrm{M} \geq 16$ with pr = 1.07, M = 23 absolutely minimizing the solution energy norm error. This coincides with the point of inflection in the solid symbol graph, Figure 4.2 right, whence the energy norm *error* changes sign, denoted (−), monotonically increasing thereafter with progressive mesh coarsening.

All solution-adapted non-uniform mesh solutions possess error in $\left\|u^h(x)\right\|_E$ *smaller* than any uniform mesh solution! This implies existence of an optimal non-uniform mesh for laminar isothermal BL GWS^h + θTS $k=1$ basis algorithm. It is additionally noteworthy in being essentially independent of the *data* $\left\|\mathrm{d}P^I/\mathrm{d}x\right\|_{\mathrm{L}2,\Omega}$!

In generating these convergence data the referenced authors were unaware of the linear weak form theory quantifying prediction of the *optimal mesh* solution. This theory, Babuska and Rheinboldt (1976), predicts that the optimal mesh solution – that possessing the extremum $\left\|u^h(x)\right\|_E$ – is supported by the solution adapted mesh M which *equi-distributes* the element energy norm $\|u_e(x)\|_E$ over the entire domain. The global norm (4.47) is computed in the summation loop

$$\left\|u^h(x)\right\|_E = \frac{1}{2\mathrm{Re}}\sum_{e=1}^{\mathrm{M}}\|u_e(x)\|_E$$
$$\|u_e(x)\|_E \equiv \det_e^{-1}\{\mathrm{U}(x)\}_e^T[\mathrm{A211}k]_e\{\mathrm{U}(x)\}_e \tag{4.49}$$

and it is $\|u_e(x)\|_E$ that is equi-distributed on the optimal mesh.

Ericson (2001) reports data addressing this omission. For a uniform M = 64 and a solution adapted pr = 1.2 M = 64 laminar BL solution at Re = E+06, the element norm DOF data, Figure 4.3, confirm the wall-normal adapted mesh solution better equi-distributes

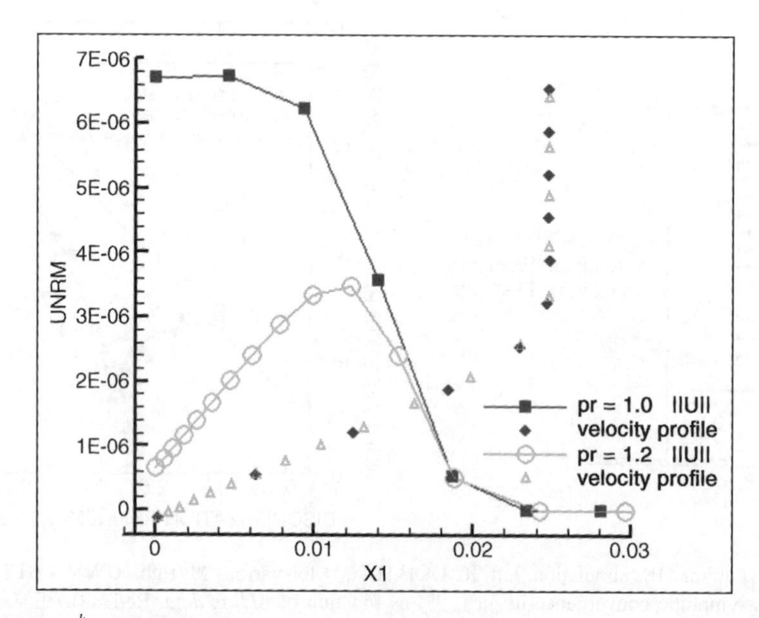

Figure 4.3 GWS^h + θTS laminar BL solution DOF {U(x)} and $\|u_e(x)\|_E$ distributions, Re = E+06: uniform and pr = 1.2 solution adapted meshes, from Ericson (2001)

$\|u_e(x)\|_E$. The associated DOF $\{U(x)\}$ distributions are visually indistinguishable on the scale of Figure 4.3, which predicts almost any "within reason" $M \geq 30$ mesh laminar BL solution will be engineering accurate!

4.6 $GWS^h + \theta TS$ Solution Optimality, Data Influence

The data in Figure 4.2 confirm that the $k = 1$ basis $GWS^h + \theta TS$ BL algorithm is "second-order accurate" as anticipated. The only *nonlinearity* in the laminar BL PDE system is the convection term $u\,\partial u/\partial x$, (4.15), leading to $GWS^h + \theta TS$ implementation via the *hyper*matrix [A3000], Table 4.1. The operational sequence for the element matrix produet $\{U_{n+1/2}\}_e^T[A3000]\{U_{n+1} - U_n\}_e$ is left product completion first, which generates a square matrix with DOF data embedding, specifically, $\{U_{n+1/2}\}_e^T[A3000] \rightarrow [A200(\{U_{n+1/2}\})]$, followed by the right matrix product with $\{U_{n+1} - U_n\}_e$.

Confirmed second-order accurate $GWS^h + \theta TS$ BL algorithm performance is directly comparable to the classic Crank–Nicolson FD (CNFD) algorithm. Moving x-station reference to superscript for clarity, the CNFD BL algorithm implementation of $u\,\partial u/\partial x$ at mesh node Y_j is $U_j^{n+1/2}(U_j^{n+1} - U_j^n)/\Delta x$. Recalling Section 3.10, the suggested exercise will verify that this CNFD stencil can be exactly recovered via assembly on the element pair sharing mesh node Y_j via alteration of the [A3000] hypermatrix, Table 4.1, to that given in (4.50)

$$S_j\left(\frac{1}{2}\{U^{n+1/2}\}_e^T\begin{bmatrix}\begin{Bmatrix}1\\0\end{Bmatrix} & \begin{Bmatrix}0\\0\end{Bmatrix}\\ \begin{Bmatrix}0\\0\end{Bmatrix} & \begin{Bmatrix}0\\1\end{Bmatrix}\end{bmatrix}\{U^{n+1} - U^n\}_e\right) \Rightarrow U_j^{n+1/2}(U_j^{n+1} - U_j^n) \tag{4.50}$$

Altering the $GWS^h + \theta TS$ BL $k = 1$ basis algorithms for (4.50) enables the sought direct comparison. Soliman and Baker (1981a) report uniform mesh refinement accuracy/convergence comparison in energy for the same range of *data* $\|dP^I/dx\|_{L2,\Omega}$. The CNFD algorithm *a posteriori* data convergence rate is quadratic, Figure 4.4 left, with error magnitude sensitive to *data* by a factor of two. For neutral $\delta(x)$ growth pressure gradient CNFD and $GWS^h + \theta TS$ solution error levels are indistinguishable. For thinning/thickening $\delta(x)$ data the CNFD error levels are uniformly larger for all M. Only for the thinning $\delta(x)$ pressure gradient does the CNFD algorithm exhibit the optimal mesh solution potential.

This direct comparison supports the assertion of *linear* weak form theory of $GWS^h + \theta TS$ algorithm solution optimality among its $k = 1$ basis peer equivalent. Ericson (2001) reports x-evolution of $GWS^h + \theta TS$ and CNFD algorithm BL solution energy norms for zero pressure gradient on the identical uniform $M = 32$ mesh. Starting from exactly identical norm maxima, the former solution uniformly exhibits an energy norm extremum during BL evolution as both decrease monotonically in advancing from the IC, Figure 4.4 right. That the extremum distinction increases *monotonically* with x validates the $GWS^h + \theta TS$ BL algorithm as comparatively *optimal*!

Soliman and Baker (1981a) report the attempt to validate linear weak form asymptotic convergence theory via uniform mesh refinement for NC $k = 2,3$ basis $GWS^h + \theta TS$ BL

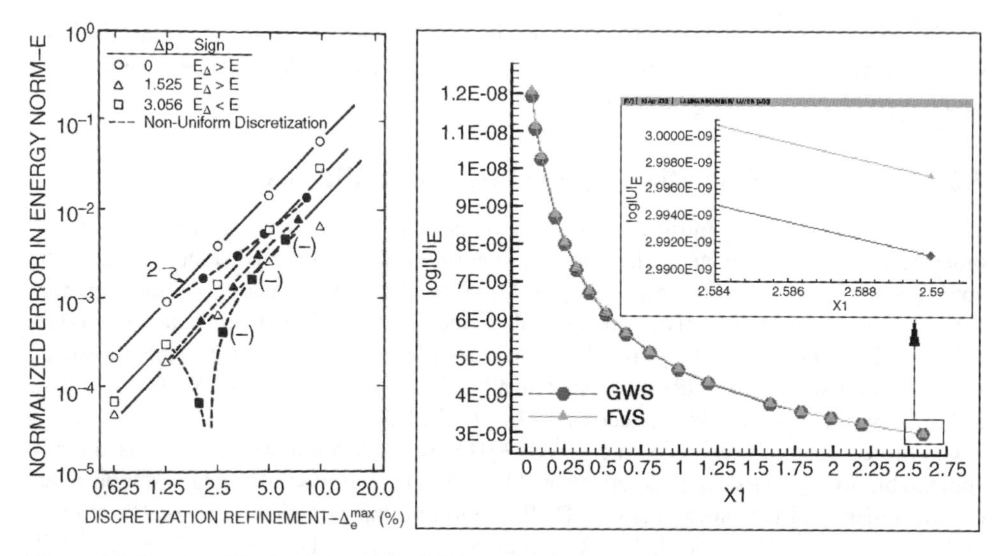

Figure 4.4 Laminar BL simulation: left, CNFD asymptotic convergence in energy, Re/L = 0.7 E+06; right: $GWS^h + \theta TS$ BL optimal solution validation, Re = E+06, uniform M = 32 mesh

algorithms. Unfortunately their matrix iteration algorithm compromised attainable precision by omission of jacobian off-diagonal blocks $[J\_UV]_e$ and $[J\_VU]_e$ and segregated $[J\_UU]_e$ and $[J\_VV]_e$ solved sequentially. Only the $k=2$ basis zero pressure gradient case *a posteriori* data converged with slope $2k=2(2)=4$, square symbols in Figure 4.5 left. The $k=2$ basis alternative pressure gradient solutions indicated convergence at suboptimal rate of 2 and none exhibited optimal mesh potential. The $k=3$ basis *a posteriori* data (diamond symbols) are thoroughly compromised by weak iterative convergence.

The influence of buoyancy on vertical surface laminar thermal BL flows for zero pressure gradient is pronounced. For a hot wall Dirichlet BC, Ericson (2001) reports $GWS^h + \theta TS$ solutions for $(\widehat{\mathbf{g}} \bullet \widehat{\mathbf{i}})$ positive, negative and zero, (4.36), corresponding to thickening, thinning and isothermal BL evolution. For uniform isothermal IC, Figure 4.5 right, compares the $\{U(x)\}$ DOF distributions generated by algorithm integration over vertical wall segments of identical length. In comparison to the isothermal profile, the BL solution approaches separation for $(\widehat{\mathbf{g}} \bullet \widehat{\mathbf{i}}) > 0$ while that for $(\widehat{\mathbf{g}} \bullet \widehat{\mathbf{i}}) < 0$ transitions to a wall jet.

4.7 Time Averaged NS, Turbulent BL Formulation

Validity of Reynolds ordering of the steady NS system requires Re be $O(\delta^{-1/2})$. For subsonic-transonic aerodynamic flow in air, Re/L is O(E+07) or larger. Hence, selecting L in Re either wing span or chord, the requirement $\delta \ll L$ is easily satisfied. Recalling basic fluid mechanics, laminar wall bounded flow becomes unstable for Re > E+03, hence aerodynamic BLs will always be *turbulent* absent actions, for example suction, blowing, surface ribblets or a surface plasma sheath.

Therefore, practical BL flow simulation requires manipulation of (4.11–4.14) for turbulent flow prediction. The legacy filtering process is *time averaging*, Schlichting (1979),

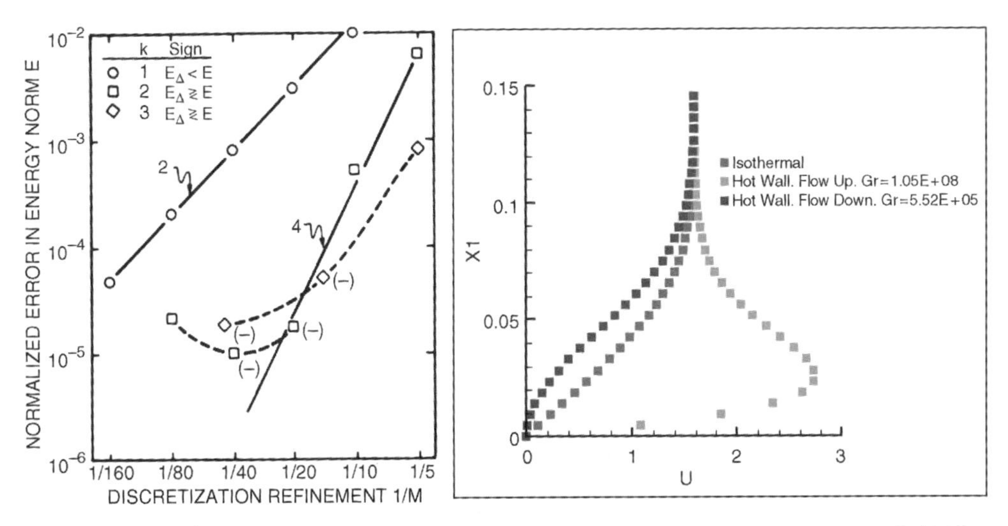

Figure 4.5 GWS^h + θTS BL solutions: left: $1 \le k \le 3$ NC basis asymptotic convergence in $\|u^h(x)\|_E$, $dp/dx = 0$; right, {U} DOF profiles for vertical BL flow up/down a Dirichlet hot wall temperature DOF

which produces what is termed in the CFD community as the *Reynolds-averaged* BL and/or PNS PDE systems. From Chapter 1 time averaging the NS PDE system requires state variable $\{q(\mathbf{x})\} = \{\mathbf{u}, p, \Theta\}^T$ resolution into a steady mean component and the unsteady fluctuation about this mean.

Switching to tensor index notation for clarity, velocity vector time average resolution is

$$\mathbf{u}(\mathbf{x}, t) \Rightarrow u_i(x_k, t) \equiv \overline{u}_i(x_k) + u'_i(x_k, t) \tag{4.51}$$

where superscript "bar" denotes steady time average with "prime" denoting the fluctuation thereabout (in space and time, realizing that turbulent flow is actually a highly unsteady phenomenon). Time averaging defines the steady component of resolution (4.51) as

$$\overline{u}_i(x_k, t) \equiv \lim_{T \to \infty} \frac{1}{T} \int_t^{t+T} u_i(x_k, \tau) d\tau \Rightarrow \overline{u}_i(x_k) \tag{4.52}$$

and by definition

$$\overline{u'}_i(x_k, t) \equiv \lim_{T \to \infty} \frac{1}{T} \int_t^{t+T} u'_i(x_k, \tau) d\tau = 0 \tag{4.53}$$

Time averaging linear terms in the BL PDE system (4.11–4.14) simply replaces each with the time average. Conversely, the time average of the nonlinear convection terms in (4.12–4.14), now expressed in the divergence form of (4.2), produces the tensor product of time averaged state variable product plus the time average of the tensor product of the fluctuations. Specifically, for $1 \le j \le 3$

$$\overline{u_j u_i} = \overline{u}_j \overline{u}_i + \overline{u'_j u'_i}, i = 1, 3 \tag{4.54}$$

$$\overline{u_j \Theta} = \overline{u}_j \overline{\Theta} + \overline{u'_j \Theta'} \tag{4.55}$$

The second terms appearing in (4.54–4.55) are called the *Reynolds stress tensor* and Reynolds *thermal vector* respectively. Reynolds ordering of the (4.54) addition to $\mathbf{DP} \bullet \widehat{\mathbf{i}}$, (4.12), yields

$$O(\mathrm{D}\mathbf{P} \bullet \widehat{\mathbf{i}}) \to O\left(\frac{\partial \overline{u'_i u'_j}}{\partial x_j}\right) = O\left(\frac{\overline{u'u'}}{1}\right) + O\left(\frac{\overline{v'u'}}{\delta}\right) + O\left(\frac{\overline{w'u'}}{1}\right) \tag{4.56}$$

confirming that Reynolds stress tensor significance cannot exceed $O(\delta)$, hence *only* the Reynolds *shear stress* $\overline{v'u'}$ need be retained. The similar operation on time averaged DE confirms that the sole BL-significant Reynolds thermal vector is $\overline{v'\Theta'}$.

A modest pressure distribution in a turbulent BL is admitted on time averaging the transverse momentum equation. Adding the pertinent Reynolds stress tensor terms into (4.5), then Reynolds ordering yields

$$O(\mathrm{D}\overline{\mathbf{P}} \bullet \widehat{\mathbf{j}}) \to O\left(\frac{\overline{vv}}{\delta}\right) + O\left(\frac{\overline{v'v'}}{\delta}\right) + O\left(\frac{\overline{P}}{\delta}\right) + \ldots \tag{4.57}$$

To order unity, time averaged ordered $\mathrm{D}\overline{\mathbf{P}} \bullet \widehat{\mathbf{j}}$ generates the ODE

$$\mathrm{D}\overline{\mathbf{P}} \bullet \widehat{\mathbf{j}} : \frac{\partial}{\partial y}\left[\overline{v'v'} + \overline{P}\right] = 0 \tag{4.58}$$

Being linear, (4.58) can be immediately integrated to distribute pressure variation through the BL. This alteration to the farfield potential flow imposed pressure distribution $P^I(x,z)$ is typically ignored.

Time averaging generated Reynolds shear stress and thermal vector constitute two *unknowns*, which must be expressed in terms of the BL state variable (and data) to achieve time averaged BL PDE system *closure*. Basically *all models* for the Reynolds shear stress tensor emulate the force-flux Stokes constitutive equation (1.14) as

$$\overline{v'u'} \equiv \tau_{xy} \equiv -\nu^t \frac{\partial \overline{u}}{\partial y} \tag{4.59}$$

The correlation function $\nu^t(x,y)$ is the *turbulent eddy viscosity*. Closure model (4.59) is extended to the Reynolds thermal vector, (4.55), by definition of a turbulent Prandtl number Pr^t. The elementary model for ν^t is via algebraic *mixing length theory* (MLT), while those utilizing the kinetic energy of turbulence, termed TKE models, end up introducing additional nonlinear PDEs. Dimensional homogeneity requires ν^t possess the units of kinematic viscosity, specifically, $D(\nu^t) = \mathrm{L}^2\tau^{-1}$.

The literature evolved Prandtl MLT model involves the following definitions

$$\nu^t \equiv (\omega l_m)^2 \left|\frac{\partial \bar{u}}{\partial y}\right| f(\delta), \quad l_m \equiv \begin{cases} \kappa y, & 0 < y \le \lambda\delta/\kappa \\ \lambda\delta, & y > \lambda\delta/\kappa \end{cases} \tag{4.60}$$

$$\omega \equiv 1 - e^{-y^+/A}, \qquad f(\delta) \equiv \left[1 + 5.5(y/\delta)^6\right]^{-1}$$

Thereby, L corresponds to ωl_m, the damped mixing length, while time τ correlates with time averaged steady velocity shear strain rate magnitude. The mixing length l_m is linear with distance y from the wall to the cutoff $\lambda\delta/\kappa$, then transitions to a fraction of the boundary layer thickness δ. The van Driest *damping function* ω is zero at the wall, and approaches unity at the top of the viscous sublayer. The Klebanoff *intermittancy function* $f(\delta)$ sharply decreases l_m to zero in the immediate vicinity of the boundary layer edge δ. The standard model coefficients are $\lambda = 0.09$, $\kappa = 0.435$, the Karman constant, and $A = 26$. Many refinements to (4.60) exist to handle wall blowing, suction, wall roughness, pressure gradients, and so on, Cebeci and Smith (1974).

TKE closure models are formed using the kinetic energy k of the velocity fluctuations, a diagonal tensor product, and a dissipation length scale l_d. The base TKE closure model definition is

$$\nu^t \equiv k^{1/2} l_d, \quad k \equiv \frac{1}{2}\left[\overline{u'u'} + \overline{v'v'} + \overline{w'w'}\right] \tag{4.61}$$

which does not admit explicit delineation of L^2 and τ dimensionality. The most utilized TKE closure models, called "k-ε" and "k-ω," are distinguished by the definition of the dissipation length scale in (4.61). For the TKE k-ε model

$$l_d \equiv \frac{C_\nu k^{3/2}}{\varepsilon}, \quad \nu^t \equiv f_\nu \frac{C_\nu k^2}{\varepsilon}, \quad \varepsilon \equiv \frac{2\nu}{3}\left[\frac{\partial u_i' \partial u_i'}{\partial x_j \partial x_k}\right]\delta_{jk} \tag{4.62}$$

with $C_\nu = 0.09$ the standard specification. The function $f_\nu(y)$ accounts for turbulence damping due to wall proximity, analogous to the MLT van Driest function, and numerous variations exist.

The definition for the k-ω model is

$$l_d \equiv k^{1/2}/\omega, \quad \nu^t \equiv k/\omega, \quad \omega = \varepsilon/\beta^* k \tag{4.63}$$

The function β^* depends in a complicated way on ω and k and their gradients. Wilcox (2006) fully describes the numerous caveats intrinsic to the closure models (4.61–4.63).

TKE closure models introduce an additional pair of nonlinear PDEs, the details on which are deferred to a following section. The *turbulent Reynolds number* is the non-D eddy viscosity

$$\mathrm{Re}^t \equiv \nu^t/\nu \tag{4.64}$$

The non-D time averaged NS PDE system describing turbulent incompressible-thermal BL flow is

$$\mathrm{D}\overline{M}\colon \frac{\partial \overline{v}}{\partial y} + \left(\frac{\partial \overline{u}}{\partial x} + \frac{\partial \overline{w}}{\partial z}\right) = 0 \tag{4.65}$$

$$\mathrm{D}\overline{\mathbf{P}} \bullet \widehat{\mathbf{i}} : \overline{\mathbf{u}} \cdot \nabla \overline{u} - \frac{1}{\mathrm{Re}} \frac{\partial}{\partial y}\left[(1 + \mathrm{Re}^t)\frac{\partial \overline{u}}{\partial y}\right] + \frac{\partial \overline{P}}{\partial x} + \frac{\mathrm{Gr}}{\mathrm{Re}^2}\overline{\Theta}(\widehat{\mathbf{g}} \bullet \widehat{\mathbf{i}}) = 0 \tag{4.66}$$

$$\mathrm{D}\overline{\mathbf{P}} \bullet \widehat{\mathbf{j}} : \overline{P}(y\colon x, z) = P^I(x, z) - \overline{v'v'}(y) \tag{4.67}$$

$$\mathrm{D}\overline{\mathbf{P}} \bullet \widehat{\mathbf{k}} : \overline{\mathbf{u}} \cdot \nabla \overline{w} - \frac{1}{\mathrm{Re}} \frac{\partial}{\partial y}\left[(1 + \mathrm{Re}^t)\frac{\partial \overline{w}}{\partial y}\right] + \frac{\partial \overline{P}}{\partial z} = 0 \tag{4.68}$$

$$D\overline{E}\colon \overline{\mathbf{u}} \cdot \nabla \overline{\Theta} - \frac{1}{\mathrm{Re}} \frac{\partial}{\partial y}\left[\left(\frac{1}{\mathrm{Pr}} + \frac{\mathrm{Re}^t}{\mathrm{Pr}^t}\right)\frac{\partial \overline{\Theta}}{\partial y}\right] - \frac{\mathrm{Ec}}{\mathrm{Re}}\left(\frac{\partial \overline{u}}{\partial y}\right)^2 = 0 \tag{4.69}$$

Having served the requirement, the overbar notation is hereon omitted. The *state variable* for the turbulent BL PDE system (4.60–4.69) remains $\{q(\mathbf{x})\} = \{\mathbf{u}, \Theta\}^T$, with turbulent Reynolds number Re^t a *nonlinear* solution dependent *parameter* with laminar flow recovered for Re^t vanishing.

4.8 Turbulent BL GWS^h + θTS, Accuracy, Convergence

The distinction between laminar and turbulent BL PDE systems, (4.11–4.14) versus (4.60–4.69), is *solely* the Reynolds shear stress tensor model Re^t, also Pr^t. This nonlinearity is a distribution, hence a GWS^h + θTS algorithm employs an appropriate degree k NC basis involving nodal DOF $\{\mathrm{RET}\}_e$. The resultant implementation alters the (linear) diffusion term matrix [A211], (4.36–4.37), to the hypermatrix expressions

$$\begin{aligned} &\frac{1}{\det_e \mathrm{Re}}[\mathrm{A211}]\{\mathrm{U}\}_e \Rightarrow \frac{1}{\det_e \mathrm{Re}}\{1 + \mathrm{RET}\}_e^T[\mathrm{A3011}]\{\mathrm{U}\}_e \\ &\frac{1}{\det_e \mathrm{RePr}}[\mathrm{A211}]\{\mathrm{T}\}_e \Rightarrow \frac{1}{\det_e \mathrm{RePr}}\left\{1 + \left(\frac{\mathrm{Pr}}{\mathrm{Pr}^t}\right)\mathrm{RET}\right\}_e^T[\mathrm{A3011}]\{\mathrm{T}\}_e \end{aligned} \tag{4.70}$$

Herein the pre-multiplier matrix $\{\bullet\}_e^T$ sums the laminar term 1 with nodal DOF of the interpolation of (4.64) multiplied by the scalar $\mathrm{Pr}/\mathrm{Pr}^t$ for DE. These terms in the jacobian matrices $[\mathrm{J\_UU}]_e$ and $[\mathrm{J\_TT}]_e$, (4.42), are accordingly altered.

Table 4.3 details the NC $k = 1,2$ basis [A3011] hypermatrices. For $k = 1$, the row matrix product defined in (4.70) produces the element average of $\{\bullet\}^T{}_e$ as the scalar multiplier on [A211] which theory predicts is the optimal form. Alternatively, the $k = 2$ basis implementation distributes $\{\bullet\}_e{}^T$ according to its specific DOF distribution.

The BL u^h energy norm definition (4.49) alteration for GWS^h + θTS algorithm turbulent flow asymptotic convergence is

Table 4.3 NC basis diffusion hypermatrix library for $k = 1,2$

$$[\mathrm{A3011}]_{k=1} = \frac{1}{2}\begin{bmatrix} \begin{Bmatrix} 1 \\ 1 \end{Bmatrix} & \begin{Bmatrix} -1 \\ -1 \end{Bmatrix} \\ \begin{Bmatrix} -1 \\ -1 \end{Bmatrix} & \begin{Bmatrix} 1 \\ 1 \end{Bmatrix} \end{bmatrix}, [\mathrm{A3011}]_{k=2} = \frac{1}{30}\begin{bmatrix} \begin{Bmatrix} 37 \\ 36 \\ -3 \end{Bmatrix} & \begin{Bmatrix} -44 \\ -32 \\ -4 \end{Bmatrix} & \begin{Bmatrix} 7 \\ -4 \\ 7 \end{Bmatrix} \\ \begin{Bmatrix} -44 \\ -32 \\ -4 \end{Bmatrix} & \begin{Bmatrix} 48 \\ 64 \\ 48 \end{Bmatrix} & \begin{Bmatrix} -4 \\ -32 \\ -44 \end{Bmatrix} \\ \begin{Bmatrix} 7 \\ -4 \\ 7 \end{Bmatrix} & \begin{Bmatrix} -4 \\ -32 \\ -44 \end{Bmatrix} & \begin{Bmatrix} -3 \\ 36 \\ 37 \end{Bmatrix} \end{bmatrix}$$

$$\|u(x)\|_E \equiv \frac{1}{2\mathrm{Re}}\int_\Omega \left[(1+\mathrm{Re}^t(x,y))\frac{\partial u(x,y)}{\partial y}\frac{\partial u(x,y)}{\partial y}\right]\mathrm{d}y \tag{4.71}$$

and implementation for arbitrary degree k NC basis is

$$\begin{aligned} \left\|u^h(x)\right\|_E &\equiv \frac{1}{2\mathrm{Re}}\sum_{e=1}^{\mathrm{M}}\left[\int_{\Omega_e}\{1+\mathrm{RET}\}_e^T\{N_k\}\{\mathrm{U}\}_e^T\frac{\mathrm{d}\{N_k\}}{\mathrm{d}y}\frac{\mathrm{d}\{N_k\}^T}{\mathrm{d}y}\{\mathrm{U}\}_e\mathrm{d}y\right] \\ &= \frac{1}{2\mathrm{Re}}\sum_{e=1}^{\mathrm{M}}\left[\det_e^{-1}\{\mathrm{U}(x)\}_e^T\{1+\mathrm{RET}\}_e^T[\mathrm{A3011}]\{\mathrm{U}(x)\}_e\right] \\ &= \frac{1}{2\mathrm{Re}}\sum_{e=1}^{\mathrm{M}}\|u_e(x)\|_E \end{aligned} \tag{4.72}$$

$\mathrm{GWS}^h + \theta\mathrm{TS}$ turbulent BL algorithm with MLT closure *a posteriori* data for $k = 1,2$ NC basis and zero imposed pressure gradient is reported, Soliman and Baker (1981b). PDE system nonlinearity requires computing and updating $\{\mathrm{Re}^t(x)\}_e$ DOF data during the matrix iteration process, (4.34). Non-uniform, *regular* solution adapted meshes are employed, which maintain the wall element span constant at that of the finest uniform $\mathrm{M} = 80$ mesh, then progressively increase element spans via geometric progression $l_{e+1} = \mathrm{pr}\, l_e$, $\mathrm{pr} > 1$.

For $1.06 \le \mathrm{pr} \le 1.63$, the $k = 1$ basis produces $60 \ge \mathrm{M} \ge 12$ elements. For direct comparison, $k = 2$ basis utilizes $1.11 \le \mathrm{pr} \le 2.81$ which proportionally halves elements to $30 \ge \mathrm{M} \ge 6$ while maintaining vertex DOF coordinates nominally identical to those for $k = 1$ basis. The BCs are no slip at the wall and homogeneous Neumann for u^h at farfield for domain span well beyond the boundary layer thickness $\delta(x)$. The ICs are $\{\mathrm{V}(x_0)\} = \{0\}$, $\{\mathrm{Re}^t(x_0)\} = \{0\}$and $\{\mathrm{U}(x_0)\}$ the slug profile in Figure 4.2 left. For each mesh M the identical number of DOF between the wall and slug profile knee maintains $\left\|q^h(x_0)\right\|^2_{H^m(\Omega)}$ a constant in (4.43), hence (4.45) does quantify error.

$GWS^h + \theta TS$ algorithm solution start up is laminar with $\{V(x)\}$ DOF computation initiated after five Δx steps. This continues until the BL shape factor achieves 90% of that for a laminar BL, whence computation of $\{Re^t(x)\}$ is initialized. The turbulent BL simulation continues until the displacement thickness δ^* achieves ~90% of that for a zero pressure gradient fully developed turbulent BL. The mesh measure h for (4.43) is the farfield element span with norm computed via (4.72).

These *regular* mesh refinement *a posteriori* data generate the $k = 1,2$ NC basis algorithm energy norm *error estimates* plotted as round/square symbols in Figure 4.6 top. The drawn straight lines are k-dependent theory slopes predicted by (4.43). The solution data are well interpolated by straight lines of slope 2 and 4, hence quantitatively *validates* (4.43)

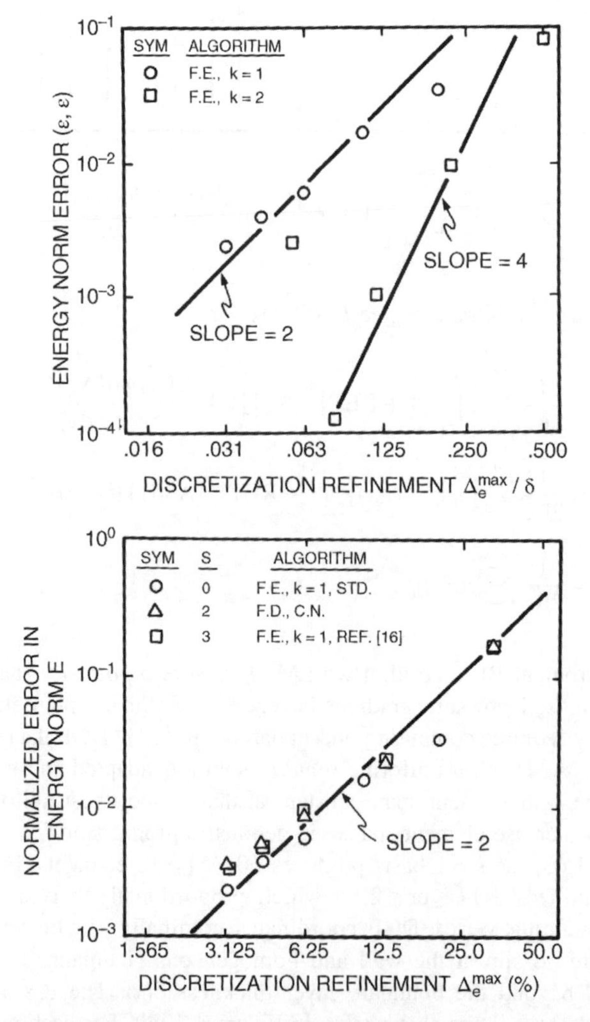

Figure 4.6 Turbulent flow BL asymptotic convergence, MLT closure, flat plate: top, $k = 1,2$ NC basis $GWS^h + \theta TS$ *a posteriori* data; bottom, identical mesh M-dependent direct comparison between $k = 1$ NC basis $GWS^h + \theta TS$, CNFD and an alternative FD algorithm

accuracy prediction as k basis degree dominated. Modest coarse mesh M "super" accuracy is apparent in both basis data, evidenced by their symbols located below fine mesh data straight lines.

Soliman and Baker (1981b) report *a posteriori* data for identical mesh $k=1$ NC basis $\text{GWS}^h + \theta\text{TS}$ and CNFD algorithm asymptotic convergence comparison prediction by (4.43). CNFD/GWS^h data are the triangle/circular symbols in Figure 4.6 bottom compared with the theory slope of two. Both algorithms generate solutions exhibiting quadratic convergence in agreement with smooth data TS second-order accuracy interpretation. The square symbols are data for a literature alternative FD [A3000] matrix. Of *prime* theoretical significance, the $\text{GWS}^h + \theta\text{TS}$ solution data exhibit *minimum* error in energy across the identical mesh M DOF spectrum. This quantitatively *validates* linear weak form theory *optimality* assertion for an explicitly *nonlinear* BL PDE with Re^t *nonlinearity* embedding in the energy norm.

The $k=1$ NC basis $\text{GWS}^h + \theta\text{TS}$ algorithm turbulent BL MLT closure accuracy assessment, Baker (1983, Chapter 6), results for $\{\text{U}(x)\}$ DOF distribution evolution comparison with the Bradshaw IDENT 2400 experiment over its $0 \le x \le 4$ ft span, Kline *et al.* (1969). The simulation IC is $\{\text{V}(x_0)\} = \{0\}$, DOF $\{\text{U}(x_0)\}$ are interpolated from the experiment first data station, $\{\text{Re}^t(x_0)\}$ is computed from these data via (4.60) and (4.64) and $\{\text{V}(x)\}$ is initialized following five Δx integration steps.

The $k=1$ basis solution DOF $\{\text{U}(x)\}$ distributions, Figure 4.7 top, exhibit *quantitative* agreement with experiment, clearly capturing the three semi-linear segments of this *non-equilibrium* turbulent BL experiment. The significance range for $\{\text{RET}(x)\}$ DOF distributions, Figure 4.7 bottom, spans 0 to E+03 while that for non-D $\{\text{U}(x)\}$ DOF is but E+00. The evolution of solution-dependent $\{\text{RET}(x,\{\text{U}(x)\})\}$ *nonlinearity* directly impacts accuracy, hence so should its assessment of error. Thus the decision herein for solution-dependent *nonlinearity* direct embedding in norm (4.72), in distinction to extraction of solution-independent *data* via some representative scalar bound, Section 3.6.

4.9 $\text{GWS}^h + \theta\text{TS}$ BL Algorithm, TKE Closure Models

The k-ε and k-ω TKE turbulence closure models each add a pair of *nonlinear* I-EBV PDEs to the turbulent BL PDE statement, (4.65–4.69). For the former and for $n=2$ the added non-D PDE pair plus differential definition system is

$$\begin{aligned}\mathcal{L}(k) = u\frac{\partial k}{\partial x} + v\frac{\partial k}{\partial y} - \frac{1}{\text{Pe}}\frac{\partial}{\partial y}\left[\left(1 + \frac{\text{Re}^t}{\text{C}_k}\right)\frac{\partial k}{\partial y}\right]&\\ - \tau_{xy}\frac{\partial u}{\partial y} + \varepsilon + \frac{\text{Gr}\,\text{Re}^t}{\text{Re}^3}\frac{\partial \Theta}{\partial y} = 0&\end{aligned} \tag{4.73}$$

$$\begin{aligned}\mathcal{L}(\varepsilon) = u\frac{\partial \varepsilon}{\partial x} + v\frac{\partial \varepsilon}{\partial y} - \frac{1}{\text{Pe}}\frac{\partial}{\partial y}\left[\frac{\text{Re}^t}{\text{C}_\varepsilon}\frac{\partial \varepsilon}{\partial y}\right]&\\ - C_\varepsilon^1 f^1 \tau_{xy}\frac{\partial u}{\partial y}\frac{\varepsilon}{k} + C_\varepsilon^2 f^2\frac{\varepsilon^2}{k} + \text{C}_\varepsilon^3\frac{\text{Gr}\,\text{Re}^t}{\text{Re}^3}\frac{\varepsilon}{k}\frac{\partial \Theta}{\partial y} = 0&\end{aligned} \tag{4.74}$$

$$\mathcal{D}(\tau_{xy}) = \tau_{xy} + f_\nu\frac{C_\nu k^2}{\varepsilon}\frac{\partial u}{\partial y} = \tau_{xy} + \text{Re}^t\frac{\partial u}{\partial y} = 0 \tag{4.75}$$

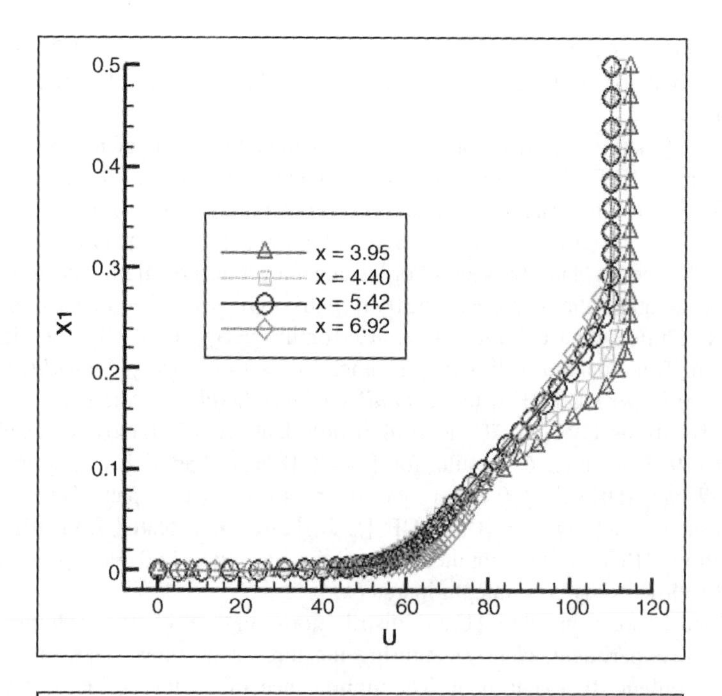

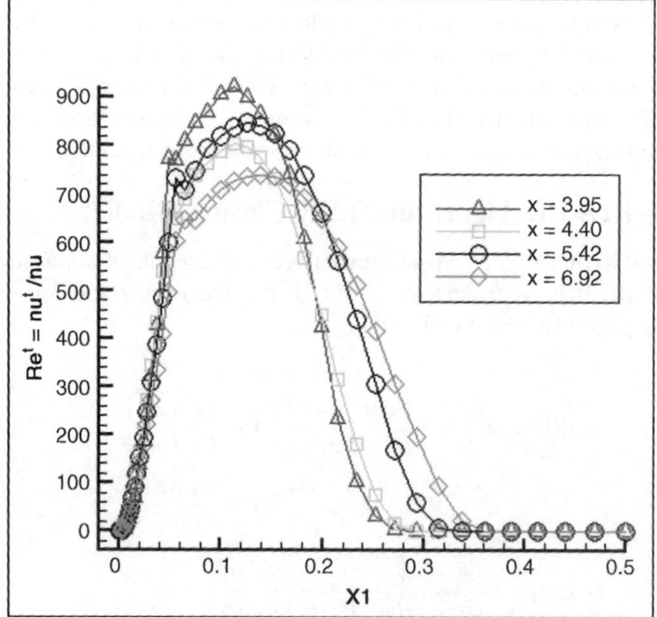

Figure 4.7 GWS^h + θTS $k = 1$ NC basis solution DOF evolution distributions, IDENT 2400 non-equilibrium turbulent BL experiment: top, velocity {U(x)}; bottom, turbulent Reynolds number {RET(x)}

The standard model constants for (4.73–4.74) are $C_k = 1.0$, $C_\varepsilon = 1.3$, $C_\varepsilon{}^1 = 1.44$, $C_\varepsilon{}^2 = 1.92$, $C_\varepsilon{}^3 = 1.0$ and Pe = RePr. The functions $f^1(y)$ and $f^2(y)$ in (4.74) are wall proximity damping functions and (4.75) is the k-ε eddy viscosity model (4.61) differential definition with wall damping function $f_\nu(y)$ and $C_\nu = 0.09$. The Lam–Bremhorst (1981) theory, representative of the range of available low-Re^t wall region models, has definitions

$$
\begin{aligned}
f_v &= \left(1 - \exp(-0.0165R_y)\right)^2(1 + 20.5/\mathrm{Re}^t) \\
f^1 &= 1 + (0.05/f_v)^3, \quad R_y = k^{1/2}y/v \\
f^2 &= 1 - \exp(-\mathrm{Re}^t)^2
\end{aligned}
\tag{4.76}
$$

The PDEs (4.73–4.74) with definitions (4.75–4.76) are nonlinear parabolic, that is, two-point boundary value and initial value, as are all BL PDE systems. Hence, encompassing BC specifications are required. For the BL solution domain extending beyond $\delta(x)$ the farfield BCs for k and ε are vanishing Neumann. At the wall the BCs are $k = 0$ and $\varepsilon = \nu\partial^2k/\partial y^2$; for computational tractability Lam–Bremhorst suggest replacing the latter with $\partial\varepsilon/\partial y = 0$.

For k-ω closure model for $n = 2$ isothermal flow, the added nonlinear non-D I-EBV PDE pair is

$$
\mathcal{L}(k) = u\frac{\partial k}{\partial x} + v\frac{\partial k}{\partial y} - \frac{1}{\mathrm{Re}}\frac{\partial}{\partial y}\left[(1 + \sigma^*\mathrm{Re}^t)\frac{\partial k}{\partial y}\right] - \tau_{xy}\frac{\partial u}{\partial y} + \beta^*\omega k = 0 \tag{4.77}
$$

$$
\mathcal{L}(\omega) = u\frac{\partial \omega}{\partial x} + v\frac{\partial \omega}{\partial y} - \frac{1}{\mathrm{Re}}\frac{\partial}{\partial y}\left[1 + \sigma\mathrm{Re}^t\frac{\partial \omega}{\partial y}\right] - \alpha\frac{\omega}{k}\tau_{xy}\frac{\partial u}{\partial y} + \beta\omega^2 = 0 \tag{4.78}
$$

This system with the second expression in (4.75) is parabolic, so farfield homogeneous Neumann BCs remain appropriate with wall BCs $k = 0 = \omega$. The auxiliary functions are $\beta \equiv \beta_0 f_\beta$ and $\beta^* \equiv \beta_0{}^* f_{\beta^*}$ and standard values for model constants are $\alpha = 13/25$, $\sigma = ½ = \sigma^*$, $\beta_0 = 9/125$ and $\beta_0{}^* = 9/100$. The functions f_β and f_{β^*} depend nonlinearly on the state variable solution, Wilcox (2006). Low-Re^t modifications are not required for k-ω closure applied to fully turbulent BL flow. The alteration of (4.78) addressing transitional $\mathrm{Re}^t \Rightarrow 0$ pertinence is derived, Wilcox (2006, (4.264)).

In the (typical) absence of experimental data, ICs for a turbulent BL TKE closure model simulation can be established via a correlation and/or BL *similarity theory*. An acceptable approximation to a turbulent BL velocity profile is via the power law, Schlichting (1979)

$$
u(y)/U_e = (y/\delta)^{1/\mathrm{n}} \tag{4.79}
$$

where U_e is the BL edge velocity, δ is BL thickness and n = 7 is typical. The similarity solution in the *log layer* of an isothermal incompressible BL is, Wilcox (2006)

$$
\begin{aligned}
U^+ &\equiv u/u_\tau = \kappa^{-1}\log(y^+) + \mathrm{B} + (2\Pi/\kappa)\sin^2\left(\frac{\pi}{2}\frac{y}{\delta}\right) \\
y^+ &\equiv u_\tau y/\nu
\end{aligned}
\tag{4.80}
$$

for κ the Karman constant, B ~ 5.0 for a smooth wall, Π ~ 0.6 for zero pressure gradient and *y*+ is Reynolds number based on distance *y* from the wall with u_τ the *friction velocity*. Figure 4.8 graphs the turbulent BL velocity profile (4.80) in similarity coordinates clearly illustrating the *log layer* sandwiched between the *sublayer*, where viscosity effects are important, and the *defect layer*.

Assuming in (4.73) and/or (4.77) that near wall production and dissipation of turbulent kinetic energy is in balance, the similarity solution (4.80) admits prediction of TKE closure model state variable distributions in the log layer (*only*) as

$$\begin{aligned} \nu^t &= \kappa y u_\tau / C_\nu, \quad k = u_\tau^2 / C_\nu^{1/2}, \quad \tau_w = u_\tau^2 / C_\nu \\ \varepsilon &= \left| u_\tau{}^3 \right| / \kappa y, \quad \omega = u_\tau / \kappa y (C_\nu)^{1/2} \end{aligned} \tag{4.81}$$

which serve reasonably for DOF IC generation. Prediction of closure model variable distributions in the laminar sublayer is possible via correlation $U^+ = y^+$, Figure 4.8.

Starting with the $\text{GWS}^h + \theta\text{TS}$ laminar BL algorithm, (4.34–4.42), implementing a TKE closure system requires computing the TKE model selected nonlinear $\{\text{Re}^t(x)\}_e$ DOF distribution via (4.62–4.63), and augmenting the turbulent BL PDE system (4.64–4.69) with (4.73–4.78). Selecting the *k*-ε model for development the TKE BL state variable is $\{q(y,x)\} = \{u, v, \Theta, k, \varepsilon, \tau_{xy}\}^T$. Formation of $\{\text{F\_}\{Q\}\}_e$, (4.35), must correspondingly be extended to DOF system $\{Q(x)\}_e = \{\text{U, V, T, K, E, TXY}\}_e{}^T$.

Retaining the NC basis, the initial-value term in the $\{\text{F\_}\{Q\}\}_e$, (4.35), is unaltered as is (4.38) for $\{\text{F\_V}\}_e$. For the typical assumption $\text{Pr}^t \equiv \text{Pr}$ the $\{\text{RES\_}(Q)\}_e$ replacements for (4.36–4.37) are

$$\begin{aligned} \{\text{RES\_U}\}_e = {} & \{\text{V}\}_e^T[\text{A3001}]\{\text{U}\}_e + \frac{1}{\det_e \text{Re}}\{1 + \text{RET}\}_e^T[\text{A3011}]\{\text{U}\}_e \\ & + \frac{\text{d}P^I}{\text{d}x}\det_e\{\text{A10}\} + \frac{\text{Gr}}{\text{Re}^2}\det_e[\text{A200}]\{\text{T}\}_e(\widehat{\mathbf{g}} \bullet \widehat{\mathbf{i}}) \end{aligned} \tag{4.82}$$

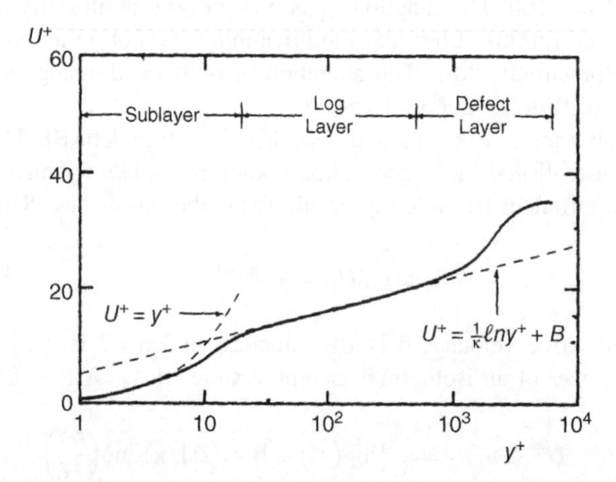

Figure 4.8 Isothermal fully turbulent BL velocity profile plotted in similarity coordinates, from Wilcox (2006) with permission

$$
\begin{aligned}
\{\mathrm{RES\_T}\}_e &= \{\mathrm{V}\}_e^T[\mathrm{A3001}]\{\mathrm{T}\}_e + \frac{1}{\det_e \mathrm{RePe}}\{1+\mathrm{RET}\}_e^T[\mathrm{A3011}]\{\mathrm{T}\}_e \\
&+ \frac{\mathrm{Ec}}{\det_e \mathrm{Re}}\{\mathrm{U}\}^T[\mathrm{A3101}]\{\mathrm{U}\}_e
\end{aligned}
\tag{4.83}
$$

The companion $\{\mathrm{RES\_}\{Q\}\}_e$ statements for $\{Q\}_e = \{\mathrm{K,E}\}_e{}^T$ are

$$
\begin{aligned}
\{\mathrm{RES\_K}\}_e &= \{\mathrm{V}\}^T[\mathrm{A3001}]\{\mathrm{K}\}_e + \frac{1}{\det_e \mathrm{Pe}}\{1 + C_k^{-1}\mathrm{RET}\}_e^T[\mathrm{A3011}]\{\mathrm{K}\}_e \\
&+ \{\mathrm{TXY}\}^T[\mathrm{A3001}]\{\mathrm{U}\}_e \\
&+ \det_e[\mathrm{A200}]\{\mathrm{E}\}_e - \frac{\mathrm{Gr}}{\mathrm{Re}^3}\{\mathrm{RET}\}_e^T[\mathrm{A3001}]\{\mathrm{T}\}_e
\end{aligned}
\tag{4.84}
$$

$$
\begin{aligned}
\{\mathrm{RES\_E}\}_e &= \{\mathrm{V}\}^T[\mathrm{A3001}]\{\mathrm{E}\}_e + \frac{1}{\det_e \mathrm{C}_\varepsilon \mathrm{Pe}}\{\mathrm{RET}\}_e^T[\mathrm{A3011}]\{\mathrm{E}\}_e \\
&+ \mathrm{C}_\varepsilon^1\{\mathrm{FE1}, \mathrm{TXY}, \mathrm{E/K}\}_e^T[\mathrm{A3001}]\{\mathrm{E}\}_e \\
&+ \mathrm{C}_\varepsilon^2 \det_e\{\mathrm{FE2}, \mathrm{E/K}\}^T[\mathrm{A3000}]\{\mathrm{E}\}_e \\
&+ \frac{\mathrm{C}_\varepsilon^3 \mathrm{Gr}}{\mathrm{Re}^3}\{\mathrm{E/K}, \mathrm{RET}\}^T[\mathrm{A3001}]\{\mathrm{T}\}_e
\end{aligned}
\tag{4.85}
$$

The $\mathrm{GWS}^N \Rightarrow \mathrm{GWS}^h$ sequence for (4.75) generates

$$
\{\mathrm{F\_TXY}\}_e = \det_e[\mathrm{A200}]\{\mathrm{TXY}\}_e + \frac{1}{\mathrm{Re}}\{\mathrm{F}_\nu, \mathrm{RET}\}^T[\mathrm{A3001}]\{\mathrm{U}\}_e = 0\}
\tag{4.86}
$$

The numerous DOF nonlinear products in (4.85–4.86), including the PDE *rational* nonlinearity ε/k, are implemented as nodal DOF products in the hypermatrix pre-multiplier, alternatively the DOF ratio $\{\mathrm{E/K}\}_e$ The wall node DOF in $\{\mathrm{E/K}\}_{e=1}$ requires suppression when using the ε BC approximation $\partial\varepsilon/\partial y = 0$, as $k = 0$ is the wall node BC. The convenient alternative replaces this DOF with the first off-wall DOF in $\{\mathrm{E/K}\}_{e=1}$, as all other wall DOF are homogeneous Dirichlet, hence are not influenced by this approximation.

The terminal $\mathrm{GWS}^h + \theta\mathrm{TS}$ turbulent BL algorithm formulation step is to expand the Newton jacobian, (4.34), for the TKE k-ε closure BL state variable $\{Q(x)\} = \{\mathrm{U, V, T, K, E, TXY}\}^T$ with $\{\mathrm{RET}\}$ the explicitly nonlinear parameter. Iterative convergence robustness results from explicit recognition of $\{\mathrm{RET}\}_e$ nonlinearity, hence for the diffusion terms in (4.82–4.85), via the chain rule

$$
\begin{aligned}
&\partial\left(\frac{1}{\det_e \mathrm{Pa}}\{1+\mathrm{RET}\}_e^T[\mathrm{A3011}]\{\mathrm{Q}\}_e\right)\Big/\partial\{Q\}_e \\
&= \frac{1}{\det_e \mathrm{Pa}}\begin{bmatrix}\{1+\mathrm{RET}\}_e^T[\mathrm{A3011}][I] \\ +\{Q\}_e^T[\mathrm{A3110}]\partial\{\mathrm{RET}\}_e/\partial\{Q\}_e\end{bmatrix}
\end{aligned}
\tag{4.87}
$$

Recall [I] is the identity matrix with non-D *parameter* Pa the placeholder for Re, Pe and C_εPe. From (4.75), via calculus

$$\begin{aligned}\partial\{\mathrm{RET}\}_e/\partial\{Q\}_e &= \partial\{\mathrm{RET}\}_e/\partial\{\mathrm{K,E}\}_e\\ &\Rightarrow \mathrm{diag}\left[2\mathrm{RET/K}\right]_e, \mathrm{diag}\left[-\mathrm{RET/E}\right]_e\end{aligned} \tag{4.88}$$

which generates element-dependent *diagonal* square matrices.

The Newton jacobian element matrix additions to (4.42) for DOF {U} are

$$\begin{aligned}[\mathrm{J\_UK}]_e &= \frac{\Delta x}{\mathrm{det}_e\,\mathrm{Re}}\{\mathrm{U}\}_e^T[\mathrm{A3110}]\mathrm{diag}\left[\mathrm{RET/K}\right]_e\\ [\mathrm{J\_UE}]_e &= \frac{-\Delta x}{2\mathrm{det}_e\,\mathrm{Re}}\{\mathrm{U}\}_e^T[\mathrm{A3110}]\mathrm{diag}\left[\mathrm{RET/E}\right]_e\end{aligned} \tag{4.89}$$

which hold as well for $[\mathrm{J\_TK}]_e$ and $[\mathrm{J\_TE}]_e$, substituting $\{\mathrm{T}\}_e^T$ for $\{\mathrm{U}\}_e^T$.

The jacobians for DOF $\{\mathrm{K,E}\}_e^T$ are

$$\begin{aligned}[\mathrm{J\_KK}]_e \equiv\ & \mathrm{det}_e\{U_{n+1/2}\}^T[\mathrm{A3000}]\\ &+\frac{\Delta x}{2}\begin{bmatrix}\{\mathrm{V}\}^T[\mathrm{A3001}]+\dfrac{1}{\mathrm{det}_e\,\mathrm{Pe}}\left\{1+\mathrm{C}_k^{-1}\mathrm{RET}\right\}_e^T[\mathrm{A3011}]\\ +\dfrac{2}{\mathrm{det}_e\,\mathrm{Pe}\,\mathrm{C}_k}\{\mathrm{K}\}_e^T[\mathrm{A3110}]\mathrm{diag}\left[\mathrm{RET/K}\right]_e\\ +2\{\mathrm{U}\}_e^T[\mathrm{A3100}]\mathrm{diag}[\mathrm{TXY/K}]_e\end{bmatrix}\end{aligned} \tag{4.90}$$

$$[\mathrm{J\_KE}]_e \equiv \frac{\Delta x}{2}\begin{bmatrix}\dfrac{1}{\mathrm{det}_e\,\mathrm{Pe}\,\mathrm{C}_k}\{\mathrm{K}\}_e^T[\mathrm{A3110}]\mathrm{diag}\left[\mathrm{RET/E}\right]_e\\ +\{\mathrm{U}\}_e^T[\mathrm{A3100}]\mathrm{diag}[\mathrm{TXY/E}]_e+\mathrm{det}_e[\mathrm{A200}]\end{bmatrix} \tag{4.91}$$

$$\begin{aligned}[\mathrm{J\_EE}]_e \equiv\ & \mathrm{det}_e\{\mathrm{U}_{n+1/2}\}_e^T[\mathrm{A3000}]\\ &+\frac{\Delta x}{2}\begin{bmatrix}\{\mathrm{V}\}_e^T[\mathrm{A3001}]+\dfrac{1}{\mathrm{det}_e\,\mathrm{Pe}}\left\{1+\mathrm{C}_\varepsilon^{-1}\mathrm{RET}\right\}_e^T[\mathrm{A3011}]\\ -\dfrac{1}{\mathrm{det}_e\,\mathrm{Pe}\,\mathrm{C}_\varepsilon}\{\mathrm{E}\}_e^T[\mathrm{A3110}]\mathrm{diag}\left[\mathrm{RET/E}\right]\\ +\mathrm{det}_e\mathrm{C}_\varepsilon^2\left(\{f^1,\mathrm{E/K}\}_e^T[\mathrm{A3000}]+\{f^1,\mathrm{E}\}_e^T[\mathrm{A3000}]\mathrm{diag}\left[1/\mathrm{K}\right]_e\right)\\ +\dfrac{\mathrm{C}_\varepsilon^3\mathrm{Gr}}{\mathrm{Re}^3}\{\mathrm{T}\}_e^T[\mathrm{A3100}]\mathrm{diag}\left[\mathrm{RET/K}^2\right]_e\end{bmatrix}\end{aligned} \tag{4.92}$$

$$[\mathrm{J\_EK}]_e \equiv \frac{\Delta x}{2}\left[\begin{array}{l} \frac{2}{\det_e \mathrm{ReC}_\varepsilon}\{\mathrm{E}\}_e^T[\mathrm{A3110}]\mathrm{diag}[\mathrm{RET/K}]_e \\ -\frac{\mathrm{C}_\varepsilon^1}{\det_e \mathrm{Pe\,C}_\varepsilon}\{f^1\mathrm{U}\}_e^T[\mathrm{A3100}]\mathrm{diag}[\mathrm{TXY, E/K}]_e \\ +2\frac{\mathrm{C}_\varepsilon^1}{\det_e \mathrm{Pe\,C}_\varepsilon}\{f^1,\mathrm{U}\}_e^T[\mathrm{A3100}]\mathrm{diag}\left[\mathrm{TXY}, (\mathrm{E/K})^2\right]_e \\ +\det_e \mathrm{C}_\varepsilon^2\left(\{f^2,\mathrm{E/K}\}_e^T[\mathrm{A3000}] + \{f^2,\mathrm{E}\}_e^T[\mathrm{A3000}]\mathrm{diag}\left[1/\mathrm{K}\right]_e\right) \\ +\frac{\mathrm{C}_\varepsilon^3\mathrm{Gr}}{\mathrm{Re}^3}\{\mathrm{T}\}_e^T[\mathrm{A3100}]\mathrm{diag}[\mathrm{RET, E/K^2}]_e \end{array}\right] \tag{4.93}$$

The MLT and k-ε TKE closure model GWS^h + θTS $k = 1$ NC basis algorithm turbulent BL solution {U} and {RET} DOF distributions at the Bradshaw IDENT 2400 non-equilibrium BL experiment last data station are compared in Figure 4.9 top. The TKE solution level of {RET} is about double that of the MLT solution, from the defect layer edge into the farfield wake. This is due to TKE closure model (4.75) lacking a Klebanoff-type *intermittancy function* $f(\delta)$, (4.60). This contributes to the profile diffusion evident in the TKE {U} distribution near δ. Elsewhere the {U} profiles are essentially identical and both exhibit quantitative agreement with experiment.

The TKE GWS^h + θTS $k = 1$ NC basis algorithm solution generated distributions of the classic BL integral parameters *skin friction* (CF/2), *shape factor* (H), *momentum thickness* (θ) and *displacement thickness* (δ*), the open symbols in Figure 4.9 bottom, are compared to experimental data computations, the solid lines. Good quantitative agreement is evident in these BL parameter distributions for this *very* demanding non-equilibrium turbulent BL experimental data set.

4.10 The Parabolic Navier–Stokes PDE System

Early in aerodynamics CFD history a variety of flowfields were simulated using the *parabolic Navier–Stokes* (PNS) PDE system. The PNS key attribute was that 3-D turbulent flow steady solutions could be generated that required *only* 2-D storage, a truly significant(!) incentive realizing how primitive compute capability was in the 1970s. Pertinent incompressible aerodynamic geometries include ducted flow, wing–body juncture region flow, jets and wakes. External supersonic–hypersonic missile shock layer geometries are also pertinent wherein closure for reacting, real gas variable gamma aerothermodynamics is required.

The fundamental assumption remains unidirectional velocity vector field admitting Reynolds ordering of the parent NS PDE system. A significant generalization beyond BL forms is required, however, as applicable aerodynamic geometries are not characterized by a unique wall normal coordinate direction. This precludes discrete solution of D*M* via the BL system ODE (4.32). The inaugural 3-D PNS CFD staggered mesh FD formulation for internal ducted flows, Patankar and Spalding (1972), introduced a Poisson PDE to approximately enforce D*M* coupled with a mass flow conserving axial pressure gradient. A summary of legacy PNS CFD algorithms exists, Baker (1983, Ch. 7).

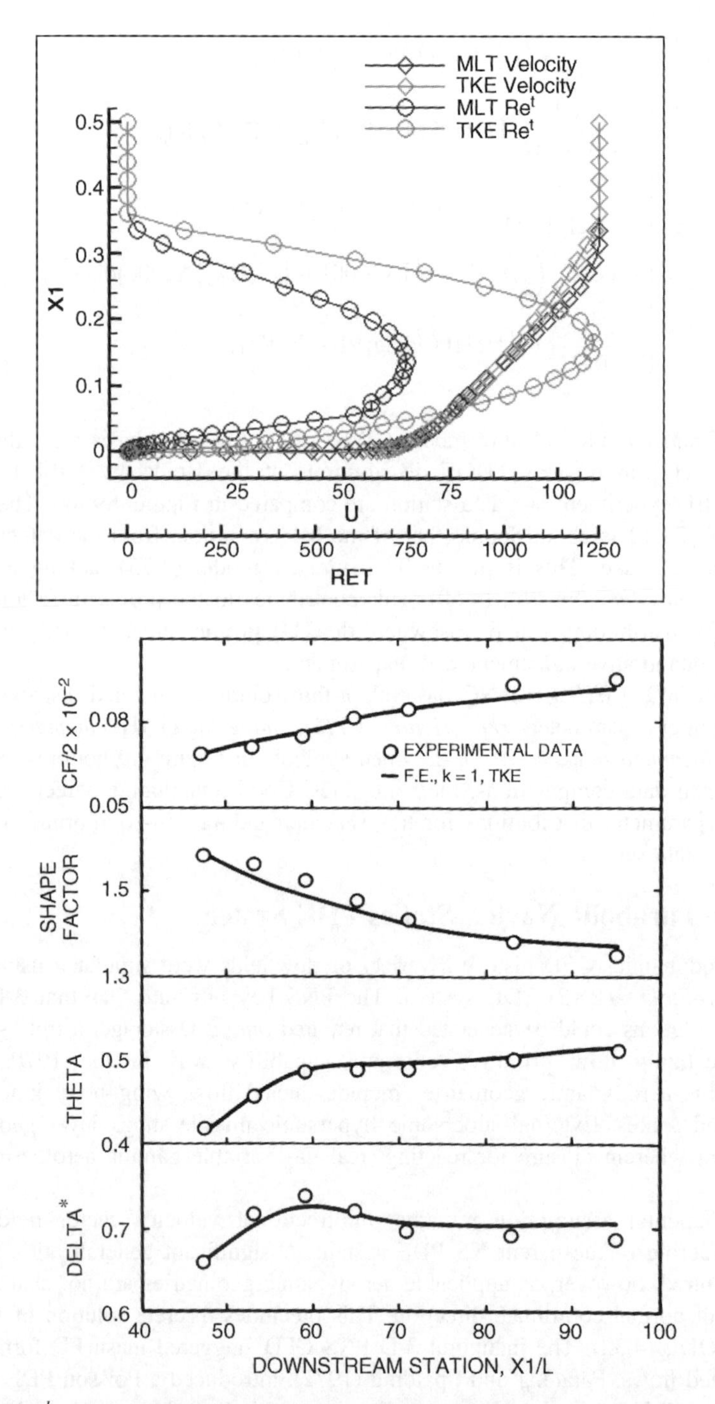

Figure 4.9 GWS^h + θTS algorithm $k = 1$ NC basis solutions, IDENT 2400 non-equilibrium turbulent BL experiment: top, MLT and TKE k-ε closure model {RET} and {U} DOF distributions at last data station; bottom, k-ε solution BL integral parameter comparisons with data

The essence of unidirectional flow in an aerodynamic juncture region is illustrated in Figure 4.10. Dual surface boundary layer growth interacts in the juncture generating multi-dimensional velocity and pressure distributions in the transverse plane spanned by x_2-x_3. Turbulence mechanisms expand to planar EBV PDE boundary value dimensionality requiring a more complete Reynolds stress tensor closure model. Of prime well-posed PDE significance, pressure effects are fully 3-D, requiring a thorough EBV weak interaction formulation.

The external aerodynamics weak interaction theory-compliant PNS PDE system formulation, including full EBV pressure resolution and complete Reynolds stress tensor, is reported in Baker and Orzechowski (1982). The theory adapts the cited ducted flow Poisson PDE for D*M* approximate enforcement and replaces the axial pressure gradient scalar computation with a 3-D pressure Poisson PDE + BCs statement. Theory resolution into a *complementary* plus *particular* Poisson solution process enables BC weak interaction communication with farfield potential theory determined 3-D pressure distributions.

The parabolic NS theory starting point for deriving the incompressible isothermal time averaged 3-D PNS PDE system remains steady (1.27–1.28)

$$\mathrm{D}M\colon \nabla \cdot \bar{\mathbf{u}} = 0 \tag{4.94}$$

$$\mathrm{D}\mathbf{P}\colon \nabla \cdot \overline{\mathbf{u}\mathbf{u}} + \nabla\bar{P} - \frac{1}{\mathrm{Re}}\nabla \cdot \left[(1 + \mathrm{Re}^t)\nabla\bar{\mathbf{u}}\right] = 0 \tag{4.95}$$

For x_1 spanning the predominant flow direction, Figure 4.10, the PNS PDE + BCs system alteration to (4.94–4.95) requires that:

- velocity vector component u_1 suffers no reversal
- diffusive transport processes in the x_1 direction are negligible in comparison to convection, and
- the overall EBV character of subsonic NS must be enforced by identification of a suitable 3-D pressure field.

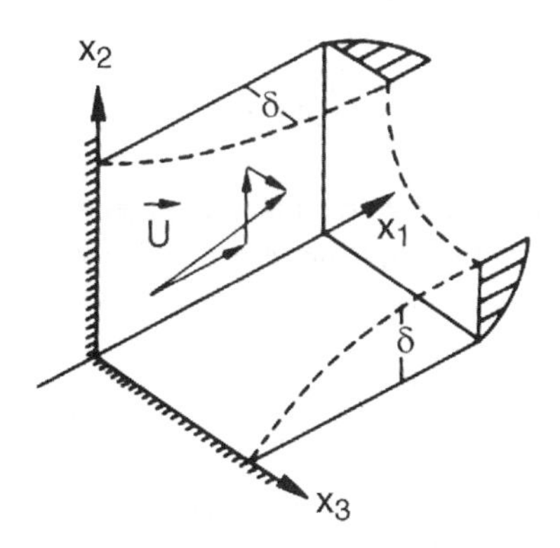

Figure 4.10 Illustration of the generic aerodynamic juncture region flowfield

Guided by the BL PDE Reynolds ordering, for $O(u_1/\mathrm{U})$ unity then $O(u_2/\mathrm{U},\ u_3/\mathrm{U})$ are assumed (must be!) $O(\delta) < O(1)$. Considering D*M*, (4.94), x_1 direction derivative is $O(1)$, hence planar derivatives transverse to x_1 must be $O(\delta^{-1})$. For **DP,** $O(1/\mathrm{Re}) = O(\delta^{-2})$ hence the order of significant Reynolds stress tensor resolutions cannot exceed $O(\delta)$, to match transverse derivative order. Thereby Reynolds stress tensor x_1 derivatives are of $O(\delta)$ and can be discarded.

In summary, replacing Re^t with the Reynolds stress tensor $\overline{u_i'u_j'}$, inserting a Poisson PDE for D*M* and for PNS theory tensor index summation conventions $1 \le (i,\ j) \le 3$ and $2 \le (k,\ l) \le 3$, the incompressible PNS PDE system with k-ε TKE closure is

$$\mathcal{L}(\phi) = -\frac{\partial^2 \phi}{\partial x_k^2} - \frac{\partial u_i}{\partial x_i} = 0 \tag{4.96}$$

$$\mathcal{L}(u_i) = u_j \frac{\partial u_i}{\partial x_j} + \delta_{ij}\frac{\partial P}{\partial x_j} + \frac{\partial}{\partial x_k}\left[\overline{u_i'u_k'} - \frac{1}{\mathrm{Re}}\frac{\partial u_i}{\partial x_k}\right] = 0 \tag{4.97}$$

$$\mathcal{L}(k) = u_j \frac{\partial k}{\partial x_j} + \frac{\partial}{\partial x_k}\left[\left(C_k^{-1}\frac{k}{\varepsilon}\overline{u_l'u_k'} - \frac{\delta_{lk}}{\mathrm{Pe}}\right)\frac{\partial k}{\partial x_l}\right] + \overline{u_1'u_k'}\frac{\partial u_1}{\partial x_k} + \varepsilon = 0 \tag{4.98}$$

$$\mathcal{L}(\varepsilon) = u_j \frac{\partial \varepsilon}{\partial x_j} + \frac{\partial}{\partial x_k}\left[C_\varepsilon^{-1}\frac{k}{\varepsilon}\overline{u_l'u_k'}\frac{\partial \varepsilon}{\partial x_l}\right] + C_\varepsilon^1\overline{u_1'u_k'}\frac{\varepsilon}{k}\frac{\partial u_1}{\partial x_k} + C_\varepsilon^2\frac{\varepsilon^2}{k} = 0 \tag{4.99}$$

The key NS constraint of D*M*, (4.94), is PNS-approximately enforced via (4.96), for $\phi(x_k)$ the transverse plane (only!) velocity *perturbation potential function*. The rigorous theory derivation is deferred to Chapter 5 when addressing the complete unsteady NS PDE system. Assuming for now that the theory is sound, the energy norm for ϕ^h is the intrinsic measure of PNS algorithm mass conservation discrete *approximation error* $\mathrm{D}M^h \neq 0$. PNS weak form algorithm design typically yields $|\phi^h|_E \le \varepsilon^2$ at convergence, for ε the terminal algebraic statement iterative convergence requirement for DOF $\{Q(x_1)\}$ determination. The well posed BCs for (4.96) are homogeneous Neumann on all domain closure segments with velocity efflux specified, otherwise $\phi = 0$, also derived in Chapter 5.

The PNS algorithm requires an EBV PDE statement for kinematic pressure *P*, the derivation of which relies on Reynolds ordering. Recalling Chapter 3, post-processing a far-field potential flow solution produces an aerodynamic surface pressure distribution. Via weak interaction theory these data are imposed on the displacement thickness distribution $\delta^*(x_i)$ lying in the PNS PDE solution domain. Select Dirichlet BC data $P^I(\delta^*(x_i))$ are thus known.

Forming the divergence of the transverse plane momentum PDE pair in (4.97), followed by Reynolds ordering generates the requisite pressure Poisson PDE with non-homogeneous Neumann (Robin) BC

$$\begin{aligned}\mathcal{L}(P) &= -\frac{\partial^2 P}{\partial x_k^2} - \frac{\partial^2}{\partial x_l \partial x_k}\left[u_l u_k + \overline{u_l'u_k'}\right] = 0\\ l(P) &= \frac{\partial P}{\partial x_k}\widehat{n}_k + \frac{1}{\mathrm{Re}}\frac{\partial^2 u_k \widehat{n}_k}{\partial x_s^2} = 0\end{aligned} \tag{4.100}$$

In (4.100) $\widehat{n}_k$ is the outwards pointing unit normal vector to domain boundary segments with throughflow with s the coordinate spanning each segment. Homogeneous (4.100), specifically, $\mathcal{L}(P) = 0$, is first solved on transverse planes for which the farfield potential solution has generated Dirichlet BC data $P^I(\delta^*(x_k, x_1))$. Homogeneous Neumann BCs are applied elsewhere.

This yields PNS algorithm *complementary pressure* distribution $P_c(x_k, x_1)$ *coupling* with the farfield. The *particular pressure* solution $P_p(x_k, x_1)$ is determined by addressing (4.100) as written with Robin BC enforcement, except for imposing Dirichlet BC $P_p = 0$ on boundaries, where $P^I(\delta^*(x_k, x_1))$ is known. The resultant PNS algorithm fully 3-D elliptic pressure distribution is $P(x_i) \equiv P_c(x_i) + P_p(x_k, x_1)$, achieved at convergence of the *weak interaction* solution sequence.

PNS PDE system closure is required for the full Reynolds stress tensor $\overline{u_i' u_j'}$ permeating (4.97–4.99). Existence of a nonlinear stress–strain rate constitutive closure is guaranteed with restrictions, Lumley (1970). Generalizing to a *continuum* tensor field the component PNS algebraic model, Gessner (1973), Gessner and Emery (1976), Baker *et al.* (1979) identify the *nonlinear* tensor field algebraic constitutive closure

$$\begin{aligned} \overline{u_i' u_j'} &= k\alpha_{ij} - C_4 \frac{k^2}{\varepsilon} S_{ij} - C_2 C_4 \frac{k^3}{\varepsilon^2} S_{ik} S_{kj} + \cdots \\ S_{ij} &\equiv 1/2\left[\partial u_i/\partial x_j + \partial u_j/\partial x_i\right] \end{aligned} \tag{4.101}$$

for $k = 1/2(\overline{u_j' u_j'})^{1/2}$ turbulent kinetic energy and S_{ij} the time averaged velocity strain rate tensor. In principal coordinates and for x_1 spanning the dominant flow direction, tensor α_{ij} definition diag$[C_1, C_3, C_3]$ recognizes normal stress *anisotropy* intrinsic to the Reynolds stress tensor for unidirectional flows, Wilcox (2006, Figure 2.5). The coefficient set for (4.101) is $C_\alpha = \{0.94, 0.067, 0.56, 0.068\}$, Launder *et al.* (1975).

Reynolds ordering (4.101) identifies stress tensor expansion terms significant in PNS closure of (4.97–4.100). Since (4.101) must be pertinent for an $n = 2$ turbulent BL, dominant shear stress *order* is

$$O(\overline{u_1' u_2'}) = O\left(C_4 \frac{k^2}{\varepsilon} \frac{\partial u_1}{\partial x_2}\right) = O(\delta) \tag{4.102}$$

Since the derivative in (4.102) is $O(\delta^{-1})$ then $O(C_4 k^2/\varepsilon) = O(\delta^2)$. As k is Reynolds tensor trace then $O(k) = O(\delta)$ and thereby $O(\varepsilon) = O(1)$ and $O(k^3/\varepsilon^2) = O(k^3) = O(\delta^3)$. These observations quantitatively confirm the PNS PDE system $O(\delta)$ Reynolds stress tensor cartesian resolution is

$$\begin{aligned} \overline{u_1' u_1'} &= C_1 k - C_2 C_4 \frac{k^3}{\varepsilon^2}\left[\left(\frac{\partial u_1}{\partial x_2}\right)^2 + \left(\frac{\partial u_1}{\partial x_3}\right)^2\right] \\ \overline{u_2' u_2'} &= C_3 k - C_2 C_4 \frac{k^3}{\varepsilon^2}\left(\frac{\partial u_1}{\partial x_2}\right)^2, \quad \overline{u_3' u_3'} = C_3 k - C_2 C_4 \frac{k^3}{\varepsilon^2}\left(\frac{\partial u_1}{\partial x_3}\right)^2 \\ \overline{u_1' u_2'} &= -C_4 \frac{k^2}{\varepsilon}\frac{\partial u_1}{\partial x_2}, \quad \overline{u_1' u_3'} = -C_4 \frac{k^2}{\varepsilon}\frac{\partial u_1}{\partial x_3}, \quad \overline{u_2' u_3'} = -C_2 C_4 \frac{k^3}{\varepsilon^2}\left(\frac{\partial u_1}{\partial x_2}\right)\left(\frac{\partial u_1}{\partial x_3}\right) \end{aligned} \tag{4.103}$$

With this determination (4.97) expanded for the transverse plane velocity vector pair $\mathcal{L}(u_k)$ confirms all terms in both PDEs are $O(\delta)$. Retaining the requisite EBV character thus requires inclusion of the pertinent $O(\delta^2)$ Reynolds stress tensor terms therein, which are

$$\begin{aligned}\overline{u_1'u_1'} &= -C_4\frac{k^2}{\varepsilon}\frac{\partial u_1}{\partial x_1}, \quad \overline{u_2'u_2'} = -C_4\frac{k^2}{\varepsilon}\frac{\partial u_2}{\partial x_2}\\ \overline{u_3'u_3'} &= -C_4\frac{k^2}{\varepsilon}\frac{\partial u_3}{\partial x_3}, \quad \overline{u_2'u_3'} = -C_4\frac{k^2}{\varepsilon}\left(\frac{\partial u_2}{\partial x_3}+\frac{\partial u_3}{\partial x_2}\right)\end{aligned} \tag{4.104}$$

4.11 GWS^h + θTS Algorithm for PNS PDE System

The subsonic external flow PNS PDE system for TKE k-ε coupled full Reynolds stress tensor closure with mass conservation and 3-D EBV pressure resolution is derived. It is constituted of I-EBV PDE system (4.97–4.99) with (4.103–4.104), plus the pure EBV PDEs (4.96) and (4.100). The PNS system state variable is $\{q(x_1, x_k\} = \{u_1, u_2, u_3, k, \varepsilon; \overline{u_i'u_j'}; \phi, P\}^T$. The PNS GWS^h + θTS algorithm corresponding DOF is labeled $\{Q(x_1)\} = \{\mathrm{U1, U2, U3, K, E; R}IJ\mathrm{; F, P}\}^T$.

Confirmation that the identified PNS PDE system is *well posed* accrues to embedding the Reynolds stress tensor resolution (4.103–4.104) into (4.97–4.98). Recalling PNS index summation conventions $1 \le i, j \le 3$ and $2 \le k, l \le 3$, also $C_4k^2/\varepsilon = \mathrm{Re}^t/\mathrm{Re}$ in (4.103), (4.96) for velocity component u_1 is the easily interpreted generalization of isothermal time averaged BL PDE system (4.66–4.68)

$$\mathcal{L}(u_1) = u_j\frac{\partial u_1}{\partial x_j} + \delta_{1j}\frac{\partial P}{\partial x_j} - \frac{1}{\mathrm{Re}}\frac{\partial}{\partial x_k}\left[(1+\mathrm{Re}^t)\frac{\partial u_1}{\partial x_k}\right] = 0 \tag{4.105}$$

Substituting *only* the $O(\delta^2)$ Reynolds stress tensors (4.104) into (4.96) for the u_k velocity components also produces the BL generalization

$$\mathcal{L}(u_k) = u_j\frac{\partial u_k}{\partial x_j} + \delta_{kj}\frac{\partial P}{\partial x_j} - \frac{1}{\mathrm{Re}}\frac{\partial}{\partial x_l}\left[(1+\mathrm{Re}^t)\frac{\partial u_k}{\partial x_l}\right] + \left(\frac{\partial\, \overline{u_k'u_l'}}{\partial x_l}\right) = 0 \tag{4.106}$$

which contains the PNS theory consistent $O(\delta)$ Reynolds stress tensor as a source term. Thereby, both (4.105) and (4.106) are verified well posed as I-EBV requiring encompassing BCs.

An exercise suggests substituting only the $O(\delta^2)$ Reynolds stress tensor contributions into the TKE closure PDEs (4.98) and (4.99). The results are, respectively

$$\frac{\partial}{\partial x_k}\left[\left(C_k^{-1}\frac{k}{\varepsilon}\overline{u_l'u_k'}\Big|_{O(\delta^2)} - \frac{\delta_{lk}}{\mathrm{Pe}}\right)\frac{\partial k}{\partial x_l}\right] = \frac{-1}{\mathrm{Re}}\frac{\partial}{\partial x_k}\left[\left(\frac{\delta_{lk}}{\mathrm{Pr}} + \frac{k}{\varepsilon}\frac{\mathrm{Re}^t}{C_k}\left(\frac{\partial u_k}{\partial x_l}\right)\right)\frac{\partial k}{\partial x_l}\right] \tag{4.107}$$

$$\frac{\partial}{\partial x_k}\left[\left(C_\varepsilon^{-1}\frac{k}{\varepsilon}\overline{u_l'u_k'}\Big|_{O(\delta^2)}\right)\frac{\partial \varepsilon}{\partial x_l}\right] = \frac{-1}{\mathrm{Re}}\frac{\partial}{\partial x_k}\left[\left(\frac{k}{\varepsilon}\frac{\mathrm{Re}^t}{C_\varepsilon}\left(\frac{\partial u_k}{\partial x_l}\right)\right)\frac{\partial \varepsilon}{\partial x_l}\right] \tag{4.108}$$

hence both PNS TKE closure PDEs are also confirmed I-EBV. The *significant* $O(\delta)$ Reynolds stress tensors are of prime importance hence, as in the BL systems, hence must appear as source terms. Finally, note the pervasive presence of Re^t which remains handled as a solution dependent *nonlinear* parameter.

The PNS state variable members pertinent to the verified I-EBV PDE system (4.97–4.99), now indexed by subscript "I," leads to definition $\{q_I(x_1, x_k\} = \{u_1, u_2, u_3, k, \varepsilon\}^T$. For $1 \leq I \leq 5$, the I-EBV PDE system is

$$\begin{aligned}\mathcal{L}(\{q_I\}) &= u_1 \frac{\partial\{q_I\}}{\partial x_1} + u_k \frac{\partial\{q_I\}}{\partial x_k} + \delta_{Ij}\frac{\partial P}{\partial x_j} \\ &+ \frac{\partial}{\partial x_k}\left[\left(f_I\left(k, \varepsilon, \overline{u_i' u_k'}\right) - \frac{1}{\mathrm{Pa}}\right)\frac{\partial\{q_I\}}{\partial x_k}\right] - s(\{q\}) = 0\end{aligned} \tag{4.109}$$

for Pa the *placeholder* for Re and Pe, $f_I(\bullet)$ the I-pertinent function of its argument and $s(\{q\})$ the source term. The statement of *any* approximation $\{q_I^N(x_1, x_k)\}$ to PNS state variable $\{q_I(x_1, x_k)\}$ leads to the discrete approximation $\{q_I^h(x_1, x_k)\}$

$$\begin{aligned}\{q_I^N(x_1, x_k)\} &\equiv \mathrm{diag}\left[\sum_{\alpha=1}^{N}\Psi_\alpha(x_k)\right]\{Q_\alpha(x_1)\} \\ &\Rightarrow \{q_I^h(x_1, x_k)\} \equiv \cup_e^M \{q_I(x_1, x_k)\}_e \\ \{q_I(x_1, x_k)\}_e &\equiv \mathrm{diag}\left[\{N_k(\xi_\alpha(x_2, x_3))\}^T\right]\{\mathrm{Q}I(x_1)\}_e\end{aligned} \tag{4.110}$$

In the last line of (4.110) diag[•] denotes diagonal matrix with elements the NC degree k trial space basis spanning $n = 2$. The column matrix $\{\mathrm{Q}I(x_1)\}$, $1 \leq I \leq 5$, is the approximation DOF for I-EBV PDE state variable $\{q_I(x_1, x_k\}$.

With (4.110), the continuum Galerkin weak statement for (4.109) is

$$\mathrm{GWS}^N = \int_\Omega \mathrm{diag}\left[\Psi_\beta(x_k)\right]\mathcal{L}\left(q_I^N(x_1, x_k)\right)\mathrm{d}\tau = \{0\}, \forall\beta \tag{4.111}$$

with $\mathrm{d}\tau = \mathrm{d}x_2\mathrm{d}x_3$ the differential element. Implementing (4.111) on the discretization Ω^h of the PNS transverse plane domain, and recalling the arguments leading to (4.27), produces the matrix ODE system

$$\begin{aligned}\mathrm{GWS}^h &= \mathrm{S}_e\int_{\Omega_e}\mathrm{diag}[N_k(\xi_\alpha)]\mathcal{L}(\{q_I(x_1, x_k)\}_e)\mathrm{d}x_k = \{0\} \\ &= \left[\mathrm{M3000}(\{\mathrm{U1}\})\right]\frac{\mathrm{d}\{\mathrm{Q}I\}}{\mathrm{d}x_1} + \{\mathrm{RES\_Q}I(\{\mathrm{Q}\})\}\end{aligned} \tag{4.112}$$

Then via (4.28–4.30) and defining $\theta \equiv 0.5$, the *terminal* nonlinear algebraic matrix statement replacing (4.31) for the $\mathrm{GWS}^h + \theta\mathrm{TS}$ time averaged TKE closed PNS algorithm for

I-EBV state variable members $\{q_I(x_1, x_k\} = \{u_1, u_2, u_3, k, \varepsilon\}^T$ is

$$\begin{aligned}\{\mathrm{F\_Q}I\} &\equiv [\mathrm{M3000}(\{\mathrm{U1}\})]_{n+1/2}\{\mathrm{Q}I_{n+1} - \mathrm{Q}I_n\} \\ &+ \frac{\Delta x_1}{2}\left[\{\mathrm{RES\_Q}I(\{Q\})\}_{n+1}\right] + \{\mathrm{RES\_Q}I(\{Q\})\}|_n = 0\}\end{aligned} \tag{4.113}$$

with iterative solution generating the DOF $\{\mathrm{Q}I(x_1)\}$ at station $n+1$.

The GWS^h algorithm for state variable members $\overline{u_i'u_j'}$ is formed on the *differential definition*

$$\mathcal{D}(\overline{u_i'u_j'}) \equiv \overline{u_i'u_j'} - k\alpha_{ij} + \mathrm{C}_4\frac{k^2}{\varepsilon}S_{ij} + \mathrm{C}_2\mathrm{C}_4\frac{k^3}{\varepsilon^2}S_{ik}S_{kj} = 0 \tag{4.114}$$

Conceptually replacing $\{q_I^N(x_1, x_k)\}$ in (4.111) with $\{\overline{u_i'u_j'}^N(x_1, x_k)\}$ and proceeding through GWS$^N \Rightarrow$ GWS^h the terminal nonlinear algebraic matrix statement for PNS algorithm DOF $\{\mathrm{R}IJ(x_1)\}$ for Reynolds stress tensor state variable members $\{q_R(x_1, x_k\} = \{\overline{u_i'u_j'}\}$ is

$$\{\mathrm{F\_R}IJ\} \equiv [\mathrm{M200}]\{RIJ\} + \{\mathrm{RES\_R}IJ(\{Q\}, S_{ij})\} = \{0\} \tag{4.115}$$

for $1 \leq (I, J) \leq 3$ and recalling the strain-rate tensor S_{ij} definition (4.101). The residuals $\{\mathrm{RES\_R}IJ(\bullet)\}$ defined in (4.115) are the GWS^h statement of stress tensor resolution (4.103–4.104). The sole admissible DOF constraint in solving (4.115) is homogeneous Dirichlet.

The PNS algorithm is completed with linear Poisson PDEs (4.96) and (4.100). Denoting these EBV *Poisson* state variable members as $\{q_P(x_1, x_k\} = \{\phi, P\}^T$, and via (4.110–4.111), GWS$^N \Rightarrow$ GWS^h produces the *linear* matrix statements

$$\{\mathrm{F\_Q}P\} \equiv [\mathrm{DIFF}]\{\mathrm{Q}P\} + \{\mathrm{b}(\{Q\}, \mathrm{data})\} = \{0\} \tag{4.116}$$

the solution of which determines DOF $\{\mathrm{Q}P\} = \{\mathrm{F}(x_1), \mathrm{P}(x_1)\}^T$.

PNS algorithm weak interaction completion is via the subsonic farfield potential flow PDE, (3.17–3.22), GWS^h implemented in three dimensional constant density form. The resultant $P^I(\delta^*(x_i))$ post-process via GWS^h (3.119) generates the state variable surface pressure DOF $\{\mathrm{P3}(x_1)\}$, the BC data required for complementary pressure solution.

The GWS^h + θTS PNS algorithm statement is complete. The matrix iterative algorithm definition remains (4.34), implemented via assembly of component constituents detailed in the next section. The algorithm iterative sequence first generates $\{\mathrm{P3}(x_1)\}$, hence the x_1-parameterized complementary pressure distribution $P_c^h(x_k, x_1)$. Then space marching over intervals $x_{n+1} = x_n + \Delta x$ generates updated PNS DOF $\{\mathrm{Q}I(x_{n+1})\} = \{\mathrm{U1, U2, U3, K, E}\}^T$, with iteration embedded updating of DOF $\{Q(x_{n+1})\}$ for $\{\mathrm{R}IJ, \mathrm{F}, \mathrm{PP}\}^T$, the latter the particular pressure solution. Completion of the x_1 space march generates the PNS algorithm solution $\{q^h(x_1, x_k\}$ from which the revised surface distribution $\delta^*(x_i)$ is generated to update $P_c^h(x_k, x_1)$ determination if necessary.

4.12 GWS^h + θTS $k = 1$ NC Basis PNS Algorithm

For $k = 1$ NC basis the GWS^h $n = 2$ integrals in (4.112–4.116) are evaluated on triangular Ω_e spanning the x_2-x_3 plane mesh Ω^h. The generated element matrices have prefix "B" and the triangle measure is $A_e = \det_e/2$, recall (3.63). The DM^h algorithm *sums* x_1 station converged perturbation potential $\{F(x_1)\}$ DOF into the accumulation $\{\Sigma F_n(x_1)\}$ over all n, which constitutes *data* when iterating at x_1 step $n+1$. Compute experience confirms DM^h algorithm performance at step $n+1$ is enhanced upon pre-multiplying both terms by step n solution diag[U1$_n$].

The GWS^h + θTS PNS algorithm element level replacement for BL (4.35) for DOF $\{QI\}_e \equiv \{UI\}_e = \{U1, U2, U3\}_e^T$, $1 \le I \le 3$, is

$$
\begin{aligned}
\{F\_QI\}_e \equiv & (\det_e/2)\{U1_{n+1/2}\}_e^T[B3000]\{QI_{n+1} - QI_n\}_e \\
& - \text{diag}[U1_n]_e[B20K]_e\left(\{F\}_e + \{\Sigma F_n\}_e\right)\delta_{IK} \\
& + \frac{\Delta x_1}{2}\left[\{RES\_QI(\{Q\})\}_e\big|_{n+1} + \{RES\_QI(\{Q\})\}_e\big|_n\right]
\end{aligned}
\tag{4.117}
$$

This is also valid for DOF $\{QI\}_e = \{K, E\}_e^T$, $4 \le I \le 5$, upon omitting the DM^h algorithm term, the second line in (4.117). The GWS^h + θTS algorithm residuals for DOF $\{UI\}_e$, (4.105–4.107), for repeated index *summation* convention $2 \le (L, K) \le 3$, are

$$
\begin{aligned}
\{RES\_UI\}_e = & \{UL\}_e^T[B300L]_e\{UI\}_e + \{b_{UI}\_\{Q\}\}_e \\
& + \frac{1}{\det_e \text{Re}}\{1 + \text{RET}\}_e^T[B30LL]_e\{UI\}_e, \quad for\ 1 \le I \le 3
\end{aligned}
\tag{4.118}
$$

The contribution $\{b_{UI}\_\{Q\}\}_e$ contain I-dependent pressure gradient terms. For DOF $\{U1\}_e$

$$
\{b_{U1}\_\{PC\}\}_e = \frac{\det_e}{2}[B200]\frac{d\{PC\}_e}{dx_1}
\tag{4.119}
$$

while for DOF $\{UK\}_e$, $2 \le (K, L) \le 3$

$$
\{b_{UK}\_\{Q\}\}_e = [B20L]_e\left((\{PP\}_e + \{PC\}_e)\delta_{KL} + \{RKL\}_e\right)
\tag{4.120}
$$

for $\{PC\}_e$ and $\{PP\}_e$ the GWS^h algorithm complementary and particular pressure solution DOF. The $\{RKL\}_e$ are GWS^h algorithm (4.115) Reynolds stress tensor transverse plane DOF.

The residual matrices for DOF $\{QI\}_e = \{K,E\}_e$ possess distinct diffusion coefficients containing very nonlinear hypermatrix pre-multiplier strings. For conceptual clarity continuum notation is employed in the following definitions

$$
\begin{aligned}
\{RES\_K\}_e = & \{UL\}_e^T[B300L]_e\{K\}_e + \{b_K\_\{Q\}\}_e \\
& + \frac{1}{\det_e \text{Re}}\left\{\frac{\delta_{lk}}{\text{Pr}} + \frac{\text{Re}^t}{C_k}\frac{k}{\varepsilon}\left(\frac{\partial u_k}{\partial x_l}\right)\right\}_e^T[B30KL]_e\{K\}_e
\end{aligned}
\tag{4.121}
$$

$$\{\text{RES\_E}\}_e = \{\text{U}L\}_e^T[\text{B300}L]_e\{\text{E}\}_e + \{\text{b}_{\text{E}}\_\{Q\}\}_e + \frac{1}{\det_e \text{Re}}\left\{\frac{\text{Re}^t}{\text{C}_\varepsilon}\frac{k}{\varepsilon}\left(\frac{\partial u_k}{\partial x_l}\right)\right\}_e^T[\text{B30}LK]_e\{\text{E}\}_e \tag{4.122}$$

The source terms for (4.121–4.122) are

$$\{b_{\text{K}}\_\{Q\}\}_e = \frac{1}{\det_e \text{C}_k}\left\{\frac{k}{\varepsilon}\overline{u_l'u_k'}\right\}_e^T[\text{B30}KL]_e\{\text{K}\}_e + \left\{\overline{u_1'u_k'}\right\}_e^T[\text{B300}K]_e\{\text{U1}\}_e + \det_e[\text{B200}]_e\{\text{E}\}_e \tag{4.123}$$

$$\{\text{b}_{\text{E}}\_\{Q\}\}_e = \frac{1}{\det_e \text{C}_\varepsilon}\left\{\frac{k}{\varepsilon}\overline{u_l'u_k'}\right\}_e^T[\text{B30}KL]_e\{\text{E}\}_e + \text{C}_\varepsilon^1\left\{\overline{u_1'u_k'}\frac{\varepsilon}{k}\right\}_e^T[\text{B300}K]_e\{\text{U1}\}_e + \text{C}_\varepsilon^2\det_e[\text{B200}]_e\left\{\text{E}^2/\text{K}\right\}_e \tag{4.124}$$

The *italicized* indices in each element matrix $[\text{B} \ldots K]_e$, $[\text{B} \ldots K \ldots]_e$, and $[\text{B} \ldots KL \ldots]_e$, (4.118–4.124), denote spatial derivative, hence are element-dependent (subscript e) via coordinate transformation(s) embedding. As detailed in Section 3.8, the library of element-independent integer array matrices emerges by extracting the $k=1$ NC basis metric data array(s) of constants. For one and two spatial derivatives, the array notations are

$$\left(\frac{\partial \zeta_\alpha}{\partial x_i}\right)_e \quad \text{and} \quad \left(\frac{\partial \zeta_\alpha}{\partial x_i}\frac{\partial \zeta_\beta}{\partial x_i}\right)_e \tag{4.125}$$

with each multiplied by $(\det_e)^{-1}$, (3.75). Extracting these data the *element-independent* library matrix *computable* form of (4.118) is

$$\begin{aligned}\{\text{RES\_U}I\}_e &= \frac{\det_e}{2}\left(\frac{\partial \zeta_\alpha}{\partial x_l}\right)_e\{\text{U}l\}^T[\text{B300}\alpha]\{\text{U}I\}_e + \{b_{\text{U}I}\_\{Q\}\}_e \\ &+ \frac{\det_e}{2\text{Re}}\left(\frac{\partial \zeta_\alpha}{\partial x_l}\frac{\partial \zeta_\beta}{\partial x_l}\right)_e\{1+\text{RET}\}_e^T[\text{B30}\alpha\beta]\{\text{U}I\}_e, \\ &\text{for } 1 \le I \le 3, 2 \le l \le 3, 1 \le \alpha, \beta \le 3\end{aligned} \tag{4.126}$$

with repeated index summation conventions clearly stated. Computable replacements for (4.117), (4.118–4.124) are similarly formed, and all *element-independent* matrices are readily established using Gauss quadrature, Section 3.9.

Continuing, the transverse velocity derivative in the cubically nonlinear turbulent diffusion hypermatrix pre-multipliers in (4.121–4.122) are element-dependent scalars. Specifically

$$\begin{aligned}\left(\frac{\partial u_k}{\partial x_l}\right)_e &= \frac{\partial}{\partial x_l}\{N_1\}^T\{\text{U}K\}_e = \left(\frac{\partial \zeta_\alpha}{\partial x_l}\right)_e\frac{\partial}{\partial \zeta_\alpha}\{N_1\}^T\{\text{U}K\}_e \\ &= \left\{\frac{\partial \zeta_1}{\partial x_l}, \frac{\partial \zeta_2}{\partial x_l}, \frac{\partial \zeta_3}{\partial x_l}\right\}_e\{\text{U}K\}_e\end{aligned} \tag{4.127}$$

via the Kronecker delta property of the $k=1$ NC basis derivative, (3.72). Therefore, these terms become scalar multipliers on element DOF ratio conversion $\left(\mathrm{Re}^t \dfrac{k}{\varepsilon}\right) \Rightarrow$ $\{\mathrm{RET,(K/E)}\}_e$ normalized by C_k or C_ε. In (4.123–4.124) the hypermatrix pre-multiplier product of $\overline{u_l' u_k'}$ with k/ε or ε/k is element implemented as DOF ratio $\{\mathrm{R}LK\mathrm{,(K/E)}\}_e$, and so on.

The $O(\delta)$ Reynolds stress tensor GWS^h algorithm (4.115), using (4.103), involves index-restricted forms of (4.127) and companion scalar expressions for velocity gradient quadratic products. Selecting $I=2=J$ for illustration, and recalling the Kronecker delta matrix $[\mathrm{M}\delta_{\alpha\beta}]$, (3.73), term evaluation produces the element scalar

$$\begin{aligned}\left(\frac{\partial u_1}{\partial x_2}\right)_e^2 &= \{\mathrm{U1}\}_e^T \frac{\partial\{N\}}{\partial x_2}\frac{\partial\{N\}^T}{\partial x_2}\{\mathrm{U1}\}_e^T \\ &= \left(\frac{\partial\zeta_\alpha}{\partial x_2}\frac{\partial\zeta_\beta}{\partial x_2}\right)_e \{\mathrm{U1}\}_e^T\left[\mathrm{M}\delta_{\alpha\beta}\right]\{\mathrm{U1}\}_e\end{aligned} \tag{4.128}$$

Then noting $\mathrm{Re}^t \equiv \mathrm{C}_4 k^2/\varepsilon$ is implemented as $\{\mathrm{RET}\}_e$ the element matrix statements for (4.115) are

$$\begin{aligned}\{\mathrm{F\_R}KL\}_e &\equiv (\det_e/2)[\mathrm{B200}]\{\mathrm{R}KL\}_e - (\det_e/2)\mathrm{C}_3\delta_{KL}[\mathrm{B200}]\{\mathrm{K}\}_e \\ &+ \frac{\det_e \mathrm{C}_2}{2}\left(\frac{\partial\zeta_\alpha}{\partial x_k}\frac{\partial\zeta_\beta}{\partial x_l}\right)_e \{\mathrm{U1}\}_e^T\left[\mathrm{M}\delta_{\alpha\beta}\right]\{\mathrm{U1}\}_e\{\mathrm{RET}\}_e^T[\mathrm{B3000}]\{\mathrm{K/E}\}_e\end{aligned} \tag{4.129}$$

$$\begin{aligned}\{\mathrm{F\_R1}L\}_e &\equiv (\det_e/2)[\mathrm{B200}]\{\mathrm{R1}L\}_e \\ &+ (\det_e/2)\left\{\frac{\partial\zeta_1}{\partial x_l}, \frac{\partial\zeta_2}{\partial x_l}, \frac{\partial\zeta_3}{\partial x_l}\right\}_e \{\mathrm{U1}\}_e[\mathrm{B200}]\{\mathrm{RET}\}_e\end{aligned} \tag{4.130}$$

The PNS algorithm is completed by GWS^h element matrix statements for (4.116). The DM^h perturbation potential statement is

$$\begin{aligned}\{\mathrm{F\_F}\}_e &\equiv [\mathrm{DIFF}]_e\{\mathrm{F}\}_e + \{\mathrm{b_F}\_\{\mathrm{U}I\}\}_e \\ &= \int_{\Omega_e}\frac{\partial}{\partial x_k}\{N_1\}\frac{\partial}{\partial x_k}\{N_1\}^T \mathrm{d}\tau\{\mathrm{F}\}_e \\ &- \int_{\Omega_e}\{N_1\}\left(\{N_1\}^T\frac{\mathrm{d}\{\mathrm{U1}\}_e}{\mathrm{d}x_1} + \frac{\partial\{N_1\}^T}{\partial x_k}\{\mathrm{U}K\}_e\right)\mathrm{d}\tau \\ &= \frac{\det_e}{2}\left(\frac{\partial\zeta_\alpha}{\partial x_k}\frac{\partial\zeta_\beta}{\partial x_k}\right)_e\left[\mathrm{M}\delta_{\alpha\beta}\right]\{\mathrm{F}\}_e \\ &- \frac{\det_e}{2}[\mathrm{B200}]\frac{\mathrm{d}\{\mathrm{U1}\}_e}{\mathrm{d}x_1} - \frac{\det_e}{2}\left(\frac{\partial\zeta_\alpha}{\partial x_k}\right)_e[\mathrm{B20}\alpha]\{\mathrm{U}K\}_e\end{aligned} \tag{4.131}$$

The fully expanded expression for $\dfrac{\det_e}{2}\left(\dfrac{\partial\zeta_\alpha}{\partial x_k}\dfrac{\partial\zeta_\beta}{\partial x_k}\right)_e\left[\mathrm{M}\delta_{\alpha\beta}\right]$ is (3.76). Estimating $\mathrm{d}\{\mathrm{U1}\}_e/\mathrm{d}x_1$ is via the BL algorithm second-order accurate TS, (4.33). In mixed notation for clarity, the *particular* pressure GWS^h algorithm element matrix statement is

$$
\begin{aligned}
\{\mathrm{F\_PP}\}_e &= \int_{\Omega_e} \frac{\partial}{\partial x_k}\{N_1\}\frac{\partial}{\partial x_k}\{N_1\}^T \mathrm{d}x_k\{\mathrm{PP}\}_e - \int_{\partial\Omega_e}\{N_1\}\frac{\partial P_P}{\partial x_k}\widehat{n}_k\,\mathrm{d}x_s \\
&+ \int_{\Omega_e}\frac{\partial\{N_1\}}{\partial x_l}\frac{\partial}{\partial x_k}\left[u_l u_k + \overline{u_l' u_k'}\right]_e \mathrm{d}x_k - \int_{\partial\Omega_e}\{N_1\}\frac{\partial}{\partial x_k}\left[u_l u_k + \overline{u_l' u_k'}\right]_e \widehat{n}_l \mathrm{d}x_s \\
&= \frac{\det_e}{2}\left(\frac{\partial\zeta_\alpha}{\partial x_k}\frac{\partial\zeta_\beta}{\partial x_k}\right)_e \left[\mathrm{M}\delta_{\alpha\beta}\right]\{\mathrm{PP}\}_e - \frac{\det_e}{\mathrm{Re}}[\mathrm{A211}]\{\mathrm{U}K\}_e \widehat{n}_k \\
&+ \frac{\det_e}{2}\left(\{\mathrm{U}K\}_e^T[\mathrm{B30}LK]_e\{\mathrm{U}L\}_e + \{\mathrm{U}L\}_e^T[\mathrm{B30}LK]_e\{\mathrm{U}K\}_e\right) \\
&+ \frac{\det_e}{2}[\mathrm{B2}LK]_e\{\mathrm{R}LK\}_e - \int_{\partial\Omega_e}\{N_1\}\frac{\partial}{\partial x_k}\left[u_l u_k + \overline{u_l' u_k'}\right]\widehat{n}_l \mathrm{d}x_s
\end{aligned}
\tag{4.132}
$$

wherein the Green–Gauss divergence theorem has been employed twice. The last surface integral vanishes identically via the homogeneous Dirichlet BC application, leading to the computable element matrix statement

$$
\begin{aligned}
\{\mathrm{F\_PP}\}_e &= \frac{\det_e}{2}\left(\frac{\partial\zeta_\alpha}{\partial x_k}\frac{\partial\zeta_\beta}{\partial x_k}\right)_e \left[\mathrm{M}\delta_{\alpha\beta}\right]\{\mathrm{PP}\}_e - \frac{\det_e}{\mathrm{Re}}[\mathrm{A211}]\{\mathrm{U}K\}_e \widehat{n}_k \\
&+ \frac{\det_e}{2}\left(\frac{\partial\zeta_\alpha}{\partial x_l}\frac{\partial\zeta_\beta}{\partial x_k}\right)_e \begin{pmatrix} \{\mathrm{U}K\}_e^T[\mathrm{B30}lk]\{\mathrm{U}L\}_e \\ +\{\mathrm{U}L\}_e^T[\mathrm{B30}lk]\{\mathrm{U}K\}_e + \left[\mathrm{M}\delta_{\alpha\beta}\right]\{\mathrm{R}LK\}_e \end{pmatrix}
\end{aligned}
\tag{4.133}
$$

The second term in (4.133) results from Green–Gauss divergence theorem implementation of the Robin BC (4.100).

The final organizing step for the $\mathrm{GWS}^h + \theta\mathrm{TS}$ PNS algorithm is matrix iteration jacobian derivation. The modest compute resource available in the late 1970s, coupled with the PNS algorithm DOF size $\{\mathrm{Q}I(x_1)\} = \{\mathrm{U1, U2, U3, K, E, R}IJ\mathrm{, F, PP, PC}\}^T$, $1 \le I \le 14$, precluded a Newton algorithm. The *a posteriori* data detailed in the following sections are generated using only the diagonal block jacobian matrices $[\mathrm{J\_}II]_e$ with element definitions

$$
\begin{aligned}
[\mathrm{JAC\_Q}I]_e &\cong \frac{\partial\{\mathrm{F\_U1}\}_e}{\partial\{\mathrm{U1}\}_e}, \quad for\ 1 \le I \le 5 \\
&= (\det_e/2)\left\{\mathrm{U1}_{n+1/2}\right\}_e^T[\mathrm{B3000}] \\
&+ \frac{\Delta x_1}{2}\left(\{\mathrm{U}L\}_e^T[\mathrm{B300}L]_e + \frac{1}{\det_e \mathrm{Re}}\{1+\mathrm{RET}\}_e^T[\mathrm{B30}LL]_e\right)
\end{aligned}
\tag{4.134}
$$

$$
[\mathrm{JAC\_R}IJ]_e \equiv \frac{\partial\{\mathrm{F\_R}IJ\}_e}{\partial\{\mathrm{R}IJ\}_e} = (\det_e/2)[\mathrm{B200}], \quad 1 \le I, J \le 3 \tag{4.135}
$$

$$
\begin{aligned}
[\mathrm{JAC\_F}]_e &\equiv \frac{\partial\{\mathrm{F\_F}\}_e}{\partial\{\mathrm{F}\}_e} = [\mathrm{DIFF}]_e = \frac{\det_e}{2}\left(\frac{\partial\zeta_\alpha}{\partial x_k}\frac{\partial\zeta_\beta}{\partial x_k}\right)_e \left[\mathrm{M}\delta_{\alpha\beta}\right] \\
&= [\mathrm{JAC\_PP}]_e = [\mathrm{JAC\_PC}]_e
\end{aligned}
\tag{4.136}
$$

The partitioned jacobian solution process starts using (4.136) to compute the complementary pressure solution DOF $\{PC(x_{n+1})\}$ with BCs provided by the 3-D farfield potential solution DOF $\{P3(x_i)\}$. For these data and using DOF {K, E, PP, RET} from station x_n, the algorithm next step iteratively couples solution for DOF {U1, U2, U3, F} at x_{n+1} using (4.134) and (4.136). Upon convergence $\{UI(x_{n+1})\}$ is input as data in iteratively solving for DOF {K,E} at station x_{n+1}, also $\{RIJ(x_{n+1})\}$, via (4.134) coupled with (4.135) with embedded updating of algebraic $\{RET(x_{n+1})\}$. Following iterative convergence of this partition, the particular pressure DOF $\{PP(x_{n+1})\}$ are post-processed using (4.136).

Once the x_1 span of the PNS solution domain is swept, interaction theory dictates the farfield 3-D potential solution DOF $\{P3(x_i)\}$ be regenerated using the PNS solution $q^h(x_1, x_k)$ to revise the displacement thickness $\delta^*(x_i)$ distribution. The PNS x_1 integration sweep sequence is then reinitialized at $x_{n=1}$ using the complementary pressure solution modification.

4.13 Weak Interaction PNS Algorithm Validation

The 3D PNS GWS^h + θTS algorithm prediction of turbulent aerodynamic juncture region flow is reported, Baker and Orzechowski (1982). Figure 4.11 left illustrates the geometry, the right rectangular juncture of two 10% thick parabolic arc surfaces. The onset flow is at $Ma_\infty = 0.08$ ($U_\infty = 100$ f/s), zero angle of attack, hence based on chord C Re/C = 0.6 E+06. Figure 4.11 right graphs the farfield complementary pressure Dirichlet BC DOF $\{PC(x_1, x_2)\}$ data, non-D as pressure coefficient C_p on $0.01 \le x_1/C \le 0.46$, mirror symmetric on $0.54 \le x_1/C \le 0.99$, generated external to PNS by a 3-D potential flow solver.

The PNS transverse plane domain of span is $0 \le x_k/C \le 0.1$ is discretized into M = 361 $\{N_1\}$ basis triangles, Figure 4.12 (most triangle diagonals omitted). For domain boundary segments keyed to lettered corners, Dirichlet BCs are imposed for DOF {U1, U2, U3, K, E, PP} on closure segment A-B-C and for DOF {F, PC} on segment D-E-F. Homogeneous Neumann BCs are applied for DOF {U1, U2, U3, K, E} on segment C-D-E-F-A, and for

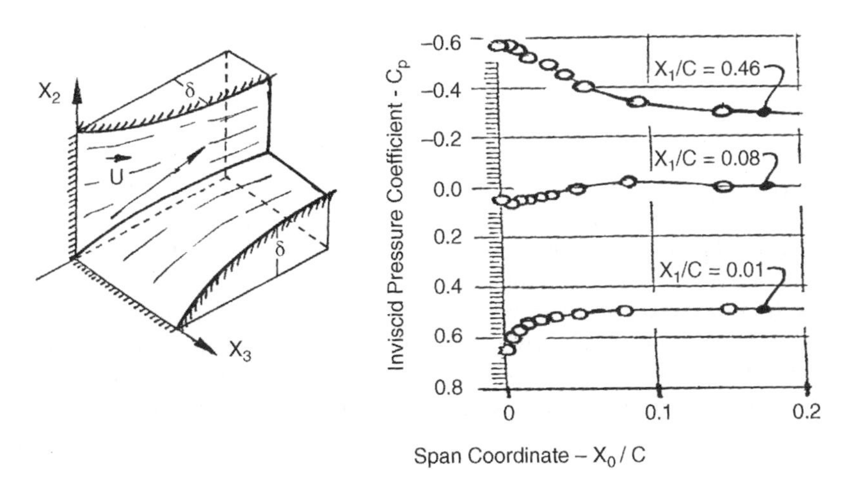

Figure 4.11 Parabolic arc aerodynamic juncture region, onset flow at $Ma_\infty = 0.08$, zero angle of attack: left, essential geometry: right, Dirichlet BC DOF $\{PC(x_1, x_2)\}$ distributions as $f(x_1/C)$

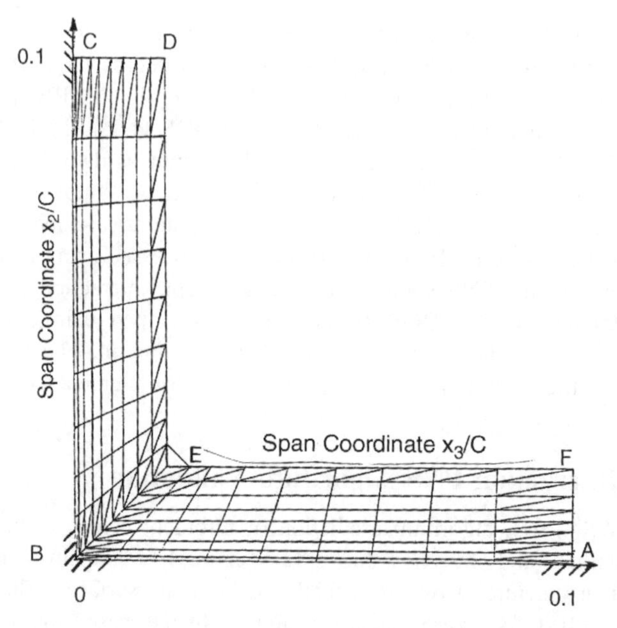

Figure 4.12 Aerodynamic juncture region turbulent PNS flowfield simulation, $\{N_1\}$ basis mesh, M = 361, BC surface corners labeled

DOF {F, PC} on segment F-A-B-C-D. The BC statement for DOF {PP} on segment C-D-E-F-A is (4.100).

The PNS IC specification for DOF {U1, K, E, RET} is via interrogation of the BL similarity solution, Figure 4.8 and (4.81), with each column of DOF data matched to free-stream velocity associated with farfield potential DOF $\{PC(x_1/C = 0.01)\}$, Figure 4.11. The IC for DOF {U2, U3} is zero with PNS algorithm solution delayed to the fourth x_1 station where the first *decent* FD estimate of $d\{U1\}/dx_1$ for (4.131) is possible, in accord with the defined BL IC procedure.

Using jacobian diagonal blocks (only), iterative procedure (4.134–4.136) required three complete PNS domain x_1-sweeps for the pressure DOF $\{P(x_1)\}=\{PC(x_1)+PP(x_1)\}$ to become stationary to *O*(E-03). The algorithm averaged four iterations per step to converge to $\max|\{\delta Q\}| < \varepsilon = 3.0$ E–04 at each x_1 station. At convergence, mass conservation satisfaction via the DM^h algorithm, (4.96) and (4.117), was $O(\varepsilon^2)$ in the potential energy norm $\|\phi^h(x_1)\|_E$. Thus, the DM^h algorithm *velocity error* is $\partial\phi^h/\partial x_k) < O(\varepsilon)$, an insignificant factor on PNS algorithm converged solution DOF $\{UI(x_1)\}$ accuracy.

The PNS third sweep solution symmetric half-plane transverse velocity vector field DOF $\{UL(x_1)\}$ distributions on $0.01 \le x_1/C \le 0.70$ are graphed in Figure 4.13. A substantial mass flux into the juncture corner is created by complementary pressure induced favorable gradient distributions on $x_1/C \le 0.50$. The influx is progressively diminished in approaching mid-chord. The plane velocity $\{UL(x_1)\}$ non-D extrema ranges $0.17 > \max|\{UL(x_1)\}|/U_\infty > 0.10$, noted in each x_1/C graphic.

A significant juncture span-parallel *wall jet* distribution is thus generated with x_1/C-dependent profiles, leading to generation of a persistent very modest magnitude

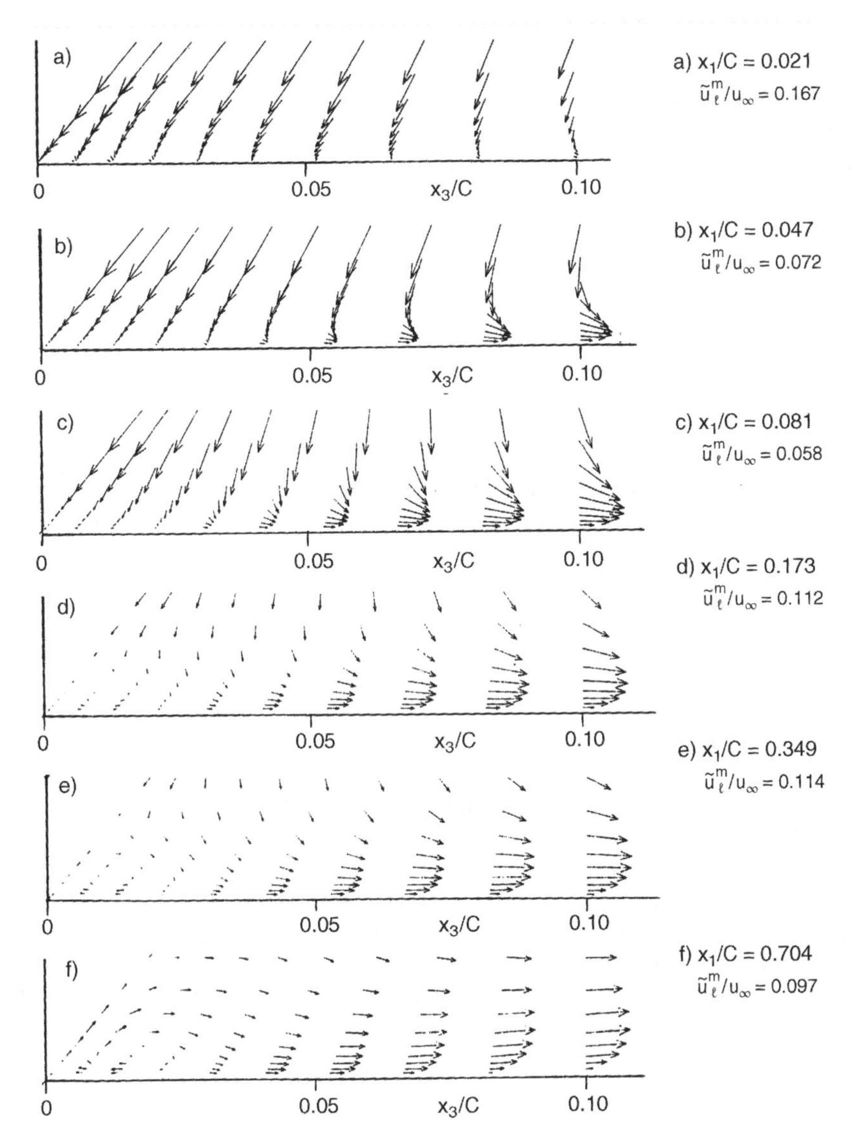

Figure 4.13 $GWS^h + \theta TS$ PNS algorithm prediction of aerodynamic juncture region transverse plane velocity vector $\{UL(x_1)\}$ DOF distributions, $Ma_\infty = 0.08$, Re/C = 0.6 E+06, k-ε closure model, full Reynolds stress tensor, symmetric half-plane data on $0.01 \le x_1/C \le 0.70$

$(\max|\{UL(x_1)\}|/U_\infty \approx 0.01)$ juncture corner axial vortex pair for all $x_1/C > 0.17$. At coordinate $x_1/C \approx 0.71$ the juncture corner first off wall DOF $\{U1(x_1)\}$ is computed negative, predicting separation, whence the PNS solution process terminates.

Notwithstanding *very* modest mesh M, this juncture region *a posteriori* data set solidly *verifies* $GWS^h + \theta TS$ 3D PNS algorithm robustness. No experimental data exist to support validation, hence PNS solution $\delta^*(x_i)$ alteration to the farfield potential pressure surface distribution is not reported.

PNS algorithm validation is reported, Baker *et al.* (1982), for *a posteriori* data from a closely coordinated CFD wind tunnel experiment for non-symmetric turbulent boundary layers merging into the wake of a NACA 63-012 airfoil, Figure 4.14. This $n = 2$ aerodynamic statement *requires* BL *and* PNS algorithm combined use in weak interaction with farfield potential flow pressure distributions computed on *displacement thickness* δ^* distribution from the combined algorithms solution.

The δ^* distribution is extracted from solution of the BL momentum thickness θ integral ODE, Schlichting (1979)

$$\mathcal{L}(\theta) = \frac{d\theta}{d\varsigma} + \frac{\theta}{u^I}(H + 2)\frac{du^I}{d\varsigma} - C_f/2 = 0 \tag{4.135}$$

The ξ coordinate is tangent to the airfoil, $H \equiv \delta^*/\theta$ is the BL shape factor, C_f is skin friction coefficient with u^I the farfield potential solution velocity component parallel to ξ. In the airfoil wake, PNS interaction with the farfield solution is via the potential flow Neumann BC (3.20), implemented using the PNS predicted efflux velocity from the PNS domain farfield boundaries.

This coordinated project rigorously identified it as imperative that the BL algorithm DOF distribution computed at the airfoil trailing edge be in agreement with that entering the experiment wake, as these BL data constitute the PNS DOF distribution IC. The test protocol was thus altered to measure u_1 and $\overline{u_i'u_j'}$ on the airfoil terminal chord C segment $0.9 \le x_1/C \le 0.998$. Due to hot wire physical size *only* the BL defect layer-far wake (see Figure 4.8) could be probed on this span.

To establish BL algorithm DOF IC, similarity theory was used to fill in airfoil surface to defect layer {U} DOF which completed the $\{U(x_1/C = 0.9)\}$ DOF IC. These data enabled establishing the $\{RET(x_1/C = 0.9)\}$ IC distribution via MLT closure model, (4.60). Thereafter, MLT–TKE closure duality coupled with matching the experiment defect layer data for $\{RIJ\}$ admitted completion of the requisite IC DOF distributions $\{K, E, RIJ(x_1/C = 0.9)\}$.

Via established BL *optimal solution adapted regular mesh* guidance, Section 4.6, a $pr = 1.06$, $M = 60$ non-uniform mesh extending well past $\delta(x_1 \ge 0.9)$ is defined. Following ~ten miniscule Δx_1 integration steps the algorithm self-initiated {V} completing the BL

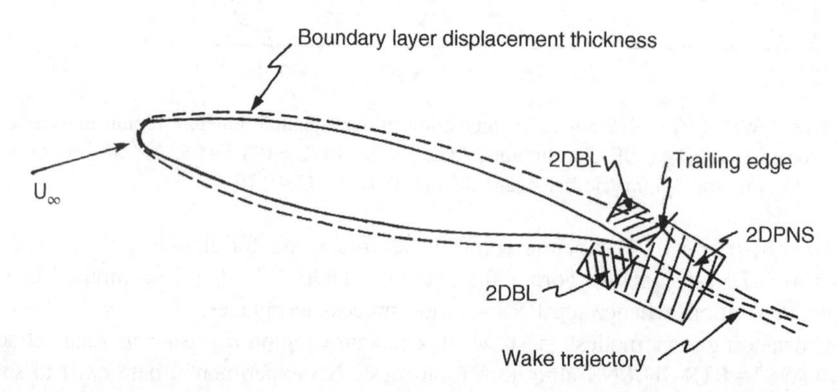

Figure 4.14 Aerodynamic weak interaction theory BL-PNS prediction of non-symmetric turbulent BL merging into a NACA 63-012 airfoil wake

state variable IC $\{QI(x_1/C=0.902)\}$. Each first left data graph in Figure 4.15 compares experiment (open symbols) with this BL algorithm IC *a posteriori* data presented as DOF interpolation. Excellent *quantitative* agreement results for BL velocity u_1 and axial normal stress $\overline{u_1'u_1'}$. The shear $\overline{u_1'u_2'}$ and transverse normal $\overline{u_2'u_2'}$ stress comparisons exhibit notable disagreement in the defect layer only.

The right three data graphs in Figure 4.15 compare experiment (open symbols) at $x_1/C=0.990$, 0.995, 0.998 with BL solution $\{QI(x_1/C)\}$ interpolations at $x_1/C=0.990$, 0.996, 1.000. Excellent *quantitative* agreement is accorded throughout, yielding solid validation that the BL $\{QI(x_1/C=1.0)\}$ DOF is the *accurate* wake PNS algorithm DOF IC. As the experiment is for zero angle of attack, these BL DOF are mirror reflected to the airfoil lower surface at $x_1/C=1.0$ to complete PNS IC.

The PNS domain transverse span, Figure 4.14, is sufficient to admit farfield homogeneous Neumann BCs for all PNS state variable DOF. Sixty progressively increasing Δx_1

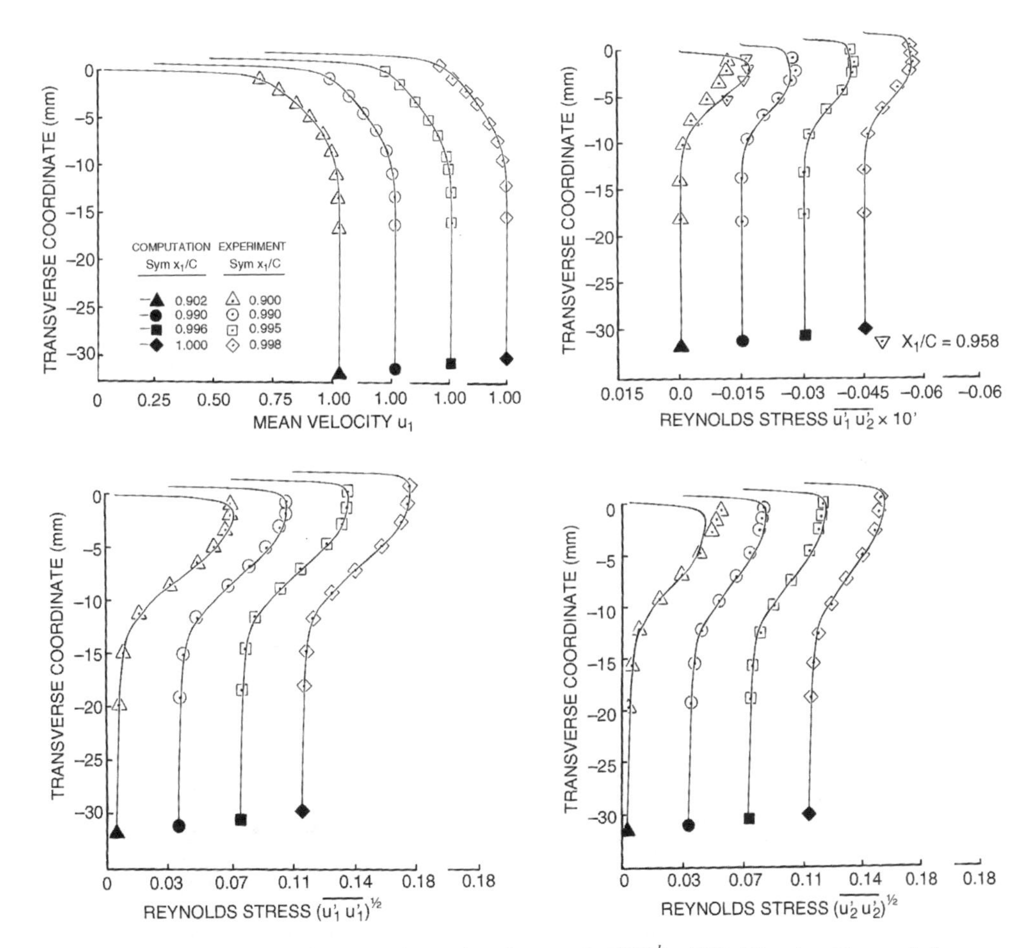

Figure 4.15 Comparison between hot wire data and $\text{GWS}^h+\theta\text{TS}$ BL algorithm solution on $0.9<x_1/C\leq 1.0$, k-ε closure model, full Reynolds stress, NACA 63-012 airfoil, zero angle of attack, $\text{Ma}_\infty=0.08$: top left, u_1; top right, $\overline{u_1'u_2'}$; bottom left, $\overline{u_1'u_1'}$; bottom right, $\overline{u_2'u_2'}$

steps generate the PNS DOF *a posteriori* data at $x_1/C = 1.0027$, the first wake data station. Thereafter 130 uniform Δx_1 steps marched the PNS solution to $x_1/C = 1.10$, the terminal data station. During PNS-external potential flow algorithm interaction post-processing the particular pressure via defined protocol ensures no violation of the parabolic assumption. The PNS second and third x_1 sweep $\{PP(x_1)\}$ DOF distributions are essentially identical, confirming weak interaction algorithm convergence in the wake.

PNS algorithm DOF data interpolations compared to data (open symbols), Figure 4.16, solidly *validate* prediction accuracy for velocity *and* the $n = 2$ parabolic-ordered Reynolds stress tensor constitutive closure (4.103–4.104) over the entire wake span $1.0 < x_1/C \leq 1.10$. Each left graph is PNS DOF $\{QI(x_1/C = 1.00002)\}$ interpolation, after the *first* Δx_1 PNS step, compared to experimental data at $x_1/C = 0.9979$. The requirement for PNS algorithm DOF IC accuracy is clearly accomplished.

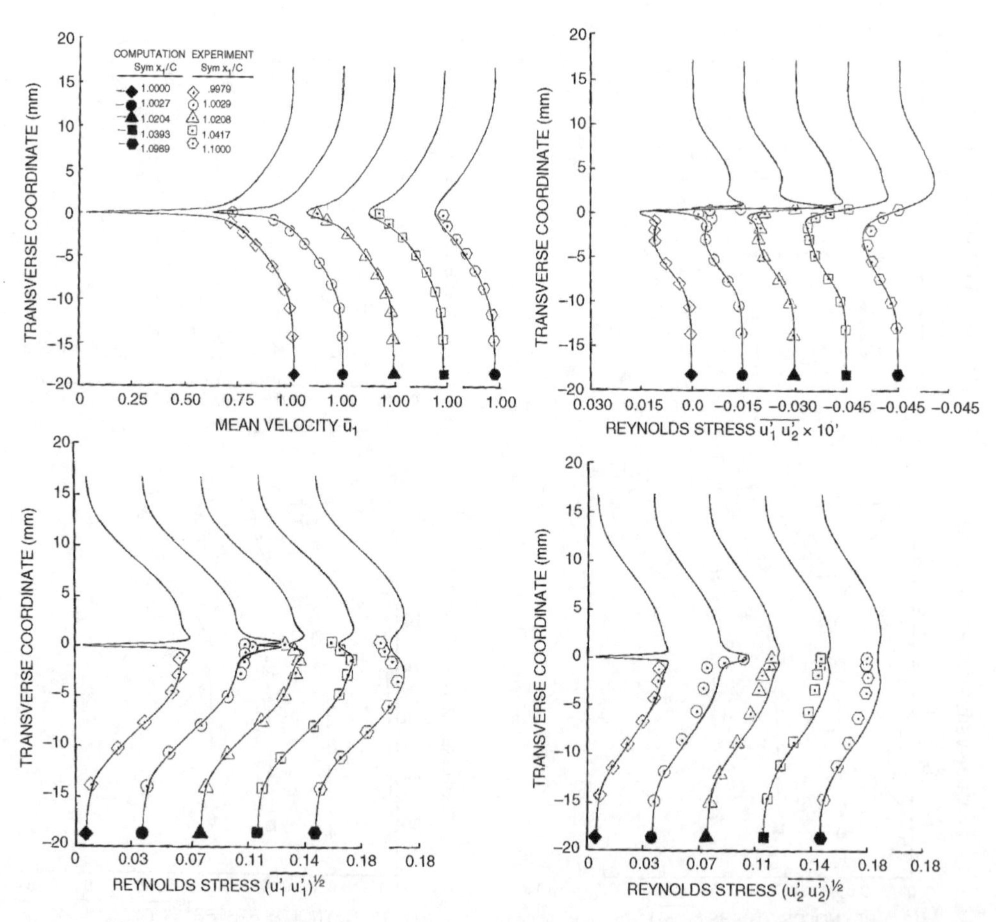

Figure 4.16 Comparison between hot wire data and $\text{GWS}^h + \theta\text{TS}$ PNS aerodynamic weak interaction algorithm solution, $x_1/C = 1.00002$, 1.0029, 1.02, 1.04, 1.10, *k*-ε closure model, full Reynolds stress tensor, NACA 63-012 airfoil, zero angle of attack, $\text{Ma}_\infty = 0.08$: top left, u_1; top right, $\overline{u_1'u_2'}$; bottom left, $\overline{u_1'u_1'}$; bottom right, $\overline{u_2'u_2'}$

The truly *fluid dynamic* feature captured in the first PNS Δx_1 step interpolation of DOF $\{QI(x_1/C = 1.00002)\}$ is ~50% recovery of the BL axial *velocity defect* following airfoil no-slip wall BC release. All PNS DOF data interpolations exhibit much sharper extrema than do the data, the result of hot wire physical size. Excellent *quantitative* agreement for all Reynolds stress distributions exists between experiment and PNS. The graphed PNS normal stresses sum the $O(\delta)$ and $O(\delta^2)$ theory contributions, (4.103–4.104). Excellent quantitative agreement exists between PNS solution and experiment wake centerline depressions in the normal stress $\overline{u'_1 u'_1}$ profiles, clearly distinct from the $\overline{u'_2 u'_2}$ profile bulges, thoroughly *validating* the Reynolds stress constitutive closure.

In summary, overall *excellent* quantitative agreement between experiment and weak form theory PNS algorithm prediction is engendered in this closely coordinated experiment. The end result is solid validation of the weak interaction PNS theory, global EBV pressure accounting, and confirmation of ordered Reynolds stress tensor closure. The defined BL algorithm IC generation protocol handles experiment data acquisition limitations. The solution adapted mesh space marching PNS algorithm DOF interpolations generate sharp *monotone* extrema as well as requisite mirror symmetries. Non-zero angle of attack *a posteriori* CFD data exist, Baker *et al.* (1982).

4.14 Square Duct PNS Algorithm Validation

Validation of $\mathrm{GWS}^h + \theta\mathrm{TS}$ PNS algorithm with ordered Reynolds stress tensor constitutive closure (4.103–4.104) is documented for bounded flow, Baker and Orzechowski (1981). The configuration is $n = 3$ steady turbulent flow in a square cross-section duct for which experimental data verify creation of transverse plane axial vortex pairs in *each* duct corner, Melling and Whitelaw (1976). Figure 4.17 illustrates the quarter plane geometry used for the simulation, as again necessitated by limited compute resource available at the time.

For wall-bounded applications weak interaction theory complementary pressure coupling is replaced with an algorithm to iteratively promote axial mass flux conservation during x_1 direction integration. The cross-section pressure correction estimate, adapted from Patankar and Spalding (1972), is

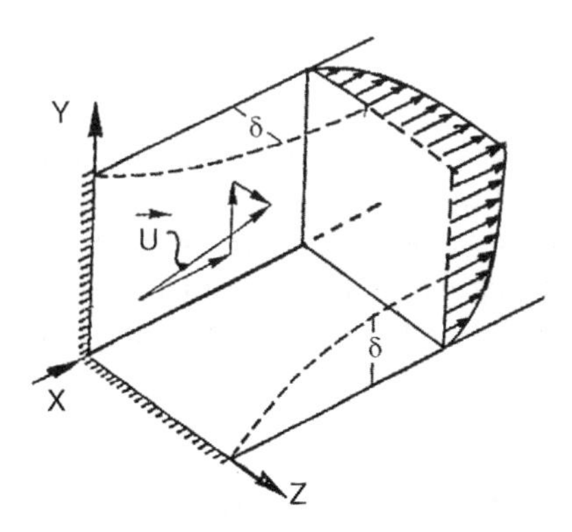

Figure 4.17 Steady turbulent flow in a square cross-section duct, symmetric quarter plane PNS algorithm solution domain

$$\mathrm{PC}(x_{n+1}) = \mathrm{PC}(x_n) + \mathrm{C}\,\Delta\left[\sum_{e=1}^{\mathrm{M}}\int_{\Omega_e}\{N_1\}^T\{\mathrm{U1}(x_1)\}_e \mathrm{d}x_2\mathrm{d}x_3\right]_{n+1,n}\Bigg/\Delta x_1 \tag{4.136}$$

where C scales the correction magnitude of computed axial mass flux difference $\Delta[\bullet]/\Delta x_1$, the integral in (4.136). Since (4.98) is homogeneous with this the sole BC, the complementary pressure DOF distribution $\{\mathrm{PC}(x_{n+1})\}$ is uniform.

Both laminar and turbulent flow PNS *a posteriori* data are reported for modestly wall-graded $\mathrm{M}=1052$ $\{N_1\}$ basis triangle mesh. Homogeneous Neumann BCs are applied to state variable DOF on both duct symmetry planes, the wall BC is no slip and Re = 4.2 E+04 matches the experiment. For the laminar specification, also the turbulent closure specification using the *standard* TKE k-ε eddy viscosity model (4.62), the PNS steady solutions were devoid of a transverse plane vortex pair. Conversely, using the ordered Reynolds stress tensor constitutive closure (4.103–4.104), the turbulent PNS simulation *did* generate the requisite vortex pair.

The top four graphics in Figure 4.18 compare the experiment, left, with the $n=3$ PNS steady solution, right, for transverse plane velocity vector $\{UL\}$ DOF distribution and axial velocity isovel contours. In comparison, false local additional vortex pairs in the PNS $\{UL\}$ distribution are observed. This results from rendering the assembled particular pressure jacobian *non-singular* via imposing Dirichlet BCs at the wall node on each symmetry plane. Recall the precise BC is uniformly homogeneous Neumann, (4.100). Krylov solvers, which can handle this singularity, were not yet invented(!) hence the authors' elected BC alteration.

Thus are generated the symmetric false local vortex pairs in the PNS $\{UL\}$ distribution. The generated vortex pair is responsible for core axial velocity penetration into the corner, leading to the comparative disparity in $\{\mathrm{U1}\}$ isovel levels. The false vortices also contribute to PNS non-D $\max|\{UL\}|=0.0043$ being half that of the experiment.

Notwithstanding, these PNS algorithm experiments *verify* the *sole cause* for vortex pair generation is *anisotropy* of the Reynolds normal stress tensor, the $O(\delta)$ $\overline{u_k' u_k'}$, $k=2$, 3 terms with $\mathrm{C_2C_4}$ multiplier in (4.103). The Figure 4.18 bottom graphs compare experiment and PNS $\overline{u_2' u_2'}$. Coarse mesh essential quantitative agreement is evident.

4.15 Summary

Prediction of aerodynamic flowfields via weak interaction theory couples farfield potential flow pressure distributions with boundary layer (BL) and parabolic Navier–Stokes (PNS) approximations to steady Navier–Stokes (NS) conservation principles PDE + BCs systems. These topics were *the* pioneering focus of fledgling CFD developments in the early 1970s, leading to the first consequentially detailed characterization of aerodynamic flowfields, both potential and viscous/turbulent.

The chapter focus is on five *key* objectives:

- establishing a thorough foundation for *nonlinear* weak form algorithm BL and PNS derivations validating permeation of *analytical vector calculus* from theorization to terminal matrix statements *and* jacobians

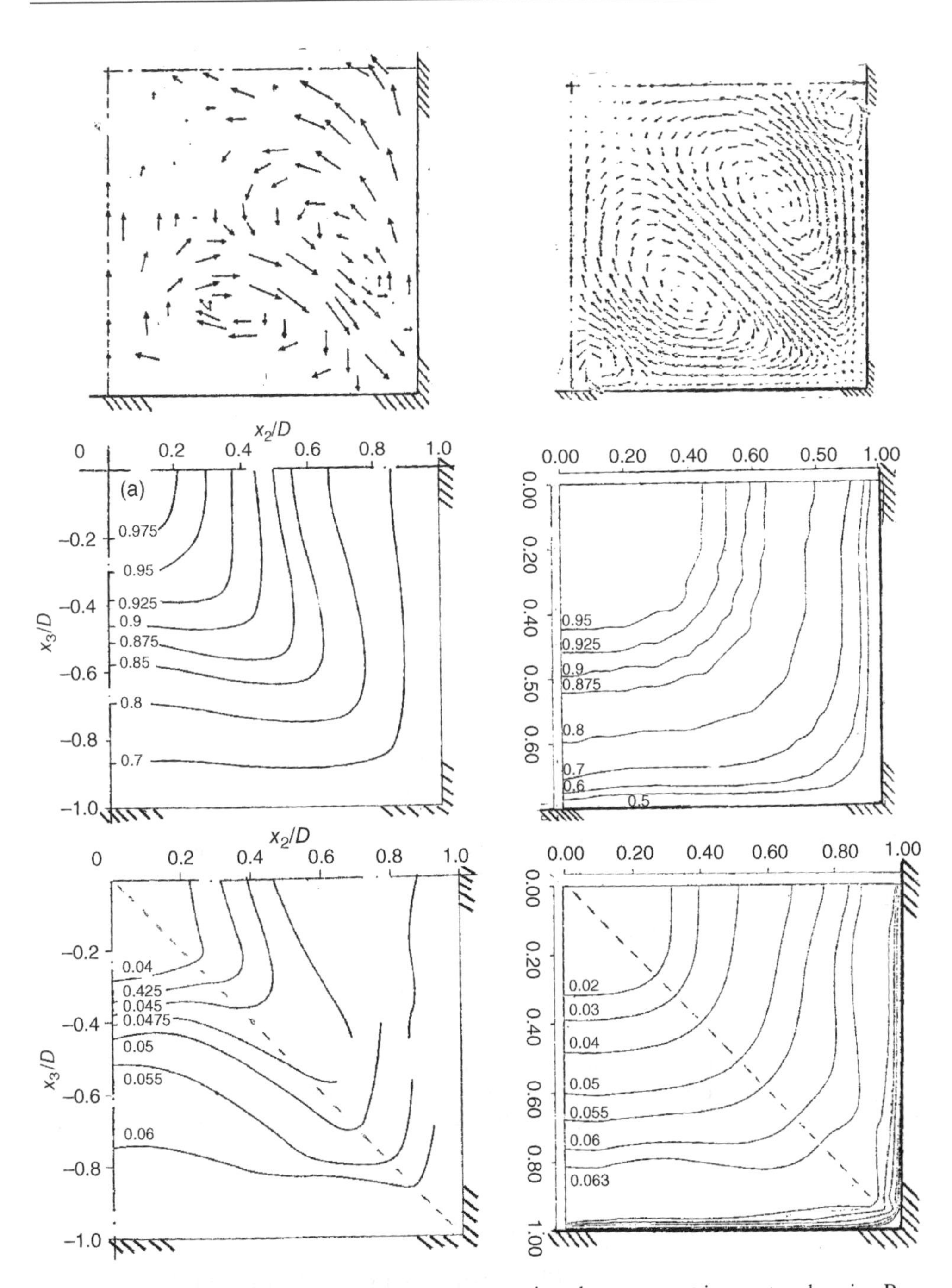

Figure 4.18 Steady turbulent flow, square cross-section duct, symmetric quarter domain, Re = 4.2 E+04: left column, experiment; right column, $n=3$ full Reynolds stress tensor PNS *a posteriori* data. *Experimental data source:* Melling, A. and Whitelaw, J.H. 1976. Reproduced with permission of Cambridge University Press

- quantitative *verification* that available *linear* weak form theory for *error estimation* and *asymptotic convergence* accurately describes *a posteriori* data generated via *uniform* and *regular solution adapted* mesh refinement for *nonlinear* PDE + BCs systems
- validation of the *linear* weak form theory assertion that the Galerkin criterion for nonlinear PDE systems generates the *optimal* solution among its peers
- establishing confidence in use of the *energy norm* and its distribution for *rigorous* prediction of $\text{GWS}^N \Rightarrow \text{GWS}^h + \theta\text{TS}$ algorithm *optimal mesh* solution existence
- highlighting fundamental identifications and performance aspects of *Reynolds stress closure models* developed for BL and PNS PDE + BCs systems.

These objectives have been solidly accomplished. The analysis fundamental to the aerodynamics problem class is *Reynolds ordering*, first developed with precision for both laminar and turbulent BL PDE + BCs statements. The resultant laminar BL $n = 3$ statement is (4.11–4.14)

$$\mathrm{D}M{:}\, \frac{\partial v}{\partial y} + \left(\frac{\partial u}{\partial x} + \frac{\partial w}{\partial z}\right) = 0$$

$$\mathbf{DP} \bullet \widehat{\mathbf{i}}\,{:}\mathbf{u}\cdot\nabla u - \frac{1}{\mathrm{Re}}\frac{\partial^2 u}{\partial y^2} + \frac{\partial P^I}{\partial x} + \frac{\mathrm{Gr}}{\mathrm{Re}^2}\Theta(\widehat{\mathbf{g}} \bullet \widehat{\mathbf{i}}) = 0$$

$$\mathbf{DP} \bullet \widehat{\mathbf{k}}\,{:}\mathbf{u}\cdot\nabla w - \frac{1}{\mathrm{Re}}\frac{\partial^2 w}{\partial y^2} + \frac{\partial P^I}{\partial z} = 0$$

$$\mathrm{D}E{:}\mathbf{u}\cdot\nabla\Theta - \frac{1}{\mathrm{Re\,Pr}}\frac{\partial^2 \Theta}{\partial y^2} - \frac{\mathrm{Ec}}{\mathrm{Re}}\left(\frac{\partial u}{\partial y}\right)^2 = 0$$

The state variable for the BL PDE system ordered by solution strategy is $\{q(x_i)\} = \{u, w, \Theta, v\}^T$. The GWS^h algorithm statement (4.27) for (4.12–4.14) is

$$\begin{aligned}\text{GWS}^h\left(\mathcal{L}(q^h)\right) &= \{0\}\\ &= \left[\text{M3000}(\{\text{U}\})\right]\frac{\mathrm{d}\{Q\}}{\mathrm{d}x} + \left\{\text{RES\_}\{Q(P^I, \text{Re}, \text{Gr}, \text{Pr}, \text{Ec})\}\right\}\end{aligned}$$

Solving (4.27) via the *nonlinear* matrix [M3000(U)] symbolic inverse identifies the x-direction derivative data required for a Taylor series (TS) ODE algorithm with implicitness parameter θ. The end result is the $\text{GWS}^h + (\theta\ 0.5)$ TS algorithm *nonlinear* algebraic matrix statement (4.31)

$$\begin{aligned}\{\text{F\_}\{Q\}\} \equiv\ &[\text{M3000}(\{\text{U}\})]_{n+1/2}\{Q_{n+1} - Q_n\}\\ &+ \frac{\Delta x}{2}\left(\{\text{RES\_}\{Q(\bullet)\}|_{n+1} + \{\text{RES\_}\{Q(\bullet)\}|_n\right) = \{0\}\end{aligned}$$

As written (4.11) is an ODE in the wall-normal direction directly amenable to a $\theta = 0.5$ TS. The nonlinear matrix iteration statement coupling the weak form restatement of the

D*M* TS with (4.31) is (4.34)

$$\begin{aligned}
\left[\text{JAC_}\{Q\}\right]\{\delta Q\}^{p+1} &= -\{\text{F_}\{Q\}\}_{n+1}^{p} \\
\{Q\}_{n+1}^{p+1} &= \{Q\}_{n+1}^{p} + \{\delta Q\}^{p+1} = \{Q\}_n + \sum_{\alpha=0}^{p}\{\delta Q\}^{\alpha+1} \\
\left[\text{JAC_}\{Q\}\right] &\equiv \frac{\partial\{\text{F_}\{Q\}\}}{\partial\{Q\}}
\end{aligned}$$

The jacobian [JAC_{Q}] is the derivative of {F_{Q}} with respect to the algorithm DOF $\{Q\}=\{\text{U, T, W, V}\}^T$ with p the iteration index.

For a laminar BL the *sole* nonlinearity is the axial convection term in (4.12), hence algorithm statement $[\text{M3000(U)}]_{n+1/2}\{Q_{n+1}-Q_n\}$ in (4.31). The key sought resolution is confirmed pertinence of *linear* weak form theory *asymptotic error* estimate (4.43)

$$\begin{gathered}
\left\|e^h(x)\right\|_E \leq Ch^{2\gamma}\left[\|\text{data}\|_{L2,\Omega} + \|\text{data}\|_{L2,\partial\Omega}\right] + C_2\Delta x^{f(\theta)}\left\|q^h(x_0)\right\|_{H^m(\Omega)}^2 \\
for\!: \gamma \equiv \min(k+1-m, r-m),\ f(\theta) = (2,3)
\end{gathered}$$

to solutions (4.34). In (4.43), theory extension to initial-value character introduces another constant C_2, $f(\theta)=2$ for $\theta=0,\ 1$, $f(\theta)=3$ for $\theta=0.5$, with $q^h(x_0)$ interpolation of the state variable distribution IC onto the DOF in the mesh. Recall $m=1$ for a second-order EBV PDE.

The $n=2$ energy norm definition for axial velocity u is (4.46)

$$\|u(x)\|_E \equiv \frac{1}{2\text{Re}}\int_{\Omega}\left[\frac{\partial u(x,y)}{\partial y}\frac{\partial u(x,y)}{\partial y}\right]\text{d}y$$

with discrete evaluation detailed in (4.47). The *data* norms bounding error *size* in (4.43) for isothermal (4.12) are defined in (4.48)

$$\left\|e^h(x)\right\|_E \leq Ch^{2k}\left\|\text{d}P^I/\text{d}x\right\|_{L2,\Omega} + C_2\Delta x^3\left\|q^h(x_0)\right\|_{H^1(\Omega)}^2$$

For DOF IC $\{\text{U}(x_0),\text{V}(x_0)\}$ ensuring $\left\|q^h(x_0)\right\|_{H^1(\Omega)}^2$ in (4.43) is M-independent, *uniform* mesh refinement *a posteriori* data for [M3000({U})] discrete variations in (4.31) quantitatively *validate* that the $\text{GWS}^h+\theta\text{TS}$ $k=1$ basis implementation is *optimally accurate* among the evaluated FD peers. The negligible impact of *data* $\left\|\text{d}P^I/\text{d}x\right\|_{L2,\Omega}$ in bounding error further confirms $\text{GWS}^h+\theta\text{TS}$ optimality in comparisons. Detailed *a posteriori* data generated by the *regular solution adapted* mesh refinement protocol quantitatively validate the linear weak form theory assertion of *optimal mesh solution* existence for fixed DOF.

Of truly fundamental significance to theoretical fluid mechanics, BL and PNS PDE + BCs systems support essentially *all* theory musings on Reynolds stress tensor *closure modeling* for CFD turbulent flow prediction. The *time averaging* statement for the NS system **DP** state variable is (4.52)

$$\overline{u}_i(x_k,t) \equiv \lim_{T\to\infty}\frac{1}{T}\int_t^{t+T} u_i(x_k,\tau)\mathrm{d}\tau \Rightarrow \overline{u}_i(x_k)$$

hence by definition (4.53)

$$\overline{u_i'}(x_k,t) \equiv \lim_{T\to\infty}\frac{1}{T}\int_t^{t+T} u_i'(x_k,\tau)\mathrm{d}\tau = 0$$

Reynolds ordering of time averaged BL $\mathrm{D}\mathbf{P} \bullet \widehat{\mathbf{i}}$ determines that the sole significant Reynolds stress tensor component is $\overline{v'u'}$, the axial shear stress. Closure models for this tensor are universally of the form (4.59)

$$\overline{v'u'} \equiv \tau_{xy} \equiv -\nu^t\frac{\partial\overline{u}}{\partial y}$$

with *nonlinear* functional $\nu^t(\{q(x_i)\})$ termed a *turbulent eddy viscosity* of *dimensionality* $D(\mathrm{L}^2\tau^{-1})$. Extension to non-isothermal aerodynamics uses (4.59) with a "turbulent" Prandtl number Pr^t model. Closure theoretical essence for eddy viscosity is detailed in Section 4.7 for a range of algebraic and PDE system models applicable to two- and three-dimensional aerodynamics.

Non-dimensionalizing (4.59) leads to definition of the *turbulent Reynolds number* (4.64)

$$\mathrm{Re}^t \equiv \nu^t/\nu$$

The non-D time averaged NS PDE system for turbulent incompressible-thermal BL flow is (4.65–4.69)

$$\mathrm{D}\overline{M}\colon \frac{\partial\overline{v}}{\partial y} + \left(\frac{\partial\overline{u}}{\partial x} + \frac{\partial\overline{w}}{\partial z}\right) = 0$$

$$\mathrm{D}\overline{\mathbf{P}} \bullet \widehat{\mathbf{i}} : \overline{\mathbf{u}}\cdot\nabla\overline{u} - \frac{1}{\mathrm{Re}}\frac{\partial}{\partial y}\left[(1+\mathrm{Re}^t)\frac{\partial\overline{u}}{\partial y}\right] + \frac{\partial\overline{P}}{\partial x} + \frac{\mathrm{Gr}}{\mathrm{Re}^2}\overline{\Theta}(\widehat{\mathbf{g}} \bullet \widehat{\mathbf{i}}) = 0$$

$$\mathrm{D}\overline{\mathbf{P}} \bullet \widehat{\mathbf{j}} : \overline{P}(y\colon x,z) = P^I(x,z) - \overline{v'v'}(y)$$

$$\mathrm{D}\overline{\mathbf{P}} \bullet \widehat{\mathbf{k}} : \overline{\mathbf{u}}\cdot\nabla\overline{w} - \frac{1}{\mathrm{Re}}\frac{\partial}{\partial y}\left[(1+\mathrm{Re}^t)\frac{\partial\overline{w}}{\partial y}\right] + \frac{\partial\overline{P}}{\partial z} = 0$$

$$\mathrm{D}\overline{E}\colon\overline{\mathbf{u}}\cdot\nabla\overline{\Theta} - \frac{1}{\mathrm{Re}}\frac{\partial}{\partial y}\left[\left(\frac{1}{\mathrm{Pr}} + \frac{\mathrm{Re}^t}{\mathrm{Pr}^t}\right)\frac{\partial\overline{\Theta}}{\partial y}\right] - \frac{\mathrm{Ec}}{\mathrm{Re}}\left(\frac{\partial\overline{u}}{\partial y}\right)^2 = 0$$

The key distinction between the laminar and turbulent BL PDE systems is the Reynolds stress closure model added *nonlinearity*. The $n=2$ $\mathrm{GWS}^h + \theta\mathrm{TS}$ implementation is via alteration of linear diffusion matrix statements to the *hypermatrix* statements (4.70)

$$\frac{1}{\det_e \text{Re}}[\text{A211}]\{\text{U}\}_e \Rightarrow \frac{1}{\det_e \text{Re}}\{1+\text{Re}^t\}_e^T[\text{A3011}]\{\text{U}\}_e$$

$$\frac{1}{\det_e \text{RePr}}[\text{A211}]\{\text{T}\}_e \Rightarrow \frac{1}{\det_e \text{Re}}\left\{\frac{1}{\text{Pr}}+\frac{\text{Re}^t}{\text{Pr}^t}\right\}_e^T[\text{A3011}]\{\text{T}\}_e$$

The linear theory asymptotic error estimate for this significantly *nonlinear* PDE system remains (4.43) for energy norm (4.47) definition altered to (4.71)

$$\|u(x)\|_E \equiv \frac{1}{2\text{Re}}\int_\Omega \left[(1+\text{Re}^t(x,y))\frac{\partial u(x,y)}{\partial y}\frac{\partial u(x,y)}{\partial y}\right]\text{d}y$$

with discrete implementation defined in (4.72). *A posteriori* data for precisely designed *regular solution adapted* mesh refinement *validate* adherence to the linear weak form theory predicted second- and fourth-order convergence rates for $\text{GWS}^h + \theta\text{TS}$ algorithm $k = 1$, 2 basis implementations. Additional data verify the $k = 1$ basis algorithm yields solution accuracy superior to the equivalent order Crank–Nicolson FD DOF construction measured in the energy norm (4.71).

The chapter on turbulent the BL subject is closed with dual I-EBV PDE systems with principal requirement to identify encompassing BCs. A low Re^t closure model is required when resolving the transitional laminar–turbulent reach of the BL. The BC alternative is via the isothermal BL *similarity solution*, assuming that near wall production and dissipation of turbulent kinetic energy is in balance. This yields dual-PDE closure model state variable DOF specifications (4.81) at the first off-wall node

$$\nu^t = \kappa y u_\tau / C_\nu, \quad k = u_\tau^2 / C_\nu^{1/2}, \quad \tau_w = u_\tau^2 / \text{C}_\nu$$

$$\varepsilon = \left|u_\tau^3\right| / \kappa y, \quad \omega = u_\tau / \kappa y (\text{C}_\nu)^{1/2}$$

The restriction is that the computed nodal y^+ lie in the similarity log layer region, typically requiring near-wall solution adapted mesh alteration.

The $n = 3$ time averaged *parabolic* Navier–Stokes PDE system is appropriate for internal/external aerodynamic *unidirectional* flowfield prediction. For x_1 spanning the dominant flow direction the PNS simplification requirements are:

- velocity vector component u_1 suffers no reversal
- diffusive and turbulent transport processes in the x_1 direction are negligible in comparison to convection, and
- the overall EBV character of subsonic NS must be enforced by identification of a suitable 3-D pressure field.

In tensor notation, replacing Re^t with Reynolds stress tensor $\overline{u_i' u_j'}$ and for index summation conventions $1 \le (i, j) \le 3$ and $2 \le (k, l) \le 3$, the $n = 3$ time averaged PNS PDE system with k-ε dual PDE closure is (4.96–4.99)

$$\mathcal{L}(\phi) = -\frac{\partial^2\phi}{\partial x_k^2} - \frac{\partial u_i}{\partial x_i} = 0$$

$$\mathcal{L}(u_i) = u_j\frac{\partial u_i}{\partial x_j} + \delta_{ij}\frac{\partial P}{\partial x_j} + \frac{\partial}{\partial x_k}\left[\overline{u_i'u_k'} - \frac{1}{\text{Re}}\frac{\partial u_i}{\partial x_k}\right] = 0$$

$$\mathcal{L}(k) = u_j\frac{\partial k}{\partial x_j} + \frac{\partial}{\partial x_k}\left[\left(\text{C}_k^{-1}\frac{k}{\varepsilon}\overline{u_l'u_k'} - \frac{\delta_{lk}}{\text{Pe}}\right)\frac{\partial k}{\partial x_l}\right] + \overline{u_1'u_k'}\frac{\partial u_1}{\partial x_k} + \varepsilon = 0$$

$$\mathcal{L}(\varepsilon) = u_j\frac{\partial \varepsilon}{\partial x_j} + \frac{\partial}{\partial x_k}\left[\text{C}_\varepsilon^{-1}\frac{k}{\varepsilon}\overline{u_l'u_k'}\frac{\partial \varepsilon}{\partial x_l}\right] + \text{C}_\varepsilon^1\overline{u_1'u_k'}\frac{\varepsilon}{k}\frac{\partial u_1}{\partial x_k} + \text{C}_\varepsilon^2\frac{\varepsilon^2}{k} = 0$$

The constraint of D*M*, (4.65), is pressure projection theory *kinematically* enforced via the transverse plane velocity potential function $\phi(x_k)$. During PNS algorithm convergence the associated *approximation error* $\text{D}M^h \neq 0$ is driven to $|\phi^h|_E \leq \varepsilon^2$, for ε the iteration convergence requirement. Via a *complementary-particular* solution strategy with pertinent axial pressure gradient BC identifications, genuine pressure determination is via (4.100)

$$\mathcal{L}(P) = -\frac{\partial^2 P}{\partial x_k^2} - \frac{\partial^2}{\partial x_l \partial x_k}\left[u_l u_k + \overline{u_l'u_k'}\right] = 0$$

$$l(P) = \frac{\partial P}{\partial x_k}\widehat{n}_k + \frac{1}{\text{Re}}\frac{\partial^2 u_k \widehat{n}_k}{\partial x_s^2} = 0$$

A Reynolds stress tensor algebraic closure is derived using Reynolds ordered tensor field theory. For coefficient set $\text{C}_\alpha = \{0.94, 0.067, 0.56, 0.068\}$, the PNS-significant $O(\delta)$ closure terms are (4.103)

$$\overline{u_1'u_1'} = \text{C}_1 k - \text{C}_2\text{C}_4\frac{k^3}{\varepsilon^2}\left[\left(\frac{\partial u_1}{\partial x_2}\right)^2 + \left(\frac{\partial u_1}{\partial x_3}\right)^2\right]$$

$$\overline{u_2'u_2'} = \text{C}_3 k - \text{C}_2\text{C}_4\frac{k^3}{\varepsilon^2}\left(\frac{\partial u_1}{\partial x_2}\right)^2, \quad \overline{u_3'u_3'} = \text{C}_3 k - \text{C}_2\text{C}_4\frac{k^3}{\varepsilon^2}\left(\frac{\partial u_1}{\partial x_3}\right)^2$$

$$\overline{u_1'u_2'} = -\text{C}_4\frac{k^2}{\varepsilon}\frac{\partial u_1}{\partial x_2}, \quad \overline{u_1'u_3'} = -\text{C}_4\frac{k^2}{\varepsilon}\frac{\partial u_1}{\partial x_3}, \quad \overline{u_2'u_3'} = -\text{C}_2\text{C}_4\frac{k^3}{\varepsilon^2}\left(\frac{\partial u_1}{\partial x_2}\right)\left(\frac{\partial u_1}{\partial x_3}\right)$$

Preservation of the I-EBV character for state variable members $u_k(x_i)$ in (4.97) requires $O(\delta^2)$ Reynolds stress tensor closure term (4.104) inclusions

$$\overline{u_1'u_1'} = -\text{C}_4\frac{k^2}{\varepsilon}\frac{\partial u_1}{\partial x_1}, \quad \overline{u_2'u_2'} = -\text{C}_4\frac{k^2}{\varepsilon}\frac{\partial u_2}{\partial x_2}$$

$$\overline{u_3'u_3'} = -\text{C}_4\frac{k^2}{\varepsilon}\frac{\partial u_3}{\partial x_3}, \quad \overline{u_2'u_3'} = -\text{C}_4\frac{k^2}{\varepsilon}\left(\frac{\partial u_2}{\partial x_3} + \frac{\partial u_3}{\partial x_2}\right)$$

The time averaged NS alteration to PNS generates the state variable $\{q(x_1, x_k\} = \{u_1, u_2, u_3, k, \varepsilon; \overline{u_i'u_j'}; \phi, P\}^T$. Direct substitution of (4.103) as appropriate confirms that PDE system (4.97–9.99) is I-EBV. The $\text{GWS}^h + \theta\text{TS}$ PNS algorithm DOF is $\{Q(x_1)\} = \{\text{U1, U2, U3, K, E; R}IJ\text{; F, P}\}^T$. Ordered as $\text{Q}I(x_1)\}$, $1 \leq I \leq 5$, the $\text{GWS}^h + \theta\text{TS}$ PNS algorithm for members $\{q_I\ (x_1, x_k\} = \{u_1, u_2, u_3, k, \varepsilon\}^T$ is the modest alteration (4.113) on the BL algorithm statement

$$\{\text{F\_Q}I\} \equiv [\text{M3000}(\{\text{U1}\})]_{n+1/2}\{\text{Q}I_{n+1} - \text{Q}I_n\}$$

$$+\frac{\Delta x_1}{2}\left[\{\text{RES\_Q}I(\{Q\})\}|_{n+1} + \{\text{RES\_Q}I(\{Q\})\}|_n\right] = \{0\}$$

The continuum GWS^N algorithm for state variable members $\{q_R\ (x_1{,}x_k\} = \{\overline{u_i'u_j'}\}$ is formed on *differential definition* (4.114)

$$\mathcal{D}(\overline{u_i'u_j'}) \equiv \overline{u_i'u_j'} - k\alpha_{ij} + \text{C}_4\frac{k^2}{\varepsilon}S_{ij} + \text{C}_2\text{C}_4\frac{k^3}{\varepsilon^2}S_{ik}S_{kj} = 0$$

Implementing GWS$^N \Rightarrow$ GWS^h the nonlinear algebraic matrix statement for DOF array $\{\text{R}IJ(x_1)\}$ is defined in (4.115)

$$\{\text{F\_R}IJ\} \equiv [\text{M200}]\{\text{R}IJ\} + \{\text{RES\_R}IJ(\{Q\}, S_{ij})\} = \{0\}$$

for $1 \leq (I, J) \leq 3$ and recalling the strain-rate tensor S_{ij} definition (4.101). Admissible DOF constraints in solving (4.115) are solely homogeneous Dirichlet.

The GWS$^N \Rightarrow$ GWS^h process for DM^h state variable members $\{q_P(x_1, x_k\} = \{\phi, P\}^T$ is routine, leading to the *linear* matrix statements (4.116)

$$\{\text{F\_Q}P\} \equiv [\text{DIFF}]\{\text{Q}P\} + \{\text{b}(\{Q\}, data)\} = \{0\}$$

for PNS algorithm DOF $\{\text{Q}P\} = \{\text{F}(x_1), \text{P}(x_1)\}^T$.

The $k = 1$ basis implementation of the GWS$^h + \theta$TS PNS algorithm including quasi-Newton jacobian is completed in Section 4.13. For a range of $n = 2$, 3 aerodynamics statements, the PNS pressure resolution theory, also the ordered full Reynolds stress algebraic closure, are *quantitatively validated* via detailed comparison of generated *a posteriori* data. with quality experimental data. In addition to weak form linear theory applicability *validation* for BL-PNS nonlinear PDE systems, particular significance accrues to Reynolds ordered algebraic stress tensor closure theory validation.

The *aerodynamics* class spans subsonic through hypersonic universally characterized by a *unidirectional* flowfield. Chapter content is restricted to *parabolic* (PNS) approximations to subsonic incompressible NS-RaNS PDE systems for internal and external aerodynamics geometries. Appendix B details parabolic *m*GWS$^h + \theta$TS modified algorithm theorization/validation for the compressible $n = 3$ RaNS (PRaNS) PDE system applicable to external *aerothermodynamic* shock layer flowfields at hypersonic free-stream Mach number.

Exercises

4.2.1 Perform Reynolds ordering of the lateral momentum and energy PDEs, hence confirm the correctness of (4.13–4.14).

4.3.1 For the Euler family of integration algorithms, (4.29), confirm the stated TS orders of accuracy for $\theta = 0$, 1, 0.5.

4.3.2 Confirm that (4.33), for the stated coefficient set, is indeed a second-order accurate TS for non-uniform spacing in the initial-value x coordinate direction.

4.4.1 Starting with (4.15–4.16), verify the accuracy of the element-level expressions (4.35–4.37) for the $\text{GWS}^h + \theta\text{TS}$ algorithm algebraic statement (4.31).

4.4.2 Confirm that (4.38) is indeed the DOF matrix form of the D*M* trapezoidal rule TS, (4.32).

4.4.3 Using the NC basis integration formula (3.81), confirm select NC $k = 1$, 2 basis matrices listed in Table 4.1 and Table 4.2.

4.4.4 Verify that the element level jacobian contributions (4.42) are correct, especially the $\{U\}_e^T$ TS levels in $[J\_UU]_e$ and $[J\_TT]_e$.

4.5.1 Verify assembly at node Y_j of the $\text{GWS}^h + \theta\text{TS}$ hypermatrix term alteration (4.50) reproduces the CNFD stencil.

4.9.1 Assuming that production and dissipation terms in isothermal (4.73) and (4.77) balance directly adjacent to a wall, confirm determinations (4.81) via the similarity solution (4.80).

4.9.2 Derive a Newton iteration algorithm to solve the nonlinear similarity solution (4.80) for the shear velocity u_τ.

4.9.3 Confirm the accuracy of the $\{RES\_\{Q\}\}_e$ matrix statements (4.82–4.85).

4.9.4 Verify the chain rule operations (4.87–4.88).

4.9.5 Confirm correctness of the TKE closure model implementation jacobians (4.90–4.93).

4.10.1 For $1 \le i \le 3$, expand (4.97) for repeated indices, hence confirm that this PDE system is initial-value in x_1 and boundary value in the x_2-x_3 plane.

4.10.2 Expand (4.98–4.99) for repeated indices, hence confirm that the TKE closure PDE system is initial-value in x_1 and boundary value in the $x_2 - x_3$ plane.

4.10.3 Using the ordering results detailed below (4.102), expand one or two Reynolds stress tensor components in (4.101), hence verify that the tensor resolution in (4.103) is uniformly $O(\delta)$.

4.10.4 Verify that the Reynolds stress tensor resolution in (4.104) is indeed uniformly $O(\delta^2)$.

4.11.1 Confirm the PNS theory PDEs (4.106–4.108).

4.12.1 Confirm the diffusion coefficient terms in (4.121–4.122).

4.12.2 Confirm the source terms $\{b_L\_\{Q\}\}_e$ in (4.119–4.120) and (4.123–4.124).

4.12.3 Verify derivations (4.127) and (4.128) and expand the final expressions to confirm they are element scalars.

4.12.4 Confirm the derivation of the DM^h GWS^h statement (4.131).

4.12.5 Confirm the derivation of the GWS^h statement (4.132–4.133) for particular pressure.

References

Babuska, I. and Rheinboldt, W.C. (1976). "*A-posteriori* error estimates for the finite element method," *J. Numerical Methods Engineering*, V.12.

Baker, A.J., Manhardt, P.D. and Orzechowski, J.A. (1979). "A numerical solution algorithm for prediction of turbulent aerodynamic corner flows," Technical Paper AIAA79:0073.

Baker, A.J. and Orzechowski, J.A. (1981). "A continuity constraint finite element algorithm for three-dimensional parabolic flow prediction," in ASME *Computers in Flow Predictions and Fluid Dynamics Experiments*, pp. 103–118.

Baker, A.J. and Orzechowski, J.A. (1982). "A numerical interaction algorithm for three dimensional subsonic turbulent juncture region flow," Technical Paper AIAA82:0100.

Baker, A.J., Yu, J.C., Orzechowski, J.A. and Gatski, T.B. (1982). "Prediction and measurement of incompressible turbulent aerodynamic trailing edge flows," *AIAA Journal*, V.20.

Baker, A.J. (1983). *Finite Element Computational Fluid Mechanics*, McGraw-Hill/Hemisphere, NY.

Cebeci, T. and Smith, A.M.O. (1974). *Analysis of Turbulent Boundary Layers*, Academic Press, NY.

Ericson, S.E. (2001). "A CFD Laboratory archive supporting the academic process," MSc thesis, University of Tennessee.

Gessner, F.B. (1973). "The origin of secondary flow in turbulent flow in a corner," *J. Fluid Mechanics*, V.58.

Gessner, F.B. and Emery, A.F. (1976). "A Reynolds stress model for turbulent corner flows, I. Model development," *ASME J. Fluids Engineering Transactions*, 1976.

Kline, S.J., Morkovin, M.V., Sovran, G. and Cockrell, D.J. (1969). "*Computation of Turbulent Boundary Layers – Volume I*," 1968 AFOSR-IFP-Stanford Conference, Thermosciences Division, Stanford University, CA.

Lam, C.K.C. and Bremhorst, K. (1981). "A modified form of k-ε model for predicting wall turbulence," *ASME Transactions*, V.103.

Launder, B.E., Reece, G.J. and Rodi, W. (1975). "Progress in the development of a Reynolds-stress turbulence closure," *J. Fluid Mechanics*, V.68.

Lumley, J.L. (1970). "Towards a turbulent constitutive relation," *J. Fluid Mechanics*, V.41.

Melling, A. and Whitelaw, J.H. (1976). "Turbulent flow in a rectangular duct," *J. Fluid Mechanics*, V.78.

Oden, J.T. and Reddy, J.N. (1976). *An Introduction to the Mathematical Theory of Finite Elements*, Wiley-Interscience, NY.

Patankar, S.V. and Spalding. D.B. (1972). "A calculation procedure for heat, mass and momentum transfer in three-dimensional parabolic flows," *J. Heat and Mass Transfer*, V.15.

Popinski, Z. and Baker, A.J. (1974). "An implicit finite element algorithm for the boundary layer equations," *J. Computational Physics*, V.21.

Schlichting, H. (1979). *Boundary Layer Theory*, McGraw-Hill, NY.

Soliman, M.O. (1978). "Accuracy and Convergence of a Finite Element Algorithm for Computational Fluid Dynamics," PhD dissertation, University of Tennessee, Knoxville.

Soliman, M.O. and Baker, A.J. (1981a). "Accuracy and convergence of a finite element algorithm for laminar boundary layer flow," *J. Computers and Fluids*, V.9.

Soliman, M.O. and Baker, A.J. (1981b). "Accuracy and convergence of a finite element algorithm for turbulent boundary layer flow," *Computer Methods Applied Mechanics and Engineering*, V.28.

Wilcox, D.C. (2006). *Turbulence Modeling for CFD*, 3rd Edition, DCW Industries, La Canada, CA.

5

The Navier–Stokes Equations: theoretical fundamentals; constraint, spectral analyses, *m*PDE theory, optimal Galerkin weak forms

5.1 The Incompressible Navier–Stokes PDE System

To this point, the focus has been on weak form algorithms for reduced forms of the incompressible-thermal Navier–Stokes conservation principles. To avoid continuously referring to previous chapters, the subject incompressible laminar-thermal NS PDE system is

$$\mathrm{D}M: \quad \nabla \cdot \mathbf{u} = 0 \tag{5.1}$$

$$\mathrm{DP}: \quad \mathrm{St}\frac{\partial \mathbf{u}}{\partial t} + \nabla \cdot \mathbf{u}\mathbf{u} + \frac{1}{\rho_0}\nabla P - \frac{1}{\mathrm{Re}}\nabla \cdot \nabla \mathbf{u} + \frac{\mathrm{Gr}}{\mathrm{Re}^2}\Theta\,\widehat{\mathbf{g}} = 0 \tag{5.2}$$

$$\mathrm{D}E: \quad \mathrm{St}\frac{\partial \Theta}{\partial t} + \nabla \cdot \mathbf{u}\Theta - \frac{1}{\mathrm{Re\ Pr}}\nabla \cdot \nabla \Theta - s_\Theta = 0 \tag{5.3}$$

$$\mathrm{D}Y: \quad \mathrm{St}\frac{\partial Y}{\partial t} + \nabla \cdot \mathbf{u}Y - \frac{1}{\mathrm{Re\ Sc}}\nabla \cdot \nabla Y - s_Y = 0 \tag{5.4}$$

The state variable for (5.1–5.4) contains *non-dimensional* (non-D) velocity vector, pressure, temperature and mass fraction: $\{q(\mathbf{x}, t)\} = \{\mathbf{u}, p, \Theta, Y\}^T$. The definitions for Stanton, Reynolds, Grashoff, Prandtl and Schmidt numbers are conventional: $\mathrm{St} \equiv \tau\mathrm{U/L}$, $\mathrm{Re} \equiv \mathrm{UL}/\nu$, $\mathrm{Gr} \equiv \mathrm{g}\beta\Delta\mathrm{TL}^3/\nu^2$, $\mathrm{Pr} \equiv \rho_0 \nu c_p/k$ and $\mathrm{Sc} = \mathrm{D}/\nu$, for D the binary diffusion coefficient. In the Boussinesq buoyancy hypothesis, $\Delta\mathrm{T} \equiv (T_{\max} - T_{\min})$ with $\beta \equiv 1/\mathrm{T_{abs}}$ and $P \equiv p/\rho_0$ is *kinematic* pressure. Note that St is typically defined unity, hence $\tau \equiv \mathrm{L/U}$, and in DE the Peclet number Pe commonly replaces RePr.

Optimal MODIFIED CONTINUOUS Galerkin CFD, First Edition. A. J. Baker.

Companion Website: www.wiley.com/go/baker/GalerkinCFD

The mathematical characterization of the NS PDE system (5.2–5.4) is constrained initial value-elliptic boundary value (I-EBV), as (5.1) constitutes a *differential constraint*. Specifically, D*M* requires that any NS state variable solution $\{q(\mathbf{x},t)\}$ possesses a velocity vector field that is *divergence-free*. Additionally, no PDE exists for kinematic pressure prediction, which "in nature" intrinsically accrues to satisfaction of the constraint of continuity, (5.1). The various procedures for enforcement of D*M*, via *exact* and *inexact* approaches, are mathematically characterized with precision in this chapter.

Thereafter, the chapter focus turns to analyses developed to characterize the *dominating error modes* introduced by the discrete implementation of a weak form (or any other!) CFD algorithm generating an approximate solution to the NS system (5.1–5.4) for practical (large!) Re. This leads to precise theoretical characterization of the topics of *phase velocity error, dispersion error* and *artificial diffusion*. The developed spectral characterizations lead to a theory for predicting *modified* $\text{GWS}^h + \theta\text{TS}$ algorithm performance using the efficient $k = 1$ basis, eventually identifying the *m*odified NS PDE (*m*PDE) system for *optimal* $m\text{GWS}^h + \theta\text{TS}$ implementation. As usual, the theoretical analyses can only be completed for linearized problem statements, with various models of the laminar NS PDE system serving this requirement.

5.2 Continuity Constraint, Exact Enforcement

Robust and efficient enforcement of D*M*, (5.1), is *the* fundamental challenge in design of incompressible NS CFD algorithms. The mathematical *constraint* is that for a velocity vector field to be admissible as a solution to **DP**, (5.2), it must be *solenoidal*, that is, divergence-free. Satisfaction of this requirement involves both exact and inexact procedures, with historical taxonomy given in Table 5.1.

Rigorous satisfaction of D*M* is trivial via vector field theory, but the acquired constraints limit general utility of these forms as CFD algorithm foundations. Any vector that is solenoidal can be expressed unequivocally, to within a constant vector, as the curl of another vector field termed a *vector potential function*. In fluid mechanics, this vector potential is called *streamfunction* and is traditionally symbolized as $\mathbf{\Psi}(\mathbf{x}, t)$. The curl of velocity is called *vorticity*, a vector usually denoted $\mathbf{\Omega}(\mathbf{x}, t)$, hence forming curl **DP** replaces the linear terms in velocity with vorticity and eliminates appearance of pressure because curl grad = 0 is a vector identity.

For the vector identity $\mathbf{\Omega} \equiv \text{curl}\ \mathbf{u} = \text{curl}\,\text{curl}\,\mathbf{\Psi}$, the resultant *vorticity-streamfunction* form of the NS PDE system (5.1–5.2) is

$$\mathcal{L}(\mathbf{\Omega}) = \text{St}\frac{\partial \mathbf{\Omega}}{\partial t} + \nabla \times \mathbf{\Psi} \cdot \nabla \mathbf{\Omega} - (\mathbf{\Omega} \cdot \nabla)\nabla \times \mathbf{\Psi} - \frac{1}{\text{Re}}\nabla^2 \mathbf{\Omega} + \frac{\text{Gr}}{\text{Re}^2}\nabla \times \Theta\,\widehat{\mathbf{g}} = \mathbf{0} \tag{5.5}$$

$$\mathcal{L}(\mathbf{\Psi}) = -\nabla^2 \mathbf{\Psi} + \nabla(\nabla \cdot \mathbf{\Psi}) - \mathbf{\Omega} = \mathbf{0} \tag{5.6}$$

While the PDE system (5.5–5.6) is mathematically precise there exist practical issues limiting its utility for CFD algorithm design. Both vector valued PDEs are elliptic boundary value (EBV), with (5.5) also initial-value, hence boundary conditions (BCs) are required defined on the *entire* closure of a solution domain. The known BCs for velocity are Dirichlet at inflow, no-slip at solid surfaces and vanishing Neumann at outflow. These data transform directly to BCs for both $\mathbf{\Psi}$ and $\nabla \mathbf{\Psi} \cdot \hat{\mathbf{n}}$, for $\hat{\mathbf{n}}$ the surface outwards pointing

Table 5.1 Taxonomy of Historical D*M* CFD formulations (from Williams and Baker, 1996)

CONTINUITY ENFORCEMENT	METHOD	ORIGINS	ISSUES/ DETRACTIONS
EXACT WITH VORTICITY	VORTICITY/ STREAMFUNCTION	Fromm [59] (FDM) Baker [61] (FEM)	Practical for 2D only. Vorticity BC at walls
	VORTICITY/VECTOR POTENTIAL	Aziz and Hellums [69]	BCs for vector potential vorticity BC
	VORTICITY/VECTOR SCALAR POTENTIALS	Aregbesola and Burley [72]	6 DOF/node in 3D vorticity BC
	VORTICITY/ VELOCITY	Fasel [76] (2D) Dennis *et al.* [77] (3D)	6 DOF/node in 3D vorticity BC
EXACT	$\boldsymbol{u}$-P Direct (mixed finite elements)	Ladyzhenskaya [52] Babuska [53] Brezzi [55]	Ill-conditioned numerical diffusion
INEXACT-ALGEBRAIC	PENALTY	Temam [86] Zienkiewicz and Godbole [49]	Ill-conditioned Reduced integrations
INEXACT-INITIAL VALUE	PSEUDO-COMPRESSIBILITY	Chorin [91]	Steady-state only numerical diffusion
INEXACT-BOUNDARY VALUE	MAC/SMAC	Harlow and Welch [99]	Staggered mesh velocity BC
	PROJECTION	Chorin [106] Temam [107]	Staggered or non-staggered meshes BC implementation
	SIMPLE, SIMPLER, SIMPLEC, SIMPLEST	Patankar and Spalding [116]	Staggered mesh, slow convergence, BC implementation
	VELOCITY CORRECTION	Schneider *et al.* [50]	Equal-order finite element, explicit with lumped mass matrix
	PISO	Issa [124, 125]	2-step predictor/corrector, staggered mesh, BC, pressure
	OPERATOR SPLITTING	Glowinski [51]	decouples nonlineartiy from incompressibility
	PRESENT	Williams and Baker [143]	implicit, equal-order, time-accurate, finite element Galerkin weak statement

unit normal vector. The BC for $\boldsymbol{\Omega}$ on inflow and outflow surfaces is thus also known, but that for a no-slip surface is not *a priori* given.

A second detraction to (5.5–5.6) is that $\boldsymbol{\Psi}$ solutions are fully coupled in (5.6) via the $(\nabla \cdot \boldsymbol{\Psi})$ term, unless $\boldsymbol{\Psi}$ itself is divergence-free. This complication is as difficult to resolve as is generation of a velocity vector $\mathbf{u}$ that is solenoidal! Finally, specified data must be sufficiently smooth such that the added differentiability of curl $\mathbf{u}$ is admissable. For incompressible NS this is the usual case with select exceptions, to be examined.

One precise mathematical resolution of these impediments accrues to considering (5.6) as the *differential definition* for $\boldsymbol{\Omega}$, followed by its direct substitution into (5.5). This produces a *biharmonic* PDE in which the sole state variable member is the vector $\boldsymbol{\Psi}$ (isothermal). Even though fourth order, it is completely well-posed as I-EBV due to the detailed available Dirichlet and Neumann BCs.

The practical detraction is the biharmonic operator $\mathrm{Re}^{-1}(\nabla^4\boldsymbol{\Psi})$, which requires the Galerkin weak form trial space reside in H^2, recall Section 3.4. This is a smoothness requirement relatively impractical for discrete implementation. Additionally, genuine NS statements possess large Re, which yields a miniscule multiplier on $\nabla^4\boldsymbol{\Psi}$. This renders the discretely implemented ∇^4 matrix ill-conditioned due to vanishingly small diagonal elements.

A proven resolution for the $\boldsymbol{\Psi}$ solenoidal issue is to transform the NS system (5.1–5.2) into *vorticity-velocity* form, generated by the vector identity $\boldsymbol{\Omega} \equiv \mathrm{curl}\, \mathbf{u}$. The resulting vector PDE system is

$$\mathcal{L}(\boldsymbol{\Omega}) = \mathrm{St}\frac{\partial \boldsymbol{\Omega}}{\partial t} + (\mathbf{u} \cdot \nabla)\boldsymbol{\Omega} - (\boldsymbol{\Omega} \cdot \nabla)\mathbf{u} - \frac{1}{\mathrm{Re}}\nabla^2\boldsymbol{\Omega} + \frac{\mathrm{Gr}}{\mathrm{Re}^2}\nabla \times \Theta\widehat{\mathbf{g}} = \mathbf{0} \tag{5.7}$$

$$\mathcal{L}(\mathbf{u}) = -\nabla^2\mathbf{u} - \nabla \times \boldsymbol{\Omega} = \mathbf{0} \tag{5.8}$$

and well-posed BCs for EBV (5.8) are known. Exact satisfaction of D*M* via definition of $\boldsymbol{\Psi}$ is relinquished, which is then required approximately enforced via Taylor series (TS) discrete BCs for $\boldsymbol{\Omega}$ on a no-slip surface. A Galerkin weak form CFD algorithm for $n = 3$ (5.7–5.8) exhibits quality performance, as detailed in the following chapter.

A second resolution for the $\boldsymbol{\Psi}$ solenoidal issue is to restrict application to $n = 2$ NS statements. This PDE system has proven highly useful for examining NS algorithm error mode characterization, the presentation on which is fully amplified in the following chapter. For $n = 2$ the streamfunction and vorticity vectors reduce to single scalar components $\boldsymbol{\Psi} \Rightarrow \psi\widehat{\mathbf{k}}$, $\boldsymbol{\Omega} \Rightarrow \omega\widehat{\mathbf{k}}$, and (5.5–5.6) simplify to the *scalar* I-EBV PDE system

$$\mathcal{L}(\omega) = \mathrm{St}\frac{\partial \omega}{\partial t} + \nabla \times \psi\,\widehat{\mathbf{k}} \cdot \nabla\omega - \frac{1}{\mathrm{Re}}\nabla^2\omega + \frac{\mathrm{Gr}}{\mathrm{Re}^2}\nabla \times \Theta\,\widehat{\mathbf{g}} = 0 \tag{5.9}$$

$$\mathcal{L}(\psi) = -\nabla^2\psi - \omega = 0 \tag{5.10}$$

A non-homogeneous Neumann BC for ω on a no-slip boundary is readily determined. Since this boundary is a surface $\psi = \mathrm{constant}$, the strictly kinematic PDE (5.10) is reduced to the *ordinary* differential equation (ODE) $\mathrm{d}^2\psi/\mathrm{dn}^2 = -\omega$, for n the wall normal coordinate. This derivative, when inserted into a TS expansion on $\psi(\mathrm{n})$, generates the Robin BC for ω, hence I-EBV PDE system (5.9–5.10) is guaranteed well-posed.

Vorticity and streamfunction are but computational variables, as the solution sought is the NS classic state variable $\{q(\mathbf{x},t)\} = \{\mathbf{u},\ P,\ \Theta\}^T$. The first member therein admits a weak form algorithm written on the curl of vector identity $\omega\widehat{\mathbf{k}} \equiv \text{curl}\,\mathbf{u}$, which produces the well-posed EBV PDE system

$$\mathcal{L}(\mathbf{u}) = -\nabla^2\mathbf{u} - \nabla\times\omega\widehat{\mathbf{k}} = \mathbf{0} \tag{5.11}$$

Pressure is determinable from the Poisson PDE formed via div $\mathbf{DP}$, which generates the well-posed EBV PDE + BCs system

$$\begin{aligned} \mathcal{L}(P) &= -\nabla^2 P - \frac{\partial}{\partial x_j}\left[u_i\frac{\partial u_j}{\partial x_i} + \frac{\text{Gr}}{\text{Re}^2}\Theta\widehat{g}_j\right] = 0 \\ \ell(P) &= \nabla P\cdot\hat{\mathbf{n}} - \left[\text{St}\frac{\partial\mathbf{u}}{\partial t} - \frac{1}{\text{Re}}\nabla^2\mathbf{u}\right]\cdot\hat{\mathbf{n}} = 0 \end{aligned} \tag{5.12}$$

Weak form algorithm solutions of (5.11–5.12) are always a post-process operation at any point in the (5.9–5.10) plus (5.3) solution sequence. Note that (5.12), recall (4.100), is valid in n dimensions.

Finally, there exists a specific weak form theory generating analytical closure for the constraint of DM for steady isothermal (5.1–5.2). Termed the "u-P direct method," and recalling the concepts and notation in Section 3.5, one seeks the discrete velocity vector function $u_i^h \in V_0^h$ and the discrete kinematic pressure scalar function $P^h \in S_0^h$ such that

$$\begin{aligned} \int_\Omega v_i^h \frac{\partial}{\partial x_j}\left[u_i^h u_j^h + P^h\delta_{ij} - \frac{1}{\text{Re}}\frac{\partial u_i^h}{\partial x_j}\right] d\tau = 0, \quad \textit{for all}\ v_i^h \in V_0^h \\ \int_\Omega q^h \frac{\partial u_j^h}{\partial x_j}\, d\tau = 0, \quad \textit{for all}\ q^h \in S_0^h \end{aligned} \tag{5.13}$$

In (5.13) $V_0^h \subset H_0^1$ and $S_0^h \subset L_0^2$ are trial function spaces of appropriate differentiability, Table 3.1, that also satisfy Dirichlet BCs (subscript 0).

In this setting, the fundamental problems are that $V_0^h \subset H_0^1$ and $S_0^h \subset L_0^2$ alone are *not sufficient* conditions to produce stable approximate solutions, the missing ingredient being the continuity constraint (5.1). Therefore let Z^h denote the subspace of *discretely divergence-free* functions

$$\begin{aligned} Z^h &= \left\{v_i^h \in V_0^h;\, \text{b}(v_i^h, q^h) = 0\ \forall q^h \in S_0^h\right\}, \\ \text{where}\quad \text{b}(v_i^h, q^h) &\equiv -\int_\Omega q^h \frac{\partial v_i^h}{\partial x_i} d\tau \end{aligned} \tag{5.14}$$

where $\forall$ denotes *for all*. The computational dilemma arises from the fact that in general $Z^h \notin Z$, where Z contains *all* divergence-free functions, which admits that discretely solenoidal functions do not necessarily remain solenoidal in the limit as $h \to 0$.

The theoretical work of Ladyzhenskaya (1969), Babuska (1973) and Brezzi (1974) led to identification of a div-stability requirement, called the LBB or *inf sup* condition. It states that, for a given pair of approximation function spaces S_0^h and V_0^h, if there exists a positive real number $\gamma > 0$ such that

$$\frac{inf}{0 \neq q^h \in S_0^h} \frac{sup}{0 \neq v_i^h \in V_0^h} \left[\frac{\mathrm{b}(v_i^h, q^h)}{\left| v_i^h \right|_1 \| q^h \|_0} \right] \geq \gamma \tag{5.15}$$

then the discrete solutions $u_i^h \in V_0^h$ and $P^h \in S_0^h$ to (5.13) will be stable.

The identification of u_i^h and P^h *trial space basis* pairs that meet the LBB condition is exhaustively developed in Gresho and Sani (1998). Summarily, the bottom line is that the FE trial space basis for velocity vector must contain polynomials at least one completeness degree k higher than that for pressure. The div-stability requirement also extends to finite difference and finite volume incompressible NS CFD methods, of course, with the well-known *staggered-mesh* technique being one method for producing a stable solution, Gunzburger (1989).

5.3 Continuity Constraint, Inexact Enforcement

There are essentially four classes of CFD algorithm *inexact* procedures for *encouraging* satisfaction of the constraint of continuity, (5.1). The adaptation from classical isotropic incompressible elasticity theory is the *penalty method*, which constitutes a constrained minimization problem in which the penalty term is an approximation to a Lagrange multiplier, Hughes *et al.* (1979).

Proof of convergence of the penalty algorithm, Temam (1968), is limited to the steady isothermal Stokes form of (5.2), the linear PDE remaining upon discarding the convection term. Noting that $\mathrm{Re} \Rightarrow 1$ in the Stokes limit, and recalling (3.34), the resultant Stokes vector bilinear weak form for encompassing Dirichlet BCs is

$$\mathrm{B}(\mathbf{u}_\alpha, \mathbf{v}) \equiv \int_\Omega \nabla \mathbf{u}_\alpha \cdot \nabla \mathbf{v} \, \mathrm{d}\tau + \lambda_\alpha / 2 \int_{\partial\Omega_n} (\nabla \cdot \mathbf{u}_\alpha)^2 \mathrm{d}\tau, \forall \, \mathbf{v} \subset H_0^1 \tag{5.16}$$

The subscript α denotes that solution $\mathbf{u}_\alpha$ depends on the magnitude of the Lagrange multiplier λ_α. The extremum of (5.16) coincides with the steady Stokes PDE form of (5.2) for the definition $\lambda_\alpha(\nabla \cdot \mathbf{u}_\alpha) \equiv -P_\alpha$. Thereby, in the penalty method the pressure distribution is equated to λ_α times the *error* in exact satisfaction of D*M*, (5.1). The proof of convergence, Temam (1979), starts with subtracting the Stokes PDE from the penalized form of Stokes yielding

$$\left[\nabla^2 \mathbf{u}_\alpha - \nabla^2 \mathbf{u} \right] + \nabla [P_\alpha - P] = \mathbf{0} \tag{5.17}$$

Multiplying (5.17) through by $(\mathbf{u}_\alpha - \mathbf{u})$, then integrating over Ω, using the divergence theorem and integration by parts, followed by the triangle inequality, produces the error bound

$$\| \mathbf{u}_\alpha - \mathbf{u} \|_{H^1}^2 + \frac{1}{2\lambda_\alpha} \| P_\alpha \|_{L2}^2 < \frac{1}{2\lambda_\alpha} \| P \|_{L2}^2 \tag{5.18}$$

From (5.18) it follows that as $\lambda_\alpha \to \infty$, then $\mathbf{u}_\alpha \to \mathbf{u}$ in H^1. Consequently, (5.18) is also used to argue that $P_\alpha \to P$ in L2.

The size of λ_α required for full NS applications at practical Re is $\lambda_\alpha \approx \mathrm{C(Re)}$ for C $O(\mathrm{E}+07)$, Hughes *et al.* (1979). The penalty algorithm main feature is that the solution for pressure is decoupled from that for the velocity vector. Its computational detraction, upon discrete implementation, is that the resulting nonlinear matrix statement for velocity is *very ill-conditioned* due to 1/Re producing small diagonal elements coupled with λ_α generating very large off-diagonal elements. Of note, the finite element CFD algorithm in the 1980s FIDAP commercial code, Engleman (1982), employed the penalty weak formulation.

A second inexact methodology, called *pseudo-compressibility*, was developed principally as an iterative method for converging to a steady state solution for (5.1–5.4), Chorin (1967). It involves proceeding through a non-physical transient via alteration of (5.1) to

$$\frac{1}{c^2}\frac{\partial P}{\partial t} + \nabla \cdot \mathbf{u} = 0 \tag{5.19}$$

where c is an artificial speed of sound, conversely an artificial compressibility. Replacing (5.1) with (5.19) renders the NS PDE system completely I-EBV hence admits using pseudo-time marching as a matrix conditioner to enhance convergence to a steady state.

A large literature was developed in the 1980s detailing adaptations to improve discrete implementation performance; a summary is provided in Williams and Baker (1996). A critical theoretical issue is sizing c for convergence optimization with significant documentation in Rogers and Kwak (1991).

The third category of inexact enforcement of DM is termed *pressure relaxation*. The pioneering formulations were named *marker and cell* (MAC), Harlow and Welch (1965), followed by *simplified marker and cell* (SMAC), Amsden and Harlow (1970), developed to circumvent homogeneous Neumann BC enforcement problems with MAC. MAC and SMAC were finite difference implemented using staggered meshes and phantom cells (outside the domain boundary) to enforce BCs. Many variations became formulated with a comprehensive summary available, Williams and Baker (1996).

The final category for inexact DM iterative procedures has become termed *pressure projection* methods. Familiar names for algorithms belonging in this category include SIMPLE (*S*emi-*I*mplicit *M*ethod for *P*ressure *L*inked *E*quations), Patankar and Spalding (1972), with the SIMPLER (SIMPLE *R*evised, Patankar, 1980) and SIMPLEC (van Doormaal and Raithby, 1984) following modifications to improve pressure iterative convergence to steady state.

None of the SIMPLE class algorithms involve solution of the genuine pressure Poisson PDE + BCs system (5.12). This operation did become included in the PISO (*P*ressure-*I*mplicit with *S*plitting of *O*perators) algorithm, Issa (1984). All SIMPLE → PISO class formulations were derived in a *discretized* finite volume form on staggered meshes, with the pressure DOF located at the cell centroid with velocity vector DOF defined at cell vertices.

5.4 The CCM Pressure Projection Algorithm

In his dissertation, Williams (1993) undertook a fundamental theoretical analysis of the concept of a pressure projection algorithm, the goal being to clearly theorize, *in the continuum*, the mathematical fundamentals. The resultant theory precisely identifies formulation and matrix iterative issues *independent* of any discretization. Of fundamental pertinence, the developed continuum theory encompasses the predecessor discretely based algorithms of the SIMPLE → PISO class by identification of specific simplifying assumptions.

Williams labeled the theory the *continuity constraint method* (CCM) algorithm. The starting point is an *explicit* time Taylor series implementing (5.2). Denote the velocity approximation by superscript tilde, set St = 1 and assume isothermal (inconsequential to the process). Switching to tensor index notation the TS is

$$
\begin{aligned}
\tilde{u}_i^{n+1} &= u_i^n + \Delta t \frac{\partial u_i^n}{\partial t} + O(\Delta t^2) \\
&= u_i^n - \Delta t \frac{\partial}{\partial x_j}\left[u_j^n u_i^n + P^n \delta_{ij} - \frac{1}{\text{Re}}\frac{\partial u_i^n}{\partial x_j}\right] + O(\Delta t^2)
\end{aligned}
\tag{5.20}
$$

No effort is made in (5.20) for the velocity vector estimate $\tilde{u}_i^{n+1} \equiv \tilde{u}_i(t+\Delta t)$ to satisfy D*M*, and in all probability it does not! Assuming that some means exists to enforce D*M* by replacing P^n with an alternative pressure P^*, the companion TS predicting the *mass-conserving* velocity vector field at time $t + \Delta t$ is

$$
\begin{aligned}
u_i^{n+1} &= u_i^n + \Delta t \frac{\partial u_i^n}{\partial t} + O(\Delta t^2) \\
&= u_i^n - \Delta t \frac{\partial}{\partial x_j}\left[u_j^n u_i^n + P^* \delta_{ij} - \frac{1}{\text{Re}}\frac{\partial u_i^n}{\partial x_j}\right] + O(\Delta t^2)
\end{aligned}
\tag{5.21}
$$

The difference between the two velocity vector predictions is simple to form, and the curl of this difference vanishes via the curl grad vector identity. Therefore, the *error* in predicting a mass-conserving velocity vector by an explicit TS must be the gradient of a scalar potential function, that is,

$$
u_i^{n+1} - \tilde{u}_i^{n+1} = -\partial\phi/\partial x_i, \quad \textit{for any}\ \tilde{u}_i^{n+1}! \tag{5.22}
$$

The fatal flaw with this analysis is that the explicit TS is theoretically inappropriate, since information at time t^{n+1} is required to estimate a P^* that satisfies D*M*. Thus an *implicit* time TS is required, and repeating the (5.21) exercise for the θ-implicit Euler algorithm yields

$$
\tilde{u}_i^{n+1} = u_i^n + \Delta t\left[\theta\frac{\partial \tilde{u}_i^{n+1}}{\partial t} + (1-\theta)\frac{\partial u_i^n}{\partial t}\right] + O\left(\Delta t^{f(\theta)}\right) \tag{5.23}
$$

Substituting (5.2) into (5.23), and then repeating the TS operation for the DM-satisfying velocity vector u^{n+1}, the difference in the TS predictions becomes

$$u_i^{n+1} - \tilde{u}_i^{n+1} = \theta \Delta t \begin{bmatrix} \tilde{u}_j \dfrac{\partial \tilde{u}_i}{\partial x_j} - \dfrac{\partial u_j u_i}{\partial x_j} + \dfrac{\partial (P^* - P)}{\partial x_i} \\ + \dfrac{1}{\text{Re}} \dfrac{\partial^2}{\partial x_j^2} (u_i - \tilde{u}_i) \end{bmatrix}_{n+1} \tag{5.24}$$

The curl of the right hand side of (5.24) *does not* vanish, hence it cannot be equated to a scalar potential function.

The *implication* is that the TS must not only be implicit but must be embedded within an iteration procedure as well. Recalling the basic Newton iteration algorithm, (4.34), near convergence at iteration $p+1$ at time t^{n+1} the error in DM will approach becoming solenoidal, hence

$$\left[u_i^{p+1} - \tilde{u}_i^{p+1} \right]_{n+1} \approx - \left[\frac{\partial \phi}{\partial x_i} \right]_{n+1}^{p+1} \tag{5.25}$$

Substituting (5.25) into (5.24) produces

$$\left[\frac{\partial \phi}{\partial x_i} \right]_{n+1}^{p+1} \approx -\theta \Delta t \left[\frac{1}{\text{Re}} \frac{\partial^3 \phi}{\partial x_i^3} - \tilde{u}_j \frac{\partial^2 \phi}{\partial x_j \partial x_i} - \frac{\partial (P^* - P)}{\partial x_i} \right]_{n+1}^{p} \tag{5.26}$$

the solution of which is intractable.

Williams therefore discarded the first two right side terms in (5.26), leaving the approximation

$$\left[\frac{\partial \phi}{\partial x_i} \right]_{n+1}^{p+1} \approx \theta \Delta t \left[\frac{\partial (P^* - P)}{\partial x_i} \right]_{n+1}^{p} \tag{5.27}$$

which can be directly integrated to produce the pressure estimate

$$P_{n+1}^{p} \approx P^{*p}_{n+1} + \frac{1}{\theta \Delta t} (\phi)_{n+1}^{p+1} \tag{5.28}$$

Realizing the *impossibility* of data generated at iteration $p+1$ affecting operations at iteration p, and further that (5.28) must be embedded into the entire matrix iteration cycle, the final form defined by Williams for pressure *estimation* at time t^{n+1} is

$$P^*|_{n+1}^{p} = P_n + \frac{1}{\theta \Delta t} \sum_{\alpha=1}^{p} \delta \phi|_{n+1}^{\alpha} \tag{5.29}$$

In (5.29), P_n is the DM-satisfying pressure post-processed via the genuine pressure PDE + BCs system (5.12) using the weak form algorithm converged DM-satisfying velocity vector solution at time t_n. Further, $\delta\phi^\alpha$ is the sequence of potential iterates generated during the CCM algorithm convergent iteration process at time t_{n+1}.

The PDE system for velocity potential function ϕ is generated by forming the divergence of (5.25). Realizing that u_i by definition satisfies D*M*, its divergence vanishes and the resultant EBV PDE + BCs statement at any iteration p is

$$\begin{aligned} \mathcal{L}(\phi) &= -\nabla^2\phi + \frac{\partial \tilde{u}_i}{\partial x_i} = 0 \\ \ell(\phi) &= -\nabla\phi \cdot \hat{\mathbf{n}} - (u_i - \tilde{u}_i)\hat{n}_i = 0 \end{aligned} \tag{5.30}$$

Realizing that the typical velocity BCs are Dirichlet at inflow and on a no-slip wall, from (5.30) the ϕ BCs thereon are homogeneous Neumann. Conversely, at an outflow the D*M*-satisfying velocity vector is unknown, hence also its difference with any velocity approximation. Therefore, the sole admissible outflow BC must be Dirichlet.

Since in computational practice the intrinsic measure of DM^h satisfaction at convergence is the potential function *energy norm* $\left\|\nabla^h \cdot \mathbf{u}^h\right\| \Rightarrow \left\|\phi^h\right\|_E$, the requirement is to *minimize this error*. Therefore, the appropriate outflow Dirichlet BC is $\phi = 0$.

It is informative to interpret the SIMPLE class of DM^h algorithms in the context of the CCM theory. The P' *pressure correction* PDE employed in the SIMPLE, SIMPLER and SIMPLEC algorithms, written in homogenous vector form, is

$$\mathcal{L}(P') = -\nabla^2 P' - \frac{1}{\Delta t}\nabla \cdot \tilde{\mathbf{u}} = 0 \tag{5.31}$$

This *is not the genuine pressure* PDE (5.12), subject to the exact BCs, but rather the DM^h velocity potential function PDE system, (5.30), with $1/\Delta t$ (for $\theta \equiv 1$) borrowed from the pressure estimator (5.28). The missing theoretical component of the SIMPLE class is the iteration statement (5.29), replaced by a sequence of *heuristic* manipulations aimed at improving the rate of convergence of pressure correction.

In distinction, PISO does include a genuine pressure Poisson PDE but erroneously inserts the $\mathrm{Re}^{-1}\nabla^2\mathbf{u}$ BC term into the $\mathcal{L}(P)$ non-homogeneity, (5.12), as a third spatial derivative. Since PISO implementations typically employ $O(\Delta x^2)$ finite volume numerics, this term TS $O(1)$ truncation error is a serious compromise to accuracy. PISO employs a P' iteration essentially identical to the SIMPLE class.

As a concluding point, it is the CCM theory that underlies the PNS algorithm DM^h formulation presented without proof in Chapter 4. Observe that (5.30) is functionally identical to (4.96); (5.29) details the PNS iterative accumulation process and the BCs match exactly. The $\mathrm{GWS}^h + \theta\mathrm{TS}$ algorithm discrete implementation of the classic state variable incompressible NS CCM theory is detailed in Chapter 7.

5.5 Convective Transport, Phase Velocity

Theorizing an accurate approximate solution algorithm for the NS conservation principle system (5.1–5.4) is significantly challenged by the fact that Re is a large number in genuine applications. NS solution *spectral content* is typically well distributed throughout wavenumber space, and content on the scale of mesh measure h cannot be resolved.

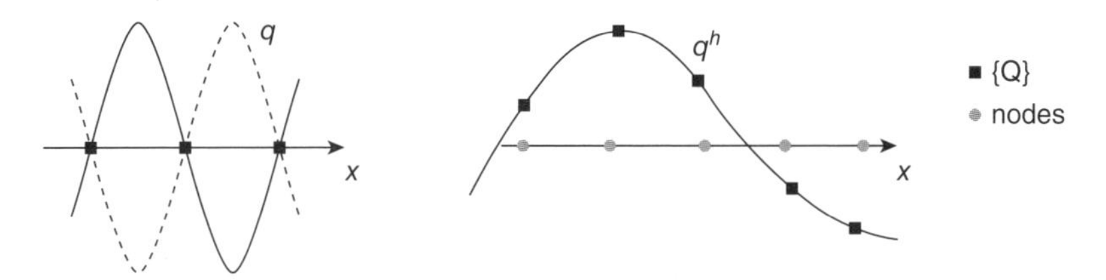

Figure 5.1 Mesh resolution of Fourier content

Figure 5.1, left and right, graphically confirms this for the example Fourier content $q = \pm\sin(2\pi x/\lambda + \phi)$ projected onto meshes of span $2h$ and $6h$. In the former, the nodal DOF $\{Q\}$ of $q^h(x)$ are always *zero*(!) for any phase and amplitude, while the latter supports resolution.

Large Re diminishes the fluid viscous (*physical*) diffusive contribution in the NS PDE system, hence the substantial derivative $\mathrm{D}(q)/\mathrm{D}t$ in (5.2–5.4), recall (1.6), totally dominates. In concert with the NS convection term explicit nonlinearity, the net result is a pathological *destabilizing* process caused by creation of a *dispersive error* mode. The theoretical requirement is to precisely characterize this phenomenon, to establish a clear rationale for constructing a viable *modified* $\mathrm{GWS}^h + \theta\mathrm{TS}$ algorithm for (5.1–5.4).

NS approximate solution spectral content is space-time domain propagated during the computational process. The *phase velocity*, usually denoted $\mathbf{c}$, is the velocity of propagation of a solution Fourier component, called a wave, by the fluid velocity $\mathbf{u}$ in the direction of the wavenumber vector $\boldsymbol{\kappa}$, Gresho and Sani (2000). Mathematically

$$\mathbf{c} \equiv \frac{(\mathbf{u} \bullet \boldsymbol{\kappa})\boldsymbol{\kappa}}{\kappa^2} \tag{5.32}$$

where $\kappa = |\boldsymbol{\kappa}|$ is wavenumber magnitude, equal to the number of wave crests existing in the interval 2π in the direction of the wave vector angle η, Haberman (1998), see Figure 5.2. Thereby, $\kappa = 2\pi/\lambda$, where λ is the crest wavelength, and $\boldsymbol{\kappa}$ is orthogonal to the wave crest. In the continuum in one dimension, the phase velocity magnitude (speed) is the scalar c which is identical with the imposed fluid speed u.

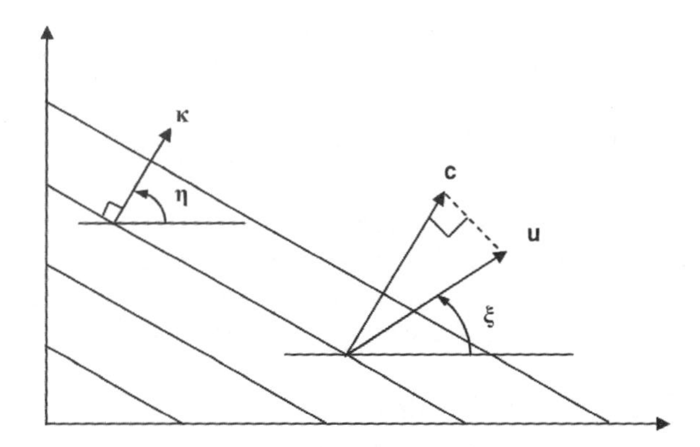

Figure 5.2 Wavenumber vector and phase velocity in the continuum

The *group velocity* $\mathbf{v}_g(\boldsymbol{\kappa})$ is the velocity vector with which smooth waves propagate energy in a dispersive medium. In rectangular cartesian coordinates the definition is

$$\mathbf{v}_g(\boldsymbol{\kappa}) \equiv \nabla_\kappa \omega = \frac{\partial \omega}{\partial \kappa_1}\hat{\mathrm{i}} + \frac{\partial \omega}{\partial \kappa_2}\hat{\mathrm{j}} + \frac{\partial \omega}{\partial \kappa_2}\widehat{\mathrm{k}} \tag{5.33}$$

where ω is the cyclic frequency and $\boldsymbol{\kappa} \Rightarrow \kappa_i$ is wavenumber resolution. In the continuum, the group velocity is independent of $\boldsymbol{\kappa}$, hence is equal to the fluid velocity $\mathbf{u}$, (5.1–5.4).

In an algorithm spatially discrete implementation, the phase and group velocities differ from their continuum values due to discretization-induced wavenumber dependence. Further, the group velocity is not always aligned with the wave vector. Reverting to one dimension for illustration, from (5.32–5.33)

$$v_g(\kappa) \equiv \frac{\partial \omega(\kappa)}{\partial \kappa} = \frac{\partial (c(\kappa)\kappa)}{\partial \kappa} = c(\kappa) + \kappa \frac{\partial (c(\kappa))}{\partial \kappa} \tag{5.34}$$

hence only if phase speed c is independent of wavenumber κ are the group and phase speeds equal. Therefore, determination of the wavenumber dependence of phase speed (velocity in n dimensions) is the key theoretical assessment required developed.

5.6 Convection-Diffusion, Phase Speed Characterization

The linear, one-dimensional unsteady convection-diffusion PDE supports phase velocity characterization of an NS CFD algorithm. The $\mathrm{St} = 1 = n$ simplification of an NS scalar transport principle PDE, for example, $\mathrm{D}\Theta$, $\mathrm{D}Y$, (5.3–5.4), is

$$\mathcal{L}(q) = \frac{\partial q}{\partial t} + u\frac{\partial q}{\partial x} - \frac{1}{\mathrm{Pa}}\frac{\partial^2 q}{\partial x^2} = 0 \tag{5.35}$$

where Pa is the non-D parameter pertinent to state variable $q(x, t)$.

Recall the Section 4.3 development leading to (4.27). Then exchanging Δx therein with Δt, the $k = 1$ NC basis implementation of the GWS^h algorithm for (5.35) generates the ordinary differential equation (ODE) system

$$\begin{aligned} \mathrm{GWS}^h\left(\mathcal{L}(q^h)\right) &= \mathrm{S}_e\left([\mathrm{A200}]_e \frac{\mathrm{d}\{Q\}_e}{\mathrm{d}t} + \{\mathrm{RES\_}Q\}_e\right) = \{0\} \\ \{\mathrm{RES\_}Q\}_e &= \mathrm{U}[\mathrm{A201}]\{Q\}_e + \frac{1}{\det_e \mathrm{Pa}}[\mathrm{A211}]\{Q\}_e \end{aligned} \tag{5.36}$$

where S_e denotes assembly, illustrated in Section 2.3.

The time derivative form of (5.36) is *data* for the Euler family of θ-implicit Taylor series, (4.29–4.31). Now assuming the given velocity vector speed $u \Rightarrow \mathrm{U}$ is constant, supplied as *data*, the terminal $\mathrm{GWS}^h + \theta\mathrm{TS}$ element matrix statement is

$$\begin{aligned} \{\mathrm{F\_}Q\}_e \equiv {} & \det_e[\mathrm{A200}]\{Q_{n+1} - Q_n\}_e \\ & + \Delta t\left[\theta\{\mathrm{RES\_}Q\}_e\big|_{n+1} + (1-\theta)\{\mathrm{RES\_}Q\}_e\big|_n\right] \end{aligned} \tag{5.37}$$

The assembly of (5.37) over all FE domains Ω_e generates the global algebraic statement (4.34). From (5.37), the jacobian is

$$\begin{aligned}[\mathrm{JAC\_}Q]_e &\equiv \frac{\partial\{\mathrm{F\_}Q\}_e}{\partial\{Q\}_e} \\ &= \det_e[\mathrm{A200}] + \theta\Delta t\left(\mathrm{U}[\mathrm{A201}] + \frac{1}{\det_e \mathrm{Pa}}[\mathrm{A211}]\right)\end{aligned} \tag{5.38}$$

Finally, since (5.35) is linear, define $\{\Delta Q\} \equiv \{Q_{n+1} - Q_n\}$ as the DOF, the change in $\{Q(t)\}$ over Δt. The resultant $\mathrm{GWS}^h + \theta\mathrm{TS}$ algorithm terminal algebraic matrix statement is

$$\begin{aligned}\mathrm{S}_e\left[\det_e[\mathrm{A200}] + \theta\Delta t\left(\mathrm{U}[\mathrm{A201}] + \frac{1}{\det_e \mathrm{Pa}}[\mathrm{A211}]\right)\right]\{\Delta Q\} \\ = -\Delta t\, \mathrm{S}_e\left(\{\mathrm{RES\_}Q\}_e\right)\Big|_n\end{aligned} \tag{5.39}$$

The characterization of $\mathrm{GWS}^h + \theta\mathrm{TS}$ algorithm phase velocity distribution results upon assembly of (5.39) into a *stencil* at the generic mesh node X_j, recall Section 3.10. For $n = 1$, this involves only the two domains Ω_e and Ω_{e+1} sharing node X_j, Figure 5.3, and $k = 1$ basis DOF labels correspond one to one with the node labels. The result of generic node assembly of the three matrix terms in (5.37–5.39) is

$$\begin{aligned}\mathrm{S}_j\left[\frac{1}{\det_e \mathrm{Pa}}[\mathrm{A211}]\{Q\}_e\right]_{\mathrm{X}_j} &= \frac{1}{\Delta x\,\mathrm{Pa}}\left[-Q_{j-1} + 2Q_j - Q_{j+1}\right] \\ \mathrm{S}_j\left[\mathrm{U}[\mathrm{A201}]\{Q\}_e\right]_{\mathrm{X}_j} &= \frac{\mathrm{U}}{2}\left[-Q_{j-1} + Q_{j+1}\right] \\ \mathrm{S}_j\left[\det_e[\mathrm{A200}]\{\Delta Q\}_e\right]_{\mathrm{X}_j} &= \frac{\Delta x}{6}\left[\Delta Q_{j-1} + 4\Delta Q_j + \Delta Q_{j+1}\right]\end{aligned} \tag{5.40}$$

and note the first expression in (5.40) is identically (3.104).

First characterize the base GWS^h algorithm (5.36) independent of θTS time discretization. The node assembly of the time derivative term matrix is the last line in (5.40) upon replacing ΔQ therein by $\mathrm{d}Q/\mathrm{d}t$. Then dividing through by Δx and setting $1/\mathrm{Pa} \equiv 0$ yields the $k = 1$ basis stencil for the *diffusion-free* GWS^h algorithm (5.36) is

$$\frac{1}{6}\frac{\mathrm{d}}{\mathrm{d}t}\left[Q_{j-1} + 4Q_j + Q_{j+1}\right] + \frac{\mathrm{U}}{2\Delta x}\left[-Q_{j-1} + Q_{j+1}\right] = 0 \tag{5.41}$$

Sought is the *continuum* PDE satisfied by the spatially discrete DOF $\{Q(t)\}$ of GWS^h algorithm, (5.41). This is accomplished by performing TS expansions about X_j for all

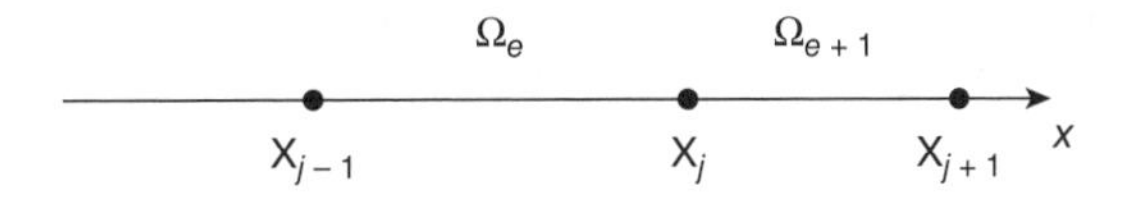

Figure 5.3 Generic element pair sharing geometric node X_j

terms in (5.41) not evaluated at X_j. Retaining a sufficient order TS expansion, the result is the *difference-differential* PDE addressed by the GWS^h algorithm generic nodal DOF $Q(t)$

$$\mathcal{L}(Q) = \frac{\partial Q}{\partial t} + \mathrm{U}\frac{\partial Q}{\partial x} + \left(\frac{1}{120} - \frac{1}{72}\right)\mathrm{U}\Delta x^4 \frac{\partial^5 Q}{\partial x^5} + \cdots = 0 \tag{5.42}$$

Comparing (5.42) to (5.35), for $1/\mathrm{Pa} = 0$ the GWS^h $k = 1$ basis algorithm generates a DOF PDE that matches the continuum PDE to $O(\Delta x^4)$, a *third-order* approximation for smooth data.

The linear weak form theory assertion that the $k = 1$ basis GWS^h algorithm is *optimal* for (5.35) within its discretization peer group is required validated. This is readily accomplished by repeating the TS operations for the DOF-comparable Crank–Nicolson FD algorithm. Recalling the matrix manipulation leading to (4.49), the CNFD algorithm stencil for (5.35) is

$$\frac{\mathrm{d}Q_j}{\mathrm{d}t} + \frac{\mathrm{U}}{2\Delta x}\left[-Q_{j-1} + Q_{j+1}\right] = 0 \tag{5.43}$$

The ensuing TS exercise generates the CNFD DOF difference-differential PDE

$$\mathcal{L}(Q) = \frac{\partial Q}{\partial t} + \mathrm{U}\frac{\partial Q}{\partial x} + \frac{1}{6}\mathrm{U}\Delta x^2 \frac{\partial^3 Q}{\partial x^3} + \cdots = 0 \tag{5.44}$$

which is but a *first-order* approximation to (5.35) for smooth data.

Clearly, this exercise verifies that the GWS^h algorithm is the optimal construction in this peer group. This occurrence is directly traced to the "mass matrix" $[\mathrm{A200}]_e$ in (5.36), the linear equivalent of the BL algorithm nonlinear $[\mathrm{A3000}]_e$ hypermatrix, assessed in Chapter 4 as responsible for the documented *optimal* performance.

That the Galerkin weak form process in fact generates optimal constructions in the large is the sought verification. Analysis of the fully discrete GWS^h + θTS algorithm, (5.41), requires identification of a Fourier mode representation of the GWS^h algorithm solution (5.36). The candidate functional form results upon realizing that the *analytical solution* to the linear PDE (5.35) for $1/\mathrm{Pa} \equiv 0$ is

$$q(x, t) = q(x, t_0)\,\mathrm{e}^{\mathrm{i}\kappa\,(x - ut)} \tag{5.45}$$

where $q(x, t_0)$ is the IC, $\mathrm{i} \equiv \sqrt{-1}$ is the imaginary unit and $\kappa \equiv 2\pi/\lambda$ is the Fourier mode wavenumber with wavelength λ.

The solution (5.45) states the IC is preserved for all time and space, and the speed of propagation of information (i.e., the phase speed c) is identically the input data u. Since only space is discretized in (5.36), the GWS^h Fourier mode representation retains time continuous. Therefore, viewing (5.45) the candidate is

$$q^h(j\Delta x, t) = q(x, t_0)\,\mathrm{e}^{\mathrm{i}\kappa\,(j\Delta x - \tilde{\mathrm{U}}t)} \tag{5.46}$$

where $\tilde{U}(\Delta x) \equiv \tilde{U}_{real} + i\tilde{U}_{imag}$ is the GWS^h *complex* approximation to real speed u in (5.35). Its departure from u, the prediction sought, is the *phase velocity error* associated with GWS^h algorithm solution (5.36).

Now including the genuine diffusion term multiplied by 1/Pa and dividing through by Δx, the stencil representation for (5.36) is

$$\begin{aligned}\frac{1}{6}\frac{d}{dt}\left[Q_{j-1} + 4Q_j + Q_{j+1}\right] &+ \frac{U}{2\Delta x}\left[-Q_{j-1} + Q_{j+1}\right] \\ &+ \frac{1}{Pa\Delta x^2}\left[-Q_{j-1} + 2Q_j - Q_{j+1}\right] = 0\end{aligned} \tag{5.47}$$

Substitution of (5.46) into (5.47) produces a statement of the form $[\bullet]\, q(x, t_0)\, e^{i\kappa\,(j\Delta x - \tilde{U}t)} = 0$. Since the multipliers exterior to the bracket do not vanish, the fact that the *real* and *imaginary* arguments in $[\bullet]$ are linearly independent enables solution for GWS^h algorithm complex phase speed resolution

$$\begin{aligned}\tilde{U}_{real} &= U\left[1 - \frac{1}{180}(\kappa\Delta x)^4 + O(\kappa\Delta x)^6\right] \\ \tilde{U}_{imag} &= U\left[-\frac{\kappa^2}{Pa} + O(\kappa\Delta x)^5\right]\end{aligned} \tag{5.48}$$

Inserting (5.48) into (5.46) generates the Fourier mode representation of the GWS^h algorithm $k=1$ basis solution for (5.35)

$$q^h(j\Delta x, t) = q(x, t_0)e^{i\kappa(j\Delta x - (U + O(\Delta x^4))t)}e^{-(\kappa^2 U/Pa + O(\Delta x^5))t} \tag{5.49}$$

Note the $O(\Delta x^4)$ phase speed alteration to $u \Rightarrow U$ in the imaginary factor exactly matches the TS analysis prediction, (5.42), a *consistency* check. The real factor in (5.49), being negative and linear in time t, generates a progressive diminution of signal during $n\Delta t$ propagation for Pa^{-1} finite.

The Fourier mode analysis of fully discrete GWS^h + θTS statement (5.37) yields *quantitative* prediction of wavenumber-dependent *spectral resolution* of the algorithm. Again dividing through by Δx and defining $1/Pa \equiv 0$, the stencil equivalent of (5.37) is

$$\begin{aligned}F\_Q_j = \frac{1}{6}&\left[\left(Q_{j-1}^{n+1} - Q_{j-1}^n\right) + 4\left(Q_j^{n+1} - Q_j^n\right) + \left(Q_{j+1}^{n+1} - Q_{j+1}^n\right)\right] \\ &+ C\left[\theta\left(-Q_{j-1}^{n+1} + Q_{j+1}^{n+1}\right) + (1-\theta)\left(-Q_{j-1}^n + Q_{j+1}^n\right)\right] = 0\end{aligned} \tag{5.50}$$

where $C \equiv U\Delta t/\Delta x$ is the *Courant number*, the local non-D time step.

From (5.45–5.46), the candidate solution is the projection of the generic node DOF Q_j at time t_n onto the spatially discrete Fourier mode representation

$$Q(j\Delta x, n\Delta t) \equiv Q_j^n = (g^h)^n\, q^h(j\Delta x) = (g^h)^n\, e^{i\kappa(j\Delta x)} \tag{5.51}$$

where g^h is the fully discrete algorithm *amplification factor*. Substituting (5.51) into (5.50) leads to exponent integer shifts for $j \pm 1$ and $n+1$. Recalling the complex variable formulae for sin(•) and cos(•), and projecting all operands to j and n, the end result is an equation of the familiar form $[\bullet]\,(g^h)^n\, e^{i\kappa(j\Delta x)} = 0$. The multiplier of the square bracket term again does not vanish, which leads to

$$g^h = \frac{1 + (1/2)\cos(\kappa\Delta x) - \mathrm{i}\,(3\mathrm{C}(1-\theta)/2)\sin(\kappa\Delta x)}{1 + (1/2)\cos(\kappa\Delta x) + \mathrm{i}\,(3\mathrm{C}\theta/2)\sin(\kappa\Delta x)} \tag{5.52}$$

as the $k=1$ basis $\mathrm{GWS}^h + \theta\mathrm{TS}$ algorithm amplification factor for all C, implicitness factor θ and Fourier mode wavenumber κ.

Note that *only* for $\theta = 0.5$ is (5.52) a rational polynomial of *complex conjugates*, whence amplification factor magnitude is unity, that is, $g^{h*}g^h = 1$, complex conjugate multiplication denoted *. For this selection, and recalling $1/\mathrm{Pa} \equiv 0$, solutions generated by the $k=1$ basis $\mathrm{GWS}^h + \theta\mathrm{TS}$ algorithm (5.37) propagate all wavenumber content devoid of *artificial diffusion*. Conversely, definitions $\theta = 0$ and $0.5 < \theta \le 1.0$ induce an artificial diffusion mechanism since $g^{h*}g^h < 1$. Finally, the algorithm is unstable for all $0 < \theta < 0.5$ since $g^{h*}g^h > 1$.

The CNFD algorithm direct comparison is required. The CNFD specification is $\theta = 0.5$ and formation of the stencil equivalent of (5.43) is a minor modification on (5.50). Retaining $1/\mathrm{Pa} \equiv 0$, the CNFD algorithm amplification factor solution is determined the rational polynomial of complex conjugates

$$g^h = \frac{1 - \mathrm{i}\,(\mathrm{C}/2)\sin(\omega\Delta x)}{1 + \mathrm{i}\,(\mathrm{C}/2)\sin(\omega\Delta x)} \tag{5.53}$$

hence is intrinsically devoid of artificial diffusion.

Determination of amplification factor enables prediction of fully discrete algorithm *phase speed* c^h distribution as a function of Courant number, implicitness factor θ and mode wavenumber κ, equivalently wavelength λ, via its definition

$$c^h(\mathrm{C}, \theta, \kappa) \equiv \frac{1}{\kappa\,\Delta x}\tan^{-1}\left[\frac{g^h_{imag}}{g^h_{real}}\right] \tag{5.54}$$

In (5.54), the inverse tangent argument is the rational polynomial of *imaginary* and *real* components of algorithm amplification factor.

Figure 5.4 compares $\mathrm{GWS}^h + \theta\mathrm{TS}$ and CNFD algorithm phase speed spectral distributions, the solid and dashed curves respectively, for Courant numbers $\mathrm{C} = 0.01$, 0.5 and 1.0. The abscissa is logarithmically scaled in mesh integer wavelength multiples $\lambda \equiv \mathrm{n}\Delta x$, $2 \le \mathrm{n} \le 50$, for short wavelength resolution. The ordinate (labeled $\Theta(\mathrm{n})$) is phase speed c^h, (5.54), with the exact solution to (5.35) the horizontal line at unity to which algorithm data asymptote for large n. For all C the CNFD phase speed distributions are clearly less accurate approaching $c^h \approx 1$ only for $\mathrm{n} \ge 20$. Conversely, for all $\mathrm{C} \le 0.5$ the $\mathrm{GWS}^h + \theta\mathrm{TS}$ algorithm phase speed is $c^h \approx 1$ for all $\mathrm{n} \ge 5$, degrading only to $\mathrm{n} \ge 10$ for $\mathrm{C} = 1.0$.

The impact of algorithm phase speed *error* is readily verified by evaluating propagation of a gaussian distribution (smooth full Fourier spectrum *data*) initial condition (IC) parallel

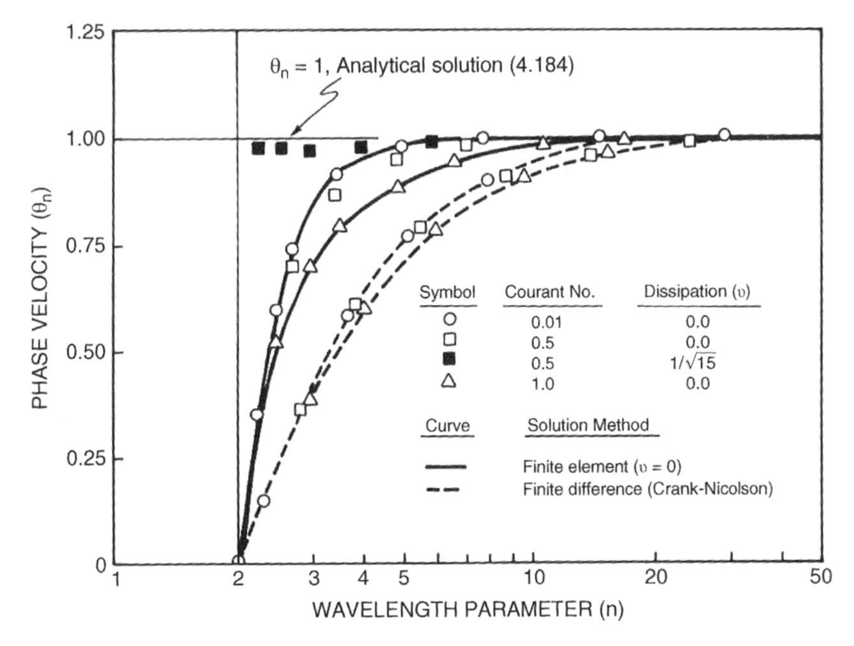

Figure 5.4 Phase speed spectral distributions for $\mathrm{GWS}^h + \theta\mathrm{TS}$ and CNFD algorithms, $\theta = 0.5$, $\mathrm{C} = 0.01$, 0.5, 1.0, from Baker (1983)

to the x axis by fixed input u. Figure 5.5 top/bottom compares $k = 1$ basis $\mathrm{GWS}^h + \theta\mathrm{TS}$ and CNFD algorithm DOF distributions for $\mathrm{C} = 0.5 = \theta$. The IC and the exact solution at the final time station are the solid curves.

The CNFD algorithm solution DOF final distribution fully quantifies *dispersion* of the gaussian IC short wavelength spectral content into longer wavelength error, which appears as a trailing oscillatory wake. The dispersive error wake magnitude is much reduced,

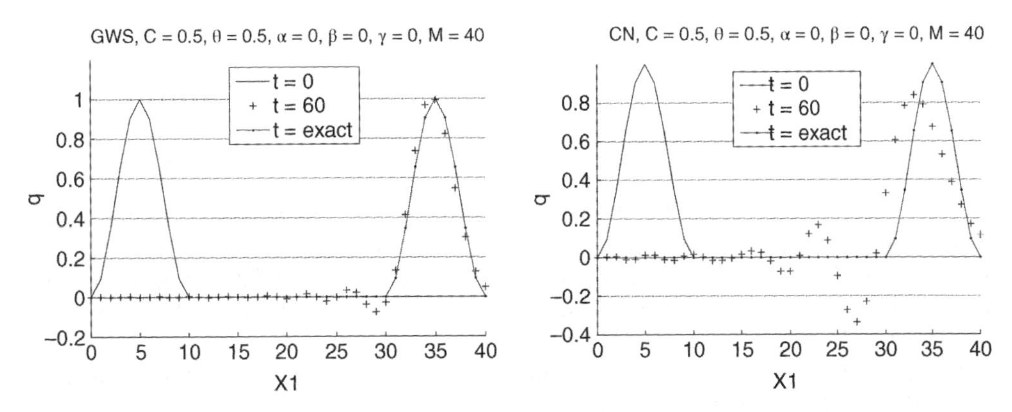

Figure 5.5 Discrete algorithm DOF data distribution, pure convection of a gaussian, 3 IC wavelength translation, $\theta = 0.5$, $\mathrm{C} = 0.5$, solid curves are IC and exact solution displaced: left, $\mathrm{GWS}^h +$ $\theta\mathrm{TS}$; right, CNFD

comparatively, in the $\mathrm{GWS}^h + \theta\mathrm{TS}$ algorithm DOF distribution, the result of the significantly more accurate phase speed distribution in mesh scale wavelengths λ, Figure 5.4, hence large wave number κ.

Numerous theories exist with goal to improve phase speed accuracy via modifications to a $\mathrm{WS}^h + \theta\mathrm{TS}$ algorithm. The majority center on definition of a test space distinct from the trial space, a *non-Galerkin* criterion algorithm. The pioneering weighted residuals (WR) algorithm, Raymond and Garder (1976), can be expressed as the modified test function spatially discrete WS^h algorithm

$$\mathrm{WS}^h\left(\mathcal{L}(q^h)\right) \equiv \mathrm{S}_e \int_{\Omega_e} \left[\{N_1(x)\} + \beta h \widehat{u} \frac{\mathrm{d}\{N_1(x)\}}{\mathrm{d}x}\right] \mathcal{L}(q^h)\mathrm{d}x = \{0\} \tag{5.55}$$

for h the mesh measure, $\widehat{\mathrm{u}}$ the velocity U unit vector and β a coefficient to be determined.

For $1/\mathrm{Pa} \equiv 0$ and substituting (5.35) into (5.55), the resultant $k = 1$ basis implementation generates the WS^h matrix statement alternative to (5.36)

$$\mathrm{WS}^h\left(\mathcal{L}(q^h)\right) = \mathrm{S}_e \left(\begin{array}{c} (l_e[\mathrm{A200}] + \beta l_e \widehat{\mathrm{u}}[\mathrm{A210}]) \dfrac{\mathrm{d}\{Q\}_e}{\mathrm{d}t} \\ + (\mathrm{U}[\mathrm{A201}] + \beta|\mathrm{U}|[\mathrm{A211}])\{Q\}_e \end{array} \right) = \{0\} \tag{5.56}$$

All matrices in (5.56) are element-independent and $|\mathrm{U}|$ denotes velocity (data) magnitude to guarantee the coefficient multiplying the diffusion matrix [A211] is positive definite.

The stencil for the $k = 1$ basis matrix product $[\mathrm{A210}]\{Q\}_e$

$$\mathrm{S}_j\left[[\mathrm{A210}]\{Q\}_e\right]_{\mathrm{X}_j} = \frac{1}{2}\left[Q_{j-1} - Q_{j+1}\right] \tag{5.57}$$

Inserting Fourier mode solution representation (5.46) into (5.56) again generates the analysis statement, which in concert with stencils in (5.47) generates $[\bullet]\, q(x, t_0)\, \mathrm{e}^{\mathrm{i}\kappa(j\Delta x - \tilde{\mathrm{U}}t)} = 0$ as terminal expression. The real and imaginary parts in $[\bullet]$ remain linearly independent, which yields the phase speed solution comparative to GWS^h (5.48)

$$\begin{aligned} \tilde{\mathrm{U}}_{real} &= \mathrm{U}\left[1 + \left(-\frac{1}{180} + \frac{\beta^2}{12}\right)(\kappa\,\Delta x)^4 + O(\kappa\,\Delta x)^6\right] \\ \tilde{\mathrm{U}}_{imag} &= |\mathrm{U}|\left[-\frac{\beta(\kappa\,\Delta x)^3}{12} + O(\kappa\,\Delta x)^5\right] \end{aligned} \tag{5.58}$$

The truncation error order for this non-Galerkin test function WS^h algorithm is minimized for $\beta \equiv 1/\sqrt{15}$, whereupon $\tilde{\mathrm{U}}_{real} \approx \mathrm{U}$ to $O(\Delta x^6)$, an $O(\Delta x^2)$ improvement over GWS^h algorithm (5.48). This definition for β renders the imaginary velocity component in (5.58) non-zero, introducing an *artificial diffusion* mechanism with "viscosity" coefficient $\nu = |\mathrm{U}|/12\sqrt{15}$.

Substituting the TS order optimal β solution into (5.46), the Fourier mode representation of the Raymond–Garder WR algorithm is

$$q^h(j\Delta x, t) = q(x, t_0)\, \mathrm{e}^{\mathrm{i}\kappa\,(j\Delta x - (\mathrm{U} + O(\Delta x^6))t)}\, \mathrm{e}^{-(\kappa^4 \nu(\Delta x^3) + O(\Delta x^5))t} \tag{5.59}$$

Of note, the artificial diffusion contribution in (5.59) is two orders more sensitive to wavenumber κ than is the physical diffusion term for $1/\mathrm{Pa} > 0$, (5.49). Hence, the impact of artificial diffusion is focused at large κ, hence smaller λ, precisely where the dispersive error mode is dominant.

The theory endpoint is amplification factor determination for the fully discrete $\mathrm{WS}^h + \theta\mathrm{TS}$ algorithm. It is an elementary exercise to substitute (5.56) into the θ-implicit TS, hence generate the alteration to (5.37). Then substituting the Fourier mode solution (5.51), the Raymond–Garder $\mathrm{WS}^h + \theta\mathrm{TS}$ algorithm amplification factor is, for their derivation selection $\theta \equiv 0.5$,

$$g^h = \frac{1 + (1/2)\cos(\kappa\Delta x) - 3\mathrm{C}\beta\sin^2(\kappa\Delta x/2) - \mathrm{i}\,(3/4)(\mathrm{C} + 2\beta)\sin(\kappa\Delta x)}{1 + (1/2)\cos(\kappa\Delta x) + 3\mathrm{C}\beta\sin^2(\kappa\Delta x/2) + \mathrm{i}\,(3/4)(\mathrm{C} - 2\beta)\sin(\kappa\Delta x)} \tag{5.60}$$

For $\mathrm{C} = 0.5$, the significant impact of this WR algorithm on short wavelength phase speed accuracy is documented by the solid square symbols in Figure 5.4. Note that c^h now approximates unity for all $\mathrm{n} > 2$! Unfortunately, $g^{h*}g^h$ is always significantly less than unity for *any* $\beta > 0$, so this phase accuracy improvement comes at the price of excessive artificial diffusion.

5.7 Theory for *Optimal* $m\mathrm{GWS}^h + \theta\mathrm{TS}$ Phase Accuracy

Minimization of phase velocity (speed) *inaccuracy* is critical to CFD algorithm performance acceptability regarding *dispersion error* pollution, the control of which has typically required infusion of artificial diffusion. The CFD literature is rich in theorizations covering the full range of discrete algorithms that improve *dispersion error control* for large Re. Legacy heuristic weak formulations proceed through Wahlbin (1974), Dendy (1974), Dendy and Fairweather (1975), Raymond and Garder (1976), leading to the widely referenced Brooks and Hughes (1982) *streamline upwind Petrov–Galerkin* (SUPG) algorithm. Each theorization employs a space of test functions distinct from the trial space, hence discarding the Galerkin criterion error extremization opportunity, Section 1.4, Figure 3.3, each with embedded coefficients required determined.

The alternative, a Galerkin form totally devoid of undefined coefficients, employs manipulations on evolutionary Taylor series to identify *continuum nonlinear* differential term additions to the textbook PDE. Thus is generated a *modified* PDE (*m*PDE) which, upon GWS^h implementation, enables *annihilation* of the *significant* order discrete approximation error. The inaugural theory is the *Taylor–Galerkin* (TG) algorithm, Donea (1984). The *Taylor weak statement* algorithm sequence follows, Baker and Kim (1987), Chaffin and Baker (1995), Sahu and Baker (2007), which summarily identifies the fully general, *n*-dimensional theory for *genuine* Galerkin weak form *optimal phase velocity* accuracy.

The theory, which widely encompasses formulations generated by diverse alternative approaches, starts with (5.35) written in divergence form. Define the *kinetic* flux vector $f \equiv uq$, then for $1/\mathrm{Pa} \equiv 0$ re-express (5.35) in flux vector form

$$\mathcal{L}(q) = \frac{\partial q}{\partial t} + \frac{\partial f}{\partial x} \Rightarrow q_t + f_x = 0, \quad \text{and } f \equiv uq \tag{5.61}$$

introducing subscript indices to denote partial differentiation. The generic TS with superscripts n and $n+1$ denoting time is

$$q^{n+1} = q^n + \Delta t q_t^n + \frac{1}{2}\Delta t^2 q_{tt}^n + \frac{1}{6}\Delta t^3 q_{ttt}^n + O(\Delta t^4) \tag{5.62}$$

Alternative forms for the second time derivative in (5.62) can be generated by multiple substitutions of (5.61), specifically

$$q_{tt} = (q_t)_t = -(f_x)_t = -(f_t)_x = -(f_q q_t)_x = (f_q f_q q_x)_x \tag{5.63}$$

assuming space/time differentiation interchangeable (euclidean space). The similar operation on the third time derivative in (5.62) produces

$$q_{ttt} = \ldots = (f_q f_q q_{t\,x})_x = -(f_q f_q (f_q q_x)_x)_x \tag{5.64}$$

From (5.61), the *jacobian* of the kinetic flux vector is $f_q \equiv u$. For u assumed constant (only for the moment), substitute $f_q \equiv u$ into (5.63–5.64) then introduce the coefficient set $\{\alpha, \beta, \gamma, \mu\}$ to account for the arbitrariness in (5.62–5.63). Absorbing the signs therein into these coefficients, collecting terms and dividing through by Δt, (5.62) takes the form

$$\begin{aligned} \frac{q^{n+1} - q^n}{\Delta t} + f_x^n - \frac{\Delta t}{2}\left(\alpha u q_t + \beta u^2 q_x\right)_x \\ - \frac{\Delta t^2}{6}\left(\gamma u^2 q_{t\,x} + \mu u^3 q_{x\,x}\right)_x + O(\Delta t^3) = 0 \end{aligned} \tag{5.65}$$

Formally take the limit of (5.65) as $\Delta t \Rightarrow \varepsilon > 0$, but retain the higher order terms. Noting that the first two terms in (5.65) combine to form $\mathcal{L}(q)$, and now removing the assumption $1/\mathrm{Pa} = 0$, the resultant conservation principle TS-modified mPDE replacing (5.61) is

$$\begin{aligned} \mathcal{L}^m(q) = \mathcal{L}(q) - \frac{\Delta t}{2}\frac{\partial}{\partial x}\left(\alpha u \frac{\partial q}{\partial t} + \beta u^2 \frac{\partial q}{\partial x}\right) \\ - \frac{\Delta t^2}{6}\frac{\partial}{\partial x}\left(\gamma u^2 \frac{\partial^2 q}{\partial x \partial t} + \mu u^3 \frac{\partial^2 q}{\partial x^2}\right) + O(\Delta t^3) = 0 \end{aligned} \tag{5.66}$$

The superscript m notation $\mathcal{L}^m(q)$ signifies addressing an mPDE throughout this text. Forming the GWS^h + θTS algorithm for (5.66) for arbitrary $\{\alpha, \beta, \gamma, \mu\}$, θ and basis degree k is straightforward. The elementary implementation is $k=1$, for which the μ term will vanish as three spatial derivatives cannot be supported. Baker and Kim (1987) resolved this $k=1$ basis limitation by introducing the alternative DOF $\Delta Q_j \equiv (Q_j - Q_{j\pm1})$ for the μ term, with sign coincident with that of u.

The $k=1$ basis GWS^h + θTS algorithm formed on mPDE (5.66) encompasses a wide range of independently derived CFD algorithms in both WS^h and FD classes, strictly dependent on definition of coefficient set $\{\alpha, \beta, \gamma, \mu)\}$ and θ. Table 5.2 details their summary wherein statement mPDE Galerkin denotes the Galerkin weak form algorithm written on (5.66).

Table 5.2 TS-generated *m*PDE theory categorization of independently derived WS^h, GWS^h and FD algorithms, from Baker and Kim (1987)

Algorithm name	θ	α	β	γ	μ
*m*PDE Galerkin	all	arbitrary	arbitrary	arb	arb
Galerkin	all	0	0	0	0
Donor cell FD	0	0	u/C	$1/C^2$	0
Lax–Wendroff FD	0	0	sgn(u)	0	0
Taylor Galerkin TG	0	0	1	1	0
Crank–Nicolson	0.5	0	0.5	1	0
Characteristic Galerkin	0	0	1	0	1
Swansea TG	0	0	1	0	0
Wahlbin	0	sgn(u)	2sgn(u)	0	0
Dendy	0	h·sgn(u)	h·sgn(u)	0	0
Raymond–Garder	0.5	$2v_o$sgn(u)/C	$2v_o$sgn(u)/C	0	0
Hughes SUPG (steady state)	—	0	sgn(u)	0	0
Euler Petrov-Galerkin	0	0	0	$(1-\nu)$	0
SUPG Petrov-Galerkin	0.5	sgn(u)	v·sgn(u)	$-v/2$	0
Warming-Beam FD	0	0	1	0	$-3(1-C)$
van Leer MUSCL	1	0	sgn(u)	0	-3
Galerkin least squares	all	2θ	2θ	0	0

Note: sgn(u) is the sign of u, $v_o = 1/\sqrt{15}$, $C \le v \le 1$, C is Courant number.

Realizing the very broad range of CFD formulation theoretical roots therein expressed, in concert with the *pervasive uniformity* visible, raises the fundamental query about existence of an *optimal* $GWS^h + \theta TS$ algorithm formed on a conservation principle TS-*m*odified *m*PDE. Sahu and Baker (2007) detail the required theory in one and n dimensions. Retaining the $k = 1$ basis choice and for arbitrary $\{\alpha, \beta, \gamma, \mu \equiv 0\}$ and θ, then substituting the Fourier mode solution (5.51) into the stencil of $GWS^h + \theta TS$ for (5.66) produces the amplification factor

$$g^h = \frac{\left(2+\gamma C^2 - 3(1-\theta)\beta C^2 - 6(1-\theta)D\right) + \left(1-\gamma C^2 + 3(1-\theta)\beta C^2 + 6(1-\theta)D\right)\cos m - i3C\left(\tfrac{1}{2}\alpha + (1-\theta)\right)\sin m}{\left(2+\gamma C^2 + 3\theta\beta C^2 + 6D\theta\right) + \left(1-\gamma C^2 - 3\theta\beta C^2 - 6D\theta\right)\cos m - i3C\left(\tfrac{1}{2}\alpha - \theta\right)\sin m} \tag{5.67}$$

In (5.67) $m \equiv \kappa h$ is non-dimensional wavenumber, the (uniform) mesh measure is h, Courant number is $C \equiv U\Delta t/h$ and $D \equiv \Delta t/Pah^2$ is the non-D diffusion term coefficient.

The precise statement of *m*PDE $GWS^h + \theta TS$ algorithm *phase speed error* in wavenumber space requires TS manipulations on amplification factor solutions. For D non-vanishing the amplification factor for the exact solution to (5.35) is

$$g_{exact} = \frac{q(x,(n+1)t)}{q(x,n\,t)} = e^{-\left(iCm + Dm^2\right)} \tag{5.68}$$

Expanding (5.68) in a TS to $O(m^7)$ in wave number space, $m \equiv \kappa h$, produces the complex function

$$
\begin{aligned}
g_{exact} = {} & 1 - \mathrm{i}Cm - \left(\frac{C^2}{2} + D\right)m^2 + \mathrm{i}C\left(\frac{C^2}{6} + D\right)m^3 + \left[C^2\left(\frac{C^2}{24} + \frac{D}{2}\right) + \frac{D^2}{2}\right]m^4 \\
& - \mathrm{i}C\left(\frac{C^4}{120} + \frac{DC^2}{6} + \frac{D^2}{2}\right)m^5 - \left[C^2\left(\frac{C^4}{720} + \frac{C^2D}{24} + \frac{D^2}{4}\right) + \frac{D^3}{6}\right]m^6 + O(m^7)
\end{aligned}
\tag{5.69}
$$

Generating the TS for (5.67), to compare to (5.69), requires it be multiplied through by the complex conjugate g^{h*} with the generated denominator then cleared via a TS of sufficiently high order. The resultant TS to $O(m^7)$ is (5.70), detailed on the following page.

Phase speed *error* for the *m*PDE $\mathrm{GWS}^h + \theta\mathrm{TS}$ algorithm is obviously the difference between g_{exact} and g^h which can be estimated using the TS expansions (5.69) and (5.70). As the theoretical focus is impact of parameter set $\{\alpha, \beta, \gamma\}$ along with C and θ in the *absence* of physical diffusion, the D terms in both expressions are now neglected. The resultant $O(m^7)$ TS for the $k = 1$ basis *m*PDE $\mathrm{GWS}^h + \theta\mathrm{TS}$ algorithm *phase speed error* is (5.71), also detailed on the next page. This establishes the theoretical basis for phase accuracy *optimization*.

Since the m^1 term is missing in (5.71), any (α, β, γ, θ, C) yields a discrete solution exhibiting $O(m^1)$ phase speed accuracy, which is also valid for any $D > 0$. For securing $O(m^2)$ accuracy the TS coefficient $[1 - 2\theta + (\alpha - \beta)]$ must vanish. Restricted to $\theta = 0.5$, the optimal $O(\Delta t^3)$ non-diffusive selection, it will vanish for any $\alpha = \beta$ and all C. Then defining $-\gamma/6 - 1/12 = 0$ causes the $O(m^3)$ TS term to vanish, hence $\gamma \equiv -0.5$ is *optimal*. For these θ and γ selections, the $O(m^4)$ TS coefficient $\alpha(1 - C^2)/24 = 0$ already vanishes for $\alpha = 0 = \beta$, also for $C = 1$, the exact equating of Δt and Δx on a uniform mesh.

In summary, the TS-modified *m*PDE weak form theory for $n = 1$ predicts the $k = 1$ basis *optimal* phase accuracy *m*odified Galerkin algorithm, hereafter denoted $m\mathrm{GWS}^h + \theta\mathrm{TS}$, results for the *m*PDE selections $\alpha = 0$, $\beta = 0$, $\gamma = -0.5$, $\theta = 0.5$ in (5.66). Algorithm *a posteriori* data thus generated will exhibit $O(m^3)$ phase accuracy for a uniform mesh independent of C. Realized solution accuracy of course degrades for larger C, and mesh non-uniformity will decrease phase accuracy by nominally one order in *m* for any algorithm.

Validation of theoretical prediction of optimal phase accuracy accrues to full Fourier content traveling gaussian wave *a posteriori* data analysis, recall Figure 5.5, for various *m*PDE parameter selections, Table 5.2. Besides standard GWS^h and Crank–Nicolson finite difference (CNFD), evaluated weak form algorithms include Raymond–Garder (RG, identical to SUPG), Figure 5.5, Taylor–Galerkin (TG), Galerkin Least Squares (GLS) and optimal $\gamma = -0.5\, m\mathrm{GWS}^h + \theta\mathrm{TS}$ (TWS). Following a 3 gaussian IC wavelength translation solution DOF distributions are compared in Figure 5.6 to the exact solution, the solid line.

All weak form algorithms are improvements over GWS^h and CNFD, Figure 5.5. However, only the $O(m^3)$ *optimal* γ mGWS^h (labeled TWS) solution maintains the gaussian peak and IC symmetry at the final time station. This performance is precisely due to *minimization* of phase dispersion error induced wake among the six algorithms tested.

$$
\begin{aligned}
g^h = {} & 1 - \mathrm{i}Cm + \left[\frac{1}{2}(\alpha-\beta-D)-\theta\right]C^2m^2 + \mathrm{i}\left[\begin{array}{l}\left(\frac{1}{4}\alpha(\alpha-\beta)+\frac{\gamma}{6}+(-\alpha+\beta)\theta+\theta^2\right)C^3\\ +CD\left(-\frac{\alpha}{2}+2\theta\right)\end{array}\right]m^3\\
& +C^2\left[\begin{array}{l} -\frac{\beta}{24}+C^2\left(\frac{\alpha^2}{8}(-\alpha+\beta)+\frac{\gamma}{6}\left(-\alpha+\frac{\beta}{2}\right)+\left(\alpha\left(\frac{3}{4}\alpha-\beta\right)+\frac{1}{4}\beta^2+\frac{1}{3}\gamma\right)\theta+\frac{3}{2}(-\alpha+\beta)\theta^2+\theta^3\right)\\ +D\left(\frac{-1}{12C^2}+\frac{\theta D}{C^2}+\left(\frac{\alpha^2}{4}+\frac{\gamma}{6}-2\alpha\theta+\beta\theta+3\theta^2\right)\right)\end{array}\right]m^4\\
& +\mathrm{i}C\left[\begin{array}{l} \frac{1}{180}+C^2\left(\frac{-\alpha\beta}{48}+\frac{\gamma}{72}+\frac{\beta\theta}{12}\right)+C^4\left\{\frac{-\alpha^3}{16}(\alpha-\beta)-\frac{\alpha\gamma}{4}\left(\frac{\alpha}{2}-\frac{\beta}{3}\right)-\frac{\gamma^2}{36}\right\}\\ +C^4\left\{\theta\left(\frac{\alpha^3}{2}-\frac{\alpha\beta}{4}(3\alpha-\beta)+\gamma\left(\frac{\alpha}{2}-\frac{\beta}{3}\right)\right)\right\}+C^4\theta^2\left\{\frac{-3\alpha}{2}\left(\alpha-\frac{3\beta}{2}\right)-\frac{3\beta^2}{4}-\frac{\gamma}{2}\right\}\\ +C^4\left(2\theta^3(\alpha-\beta)-\theta^4\right)\\ +D\left(\frac{-\alpha}{24}+\frac{\theta}{6}+(\alpha-3\theta)\theta D+C^2\left(\frac{\alpha^3}{8}+\frac{\alpha\gamma}{6}+\left(-\frac{3}{2}\alpha+\beta\right)\alpha\theta-\frac{2}{3}\gamma\theta+\left(\frac{9}{2}\alpha-3\beta-4\theta\right)\theta^2\right)\right)\end{array}\right]m^5\\
& +C^2\left[\begin{array}{l} \frac{-\alpha}{180}-\frac{\beta}{720}+\frac{\theta}{90}+C^2\left\{\frac{\alpha\beta}{12}\left(\frac{\alpha}{8}-\theta\right)-\frac{\gamma}{72}(\alpha-\beta)+\frac{\gamma\theta}{36}+\frac{\beta\theta}{8}\left(\frac{\beta}{3}+\theta\right)\right\}\\ +C^4\left\{\frac{\alpha^5}{32}-\frac{\alpha^4\beta}{32}+\frac{\alpha^2\gamma}{4}\left(\frac{\alpha}{3}-\frac{\beta}{4}\right)+\frac{\gamma^2}{24}\left(\alpha-\frac{\beta}{3}\right)\right\}\\ +C^4\theta\left\{\frac{-5\alpha^4}{16}+\frac{\alpha^2\beta}{2}\left(\alpha-\frac{3\beta}{16}\right)+\frac{\gamma}{2}\left(-\alpha(\alpha-\beta)-\frac{1}{6}(\beta^2+\gamma)\right)\right\}\\ +\theta^2C^4\left\{\frac{5\alpha^3}{4}+9\alpha\beta\left(-\frac{\alpha}{4}+\frac{\beta}{8}\right)-\frac{\beta^3}{8}+\gamma\left(\alpha-\frac{3\beta}{4}\right)\right\}\\ +C^4\theta^3\left\{\frac{-1}{2}(5\alpha^2+3\beta^2)+4\alpha\beta-\frac{2}{3}\gamma+\frac{5}{2}\theta(\alpha-\beta)\right\}-C^4\theta^5\\ +D\left(\begin{array}{l}\frac{1}{C^2}\left(\frac{-1}{360}+\frac{1}{6}D\theta-D^2\theta^2\right)+\frac{\alpha^2}{48}+\frac{\gamma}{36}-\frac{1}{6}\alpha\theta-\frac{3}{4}D\alpha^2\theta+\frac{1}{6}\beta\theta-\frac{1}{3}D\gamma\theta\\ +\frac{\theta^2}{4}+\frac{9}{2}D\alpha\theta^2-\frac{3}{2}D\beta\theta^2-6D\theta^3\end{array}\right)\end{array}\right]m^6\\
& +C^2\left[DC^2\left(\begin{array}{l}-\frac{\alpha^4}{16}-\frac{1}{8}\alpha^2\gamma-\frac{\gamma^2}{36}+\alpha^3\theta-\frac{3}{4}\alpha^2\beta\theta+\alpha\gamma\theta-\frac{1}{3}\beta\gamma\theta-\frac{9}{2}\alpha^2\theta^2+\frac{9}{2}\alpha\beta\theta^2-\frac{3}{4}\beta^2\theta^2\\ -\frac{3}{2}\gamma\theta^2+8\alpha\theta^3-6\beta\theta^3-5\theta^4\end{array}\right)\right]m^6+O(m^7)
\end{aligned}
\tag{5.70}
$$

$$
\begin{aligned}
c^h = & -[1+2\theta+(\alpha-\beta)]\frac{\mathrm{C}^2 m^2}{2} + \mathrm{i}\left[\frac{1}{6}-\frac{1}{4}\alpha(\alpha-\beta)-\frac{\gamma}{6}-(-\alpha+\beta)\theta-\theta^2\right]\mathrm{C}^3 m^3 \\
& + \mathrm{C}^2\left[\frac{\mathrm{C}^2}{24}+\frac{\beta}{24}-\mathrm{C}^2\left(\frac{\alpha^2}{8}(-\alpha+\beta)+\frac{\gamma}{6}\left(-\alpha+\frac{\beta}{2}\right)+\left(\alpha\left(\frac{3}{4}\alpha-\beta\right)+\frac{1}{4}\beta^2+\frac{1}{3}\gamma\right)\theta+\frac{3}{2}(-\alpha+\beta)\theta^2+\theta^3\right)\right]m^4 \\
& + \mathrm{iC}\left[\begin{array}{l}
-\dfrac{\mathrm{C}^4}{120}-\dfrac{1}{180}-\mathrm{C}^2\left(\dfrac{-\alpha\beta}{48}+\dfrac{\gamma}{72}+\dfrac{\beta\theta}{12}\right)-\mathrm{C}^4\left\{\dfrac{-\alpha^3}{16}(\alpha-\beta)-\dfrac{\alpha\gamma}{4}\left(\dfrac{\alpha}{2}-\dfrac{\beta}{3}\right)-\dfrac{\gamma^2}{36}\right\} \\
-\mathrm{C}^4\left\{\theta\left(\dfrac{\alpha^3}{2}-\dfrac{\alpha\beta}{4}(3\alpha-\beta)+\gamma\left(\dfrac{\alpha}{2}-\dfrac{\beta}{3}\right)\right)\right\}-\mathrm{C}^4\theta^2\left\{\dfrac{-3\alpha}{2}\left(\alpha-\dfrac{3\beta}{2}\right)-\dfrac{3\beta^2}{4}-\dfrac{\gamma}{2}\right\} \\
-\mathrm{C}^4\left(2\theta^3(\alpha-\beta)-\theta^4\right)
\end{array}\right]m^5 \\
& + \mathrm{C}^2\left[\begin{array}{l}
-\dfrac{\mathrm{C}^4}{720}+\dfrac{\alpha}{180}+\dfrac{\beta}{720}-\dfrac{\theta}{90}-\mathrm{C}^2\left\{\dfrac{\alpha\beta}{12}\left(\dfrac{\alpha}{8}-\theta\right)-\dfrac{\gamma}{72}(\alpha-\beta)+\dfrac{\gamma\theta}{36}+\dfrac{\beta\theta}{8}\left(\dfrac{\beta}{3}+\theta\right)\right\} \\
-\mathrm{C}^4\left\{\dfrac{\alpha^5}{32}-\dfrac{\alpha^4\beta}{32}+\dfrac{\alpha^2\gamma}{4}\left(\dfrac{\alpha}{3}-\dfrac{\beta}{4}\right)+\dfrac{\gamma^2}{24}\left(\alpha-\dfrac{\beta}{3}\right)\right\} \\
-\mathrm{C}^4\theta\left\{\dfrac{-5\alpha^4}{16}+\dfrac{\alpha^2\beta}{2}\left(\alpha-\dfrac{3\beta}{16}\right)+\dfrac{\gamma}{2}\left(-\alpha(\alpha-\beta)-\dfrac{1}{6}\left(\beta^2+\gamma\right)\right)\right\} \\
-\theta^2\mathrm{C}^4\left\{\dfrac{5\alpha^3}{4}+9\alpha\beta\left(-\dfrac{\alpha}{4}+\dfrac{\beta}{8}\right)-\dfrac{\beta^3}{8}+\gamma\left(\alpha-\dfrac{3\beta}{4}\right)\right\} \\
-\mathrm{C}^4\theta^3\left\{\dfrac{-1}{2}\left(5\alpha^2+3\beta^2\right)+4\alpha\beta-\dfrac{2}{3}\gamma+\dfrac{5}{2}\theta(\alpha-\beta)\right\}-\mathrm{C}^4\theta^5
\end{array}\right]m^6+O(m^7)
\end{aligned}
\tag{5.71}
$$

The theory for *optimal* $m\mathrm{GWS}^h+\theta\mathrm{TS}$ algorithms is extended to $k=2$, 3 basis implementations, Chaffin and Baker (1995). The theory $k=2$, 3 predictions remain $\alpha=0=\beta$ and $\theta=0.5$, and determine optimal $\gamma=-0.4$ for $k=2$ and $\gamma=-0.333$ (i.e., 1/3) for $k=3$. Stencil quantification of phase accuracy order in m is not possible due to $k=2$, 3 basis non-vertex DOFs not connecting across element boundaries.

The *m*PDE theory enables quantified prediction of algorithm spectral *phase speed error distribution*. Selecting $\theta=0.5=\mathrm{C}$, theory predicted distributions are abscissa logarithm scaled in *integer mesh* units $\mathrm{n}\Delta x$ to expose distinguishing features in the *all important* short wavelength (n small) spectrum, Figure 5.7. The non-Galerkin algorithms evaluated (with graph labels) are $O(\Delta x^2)$ Crank–Nicolson finite difference (FD), $O(\Delta x^1)$) upwind finite difference (UW), $O(\Delta x^3)$ upwind FV QUICK3 (Q3) and $O(\Delta x^5)$ upwind FV QUICK5 (Q5). The graphed distribution comparisons are completed with $1\le k\le 3$ $\mathrm{GWS}^h+\theta\mathrm{TS}$ (labeled GL, GQ, GC), and $1\le k\le 3$ optimal $\gamma=-(0.5, 0.4, 0.333)$ $m\mathrm{GWS}^h+\theta\mathrm{TS}$ (labeled TL, TQ, TC).

The theory prediction of error incurred in *one time step*, Figure 5.7, confirms *all*(!) algorithms possess 100% phase speed error at wavelength $\lambda=2\Delta x$, Figure 5.1 left anticipated. Further, each asymptotes to negligible phase speed error for $\lambda>10\Delta x$, recall Figure 5.1 right. Focusing on the algorithm label column at abscissa coordinate $\sim 2.5\Delta x$, poorest phase speed accuracy is shared by the FD and upwind algorithms with each Δt step generating ~70% error. The two FV QUICK algorithms are proportional improvements with order, error ~50%. The $k=1$, 2, 3 GWS^h algorithms generate proportional improvement with order error level reductions to ~30, 15, 5%. The *key* observable is $k=1$, 2 basis *optimal* γ $m\mathrm{GWS}^h$ algorithms exhibit the phase accuracy distributions of the next *higher* degree k basis GWS^h algorithm, with $k=3$ basis $m\mathrm{GWS}^h$ error nominally zero! (Note: the $m\mathrm{GWS}^h$ optimal γ magnitudes table listed in Figure 5.7 have the 1/6 multiplier in (5.66).)

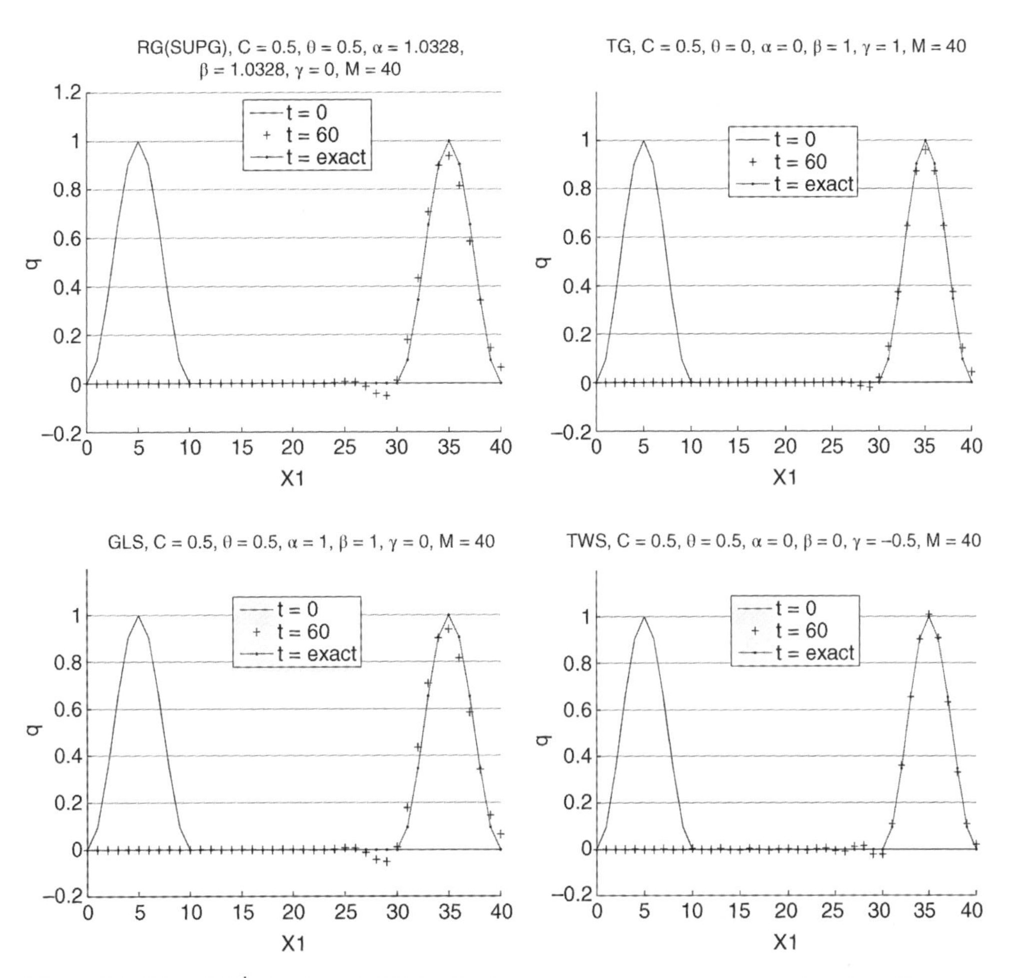

Figure 5.6 Select WS^h algorithm DOF distributions, pure convection of a gaussian, three IC wavelength translation, solid line is IC and exact solution, $\theta = 0.5$, $C = 0.5$: top left RG(SUPG); top right TG; bottom left GLS; bottom right *optimal* γ mGWS$^h + \theta$TS (TWS)

The *m*PDE theory phase speed spectral resolution data, Figure 5.7, predict *precisely* the algorithm comparative *a posteriori* DOF distributions for the traveling gaussian verification, Figures 5.5 and 5.6. The theory is readily extended to prediction of amplification factor departure from unity, recall (5.67), which quantifies algorithm spectral distribution of generated *artificial diffusion error*. For the selected algorithms, the *m*PDE theory prediction results are graphed in Figure 5.8 for abscissa logarithmic scaled in mesh units nΔx for n small resolution. Theory predictions again correspond to error incurred in only *one* time step of an algorithm for $\theta = 0.5 = C$.

The Crank–Nicolson (FD) and $k = 1$, 2, 3 basis GWS^h algorithms (labeled centered fd/fe) generate no artificial diffusion since $g^{h*}g^h = 1$ for all λ. The upwind FD algorithm (UW, always commercial CFD code available) is *grossly* diffusive leaving less than 20% of the original amplitude in the $2\Delta x \le \lambda \le 3\Delta x$ range after only one time step! The optimal γ, $\alpha = 0$, $\beta > 0$ mGWS$^h k = 1$ basis algorithm generates this dissipation level at $2\Delta x$ *only* for

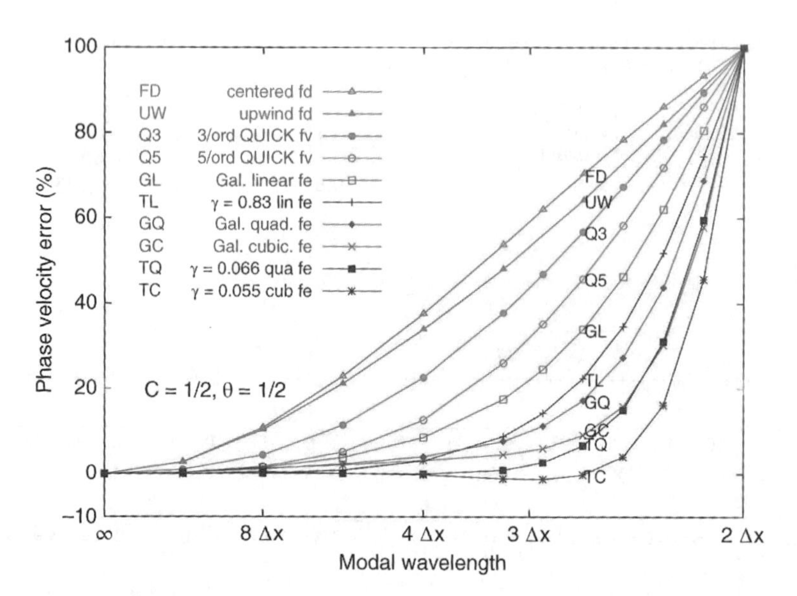

Figure 5.7 Spectral distributions of $n=1$ phase velocity (speed) error per Δt step, selected FD, FV, $1 \leq k \leq 3$ basis GWS^h and optimal γ mGWS^h $1 \leq k \leq 3$ basis algorithms, $\theta=0.5$, $C=0.5$, from Chaffin and Baker (1995). *Source:* Chaffin, D.J. and Baker, A.J. 1995. Reproduced with permission of Wiley

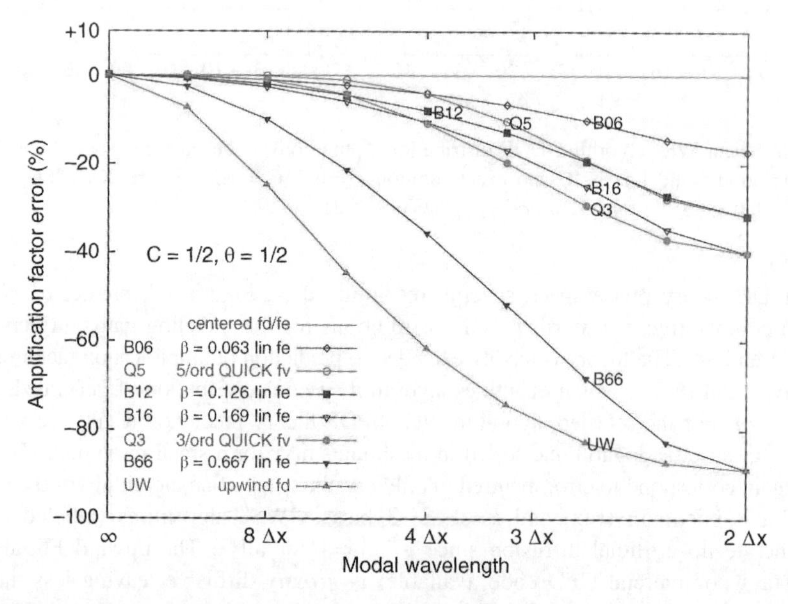

Figure 5.8 Amplification factor error spectral distribution per Δt step for select FD, FV, $1 \leq k \leq 3$ basis GWS^h and optimal γ mGWS^h $1 \leq k \leq 3$ basis algorithms, $\theta=0.5$, $C=0.5$, from Chaffin and Baker (1995). *Source:* Chaffin, D.J. and Baker, A.J. 1995. Reproduced with permission of Wiley

$\beta = 2(0.667) = 4/3$, five times larger(!) than Raymond–Garder theory parameter $\beta \equiv 1/\sqrt{15}$. For this solution, labeled B66, the 2 multiplier comes from the definition in (5.66).

The FV upwind QUICK third- and fifth-order schemes (labeled Q3, Q5) were derived to overcome the UW FD algorithm totally excess artificial diffusion. The dissipation level of Q3 and Q5 FV algorithms at $2\Delta x$ are emulated by $k = 1$ basis $m\text{GWS}^h$ algorithm for $\beta = 2(0.169) = 0.338$ and $\beta = 2(0.126) = 0.252$, the latter essentially identical to Raymond–Garder optimal parameter $\beta \equiv 1/\sqrt{15}$. Finally, defining $\beta = 2(0.063) = 0.126$ generates $< 20\%$ signal dissipation per Δt at $2\Delta x$, decreasing to $\sim 5\%$ at $3\Delta x$ and vanishes for all $\lambda > 4\Delta x$.

5.8 Optimally Phase Accurate $m\text{GWS}^h + \theta\text{TS}$ in n Dimensions

Multi-dimensional mPDE theory predicts the sought $k = 1$ basis *optimally phase accurate* $m\text{GWS}^h + \theta\text{TS}$ algorithm. In tensor index notation, the scalar transport conservation principle departure point is n-D (5.61)

$$\mathcal{L}(q) = \frac{\partial q}{\partial t} + u_j \frac{\partial q}{\partial x_j} \equiv q_t + \frac{\partial f_j}{\partial x_j} = 0, \quad f_j \equiv u_j q \tag{5.72}$$

For TE denoting truncation error, the fourth-order TS remains (5.62), the operations paralleling (5.63–5.64) are direct extensions, and in concert generate the replacement for (5.65)

$$\begin{aligned} &\frac{q^{n+1} - q^n}{\Delta t} + \frac{\partial f_j}{\partial x_j} - \frac{\Delta t}{2} \frac{\partial}{\partial x_j} \left(\alpha \frac{\partial f_j}{\partial q} \frac{\partial q}{\partial t} + \beta \frac{\partial f_j}{\partial q} \frac{\partial f_k}{\partial x_k} \right) \\ &- \frac{\Delta t^2}{6} \frac{\partial}{\partial x_j} \left(\gamma \frac{\partial f_j}{\partial q} \frac{\partial}{\partial x_k} \left(\frac{\partial f_k}{\partial q} \frac{\partial q}{\partial t} \right) + \mu \frac{\partial f_j}{\partial q} \frac{\partial}{\partial x_k} \left(\frac{\partial f_k}{\partial q} \frac{\partial f_l}{\partial x_l} \right) \right) + \text{TE} = 0 \end{aligned} \tag{5.73}$$

Taking the limit $\Delta t \Rightarrow \varepsilon > 0$ and retaining the higher-order terms leads to the TS *m*odified n-dimensional scalar transport conservation principle mPDE replacement for (5.72) as

$$\begin{aligned} \mathcal{L}^m(q) = \mathcal{L}(q) &- \frac{\Delta t}{2} \frac{\partial}{\partial x_j} \left[\alpha u_j \frac{\partial q}{\partial t} + \beta u_j u_k \frac{\partial q}{\partial x_k} \right] \\ &- \frac{\Delta t^2}{6} \frac{\partial}{\partial x_j} \left[\gamma u_j u_k \frac{\partial^2 q}{\partial x_k \partial t} + \mu u_j u_k \frac{\partial}{\partial x_k} \left(u_l \frac{\partial q}{\partial x_l} \right) \right] + O(\Delta t^3) \end{aligned} \tag{5.74}$$

on which the theoretical analysis is based.

Unfortunately, the theory algebraic operational details are tractable *only* for the restriction $n = 2$, $1/\text{Pa} = 0$ and u_j a constant. Therefore, the mPDE rectangular cartesian resolution of (5.74) admitting Fourier modal analysis completion is

$$\begin{aligned} \mathcal{L}^m(q) = &\frac{\partial q}{\partial t} + u \frac{\partial q}{\partial x} + v \frac{\partial q}{\partial y} - \frac{\alpha \Delta t}{2} \left[\frac{\partial}{\partial x} \left(u \frac{\partial q}{\partial t} \right) + \frac{\partial}{\partial y} \left(v \frac{\partial q}{\partial t} \right) \right] \\ &- \frac{\beta \Delta t}{2} \left[\frac{\partial}{\partial x} \left(u^2 \frac{\partial q}{\partial x} \right) + \frac{\partial}{\partial x} \left(uv \frac{\partial q}{\partial y} \right) + \frac{\partial}{\partial y} \left(vu \frac{\partial q}{\partial x} \right) + \frac{\partial}{\partial y} \left(v^2 \frac{\partial q}{\partial y} \right) \right] \\ &- \frac{\gamma \Delta t^2}{6} \left[\begin{array}{l} \dfrac{\partial}{\partial x} \left(u^2 \dfrac{\partial}{\partial x} \dfrac{\partial q}{\partial t} \right) + \dfrac{\partial}{\partial x} \left(uv \dfrac{\partial}{\partial y} \dfrac{\partial q}{\partial t} \right) \\ + \dfrac{\partial}{\partial y} \left(vu \dfrac{\partial}{\partial x} \dfrac{\partial q}{\partial t} \right) + \dfrac{\partial}{\partial y} \left(v^2 \dfrac{\partial}{\partial y} \dfrac{\partial q}{\partial t} \right) \end{array} \right] + O(\Delta t^3) = 0 \end{aligned} \tag{5.75}$$

For cited restrictions the exact solution to (5.72) is

$$q(x, y, t) = q_0(x, y, t_0)e^{-i(\kappa_1(x-ut)+\kappa_2(y-vt))} \tag{5.76}$$

and the analytical amplification factor definition remains (5.68). Since $\kappa_1 = \kappa\cos(\eta)$ and $\kappa_2 = \kappa\sin(\eta)$, the preferred formal definition is

$$g_{exact} = e^{-im\left(C_x \cos\eta + C_y \sin\eta\right)} \tag{5.77}$$

where (C_x, C_y) denotes the cartesian resolution of the *Courant vector* **C**, the magnitude of which remains the non-D time step. The resultant analytical solution TS expansion complete to $O(m^4)$ is

$$\begin{aligned} g_{exact} &= 1 - i[C_x\cos(\eta) + C_y\sin(\eta)]m \\ &\quad - \frac{1}{2}[C_x\cos(\eta) + C_y\sin(\eta)]^2 m^2 \\ &\quad + \frac{i}{6}[C_x\cos(\eta) + C_y\sin(\eta)]^3 m^3 + O(m^4) \end{aligned} \tag{5.78}$$

The amplification factor g^h for the discrete approximate solution $q^h(x, y, t)$ to *m*PDE (5.75) starts with assembly of $m\text{GWS}^h + \theta\text{TS}$ at the generic node (X_j, Y_k). This produces a stencil technically similar to the $n = 1$ illustrations via the analogous Fourier representations

$$\begin{aligned} Q_{j-1,k-1} &= Q(x - \Delta x, y - \Delta y) \equiv Q_{j,k}e^{-i(\kappa_1\Delta x)}e^{-i(\kappa_2\Delta y)} \\ Q_{j+1,k+1} &= Q(x + \Delta x, y + \Delta y) \equiv Q_{j,k}e^{i(\kappa_1\Delta x)}e^{i(\kappa_2\Delta y)} \end{aligned} \tag{5.79}$$

for the wavenumber definitions $\kappa_1 = \kappa\cos(\eta)$ and $\kappa_2 = \kappa\sin(\eta)$, recall Figure 5.2. For $k = 1$ bilinear tensor product (TP) basis implementation, (3.26), the $m\text{GWS}^h + \theta\text{TS}$ algorithm amplification factor solution is

$$g^h = \frac{\begin{array}{c}\left(b_{j-1}e^{-i\kappa_1} + b_j + b_{j+1}e^{+i\kappa_1}\right)e^{-i\kappa_2} + \left(b_{j-1}e^{-i\kappa_1} + b_j + b_{j+1}e^{+i\kappa_1}\right) \\ +\left(b_{j-1}e^{-i\kappa_1} + b_j + b_{j+1}e^{+i\kappa_1}\right)e^{+i\kappa_2}\end{array}}{\begin{array}{c}\left(a_{j-1}e^{-i\kappa_1} + a_j + a_{j+1}e^{+i\kappa_1}\right)e^{-i\kappa_2} + \left(a_{j-1}e^{-i\kappa_1} + a_j + a_{j+1}e^{+i\kappa_1}\right) \\ +\left(b_{j-1}e^{-i\kappa_1} + b_j + b_{j+1}e^{+i\kappa_1}\right)e^{+i\kappa_2}\end{array}} \tag{5.80}$$

Theoretical analysis completion for (5.80) is tractable *only* for a uniform mesh, hence set $\kappa_1\Delta x \equiv m \equiv \kappa_2\Delta y$, where m remains the non-D wavenumber and $\Delta x \equiv h \equiv \Delta y$ in (5.79). The resultant TS expansion of (5.80) to $O(m^4)$ is archived in Sahu and Baker (2007, Appendix A). The analysis endpoint is the spectral distribution in wavenumber space of the $m\text{GWS}^h + \theta\text{TS}$ algorithm phase velocity *vector* error, the *n*-dimensional complement to (5.71).

The error definition remains $c^h \equiv g_{exact} - g^h$, again approximately quantified via TS expansions. This solution to $O(m^4)$ is archived, Sahu and Baker (2007, Appendix B). The significant non-vanishing term is with $O(m^2)$ multiplier, hence optimizing *phase velocity error* for an $m\text{GWS}^h + \theta\text{TS}$ algorithm requires

$$\left|c^h\right| = \begin{pmatrix} \dfrac{C_x^2}{2} + C_xC_y + \dfrac{C_y^2}{2} - \dfrac{1}{2}(\alpha + \beta)\left(C_x^2 + C_y^2\right) \\ -\alpha C_xC_y - C_x^2\theta - 2C_xC_y\theta - C_y^2\theta \end{pmatrix} m^2 + O(m^3) \tag{5.81}$$

vanish identically. This occurs for $\alpha \equiv 0 \equiv \beta$, $\theta \equiv ½$, and γ remains arbitrary. Thereby, in n dimensions, the $k = 1$ TP basis $\text{GWS}^h + \theta\text{TS}$ algorithm for $\theta \equiv 0.5$ in (5.72) is phase velocity accurate to $O(m^3)$ in non-D wavenumber space.

The optimizing opportunity rests on determination of whether the phase velocity wavenumber order can be improved for an alternative (α, β, γ, θ) selection $m\text{GWS}^h + \theta\text{TS}$ algorithm. This requires probing candidates for phase velocity dependence on phase angle η, recall Figure 5.2, as well as Courant vector $\mathbf{C}$ resolution and magnitude. A manipulation of the (α, β) coefficient set is required to recover multi-dimensional Brooks–Hughes SUPG algorithm, also Raymond–Garder, to compensate for these weak form theories not derived in n-D.

Since velocity is a vector and for $n > 1$, $k = 1$ basis domains possess at least two measures, recall Figure 3.4. This requires definition of a *local time scale* selected as $\Delta t \equiv |u_i|h_i/u^2 = |\hat{u}_i|h_i/|\mathbf{u}|$, a non-negative number, summation on the repeated index with h_i the mesh measure in the direction of u_i. The resulting modifications to the (α, β) terms in (5.75), for superscript "hat" denoting unit vector, are

$$\begin{aligned} \frac{\Delta t}{2}(\alpha u_j) &\equiv \frac{\alpha h_i|u_i|u_j}{2u^2} = \frac{\alpha h_i|\hat{u}_i|\hat{u}_j}{2} \\ \frac{\Delta t}{2}(\beta u_j u_k) &\equiv \frac{\beta h_i|u_i|u_j u_k}{2u^2} = \frac{\beta h_i|\hat{u}_i|\hat{u}_j u_k}{2} \end{aligned} \tag{5.82}$$

The β term dimensionality is $D(L^2\tau^{-1})$, dissipative *only* for tensor product $u_j u_k$ non-negative, with that for the α term $D(L)$.

Sahu and Baker (2007) define the wave vector angle sample space as $\eta = \pi/2$, $5\pi/8$, $3\pi/4$, π, $5\pi/4$ and $11\pi/8$, Figure 5.9. Included thereon are the coordinate pairs producing Courant vector magnitude variation $0.19 \leq |\mathbf{C}| \leq 0.41$ for an imposed solid body rotation velocity vector field $\mathbf{u} = (r\omega\Delta t/h)\hat{\mathbf{e}}$, where ω is angular velocity with $\hat{\mathbf{e}}$ the unit vector tangent to the η coordinate. For these definitions, the theory predicts the range of $m\text{GWS}^h + \theta\text{TS}$ algorithm phase velocity *error* is angular quadrant independent, as do comparative theories, Christon (1999), Christon *et al.* (2004).

Along with Crank–Nicolson (CN), the WS^h, GWS^h and $m\text{GWS}^h + \theta\text{TS}$ algorithm formulations evaluated, with graph label and α, β, γ, θ selection denoted within (•), are:

Galerkin (GWS: 0, 0, 0, 1/2)
Taylor–Galerkin (TG: 0, 1, 1, 0)
SUPG/Raymond–Garder(SUPG(RG)); $1/\sqrt{15}$, $1/\sqrt{15}$, 0, 1/2)
Galerkin Least-Squares (GLS: 2θ, 2θ, 0, 1/2)
optimal γ $m\text{GWS} + \theta\text{TS}$ (TWS: 0, 0, −1/2, 1/2)

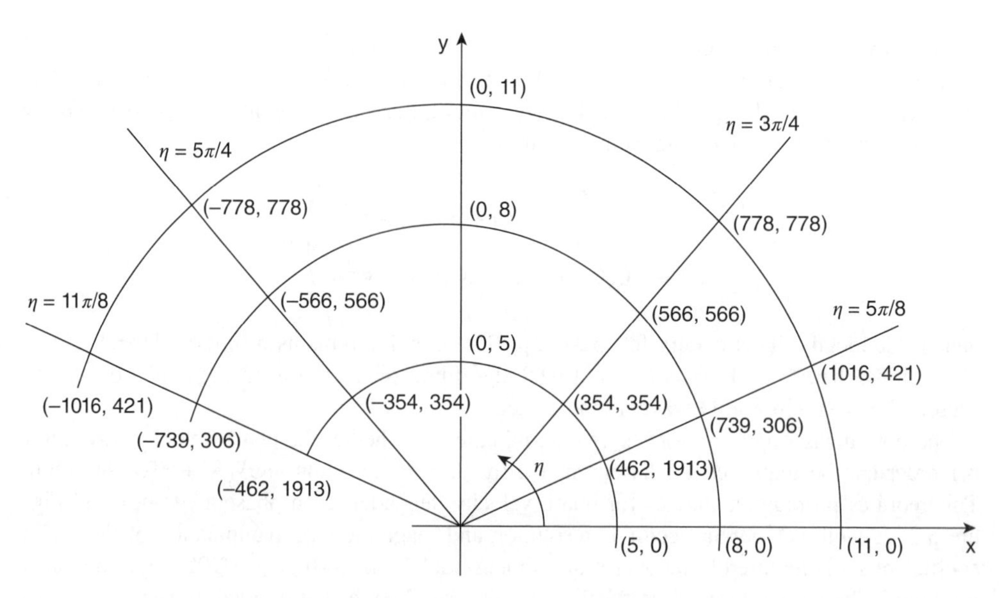

Figure 5.9 Sample space of wave vector angles and coordinates for theoretical prediction of phase velocity error dependence on m and **C**

Figure 5.10 graphs computed solution spectrum essence for a single wave vector angle η and |**C**|; Sahu and Baker (2007) report solutions for the complete range of η and Courant vector magnitude |**C**|. The integers scaling the abscissa are TS truncation error order exponent $O(m^n)$, $1 \le n \le 5$, in non-D wavenumber $m = \kappa\Delta x$. The ordinate is the TS $O(m^n)$ term exponent n.

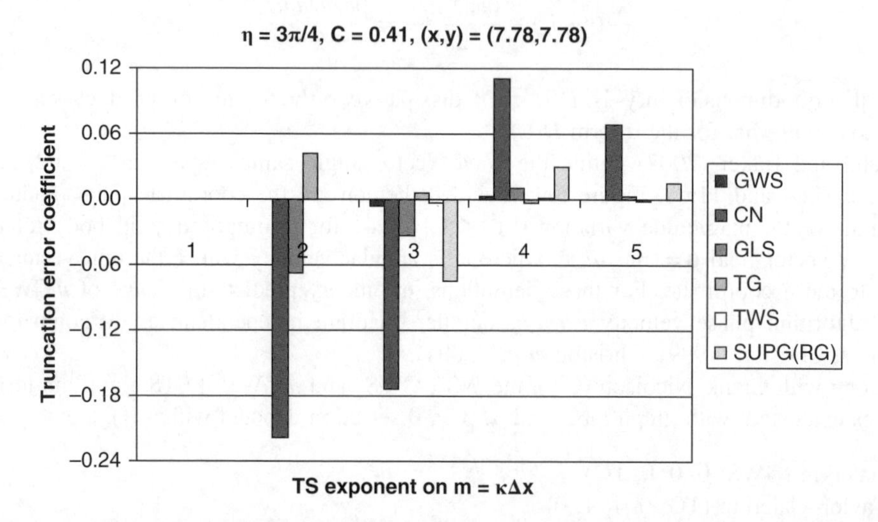

Figure 5.10 Theory predicted Taylor series wavenumber order dependence, CN and selected WS^h, GWS^h, $mGWS^h + \theta TS$ algorithms, $n = 2$ pure convection, $\eta = 5\pi/8$, $|\mathbf{C}| = 0.41$, from Sahu and Baker (2007). *Source:* Sahu, S. and Baker, A.J. 2007. Reproduced with permission of Wiley

The CN, GLS and TG algorithms each possess non-vanishing $O(m^2)$ TS coefficients, hence their phase velocity error is one order larger than that of the GWS^h algorithm. Among those evaluated, the CN algorithm possesses the extremum TS coefficient magnitude across the order spectrum. In distinction, the TS $O(m^2)$ coefficients vanish for mGWS^h (TWS) and SUPG(RG), with mGWS^h uniformly exhibiting minimal coefficient magnitudes throughout $O(m^3) - O(m^5)$.

The large size of the CN, SUPG/RG and TG $O(m^2) - O(m^3)$ coefficients, Figure 5.10, clouds distinctions between the GWS^h and *optimal* γ mGWS^h algorithms. Figure 5.11 clarifies this comparison for $|\mathbf{C}| = 0.41$ for all wave vector angles tested. The $n = 1$ prediction of $\gamma = -0.5$ optimal for the $k = 1$ basis implementation of mGWS$^h + \theta$TS transcends to the n-dimensional algorithm, generating minimal TS truncation error coefficients across the entire wavenumber TS order spectrum. The mPDE theory n-dimensional consistency is thus firmly quantified via the $\gamma = -0.5$ mGWS$^h + \theta$TS algorithm phase velocity predicted $O(m^4)$ parallel to principal coordinates in cartesian space (the $\pi/2$ data).

This analytical proof in concert with the *a posteriori* data in Figure 5.7 confirm that $1 \leq k \leq 3$ basis mGWS$^h + \theta$TS algorithms applied to the TS-theory mPDE replacement for (5.72)

$$\mathcal{L}^m(q) \equiv [m]\frac{\partial q}{\partial t} + u_j\frac{\partial q}{\partial x_j} = 0, [m] \equiv \left[1 - \frac{\gamma\, \Delta t^2}{6}\frac{\partial}{\partial x_j}\left(u_j u_k \frac{\partial}{\partial x_k}\right)\right] \tag{5.83}$$

will generate mGWS^h weak form solutions exhibiting *optimal phase velocity accuracy* in n dimensions. The mPDE theory predicts that $\gamma = -(1/2,\ 1/2.5,\ 1/3)$ are the *optimal* definitions for $k = 1,\ 2,\ 3$ basis implementation, hence (5.83) contains *no* arbitrary coefficients.

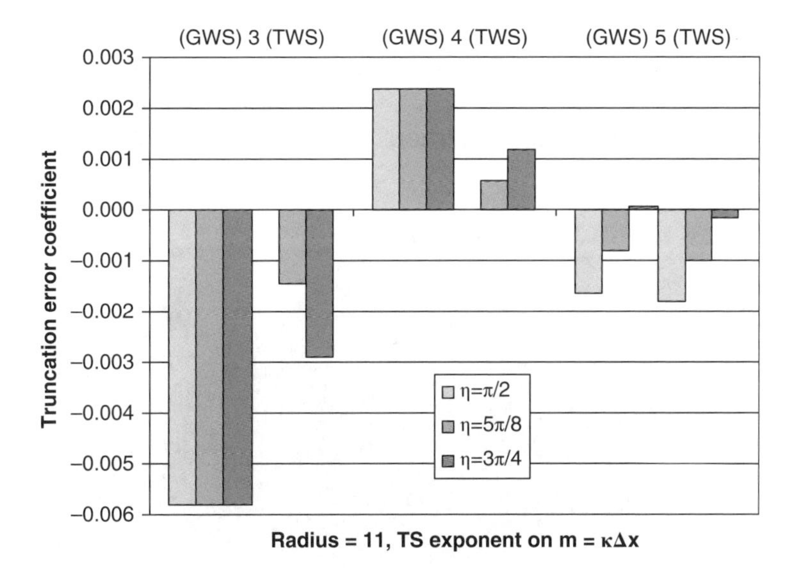

Figure 5.11 Theoretical TS wavenumber error order for GWS^h and $\gamma = -0.5$ mGWS^h $k = 1$ basis, $\theta = 0.5$ algorithms, $n = 2$ pure convection, various wave angles η, $|\mathbf{C}| = 0.41$, from Sahu and Baker (2007). *Source:* Sahu, S. and Baker, A.J. 2007. Reproduced with permission of Wiley

The definitive $m\text{GWS}^h + \theta\text{TS}$ theory *validation* is the $n = 2$ convected gaussian IC termed the *rotating cone*. The IC is the gaussian, Figure 5.5, rotated around its centroidal axis, perspective graphed in Figure 5.12. The analytical solution is *exact* preservation of the IC distribution as it is convected around the horizontal plane by the imposed *rotational* velocity vector field $\mathbf{u}(x,y) = (r\omega\Delta t/h)\hat{\mathbf{e}}$ which is identical with the Courant vector distribution $\mathbf{C}(x,y)$.

The rotated gaussian IC spectrum possesses full Fourier content, the wavenumber requisite for acquiring precisely definitive *a posteriori* data. Solutions generated by weak form WS^h, GWS^h and $m\text{GWS}^h$ algorithms, also CNFD, for elapsed time $n\Delta t$ execution required to return the IC to its original location, Figure 5.12, are perspective graphed in Figure 5.13. Each plot lists solution peak mass fraction at the final time in percent (the IC is 100%), also the extremum *negative* mass fraction generated by *phase velocity error*. This is visualized by the *dispersion error*-induced dark contoured wake in the horizontal plane for clockwise travel of the IC. The final solution distortion from axisymmetric is quantified by contour decile bands.

The CNFD solution is totally distorted by phase velocity error, the consequence of being an $O(m^1)$ algorithm. The peak mass fraction is less than half the IC and the wake extremum is huge (−28%). The GWS^h algorithm solution is moderately accurate, with dispersion error distorting the peak to 94% and inducing a wake extremum (−15%) half that of CNFD.

The SUPG/RG, GLS and TG algorithms in this order reduce the dispersion wake extrema to −12, −5, −3%, each an improvement on the GWS^h solution. However, each artificially diffuses the peak to 89, 83, 82% respectively. The *m*PDE theory optimal γ $m\text{GWS}^h + \theta\text{TS}$ algorithm (labeled TWS) solution preserves the peak at 100%, and essentially retains the IC axisymmetry while generating a comparable dispersion error wake extremum of −6%.

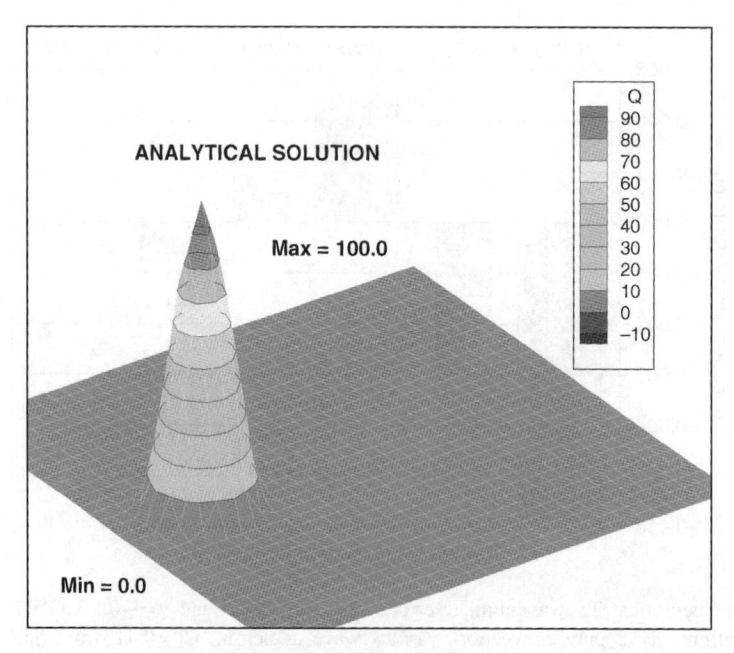

Figure 5.12 The rotated gaussian IC and analytical solution for the *rotating cone* verification problem

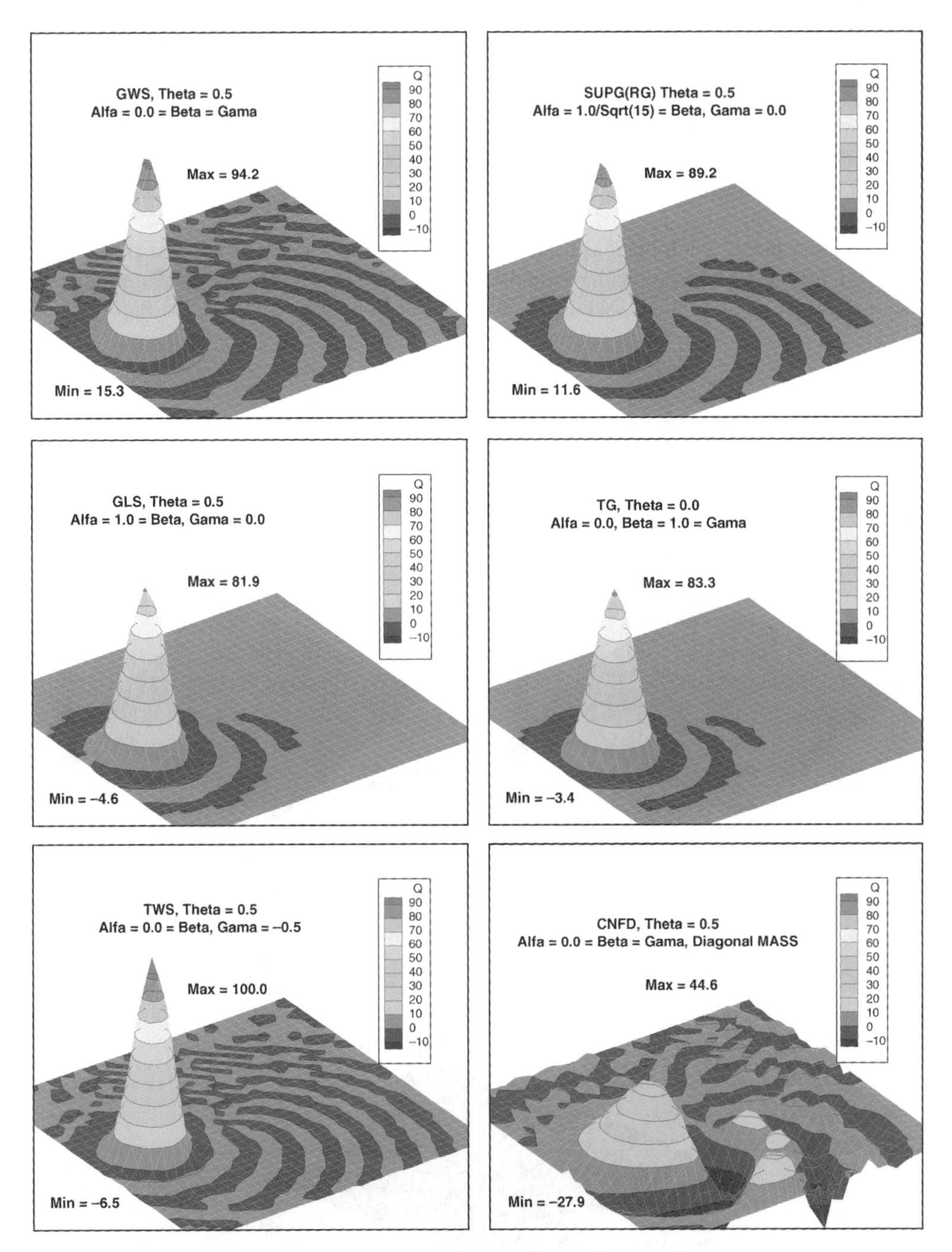

Figure 5.13 Rotating cone *a posteriori* data for select WS^h, GWS^h, $m\mathrm{GWS}^h$ and CNFD algorithms, one clockwise translation around the horizontal plane, $k = 1$ TP basis uniform $\mathrm{M} = 32 \times 32$ cartesian mesh, Courant vector magnitude $|\mathbf{C}| = 0.3$ at the IC centroid, $|\mathbf{C}| = 0.41$ at the outboard IC edge: upper left GWS; upper right SUPG/RG; middle left GLS; middle right TG; lower left *optimal* γ GWS; lower right CNFD, from Sahu and Baker (2007). *Source:* Sahu, S. and Baker, A.J. 2007. Reproduced with permission of Wiley

Recalling Figure 5.7, the *m*PDE theory predicts, and *a posteriori* data confirm, Baker (1983, Figure 4.21), that the $k=2$ TP basis GWS^h algorithm rotating cone solution is indistinguishable from that of $k=1$ optimal γ *m*GWS$^h + \theta$TS, Figure 5.13. The detraction of $k=2$ optimal $\gamma = -0.4$ *m*GWS$^h + \theta$TS is significantly increased jacobian bandwidth, a severe compute penalty in transition to genuine NS *m*PDE systems. The $k=2$ TP basis algorithm does generate improved accuracy regarding dispersion error extrema *only*.

Additional theory confirmation *a posteriori* data for the rotating cone verification exist for $k=1$ natural coordinate (NC) basis GWS, TG and optimal γ *m*GWS$^h + \theta$TS implementations, Figure 5.14. The IC is located diametrically opposite to those for Figure 5.12.

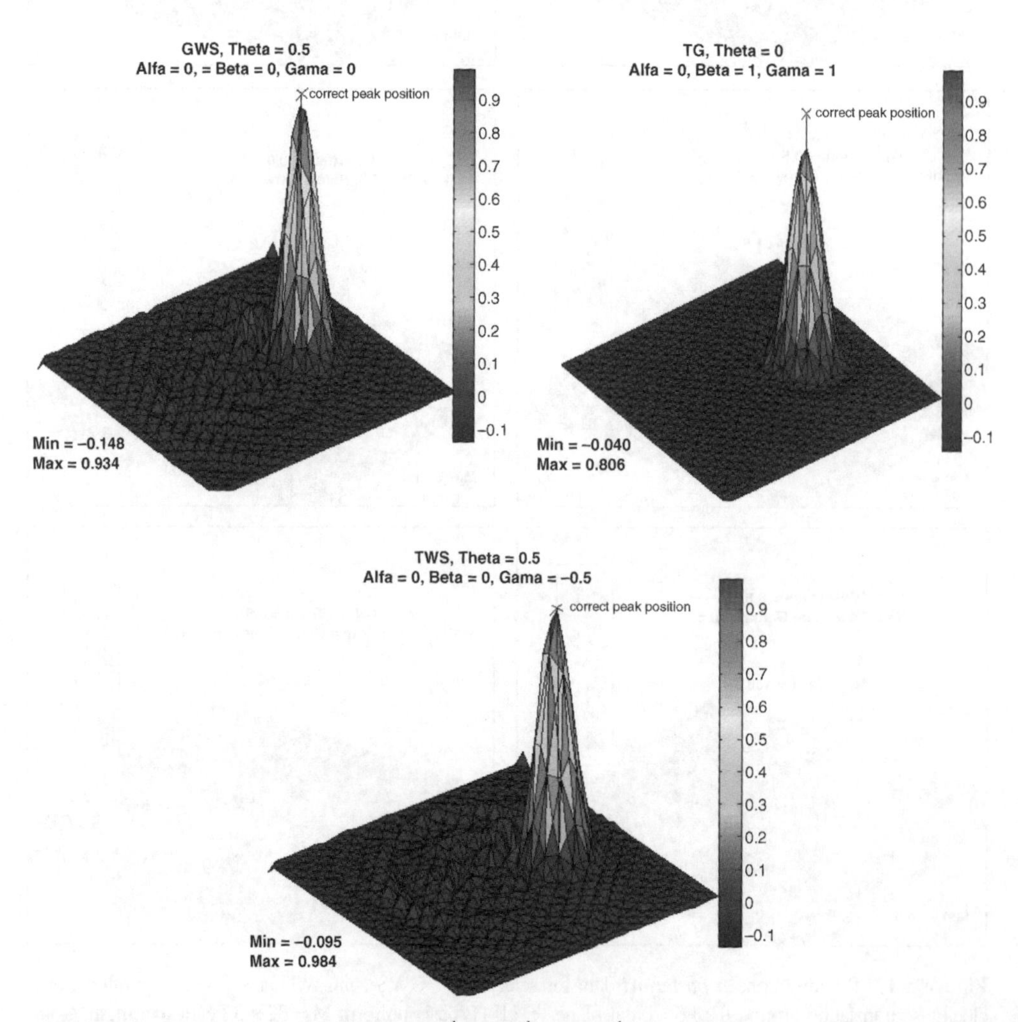

Figure 5.14 Rotating cone verification, WS^h, GWS^h, *m*GWS^h algorithms, one translation clockwise from IC, $M = 128 \times 128$ $k=1$ NC basis implementations, $|\mathbf{C}| = 0.3$ at IC centroid: upper left GWS^h; upper right TG, lower left optimal γ *m*GWS^h.

The mesh is a triangulation of the uniform $M = 32 \times 32$ TP basis mesh yielding identical DOF number and coordinates. A slight decrease in solution peak accuracy results for each algorithm, with a comparable modest increase in dispersion error induced wake extrema. These data are: GWS^h (93, −15), TG (81, −4), optimal γ mGWS^h (98, −9).

5.9 Theory for *Optimal* mGWS^h Asymptotic Convergence

The *m*PDE theory developed to here neglects the NS PDE system viscous (turbulent) contributions by defining the diffusion term parameter $D \equiv \Delta t/\mathrm{Pa}h^2$, (5.67–5.70), to vanish. This enables focusing on mGWS^h algorithm optimization of Fourier content propagation. In practical CFD genuine diffusion-turbulence effects are always present, Chapter 4. Therefore, the concluding theoretical requirement is to derive a complete NS *m*PDE system for which the mGWS$^h + \theta$TS algorithm exhibits optimal performance opportunities.

Classic weak form theory predicts that improving accuracy, via an enhanced asymptotic convergence rate in the energy norm (3.60), (4.43), accrues to trial space basis completeness degree k enrichment. The theoretical foundation is bounding approximation error by interpolation error for a suitable *bilinear form*, Sections 3.5, 3.6. This requires existence of second-order spatial derivatives in the analyzed NS PDE system, that is, the diffusion terms.

The NS **DP** PDE viscous term parameter Pa is Re, hence $1/\mathrm{Pa} \Rightarrow 1/\mathrm{Re}$ which, in practice, tends to be very small, for example $O(\mathrm{E}-06)$. Manipulation of NS **DP** PDE for turbulent flow pertinence does introduce Re^t, a *distribution* with non-D extrema $O(\mathrm{E}+03)$ in highly turbulent flow reaches, for example boundary/shear layers. The extremum ratio hardly exceeds $O(\mathrm{E}-03)$, hence compared to convection the influence is minimal.

Hence *m*PDE theory must focus on NS **DP** for $D \Rightarrow \varepsilon > 0$, vanishingly small. Exchanging initial-value coordinate x with time t, the asymptotic error estimate (4.43) remains appropriate for NS **DP** with D finite. However, in the limit $D = 0$, the asymptotic error estimate is fundamentally altered to, Oden and Reddy (1976),

$$\left\| e^h(n\Delta t) \right\|_E \leq Ch^2 \int_0^{n\Delta t} \| q(\mathbf{x}, \tau \|^2_{H^{k+1}} \mathrm{d}\tau + C_2 \Delta t^{f(\theta)} \left\| q^h(x_0) \right\|^2_{H^0(\Omega)} \tag{5.84}$$

The error L2 norm bound $\left[\|\mathrm{data}\|_{L2,\Omega} + \|\mathrm{data}\|_{L2,\partial\Omega} \right]$ in (4.43) is *replaced* by evolution over time $t = 0 \Rightarrow n\Delta t$ of the H^{k+1} Sobolev norm of the (unknown!) *analytical* solution $q(\mathbf{x}, t)$. Weak form algorithm performance improvement via a degree $k > 1$ basis thus requires extra solution *smoothness*. But of true *significance* to CFD, the theory predicts asymptotic convergence in (5.84) is *independent* of basis degree k, totally negating the classic theory advantage of a $k > 1$ basis!

Hence the fundamental question, "Does a $k > 1$ basis mGWS$^h + \theta$TS algorithm offer true practice value in CFD?" Factually, seeking the answer is the genesis of *m*PDE theory. The conclusion identifies an NS **DP** *m*PDE for $1/\mathrm{Re} \Rightarrow \varepsilon > 0$ leading to derivation of the $k = 1$ basis *optimal m*odified mGWS$^h + \theta$TS algorithm which, for *sufficiently large* Re, exhibits the asymptotic convergence of a $k = 2$ basis $GWS^h + \theta$TS algorithm!

The theory, Kolesnikov and Baker (2001), employs extensions on the TS *m*PDE process utilized to this juncture. The inaugural NS **DP** PDE analysis model is the steady $n=1$ simplification of (5.2)

$$\mathcal{L}(q) = f(q,x)\frac{\mathrm{d}q}{\mathrm{d}x} - \varepsilon(x)\frac{\mathrm{d}^2 q}{\mathrm{d}x^2} = 0 \tag{5.85}$$

The $k=1$ basis stencil for DOF $\{Q\}$ of the GWS^h algorithm for (5.85), replacing interpolation of the functional $f(q, x)$ and the data $\varepsilon(x)$ with the values taken at node X_j is

$$f_j\left(a_1 Q_{j-1} + a_2 Q_j + a_3 Q_{j+1}\right) - \varepsilon_j\left(a_4 Q_{j-1} + a_5 Q_j + a_6 Q_{j+1}\right) = 0 \tag{5.86}$$

The coefficient array a_i admits full generality, which for uniform mesh is subject to constraints $(a_1+a_2+a_3)=0=(a_4+a_5+a_6)$, $(a_3+a_1)=0=(a_6-a_4)$, $(a_3-a_1)=h^{-1}$ and $(a_6+a_4)=2h^{-2}$. One can readily verify the $k=1$ basis array definition is $a_i=(-1/2h,\ 0,\ 1/2h,\ h^{-2},\ -2h^{-2},\ h^{-2})$.

Forming the TS about X_j for all terms in (5.86), using subscript notation for spatial derivatives, then dropping the j subscript for clarity, recall (5.42), the $O(h^\infty)$ *infinite*(!) order accurate TS is

$$\begin{aligned} &f\begin{bmatrix} (a_1+a_2+a_3)Q + (a_3-a_1)hQ_x + (a_3+a_1)\dfrac{h^2}{2}Q_{xx} \\ + (a_3-a_1)\displaystyle\sum_{n=3,2}^{\infty}\frac{h^n}{n!}Q_{(n)} + (a_3+a_1)\sum_{n=4,2}^{\infty}\frac{h^n}{n!}Q_{(n)} \end{bmatrix} \\ &-\varepsilon\begin{bmatrix} (a_4+a_5+a_6)Q + (a_6-a_4)hQ_x + (a_6+a_4)\dfrac{h^2}{2}Q_{xx} \\ + (a_6-a_4)\displaystyle\sum_{n=3,2}^{\infty}\frac{h^n}{n!}Q_{(n)} + (a_6+a_4)\sum_{n=4,2}^{\infty}\frac{h^n}{n!}Q_{(n)} \end{bmatrix} = 0 \end{aligned} \tag{5.87}$$

As (5.87) is infinite-order accurate, hence equated to zero, it may be used to accurately replace higher-order derivatives leading to coefficient arrays a_i alternative to those for the $k=1$ basis GWS^h. The *key requirement* for these operations is retaining a tri-diagonal *stencil* to assure that the final form is amenable to an eventual $k=1$ basis GWS^h restatement for (5.85).

Retaining the first three derivatives in each term in (5.87) and denoting the truncated higher-order terms as *H.O.T.* yields

$$fQ_x - \varepsilon Q_{xx} + h^2(fQ_{xxx}/6 - \varepsilon Q_{xxxx}/12) + H.O.T. = 0 \tag{5.88}$$

This DOF equation is easily rearranged to

$$\varepsilon Q_{xx} = fQ_x - h^2(fQ_{xxx}/6 - \varepsilon Q_{xxxx}/12) + H.O.T. = 0 \tag{5.89}$$

which represents the $O(h^\infty)$ PDE (ODE in this $n=1$ case) satisfied by the model **DP** DOF in (5.86).

Being $O(h^\infty)$, (5.89) can be differentiated any number of times to establish expressions for the higher-order derivatives therein. The goal is to establish a PDE (ODE) the second-order approximation to which results in a higher-order approximation of the original equation. Now assuming ε is constant and differentiating (5.89) repeatedly while neglecting the generated higher-order terms produces

$$
\begin{aligned}
\varepsilon Q_{xxx} &= f_x Q_x + f Q_{xx} - H.O.T. \\
\varepsilon Q_{xxxx} &= f_{xx} Q_x + 2 f_x Q_{xx} + f Q_{xxx} - H.O.T.
\end{aligned}
\tag{5.90}
$$

Substituting (5.90) into (5.89) and neglecting all terms exceeding $O(h^4)$ yields

$$
fQ_x - \varepsilon Q_{xx} + h^2 Q_x \left(\frac{f f_x}{12\varepsilon} - \frac{f_{xx}}{12} \right) + h^2 Q_{xx} \left(\frac{f^2}{12\varepsilon} - \frac{f_x}{6} \right) + O(h^4) = 0 \tag{5.91}
$$

which represents the PDE (ODE) satisfied by model **DP** DOF in (5.87) *including* the $O(h^2)$ truncation error resulting from the second-order accurate approximation used in writing (5.86).

Clearly, sign reversal of the truncation error terms in (5.91) leads to their elimination. Thus is produced the sought TS *modified* replacement for (5.85)

$$
fQ_x - \varepsilon Q_{xx} - h^2 Q_x \left(\frac{f f_x}{12\varepsilon} - \frac{f_{xx}}{12} \right) - h^2 Q_{xx} \left(\frac{f^2}{12\varepsilon} - \frac{f_x}{6} \right) + O(h^4) = 0 \tag{5.92}
$$

Via the obvious connection between DOF ODE (5.92) and the original continuum expression, the **DP** model TS *m*PDE (*m*ODE in this case) expressed in (5.92) is

$$
fq_x - \varepsilon q_{xx} - h^2 q_x \left(\frac{f f_x}{12\varepsilon} - \frac{f_{xx}}{12} \right) - h^2 q_{xx} \left(\frac{f^2}{12\varepsilon} - \frac{f_x}{6} \right) + O(h^4) = 0 \tag{5.93}
$$

The required restriction is $\varepsilon > 0$, but vanishingly small, which enables deleting the less significant added terms in (5.93). Recalling the chain rule for differentiation, the *continuum* **DP** model terminal $O(h^4)$ *m*PDE (*m*ODE) replacement for (5.85) is

$$
\mathcal{L}^m(q) = f\frac{\mathrm{d}q}{\mathrm{d}x} - \varepsilon \frac{\mathrm{d}^2 q}{\mathrm{d}x^2} - \frac{h^2}{12\varepsilon} f \frac{\mathrm{d}}{\mathrm{d}x}\left(f \frac{\mathrm{d}q}{\mathrm{d}x} \right) = 0 \tag{5.94}
$$

Via this TS analysis, a $k=1$ basis GWS^h solution for (5.94) will exhibit $O(h^4)$ accuracy for smooth data *and* sufficiently *small* ε. More precisely, the *asymptotic convergence rate* in the *energy norm* for an $m\mathrm{GWS}^h$ algorithm solution to (5.94) is theoretically predicted $O(h^4)$, in clear distinction to the smooth data $2\gamma = 2(k=1) = 2$ convergence rate for (5.86) stated in (3.60).

Similar arguments generate another modification to (5.85), the $m\mathrm{GWS}^h$ algorithm solution to which will exhibit $O(h^6)$ asymptotic convergence in energy for sufficiently small ε.

Now assuming $f(q,x)$ in (5.85) is fluid speed $u(x)$, given *data*, and via the outlined procedures Kolesnikov and Baker (2001) derive the replacement for (5.93) as

$$\begin{aligned} &fq_x - \varepsilon q_{xx} - h^2 q_x\left(\frac{uu_x}{12\varepsilon} - \frac{u_{xx}}{12}\right) - h^2 q_{xx}\left(\frac{u^2}{12\varepsilon} - \frac{u_x}{6}\right) \\ &- h^4 q_x\left(\frac{uu_x^2}{180\varepsilon^2} - \frac{u^3 u_x}{720\varepsilon^3} - \frac{u_{xxxx}}{360} - \frac{u^2 u_{xx}}{720\varepsilon^2} + \frac{uu_{xxx}}{180\varepsilon}\right) \\ &- h^4 q_{xx}\left(\frac{u_{xx}^2}{180\varepsilon} - \frac{u_{xxx}}{90} + \frac{uu_{xx}}{72\varepsilon} + \frac{u^2 u_x}{360\varepsilon^2} - \frac{u^4}{720\varepsilon^3}\right) \\ &+ O(h^6) = 0 \end{aligned} \tag{5.95}$$

A key observation in (5.95) is that the TS modifications leading to the $O(h^4)$ asymptotic convergence rate secured for *m*ODE (5.94) are unaltered by manipulations to the $O(h^6)$ *m*ODE. The detraction in (5.95) is the high degree of differentiability required of $u(x)$.

Continuing, the famous nonlinear *Burgers equation* is recovered for the definition $f(q,\ x) \equiv q(x)$ in (5.85). The $O(h^4)$ TS-modified *m*ODE remains (5.94) in the appropriate nomenclature

$$\mathcal{L}^m(q) = q\frac{dq}{dx} - \varepsilon\frac{d^2q}{dx^2} - \frac{h^2}{12\varepsilon}q\frac{d}{dx}\left(q\frac{dq}{dx}\right) = 0 \tag{5.96}$$

The $k=1$ basis *m*GWS^h algorithm for (5.96) will generate smooth data (*only*) solutions exhibiting $O(h^4)$ asymptotic convergence in the energy norm. The caveat *only* is important since solutions to (5.96) can exhibit singular behavior due to the NS model nonlinearity. Factually, as a function of BCs, Burgers equation solutions can emulate the *shock* generated by a genuine NS PDE solution.

Anticipating regular solution adapted mesh refinement, the terminal analysis seeks alteration of theory predicted $O(h^4)$ convergence rate for (5.94) due to a non-uniform mesh. Recalling the BL discussion, Section 4.5, assume $n=1$ mesh generation employs the geometric *progression ratio* $l_{e+1} \equiv \mathrm{pr}(l_e)$ for pr a constant. For (5.85), the resultant alteration to the $O(h^\infty)$ TS (5.87) is

$$\begin{aligned} &f\left[\begin{aligned} &(a_1+a_2+a_3)Q + (a_3\mathrm{pr} - a_1)hQ_x + \left(a_3\mathrm{pr}^2 + a_1\right)\frac{h^2}{2}Q_{xx} \\ &+ (a_3\mathrm{pr}^n - a_1)\sum_{n=3,2}^{\infty}\frac{h^n}{n!}Q_{(n)} + (a_3\mathrm{pr}^n + a_1)\sum_{n=4,2}^{\infty}\frac{h^n}{n!}Q_{(n)} \end{aligned}\right] \\ &-\varepsilon\left[\begin{aligned} &(a_4+a_5+a_6)Q + (a_6\mathrm{pr} - a_4)hQ_x + \left(a_6\mathrm{pr}^2 + a_4\right)\frac{h^2}{2}Q_{xx} \\ &+ (a_6\mathrm{pr}^n - a_4)\sum_{n=3,2}^{\infty}\frac{h^n}{n!}Q_{(n)} + (a_6\mathrm{pr}^n + a_4)\sum_{n=4,2}^{\infty}\frac{h^n}{n!}Q_{(n)} \end{aligned}\right] = 0 \end{aligned} \tag{5.97}$$

The modified coefficient constraint set associated with (5.97) is $(a_1+a_2+a_3)=0=(a_4+a_5+a_6)$, $(a_3\,\mathrm{pr}^2+a_1)=0=(a_6\,\mathrm{pr}^2-a_4)$, also $(a_3\,\mathrm{pr}-a_1)=h^{-1}$ and $(a_6\,\mathrm{pr}^2+a_4)=2h^{-2}$.

The specific determination for this TS coefficient set for the $k=1$ basis GWS^h implementation is

$$a_1 = \frac{\text{pr}}{h(1+\text{pr})}, \quad a_2 = \frac{\text{pr}-1}{h\text{pr}}, \quad a_3 = \frac{1}{h\text{pr}(1+\text{pr})},$$
$$a_4 = \frac{2}{h^2(1+\text{pr})}, \quad a_5 = \frac{-2}{h^2\text{pr}}, \quad a_6 = \frac{2}{h^2\text{pr}(1+\text{pr})} \tag{5.98}$$

Substituting these data into (5.97), then neglecting terms of significance less than $O(h^2)$ produces the TS-modified DOF mODE replacement for (5.92)

$$fQ_x - \varepsilon Q_{xx} - \frac{\varepsilon(\text{pr}-1)h}{3}Q_{xxx} + \frac{f\,\text{pr}\,h^2}{6}Q_{xxx} + O(h^2) = 0 \tag{5.99}$$

The conclusion from (5.99) is that a tri-diagonal algorithm written on a non-uniform mesh, assuming ε sufficiently small, can at best be $O(h)$ due to the $O(h^2)$ truncation error.

The flaw in deriving (5.99) is considering only even-order truncation error terms. Adding the odd-order terms significantly complicates the analysis requiring use of symbolic manipulation software, for example Mathematica, Maple. Following this tedious process, and reversing the signs of the generated truncation error terms, Kolesnikov (2000) derives the $O(h^3)$ mODE stencil replacement for (5.99)

$$\begin{aligned} fQ_x - \varepsilon Q_{xx} &+ Q_x\left[\frac{(\text{pr}^2-\text{pr}+1)h^2 f_{xx}}{12} - \frac{(\text{pr}^2+\text{pr}+1)h^2 f f_x}{36\varepsilon}\right] \\ &+ Q_{xx}\left[\frac{(\text{pr}^2-\text{pr}+1)h^2 f_x}{6} - \frac{(\text{pr}^2+\text{pr}+1)h^2 f^2}{36\varepsilon}\right] \\ &+ O(h^3) = 0 \end{aligned} \tag{5.100}$$

Then, as detailed for the uniform mesh theory, recalling differential calculus and for ε sufficiently small the improved order *continuum* mODE replacement for (5.94) is

$$\mathcal{L}^m(q) = f\frac{dq}{dx} - \varepsilon\frac{d^2q}{dx^2} - \frac{h^2(\text{pr}^2+\text{pr}+1)}{36\varepsilon}f\frac{d}{dx}\left(f\frac{dq}{dx}\right) = 0 \tag{5.101}$$

A $k=1$ basis mGWS^h algorithm applied to (5.101) will generate solutions exhibiting $O(h^3)$ asymptotic convergence for smooth data. Of fundamental significance, as anticipated, the rigorous theory predicts non-uniform mesh degradation of asymptotic convergence rate by one order in h. The consistency check is (5.101) precisely reduces to (5.94) for $\text{pr}=1$ uniform mesh.

Select verification statement *a posteriori* data are reported comparing $O(h^2)$ GWS^h and $O(h^4)$ mGWS^h algorithm performance. For the $n=1$ steady linear convection-diffusion ODE (5.85), for any $\text{Pa} \equiv 1/\varepsilon$ the non-D analytical solution for BCs $q(0)=0$ and $q(1)=1$ is

$$q(x) = \left(1 - e^{\text{Pa}\,x}\right)/\left(1 - e^{\text{Pa}}\right) \tag{5.102}$$

This solution is essentially zero everywhere except in a thin *wall layer* adjacent to $x = 1$ with span dependent on Pa. The $O(h^4)$ *m*ODE is (5.94) now written in non-D form is

$$\mathcal{L}^m(q) = u\frac{\mathrm{d}q}{\mathrm{d}x} - \left[\frac{1}{\mathrm{Pa}} + \frac{\mathrm{Pa}h^2u^2}{12}\right]\frac{\mathrm{d}^2q}{\mathrm{d}x^2} = 0 \tag{5.103}$$

wherein non-D $f = u \equiv 1$ and 1/Pa replaces ε. Since Pa is assumed *large* (5.103) clearly confirms $O(h^4)$ asymptotic convergence results from a substantial infusion of *artificial diffusion* with magnitude explicitly dependent on mesh resolution via the h^2 multiplier.

For M = 20, 120 and 220 uniform discretizations and Pa = E+03, the $k = 1$ basis $O(h^2)$ GWS^h and $O(h^4)$ *m*GWS^h algorithm DOF data are compared in Figure 5.15. All GWS^h solutions are non-monotone, in agreement with theory predicting the $k = 1$ basis uniform mesh solution monotonicity requirement M ≥ (Pa/2), Fletcher (1991). Conversely, all *m*GWS^h solutions are *monotone* with that for M = 20 *totally wrong*(!) due to excess artificial diffusion. The M = 120 solution approaches engineering accurate, the M = 220 solution is mathematically accurate.

Defining Pa = 10 violates the small ε assumption in converting (5.93) to (5.94). Returning the differential terms deleted in (5.93) for this assumption, *m*GWS^h solution *a posteriori* data confirm asymptotic convergence is the theory-predicted $O(h^4)$, Kolesnikov and Baker (2001, Table 1). For Pa = E+03 and non-uniform *regular* solution adapted mesh refinement starting with M = 20, pr = 0.6, *m*ODE (5.96) *m*GWS^h solutions generate *a posteriori* data confirming $O(h^4)$ convergence by halving the smallest element measure in the sequence M = 20, 40, 80.

The identical process carried to larger M also generates monotone GWS^h algorithm *a posteriori* data confirming $O(h^2)$ convergence. Note any *non-monotone* GWS^h solution is useless for assessing asymptotic convergence because the energy norm, a symmetric quadratic form, accumulates the *false* energy associated with dispersion error mesh scale oscillations.

The *nonlinear* verification is the $n = 1$ viscous Burgers equation. For BCs $q(0) = 1$ and $q(1) = -1$, the steady solution is a standing *stationary wave* with *thickness* dependent on Pa. The pertinent non-D restatement of nonlinear (5.96) is

$$\mathcal{L}^m(q) = q\frac{\mathrm{d}q}{\mathrm{d}x} - \frac{1}{\mathrm{Pa}}\frac{\mathrm{d}^2q}{\mathrm{d}x^2} - \frac{\mathrm{Pa}h^2}{12}q\frac{\mathrm{d}}{\mathrm{d}x}\left(q\frac{\mathrm{d}q}{\mathrm{d}x}\right) = 0 \tag{5.104}$$

For M = 22, 122, 222 uniform discretizations and Pa = E + 03, all $k = 1$ basis GWS^h algorithm DOF distributions are non-monotone, polluted by dispersion error about the wave front. Conversely, all *m*GWS^h algorithm solutions are symmetric and *monotone*.

The coarse mesh solution again is totally inaccurate(!) due to artificial diffusion, Figure 5.16 upper left. The finest mesh DOF distribution sharply resolves the standing wave over three Ω_e, a mathematically accurate resolution. For Pa = 10, upon eliminating the small ε assumption in generating (5.96), *m*GWS^h algorithm *a posteriori* data confirm $O(h^4)$ convergence, as theoretically predicted, Kolesnikov and Baker (2001, Table 2).

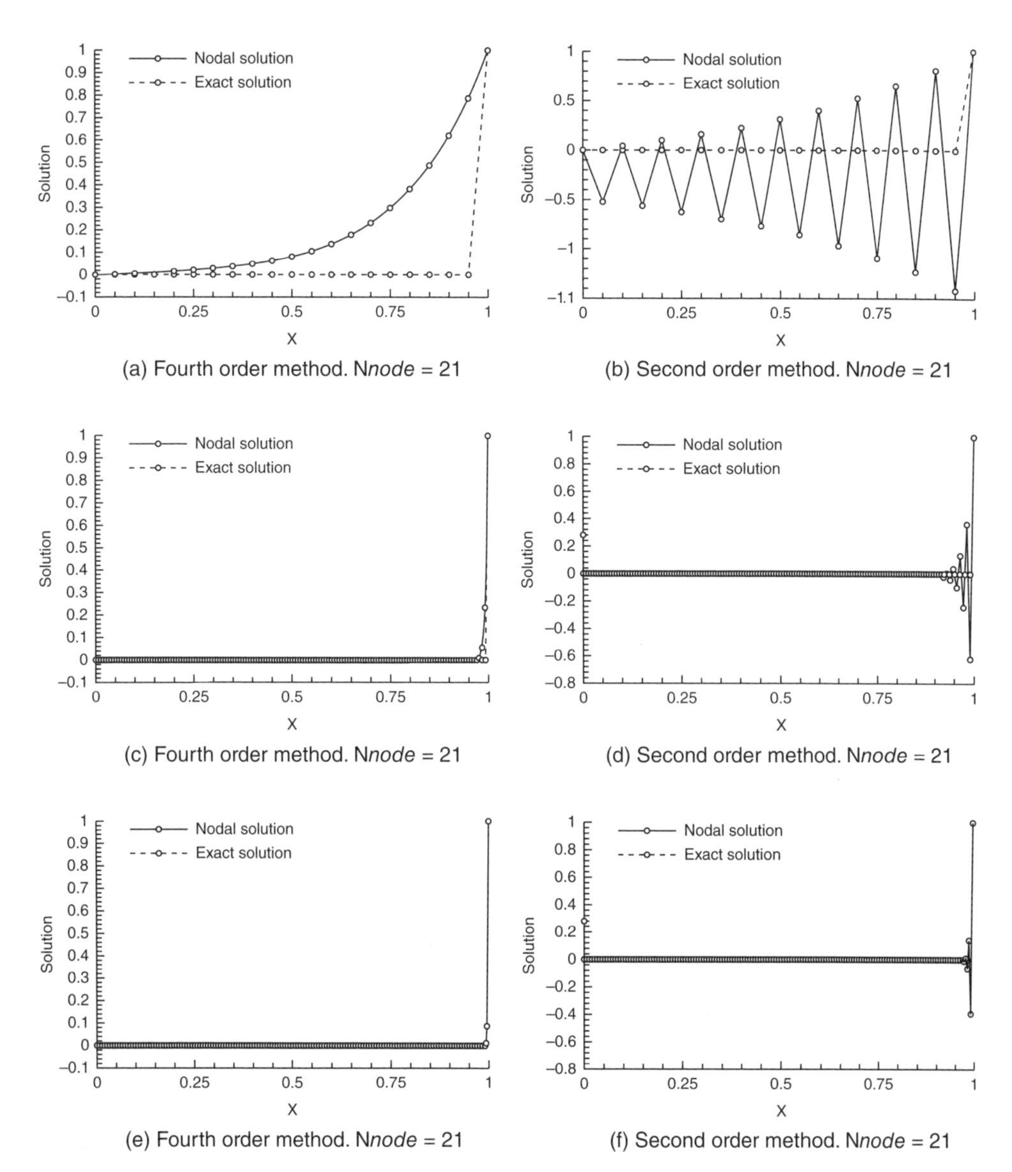

(a) Fourth order method. Nnode = 21

(b) Second order method. Nnode = 21

(c) Fourth order method. Nnode = 21

(d) Second order method. Nnode = 21

(e) Fourth order method. Nnode = 21

(f) Second order method. Nnode = 21

Figure 5.15 Wall layer verification, Pa = E+03, $k = 1$ basis *a posteriori* DOF distributions, uniform M = 20, 120, 220 meshes top to bottom: left column, $O(h^4)$ mGWS^h; right column, $O(h^2)$ GWS^h, from Kolesnikov and Baker (2001). *Source:* Kolesnikov, A. and Baker, A.J. 2001. Reproduced with permission of Elsevier

The TS-modified mPDE theory is readily generalized for an n-dimensional D**P** model PDE. For $f \Rightarrow f_j \Rightarrow u_j$ the n-D replacement for (5.85) is

$$\mathcal{L}(q) = u_j \frac{\partial q}{\partial x_j} - \varepsilon \frac{\partial^2 q}{\partial x_j^2} = 0 \tag{5.105}$$

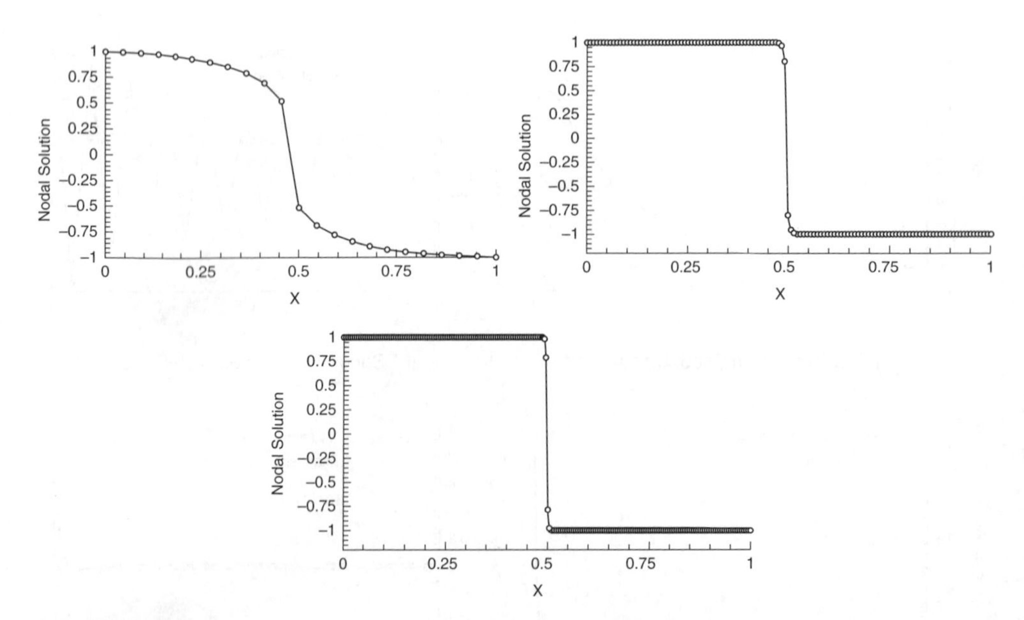

Figure 5.16 Stationary wave nonlinear verification, Pa = E+03, $k = 1$ basis mGWS^h algorithm DOF distributions, uniform mesh: M = 22 upper left, M = 122 upper right, M = 222 lower right. *Source:* Kolesnikov, A. and Baker, A.J. 2001. Reproduced with permission of Elsevier

For cartesian resolution of (5.105) on a uniform cubical mesh of measure h, and retaining the $k = 1$ trilinear TP basis, (3.26), the required tedious algebraic process is fully detailed in Kolesnikov (2000, Ch. 4). An $O(h^4)$ mGWS^h DOF stencil analogous to (5.92) results, which upon assuming ε small leads to the model **D**P n-D mPDE replacement for (5.105)

$$\mathcal{L}^m(q) = u_j \frac{\partial q}{\partial x_j} - \varepsilon \frac{\partial^2 q}{\partial x_j^2} - \frac{h^2}{12\varepsilon} u_k \frac{\partial}{\partial x_k}\left[u_j \frac{\partial q}{\partial x_j}\right] = 0 \tag{5.106}$$

Thereby the $k = 1$ basis mGWS^h algorithm for n-D mPDE (5.106), for *arbitrary data* and *sufficiently small* ε, equivalently *large* Pa = Re for **D**P, will generate uniform mesh solutions adhering to the *asymptotic error estimate*

$$\left\|\{e^h\}\right\|_E \le Ch^{2\gamma}\left[\|\mathrm{data}\|_{L2,\Omega} + \|\mathrm{data}\|_{L2,\partial\Omega}\right], \gamma \equiv \min(k+1, r-1) \tag{5.107}$$

for C a constant and r the measure of *non-smooth data* influence. The verification statement is the $n = 2$ extension of the scalar transport wall layer ODE, (5.104). For imposed velocity vector $\mathbf{u} = \hat{\mathbf{i}} + \hat{\mathbf{j}}$ the DOF at the square domain outflow corner *only* is fixed at Q = 1. All outflow boundary BCs are homogeneous Neumann, all inflow boundary DOF are fixed at {Q} = {0} and ε is variable.

For ε = 0.1, Pa = 10 and uniform M = 10^2, 20^2, 40^2, 80^2 meshes, all GWS^h and mGWS^h DOF distributions are monotone, appear visually identical and possess identical energy norms, Figure 5.17 upper left. Conversely, for ε = 0.001, Pa = E+03 and uniform $k = 1$ TP basis M = 10^2, 20^2, 40^2, 80^2, 160^2 meshes, all GWS^h solutions are non-monotone, all mGWS^h

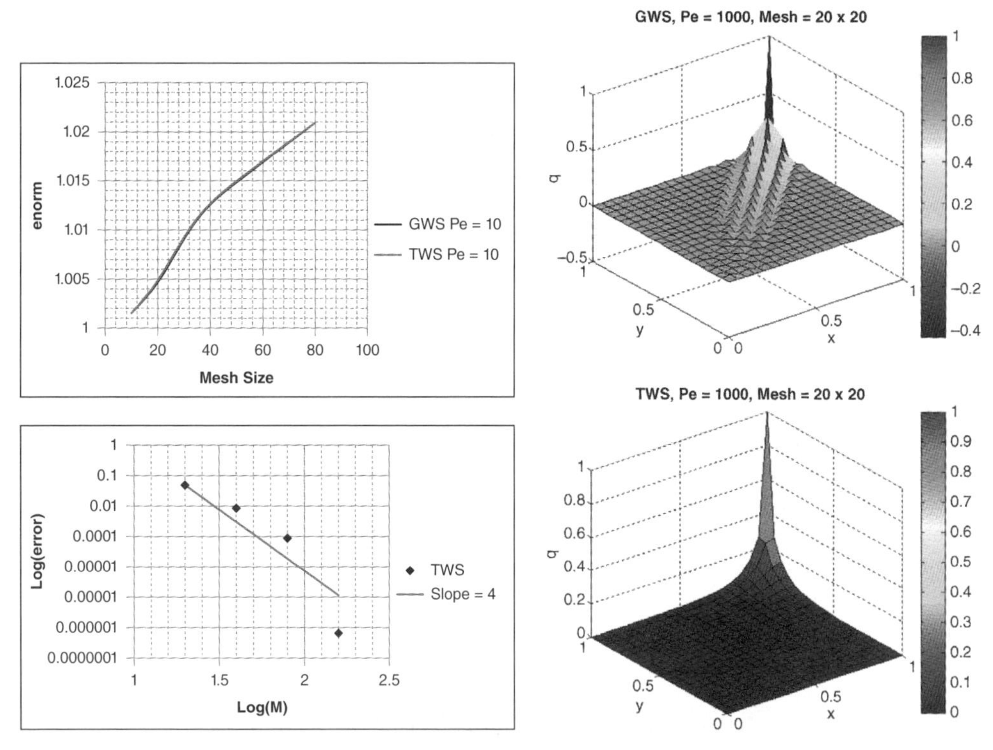

Figure 5.17 Wall layer verification problem, $n=2$, $k=1$ TP basis GWS^h and mGWS^h (labeled TWS) solutions: upper left, Pa = 10 norm overlay for all M; lower left, mGWS^h algorithm $O(h^4)$ convergence verification, Pa = E+03; upper/lower right, Pa = E+03 GWS^h/mGWS^h algorithm DOF perspective distributions, uniform $M = 20^2$ mesh

solutions are monotone with those for $M < 40^2$ suffering excess artificial diffusion, Figures 5.17 upper/lower right. The Pa = E+03 $M \geq 40^2 m$GWS^h*a posteriori* data sequence confirms theory predicted $O(h^4)$ convergence, (5.107), Figure 5.17 lower left. Therein, the exponential decrease in $M = 160^2$ solution error indicates it as the $k=1$ basis interpolation of the exact solution.

5.10 The *Optimal m*GWS^h + θTS $k = 1$ Basis NS Algorithm

Identifying the n-D NS mPDE system replacement for (5.1–5.4), the $k=1$ basis *optimal modified m*GWS^h + θTS algorithm completes the chapter. The concluding theoretical step is extension of (5.105) to the NS model I-EBV PDE

$$\mathcal{L}(q) = \frac{\partial q}{\partial t} + u_j \frac{\partial q}{\partial x_j} - \varepsilon \frac{\partial^2 q}{\partial x_j^2} = 0 \tag{5.108}$$

From the original theory of Lax–Wendroff (1960) as amended by Chaffin (1997), Kolesnikov (2000) derives θ-implicit trapezoidal rule $O(\Delta t^3)$ truncation error elimination.

For a TS on full time-step Δt, a partial forward TS at $(1-\alpha)\Delta t$ and two backwards TS from Δt, the linear combination leads to the modified θ-implicit TS for the time-derivative in (5.108)

$$
\begin{aligned}
q^{n+1} - q^n - \Delta t\left(\theta\frac{\partial q^{n+1}}{\partial t} + (1-\theta)\frac{\partial q^n}{\partial t}\right) = \\
-\frac{\Delta t^2}{6}\left[(3\theta-1)\frac{\partial^2 q^{n+1}}{\partial t^2} + (3\theta-2)\frac{\partial^2 q^n}{\partial t^2}\right] \\
-\frac{\Delta t^4}{72}\left[-(3\theta-1)^2\frac{\partial^4 q^{n+1}}{\partial t^4} + (3\theta-2)^2\frac{\partial^4 q^n}{\partial t^4}\right] + O(\Delta t^5)
\end{aligned}
\tag{5.109}
$$

Terms on the left side of the equality in (5.109) constitute the Euler single step θ-implicit TS, with the right side terms the theory-identified $O(\Delta t^3)$ truncation error correction. For the time-accurate TS definition $\theta = 0.5$, (5.108) is then used to replace the higher-order time derivatives in (5.109) with spatial derivatives, recall (5.63–5.64).

Upon tedious algebra operations completion, restricting to $k=1$ basis generates the sought tri-diagonal $m\text{GWS}^h$ DOF *stencil*. Following symmetrizing operations, taking the limit $\Delta t \Rightarrow \delta > 0$, recall (5.65–5.66), and assuming ε sufficiently small leads to the *m*PDE replacement for (5.108), Kolesnikov (2000, Ch. 4)

$$
\begin{aligned}
\mathcal{L}^m(q) = \frac{\partial q}{\partial t} + u_j\frac{\partial q}{\partial x_j} - \varepsilon\frac{\partial^2 q}{\partial x_j^2} - \frac{h^2}{12\varepsilon}u_k\frac{\partial}{\partial x_k}\left[u_j\frac{\partial q}{\partial x_j}\right] \\
+\frac{h^2}{6\Delta t\varepsilon}\left[u_j\frac{\partial q}{\partial x_j}\right] - \frac{h^2}{6\Delta t}\frac{\partial^2 q}{\partial x_j^2} = 0
\end{aligned}
\tag{5.110}
$$

The *m*PDE theory replacement for the NS model unsteady I_EBV PDE is thus derived. Note that (5.110) embeds the steady state *m*PDE (5.106) upon neglect of temporal truncation *error annihilation* terms. Thus is identified the NS I_EBV *m*PDE leading to *optimal* $k=1$ basis $m\text{GWS}^h + \theta\text{TS}$ performance. Selecting $\theta \equiv 0.5$, the generated *n*-D approximation $q^h(\mathbf{x}, n\Delta t)$ to the (5.110) exact solution $q(\mathbf{x}, t)$ will exhibit $O(\Delta t^4)$ temporal accuracy and $O(h^4)$ asymptotic convergence in energy for *smooth* data and vanishingly *small* ε, the sought theory.

Incompressible-thermal NS algorithm completion is *modulo* the pressure projection theory for iterative implementation of the differential constraint (5.1), that is, $DM^h \sim 0$ to adequate discrete precision. For the NS state variable $\{q(\mathbf{x}, t)\} = \{u_i, \Theta, Y\}^T$ and this theory-generated NS *A*uxiliary state variable $\{q_A(\mathbf{x}, t)\} = \{\phi, P\}^T$, the complete NS I-EBV plus EBV PDE system for $\text{St} \equiv 1$ is

$$
\mathcal{L}(\{q\}) = \frac{\partial\{q\}}{\partial t} + \frac{\partial}{\partial x_j}\left[u_j\{q\} - \frac{1}{\text{Pa}}\frac{\partial\{q\}}{\partial x_j}\right] - \{s(P, \{q\})\} = \{0\} \tag{5.111}
$$

$$
\mathcal{L}(q_A) = -\nabla^2 q_A - s_A(\{q\}) = 0 \tag{5.112}
$$

Recall that the *kinetic* influence of pressure projection theory, Section 5.4, is a source term contribution in (5.111), with genuine pressure distribution $P(\mathbf{x}, t)$ determination a convergence post-process operation.

The *optimal* modified *continuous* Galerkin CFD theory *m*PDE replacement for (5.110–5.111) accrues to inserting the $O(m^3)$ optimal γ *m*PDE modification (5.74)

$$\mathcal{L}^m(\{q\}) \equiv \left[1 - \frac{\gamma\,\Delta t^2}{6}\frac{\partial}{\partial x_j}\left(u_j u_k \frac{\partial}{\partial x_k}\right)\right]\frac{\partial\{q\}}{\partial t} - s(P, \{q\})$$
$$+\frac{\partial}{\partial x_j}\left[u_j\{q\} - \frac{1}{\mathrm{Pa}}\frac{\partial\{q\}}{\partial x_j} - \frac{\mathrm{Pa}h^2}{12}u_j\,u_k\frac{\partial\{q\}}{\partial x_k} + \frac{h^2}{6\Delta t}\left(\mathrm{Pa}\,u_j\{q\} - \frac{\partial\{q\}}{\partial x_j}\right)\right] = 0 \tag{5.113}$$

The $k=1$ basis $m\mathrm{GWS}^h + \theta\mathrm{TS}$ algorithm solution approximation to (5.113) is theoretically predicted to exhibit $O(\Delta t^4, h^4, m^3)$ accuracy for smooth data and $\theta \equiv 0.5$.

Recognition of iterative NS algorithm discrete satisfaction of (5.1) admits inserting both *m*PDE terms with multiplier $\mathrm{Pa}h^2$ inside the divergence operator in (5.113), compare to (5.110). This renders the velocity *nonlinear* tensor product *m*PDE addition $u_j\,u_k$ *identical* to the unsteady TS theory *m*PDE term multiplied by $\beta\Delta t/2$. This admits replacement of *a priori* unknown coefficient β and constitutes robust verification of theoretical consistency between the very distinct *m*PDE theoretical tracks.

Thus becomes identified a key distinction between 1-D and *n*-D *m*PDE (5.113). The coefficient β is non-dimensional while the dimensionality of *m*PDE term $\mathrm{Pa}h^2 u_j\,u_k$ is $D(\mathrm{L}^2\tau^{-1})$ for D**P**, identically that of kinematic viscosity. In 1-D the tensor product reduces to a positive definite scalar generating pure numerical diffusion, as amply illustrated. In *n*-D the tensor product $u_j\,u_k$ is *non-positive definite*, which constitutes a *dissipative/anti-dissipative* operator. That this indeed is the *m*PDE mechanism is quantitatively validated via *a posteriori* data detailed throughout the remaining chapters.

The final required derivation is the *m*PDE replacement for decoupled EBV PDEs (5.112). The TS error annihilation theory is identically that generating (5.106) from (5.105). As detailed in Kolesnikov (2000), the restriction to tri-diagonal DOF stencil on a uniform rectangular cartesian mesh leads to (5.112) replacement with *m*PDE

$$\mathcal{L}^m(q_A) = -\nabla^2 q_A - s_A(\{q\}) - \frac{h^2}{12}\nabla^2 s_A(\{q\}) = 0 \tag{5.114}$$

The $k=1$ basis $m\mathrm{GWS}^h + \theta\mathrm{TS}$ algorithm solution of (5.114) is theory predicted to exhibit the sought $O(h^4)$ asymptotic convergence rate.

5.11 Summary

Fundamental *theoretical issues* underlying the design of an *optimal* modified continuous *genuine* Galerkin weak form algorithm for the *infamous* Navier–Stokes equations, the explicitly nonlinear *differentially constrained* PDE system (5.1–5.4), is complete. The inaugural assessment was focused on options for handling the constraint of mass conservation, (5.1), via both exact theory and inexact methodology. The viable class of exact procedures invariably involves transformation of D**P** into a PDE system written for vorticity, the curl of the velocity vector. This operation eliminates the appearance of pressure therein, hence the need to address its resolution.

Among the class of inexact procedures, the summary conclusion is that the mathematical rigor of the *continuity constraint* pressure projection theory, Williams (1993), best meets the *robustness* requirement. It additionally encompasses as simplifying assumptions the wide range of CFD literature published algorithms. This theory adds two EBV Poisson PDEs to the NS PDE system (5.2–5.4), each of which exhibits *well-posed* boundary conditions (BCs) specifically applicable to multiply connected solution domains.

The resultant non-D NS PDE + BCs system implementing the continuity constraint theory with genuine kinematic pressure is defined in (5.30) and (5.12)

$$\mathcal{L}(\phi) = -\nabla^2\phi + \frac{\partial \tilde{u}_i}{\partial x_i} = 0$$

$$\ell(\phi) = -\nabla\phi \cdot \hat{\mathbf{n}} - (u_i - \tilde{u}_i)\hat{n}_i = 0$$

$$\mathcal{L}(P) = -\nabla^2 P - \frac{\partial}{\partial x_j}\left[u_i \frac{\partial u_j}{\partial x_i} + \frac{\mathrm{Gr}}{\mathrm{Re}^2}\Theta \widehat{g}_j\right] = 0$$

$$\ell(P) = \nabla P \cdot \hat{\mathbf{n}} - \left[\mathrm{St}\frac{\partial \mathbf{u}}{\partial t} - \frac{1}{\mathrm{Re}}\nabla^2 \mathbf{u}\right] \cdot \hat{\mathbf{n}} = 0$$

for $\tilde{u}_i$ *any* approximation to a NS velocity vector field, specifically *not* required divergence-free.

The iteration strategy for pressure *kinetic impact* on the NS PDE system solution process at forward time $t_{n+1} = t_n + \Delta t$ is (5.29)

$$P^*|_{n+1}^{p} = P_n + \frac{1}{\theta \Delta t}\sum_{\alpha=1}^{p} \delta\phi|_{n+1}^{\alpha}$$

for P_n the solution of (5.12) computed using the converged $\mathrm{D}M^h$ satisfying velocity vector field at time t_n. The *quantitative* measure of *error* in discrete approximation $\mathrm{D}M^h = \nabla^h \cdot \mathbf{u}^h$ to (5.1) at any time t_{n+1} is the velocity potential *energy norm* $\|\phi^h\|_E = \|\nabla^h \cdot \mathbf{u}^h\|$, via the theory generating (5.30). *A posteriori* data detailed in subsequent chapters confirm the "size" of this error is $O\left(\|\phi^h\|_E\right) \leq \delta \approx O(\varepsilon^2)$ for ε the algorithm algebraic statement iterative convergence requirement.

Thereby, as the $\mathrm{D}M^h$ error at convergence is guaranteed *insignificant*, (5.30) need not be augmented with the *m*PDE term in (5.114). This also holds true for (5.12), since genuine pressure enters the solution process as *data*, and data interpolation error tends to zero faster than the solution asymptotic convergence rate, Strang and Fix (1979).

The detailed TS theory manipulations leading to the terminal *m*PDE system replacement for (5.2–5.4) predict that the *optimal modified continuous Galerkin* CFD algorithm for the incompressible NS state variable $\{q(\mathbf{x}, t)\} = \{u_i, \Theta, Y\}^T$ is best expressed as a divergence form *m*PDE

$$\mathcal{L}^m(\{q(q_A)\}) = [m]\frac{\partial \{q\}}{\partial t} + \frac{\partial}{\partial x_j}\left[\{f_j\} - \{f_j^d\}\right] - \{s\} = \{0\} \tag{5.115}$$

In (5.115), the *kinetic flux vector* $\{f_j\}$ implicitly embeds the continuity constraint theory auxiliary state variable $\{q_A(\mathbf{x}, t)\} = \{\phi, P\}^T$, $\{f_j^d\}$ is the *dissipative flux vector* and $\{s\}$ remains the source term. The *m*PDE definitions are

$$\{f_j(\{q(q_A)\})\} = \begin{Bmatrix} u_j\left(1 + \dfrac{\mathrm{Re}h^2}{6\Delta t}\right)u_i + \left(P_n + \dfrac{1}{\theta\Delta t}\displaystyle\sum_{\alpha=1}^{p}\delta\phi|_{n+1}^{\alpha}\right)\delta_{ij} \\ u_j\left(1 + \dfrac{\mathrm{RePr}\,h^2}{6\Delta t}\right)\Theta \\ u_j\left(1 + \dfrac{\mathrm{ReSc}\,h^2}{6\Delta t}\right)Y \end{Bmatrix} \tag{5.116}$$

$$\left\{f_j^d(\{q\})\right\} = \begin{Bmatrix} \left(\dfrac{1}{\mathrm{Re}} + \dfrac{h^2}{6\Delta t}\right)\dfrac{\partial u_i}{\partial x_j} + \dfrac{\mathrm{Re}\,h^2}{12}u_j u_k \dfrac{\partial u_i}{\partial x_k} \\ \left(\dfrac{1}{\mathrm{RePr}} + \dfrac{h^2}{6\Delta t}\right)\dfrac{\partial \Theta}{\partial x_j} + \dfrac{\mathrm{RePr}\,h^2}{12}u_j u_k \dfrac{\partial \Theta}{\partial x_k} \\ \left(\dfrac{1}{\mathrm{ReSc}} + \dfrac{h^2}{6\Delta t}\right)\dfrac{\partial Y}{\partial x_j} + \dfrac{\mathrm{ReSc}\,h^2}{12}u_j u_k \dfrac{\partial Y}{\partial x_k} \end{Bmatrix} \tag{5.117}$$

$$[m] = \mathrm{diag}\left[1 - \frac{\gamma\,\Delta t^2}{6}\frac{\partial}{\partial x_j}\left(u_j u_k \frac{\partial}{\partial x_k}\right)\right] \tag{5.118}$$

$$\{s(\{q\})\} = \left\{-\frac{\mathrm{Gr}}{\mathrm{Re}^2}\Theta\widehat{g}_i, \frac{\mathrm{Ec}}{\mathrm{Re}}\frac{\partial}{\partial x_j}\left(u_i\frac{\partial u_j}{\partial x_i}\right), s(Y)\right\}^T \tag{5.119}$$

The I-EBV *m*PDE system (5.115–5.119), coupled with the EBV PDE systems (5.12) and (5.30) operating *modulo* (5.29), constitutes the derived NS *m*PDE system to be addressed by *m*GWS^N.

For *m*GWS^N transition to $k=1$ basis, $\theta \equiv 0.5$ *m*GWS$^h + \theta$TS, this solution to the NS *m*PDE system is *smooth* data TS predicted $O(h^4, m^3, \Delta t^4)$ accurate. For *arbitrary* data the *regular solution adapted* mesh theoretical *asymptotic error* estimate for generated approximation $\{q^h(\mathbf{x}, t)\} = \{\mathbf{u}^h, \Theta^h, \psi^h\}^T$ to NS state variable $\{q(\mathbf{x}, t)\} = \{u_i, \Theta, Y\}^T$ is

$$\begin{aligned} \left\|\{e^h(n\Delta t)\}\right\|_E &\le Ch^{2\gamma}\left[\|\mathrm{data}\|_{L2,\Omega} + \|\mathrm{data}\|_{L2,\partial\Omega}\right] + C_2\Delta t^4\left\|\{q^h(t_0)\}\right\|_{H^1(\Omega)}^2 \\ & for: \gamma \equiv \min(k(\equiv 1) + 1 = 2, r-1) \end{aligned} \tag{5.120}$$

for *sufficiently large* Re with C and C_2 constants.

The *m*GWS$^h + \theta$TS algorithm terminal form is a nonlinear matrix statement for determination of approximation DOF $\{Q(t)\}$. Recalling (4.34–4.38), the iterative Newton algorithm statement is

$$\begin{aligned} [\mathrm{JAC\_}\{Q\}]\{\delta Q\}^{p+1} &= -\{\mathrm{F\_}\{Q\}\}_{n+1}^p, \ [\mathrm{JAC\_}\{Q\}] \equiv \frac{\partial\{\mathrm{F\_}\{Q\}\}}{\partial\{Q\}} \\ \{\mathrm{F\_}\{Q\}\}_{n+1}^p &= [\mathrm{M}]\{Q_{n+1}^p - Q_n\} \\ &\quad + \Delta t\left(\theta\{\mathrm{RES\_}\{Q\}\}_{n+1}^p + (1-\theta)\{\mathrm{RES\_}\{Q\}\}_n\right) \\ \{Q\}_{n+1}^{p+1} &= \{Q\}_{n+1}^p + \{\delta Q\}^{p+1} = \{Q\}_n + \sum_{\alpha=0}^{p}\{\delta Q\}^{\alpha+1} \end{aligned} \tag{5.121}$$

Note the matrix $[m]$ in (5.115) is replaced in (5.121) with [M] which contains contributions from the mPDE terms in (5.116–5.117) with $1/\Delta t$ multiplier. This multiplier for these terms cancels the Δt multiplier on {RES_{•}} in (5.121). The functional form of [M] results by augmenting $\{\text{RES\_}\{\bullet\}\}_n$ in (5.121) with $(\pm)\theta\{m\text{PDE terms with } 1/\Delta t \text{ multiplier}\}_n$. Reversal of the divergence theorem manipulation in (5.113) then leads to mPDE-altered $[m]$ matrix definition

$$[\mathrm{M}] = \mathrm{diag}\begin{bmatrix} 1 - \dfrac{\gamma\Delta t^2}{6}\dfrac{\partial}{\partial x_j}\left(u_j u_k \dfrac{\partial}{\partial x_k}\right) + \dfrac{\theta h^2}{6}\left(u_j \mathrm{Re} - \dfrac{\partial}{\partial x_j}\right)\dfrac{\partial}{\partial x_j} \\ 1 - \dfrac{\gamma\Delta t^2}{6}\dfrac{\partial}{\partial x_j}\left(u_j u_k \dfrac{\partial}{\partial x_k}\right) + \dfrac{\theta h^2}{6}\left(u_j \mathrm{RePr} - \dfrac{\partial}{\partial x_j}\right)\dfrac{\partial}{\partial x_j} \\ 1 - \dfrac{\gamma\Delta t^2}{6}\dfrac{\partial}{\partial x_j}\left(u_j u_k \dfrac{\partial}{\partial x_k}\right) + \dfrac{\theta h^2}{6}\left(u_j \mathrm{ReSc} - \dfrac{\partial}{\partial x_j}\right)\dfrac{\partial}{\partial x_j} \end{bmatrix} \tag{5.122}$$

This operation eliminates appearance of *all* mPDE differential terms with $1/\Delta t$ multiplier in flux vectors (5.116–5.117). In concert it generates *data* augmentations to $\{s\}$, (5.119), to the revised matrix

$$\{s\} = \left\{ \begin{array}{l} -\dfrac{\mathrm{Gr}}{\mathrm{Re}^2}\Theta\widehat{g}_i \qquad\quad - \dfrac{h^2}{6}\left[u_j \mathrm{Re}\dfrac{\partial u_i}{\partial x_j} - \dfrac{\partial^2 u_i}{\partial x_j^2}\right]_n \\ \dfrac{\mathrm{Ec}}{\mathrm{Re}}\dfrac{\partial}{\partial x_j}\left(u_i \dfrac{\partial u_j}{\partial x_i}\right) - \dfrac{h^2}{6}\left[u_j \mathrm{RePr}\dfrac{\partial \Theta}{\partial x_j} - \dfrac{\partial^2 \Theta}{\partial x_j^2}\right]_n \\ s(Y) \qquad\qquad\quad - \dfrac{h^2}{6}\left[u_j \mathrm{ReSc}\dfrac{\partial Y}{\partial x_j} - \dfrac{\partial^2 Y}{\partial x_j^2}\right]_n \end{array} \right\} \tag{5.123}$$

For this terminal linear algebra statement enabled manipulation, it is of significant *weak form theory consequence* that the TS theory mPDE-identified alterations elevating $k=1$, $\theta=0.5$ GWS^h + θTS NS algorithm smooth data performance from $O(h^2, m^2, \Delta t^2)$ to $O(h^4, m^3, \Delta t^4)$ reside mainly in the diagonal matrix multiplier on the time derivative in the parent NS PDE system. Thereby, the "*mass matrix*" referenced in the earliest finite element CFD literature as affecting solution accuracy takes on a thoroughly encompassing theoretical role.

Exercises

5.2.1 Verify that the vorticity-streamfunction vector PDE system (5.5–5.6) is correct.

5.2.2 Substitute (5.6) directly into (5.5), hence derive the biharmonic PDE written on streamfunction (and temperature) only.

5.2.3 Verify the accuracy of the $n=2$ vorticity-streamfunction PDE system (5.9–5.10).

5.2.4 Form div **DP**, (5.2), hence verify the pressure Poisson PDE in (5.12). Then form the GWS^N for this PDE and substituting (5.2) derive the Neumann BC stated in (5.12).

5.3.1 Verify the extremum of (5.16) coincides with the steady Stokes PDE form of (5.2) for the definition $\lambda_\alpha(\nabla \cdot \mathbf{u}_\alpha) \equiv -P_\alpha$.

5.4.1 Form the curl of the difference between (5.20) and (5.21), hence verify that the divergence error is solenoidal, (5.22).

5.4.2 Use the implicit TS (5.23) to form the equivalent of (5.20) and (5.21), hence verify (5.24).

5.4.3 Verify the velocity vector potential PDE system (5.30).

5.6.1 Verify the $\text{GWS}^h + \theta\text{TS}$ algorithm statement sequence (5.36–5.39) for PDE (5.35).

5.6.2 Verify the algorithm recursion relations in (5.40).

5.6.3 Execute the Taylor series operations on (5.41) hence validate the $\text{GWS}^h + \theta\text{TS}$ algorithm difference-differential PDE (5.42).

5.6.4 Verify the Crank–Nicolson difference-differential PDE (5.44).

5.6.5 Substitute the Fourier mode solution (5.46) into (5.47), hence validate the GWS^h algorithm complex velocity solution (5.48), then verify (5.49).

5.6.6 Substitute the fully discrete Fourier mode solution (5.51) into (5.50), hence verify the $\text{GWS}^h + \theta\text{TS}$ algorithm amplification factor (5.52). Confirm its magnitude is unity *only* for $\theta = 0.5$.

5.6.7 Verify the WS^h algorithm statement (5.56) for (5.55).

5.6.8 Confirm the WS^h algorithm Fourier mode complex velocity solution (5.58), also the Fourier mode solution statement (5.59).

5.6.9 Starting with (5.56), form the $\text{WS}^h + \theta\text{TS}$ algorithm statement stencil. Then substitute the fully discrete Fourier mode solution, hence validate the amplification factor solution (5.60).

5.7.1 Proceed through the TS operations summarized in (5.63–5.65), hence verify the TS *m*odified conservation principle *m*PDE (5.66).

5.7.2 Simplifying to the case $1/\text{Pa} = 0 = \text{D}$, generate the amplification factor solution for the $m\text{GWS}^h + \theta\text{TS}$ algorithm (5.67).

5.7.3 Starting with (5.71), verify the optimal parameter definition for the $m\text{GWS}^h + \theta\text{TS}$ algorithm is $\alpha = 0$, $\beta = 0$, $\gamma = -0.5$, $\theta = 0.5$ independent of Courant number C.

5.8.1 Starting with (5.72), proceed through the TS operations equivalent to (5.63–5.65) to verify the TS-*m*odified conservation principle *m*PDE (5.74).

5.8.2 Substitute the Fourier representation (5.76) into one or two terms in (5.75), hence confirm a term in the amplification factor solution (5.77).

5.8.2 Verify the mGWS^h + θTS algorithm $O(m^2)$ phase velocity error vanishes identically for definitions $\alpha = 0 = \beta$, $\theta = ½$ in (5.81).

5.9.1 Confirm for (5.86) that $a_i = (-1/2h, 0, 1/2h, h^{-2}, -2h^{-2}, h^{-2})$ defines the $k = 1$ basis GWS^h stencil for (5.85).

5.9.2 Expand a few terms in the TS (5.87) hence confirm its validity.

5.9.3 Verify the DOF $O(h^4)$ algorithm stencil (5.91).

5.9.4 Verify the $O(h^4)$ mPDE (mODE) (5.94) for the assumption ε vanishingly small.

5.9.5 The uniform mesh solutions, Figures 5.15 and 5.16, obviously contain excess DOF. Viewing (5.101), estimate pr distributions for both verification statements that support mathematically accurate solutions on an $M \approx 40$ regular solution adapted mesh.

5.10.1 Verify that (5.113) includes all mPDE theory error annihilation differential terms derived in the chapter.

5.10.2 Confirm dimensionality $D(L^2\tau^{-1})$ is correct for mPDE theory term $\mathrm{Pa}h^2 u_j u_k$ for NS **DP** state variable members.

5.10.3 Determine the dimensionality of the mPDE theory term $\mathrm{Pa}h^2 u_j u_k$ for NS scalar state variable members.

5.10.4 Derive the linear Poisson EBV mPDE (5.114).

5.11.1 Verify that (5.115–5.118) is the flux vector representation of NS PDE system (5.2–5.4) upon inclusion of all TS theory-generated differential calculus terms specified in (5.113).

5.11.2 Execute the $(\pm)(\theta)\{m\text{PDE terms with } 1/\Delta t \text{ multiplier}\}_n$ operation in $\{\text{RES\_}\{\bullet\}\}_n$ in (5.121), hence confirm the correctness of (5.122) and (5.123).

References

Amsden, A.A. and Harlow, F.H. (1970). "The SMAC method: a numerical technique for calculating incompressible fluid flows," Los Alamos Scientific Laboratory Report LA-4370.

Babuska, I. (1973). "The finite element method with Lagrange multipliers," *J. Numerical Mathematics*, V.20.

Baker, A.J. (1983). *Finite Element Computational Fluid Mechanics*, Hemisphere Publishing Corp. New York.

Baker, A.J. and Kim, J.W. (1987). "A Taylor weak statement for hyperbolic conservation laws," *J. Numerical Methods Fluids*, V.7.

Brezzi, F. (1974). "On the existence, uniqueness and approximation for saddle-point problems arising from Lagrange multipliers," *RAIRO Anal. Numer.*, V.8.

Brooks, A. and Hughes, T.J.R. (1982). "Streamline upwind/Petrov-Galerkin formulations for convection dominated flows with particular emphasis on the incompressible Navier–Stokes equations," *Computer Methods Applied Mechanics and Engineering*, V.32.

Chaffin, D.J. and Baker, A.J. (1995). "On Taylor weak statement finite element methods for computational fluid dynamics," *J. Numerical Methods Fluids*, V.21.

Chaffin, D.J. (1997). "A Taylor weak statement finite element method for computational fluid dynamics," PhD dissertation, University of Tennessee.

Chorin, A.J. (1967). "A numerical method for solving incompressible viscous flow problems," *J. Computational Physics*, V.2.

Christon, M.A. (1999). "The influence of the mass matrix on the dispersive nature of the semi-discrete, second-order wave equation," *Comp. Methods Appl. Mechanics Engineering*, V.173.

Christon, M.A., Martinez, M.J. and Voth, T.E. (2004). "Generalized Fourier analysis of the advection-diffusion equation, II. one-dimensional domains," *J. Numerical Methods Fluids*, V.45.

Dendy, J.E. (1974). "Two methods of Galerkin type achieving optimum L^2 rates of convergence for first-order hyperbolics," *SIAM J. Numerical Analysis*, V.11.

Dendy, J.E. and Fairweather, G. (1975). "Alternating-direction Galerkin methods for parabolic and hyperbolic problems on rectangular polygons," *SIAM J. Numerical Analysis*, V.12.

Donea, J. (1984). "A Taylor–Galerkin algorithm for hyperbolic conservation laws," *J. Numerical Methods Engineering*, V.20.

Engleman, M.S. (1982). "FIDAP: A Fluid Dynamics Analysis Program," *Advances in Engineering Software*, V.4. pp. 163–166.

Fletcher, C.A.J. (1991). *Computational Techniques for Fluid Dynamics, 1: Fundamental and General Techniques*, Springer-Verlag, NY.

Gresho, P.M. and Sani, R.L. (1998). *Incompressible Flow and the Finite Element Method; Advection-Diffusion and Isothermal Laminar Flow*, Wiley, England.

Gresho, P.M. and Sani, R.L. (2000). *Incompressible Flow and the Finite Element Method*, Wiley, England.

Gunzburger, M.D. (1989). *Finite Element Methods for Viscous Incompressible Flows*, Academic Press, NY.

Haberman, R. (1998). *Elementary Applied Partial Differential Equations*, 3rd Edition, Prentice-Hall, NJ.

Harlow, F.H. and Welch, J.E. (1965). "Numerical calculation of time-dependent viscous incompressible flow of fluid with free surface," *Physics of Fluids*, V.8.

Hughes, T.J.R., Liu, W.K. and Brooks, A. (1979). "Finite element analysis of incompressible viscous flows by the penalty method," *J. Computational Physics*, V.30.

Issa, R.I. (1984). "Solution of the implicitly discretised fluid flow equations by operator-splitting," *J. Computational Physics*, V.62.

Kolesnikov, A. (2000). "Efficient implementation of high order methods in computational fluid dynamics," PhD dissertation, University of Tennessee, Knoxville, TN.

Kolesnikov, A. and Baker, A.J. (2001). "An efficient high order Taylor weak statement formulation for the Navier–Stokes equations," *J. Computational Physics*, V.173.

Ladyzhenskaya, O. (1969). *The Mathematical Theory of Viscous Incompressible Flow*, Gordon and Breach Publishers, NY.

Lax, P.D. and Wendroff, B. (1960). "Systems of Conservation Laws," *Comm. Pure and Appl. Math*, V.7, pp. 217–237.

Oden, J.T., and Reddy, J.N. (1976). *An Introduction to the Mathematical Theory of Finite Elements*, Wiley-Interscience, NY.

Patankar, S., and Spalding, D.B. (1972). "A calculation procedure for heat, mass and momentum transfer in three-dimensional parabolic flows," *J. Heat and Mass Transfer*, V.15.

Patankar, S. (1980). *Numerical Heat Transfer and Fluid Flow*, Hemisphere, NY.

Raymond, W.H. and Garder, A. (1976). "Selective damping in a Galerkin method for solving wave problems with variable grids," *Monthly Weather Review*, V.104.

Rogers, S.E. and Kwak, D. (1991). "Steady and unsteady solutions of the incompressible Navier–Stokes equations," *AIAA Journal*, V.29.

Sahu. S. and Baker, A.J. (2007). "A modified conservation principles theory leading to an optimal Galerkin CFD algorithm," *J. Numerical Methods Fluids*, V.55.

Strang, G. and Fix, G.J. (1979). *An Analysis of the Finite Element Method*, Prentice-Hall, NJ.

Temam, R. (1968). "Une methode d'approximation de la solution des equations de Navier–Stokes," *Bull. Soc. Math. France*, V.96.

Temam, R. (1979). *Navier–Stokes Equations, Theory and Numerical Analysis*, North Holland Press, Amsterdam.

vanDoormaal, J.P. and Raithby, G.D. (1984). "Enhancements of the SIMPLE method for predicting incompressible fluid flow," *J. Numerical Heat Transfer*, V.7.

Wahlbin, L.B. (1974). "A dissipative Galerkin method applied to some quasi-linear hyperbolic equations," *RAIRO*, V.8.

Williams, P.T. (1993). "A 3-dimensional time-accurate incompressible Navier–Stokes finite element CFD algorithm," PhD dissertation, University of Tennessee.

Williams, P.T. and Baker, A.J. (1996). "Incompressible computational fluid dynamics and the continuity constraint method for the 3-D Navier–Stokes equations," *J. Numerical Heat Transfer, Part B, Fundamentals*, V.29.

6

Vector Field Theory Implementations:

vorticity-streamfunction, vorticity-velocity formulations

6.1 Vector Field Theory NS PDE Manipulations

The incompressible laminar-thermal NS PDE system vector statement remains

$$\mathrm{D}M: \quad \nabla \cdot \mathbf{u} = 0 \tag{6.1}$$

$$\mathrm{D}\mathbf{P}: \quad \mathrm{St}\frac{\partial \mathbf{u}}{\partial t} + \nabla \cdot \mathbf{u}\mathbf{u} + \nabla P - \frac{1}{\mathrm{Re}}\nabla \cdot \nabla \mathbf{u} + \frac{\mathrm{Gr}}{\mathrm{Re}^2}\Theta\,\widehat{\mathbf{g}} = 0 \tag{6.2}$$

$$\mathrm{D}E: \quad \mathrm{St}\frac{\partial \Theta}{\partial t} + \nabla \cdot \mathbf{u}\Theta - \frac{1}{\mathrm{Re\,Pr}}\nabla \cdot \nabla \Theta - s_\Theta = 0 \tag{6.3}$$

$$\mathrm{D}Y: \quad \mathrm{St}\frac{\partial Y}{\partial t} + \nabla \cdot \mathbf{u}Y - \frac{1}{\mathrm{ReSc}}\nabla \cdot \nabla Y - s_Y = 0 \tag{6.4}$$

The state variable for (6.1–6.4) is $\{q(\mathbf{x}, t)\} = \{\mathbf{u}, P, \Theta, Y\}^T$, the *non-dimensional* (non-D) velocity vector, kinematic pressure $P \equiv p/\rho_0$, potential temperature Θ and mass fraction Y. The Stanton, Reynolds, Grashoff, Prandtl and Schmidt number definitions remain as $\mathrm{St} \equiv \tau\mathrm{U/L}$, $\mathrm{Re} \equiv \mathrm{UL}/\nu$, $\mathrm{Gr} \equiv \mathrm{g}\beta\Delta\mathrm{TL}^3/\nu^2$, $\mathrm{Pr} \equiv \rho_0 \nu c_p/k$ and $\mathrm{Sc} \equiv \mathrm{D}/\nu$. In the Boussinesq buoyancy model $\Delta\mathrm{T} \equiv (T_{max} - T_{min})$ and $\beta \equiv 1/\mathrm{T}_{abs}$. For non-periodic flows St is typically unity, hence $\tau \equiv \mathrm{L/U}$, and in DE the Peclet number Pe commonly replaces RePr.

The mathematical constraint imposed by *continuity*, (6.1), is that *any* velocity vector field solution to D**P**, (6.2), must be *solenoidal*, that is, divergence-free. Vector field theory admits rigorous satisfaction of DM since any solenoidal vector can be expressed, to within the gradient of a solution to the laplacian, as the curl of a vector potential function. In fluid

Optimal MODIFIED CONTINUOUS Galerkin CFD, First Edition. A. J. Baker.

Companion Website: www.wiley.com/go/baker/GalerkinCFD

mechanics, the vector potential is *streamfunction*, denoted $\mathbf{\Psi}(\mathbf{x}, t)$, while the curl of velocity is *vorticity*, usually symbolized as $\mathbf{\Omega}(\mathbf{x}, t)$.

Invoking the vector identities $\mathbf{\Omega} \equiv \text{curl}\, \mathbf{u}$ and $\mathbf{u} \equiv \text{curl}\, \mathbf{\Psi}$ and defining $\text{St} \equiv 1$ produces the *vorticity-streamfunction* PDE system replacement for (6.1–6.3)

$$\mathcal{L}(\mathbf{\Omega}) = \frac{\partial \mathbf{\Omega}}{\partial t} + \nabla \times \mathbf{\Psi} \cdot \nabla \mathbf{\Omega} - (\mathbf{\Omega} \cdot \nabla)\nabla \times \mathbf{\Psi} - \frac{1}{\text{Re}} \nabla^2 \mathbf{\Omega} + \frac{\text{Gr}}{\text{Re}^2} \nabla \times \Theta\, \widehat{\mathbf{g}} = \mathbf{0} \tag{6.5}$$

$$\mathcal{L}(\Theta) = \frac{\partial \Theta}{\partial t} + \nabla \times \mathbf{\Psi} \cdot \nabla \Theta - \frac{1}{\text{RePr}} \nabla \cdot \nabla \Theta - s_\Theta = 0 \tag{6.6}$$

and also generates the strictly kinematic vector identity

$$\mathcal{L}(\mathbf{\Psi}) = -\nabla^2 \mathbf{\Psi} + \nabla(\nabla \cdot \mathbf{\Psi}) - \mathbf{\Omega} = \mathbf{0} \tag{6.7}$$

The initial-elliptic boundary value (I-EBV) PDE system (6.5–6.7) state variable is $\{q(\mathbf{x},t)\} = \{\mathbf{\Omega}, \mathbf{\Psi}, \Theta\}^T$ with boundary conditions (BCs) required on the *entire* closure of the solution domain. The key attribute of (6.5) is elimination of kinematic pressure P as state variable member. The associated compromise is that problem *data* must be sufficiently smooth to admit $\mathbf{\Omega}$ as the state variable replacement for $\mathbf{u}$.

The main detraction to $m\text{GWS}^h + \theta\text{TS}$ implementation of (6.5–6.7) is the $(\nabla \cdot \mathbf{\Psi})$ term in (6.7), which fully couples the EBV solution for streamfunction $\mathbf{\Psi}$. Identifying a divergence-free $\mathbf{\Psi}$ has proven impractical.

One alternative is to transform (6.1–6.2) to the *vorticity-velocity* I-EBV/EBV PDE system

$$\mathcal{L}(\mathbf{\Omega}) = \frac{\partial \mathbf{\Omega}}{\partial t} + (\mathbf{u} \cdot \nabla)\mathbf{\Omega} - (\mathbf{\Omega} \cdot \nabla)\mathbf{u} - \frac{1}{\text{Re}} \nabla^2 \mathbf{\Omega} + \frac{\text{Gr}}{\text{Re}^2} \nabla \times \Theta\, \widehat{\mathbf{g}} = \mathbf{0} \tag{6.8}$$

$$\mathcal{L}(\mathbf{u}) = -\nabla^2 \mathbf{u} - \nabla \times \mathbf{\Omega} = \mathbf{0} \tag{6.9}$$

by invoking only the vector identity $\mathbf{\Omega} \equiv \text{curl}\, \mathbf{u}$ and retaining (6.6) with $\nabla \times \mathbf{\Psi}$ replaced by $\mathbf{u}$. The state variable is $\{q(\mathbf{x}, t)\} = \{\mathbf{\Omega}, \mathbf{u}, \Theta\}^T$, which remains absent kinematic pressure P. The main detraction is that exact satisfaction of DM via $\mathbf{\Psi}$ is relinquished. This can be compensated by approximate DM^h enforcement via precisely designed Taylor series (TS) discrete BCs for $\mathbf{\Omega}$ on no-slip boundary segments.

A second alternative that does retain exact DM satisfaction replaces $\mathbf{\Omega}$ in (6.5) via the identity $\mathbf{\Omega} \equiv \text{curl curl}\, \mathbf{\Psi}$. The state variable is $\{q(\mathbf{x}, t)\} = \{\mathbf{\Psi}, \Theta\}^T$ with result a *biharmonic* I-EBV PDE, absent $\mathbf{\Omega}$, for which well-posed BCs of both Dirichlet and Neumann type are *a priori* known. Both dominating detractions reside in the lead differential term $\text{Re}^{-1}(\nabla^4 \mathbf{\Psi})$. The biharmonic operator requires the $m\text{GWS}^h + \theta\text{TS}$ trial space basis reside in a subspace of H^2, recall Section 3.4, a computationally *impractical* completeness requirement. Secondly, the Re^{-1} multiplier becomes vanishingly small in practice, yielding a thoroughly ill-conditioned terminal matrix statement jacobian.

This chapter implements $\text{GWS}^h/m\text{GWS}^h + \theta\text{TS}$ algorithms for vector field theory *m*PDE I-EBV/EBV restatements of (6.5–6.9), restricted to $n=2$ and (6.6), (6.8–6.9) for $n=3$.

The key sought result is quantified *validation* that weak form *linear* theories accurately predict solution *a posteriori* data generated by carefully designed compute exercises addressing these *nonlinear* alterations to the full laminar-thermal NS PDE system (6.1–6.4).

6.2 Vorticity-Streamfunction PDE System, $n = 2$

The identified PDE system (6.5–6.7) detraction is eliminated by restriction to $n=2$. This I-EBV system, thoroughly examined in the legacy CFD literature, has proven particularly pertinent to *validate* $\mathrm{GWS}^h/m\mathrm{GWS}^h + \theta\mathrm{TS}$ algorithm performance with linear weak form theory. On $n=2$ the streamfunction and vorticity vectors possess but single scalar components, $\mathbf{\Psi} \Rightarrow \psi\widehat{\mathbf{k}}$ and $\mathbf{\Omega} \Rightarrow \omega\widehat{\mathbf{k}}$, hence (6.5–6.7) simplifies to

$$\mathcal{L}(\omega) = \frac{\partial\omega}{\partial t} + \nabla\times\psi\,\widehat{\mathbf{k}}\cdot\nabla\omega - \frac{1}{\mathrm{Re}}\nabla^2\omega + \frac{\mathrm{Gr}}{\mathrm{Re}^2}\nabla\times\Theta\,\widehat{\mathbf{g}} = 0 \tag{6.10}$$

$$\mathcal{L}(\Theta) = \frac{\partial\Theta}{\partial t} + \nabla\times\psi\,\widehat{\mathbf{k}}\cdot\nabla\Theta - \frac{1}{\mathrm{RePr}}\nabla^2\Theta - s_\Theta = 0 \tag{6.11}$$

$$\mathcal{L}(\psi) = -\nabla^2\psi - \omega = 0 \tag{6.12}$$

The state variable for the I-EBV/EBV PDE system (6.10–6.12) is $\{q(\mathbf{x}, t)\} = \{\omega, \Theta, \psi\}^T$ and BCs remain required on the entire domain closure. On inflow boundaries spanned by tangential coordinate s, the typical BC specification is $\mathbf{u}(s)\cdot\hat{\mathbf{n}}$ and $\Theta(s)$, the former readily transformed to $\psi(s)$ and $\omega(s)$ using vector identities. All boundaries with outflow must be sufficiently remote such that the vanishing Neumann BC is appropriate for $\{q(\mathbf{x}, t)\}$. No-slip walls are surfaces $\psi = \text{constant}$, with specific levels determined by integration of the $\mathbf{u}(s)\cdot\hat{\mathbf{n}}$ data across all inflow boundary segments. Finally, the full range of Dirichlet and Neumann temperature BCs are applicable.

A no-slip wall Robin BC for $\omega(s)$ is readily derived from (6.12). Since any wall is $\psi =$ constant, this strictly kinematic PDE transforms thereon to the ordinary differential equation (ODE) $\mathrm{d}^2\psi/\mathrm{d}n^2 + \omega = 0$ for n the wall normal coordinate. The Taylor series (TS) expansion for $\psi(n)$ into the near-wall field flow of measure Δn is

$$\psi(\Delta n) = \psi_{\mathrm{w}} + \Delta n\left.\frac{\mathrm{d}\psi}{\mathrm{d}n}\right|_{\mathrm{w}} + \frac{\Delta n^2}{2}\left.\frac{\mathrm{d}^2\psi}{\mathrm{d}n^2}\right|_{\mathrm{w}} + \frac{\Delta n^3}{3!}\left.\frac{\mathrm{d}^3\psi}{\mathrm{d}n^3}\right|_{\mathrm{w}} + O(\Delta n^4) \tag{6.13}$$

which upon substitution of $\mathrm{d}^2\psi/\mathrm{d}n^2 + \omega = 0$ becomes

$$\psi(\Delta n) = \psi_{\mathrm{w}} - \mathrm{U}_{\mathrm{w}}\Delta n - \frac{\Delta n^2}{2}\omega_{\mathrm{w}} - \frac{\Delta n^3}{6}\left.\frac{\mathrm{d}\omega}{\mathrm{d}n}\right|_{\mathrm{w}} + O(\Delta n^4) \tag{6.14}$$

The identity $\mathbf{u} \equiv \mathrm{curl}\,\psi\widehat{\mathbf{k}} \Rightarrow \mathrm{U}_{\mathrm{w}} \equiv \mathrm{d}\psi/\mathrm{d}n|_{\mathrm{w}}$ is tangential speed *data* of the wall should it be translating.

Recognizing that the TS n direction opposes the domain outward pointing unit normal $\hat{\mathbf{n}}$, rearrangement of (6.14) produces the TS $O(\Delta n^4)$ discrete vorticity no-slip wall Robin BC

$$\ell(\omega) = \nabla\omega\cdot\hat{\mathbf{n}} + \frac{3}{\Delta n}\omega_{\mathrm{w}} + \frac{6}{\Delta n^2}\mathrm{U}_{\mathrm{w}} + \frac{6}{\Delta n^3}(\psi_{\mathrm{w}+1} - \psi_{\mathrm{w}}) \tag{6.15}$$

CFD code practice has never used recently derived (6.15), instead implements Dirichlet BCs by truncating TS (6.13) to lower orders in Δn

$$\begin{aligned}\omega_{\mathrm{w}} &\equiv \frac{-\omega_{\mathrm{w+1}}}{2} - \frac{3}{\Delta n^2}(\psi_{\mathrm{w+1}} - \psi_{\mathrm{w}}) - \frac{3}{\Delta n}\mathrm{U_w} + O(\Delta n^2) \\ \omega_{\mathrm{w}} &\equiv \frac{-2}{\Delta n^2}(\psi_{\mathrm{w+1}} - \psi_{\mathrm{w}}) - \frac{2}{\Delta n}\mathrm{U_w} + O(\Delta n)\end{aligned} \tag{6.16}$$

The notation in (6.15–6.16) is literature common but herein imprecise as (6.16) expresses DOF, not continuum, constraints. Subscript "w + 1" denotes ψ^h and ω^h DOF at the node Δn distant from the wall, with "w" signifying the wall node.

With (6.15–6.16) the $n = 2$ vorticity-streamfunction I-EBV/EBV PDE system (6.10–6.12) is *well-posed.* However, these are but computational variables as the desired solution remains velocity vector and pressure. Velocity is determinable via two approaches. The curl of identity $\omega\widehat{\mathbf{k}} \equiv \mathrm{curl}\,\mathbf{u}$ yields the well-posed EBV PDE system

$$\mathcal{L}(\mathbf{u}) = -\nabla^2\mathbf{u} - \nabla \times \omega\,\widehat{\mathbf{k}} = \mathbf{0} \tag{6.17}$$

The alternative is GWS^h on *differential definition* $\mathcal{D}(\mathbf{u}) \equiv \mathbf{u} - \mathrm{curl}\,\psi\,\widehat{\mathbf{k}} = 0$ which admits homogeneous Dirichlet DOF constraints *only*.

Pressure distributions are determined from the Poisson PDE generated by forming div **DP**. The resultant well-posed EBV PDE + BCs system remains

$$\begin{aligned}\mathcal{L}(P) &= -\nabla^2 P - \frac{\partial}{\partial x_j}\left[u_i\frac{\partial u_j}{\partial x_i} + \frac{\mathrm{Gr}}{\mathrm{Re}^2}\Theta\widehat{g}_j\right] = 0 \\ \ell(P) &= \nabla P\cdot\hat{\mathbf{n}} - \left[\frac{\partial\mathbf{u}}{\partial t} - \frac{1}{\mathrm{Re}}\nabla^2\mathbf{u}\right]\cdot\hat{\mathbf{n}} = 0\end{aligned} \tag{6.18}$$

GWS^h algorithms for (6.17–6.18) are post-convergence operations anywhere in the time evolution solution sequence for (6.10–6.12).

6.3 Vorticity-Streamfunction *m*GWS^h Algorithm

The nonlinear *m*PDE replacement for (6.10–6.12) is the ideal platform supporting *m*GWS^h + θTS algorithm *validation* of *linear* weak form theory *pertinence* for prediction of accuracy, convergence, stability and *regular* solution adapted mesh *optimal* solution existence. The steady solution *a posteriori* DOF data supporting this activity are generated via time evolution from an initial condition (IC). The *m*PDE alteration to (6.10–6.12) thereby omits the $O(\Delta t^4, m^3)$ augmentation terms. Inserting the steady theory $O(h^4)$ augmentations, defined in (5.113–5.117), generates the *m*PDE system

$$\begin{aligned}\mathcal{L}^m(\omega) = &\frac{\partial\omega}{\partial t} + \nabla\times\psi\,\widehat{\mathbf{k}}\cdot\nabla\omega - \frac{1}{\mathrm{Re}}\nabla^2\omega \\ &+ \frac{\mathrm{Gr}}{\mathrm{Re}^2}\nabla\times\Theta\,\widehat{\mathbf{g}} - \frac{\mathrm{Re}h^2}{12}\frac{\partial}{\partial x_j}\left[u_j\,u_k\frac{\partial\omega}{\partial x_k}\right] = 0\end{aligned} \tag{6.19}$$

$$\mathcal{L}^m(\Theta) = \frac{\partial \Theta}{\partial t} + \nabla \times \psi \widehat{\mathbf{k}} \cdot \nabla \Theta - \frac{1}{\text{RePr}} \nabla^2 \Theta \\ - s_\Theta - \frac{\text{RePr} h^2}{12} \frac{\partial}{\partial x_j} \left[u_j u_k \frac{\partial \Theta}{\partial x_k} \right] = 0 \tag{6.20}$$

$$\mathcal{L}^m(\psi) = -\nabla^2 \psi - \omega - \frac{h^2}{12} \nabla^2 \omega = 0 \tag{6.21}$$

As always the end result of the $m\text{GWS}^h + \theta\text{TS}$ algorithm is a nonlinear iterative algebraic matrix statement of form (5.119)

$$\begin{aligned} [\text{JAC\_}\{\text{Q}I\}]\{\delta \text{Q}I\}^{p+1} &= -\{\text{F\_Q}I\}_{n+1}^{p} \\ [\text{JAC\_}\{\text{Q}I\}] &\equiv \frac{\partial \{\text{F\_Q}I\}}{\partial \{\text{Q}I\}} \\ \{\text{Q}I\}_{n+1}^{p+1} &= \{\text{Q}I\}_{n+1}^{p} + \{\delta \text{Q}I\}^{p+1} \\ &= \{\text{Q}I\}_n + \sum_{\alpha=0}^{p} \{\delta \text{Q}I\}^{\alpha+1} \end{aligned} \tag{6.22}$$

written on state variable DOF $\{\text{Q}I(t)\} = \{\text{OMG, TEM, PSI}\}^T$ of the discrete approximate solution state variable $\{q^h(\mathbf{x}, t)\} = \{\omega^h, \Theta^h, \psi^h\}^T$. The jacobian (6.22) is fully coupled in $\{\text{Q}I\}$, $1 \leq I \leq 3$, with p the iteration index and n denoting timing $t_n = t_0 + n\Delta t$.

All matrices for (6.22) are formed on the generic element of the discretization Ω^h of the domain Ω with subsequent *assembly* generating the global matrix statement (F_QI}. Recalling matrix prefix B denotes two dimensions, for NC or TP $k = 1$ basis the $m\text{GWS}^h$ + θTS element matrix statements for $\{\text{Q}I(t)\} = \{\text{OMG, TEM}\}^T$, $1 \leq I \leq 2$, are

$$\{\text{F\_Q}I\}_e \equiv [\text{B200}]_e \{\text{Q}I_{n+1} - \text{Q}I_n\}_e \\ + \Delta t \left[\theta \{\text{RES\_}\{\text{Q}I\}\}_e \big|_{n+1} + (1-\theta) \{\text{RES\_}\{\text{Q}I\}\}_e \big|_n \right] \tag{6.23}$$

$$\{\text{RES\_OMG}\}_e = \{\text{PSI}\}_e^T [\text{B3}L0K]_e \{\text{OMG}\}_e \text{e}_{KL} \\ + \frac{1}{\text{Re}} [\text{B2}KK]_e \{\text{OMG}\}_e + \frac{\text{Gr}}{\text{Re}^2} [\text{B20}K]_e \{\text{TEM}\}_e \delta_{K1} \\ + \frac{h^2\, \text{Re}}{12} \{\text{U}J\, \text{U}K\}_e^T [\text{B30}JK]_e \{\text{OMG}\}_e \tag{6.24}$$

$$\{\text{RES\_TEM})\}_e = \{\text{PSI}\}_e^T [\text{B3}L0K]_e \{\text{TEM}\}_e \text{e}_{KL} \\ + \frac{1}{\text{RePr}} [\text{B2}KK]_e \{\text{TEM}\}_e \\ + \frac{h^2\, \text{RePr}}{12} \{\text{U}J\, \text{U}K\}_e^T [\text{B30}JK]_e \{\text{TEM}\}_e \tag{6.25}$$

For DOF $\{\mathrm{Q3}\} = \{\mathrm{PSI}\}$ the element matrix statement is

$$\begin{aligned}\{\mathrm{F\_PSI}\}_e \equiv [\mathrm{B2}KK]_e\{\mathrm{PSI}\}_e - [\mathrm{B200}]_e\{\mathrm{OMG}\}_e \\ - \frac{h^2}{12}[\mathrm{B2}KK]_e\{\mathrm{OMG}\}_e\end{aligned} \tag{6.26}$$

The italicized indices sum $1 \le (J,\ K,\ L) \le 2 = n$. In (6.24–6.25), δ_{K1} is the Kronecker delta with e_{KL} the curl alternator ($\mathrm{e}_{KL} = +1, -1$ for index ordering KL, LK). Matrices with e-subscripts have embedded element-dependent metric data.

Forming the Newton jacobian, (6.22), or any approximation thereto, employs analytical calculus. Denoting the absent DOF couplings as [0], the element jacobian is

$$[\mathrm{JAC\_}\{\bullet\})]_e \equiv \frac{\partial\{\mathrm{F\_Q}I\}_e}{\partial\{\mathrm{Q}J\}_e} = \begin{bmatrix} [\mathrm{JAC\_OO}], & [\mathrm{JAC\_OT}], & [\mathrm{JAC\_OP}] \\ [0], & [\mathrm{JAC\_TT}], & [\mathrm{JAC\_TP}] \\ [\mathrm{JAC\_PO}], & [0], & [\mathrm{JAC\_PP}] \end{bmatrix}_e \tag{6.27}$$

with {O, T, P} shorthand for {OMG, TEM, PSI}. The suggested exercise will verify the element matrices populating (6.27) are

$$[\mathrm{JAC\_OO}]_e = [\mathrm{B200}]_e + \theta\Delta t\left(\begin{aligned}\{\mathrm{PSI}\}_e^T[\mathrm{B3}L0K]_e\,\mathrm{e}_{KL} + \frac{1}{\mathrm{Re}}[\mathrm{B2}KK]_e \\ + \frac{h^2\,\mathrm{Re}}{12}\{\mathrm{U}J\,\mathrm{U}K\}_e^T[\mathrm{B30}JK]_e\end{aligned}\right) \tag{6.28}$$

$$[\mathrm{JAC\_OT}]_e = \theta\Delta t\left(\frac{\mathrm{Gr}}{\mathrm{Re}^2}[\mathrm{B20}K]_e\delta_{K1}\right) \tag{6.29}$$

$$[\mathrm{JAC\_OP}]_e = \theta\Delta t\left(\{\mathrm{OMG}\}_e^T[\mathrm{B3}K0L]_e\,\mathrm{e}_{KL}\right) \tag{6.30}$$

$$[\mathrm{JAC\_TT}]_e = [\mathrm{B200}]_e + \theta\Delta t\left(\begin{aligned}\{\mathrm{PSI}\}_e^T[\mathrm{B3}L0K]_e\,\mathrm{e}_{KL} + \frac{1}{\mathrm{Pe}}[\mathrm{B2}KK]_e \\ + \frac{h^2\mathrm{Pe}}{12}\{\mathrm{U}J\,\mathrm{U}K\}_e^T[\mathrm{B30}JK]_e\end{aligned}\right) \tag{6.31}$$

$$[\mathrm{JAC\_TP}]_e = \theta\Delta t\left(\{\mathrm{TEM}\}_e^T[\mathrm{B3}K0L]_e\,\mathrm{e}_{KL}\right) \tag{6.32}$$

$$[\mathrm{JAC\_PO}]_e = -[\mathrm{B200}]_e - \frac{h^2}{12}[\mathrm{B2}KK]_e, \quad [\mathrm{JAC\_PP}]_e = [\mathrm{B2}KK]_e \tag{6.33}$$

Note the index order exchange in cross-coupling matrices $[\mathrm{JAC\_OP}]_e$ and $[\mathrm{JAC\_TP}]_e$, compare to (6.24–6.25), and $\mathrm{RePr} \Rightarrow \mathrm{Pe}$ in (6.31). This jacobian is linearized by assuming the mPDE matrix nonlinearity is retarded using the converged solution data $\{\mathrm{U}J\,\mathrm{U}K(t_n)\}_e^T$. Finally, vorticity DOF constraint (6.16) insertion requires appropriate replacements in $\{\mathrm{F\_OMG}\}_e$ and $[\mathrm{JAC\_OP}]_e$. Their omission in the jacobian severely destabilizes the iteration process.

The velocity vector DOF are generated via GWS^h on EBV PDE (6.17) yielding the element matrix statement

$$\{F\_\mathrm{U}I\}_e \equiv [\mathrm{B2}KK]_e\{\mathrm{U}I\}_e - [\mathrm{B20}L]_e\{\mathrm{OMG}\}_e\,\mathrm{e}_{IL} \tag{6.34}$$

Alternatively, recalling the alternating tensor e_{ijk}, continuum GWS^N on *differential definition* $\mathcal{D}(\mathbf{u}) = \mathbf{u} - \mathrm{curl}\,\psi\,\widehat{\mathbf{k}} = 0$

$$\mathrm{GWS}\left(\mathbf{u}^N\right) \equiv \int_\Omega \Psi_\beta(\mathbf{x})\left[u_i^N - \mathrm{e}_{ijk}\frac{\partial \psi_k}{\partial x_j}\right]\mathrm{d}\tau \equiv 0, \quad \forall\beta \tag{6.35}$$

transitioned to GWS^h generates the element matrix statement

$$\{\mathrm{F\_U}I\}_e = [\mathrm{B200}]_e\{\mathrm{U}I\}_e - [\mathrm{B20}L]_e\{\mathrm{PSI}\}_e\,\mathrm{e}_{IL} \tag{6.36}$$

Since (6.34) is an EBV PDE BCs are required specified on the entire domain closure. The sole constraint applicable in solving (6.36) is homogeneous Dirichlet for DOF $\{\mathrm{U}I\}$ constrained by a no-slip wall BC.

Following two applications of the Green–Gauss divergence theorem and imposition of D*M*, (6.1), the continuum GWS^N for (6.18) is

$$\begin{aligned}
\mathrm{GWS}\left(\mathcal{L}(P^N)\right) &\equiv \int_\Omega \Psi_\beta(\mathbf{x})\left[-\frac{\partial}{\partial x_j}\left(\frac{\partial P^N}{\partial x_j} + u_i^N\frac{\partial u_j^N}{\partial x_i} + \frac{\mathrm{Gr}}{\mathrm{Re}^2}\Theta^N\widehat{g}_j\right)\right]\mathrm{d}\tau \\
&= \int_\Omega \frac{\partial \Psi_\beta(\mathbf{x})}{\partial x_j}\left[\frac{\partial P^N}{\partial x_j} + u_i^N\frac{\partial u_j^N}{\partial x_i} + \frac{\mathrm{Gr}}{\mathrm{Re}^2}\Theta^N\widehat{g}_j\right]\mathrm{d}\tau \\
&\quad + \int_{\partial\Omega} \Psi_\beta(\mathbf{x})\left[\frac{\partial \mathbf{u}^N}{\partial t} - \frac{1}{\mathrm{Re}}\nabla^2\mathbf{u}^N\right]\cdot\hat{\mathbf{n}}\mathrm{d}\sigma \equiv 0, \quad \forall\beta
\end{aligned} \tag{6.37}$$

Proceeding through $\mathrm{GWS}^N \rightarrow \mathrm{GWS}^h$ the $n=2$ element matrix statement for pressure DOF $\{\mathrm{PRS}\}$ is

$$\begin{aligned}
\{\mathrm{F_PRS}\}_e &= [\mathrm{B2}KK]_e\{\mathrm{PRS}\}_e + \{\mathrm{U}I\}_e^T[\mathrm{B30}JI]_e\{\mathrm{U}J\}_e \\
&\quad + \frac{\mathrm{Gr}}{\mathrm{Re}^2}[\mathrm{B2}J0]_e\{\mathrm{TEM}\}_e\widehat{g}_j + [\mathrm{A200}]_e\frac{\mathrm{d}}{\mathrm{d}t}\{\mathrm{U}J \bullet n_j\}_e \\
&\quad + \frac{1}{\mathrm{Re}}\left([\mathrm{A211}]_e\{\mathrm{U}J \bullet \widehat{n}_j\}_e - \mathrm{dU}J_e/\mathrm{ds}|_\mathrm{w}\right)
\end{aligned} \tag{6.38}$$

retaining the BC unsteady term contribution for theory completeness.

The last line in (6.38) is via a second Green–Gauss divergence theorem application on the $\partial\Omega_e$ BC laplacian operator in (6.37). The last term therein is wall shear in the s coordinate spanning an outflow boundary plane. Recalling GWS^h flux determination for nodal DOF Dirichlet BC constrained, Section 2.3, the exercise will verify that last term imposition amounts to deleting the wall node contributions from $[\mathrm{A211}]_e\{\mathrm{U}J \bullet \widehat{n}_j\}_e$!

Algorithm element matrices subscripted e require extraction of embedded metric data. The $[\mathrm{B200}]_e$ transforms to $(\det_e/2)[\mathrm{B200}]$ or $(\det_e/4)[\mathrm{B200}]$ for the NC or TP $k=1$ basis. The diffusion matrix $[\mathrm{B2}KK]_e$ transformation is thoroughly detailed in Chapter 3, (3.74–3.76), (3.99–3.100). For summation conventions $1 \le (i, j, k) \le n = 2$ and $1 \le \alpha \le 3$, the NC and TP basis forms are

$$[\mathrm{B2}KK]_e \Rightarrow \frac{1}{2\det_e}(\zeta_{\alpha i}\zeta_{\beta i})_e [\mathrm{B}\delta_{\alpha\beta}] \quad or \quad \frac{1}{4\det_e}(\eta_{ji}\eta_{ki})_e [\mathrm{B2}jk] \tag{6.39}$$

The new nonlinear element hypermatrices are $[\mathrm{B3}L0K]_e \mathrm{e}_{KL}$, $[\mathrm{B3}K0L]_e \mathrm{e}_{KL}$, and $[\mathrm{B30}JK]_e$. The first two result from

$$\begin{aligned}
\int_{\Omega_e} \{N_1(\boldsymbol{\xi}, \boldsymbol{\eta})\} \nabla \times \psi_e \widehat{\mathbf{k}} \cdot \nabla q_e \mathrm{d}x\mathrm{d}y &= \int_{\Omega_e} \{N_1\} \left[\frac{\partial \psi_e}{\partial y}\frac{\partial q_e}{\partial x} - \frac{\partial \psi_e}{\partial x}\frac{\partial q_e}{\partial y} \right] \mathrm{d}x\mathrm{d}y \\
&= \{\mathrm{PSI}\}_e^T \int_{\Omega_e} \left[\frac{\partial\{N_1\}}{\partial y}\{N_1\}\frac{\partial\{N_1\}^T}{\partial x} - \frac{\partial\{N_1\}}{\partial x}\{N_1\}\frac{\partial\{N_1\}^T}{\partial y} \right] \det_e \mathrm{d}\tau \{Q\}_e \\
&\equiv \{\mathrm{PSI}\}_e^T [\mathrm{B3}L0K]_e \{Q\}_e \, \mathrm{e}_{KL}
\end{aligned} \tag{6.40}$$

Proceeding through the coordinate transformations

$$[\mathrm{B3}L0K]_e \Rightarrow \frac{1}{2\det_e}\left(\zeta_{\alpha L}\,\zeta_{\beta K}\right)_e [\mathrm{B3}\alpha 0\beta] \quad or \quad \frac{1}{4\det_e}(\eta_{iL}\,\eta_{jK})[\mathrm{B3}i0j] \tag{6.41}$$

These element-*in*dependent NC/TP $k=1$ basis hypermatrices $[\mathrm{B3}\alpha 0\beta]$ and $[\mathrm{B3}i0j]$ are skew-symmetric rational integer arrays. The matrix elements of the hypermatrix $[\mathrm{B3}K0L]_e$ are the row-column transpose of those in $[\mathrm{B3}L0K]_e$.

The *m*PDE hypermatrix $[\mathrm{B30}KL]_e$ for the $k=1$ NC basis equates to an average, that is, the scalar $\{\mathrm{B10}\}^T\{\mathrm{U}J\mathrm{U}K\}_e$, multiplier on $[\mathrm{B2}KK]_e$, recall Table 4.3 for the $n=1$ case. The TP basis distributes the velocity tensor product on Ω_e; the transformed statement is

$$\frac{h^2\mathrm{Pa}}{12}\{\mathrm{U}J\,\mathrm{U}K\}_e^T[\mathrm{B30}JK]_e \Rightarrow \frac{h^2\mathrm{Pa}}{48\det_e}(\eta_{Jl}\eta_{Km})_e\{\mathrm{U}J\,\mathrm{U}K\}_e^T[\mathrm{B30}lm] \tag{6.42}$$

for Pa ⇒ Re, RePr as appropriate, with $1 \le (l, m) \le 2$ additional summation index pairs.

The curl operation generating (6.10) rotates the buoyancy body force to an x_1 directional derivative. Extracting metric data

$$[\mathrm{B20}K]_e\delta_{K1} \Rightarrow (\zeta_{\alpha 1})_e[\mathrm{B20}\alpha] \quad or \quad \left(\eta_{k\,1}\right)_e[\mathrm{B20}k] \tag{6.43}$$

for summation conventions $1 \le \alpha \le 3$, $1 \le k \le 2$ for the NC, TP $k=1$ bases. Replacing the Kronecker delta in (6.43) with the alternator induces *only* the appropriate index alteration.

6.4 Weak Form Theory Verification, $\mathrm{GWS}^h/m\mathrm{GWS}^h$

Weak form linear theory predictions requiring validation for $k=1$ basis $\mathrm{GWS}^h/m\mathrm{GWS}^h +$ $\theta\mathrm{TS}$ $\{q^h(\mathbf{x},t)\}=\{\omega^h, \Theta^h, \psi^h\}^T$ solutions to the NS vorticity-streamfunction PDE/*m*PDE systems include:

- adherence to *linear* theory asymptotic error estimates in state variable system energy norms for explicitly *nonlinear* NS PDE systems
- weak form Galerkin criterion solution optimal compared with discrete implementation peer group for identical DOF
- fixed DOF optimal mesh solution accrues to the regular solution adapted mesh that supports energy norm equi-distributions
- *m*odified weak form theory $m\mathrm{GWS}^h$ significant-order dispersion error *annihilation* for large Re, Pe
- algebraic stability in the presence of significant dispersion error
- absence of artificial diffusion pollution
- quantification of attainable solution *accuracy*.

Theory assessments plus solution accuracy confirmation is accomplished via precisely designed compute experiments for *benchmark* and *validation* NS statements. The benchmark class possesses quality, independently generated CFD algorithm solution DOF data for comparison. Validation requires existence of quality experimental data.

The linear theory *asymptotic error estimate* for $\mathrm{GWS}^h+\theta\mathrm{TS}$ solutions to *non-modified* NS EBV PDE system (6.10–6.12) is (4.43) with $\gamma=\min(k, r-1)$ and exchanging Δx with Δt. Conversely, for $k=1$ basis $m\mathrm{GWS}^h+\theta\mathrm{TS}$ solutions to *m*PDE system (6.19–6.21), theory predicts $\gamma\rightarrow\min(k+1, r-1)$, Section 5.9.

The *unified* asymptotic error estimate for $\mathrm{GWS}^h/m\mathrm{GWS}^h+\theta\mathrm{TS}$ solutions $\{q^h(\mathbf{x},t)\}=$ $\{\omega^h, \Theta^h, \psi^h\}^T$ to the NS state variable is

$$\begin{aligned}\left\|\{e^h(n\Delta t)\}\right\|_E &\leq Ch^{2\gamma}\left[\|\mathrm{data}\|_{L2,\Omega}+\|\mathrm{data}\|_{L2,\partial\Omega}\right]+C_2\Delta t^{f(\theta)}\left\|\{q^h(t_0)\}\right\|^2_{H^1(\Omega)}\\ &\text{for}: \gamma\equiv\min(k,(k+1),r-1), f(\theta)=(2,3)\end{aligned} \tag{6.44}$$

for C and C_2 constants *and* caveat $(k+1)\Rightarrow(1+1)=2$ for *sufficiently large* Re.

A *benchmark* NS statement devised by the FD CFD community in early 1970s, subsequently transitioning to the "holy grail" test, is the *lid driven cavity*. It is the $n=2$ unit square cavity model of journal bearing lubrication with imposed lid translation velocity, Figure 6.1 top left. Contrary to reality, no flow is allowed between side walls and lid, hence $\psi=0$ is the streamfunction BC on the domain boundary entirety.

The absence of lid–wall through flow generates the benchmark definition key attribute, specifically the vorticity solution is *singular*! This vorticity solution aspect is *not* captured using FD second order accurate laplacian stencil (3.103), which *omits* domain corner DOF. Conversely, $\mathrm{GWS}^h/m\mathrm{GWS}^h$ laplacian stencils omit no DOF, (3.111), thus leading to *highly* informative benchmark driven cavity solutions.

The $k=1$ basis $\mathrm{M}=16^2$ uniform mesh *modest* $\mathrm{Re}=100$ GWS^h steady solution DOF interpolation is Figure 6.1 top right, the arrow denoting lid motion. The mesh overlay on

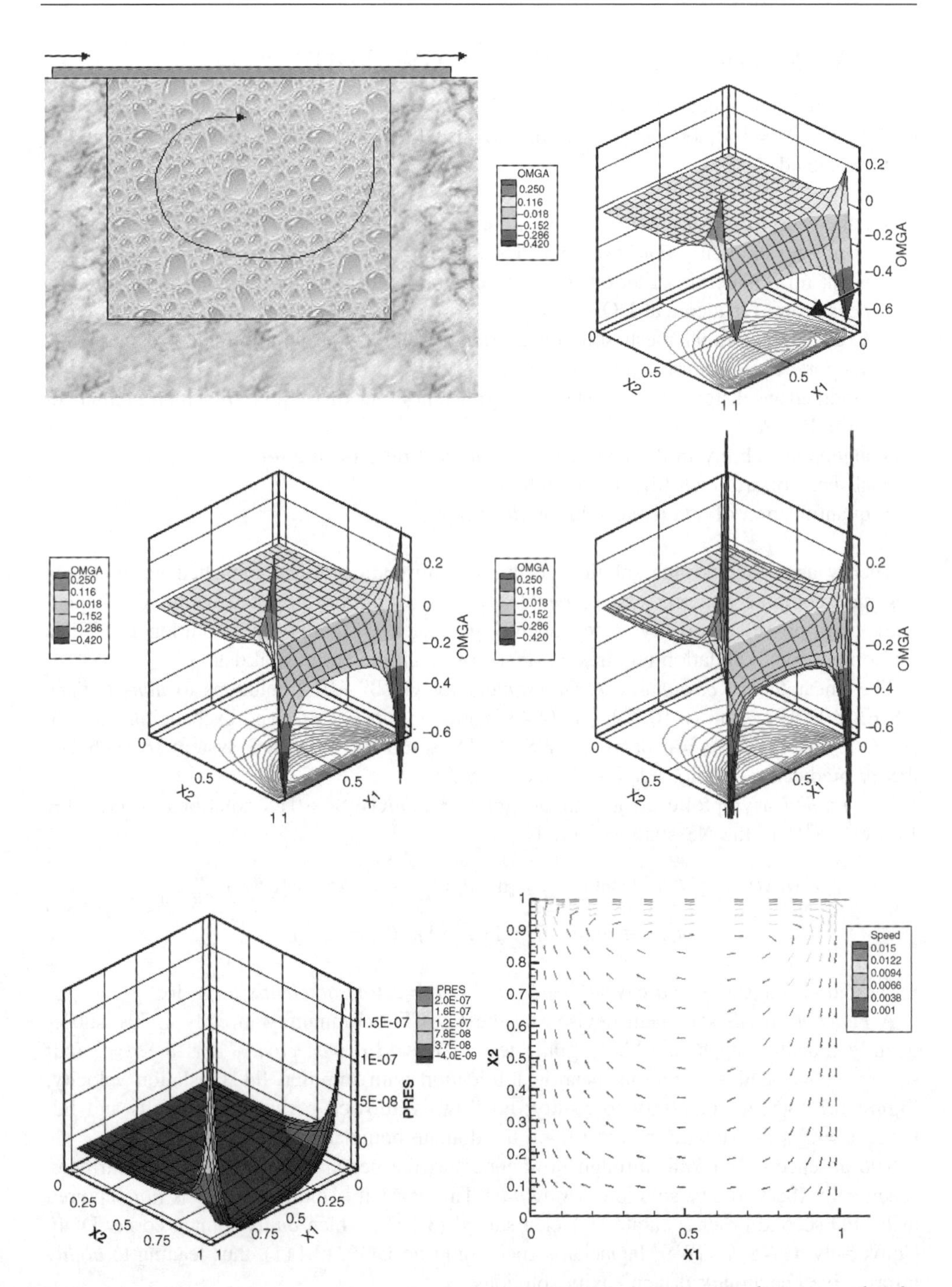

Figure 6.1 Legacy driven cavity benchmark, top left, motion geometry: GWS^h steady $\{\omega^h, \psi^h\}$ DOF distributions, Re = 100, $k=1$ TP M = 16^2 cartesian mesh: top right, uniform; middle left/right, pr = 0.9/0.7; bottom, for pr = 0.7: left, P^h; right, $\mathbf{u}^h$ unit vectors colored by speed

ω^h perspective clearly visualizes resolution, in particular the significant gradients evident near each lid–wall corner. The companion ψ^h contours are plotted on the base plane.

Resolution of vorticity solution corner singularities are markedly improved upon *regular solution adapted* mesh refinement via geometric progression ratios $l_{e+1} \equiv \mathrm{pr}(l_e)$ from cavity center to each wall. Figure 6.1, middle left, right, graphs Re = 100 GWS^h steady solution ω^h DOF interpolations for pr = 0.9, 0.7. The mesh overlay perspectives clearly quantify progressive ω^h solution lid–corner singularity mesh resolution attempts. The planar ψ^h contours are visually unaffected. (Note: ω^h off- scale extrema are truncated for uniform mesh data direct comparison.)

The pr = 0.7 ω^h DOF data are post-processed for velocity vector distribution, (6.34), which generates the {**U**} DOF data required to compute *genuine pressure* (6.38). Each cavity corner is a stagnation point and these data generate the lid–corner sharp pressure extrema, Figure 6.1 bottom left, associated with dynamic pressure recovery, (3.114). This prediction is due *strictly* to the velocity vector tensor product derivative source term in continuum Poisson PDE, (6.18).

The GWS^h laplacian global matrix assembled from (6.38) is *singular*, the result of the uniformly homogeneous Neumann BC specification for ψ^h. This previous "show-stopper," recall Section 4.14, Figure 4.18 top right, is *not a problem* using the *preconditioned conjugate gradient* Krylov solver, Section 2.12. The post-processed pr = 0.7 Re = 100 velocity *unit* vector {**U**/|**U**|} DOF distribution colored by speed, Figure 6.1 bottom right, confirms the minimal domain span of the stagnation region flowfields for Re = 100.

The Re = 100 uniform mesh refinement M = 16^2, 32^2, 64^2*a posteriori* data confirm that the $k = 1$ TP basis GWS^h algorithm *asymptotic convergence* rate in energy for both state variable {ω^h, ψ^h} members is quadratic, as theory predicted (6.44). Despite the near singular behavior the $\|\omega^h\|_E$ norm data solidly confirm theory, Figure 6.2, Ericson (2001). The M = 16^2, 32^2 data for $\|\psi^h\|_E$ is firmly quadratic; the M = 32^2 and 64^2 data being nominally equal may relate to ω^h singularity local impact. The FD CFD community-defined *legacy* point norms are extremum ψ^h DOF and co-located ω^h DOF. This comparative TS

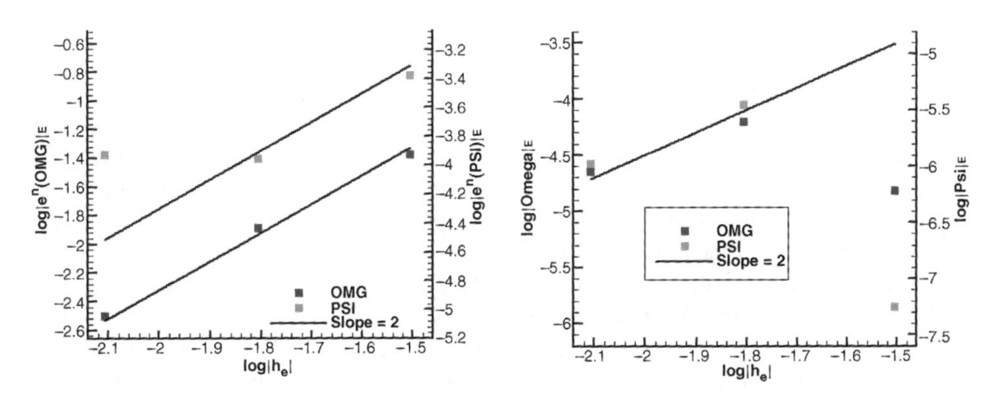

Figure 6.2 Driven cavity benchmark, asymptotic convergence, $k = 1$ TP basis GWS^h steady vorticity and streamfunction solutions: left, in energy norms $\|\omega^h\|_E$ and $\|\psi^h\|_E$; right, in FD legacy point norm DOF pair

order-of-accuracy convergence DOF assessment, Figure 6.2 right, is solidly quadratic on the finer meshes for both state variable members.

For Re = 10, 100, mGWS^h algorithm uniform mesh refinement *a posteriori* data are reported, Kolesnikov and Baker (2001, Table 4). These quantitatively confirm theory prediction of enhanced asymptotic convergence limited to large Re, as the Re = 10 rate for $\|\psi^h\|_E$ is $O(h^2)$. The $O(h^3)$ rate for Re = 100 is one order smaller than theory, likely due to Re = 100 *not being* sufficiently large. However, it does constitute theory verification because the rate quantitatively exceeds GWS^h $O(h^2)$ convergence. Supporting data for $\|\omega^h\|_E$ are not reported. Finally, their Table 3 DOF data confirm mGWS^h theory TS DOF $O(h^4)$ convergence for an $n = 2$ scalar convection-diffusion PDE at Pa = E + 03. This prediction is in agreement with quantified convergence in energy, Figure 5.17.

For substantial Re, driven cavity benchmark vorticity corner singularities pathologically induce rampant dispersion error pollution due to convection nonlinearity discrete approximation. However, the robustness of the GWS^h Newton algorithm (6.27–6.33) generates monotone convergence to steady solutions, even those grossly polluted by mesh scale oscillations Thus is enabled quantification of mGWS^h algorithm phase selective *dispersion error annihilation* absent artificial diffusion pollution.

For Re = 2000 and modestly solution adapted pr = 0.9 M = 32^2 mesh, the $k = 1$ TP basis GWS^h steady ω^h DOF interpolation is perspective graphed with mesh overlay, Figure 6.3 top left. The DOF extrema are again truncated for visualization of the mesh scale $O(2\Delta x)$ oscillatory dispersion error pollution everywhere between the extrema. The largely solution adapted pr = 0.7, M = 32^2 mesh GWS^h steady solution for ω^h nearly eliminates $O(2\Delta x)$ dispersion error, Figure 6.3 top right, at the expense of inducing coarse grid inaccuracy along the opposing wall, compare mesh overlay differences.

In distinction, the moderately solution adapted pr = 0.8, M = 32^2 mesh mGWS^h steady solution for ω^h, Figure 6.3 bottom left, retains quality global resolution while totally *annihilating*(!) the $O(2\Delta x)$ dispersive error admitted by the GWS^h algorithm. The resulting post-processed mGWS^h P^h solution DOF interpolation, Figure 6.3 bottom right, is *monotone* predicting *very sharp* peaks in both lid–corner stagnation approximations, Figure 6.3 bottom right. Extrema in all Figure 6.3 solution graphics are well off scale, the truncation compromise required for interpolated DOF resolution quantification.

mGWS^h post-processed $\{\mathbf{U}/|\mathbf{U}|\}$ DOF distribution colored by speed, Figure 6.4, details the local very complex velocity distribution in approaching the left wall–lid stagnation point. The vertical velocity attached wall jet separates well short of the corner, turning sharply into and then out of the near corner domain interior prior to sliding lid attachment. The downstream lid–wall corner juncture closely emulates classical stagnation point flow, transitioning to a high speed wall jet down that vertical surface. The truly substantial corner pressure extrema, Figure 6.3, are due solely to these local velocity field tensor product distributions residing in the Poisson PDE source term, (6.18).

The linear weak form theory assertion of GWS^h hence mGWS^h *solution optimality* is validated via *a posteriori* DOF data comparisons with the FD CFD literature defined driven cavity legacy point norm pair. These FD definitions are Ψ^h extremum DOF, BC-guaranteed cavity interior located, its coordinate pair, and co-located ω^h DOF, definitely not an extremum. The universally referenced benchmark data, Ghia *et al.* (1982), were generated for $100 \leq \mathrm{Re} \leq 3200$ using a second order FD multi-grid algorithm on uniform square meshes $16^2 \leq \mathrm{M} \leq 256^2$.

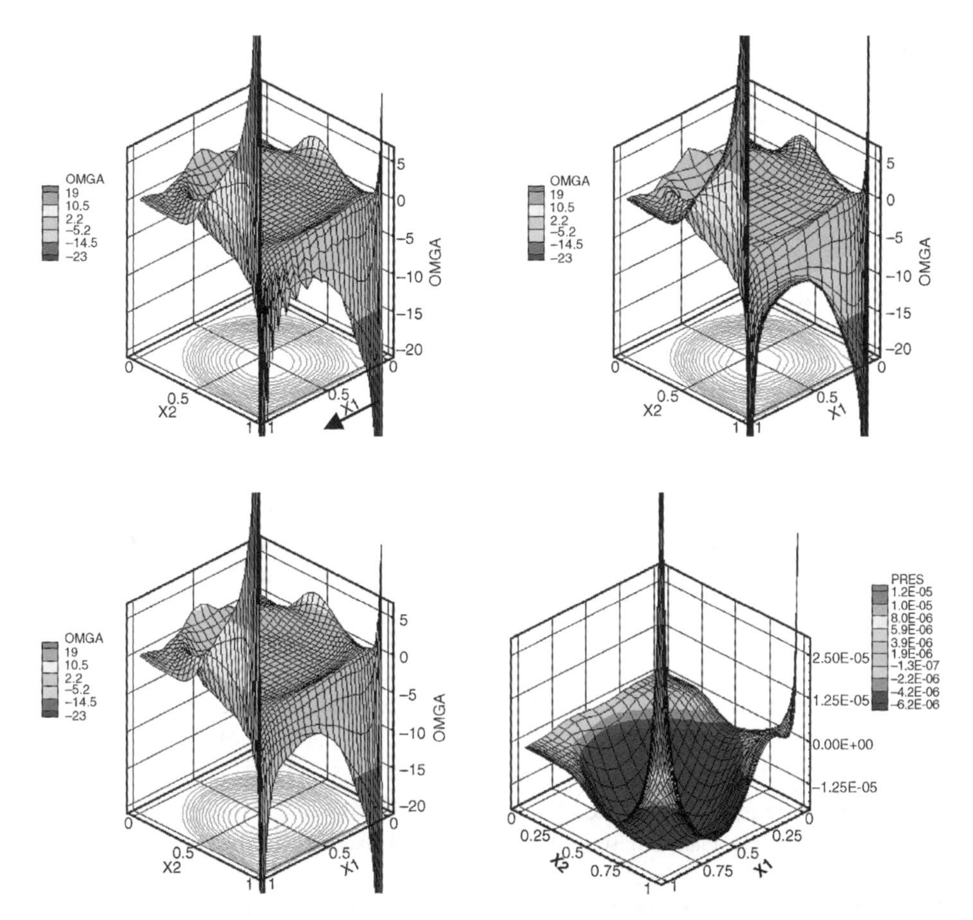

Figure 6.3 Driven cavity benchmark, $k=1$ TP basis steady vorticity-streamfunction solutions, $\mathrm{Re}=2000$, various $\mathrm{M}=32^2$ meshes: top left, GWS^h, pr = 0.9; top right, GWS^h pr = 0.7; bottom left, $m\mathrm{GWS}^h$, pr = 0.8; bottom right, $m\mathrm{GWS}^h$ solution post-processed pressure

Their $\mathrm{M}=256^2$ solution DOF data are compared to the $k=1$ TP basis GWS^h and $m\mathrm{GWS}^h$ solution extremum DOF in Table 6.1. For $\mathrm{Re}\leq 1000$, GWS^h and $m\mathrm{GWS}^h$ data are for a uniform $\mathrm{M}=64^2$ mesh, while those for $\mathrm{Re}=3200$ are for a pr = 0.8 $\mathrm{M}=64^2$ solution adapted mesh. ***Bold italic*** numerals denote DOF comparative significant digit. For all $\mathrm{Re}\leq 1000$ both GWS^h and $m\mathrm{GWS}^h$ algorithm $\mathrm{M}=64^2$ uniform mesh solutions possess Ψ^h DOF extrema exceeding the FD benchmark, hence are comparatively *optimal* in this solution measure.

This is particularly noteworthy because theory predicts optimality for *strictly identical* DOF-mesh data. This is *not* the case for Table 6.1 data, since the $\mathrm{M}=256^2$ FD benchmark contains four times the GWS^h DOF. The GWS^h and $m\mathrm{GWS}^h$ solution co-located ω^h DOF are both also larger for this Re range, but this does not constitute a quantified optimum.

For $\mathrm{Re}=3200$, neither the GWS^h nor $m\mathrm{GWS}^h$ $\mathrm{M}=64^2$ Ψ^h extremum DOF exceeds the $\mathrm{M}=256^2$ FD benchmark data. This is not unexpected realizing cavity vorticity singularity

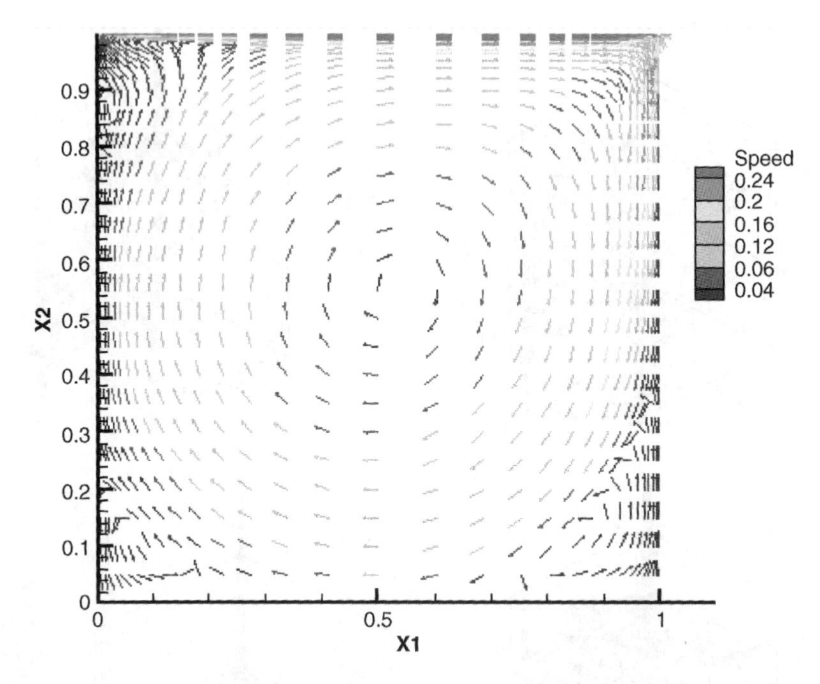

Figure 6.4 Driven cavity benchmark, $k=1$ TP basis steady $m\text{GWS}^h$ solution post-processed velocity unit vector $\{\mathbf{U}/|\mathbf{U}|\}$ DOF distribution, colored by speed, Re = 2000, pr = 0.8, $M = 32^2$ mesh

mesh resolution demand at this Re. The $m\text{GWS}^h$ ω^h DOF exceeds that for GWS^h, the result of dispersion error annihilation. That it also well exceeds the FD benchmark data is open to interpretation.

Linear weak form theory predicts that the *solution adapted* mesh for which the energy norm is *extremum* supports the *optimal* solution attainable for fixed DOF. For $100 \le \text{Re} \le 2000$, pr-adjusted $M = 32^2$ mesh GWS^h algorithm *a posteriori* norm data confirm that $\|\omega^h\|_E$ and $\|\psi^h\|_E$ trend in opposite directions with pr, Figure 6.5, Ericson (2001). Since both state variable norm extrema exist within $0.9 \le \text{pr} \le 0.8$, they bound the *optimally accurate* fixed DOF $M = 32^2$ mesh solutions attainable for $100 \le \text{Re} \le 2000$. That the pr = 0.8 $M = 64^2$ mesh also supported the determined optimal $m\text{GWS}^h$ solution for Re = 3200 adds substance to this conclusion.

Table 6.1 Optimality, solution point norm extremum DOF

Re	Ghia *et al.* (1982)		GWS^h		$m\text{GWS}^h$	
	$\lvert\{\text{PSI}\}\rvert_{max}$	{OMG}	$\lvert\{\text{PSI}\}\rvert_{max}$	{OMG}	$\lvert\{\text{PSI}\}\rvert_{max}$	{OMG}
100	0.103423	3.16646	0.10377	3.29898	0.10358	3.30388
400	0.113909	2.29469	0.11467	2.31917	0.11771	2.31648
1000	0.117929	2.04908	0.11881	2.09925	0.11835	2.10249
3200	0.120377	1.98860	0.11835	1.49245	0.11836	2.10410

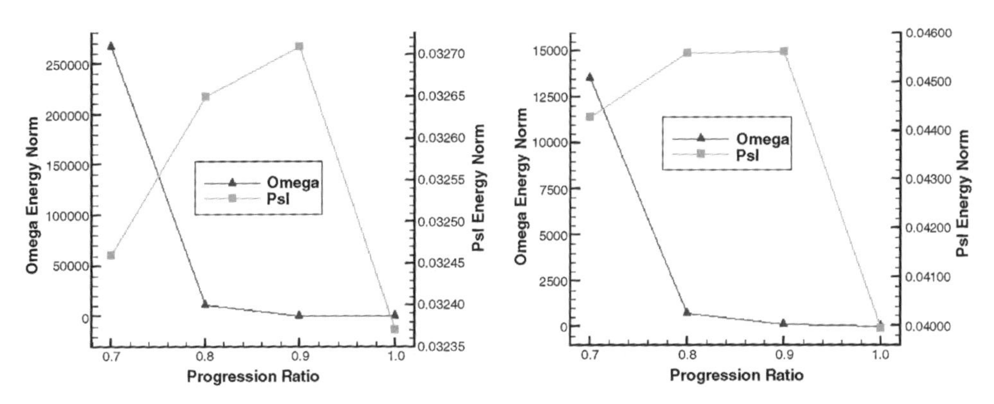

Figure 6.5 Driven cavity, $k=1$ TP basis $\mathrm{M}=32^2$, $1.0 \leq \mathrm{pr} \leq 0.7$ GWS^h steady solution $\|\omega^h\|_E$ and $\|\psi^h\|_E$ data: left, Re = 100; right, Re = 2000

To quantify these observations, weak form theory predicts that the *optimal mesh* solution – the one exhibiting $\|q^h\|_E$ extrema – results for the mesh that *equi-distributes* $\|q_e\|_{E,\Omega_e}$ over all Ω_e constituting Ω^h, Babuska and Rheinboldt (1976). Moving away from the singularity dominated driven cavity, regular solutions exist for the $n=2$ natural convection *thermal cavity* for a range of Rayleigh number Ra = GrPr. Cold/hot temperature Dirichlet BCs exist on opposite vertical walls, Figure 6.6 top left, with their difference ΔT input to Gr, (6.2). Adiabatic BCs are applied to both horizontal walls, and being a cavity the streamfunction BC remains $\psi=0$. Wall vorticity BCs are selected from (6.15–6.16).

Uniform $\mathrm{M}=32^2$ mesh $k=1$ GWS^h steady temperature DOF interpolations for Ra $\equiv$ E + 03, E + 06, Figure 6.6 top right, middle left, confirm cavity central isotherm distribution rotates from nearly vertical to horizontal with increasing Ra. Definition Ra = E + 06 generates very large wall-normal temperature gradients. The GWS^h steady isotherm distributions for this Ra range, Ericson (2001), exhibit excellent quantitative agreement with the benchmark data, de Vahl Davis (1983).

As solution gradients are extremum adjacent to vertical walls, the *optimal mesh* solution will result for near wall pr-modified *regular* mesh refinement. For Ra = E + 06, moving from uniform to wall attracted pr = 0.9 confirms that the steady ω^h solution energy norm *distribution* $\|\omega_e\|_{E,\Omega_e}$ over Ω_e trends towards equi-distribution by ~50% reduction of extrema, compare perspective graphics Figure 6.6 middle right, bottom left. Transitioning the mesh from uniform to pr = 0.9, state variable $\{q^h\}=\{\omega^h, \Theta^h, \psi^h\}^T$ global energy norms monotonically asymptote towards extrema, $\|\omega^h\|_E$ from below, $\|T^h\|_E$ and $\|\psi^h\|_E$ from above, Figure 6.6 lower right.

These pr = 0.9 $\mathrm{M}=32^2$ norm data indicate that pr is just short of the wall pr-adaptation required to support the optimal mesh solution for $\mathrm{M}=32^2$ DOF. The *m*GWS^h solution energy norms, the open symbols barely visible along abscissa coordinate pr = 1.0, Figure 6.6 lower right, confirm that *m*GWS^h solution norm extrema bias, compared to the GWS^h solution, is in the direction predicted by the theory.

These benchmark statement GWS^h/*m*GWS^h algorithm solution data *quantitatively verify* each identified aspect of *linear* weak form theory pertinence to genuine *nonlinear* NS

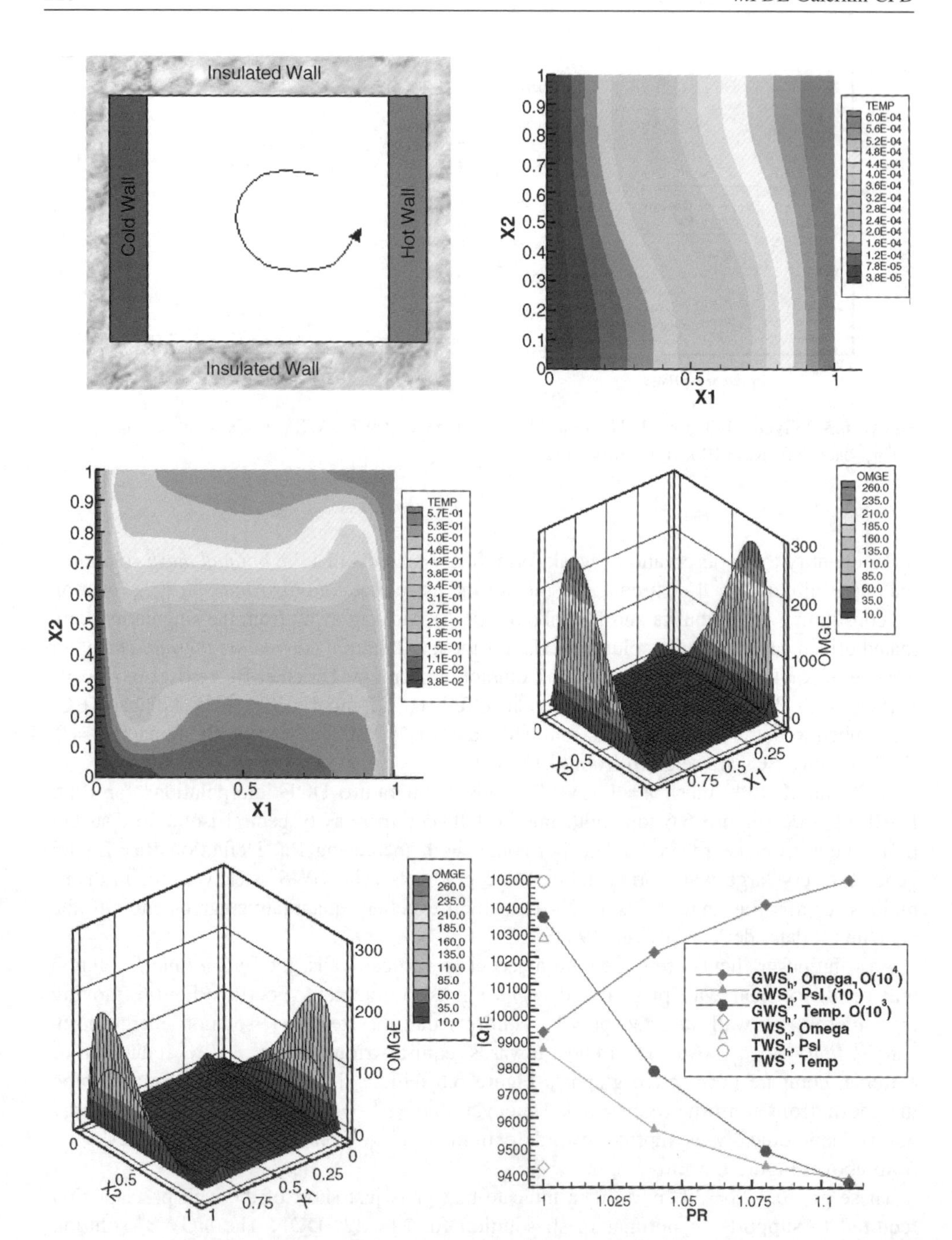

Figure 6.6 Thermal cavity benchmark, geometry with BCs top left: left to right, top to bottom, $k=1$ TP basis $\mathrm{M}=32^2$ GWS^h steady solution: temperature, Ra = E + 03, Ra = E + 06; $\|\omega_e\|_{E,\Omega_e}$ for $1.0 \le 1/\mathrm{pr} \le 1.1/0.9$, GWS^h, *m*GWS^h solution optimal mesh search in $\|T^h\|_E$, $\|\psi^h\|_E$, $\|\omega^h\|_E$

PDE/*m*PDE systems. The issue remaining is algorithm *validation*, with selected specification the close-coupled step wall diffuser. For duct entrance/diffuser cross-section area ratio 1/2, downstream coordinates where reversed flow bounding streamlines intersect the diffuser symmetry centerplane walls are experimentally measured for $100 \leq \text{Re} \leq 7500$, Armaly *et al.* (1983). For $\text{Re} \leq 400$ *only*, these data support validation of laminar flow $n=2$ CFD algorithm *a posteriori* data. (The step wall diffuser $n=3$ $m\text{GWS}^h$ algorithm validation, Chapter 7, quantifies three-dimensionality dominance for all $\text{Re} > 400$.)

Regular solution adapted mesh refinement $k=1$ NC basis GWS^h steady solution *a posteriori* data for $100 \leq \text{Re} \leq 600$ are reported, Baker (2012). The inflow BC is the fully developed flow parabolic profile $u = u(y^2)$. Via the vector identities, inflow DOF Dirichlet data are computed as $\psi = \psi(y^3)$ and $\omega = \omega(y)$. Integrating $\psi(y^3)$ across the inflow generates ψ^h DOF Dirichlet data for top and bottom boundaries. The $O(\Delta n^2)$ wall vorticity Dirichlet BC is selected, (6.16). As with *all* NS genuine statements, *a priori* absence of state variable data at outflow confirms that the sole admissable ψ^h and ω^h BC is homogeneous Neumann. This requires the outflow boundary be sufficiently distant from geometry induced interior flow complexity.

The reported GWS^h steady solution *a posteriori* data are detailed as ψ^h DOF distribution interpolations, Figure 6.7. For $\text{Re} = 100$ the primary recirculation bubble

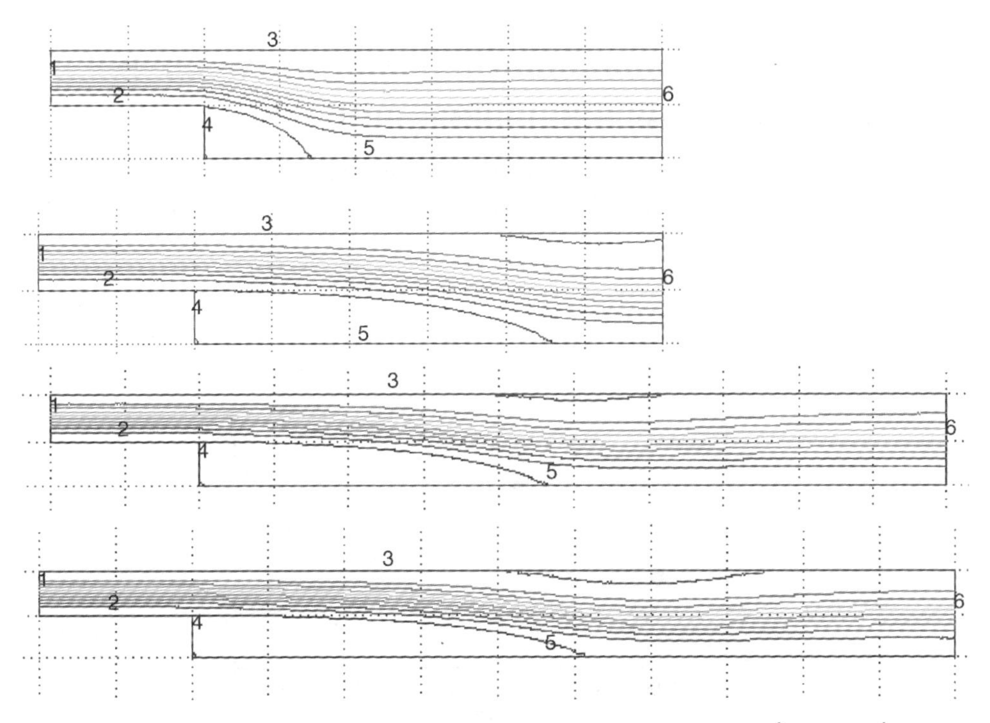

Figure 6.7 Close-coupled stepwall diffuser validation, $k=1$ NC basis GWS^h steady ψ^h solution DOF interpolations, top to bottom: $\text{Re} = 100$, 500, $\text{Re} = 500$ extended domain, $\text{Re} = 600$ extended domain

separates at the step face lip with lower wall reattachment nominally equal to step face height, top graphic. For Re = 200, 300, 400 the recirculation region lower wall attachment translates uniformly downstream, and for Re = 400 an upper wall very shallow recirculation region is generated. At Re = 500 the larger upper wall secondary recirculation bubble intersects the outflow boundary, following graphic. The ψ^h distribution no longer exits the solution domain with vanishing normal derivative and thus violates the outflow Neumann BC.

The required solution adapted mesh correction is elongating the domain. The third graphic confirms that the Re = 500 solution is indeed outflow BC-admissible. The bottom graphic confirms that the GWS^h solution Re = 600 upper wall recirculation reattachment point extends well past the original domain outflow boundary. The primary recirculation reattachment has moved further downstream with outflow solution distribution confirmed BC-appropriate.

The *a posteriori* data in Figure 6.7 are on the finer mesh of an M = 348/2573, then M = 2573/3807 regular solution adapted mesh refinement study, short/long domains. The validation criterion is primary recirculation streamline reattachment coordinate on the lower wall. Normalized by step height these GWS^h algorithm data are in excellent quantitative agreement with experiment for Re $\leq$ 400, Figure 6.8, then progressively depart the experiment for 400 < Re $\leq$ 600.

As confirmation of the CFD axiom, "*you get out what you put in, especially on the boundaries*," the Re = 600 short domain GWS^h solution reattachment coordinate agrees almost exactly(!) with experiment. This documents "*getting the right answer for the wrong reason*," the result of inattention to simulation definition detail, particularly the BCs!

6.5 Vorticity-Velocity *m*GWS^h Algorithm, $n = 3$

Weak form theory predictions *pertinence* to GWS^h/*m*GWS^h algorithm $n = 2$ vorticity-streamfunction explicitly nonlinear PDE/*m*PDE systems is *quantitatively verified*, including

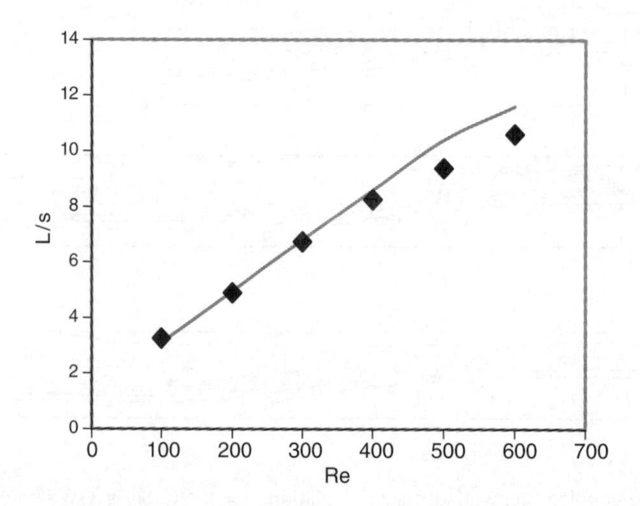

Figure 6.8 Close coupled stepwall diffuser primary recirculation intercept, 100 $\leq$ Re $\leq$ 600, symbols are $n = 2$ GWS^h solution, solid line is experiment

solution *validation*. This completed, the sole practical vector field theory CFD formulation must address the $n = 3$ *vorticity-velocity* PDE system

$$\mathcal{L}(\boldsymbol{\Omega}) = \frac{\partial \boldsymbol{\Omega}}{\partial t} + (\mathbf{u} \cdot \nabla)\boldsymbol{\Omega} - (\boldsymbol{\Omega} \cdot \nabla)\mathbf{u} - \frac{1}{\text{Re}} \nabla^2 \boldsymbol{\Omega} + \frac{\text{Gr}}{\text{Re}^2} \nabla \times \Theta \, \widehat{\mathbf{g}} = \mathbf{0} \tag{6.8}$$

$$\mathcal{L}(\mathbf{u}) = -\nabla^2 \mathbf{u} - \nabla \times \boldsymbol{\Omega} = \mathbf{0} \tag{6.9}$$

$$\mathcal{L}(\Theta) = \frac{\partial \Theta}{\partial t} + \mathbf{u} \cdot \nabla \Theta - \frac{1}{\text{RePr}} \nabla^2 \Theta - s_\Theta = 0 \tag{6.45}$$

The key formulation feature is again absence of pressure in the state variable $\{q(\mathbf{x},t)\} \equiv \{\boldsymbol{\Omega}, \Theta, \mathbf{u}\}^T$. This mixed I-EBV/EBV PDE system is initial-value for vorticity $\boldsymbol{\Omega}$ and potential temperature Θ and strictly boundary value for velocity $\mathbf{u}$.

The laminar flow NS PDE system (6.8–6.9) is verified well-posed, Gunzburger and Peterson (1998), with experience confirming (6.43) is also well-posed. The relinquishing of exact satisfaction of D*M* via $\boldsymbol{\Psi}$ is replaced by a no-slip wall strongly coupled *discrete* vorticity BC, recall (6.15–6.16), TS-derived from kinematics $\boldsymbol{\Omega} \equiv \nabla \times \mathbf{u}$ on boundaries with Dirichlet BC $\mathbf{u} = \mathbf{0}$, Daube (1992). The resultant $\text{GWS}^h + \theta\text{TS}$ matrix iteration algorithm for (6.8), (6.9), (6.45) must be fully coupled, producing a *large* jacobian!

The GWS^N algorithm state variable is $\{qN(\mathbf{x}, t)\} \equiv \{\Omega_x, \Omega_y, \Omega_z, \Theta, u, v, w\}^T$. The $k = 1$ basis $\text{GWS}^h + \theta\text{TS}$ algorithm statement for resultant DOF $\{\text{Q}I(t)\} = \{\text{OX, OY, OZ, T, U, V, W}\}^T$, $1 \le I \le 7$, remains

$$\begin{aligned}
[\text{JAC_}\{\text{Q}I\}]\{\delta\text{Q}I\}^{p+1} &= -\{\text{F_Q}I\}_{n+1}^{p} \\
[\text{JAC_}\{\bullet\}] &\equiv \frac{\partial\{\text{F_Q}I\}}{\partial\{\text{Q}J\}} \\
\{\text{Q}I\}_{n+1}^{p+1} &= \{\text{Q}I\}_{n+1}^{p} + \{\delta\text{Q}I\}^{p+1} \\
&= \{\text{Q}I\}_n + \sum_{\alpha=0}^{p} \{\delta\text{Q}I\}^{\alpha+1}
\end{aligned} \tag{6.22}$$

for iteration index p and n denoting time level $t_n = t_0 + n\Delta t$.

For algorithm implementation via parallel processing, Wong (1995), GWS^h steady solution *a posteriori* data generation employs the *conjugate gradient squared* (CGS), Sonneveld (1989), and *bi-conjugate gradient stabilized* (BiCGSTAB), van der Vorst (1992), sparse solvers for non-symmetric jacobians. The large Newton jacobian is thus never formed, instead replaced by the sequence of vector operations intrinsic to Krylov-class solvers, Section 2.12.

All matrices defined in (6.22) are formed as element integrals, then assembled over the discretization Ω^h of the domain Ω. Recalling matrix prefix C denotes $n = 3$, the $k = 1$ NC/TP basis $m\text{GWS}^h + \theta\text{TS}$ algorithm for (6.8) and (6.45) generates the $1 \le I \le 4$ element matrix statement

$$\begin{aligned}
\{\text{F_Q}I\}_e \equiv & \left([\text{C200}]_e - \frac{\gamma\,\Delta t^2}{6}\{\text{U}J\,\text{U}K\}_e^T[\text{C30}JK]_e\right)\{\text{Q}I_{n+1} - \text{Q}I_n\}_e \\
& + \Delta t\left[\theta\{\text{RES_}\{\text{Q}I\}\}_e\big|_{n+1} + (1-\theta)\{\text{RES_}\{\text{Q}I\}\}_e\big|_n\right]
\end{aligned} \tag{6.46}$$

The $\{\text{RES\_}\{\text{Q}I\}\}_e$ element matrices defined in (6.46) are

$$\begin{aligned}\{\text{RES\_O}I\}_e &= \{\text{U}J\}_e^T[\text{C300}J]_e\{\text{O}I\}_e - \{\text{O}J\}_e^T[\text{C300}J]_e\{\text{U}I\}_e \\ &+ \frac{1}{\text{Re}}[\text{C2}KK]_e\{\text{O}I\}_e - \frac{\text{Gr}}{\text{Re}^2}[\text{C20}K]_e\{\text{T}\}_e\widehat{\text{g}}_J\text{e}_{IKJ} \\ &+ \frac{h^2\text{Re}}{12}\{\text{U}J\,\text{U}K\}_e^T[\text{C30}JK]_e\{\text{O}I\}_e\end{aligned} \tag{6.47}$$

$$\begin{aligned}\{\text{RES\_T}\}_e &= \{\text{U}J\}_e^T[\text{C300}J]_e\{\text{T}\}_e + \frac{1}{\text{RePr}}[\text{C2}KK]_e\{\text{T}\}_e \\ &+ \frac{h^2\text{RePr}}{12}\{\text{U}J\,\text{U}K\}_e^T[\text{C30}JK]_e\{\text{T}\}_e\end{aligned} \tag{6.48}$$

For $\{\text{Q}I\} = \{\text{U}I\}$, $5 \le I \le 7$, the $m\text{GWS}^h$ element matrix statement is

$$\begin{aligned}\{\text{F\_U}I\}_e &\equiv [\text{C2}KK]_e\{\text{U}I\}_e - [\text{C20}K]_e\{\text{O}J\}_e\text{e}_{IKJ} \\ &- \frac{h^2}{12}[\text{C2}KK]_e\{\text{curl}\ \Omega I\}_e\end{aligned} \tag{6.49}$$

Italicized indices are summed $1 \le J,\ K \le 3 = n$, e_{IKJ} is the alternating tensor (+1, −1 for IKJ cyclic, anticyclic) with $\widehat{\text{g}}_J$ the gravity unit vector. Matrices e-subscripted contain element-dependent metric data, which upon extraction leads to the $n = 3$, $k = 1$ TP/NC basis element-independent matrix libraries.

The mPDE addition to (6.49), recall (5.113), contains three spatial derivatives, hence is $k = 1$ basis incompatible. As detailed for velocity, (6.35), $k = 1$ basis determination is via GWS^N written on the differential definition $\mathcal{D}(\text{curl}\ \Omega I) = \text{curl}\ \Omega I - \nabla \times \boldsymbol{\Omega} = \mathbf{0}$. The resultant GWS^h element matrix statement is

$$\{\text{F\_}\nabla \times \Omega\}_e \equiv [\text{C200}]_e\{\text{curl}\ \Omega I\}_e - [\text{C20}K]_e\{\text{O}J\}_e\text{e}_{IKJ} \tag{6.50}$$

with solution yielding DOF $\{\text{curl}\ \boldsymbol{\Omega}\mathbf{I}\}_e$, $1 \le I \le 3$, as required for (6.49).

Jacobian formation completes the algorithm statement. As well illustrated, element contributions are generated via calculus operations on the element matrix statements (6.46–6.49). Denoting absent DOF couplings as [0] the element block jacobian is

$$[\text{JAC\_}\{\bullet\}]_e \equiv \frac{\partial\{\text{F\_}QI\}_e}{\partial\{QJ\}_e}$$

$$= \begin{bmatrix} [\text{J\_O}_x\text{O}_x], & [\text{J\_O}_x\text{O}_y], & [\text{J\_O}_x\text{O}_z], & [\text{J\_O}_x\text{T}], & [\text{J\_O}_x\text{U}], & [\text{J\_O}_x\text{V}], & [\text{J\_O}_x\text{W}] \\ [\text{J\_O}_y\text{O}_x], & [\text{J\_O}_y\text{O}_y], & [\text{J\_O}_y\text{O}_z], & [\text{J\_O}_y\text{T}], & [\text{J\_O}_y\text{U}], & [\text{J\_O}_y\text{V}], & [\text{J\_O}_y\text{W}] \\ [\text{J\_O}_z\text{O}_x], & [\text{J\_O}_z\text{O}_y], & [\text{J\_O}_z\text{O}_z], & [0], & [\text{J\_O}_z\text{U}], & [\text{J\_O}_z\text{V}], & [\text{J\_O}_z\text{W}] \\ [0], & [0], & [0], & [\text{J\_TT}], & [\text{J\_TU}], & [\text{J\_TV}], & [\text{J\_TW}] \\ [0], & [\text{J\_UO}_y], & [\text{J\_UO}_z], & [0], & [\text{J\_UU}], & [0], & [0] \\ [\text{J\_VO}_x], & [0], & [\text{J\_VO}_z], & [0], & [0], & [\text{J\_VV}], & [0] \\ [\text{J\_WO}_x], & [\text{J\_WO}_y], & [0], & [0], & [0], & [0], & [\text{J\_WW}] \end{bmatrix}_e \tag{6.51}$$

with $\{\text{O}_x, \text{O}_y, \text{O}_z, \text{T}, \text{U}, \text{V}, \text{W}\}$ shorthand for DOF $\{\text{Q}I\}$.

The suggested exercises verify that the DOF self-coupling matrices, the diagonal entries in (6.51), with ***no*** summation on index I are

$$[\mathrm{J\_OIOI}]_e = [\mathrm{C200}]_e - \frac{\gamma\,\Delta t^2}{6}\{\mathrm{U}J\,\mathrm{U}K\}_e^T[\mathrm{C30}JK]_e + \theta\Delta t\begin{pmatrix}\{\mathrm{U}J\}_e^T[\mathrm{C300}J]_e - \{\mathrm{UI}\}_e^T[\mathrm{C3}J\mathrm{00}]_e\delta_{J\mathrm{I}} \\ +\dfrac{1}{\mathrm{Re}}[\mathrm{C2}KK]_e + \dfrac{h^2\mathrm{Re}}{12}\{\mathrm{U}J\,\mathrm{U}K\}_e^T[\mathrm{C30}JK]_e\end{pmatrix} \tag{6.52}$$

$$[\mathrm{J\_TT}]_e = [\mathrm{C200}]_e - \frac{\gamma\,\Delta t^2}{6}\{\mathrm{U}J\,\mathrm{U}K\}_e^T[\mathrm{C30}JK]_e + \theta\Delta t\begin{pmatrix}\{\mathrm{U}J\}_e^T[\mathrm{C300}J]_e + \dfrac{1}{\mathrm{Re}}[\mathrm{C2}KK]_e \\ +\dfrac{h^2\mathrm{Pe}}{12}\{\mathrm{U}J\,\mathrm{U}K\}_e^T[\mathrm{C30}JK]_e\end{pmatrix} \tag{6.53}$$

$$[\mathrm{J\_UIUI}]_e = [\mathrm{C2}KK]_e \tag{6.54}$$

The DOF off-diagonal cross-coupling jacobian matrices are

$$[\mathrm{J\_O}I\mathrm{O}K]_e = \theta\Delta t\{\mathrm{U}I\}_e^T[\mathrm{C3}K\mathrm{00}]_e \tag{6.55}$$

$$[\mathrm{J\_O}I\mathrm{T}] = -\theta\Delta t\frac{\mathrm{Gr}}{\mathrm{Re}^2}[\mathrm{C20}K]_e\widehat{\mathrm{g}}_J\mathrm{e}_{IKJ} \tag{6.56}$$

$$[\mathrm{J\_O}I\mathrm{U}J]_e = \theta\Delta t\left(\{\mathrm{O}I\}_e^T[\mathrm{C3}J\mathrm{00}]_e - \{\mathrm{O}I\}_e^T[\mathrm{C300}J]_e\delta_{IJ}\right) \tag{6.57}$$

$$[\mathrm{J\_TU}J]_e = \theta\Delta t\{\mathrm{T}\}_e^T[\mathrm{C3}J\mathrm{00}]_e \tag{6.58}$$

$$[\mathrm{J\_U}I\mathrm{O}J]_e = -[\mathrm{C20}K]_e\mathrm{e}_{IKJ} \tag{6.59}$$

For the $m\mathrm{GWS}^h + \theta\mathrm{TS}$ algorithm the nonlinear coupling due to *m*PDE terms is eliminated by $\{\mathrm{U}J\mathrm{U}K(t_n)\}_e$ evaluation with the converged solution at previous time step t_n.

The *key* to NS vorticity-velocity formulation control of $\mathrm{D}M^h \neq 0$ error is derivation of GWS^h basis degree *consistent* TS approximations to kinematic definition $\mathbf{\Omega}^h \bullet \widehat{\mathbf{n}} \equiv \nabla^h \times \mathbf{u}^h \bullet \widehat{\mathbf{n}}$ on solution domain no-slip boundaries with unit normal $\widehat{\mathbf{n}}$. Restricting to rectangular cartesian coordinates on boundary surfaces with orientations $\widehat{\mathbf{n}} = \pm\widehat{\mathbf{k}}$

$$\mathbf{\Omega}\cdot\widehat{\mathbf{k}} \equiv \Omega_z = \nabla\times\mathbf{u}\bullet\widehat{\mathbf{k}} = \frac{\partial v}{\partial x} - \frac{\partial u}{\partial y} = 0 \tag{6.60}$$

Similarly, Ω_x and Ω_y vanish respectively on bounding no-slip surfaces with normals $\widehat{\mathbf{n}} = \pm\widehat{\mathbf{i}}$ and $\widehat{\mathbf{n}} = \pm\widehat{\mathbf{j}}$.

The BCs for the remaining scalar components of $\mathbf{\Omega}$ are formed via an appropriate-order TS on velocity scalar components in wall normal coordinates, recall the $n = 2$ vorticity wall normal TS (6.14). For subscripts 0 and 1 denoting wall and first off-wall DOF, in continuum notation and suppressing h for clarity, for surface orientation $\widehat{\mathbf{n}} = \widehat{\mathbf{k}}$ the appropriate TS is

$$u_1 = u_0 + \Delta z \left.\frac{\partial u}{\partial z}\right|_0 + \left.\frac{\Delta z^2}{2}\frac{\partial^2 u}{\partial z^2}\right|_0 + O(\Delta z^3)$$
$$= u_0 + \Delta z\,\Omega_{y0} + \left.\frac{\Delta z^2}{2}\frac{\partial \Omega_y}{\partial z}\right|_0 + O(\Delta z^3), \quad \text{for } x, y \text{ constant} \tag{6.61}$$

The (6.61) order commensurate TS for Ω_y is

$$\Omega_{y1} = \Omega_{y0} + \Delta z \left.\frac{\partial \Omega_y}{\partial z}\right|_0 + O(\Delta z^2) \tag{6.62}$$

which when substituted into (6.61) produces the $O(\Delta z)$ vorticity DOF constraint at the no-slip surface

$$\Omega_{y0} = -\Omega_{y1} + \frac{2}{\Delta z}(u_1 - u_0) + O(\Delta z^2), \quad \text{for } \widehat{\mathbf{n}} = \widehat{\mathbf{k}} \tag{6.63}$$

Improving the TS in (6.61) and (6.62) one order generates the $O(\Delta z^2)$ alternative to (6.63)

$$3\Omega_{y0} = -\left(4\Omega_{y1} - \Omega_{y2}\right) - \frac{1}{\Delta z}(7u_0 - 8u_1 - u_2) + O(\Delta z^3) \tag{6.64}$$

Implementing basis degree k consistent vorticity TS BC DOF wall constraints is crucial to $\mathrm{D}M^h \neq 0$ accuracy. As confirmation, for a $k=2$ TP basis vorticity-velocity GWS^h implementation, replacing (6.63) with (6.64) proved mandatory to $\mathrm{D}M^h$ error control for an $n=2$ duct flow benchmark, Wong (1995).

For the $n=3$, $k=1$ TP basis GWS^h algorithm, vorticity wall DOF constraints of $O(\Delta z)$ proved robust in enforcing $\mathrm{D}M^h \leq O(\epsilon)$ for iterative convergence $\epsilon \leq$ E-04. Summarizing, the set of $k=1$ basis-consistent $O(\Delta n)$ no-slip wall vorticity DOF constraints for domain closure surfaces with unit normal vectors $\widehat{\mathbf{n}} = \pm\left(\widehat{\mathbf{i}}, \widehat{\mathbf{j}}, \widehat{\mathbf{k}}\right)$ is

$$\Omega_{y0} = -\Omega_{y1} \mp \frac{2}{\Delta x}(\mathrm{W}_1 - \mathrm{W}_0); \Omega_{z0} = -\Omega_{z1} \pm \frac{2}{\Delta x}(\mathrm{V}_1 - \mathrm{V}_0); \Omega_{x0} = 0 \tag{6.65}$$

$$\Omega_{x0} = -\Omega_{x1} \pm \frac{2}{\Delta y}(\mathrm{W}_1 - \mathrm{W}_0); \Omega_{z0} = -\Omega_{z1} \mp \frac{2}{\Delta y}(\mathrm{U}_1 - \mathrm{U}_0); \Omega_{y0} = 0 \tag{6.66}$$

$$\Omega_{x0} = -\Omega_{x1} \mp \frac{2}{\Delta z}(\mathrm{V}_1 - \mathrm{V}_0); \Omega_{y0} = -\Omega_{y1} \pm \frac{2}{\Delta z}(\mathrm{U}_1 - \mathrm{U}_0); \Omega_{z0} = 0 \tag{6.67}$$

As with the vorticity-streamfunction formulation, the DOF constraints (6.65–6.67) must replace those generated by GWS^h in (6.46–6.47). For iterative stability the vorticity DOF cross-coupling entries in jacobians (6.52), (6.55), (6.57) must be replaced with the results of appropriate differentiations in (6.65–6.67). Computational experience confirms that fully implicit coupling is crucial to maintaining mass conservation error $\mathrm{D}M^h \le O(\epsilon)$, Wong (1995).

Formulation completion rests on extracting element metric data from the matrices in (6.46–6.59). Matrix $[\mathrm{C200}]_e$ transforms to $(\det_e/6)[\mathrm{C200}]$ or $(\det_e/8)[\mathrm{C200}]$ for the NC or TP $k=1$ basis. The transformation of diffusion matrix $[\mathrm{C2}KK]_e$ is detailed in Chapter 3, (3.74–3.76), (3.99–3.100). For summation index conventions $1 \le i,\ j,\ k \le n = 3$ and $1 \le \alpha$, $\beta \le 4$, the NC and TP basis forms are

$$[\mathrm{C2}KK]_e \Rightarrow \frac{1}{6\det_e}(\zeta_{\alpha i}\zeta_{\beta i})_e[\mathrm{C}\delta_{\alpha\beta}] \quad \text{and} \quad \frac{1}{8\det_e}\left(\eta_{j\,i}\eta_{k\,i}\right)_e[\mathrm{C2}jk] \tag{6.68}$$

The convection hypermatrix $[\mathrm{C300}J]_e$ and directional derivative $[\mathrm{C20}K]_e$ matrix are newly generated. Both are devoid of $\det_e$, possessing only one spatial derivative. The NC and TP basis forms are

$$\begin{aligned}[\mathrm{C300}J]_e &\Rightarrow \left(\zeta_{\alpha J}\right)_e[\mathrm{C300}\alpha] \quad \text{and} \quad \left(\eta_{iJ}\right)[\mathrm{C300}i] \\ [\mathrm{C20}J]_e &\Rightarrow \left(\zeta_{\alpha J}\right)_e[\mathrm{C20}\alpha] \quad \text{and} \quad \left(\eta_{iJ}\right)[\mathrm{C20}i]\end{aligned} \tag{6.69}$$

Several jacobian matrices have the transform convection hypermatrix $[\mathrm{C3}J\mathrm{00}]$. The (6.69) alterations simply replace $[\mathrm{C300}\alpha]$ with $[\mathrm{C3}\alpha\mathrm{00}]$ and $[\mathrm{C300}i]$ with $[\mathrm{C3}i\mathrm{00}]$.

The *m*PDE hypermatrix $[\mathrm{C30}KL]_e$ for the $k=1$ NC basis equates to the element-average scalar $\{\mathrm{C10}\}^T\{\mathrm{U}J\mathrm{U}K\}_e$ multiplier on the $[\mathrm{C2}KK]_e$ matrix, recall Table 4.3. The TP basis implementation distributes the velocity tensor product nodally as

$$\frac{h^2\mathrm{Pa}}{12}\{\mathrm{U}J\,\mathrm{U}K\}_e^T[\mathrm{C30}JK]_e \Rightarrow \frac{h^2\mathrm{Pa}}{96\det_e}\left(\eta_{J\,l}\eta_{K\,m}\right)_e\{\mathrm{U}J\,\mathrm{U}K\}_e^T[\mathrm{C30}lm] \tag{6.70}$$

where $\mathrm{Pa} \Rightarrow \mathrm{Re}$, RePr and $1 \le (l,\ m) \le 3$ are a new summation index pair.

6.6 Vorticity-Velocity GWS^h + θTS Assessments, $n = 3$

Incompressible laminar-thermal NS $n=3$ benchmark/validation statements include developing flow in a rectangular cross-section duct, the close-coupled step-wall diffuser and lid-driven and thermal cavities. Inaugural benchmark driven cavity $n=3$ vorticity-velocity GWS^h solutions for $\mathrm{Re} \le 400$ are reported, Guj and Stella (1988), Guevremont *et al.* (1993). Comprehensive $k=1$ TP basis GWS^h + θTS algorithm steady solutions for $n=3$ benchmarks/validations for $\mathrm{Re} \le 2000$ and $\mathrm{Ra} \le \mathrm{E}+07$ are reported, Wong and Baker (2002). *A posteriori* data generation is via time marching the unsteady GWS^h + θTS algorithm. *m*GWS^h algorithm data are unavailable as theory derivation occurred subsequent to generation of the GWS^h data.

Validation measures for fully enclosed duct flow include semi-analytical solutions for the fully developed flow transverse plane axial velocity distribution, White (1974), also correlated entrance length to reach fully developed velocity profile as a function of Re, Han (1960). For a uniform $\mathrm{M}=16^3$, $k=1$ TP basis mesh and $\mathrm{Re}=100$, Figure 6.9 top

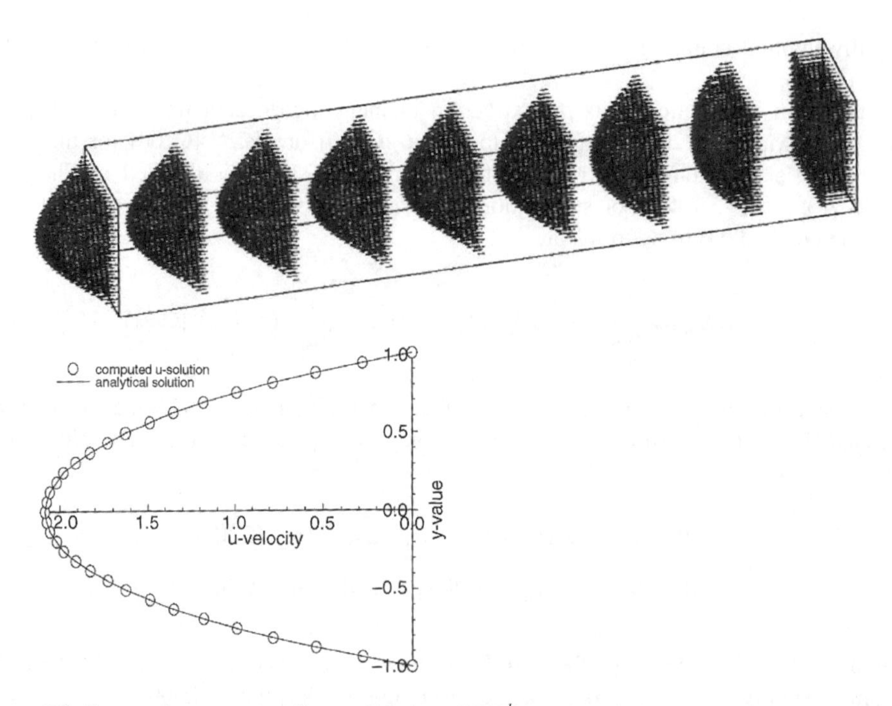

Figure 6.9 Square duct entrance flow validation, GWS^h steady solution, Re = 100, $k = 1$ TP basis uniform M = 16^3 mesh: top, $u^h(x, y, z)$ evolution from inflow slug {U} DOF BC; bottom, symmetry line $\{U(z_{outflow})\}$ DOF compared with analytical solution

graphs in perspective square duct axial velocity $u^h(x, y, z)$ development from the inflow plane slug {U} DOF BC, right, to the fully developed flow paraboloidal distribution, left. Figure 6.9 bottom graphs exit plane symmetry line {U} DOF compared to the analytical solution (solid line) confirming full *quantitative* agreement.

The $n = 3$ driven cavity is the unit cube with top wall translating parallel to the x axis at non-D velocity $u = 1$. Being a cavity, the no-slip BC dominates hence wall vorticity DOF (6.65–6.67) are universally implemented. The IC is zero vorticity everywhere; lid translation initializes solution evolution. Vorticity-velocity formulation GWS^h steady solutions at Re = 100, 400, 1000 are reported, Wong and Baker (2002). For wall adapted $k = 1$ TP basis M = 48^3 mesh, the Krylov parallel solver utilized domain decomposition into 12 blocks generating a 930/78 Mbytes global/local memory requirement.

The Re = 100 GWS^h steady solution *a posteriori* data are perspective graphics summarized in Figure 6.10. The cavity top plane translation direction is downward to the left. Vorticity vector cartesian resolution $\Omega_x/\Omega_y/\Omega_z$ distributions are plotted, right column top to bottom, on each respective $x/y/z$ axis direction cavity mid-plane, also on the back plane with normal parallel to that coordinate direction. The velocity vector DOF mid-plane distributions, Figure 6.10 left column, coincide with each $x/y/z$ axis vorticity scalar component. The contour lines superposed on each velocity vector graph are streamlines in the plane computed via GWS^h on the planar $n = 2$ differential definition $\mathcal{D}(\psi) = \nabla^2\psi + \omega = 0$ normal to each graphed plane.

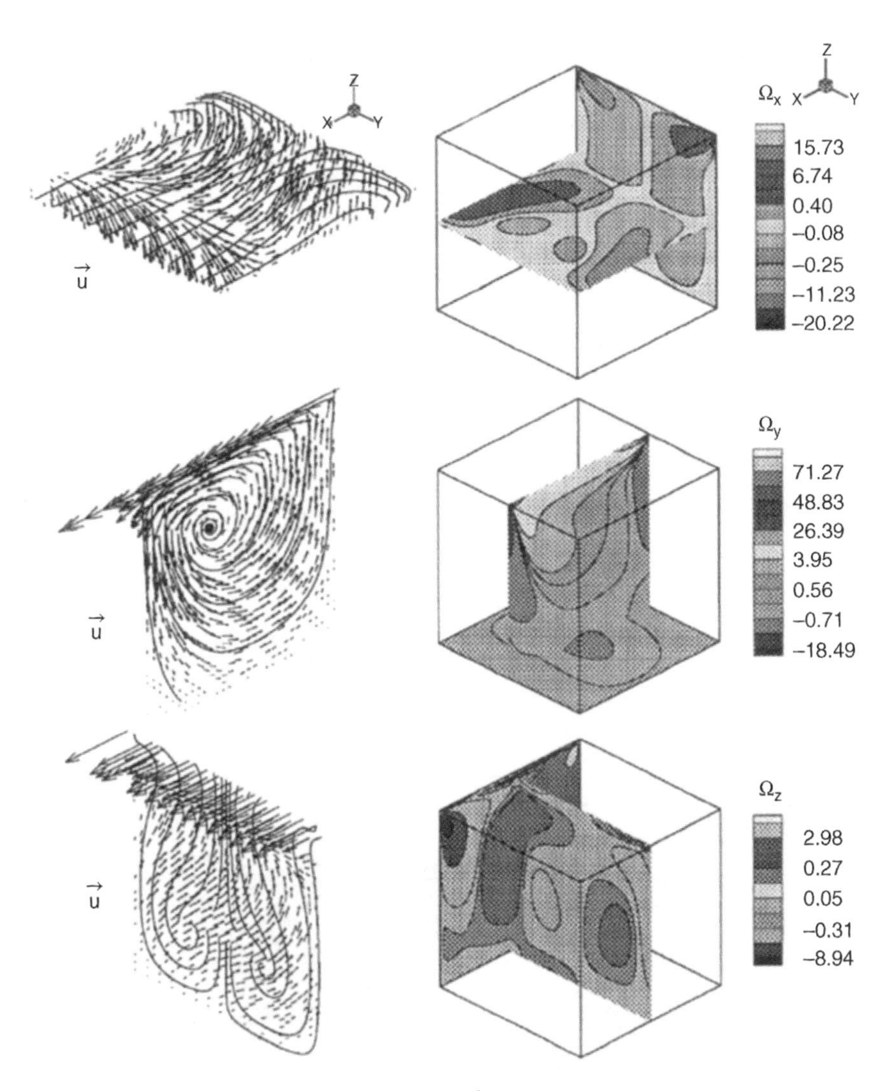

Figure 6.10 Driven cavity $n=3$ benchmark GWS^h steady solution perspective, $k=1$ TP basis $M=48^3$ mesh, Re = 100: left column, $\mathbf{u}^h(x, y, z)$ with planar streamline contours; right column, $\boldsymbol{\Omega}^h(x, y, z)$ scalar component planar floods. *Source:* Wong, K.L. and Baker, A.J. 2002. Reproduced with permission by Wiley

This solution data presentation hopefully assists in visualization of the very detailed, fully three-dimensional vortex-dominated driven cavity flowfield. The Re = 1000 GWS^h steady solution *a posteriori* data are summarized in Figure 6.11 in the identical perspective format. The solution distinctions in transition to Re = 1000 are discernible mainly on the streamfunction superpositions on the planar velocity vector DOF distribution. Instigation of measureable rear/front wall upwelling/downwelling is visible in the left top graphic. The mid-plane circulation pattern is essentially unaltered except for the vortex center

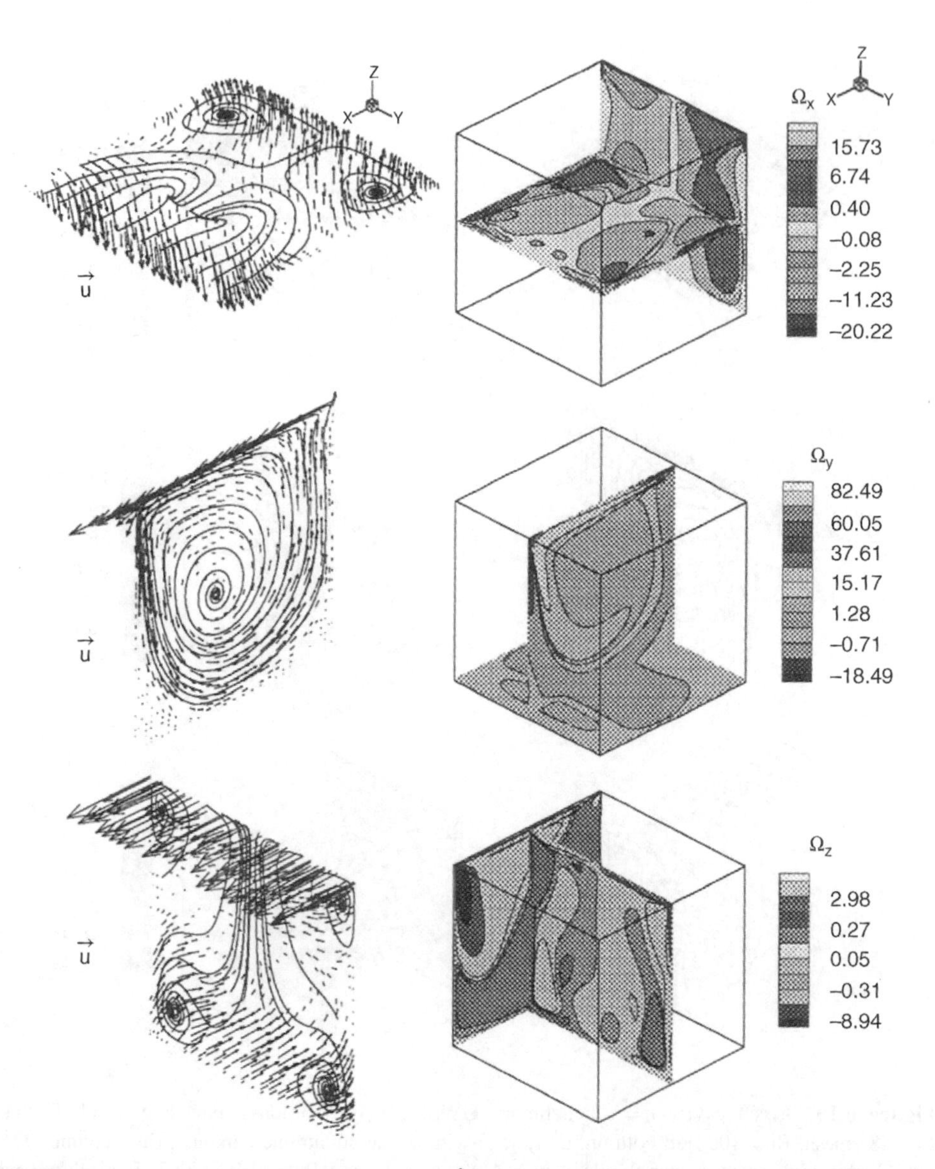

Figure 6.11 Driven cavity $n=3$ benchmark GWS^h steady solution perspective, $k=1$ TP basis solution adapted $M=48^3$ mesh, $Re=1000$: left column, $\mathbf{u}^h(x, y, z)$ with planar streamline contours; right column, $\boldsymbol{\Omega}^h(x, y, z)$ scalar component planar floods. *Source:* Wong, K.L. and Baker, A.J. 2002. Reproduced with permission by Wiley

moving to near mid-cavity. The vortex generated circulation structures in the bottom graphic are significantly more intense.

A summary of driven cavity flowfield Re dependence is detailed as planar projection of the velocity vector field onto the three centroidal cavity planes, Figure 6.12. In the y plane, center column, the primary vortex axis starts upper right and progresses towards cavity

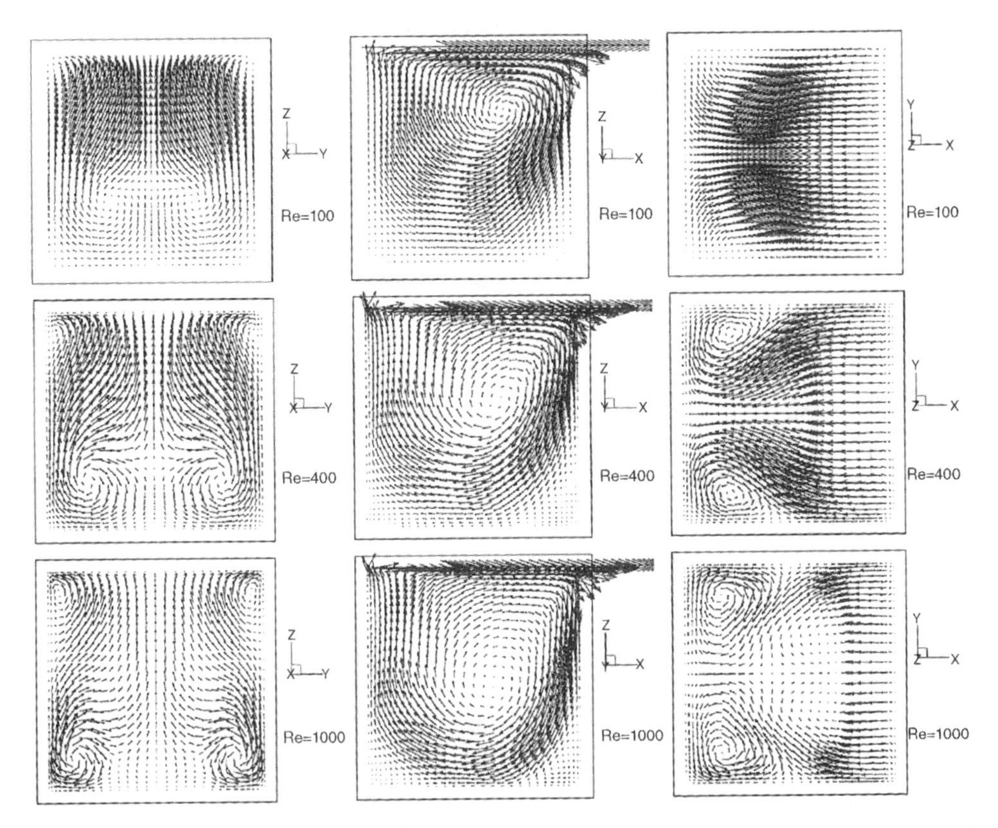

Figure 6.12 Driven cavity $n=3$ GWS^h flowfield summary, solution adapted $k=1$ TP basis $M=48^3$ mesh, velocity vector planar projection onto cavity bisection $x/y/z$ planes: top row Re = 100; middle row Re = 400; bottom row Re = 1000. *Source:* Wong, K.L. and Baker, A.J. 2002. Reproduced with permission by Wiley

center with increasing Re. The secondary vortex structure axes gravitate towards the x sidewalls, left column, and upwards towards the z sidewall, right column. Note the Re = 1000 center column velocity distribution exhibits *quantitative* similarity with the $n=2$, Re = 2000 solution, Figure 6.4.

The absolutely key theory issue with the vorticity-velocity formulation is control of DM^h error via k basis degree consistent TS order no-slip wall vorticity BCs. Wong (1995) validates $O(\Delta n)$ BCs (6.65–6.67) consistency with the $k=1$ TP basis GWS^h algorithm via $M=64^2$ mesh solution comparison with the $n=2$ driven cavity $M=256^2$ mesh benchmark data of Ghia *et al.* (1982). The generated excellent *quantitative* agreement of this $n=2$ solution multiple recirculation intercepts, Figure 6.13, quantitatively *validates* the vorticity BC selection control of DM^h error.

Returning to $n=3$, cavity centroidal plane velocity DOF *a posteriori* data {U($x=0.5$, y, $z=0.5$) are published by Guj and Stella (1993), Fujima *et al.* (1994) and Jiang *et al.* (1994). While not "certified" validation quality, these data do enable *quantitative* accuracy

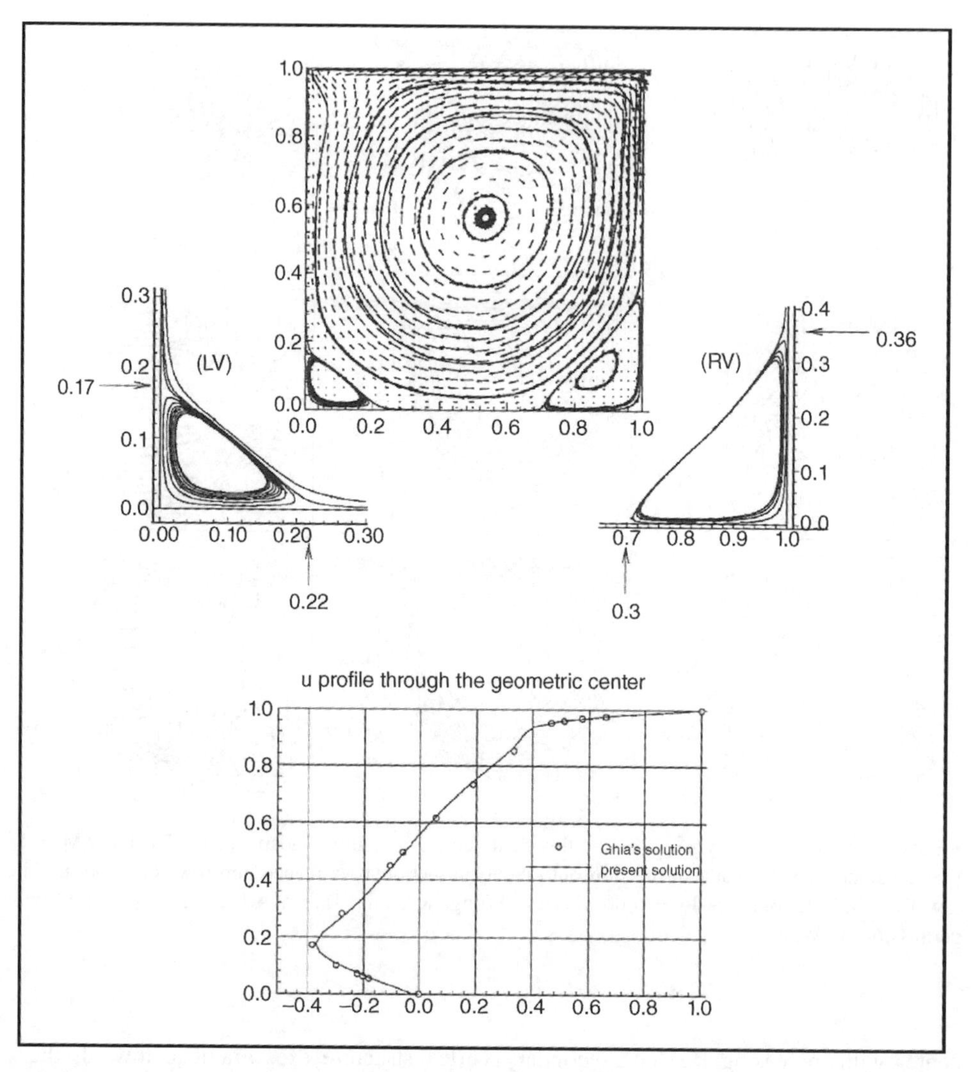

Figure 6.13 Driven cavity $n=2$ GWS^h solution recirculation zone wall intercepts comparison to benchmark data of Ghia *et al.* (1982), solution-adapted $M=64^2$, $k=1$ TP basis mesh, Re = 1000, from Wong (1995)

assessment of the GWS^h algorithm for this Re range. For Re = 400, the three solutions exhibit excellent quantitative agreement, top graphic in Figure 6.14. This holds true for the Re = 1000 comparison, bottom graphic, except in the {U} *extremum* reach where the GWS^h DOF distribution is the ~average of the comparative data. The reported maximum disparity is 0.004, ~1.5% of extremum DOF $\{U(x=0.5,\ y=0.1,\ z=0.5)\} = -0.279$. Unfortunately, as typical of $n=3$ benchmarks, solution adapted regular mesh refinement data are not reported.

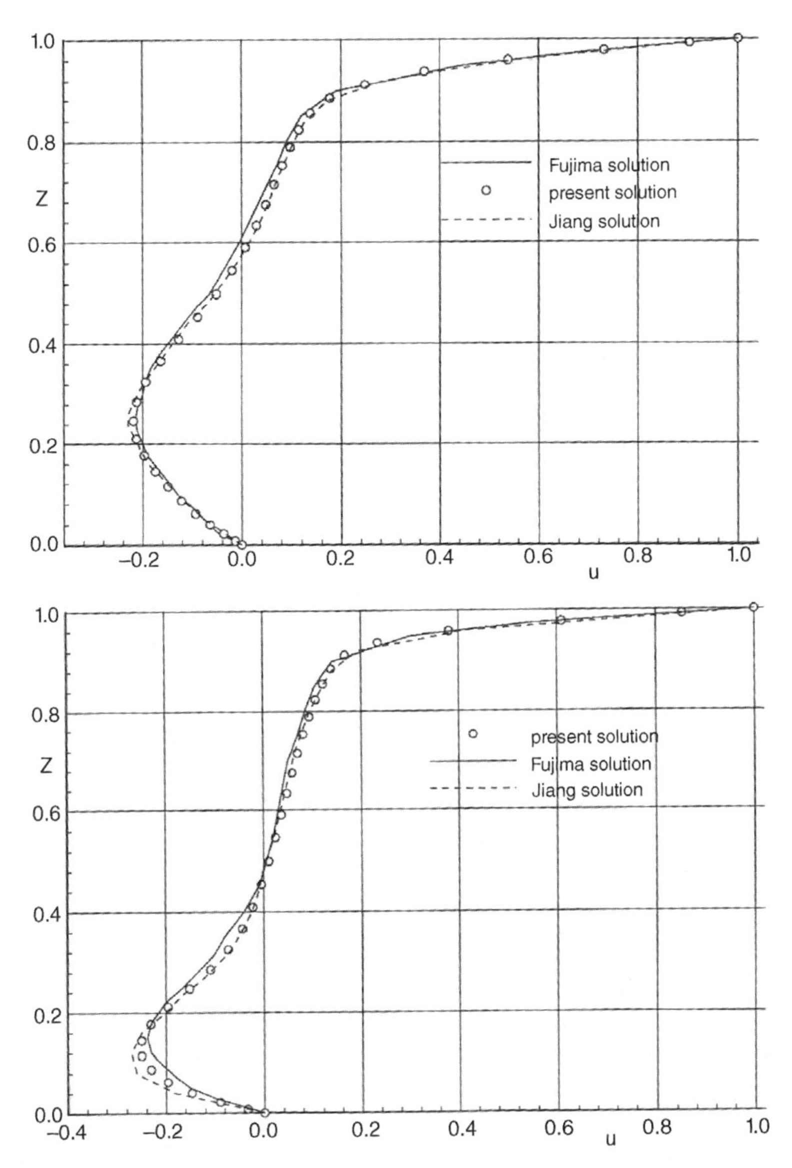

Figure 6.14 Driven cavity GWS^h solution validation, $n=3$, solution adapted $k=1$ TP basis $M=64^3$ mesh, $\{U(x=0.5,\ y,\ z=0.5)\}$ DOF comparison to benchmark data of Fujima *et al.* (1994), Jiang *et al.* (1994): top, Re = 400; bottom, Re = 1000. *Source:* Wong, K.L. and Baker, A.J. 2002. Reproduced with permission by Wiley

The $n=3$ thermal cavity benchmark geometry with BCs is Figure 6.15. Being a cavity, all domain boundaries are no-slip hence the vorticity kinematic DOF BCs (6.65–6.67) are applied for all wall DOF. The temperature IC is linear interpolation of the Dirichlet BC wall data; the velocity IC is zero. The unit cube was discretized into a solution adapted $M=48^3$, $k=1$ TP basis mesh for GWS^h steady solutions for $E+04 \leq Ra \leq 1.6E+07$. The

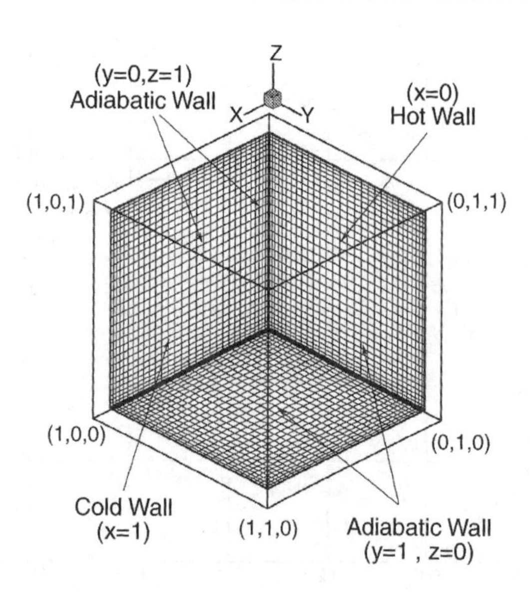

Figure 6.15 3-D cavity

domain is partitioned into eight uniform size rectangular blocks and utilizes 130 Mbytes of global memory for parallel execution.

Analogous to Figure 6.10, the cavity mid-plane perspective velocity vector and temperature flood graphic of steady *a posteriori* data essence for Ra = E + 06 is presented in Figure 6.16. The centerplane temperature flood clearly delineates very large temperature gradients adjacent to both Dirichlet BC walls, a solution character qualitatively identical to the $n = 2$ simulation, Figure 6.6. The contour lines sketched on each mid-plane velocity vector perspective are the corresponding $n = 2$ planar streamline distributions. This data presentation admits clear visualization of the multiple vortex velocity fields intrinsic to the $n = 3$ Ra = E + 06 thermal cavity.

Thermal cavity steady solution dependence on Ra is summarized in Figure 6.17. Graphed are symmetry plane centerline temperature DOF distribution, upper left, transverse (z) coordinate variation of centerline temperature, upper right, and centerline u^h and w^h resolution DOF component distributions, lower graphic pair. The sharp dependence of near wall temperature gradient on Ra is apparent, as is its moderation by proximity to the transverse plane adiabatic wall BCs for Ra = E + 06. The centerline u^h magnitude DOF distribution monotonically decreases by a factor of five with increasing Ra over the range. Conversely, velocity resolution w^h DOF parallel to the fixed temperature walls remain nominally constant while the transverse reach diminishes substantially with increasing Ra.

On the symmetry plane centerline, Figure 6.18 documents validation of the GWS^h steady solution {T} DOF distribution by comparison with a Ra = E + 05 benchmark solution generated on a non-uniform $M = 62^3$ equivalent mesh, Fusegi *et al.* (1991). Quantitative validation accrues to the centerline *extremum* {U} DOF difference being 0.0053 on 0.1468, a 3.5% distinction, Wong (1995).

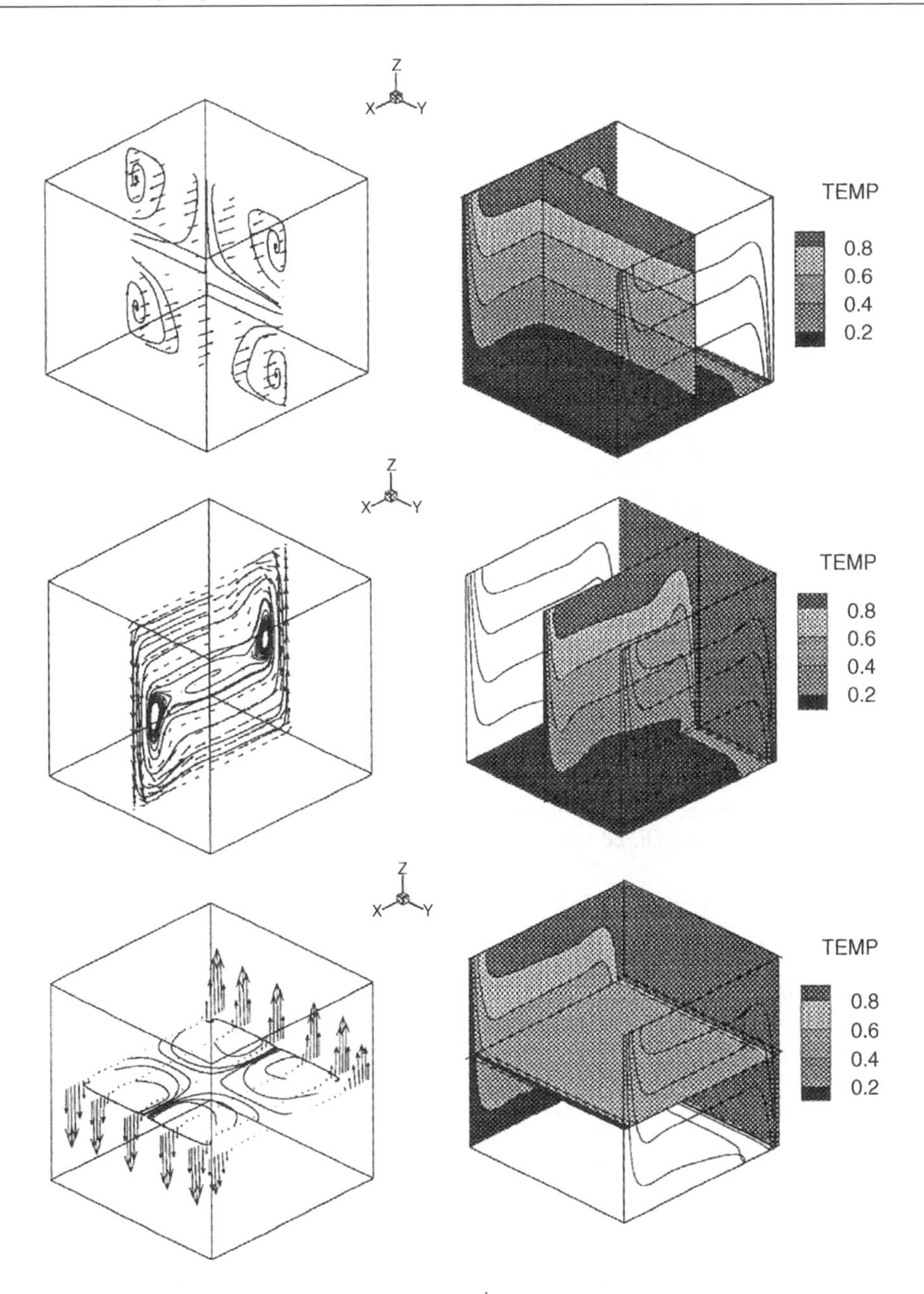

Figure 6.16 Thermal cavity $n=3$ benchmark GWS^h steady solution perspectives, solution adapted $k=1$ TP basis $M=48^3$ mesh, Ra = E + 06: left column, $\mathbf{u}^h(x, y, z)$ with planar streamline contours added; right column, $T^h(x, y, z)$ planar flood with sidewall isotherm projection. *Source:* Wong, K.L. and Baker, A.J. 2002. Reproduced with permission by Wiley

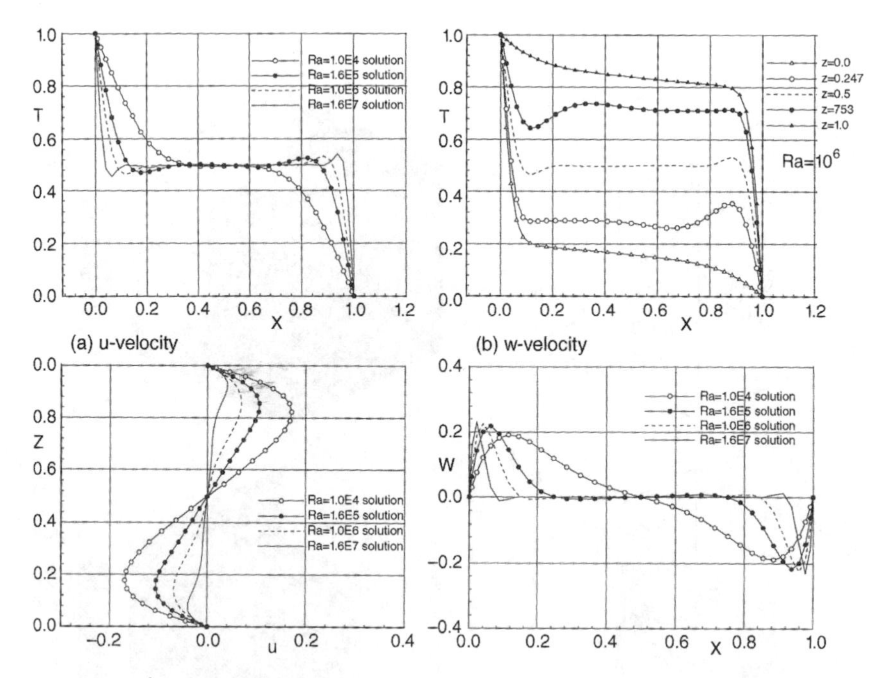

Figure 6.17 GWS^h solution DOF *a posteriori* data confirming thermal cavity dependence on Ra, $n=3$, solution adapted $k=1$ TP basis $\mathrm{M}=64^3$ mesh, $\mathrm{E}+04 \leq \mathrm{Ra} \leq 1.6\mathrm{E}+07$, from Wong and Baker (2002): top left, $\{\mathrm{T}(x, y=0.5, z=0.5)\}$; top right, $\{\mathrm{T}(x, y=0.5, z)\}$ for $\mathrm{Ra}=\mathrm{E}+06$; bottom left, $\{\mathrm{U}(x, y=0.5, z=0.5)\}$; bottom right, $\{\mathrm{W}(x, y=0.5, z=0.5)\}$. *Source:* Wong, K.L. and Baker, A.J. 2002. Reproduced with permission by Wiley

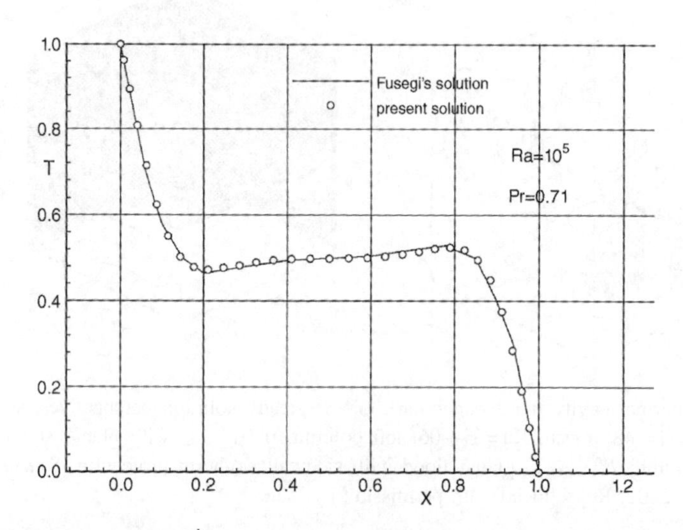

Figure 6.18 Thermal cavity GWS^h steady solution validation, $n=3$, solution adapted $k=1$ TP basis $\mathrm{M}=64^3$ mesh, $\mathrm{Ra}=\mathrm{E}+05$: $\{\mathrm{T}(x, y=0.5, z=0.5)\}$ DOF distribution compared with Fusegi *et al.* (1991), from Wong (1995)

6.7 Summary

The chapter details vector field theory manipulations rendering the incompressible-thermal NS conservation principle *constrained* I-EBV PDE + BCs system well-posed. The starting point is (6.1–6.3)

$$
\begin{aligned}
&\mathrm{D}M: && \nabla \cdot \mathbf{u} = 0\\
&\mathrm{D}\mathbf{P}: && \mathrm{St}\frac{\partial \mathbf{u}}{\partial t} + \nabla \cdot \mathbf{uu} + \nabla P - \frac{1}{\mathrm{Re}}\nabla \cdot \nabla \mathbf{u} + \frac{\mathrm{Gr}}{\mathrm{Re}^2}\Theta\,\widehat{\mathbf{g}} = 0\\
&\mathrm{D}E: && \mathrm{St}\frac{\partial \Theta}{\partial t} + \nabla \cdot \mathbf{u}\Theta - \frac{1}{\mathrm{RePr}}\nabla \cdot \nabla \Theta - s_\Theta = 0
\end{aligned}
$$

with state variable $\{q(\mathbf{x}, t)\} = \{\mathbf{u}, P, \Theta\}^T$, the non-D velocity vector, kinematic pressure $P \equiv p/\rho_0$, and potential temperature Θ. The key theoretical requirement is enforcing DM, (6.1), via manipulations which as well eliminate pressure as a state variable member.

Computationally viable vector field theory alterations universally insert vorticity (vector) into the state variable. Thus are introduced two *key theoretical issues*:

- the NS velocity vector field must be sufficiently smooth such that its curl lies in H^1, recall Table 3.1, and
- vorticity BCs on a no-slip wall are not *a priori* known, hence must be derived via TS approximation to kinematics.

Invoking $\mathbf{\Omega} \equiv \mathrm{curl}\ \mathbf{u}$ and $\mathbf{u} \equiv \mathrm{curl}\ \mathbf{\Psi}$ and setting $\mathrm{St} \equiv 1$ produces the *vorticity-streamfunction* PDE system (6.5–6.6) replacement for (6.1–6.3)

$$
\begin{aligned}
\mathcal{L}(\mathbf{\Omega}) &= \frac{\partial \mathbf{\Omega}}{\partial t} + \nabla \times \mathbf{\Psi} \cdot \nabla \mathbf{\Omega} - (\mathbf{\Omega} \cdot \nabla)\nabla \times \mathbf{\Psi} - \frac{1}{\mathrm{Re}}\nabla^2 \mathbf{\Omega} + \frac{\mathrm{Gr}}{\mathrm{Re}^2}\nabla \times \Theta\widehat{\mathbf{g}} = \mathbf{0}\\
\mathcal{L}(\Theta) &= \frac{\partial \Theta}{\partial t} + \nabla \times \mathbf{\Psi} \cdot \nabla \Theta - \frac{1}{\mathrm{RePr}}\nabla \cdot \nabla \Theta - s_\Theta = 0
\end{aligned}
$$

and appends the strictly *kinematic* vector identity (6.7)

$$
\mathcal{L}(\mathbf{\Psi}) = -\nabla^2 \mathbf{\Psi} + \nabla(\nabla \cdot \mathbf{\Psi}) - \mathbf{\Omega} = 0
$$

Algorithm implementation detraction is the $\nabla \cdot \mathbf{\Psi}$ term in (6.7) which fully couples the solution process for streamfunction $\mathbf{\Psi}$. The alternative is to identify a divergence free $\mathbf{\Psi}$ which is proven as impractical as theorizing $\mathbf{u}$ that is solenoidal.

The alternative to (6.5–6.7) is the *vorticity-velocity* PDE system. Invoking only $\mathbf{\Omega} \equiv \mathrm{curl}$ $\mathbf{u}$ generates (6.8–6.9)

$$
\begin{aligned}
\mathcal{L}(\mathbf{\Omega}) &= \frac{\partial \mathbf{\Omega}}{\partial t} + (\mathbf{u} \cdot \nabla)\mathbf{\Omega} - (\mathbf{\Omega} \cdot \nabla)\mathbf{u} - \frac{1}{\mathrm{Re}}\nabla^2 \mathbf{\Omega} + \frac{\mathrm{Gr}}{\mathrm{Re}^2}\nabla \times \Theta\,\widehat{\mathbf{g}} = \mathbf{0}\\
\mathcal{L}(\mathbf{u}) &= -\nabla^2 \mathbf{u} - \nabla \times \mathbf{\Omega} = \mathbf{0}
\end{aligned}
$$

The ability to induce exact D*M* satisfaction via $\boldsymbol{\Psi}$ is relinquished, which is approximately recovered by specifically designed differential definition discrete TS statements for $\boldsymbol{\Omega}$ on no-slip walls.

The literature favorite *legacy* computational alternative to (6.5–6.7) is reduction to $n = 2$ whence $\nabla \cdot \boldsymbol{\Psi}$ is identically zero. Streamfunction and vorticity are reduced to single scalar components $\boldsymbol{\Psi} \Rightarrow \psi \widehat{\mathbf{k}}, \boldsymbol{\Omega} \Rightarrow \omega \widehat{\mathbf{k}}$, and (6.5–6.7) simplify to (6.10–6.12)

$$\mathcal{L}(\omega) = \frac{\partial \omega}{\partial t} + \nabla \times \psi \widehat{\mathbf{k}} \cdot \nabla \omega - \frac{1}{\mathrm{Re}} \nabla^2 \omega + \frac{\mathrm{Gr}}{\mathrm{Re}^2} \nabla \times \Theta \widehat{\mathbf{g}} = 0$$

$$\mathcal{L}(\Theta) = \frac{\partial \Theta}{\partial t} + \nabla \times \psi \widehat{\mathbf{k}} \cdot \nabla \Theta - \frac{1}{\mathrm{RePr}} \nabla^2 \Theta - s_\Theta = 0$$

$$\mathcal{L}(\psi) = -\nabla^2 \psi - \omega = 0$$

The state variable for the I-EBV system (6.10–6.11) plus EBV (6.12) is $\{q(\mathbf{x}, t)\} = \{\omega, \Theta, \psi\}^T$ hence BCs are required specified on the entire domain closure. The vorticity no-slip wall BC is *a priori* unknown; however, a Robin BC is extractable from (6.12) via a TS written on ψ of $O(\Delta n^4)$ in the wall normal direction yielding (6.15)

$$\ell(\omega) = \nabla \omega \cdot \hat{\mathbf{n}} + \frac{3}{\Delta n} \omega_w + \frac{6}{\Delta n^2} \mathrm{U_w} + \frac{6}{\Delta n^3} (\psi_{\mathrm{w}+1} - \psi_{\mathrm{w}})$$

Subscripts w and w + 1 denote wall and first off-wall nodes, separated by Δn, and $\mathrm{U_w}$ is wall tangential translation speed. Code practice commonly replaces (6.15) with TS truncations of $O(\Delta n^2)$ or $O(\Delta n)$.

Vorticity is a mathematical crutch because sought output is typically velocity vector and pressure distributions. The former is predictable via curl $\omega \widehat{\mathbf{k}} \equiv$ curl curl $\mathbf{u}$ which generates the EBV system (6.17)

$$\mathcal{L}(\mathbf{u}) = -\nabla^2 \mathbf{u} - \nabla \times \omega \widehat{\mathbf{k}} = \mathbf{0}$$

well-posed via known encompassing BCs. The alternative is GWS^N on the *differential definition* $\mathcal{D}(\mathbf{u}) \equiv \mathbf{u} - \mathrm{curl}\, \psi \widehat{\mathbf{k}} = 0$ for which only homogeneous Dirichlet constraints are applicable.

The pressure distribution is post-processed via the well-posed Poisson EBV PDE + BCs (6.18) generated from div **DP** for solution converged $\nabla^h \cdot \mathbf{u}^h \approx 0$

$$\mathcal{L}(P) = -\nabla^2 P - \frac{\partial}{\partial x_j} \left[u_i \frac{\partial u_j}{\partial x_i} + \frac{\mathrm{Gr}}{\mathrm{Re}^2} \Theta \widehat{g}_j \right] = 0$$

$$\ell(P) = \nabla P \cdot \hat{\mathbf{n}} - \left[\frac{\partial \mathbf{u}}{\partial t} - \frac{1}{\mathrm{Re}} \nabla^2 \mathbf{u} \right] \cdot \hat{\mathbf{n}} = 0$$

A range of *a posteriori* data exists for precisely designed compute experiments for $\mathrm{GWS}^h/m\mathrm{GWS}^h + \theta\mathrm{TS}$ algorithms for PDE + BCs system (6.10–6.12), also their *m*PDE theory replacements (6.19–6.21)

$$\mathcal{L}^m(\omega) = \frac{\partial \omega}{\partial t} + \nabla \times \psi\,\widehat{\mathbf{k}} \cdot \nabla\omega - \frac{1}{\mathrm{Re}}\nabla^2\omega$$

$$+\frac{\mathrm{Gr}}{\mathrm{Re}^2}\nabla \times \Theta\,\widehat{\mathbf{g}} - \frac{\mathrm{Re}h^2}{12}\frac{\partial}{\partial x_j}\left[u_j u_k \frac{\partial \omega}{\partial x_k}\right] = 0$$

$$\mathcal{L}^m(\Theta) = \frac{\partial \Theta}{\partial t} + \nabla \times \psi\,\widehat{\mathbf{k}} \cdot \nabla\Theta - \frac{1}{\mathrm{RePr}}\nabla^2\Theta$$

$$- s_\Theta - \frac{\mathrm{RePr}h^2}{12}\frac{\partial}{\partial x_j}\left[u_j u_k \frac{\partial \Theta}{\partial x_k}\right] = 0$$

$$\mathcal{L}^m(\psi) = -\nabla^2\psi - \omega - \frac{h^2}{12}\nabla^2\omega = 0$$

The *asymptotic error* estimate for $\mathrm{GWS}^h/m\mathrm{GWS}^h + \theta\mathrm{TS}$ algorithm predictions for state variable $\{q^h(\mathbf{x}, t)\} = \{\omega^h, \Theta^h, \psi^h\}^T$ is (6.44)

$$\left\|\{e^h(n\Delta t)\}\right\|_E \le Ch^{2\gamma}\left[\|\mathrm{data}\|_{L2,\Omega} + \|\mathrm{data}\|_{L2,\partial\Omega}\right] + C_2\Delta t^{f(\theta)}\left\|\{q^h(t_0)\}\right\|^2_{H^1(\Omega)}$$

$$\text{for} : \gamma \equiv \min(k, (k+1), r-1), f(\theta) = (2, 3)$$

for C and C_2 constants *with* caveat $(k+1) \Rightarrow (1+1) = 2$ for $m\mathrm{GWS}^h$ algorithm $k = 1$ basis mPDE solutions for *sufficiently large* Re.

The FD CFD community-conceived lid driven cavity *benchmark* precisely supports *theory validation* regarding steady solution *asymptotic convergence*, *error estimation* and enhanced *stability* via $m\mathrm{GWS}^h$ theory moderation of dispersion error. Confirmation the GWS^h solution *optimal*ity in *quantified* for $100 \le \mathrm{Re} < 3200$, extending the linear theory caveat of DOF-equivalence well into the nonlinear range, plus validating $m\mathrm{GWS}^h$ theory improvement on GWS^h optimality.

In addition to theory validation, the driven cavity fundamental corner BC *singularity* is thoroughly characterized. For modest $\mathrm{Re} = 100$, solution adapted regular mesh refinement *a posteriori* data quantify corner vorticity *unbounded* singularities. These are *not predicted* by FD constructions which omit the corner vorticity DOF from the laplacian stencil. Resulting GWS^h post-processed velocity and pressure DOF distributions quantify the very complex velocity vector fields in lid–corner juncture regions, as well as stagnation region pressure discrete approximations of near singular behavior.

Identification of the *optimal mesh solution* via energy norm *equi-distribution* for fixed DOF is quantitatively assessed for the $n = 2$ thermal cavity benchmark. Absent corner BC singularities, solution adapted regular mesh refinement *a posteriori* data confirm the linear theory prediction. For the $n = 2$ step-wall diffuser *validation* statement, prediction comparison with experimental data clearly enforces the requirement for outflow plane homogeneous Neumann BC validity.

Chapter completion derives the $m\mathrm{GWS}^h + \theta\mathrm{TS}$ algorithm for the $n = 3$ vorticity-velocity NS PDE system (6.8–6.9) plus (6.45) for state variable $\{q(\mathbf{x}, t)\} \equiv \{\Omega_x, \Omega_y, \Omega_z, \Theta, u, v, w\}^T$. Section 6.5 details algorithm derivation, also the absolutely *key* issue of $\mathrm{GWS}^h k$ basis completeness degree TS *order consistent* approximation of the kinematic constraint $\mathbf{\Omega}^h \bullet \widehat{\mathbf{n}} \equiv \nabla^h \times \mathbf{u}^h \bullet \widehat{\mathbf{n}}$ on no-slip walls with unit normal $\widehat{\mathbf{n}}$.

A posteriori data for developing laminar square cross-section duct flow validates attainment of the requisite axial velocity fully developed paraboloidal distribution. Additional data for the cubical driven cavity and thermal cavity $n = 3$ specifications confirm excellent quantitative agreement with the available benchmark CFD data for $\mathrm{Re} \leq 1000$ and $\mathrm{Ra} \leq \mathrm{E}+07$ respectively.

Summarily, solid *quantitative validation* of the pertinence of weak form *linear* theories to prediction of very *nonlinear* $\mathrm{GWS}^h/m\mathrm{GWS}^h + \theta\mathrm{TS}$ algorithm steady solution performance exists for vector field theory alterations to the primitive NS PDE system. Validated as well is $m\mathrm{GWS}^h$ theory prediction of mPDE differential term alterations to NS and EBV Poisson PDEs generating *optimal* performance $k = 1$ basis algorithm implementations.

Exercises

6.2.1 Verify the $n = 2$ vorticity-streamfunction PDE system (6.10–6.12) starting with (6.5–6.7).

6.2.2 Starting with (6.14) confirm the lower-order TS for wall vorticity DOF determination in (6.16).

6.2.3 Form div **DP** from (6.2) to verify the pressure Poisson PDE + BCs system (6.18).

6.3.1 Substitute (6.19–6.21) into the $m\mathrm{GWS}^h + \theta\mathrm{TS}$ algorithm, recall Section 5.6, hence derive the algorithm matrix statement (6.23–6.26).

6.3.2 Using the results of 6.3.1 and the matrix iteration algorithm statement (6.22), verify the empty matrices in (6.27) as well as the accuracy of statements (6.28–6.33).

6.3.3 Form the divergence of (6.2), then proceed through the Green–Gauss process to confirm kinematic pressure GWS^N algorithm statement (6.37).

6.3.4 Confirm that (6.38) is the accurate discrete implementation of (6.37) for $n = 2$.

6.3.5 Verify that imposing the wall BC shear term in (6.38) amounts to eliminating the wall node contribution from $[\mathrm{A211}]_e\{\mathrm{U}J \bullet \widehat{\mathbf{n}}_j\}_e$.

6.5.1 Verify the vorticity-velocity $m\mathrm{GWS}^h + \theta\mathrm{TS}$ algorithm matrix statement (6.46–6.49).

6.5.2 Validate several of the $m\mathrm{GWS}^h + \theta\mathrm{TS}$ algorithm jacobian matrix contributions in (6.51–6.59).

6.5.3 Confirm the correctness of one or more of the TS-generated no-slip wall vorticity DOF constraints (6.65–6.67).

References

Armaly, B.F., Durst, F., Pereira, J.C.F. and Schonung, B., (1983). "Experimental and theoretical investigation of backward-facing step flow," *J. Fluid Mechanics*, V. 127.

Babuska, I. and Rheinboldt, W.C. (1976). "*A-posteriori* error estimates for the finite element method," *J. Numerical Methods Engineering*, V. 12.

Baker, A.J. (2012). *Finite Elements ⇔ Computational Engineering Sciences*, John Wiley and Sons, London.

Daube, O., (1992). "Resolution of the 2D Navier-Stokes equations in velocity-vorticity form by means of an influence matrix technique," *J. Comp. Physics*, V. 103.

deVahl Davis, G., (1983). "Natural convection of air in a square cavity; a benchmark numerical solution," *J. Numerical Methods Fluids*, V. 3.

Ericson, S.C., (2001). "A CFD Laboratory archive supporting the academic process," MSc thesis, University of Tennessee.

Fusegi, T., Hyun, J.M., Kuwahara, K. and Farouk, B., (1991). "A numerical study of three-dimensional natural convection in a differentially heated cubical enclosure," *J. Heat & Mass Transfer*, V. 34.

Fujima, S., Tabata, M. and Fukasawa, Y., (1994). "Extension to three-dimensional problems of the upwind finite element scheme based on the choice of up- and down-wind points," *Computer Methods Applied Mechanics & Engineering*, V. 112.

Ghia, K.N., Ghia, U. and Shin, C.T., (1982). "High-Re solutions for incompressible flow using the Navier–Stokes equations and a multi-grid method," *J. Computational Physics*, V. 48.

Goldstein, R.J. and Freid, D.K. (1967). "Measurement of laminar flow development in a square duct using a laser Doppler flowmeter," *J. Applied Mechanics*, V. 34.

Guevremont, G., Habashi, W.G., Kotiuga, P.L. and Hafez, M.M., (1993). "Finite element solution of the 3D compressible Navier-Stokes equations by a velocity-vorticity method," *J. Computational Physics*, V. 107.

Guj, G. and Stella, F., (1988). "Numerical solutions of the high-Re recirculating flows in vorticity-velocity form," *J. Numerical Methods Fluids*, V. 8.

Guj, G. and Stella, F., (1993). "A vorticity-velocity method for numerical solution of the 3D incompressible flows," *J. Computational Physics*, V. 106.

Gunzburger, M.D. and Peterson, J.S., (1998). "On finite element approximations of the streamfunction-vorticity and velocity-vorticity equations," *J. Numerical Methods Fluids*, V. 8.

Han, L.S. (1960). "Hydrodynamic entrance lengths for incompressible laminar flow in rectangular ducts," *J. Applied Mechanics*, V. 27.

Jiang, B.N., Lin, T.L. and Povinelli, L.A., (1994). "Large scale computation of incompressible viscous flows by least-squares finite element method," *Computer Methods Applied Mechanics & Engineering*, V. 114.

Kolesnikov, A. and Baker, A.J., (2001). "An efficient high order Taylor weak statement formulation for the Navier–Stokes equations," *J. Computational Physics*, V. 173.

Sonneveld, P., (1989). "CGS, a fast Lanczos-type solver for non-symmetric linear systems," *SIAM J. Scientific & Statistical Computing*, V. 10.

van derVorst, H., (1992). "Bi-CGSTAB: a fast and smoothly converging variant of the Bi-CG for the solution of non-symmetric linear systems," *SIAM J. Scientific & Statistical Computing*, V. 13.

White, F. (1974). *Viscous Fluid Flow*, McGraw-Hill, NY.

Wong, K.L., (1995). "A parallel finite element algorithm for 3-D incompressible flow in velocity-vorticity form," PhD dissertation, University of Tennessee.

Wong, K.L. and Baker, A.J., (2002). "A 3-D incompressible Navier-Stokes velocity-vorticity weak form finite element algorithm," *J. Numerical Methods Fluids*, V. 38.

7

Classic State Variable Formulations:

GWS/mGWS^h + θTS algorithms for Navier–Stokes; accuracy, convergence, validation, BCs, radiation, ALE formulation

7.1 Classic State Variable Navier–Stokes PDE System

Chapter content to this juncture has *verified* the accuracy, convergence, stability and optimality properties of standard and *optimal m*odified continuous Galerkin weak form algorithms applied to Reynolds ordered and vector field theory manipulated incompressible-thermal Navier–Stokes conservation principle PDE + BCs system. The net consequence is solid *quantitative validation* of the pertinence of *linear* weak form theory to characterizing *nonlinear* NS PDE system algorithm performance via detailed analysis of *a posteriori* data generated in precisely designed and executed compute experiments.

Attention now turns to GWS^h/mGWS^h + θTS algorithm derivation/validation for the *classic state variable* form of D*M*-constrained NS PDE system

$$\mathrm{D}M: \quad \nabla \cdot \mathbf{u} = 0 \tag{7.1}$$

$$\mathrm{D}\mathbf{P}: \quad \frac{\partial \mathbf{u}}{\partial t} + \nabla \cdot \mathbf{u}\mathbf{u} + \frac{1}{\rho_0}\nabla p - \frac{1}{\mathrm{Re}}\nabla \cdot \nabla \mathbf{u} + \frac{\mathrm{Gr}}{\mathrm{Re}^2}\Theta\,\widehat{\mathbf{g}} = 0 \tag{7.2}$$

$$\mathrm{D}E: \quad \frac{\partial \Theta}{\partial t} + \nabla \cdot \mathbf{u}\Theta - \frac{1}{\mathrm{RePr}}\nabla \cdot \nabla \Theta - s_\Theta = 0 \tag{7.3}$$

$$\mathrm{D}Y: \quad \frac{\partial Y}{\partial t} + \nabla \cdot \mathbf{u}Y - \frac{1}{\mathrm{ReSc}}\nabla \cdot \nabla Y - s_Y = 0 \tag{7.4}$$

The state variable for (7.1–7.4) is (non-D) velocity vector, pressure, temperature and mass fraction, $\{q(\mathbf{x}, t)\} \equiv \{\mathbf{u}, p, \Theta, Y\}^T$. The Reynolds, Grashoff, Prandtl and Schmidt fundamental characterizing groups remain $\mathrm{Re} \equiv \mathrm{UL}/\nu$, $\mathrm{Gr} \equiv \mathrm{g}\beta\Delta\mathrm{TL}^3/\nu^2$, $\mathrm{Pr} \equiv \rho_0 \nu c_p/k$ and $\mathrm{Sc} = \mathrm{D}/\nu$. In the Boussinesq buoyancy model, $\Delta\mathrm{T} \equiv (T_{\max} - T_{\min})$ and $\beta \equiv 1/\mathrm{T}_{\mathrm{abs}}$, and $P \equiv p/\rho_0$ defines kinematic pressure for density ρ_0 otherwise constant. The assumption is $\mathrm{St} \equiv \tau\mathrm{U/L} \equiv 1$ for Stanton number, hence $\tau \equiv \mathrm{L/U}$ is the non-D timescale, and in D*E* the Peclet number $\mathrm{Pe} = \mathrm{RePr}$ is a common replacement.

As often stated, the mathematical character of (7.2–7.4) is *constrained* initial-elliptic boundary value (I-EBV), as (7.1) requires *admissable* NS solutions $\{q(\mathbf{x}, t)\}$ possess a *divergence-free* velocity vector field. The theory generalizing the numerous published SIMPLE-type algorithms is the *continuity constraint method* (CCM), Williams (1993), Williams and Baker (1996). As detailed in Section 5.4, during nonlinear matrix iteration for the $m\mathrm{GWS}^h + \theta\mathrm{TS}$ algorithm DOF $\{Q(t_{n+1})\}$, the derivation confirms the *kinetic action* of pressure at time t_{n+1} is approximated as

$$P^*\big|_{n+1}^{p} = P_n + \frac{1}{\theta\Delta t}\sum_{\alpha=1}^{p} \delta\phi\big|_{n+1}^{\alpha} \tag{7.5}$$

In (7.5), P_n is the genuine kinematic pressure at time t_n, θ is the time integration implicitness factor and Δt the time step size.

The velocity *potential* function ϕ in (7.5) is identified from the divergence of (5.25). Therein, since u_i by definition satisfies D*M*, its divergence vanishes and the resultant EBV PDE + BCs statement at *any* iteration *p* for *any* approximation $u_i^h(x_i, t_{n+1})$ to velocity vector u_i is

$$\begin{aligned} \mathcal{L}(\phi) &= -\nabla^2\phi + \frac{\partial u_i^h}{\partial x_i} = 0 \\ \ell(\phi) &= -\nabla\phi \cdot \hat{\mathbf{n}} - \left(u_i - u_i^h\right)\hat{n}_i = 0 \end{aligned} \tag{7.6}$$

The sequence of potential solution iterates $\delta\phi^\alpha$ generated via (7.6) during CCM algorithm convergent at time t^{n+1} is the data for (7.5).

The typical inflow velocity vector BC is Dirichlet, and on no-slip walls (7.6) confirms the ϕ BC must be homogeneous Neumann. On any outflow boundary segment the D*M*-satisfying velocity vector is unknown, hence also its difference with *any* velocity approximation. Therefore, the sole admissible outflow BC is Dirichlet. In practice the intrinsic measure of the *error* in D*M*, that is, $\mathrm{D}M^h$, is the *energy norm* $\|\nabla^h \cdot \mathbf{u}^h\| \Rightarrow \|\phi^h\|_E$, with practical requirement *minimization*. Hence for all nodes on outflow boundaries the Dirichlet BC is $\phi = 0$.

With (7.5–7.6), and as thoroughly exposed in Section 6.2, the determination of the kinematic pressure $P(x_i, t_{n+1})$ involves algorithm convergence *post-process* solution of the Poisson equation generated by forming the divergence of (7.2) for velocity vector $u_i^h(x_i, t_{n+1})$ indeed *solenoidal*. Recalling (6.18), thus is generated the *well-posed* EBV PDE + BCs system

$$\begin{aligned} \mathcal{L}(P) &= -\nabla^2 P - \frac{\partial}{\partial x_j}\left[u_i \frac{\partial u_j}{\partial x_i} + \frac{\mathrm{Gr}}{\mathrm{Re}^2}\Theta\widehat{g}_j\right] = 0 \\ \ell(P) &= \nabla P \cdot \hat{\mathbf{n}} - \left[\frac{\partial \mathbf{u}}{\partial t} - \frac{1}{\mathrm{Re}}\nabla^2\mathbf{u}\right] \cdot \hat{\mathbf{n}} = 0 \end{aligned} \tag{7.7}$$

In conclusion, the mixed I-EBV/EBV PDE + BCs system (7.2–7.7) is well-posed for theorizing $m\text{GWS}^h + \theta\text{TS}$ weak form algorithms generating discrete approximate solutions to the classic state variable NS conservation principle PDE hence *m*PDE systems.

7.2 NS Classic State Variable *m*PDE System

The weak form $m\text{GWS}^h + \theta\text{TS}$ algorithm addresses the TS theory *modified* PDE (*m*PDE) system replacements for (7.2–7.7) pertinent to implementation via the $k = 1$ basis *only*. I-EBV system (7.2–7.4) is pertinent for NS state variable partition $\{q(\mathbf{x}, t)\} = \{u_i, \Theta, Y\}^T$, hence also its *m*PDE replacement. For *arbitrary* data the linear asymptotic convergence theory generates the *error estimate*

$$\left\|\{e^h(n\Delta t)\}\right\|_E \leq Ch^{2\gamma}\left[\|\text{data}\|_{L2,\Omega} + \|\text{data}\|_{L2,\partial\Omega}\right] + C_2\Delta t^{f(\theta)}\left\|\{q^h(t_0)\}\right\|^2_{H^1(\Omega)}$$
$$\text{for}: \gamma \equiv min(k, (k+1), r-1), f(\theta) = (2, 3, 4) \tag{7.8}$$

for C and C_2 constants. For *smooth data* (7.8) predicts $k = 1$ basis convergence is $O(h^2)$ for *m*PDE specification $\beta \equiv \Delta t/2$, and improves to $O(h^4)$ for $\beta \equiv h^2\text{Re}/12$ and *sufficiently large* Re. As before, $f(\theta) = (2, 3, 4)$ for θ corresponding to explicit and/or backwards Euler, the trapezoidal rule, then the $\beta \equiv h^2\text{Re}/12$ *m*PDE. Additional to (7.8), *optimal* $\gamma \equiv -0.5$ $k = 1$ basis phase accuracy is $O(m^3)$, $m \equiv \kappa h$ the non-D wave number. Hence *optimal modified continuous* Galerkin $k = 1$ basis solutions can potentially exhibit $O(h^4, m^3, \Delta t^4)$ accuracy with stability for smooth data.

The TS theory *m*PDE replacement for the I-EBV PDE system (7.2–7.4) is derived, (5.114). Upon passage through the Newton iteration algorithm (5.119), the terminal NS *m*PDE system divergence form is

$$\mathcal{L}^m(\{q\}) = [\text{M}]\frac{\partial\{q\}}{\partial t} + \frac{\partial}{\partial x_j}\left[\{f_j\} - \{f_j^d\}\right] - \{s\} = \{0\} \tag{7.9}$$

In (7.9), [M] is the diagonal *m*PDE-*augmented* "mass matrix," $\{f_j\}$ the *kinetic flux vector*, $\{f_j^d\}$ the *m*PDE-augmented *dissipative flux vector*, with $\{s\}$ the source term. The *m*PDE definitions in (7.9) for the classic state variable NS formulation are

$$[\text{M}] = \text{diag}\begin{bmatrix} 1 - \dfrac{\gamma\Delta t^2}{6}\dfrac{\partial}{\partial x_j}\left(u_j u_k \dfrac{\partial}{\partial x_k}\right) + \dfrac{\theta h^2}{6}\left(u_j\text{Re} - \dfrac{\partial}{\partial x_j}\right)\dfrac{\partial}{\partial x_j} \\ 1 - \dfrac{\gamma\Delta t^2}{6}\dfrac{\partial}{\partial x_j}\left(u_j u_k \dfrac{\partial}{\partial x_k}\right) + \dfrac{\theta h^2}{6}\left(u_j\text{RePr} - \dfrac{\partial}{\partial x_j}\right)\dfrac{\partial}{\partial x_j} \\ 1 - \dfrac{\gamma\Delta t^2}{6}\dfrac{\partial}{\partial x_j}\left(u_j u_k \dfrac{\partial}{\partial x_k}\right) + \dfrac{\theta h^2}{6}\left(u_j\text{ReSc} - \dfrac{\partial}{\partial x_j}\right)\dfrac{\partial}{\partial x_j} \end{bmatrix} \tag{7.10}$$

$$\{f_j(\{q\})\} = \begin{Bmatrix} u_j u_i + P^*|^p_{n+1}\delta_{ij} \\ u_j\Theta \\ u_j Y \end{Bmatrix}, \; for\ 1 \leq i \leq n \tag{7.11}$$

$$\{f^d(\{q\})\} = \begin{Bmatrix} \dfrac{1}{\text{Re}}\dfrac{\partial u_i}{\partial x_j} + \dfrac{\text{Re}\,h^2}{12} u_j\, u_k \dfrac{\partial u_i}{\partial x_k} \\ \dfrac{1}{\text{RePr}}\dfrac{\partial \Theta}{\partial x_j} + \dfrac{\text{RePr}\,h^2}{12} u_j\, u_k \dfrac{\partial \Theta}{\partial x_k} \\ \dfrac{1}{\text{ReSc}}\dfrac{\partial Y}{\partial x_j} + \dfrac{\text{ReSc}\,h^2}{12} u_j\, u_k \dfrac{\partial Y}{\partial x_k} \end{Bmatrix} \tag{7.12}$$

$$\{s\} = \begin{Bmatrix} -\dfrac{\text{Gr}}{\text{Re}^2}\Theta \widehat{g}_i & -\dfrac{h^2}{6}\left[u_j \text{Re}\dfrac{\partial u_i}{\partial x_j} - \dfrac{\partial^2 u_i}{\partial x_j^2}\right]_n \\ \dfrac{\text{Ec}}{\text{Re}}\dfrac{\partial}{\partial x_j}\left(u_i \dfrac{\partial u_j}{\partial x_i}\right) & -\dfrac{h^2}{6}\left[u_j \text{RePr}\dfrac{\partial \Theta}{\partial x_j} - \dfrac{\partial^2 \Theta}{\partial x_j^2}\right]_n \\ s(Y) & -\dfrac{h^2}{6}\left[u_j \text{ReSc}\dfrac{\partial Y}{\partial x_j} - \dfrac{\partial^2 Y}{\partial x_j^2}\right]_n \end{Bmatrix} \tag{7.13}$$

Note that *none* of the *m*PDE modifications to the NS PDE system reside in the kinetic flux vector, the favorite differential operator location for alternative (FD/FV) CFD manipulations. The [M] matrix time derivative multiplier is totally unique to the weak formulation. Finally, $\text{GWS}^h/m\text{GWS}^h + \theta\text{TS}$ algorithms share a completely common $k=1$ basis implementation, the latter simply possessing additional differential terms except in the kinetic flux vector.

The pressure projection theory *A*ugments the classic NS state variable with $\{q_A(\mathbf{x}, t)\} = \{\phi, P\}^T$. While anticipated not required, the *m*PDE replacement for EBV systems (7.6–7.7) is

$$\mathcal{L}^m(q_A) = -\nabla^2 q_A - s_A(\{q\}) - \frac{h^2}{12}\nabla^2 s_A(\{q\}) = 0 \tag{7.14}$$

Thus is completed the NS classic state variable *m*PDE system.

7.3 NS Classic State Variable $m\text{GWS}^h + \theta\text{TS}$ Algorithm

The *optimal m*odified continuous Galerkin weak form CFD algorithm for the NS *m*PDE system (7.9–7.14) seeks an *approximation* $\{q^N(\mathbf{x}, t)\}$ to the classic state variable $\{q(\mathbf{x}, t)\} = \{u_i, \Theta, Y\}^T$. As always, it is the matrix product of a *trial space* $\{\Psi(\mathbf{x})\}$, a set of N specified functions lying in H^1 (for $k=1$ basis implementation) with a set of time-dependent unknown coefficients. In the *continuum*

$$\{q(\mathbf{x}, t)\} \approx \{q^N(\mathbf{x}, t)\} \equiv \text{diag}\left[\{\Psi(\mathbf{x})\}^T\right]\{Q(t)\} \tag{7.15}$$

where diag[•] denotes a *diagonal* matrix and $\{Q(t)\}$ is a column matrix of time-dependent coefficients constituting the unknown *degrees-of-freedom* (DOF) of the solution approximation.

The Galerkin weak form criterion requires the *approximation error* associated with (7.15), specifically, $\{e^N(\mathbf{x}, t)\} \equiv \{q(\mathbf{x}, t)\} - \{q^N(\mathbf{x}, t)\}$, be orthogonal to the trial space $\{\Psi(\mathbf{x})\}$. Via an interpolation and *extremization*, recall Chapter 2, the weak form scalar is converted into a matrix statement of order identical to the DOF order $\{Q(t)\}$. The resultant *m*odified continuous Galerkin *weak statement* is

$$\begin{aligned}\{m\mathrm{GWS}^N(t)\} &\equiv \int_\Omega \mathrm{diag}\left[\{\Psi(\mathbf{x})\}\right]\mathcal{L}^m(\{q^N\})\mathrm{d}\tau \equiv \{0\} \\ &= [M(\bullet)]\frac{\mathrm{d}\{Q\}}{\mathrm{d}t} + \left\{\mathrm{RES}\left(\{f^N(\bullet)\} - \{f^{d\,N}(\bullet)\} - \{s(\bullet)\}\right)\right\}\end{aligned} \tag{7.16}$$

an order $N \times \mathrm{DOF}(\{q^N\})$ nonlinear matrix ordinary differential equation (ODE) system. The pervasive nonlinear functional dependencies (•) in (7.16) correlate one-to-one with the *m*PDE system (7.10–7.13).

Conceptually solving (7.16) for $\mathrm{d}\{Q(t)\}/\mathrm{d}t$ by clearing $[\mathrm{M}]^{-1}$ generates the data necessary for completing a Taylor series (TS) underlying ODE integration algorithms. Selecting the θ implicit single step Euler family, with timing $t_{n+1} = t_n + \Delta t$, the resultant terminal nonlinear algebraic equation system is

$$\begin{aligned}\{m\mathrm{GWS}^N(t)\} + \theta\mathrm{TS} &\equiv \{\mathrm{F\_}\{Q\}\} \\ &= [\mathrm{M}]\left(\{Q\}_{n+1} - \{Q\}_n\right) + \Delta t\left(\theta\{\mathrm{RES}\}_{n+1} + (1-\theta)\{\mathrm{RES}\}_n\right)\end{aligned} \tag{7.17}$$

For iteration index p, the nonlinear matrix solution process for the $\{q^N\}$ coupled system (7.17) is

$$\begin{aligned}\left[\mathrm{JAC\_}\{Q\}\right]\{\delta Q\}^{p+1} &= -\{\mathrm{F\_}\{Q\}\}_{n+1}^p \\ \{Q\}_{n+1}^{p+1} &= \{Q\}_{n+1}^p + \{\delta Q\}^{p+1} = \{Q\}_n + \sum_{\alpha=0}^{p}\{\delta Q\}^{\alpha+1} \\ \left[\mathrm{JAC\_}\{Q\}\right] &\equiv \frac{\partial\{\mathrm{F\_}\{Q\}\}}{\partial\{Q\}}\end{aligned} \tag{7.18}$$

The companion *m*odified Galerkin weak statement for the NS closure theory *A*ugmented state variable members is

$$\{m\mathrm{GWS}^N(t)\} \equiv \int_\Omega \{\Psi(\mathbf{x})\}\mathcal{L}^m(\{q_A^N\})\mathrm{d}\tau \equiv \{\mathrm{F\_}QA\} \equiv \{0\} \tag{7.19}$$

Each scalar matrix statement resulting from (7.19) is of DOF order N, and the solution DOF thereof are parametric functions of time via source term dependence on $\{q^N(\mathbf{x}, t)\}$.

The terminal step to computability is to evaluate the integrals in (7.17) and (7.19). At this juncture the spatial domain of dependence of (7.2–7.7) with boundary is *discretized* into $\Omega \cup \partial\Omega \approx \Omega^h \cup \partial\Omega^h$ by a computational mesh of measure h, for ∪ denoting union (non-overlapping sum). The global-span trial space $\{\Psi(\mathbf{x})\}$ is then projected onto the much smaller(!) *trial space basis* $\{N_k(\boldsymbol{\eta})\}$ subspace, a column matrix of completeness degree k

polynomials written on a local coordinate system $\boldsymbol{\eta}$ spanning each discrete domain (finite element).

Belonging to the *generic* domain Ω_e denoted by subscript e, and recalling $\Omega^h \equiv \cup\ \partial\Omega_e$, transition from the continuum approximation $\{q^N(\mathbf{x}, t)\}$ to the *discrete* approximate solution $\{q^h(\mathbf{x}, t)\}$ is via the symbolic sequence

$$
\begin{aligned}
\left\{q^N(\mathbf{x},t)\right\} &\equiv \mathrm{diag}\left[\{\Psi(\mathbf{x})\}^T\right]\{Q(t)\} \Rightarrow \left\{q^h(\mathbf{x},t)\right\} \equiv \cup_e \{q(\mathbf{x},t)\}_e \\
\{q(\mathbf{x},t)\}_e &\equiv \mathrm{diag}\left[\{N_k(\boldsymbol{\eta}(\mathbf{x}))\}^T\right]\{Q(t)\}_e
\end{aligned} \tag{7.20}
$$

The definition (7.20) reduces global matrix integral evaluations defined in (7.17–7.19) to those operations on the generic domain Ω_e. All locally computed matrices, each of order element DOF, are projected into the global $N\times$DOF-order algebraic matrix statement via *assembly*, the n-D coupled DOF extension on the example detailed in Section 2.3.

Thereby, formation of (7.17–7.19) is accomplished as

$$
\begin{aligned}
\left[\mathrm{JAC}^h\_\{Q\}\right] &\Rightarrow \mathrm{S}_e\left[\mathrm{JAC}\_\{Q\}\right]_e \equiv \mathrm{S}_e\left(\frac{\partial\{\mathrm{F}\_Q\}_e}{\partial\{Q\}_e}\right) \\
\{\mathrm{F}\_Q^h\}_{n+1}^p &\Rightarrow \mathrm{S}_e\{\mathrm{F}\_Q\}_{e,n+1}^p
\end{aligned} \tag{7.21}
$$

$$
\left\{\mathrm{F}\_QA^h\right\} \Rightarrow \mathrm{S}_e\left([\mathrm{DIFF}]_e\{QA\}_e - \{\mathrm{b}(\{q^h\})\}_e\right) \tag{7.22}
$$

In (7.22), $[\mathrm{DIFF}]_e$ for GWS^h is the element discrete laplacian matrix, thoroughly detailed in Chapter 3, with $\{\mathrm{b}(\bullet)\}_e$ the Poisson PDE-specific element source term augmented for *m*GWS^h, (7.14), if necessary.

The terminal formulation step is to iteratively couple solution for potential ϕ^h with the classic state variable. Hence, (7.21) is formed for $\left\{q^h(\mathbf{x},t)\right\} = \left\{u_i^h, \Theta^h, Y^h, \varphi^h\right\}^T$ which at iterative convergence generates DOF $\{QI(t_{n+1})\}$, (7.20). Then, (7.22) is post-convergence direct solved for the P^h solution DOF at t_{n+1}.

The *m*GWS^h + θTS algorithm for the n-D classic state variable NS *m*PDE system (7.9–7.14) is defined. It is noteworthy that the *analytically* derived *m*PDE nonlinear differential calculus terms responsible for elevating the $k=1$ basis traditional $O(h^2, m^2, t^2)$ convergence to *optimal* $O(h^4, m^3, t^4)$, for *smooth data*, hence error estimate (7.8), reside *mainly* in the *m*GWS^h algorithm mass matrix [M], (7.10). This weak form analytical theory guidance is totally distinct from that underlying legacy FD/FV CFD theories wherein [M] is invariably the identity matrix [I].

7.4 NS *m*GWS^h + θTS Algorithm Discrete Formation

The *m*GWS^h + θTS algorithm algebraic statement for the n-D classic state variable NS *m*PDE system is (7.17–7.18), subsequently formed via (7.21–7.22). The CCM pressure projection theory state variable is $\left\{q^h(\mathbf{x},t)\right\} = \left\{u_i^h, \Theta^h, Y^h, \phi^h; P^h\right\}^T$ with determination sought for the approximate solution DOF $\{QI(t_{n+1})\}$, $1 \le I \le n+4$. Selectively for clarity, the DOF $\{QI(\bullet)\}$, $1 \le I \le 3$, 6 will be denoted $\{UI(\bullet)\}$, $\{\mathrm{PHI}(\bullet)\}$. The solution of (7.22) for P^h is a decoupled post-convergence operation.

In the defined index notation, (7.17) is formed via evaluation of the $m\text{GWS}^h + \theta\text{TS}$ matrix statement on the generic Ω_e

$$\begin{aligned}\{\text{F\_}QI\}_e &= \left[\text{M}(\{Q\}, data)\right]_e \{QI_{n+1} - QI_n\}_e \\ &+ \Delta t\left[\theta\{\text{RES\_}QI(\{Q\})\}_e\big|_{n+1} + (1-\theta)\{\text{RES\_}QI(\{Q\})\}_e\big|_n\right]\end{aligned} \tag{7.23}$$

followed by assembly. The lead term in $[\text{M}(\bullet)]_e$ is the *optimal* γ $m\text{GWS}^h$ element matrix $[m]_e$, (5.83). For $1 \le I \le n+2$, and using the Green–Gauss theorem, the element matrix statement is

$$\begin{aligned}[m]_e &= \int_{\Omega_e} \{N_1\}\left[1 - \frac{\gamma\,\Delta t^2}{6}\frac{\partial}{\partial x_j}\left(u_j^h u_k^h \frac{\partial}{\partial x_k}\right)\right]_e \{N_1\}^T \mathrm{d}\tau \\ &= \det{}_e[\text{M200}] + \frac{\gamma\,\Delta t^2}{6}\{\text{UJUK}\}_e^T[\text{M30}JK]_e\end{aligned} \tag{7.24}$$

The theorem-generated surface integral is omitted in (7.24), the consequence of the typical state variable global domain BCs.

The matrix notation for (7.24) is $\text{M} \Rightarrow (\text{C, B, A})$ for $\Omega \subset \mathbb{R}^n$, $n = (3, 2, 1)$. If subscripted e, the dependence on Ω_e requires conversion to the element-independent library form, Sections 3.8–3.9. The coordinate transformation determinant $\det_e$ is the *measure* of Ω_e related to its volume, area or length.

While (7.24) and the following are written for arbitrary basis implementation, the mPDE theory is valid only for the $k = 1$ basis for which optimal $\gamma \equiv -0.5$. Defining Pa as the placeholder for parametric *data* Re, RePr, ReSc enables compact expression of the element matrix $[\text{M}]_e$ (7.10) as

$$\begin{aligned}[\text{M}]_e &= [m]_e + \int_{\Omega_e} \{N_1\}\left[\frac{\theta h^2}{6}\left(u_j\text{Pa} - \frac{\partial}{\partial x_j}\right)\frac{\partial}{\partial x_j}\right]_e \{N_1\}^T \mathrm{d}\tau \\ &= \det{}_e[\text{M200}] + \frac{\gamma\,\Delta t^2}{6}\{\text{UJUK}\}_e^T[\text{M30}JK]_e \\ &+ \frac{\theta h^2}{6}\left(\text{Pa}\{\text{U}J\}_e^T[\text{M300}J]_e + [\text{M2}JJ]_e\right)\end{aligned} \tag{7.25}$$

with Green–Gauss theorem employed on the last term and deleting the generated surface integral as typically BC overwritten.

The kinetic flux vector element matrix contribution to $\{\text{RES\_}QI\}_e$ for $1 \le I \le n+2$, omitting the kinetic pressure term P^*, is

$$\begin{aligned}\{\text{RES\_}QI(\{f_j\})\}_e &= \int_{\Omega_e} \{N_1\}\frac{\partial}{\partial x_j}u_j^h\{N_1\}^T \mathrm{d}\tau\{QI\}_e \\ &= \{\text{U}J\}_e^T[\text{M300}J]_e\{QI\}_e\end{aligned} \tag{7.26}$$

For $1 \le I \le n = 3$ only and via (7.5) the kinetic pressure contribution in (7.11) converts to the (7.26) augmentation pair

$$
\begin{aligned}
\{\text{RES\_U}I(\{f_j\})\}_e &= \int_{\Omega_e} \{N_1\} \frac{\partial}{\partial x_j} \{P^*|_{n+1}^p \delta_{ij}\} d\tau \\
&= \int_{\Omega_e} \{N_1\} \frac{\partial}{\partial x_i} \left\{ P_n^h + \frac{1}{\theta \Delta t} \sum_{\alpha=1}^{p} \delta \phi^h|_{n-1}^{\alpha} \right\} d\tau \\
&= \int_{\Omega_e} \{N_1\} \frac{\partial \{N_1\}^T}{\partial x_i} \left(\{\text{Q7}(n)\}_e + \frac{1}{\theta \Delta t} \sum_{\alpha=1}^{p} \delta \{\text{Q6}(n+1)\}_e^{\alpha} \right) d\tau \\
&= [\text{M20}I]_e \{\text{Q7}(t_n)\}_e + \frac{1}{\theta \Delta t} [\text{M20}I]_e \sum_{\alpha=1}^{p} \{\delta \text{Q6}\}_e^{\alpha}, 1 \le I \le 3
\end{aligned}
\tag{7.27}
$$

The lead matrix product in (7.27) is *data*, the contribution from the previous time station solution $P^h(\mathbf{x}, t_n)$. The second term tightly couples the continuity constraint via the sequence of velocity potential iterates at time t_{n+1} (*only*). Note its $1/\theta\Delta t$ multiplier cancels the residual multiplier hence this term impact is independent of time step Δt.

Via Pa notation the dissipative flux vector residual matrix is also compactly expressed for $1 \le I \le n+2$ as

$$
\begin{aligned}
\{\text{RES\_Q}I(\{f_j^d\})\}_e &= \int_{\Omega_e} \{N_1\} \frac{\partial}{\partial x_j} \left(\frac{-1}{\text{Pa}} \frac{\partial}{\partial x_j} - \frac{\text{Pa}\, h^2}{12} u_j^h u_k^h \frac{\partial}{\partial x_k} \right) \{N_1\}^T d\tau \{\text{Q}I\}_e \\
&= \frac{1}{\text{Pa}} [\text{M2}KK]_e \{\text{Q}I\}_e + \frac{\text{Pa}\, h^2}{12} \{\text{UJUK}\}_e^T [\text{M30}JK]_e \{\text{Q}I\}_e
\end{aligned}
\tag{7.28}
$$

The element matrix for the "genuine physics" contributions to the source term (7.13) is

$$
\{s\}_e \Rightarrow \left\{
\begin{array}{l}
\dfrac{-\text{Gr}}{\text{Re}^2} \displaystyle\int_{\Omega_e} \{N_1\}\{N_1\}^T d\tau \{\text{Q4}\}_e \widehat{g}_I \\
\dfrac{\text{Ec}}{\text{Re}} \displaystyle\int_{\Omega_e} \{N_1\} \frac{\partial}{\partial x_j} \left(\{N_1\}^T \{\text{U}I\}_e \frac{\partial}{\partial x_i} \{N_1\}^T \{\text{U}J\}_e \right) d\tau \\
\displaystyle\int_{\Omega_e} \{N_1\}\{N_1\}^T d\tau \{\text{SRC}Y\}_e
\end{array}
\right\}
\tag{7.29}
$$

The representative *m*PDE contribution is

$$
\{s\}_e \Rightarrow -\int_{\Omega_e} \{N_1\} \frac{h^2}{6} \left[\text{Re}\{N_1\}^T \{\text{U}J\}_e - \frac{\partial}{\partial x_j} \right]_n \frac{\partial \{N_1\}^T}{\partial x_j} d\tau \{\text{U}I\}_e
\tag{7.30}
$$

Using the Green–Gauss theorem in evaluating the D*E* and terminal *m*PDE terms and discarding the generated surface integrals yields the source term element matrix

$$
\{s\}_e = \left\{ \begin{array}{l} \dfrac{-\mathrm{Gr}\,\widehat{g}_I}{\mathrm{Re}^2}\det_e[\mathrm{M200}]\{\mathrm{Q4}\}_e - \dfrac{h^2}{6}\left[\left(\mathrm{Re}\{\mathrm{U}J\}_e^T[\mathrm{M300}J]_e + [\mathrm{M2}JJ]_e\right)\{\mathrm{U}I\}_e\right]_n \\ \dfrac{-\mathrm{Ec}}{\mathrm{Re}}\{\mathrm{U}I\}_e^T[\mathrm{M30}JI]_e\{\mathrm{U}J\}_e - \dfrac{h^2}{6}\left[\left(\mathrm{RePr}\{\mathrm{U}J\}_e^T[\mathrm{M300}J]_e + [\mathrm{M2}JJ]_e\right)\{\mathrm{Q4}\}_e\right]_n \\ \det_e[\mathrm{M200}]\{\mathrm{SRC}Y\}_e - \dfrac{h^2}{6}\left[\left(\mathrm{ReSc}\{\mathrm{U}J\}_e^T[\mathrm{M300}J]_e + [\mathrm{M2}JJ]_e\right)\{\mathrm{Q5}\}_e\right]_n \end{array} \right\}
\tag{7.31}
$$

The algorithm iterative statement is completed with the element matrix for (7.6). Since at convergence $O(\mathrm{D}M^h) \le O(\varepsilon^2)$, replacement of (7.6) with *m*PDE (7.14) is unnecessary hence

$$
\begin{aligned}
\left\{\mathrm{GWS}^h(\mathcal{L}(\phi^h))\right\}_e &\Rightarrow \int_{\Omega_e}\{N_1\}\left(-\nabla^2\phi^h + \frac{\partial u_i^h}{\partial x_i}\right)\mathrm{d}\tau \\
&= \int_{\Omega_e}\left(\frac{\partial\{N_1\}}{\partial x_j}\frac{\partial\{N_1\}^T}{\partial x_j}\{\mathrm{PHI}\}_e + \{N_1\}\frac{\partial\{N_1\}^T}{\partial x_i}\{\mathrm{U}I\}_e\right)\mathrm{d}\tau \\
&= [\mathrm{M2}JJ]_e\{\mathrm{PHI}\}_e + [\mathrm{M20}I]_e\{\mathrm{U}I\}_e
\end{aligned}
\tag{7.32}
$$

The terminal formation step is the element matrix statement for genuine kinematic pressure. For $n=2$ the GWS^h construction for (7.7) is fully detailed in (6.37–6.38). As pressure constitutes algorithm *data*, replacement of (7.7) with *m*PDE (7.14) is also considered unnecessary. Generalizing to $n=3$ is straightforward, hence

$$
\begin{aligned}
\{\mathrm{F\_Q7}\}_e &= [\mathrm{M2}KK]_e\{\mathrm{Q7}\}_e + \{\mathrm{U}I\}_e^T[\mathrm{M30}JI]_e\{\mathrm{U}J\}_e \\
&\quad + \frac{\mathrm{Gr}}{\mathrm{Re}^2}[\mathrm{M2}J0]_e\{\mathrm{Q4}\}_e\widehat{g}_j + [\mathrm{N200}]_e\frac{\mathrm{d}}{\mathrm{d}t}\{\mathrm{U}J\bullet\widehat{n}_j\}_e \\
&\quad + \frac{1}{\mathrm{Re}}\left([\mathrm{N2}KK]_e\{\mathrm{U}J\bullet\widehat{n}_j\}_e - [\mathrm{A20}S]_e\{\mathrm{U}J\bullet\widehat{n}_j\}_e\Big|_{\mathrm{w}}\right)
\end{aligned}
\tag{7.33}
$$

As introduced in Section 6.3, the last line in (7.33) results from the Green–Gauss theorem applied to the laplacian operator in the Neumann BC (7.7). The term with matrix $[\mathrm{A20}S]$ is the theorem-generated shear in coordinate direction s normal to the boundary enclosing the outflow plane. Recalling the Section 2.3 solution for GWS^h flux DOF at a Dirichlet BC-constrained node, suggested exercise completion will verify that implementing $[\mathrm{A20}S]_e\{\mathrm{U}J\bullet\widehat{n}_j\}_e\Big|_{\mathrm{w}}$ amounts to deleting the companion $[\mathrm{N2}KK]_e\{\mathrm{U}J\bullet\widehat{n}_j\}_e$ contribution at each no-slip wall node.

Except for [M200], all matrices defined in (7.23–7.33) are subscripted e indicating Ω_e dependence. Replacement with the element-independent library matrices is accomplished by implementing the coordinate transformation. These operations, amply illustrated in preceding chapters, produce

$$
\begin{aligned}
[\mathrm{M20}J]_e &= \left(\frac{\partial \eta_k}{\partial x_j}\right)_e [\mathrm{M20}k], [\mathrm{M2}J0]_e = \left(\frac{\partial \eta_k}{\partial x_j}\right)_e [\mathrm{M2}k0] \\
[\mathrm{M300}J]_e &= \left(\frac{\partial \eta_k}{\partial x_j}\right)_e [\mathrm{M300}k] \\
[\mathrm{M2}KK]_e &= \det{}_e^{-1}\left(\frac{\partial \eta_l}{\partial x_k}\frac{\partial \eta_m}{\partial x_k}\right)_e [\mathrm{M2}lm] \\
[\mathrm{M30}JK]_e &= \det{}_e^{-1}\left(\frac{\partial \eta_l}{\partial x_j}\frac{\partial \eta_m}{\partial x_k}\right)_e [\mathrm{M30}lm]
\end{aligned}
\tag{7.34}
$$

Finally, select *m*PDE operators possess an h^2 multiplier. The theory rigorously admits non-uniform mesh with progression ratio pr only for $n=1$, (5.101). To implement the h^2 multiplier on appropriate $n=3$ library element matrices (7.34) requires a relationship with element matrix multiplier $\det_e^{-1}$. For the tri-linear TP basis the essential measure relationship is $h^n \propto 2^n \det_e$, refer to (3.26) and (3.93). Replacing proportionality with equality, $h^2/\det_e = (2)^{2n/3}/\det_e{}^{1/3}$ is the relationship for $n=3$ while that for $n=2$ is simply $h^2/\det_e = 4$.

Thereby, for *m*PDE term matrices in (7.25), (7.28) and (7.30) the $n=3$ definitions in (7.34) are replaced with

$$
\begin{aligned}
h^2[\mathrm{M2}KK]_e &= \frac{(2)^{2n/3}}{\det_e^{1/3}}\left(\frac{\partial \eta_l}{\partial x_k}\frac{\partial \eta_m}{\partial x_k}\right)_e [\mathrm{M2}lm] \\
h^2[\mathrm{M300}J]_e &= (2)^{2n/3}\det{}_e^{2/3}\left(\frac{\partial \eta_k}{\partial x_j}\right)_e [\mathrm{M300}k] \\
\mathrm{Pa}\, h^2[\mathrm{M30}JK]_e &= \frac{(2)^{2n/3}\mathrm{Pa}}{\det_e^{1/3}}\left(\frac{\partial \eta_l}{\partial x_j}\frac{\partial \eta_m}{\partial x_k}\right)_e [\mathrm{M30}lm]
\end{aligned}
\tag{7.35}
$$

Viewing (7.34–7.35) a small number of library matrices are required to implement the $m\mathrm{GWS}^h + \theta\mathrm{TS}$ algorithm. For the $k=1$ NC basis choice, from (3.81) and (3.69) the tetrahedron measure relationship is $h^n \propto \det_e/(n!)$. Replacing proportionally with equality, the $k=1$ NC basis matrix multiplier replacements in (7.35) for $n=3$ are $2^{2n/3}/\det_e{}^{1/3} \Rightarrow 1/(n!)^{2/3} \det_c{}^{1/3}$ and $2^{2n/3} \det_e{}^{2/3} \Rightarrow \det_c{}^{2/3}/(n!)^{2/3}$ with $h^2/\det_e = 1/2$ for $n=2$.

7.5 $m\mathrm{GWS}^h + \theta\mathrm{TS}$ Algorithm Completion

The $m\mathrm{GWS}^h + \theta\mathrm{TS}$ algorithm definitions $\{\mathrm{F\_Q}I\}_e$ in (7.23) for construction of the algebraic jacobian, for $1 \le I \le n+2$, are

$$
\begin{aligned}
[\mathrm{M}]_e = {} & \det{}_e[\mathrm{M200}] + \frac{\gamma\,\Delta t^2}{6}\{\mathrm{U}J\mathrm{U}K\}_e^T[\mathrm{M30}JK]_e \\
& + \frac{\theta h^2}{6}\left(\mathrm{Pa}\{\mathrm{U}J\}_e^T[\mathrm{M300}J]_e + [\mathrm{M2}JJ]_e\right)
\end{aligned}
\tag{7.36}
$$

$$
\begin{aligned}
\{\text{RES\_Q}I\}_e = {} & \{\text{U}J\}_e^T[\text{M300}J]_e\{\text{Q}I\}_e + \frac{1}{\text{Pa}}[\text{M2}JJ]_e\{\text{Q}I\}_e \\
& + \frac{\text{Pa}\,h^2}{12}\{\text{U}J\text{U}K\}_e^T[\text{M30}JK]_e\{\text{Q}I\}_e - \{s\}_e
\end{aligned}
\tag{7.37}
$$

$$
\{s\}_e = \left\{
\begin{array}{l}
\dfrac{-\text{Gr}\,\widehat{g}_I}{\text{Re}^2}\det_e[\text{M200}]\{\text{Q4}\}_e - [\text{M20}I]_e\{\text{Q7}(t_n)\}_e \\
\qquad -\dfrac{h^2}{6}\left[\left(\text{Re}\{\text{U}J\}_e^T[\text{M300}J]_e + [\text{M2}JJ]_e\right)\{\text{U}I\}_e\right]_n \\
\dfrac{-\text{Ec}}{\text{Re}}\{\text{U}I\}_e^T[\text{M30}JI]_e\{\text{U}J\}_e \\
\qquad -\dfrac{h^2}{6}\left[\left(\text{RePr}\{\text{U}J\}_e^T[\text{M300}J]_e + [\text{M2}JJ]_e\right)\{\text{Q4}\}_e\right]_n \\
\det_e[\text{M200}]\{\text{SRC}Y\}_e \\
\qquad -\dfrac{h^2}{6}\left[\left(\text{ReSc}\{\text{U}J\}_e^T[\text{M300}J]_e + [\text{M2}JJ]_e\right)\{\text{Q5}\}_e\right]_n
\end{array}
\right\}
\tag{7.38}
$$

The matrix entries in (7.38) are I-distinct by rows, and the CCM theory addition to (7.37) for $1 \leq I \leq n$, $6 \leq J \leq 7$ is

$$
\{\text{RES\_U}I(\text{Q}J)\}_e = [\text{M20}I]_e\{\text{Q7}(t_n)\}_e + \frac{1}{\theta\Delta t}[\text{M20}I]_e\sum_{\alpha=1}^{p}\{\delta\text{Q6}\}_e^{\alpha}
\tag{7.39}
$$

Finally, for $I = 6$ from (7.32)

$$
\{\text{F\_Q6}\}_e = [\text{M2}JJ]_e\{\text{Q6}\}_e + [\text{M20}J]_e\{\text{U}J\}_e
\tag{7.40}
$$

Jacobian formation employs calculus as amply illustrated. DOF self-coupling generates matrices on the diagonal of the block jacobian for $\{\text{Q}I\}$, $1 \leq I \leq n+2$. From the definition (7.21), via (7.36–7.39), and neglecting nonlinearities due to the $\{\text{U}J\text{U}K\}_e^T$ multipliers on mPDE matrices

$$
\begin{aligned}
[\text{JAC\_Q}I]_e &\equiv \frac{\partial\{\text{F\_Q}I\}_e}{\partial\{\text{Q}I\}_e},\; 1 \leq I \leq n+2 \\
&= \frac{\partial}{\partial\{\text{Q}I\}_e}\left([\text{M}(\bullet)]_e\{\text{Q}I_{n+1}\}_e + \theta\Delta t\{\text{RES\_Q}I(\{Q\})\}_e\big|\right) \\
&\approx \det_e[\text{M200}] + \frac{\gamma\,\Delta t^2}{6}\{\text{U}J\text{U}K\}_e^T[\text{M30}JK]_e \\
&\quad + \frac{\theta h^2}{6}\left(\text{Pa}\{\text{U}J\}_e^T[\text{M300}J]_e + [\text{M2}JJ]_e\right) \\
&\quad + \theta\Delta t\left(
\begin{array}{l}
\{\text{U}J\}_e^T[\text{M300}J]_e + \{\text{U}I\}_e^T[\text{M3}J\text{00}]_e\delta_{IJ} \\
+\dfrac{1}{\text{Pa}}[\text{M2}KK]_e + \dfrac{\text{Pa}\,h^2}{12}\{\text{U}J\text{U}K\}_e^T[\text{M30}JK]_e
\end{array}
\right)
\end{aligned}
\tag{7.41}
$$

The key to iterative stability is coupling of the velocity potential solution; from (7.39) for $1 \le I \le n$ the jacobian off-diagonal matrices are

$$\frac{\partial\{\mathrm{F\_Q}I\}_e}{\partial\{\mathrm{Q6}\}_e} \Rightarrow \frac{\partial\{\mathrm{F\_U}I\}_e}{\partial\{\delta\mathrm{Q6}\}_e} = [\mathrm{M20}I]_e, 1 \le I \le n \tag{7.42}$$

and note cancellation of the $\theta\Delta t$ multiplier. Neglecting the term in $\{s\}_e$ with multiplier Ec/Re, added off-diagonal block jacobian matrices are

$$\begin{aligned}\frac{\partial\{\mathrm{F\_Q}I\}_e}{\partial\{\mathrm{Q4}\}_e} &= -\theta\Delta t\frac{\mathrm{Gr}\,\widehat{g}_I}{\mathrm{Re}^2}\det_e[\mathrm{M200}], 1 \le I \le n\\ \frac{\partial\{\mathrm{F\_Q4}\}_e}{\partial\{\mathrm{Q}I\}_e} &= \theta\Delta t\{\mathrm{U}I\}_e^T[\mathrm{M300}J]_e\delta_{IJ}, 1 \le I \le n\end{aligned} \tag{7.43}$$

and the coupling of $\{\mathrm{Q5}\}$ to $\{\mathrm{Q}I\}_e$ for $1 \le I \le n$ is expressed in (7.43).

For clarity, the block jacobian matrices are expressed in element dependent form, hence must be converted to library matrices via (7.34–7.35). Economy accrues to combining the distinct multipliers on common matrices in (7.41) yielding

$$\begin{aligned}[\mathrm{JAC\_Q}I]_e \approx\ & \det_e[\mathrm{M200}] + \{\mathrm{U}I\}_e^T[\mathrm{M3}J\mathrm{00}]_e\,\delta_{IJ}\\ &+\left(\frac{\gamma\,\Delta t^2}{6} + \theta\Delta t\frac{\mathrm{Pa}\,h^2}{12}\right)\{\mathrm{U}J\mathrm{U}K\}_e^T[\mathrm{M30}JK]_e\\ &+\theta\left(\frac{\mathrm{Pa}h^2}{6} + \Delta t\right)\{\mathrm{U}J\}_e^T[\mathrm{M300}J]_e\\ &+\theta\left(\frac{h^2}{6} + \frac{\Delta t}{\mathrm{Pa}}\right)[\mathrm{M2}KK]_e\end{aligned} \tag{7.44}$$

with the h^2 multipliers appropriately replaced as $f(\det_e)$, (7.35).

7.6 *m*GWS^h+θTS Algorithm Benchmarks, $n = 2$

The $n = 2$ natural convection *thermal cavity* benchmark, detailed for vorticity-streamfunction GWS^h/*m*GWS^h algorithms in Section 6.4, remains pertinent for theory validation. Classic state variable GWS^h algorithm *a posteriori* data on a uniform $\mathrm{M} = 32^2$ $k = 1$ TP basis mesh for $10^3 \le \mathrm{Ra} \le 10^6$ are reported, Ericson (2001). The temperature solution graphic distributions are in excellent qualitative agreement with benchmark data, de Vahl Davis (1983).

Also published are regular solution adapted mesh data at $\mathrm{Ra} = 10^6$ for $1.0 \le \mathrm{pr} \le 0.7$ wall-normal mesh refinements supporting linear theory validation for scalar-vector state variable mix. Figure 7.1 top left graphs the steady solution normalized global energy norm distributions for the five-member NS state variable. The scalar member data indicate extrema potential existence for $0.8 \le \mathrm{pr} \le 0.9$. Quantitative verification that this indicates the optimally accurate $\mathrm{M} = 32^2$ solution rests on energy norm equi-distribution assessment.

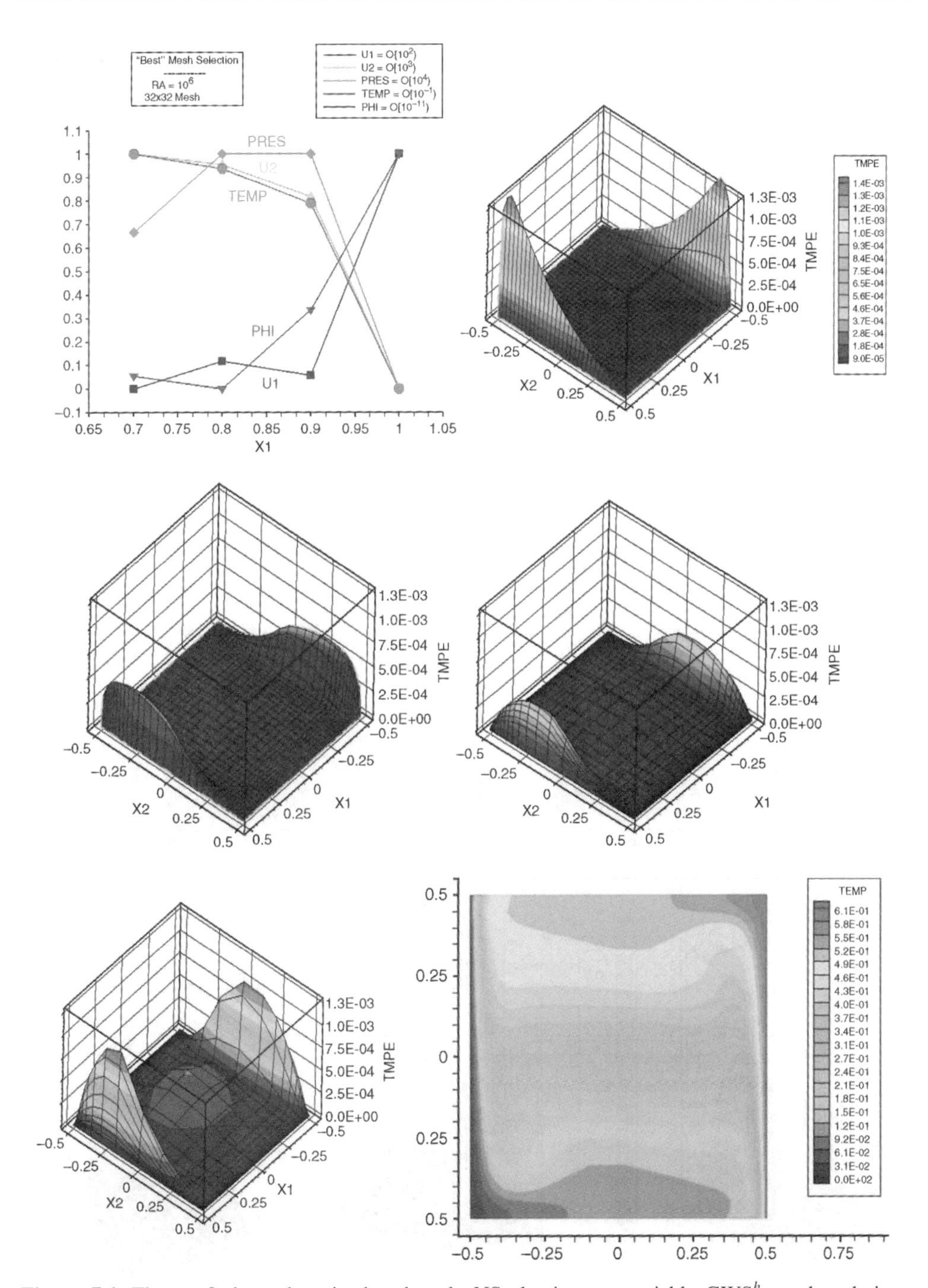

Figure 7.1 The $n = 2$ thermal cavity benchmark, NS classic state variable GWS^h steady solution, $\mathrm{Ra} = 10^6$, $\mathrm{M} = 32^2$ TP $k = 1$ basis, $1.0 \leq \mathrm{pr} \leq 0.7$: top left, NS state variable global energy norm distributions; top right-to-bottom left, $\mathrm{pr} = (1.0, 0.9, 0.8, 0.7)$ sequence norm $\left\|\{T^h(t_{converged})\}\right\|_{E,\Omega_e}$ distribution perspectives; bottom right, temperature distribution, $\mathrm{pr} = 0.8$

Temperature is the key scalar member; the next four graphics, right then down in Figure 7.1, graph elemental $\left\|T^h(t_{converged})\right\|_{E,\Omega_e}$ distributions for the sequence pr = (1.0, 0.9, 0.8, 0.7). Clearly, the middle pr solutions best approach norm equi-distribution, possessing nearly identical extrema (0.00052, 0.00056). Therefore, the *scalar equi-distribution theory* basis confirms the $k = 1$ basis $M = 32^2$ mesh supporting the *best attainable* GWS^h solution occurs for pr ~ 0.8. Figure 7.1 bottom right graphs this temperature distribution.

Confirming that this Ra = 10^6 solution-adapted $M = 32^2 k = 1$ basis solution is optimal for the NS vector state variable members is required. The large wall-normal temperature gradient distribution along both Dirichlet BC walls, Figure 7.1 bottom right, coincides with the induced wall jets. Figure 7.2 top left graphs the velocity vector distribution, scaled by vector length, confirming that the cavity central flowfield is featureless. The wall parallel velocity u_2^h at quarter wall-height quantifies the dominant hot wall jet, Figure 7.2 top right. NS velocity state variable solution significance is clearly wall adjacent with resolution pr mesh refinement sensitive. For pr = (1.0, 0.9, 0.8) the Figure 7.2 bottom graphic sequence is elemental norm $\left\|u_2^h(t_{converged})\right\|_{E,\Omega_e}$ distributions. The extremum occurring for pr = 0.9

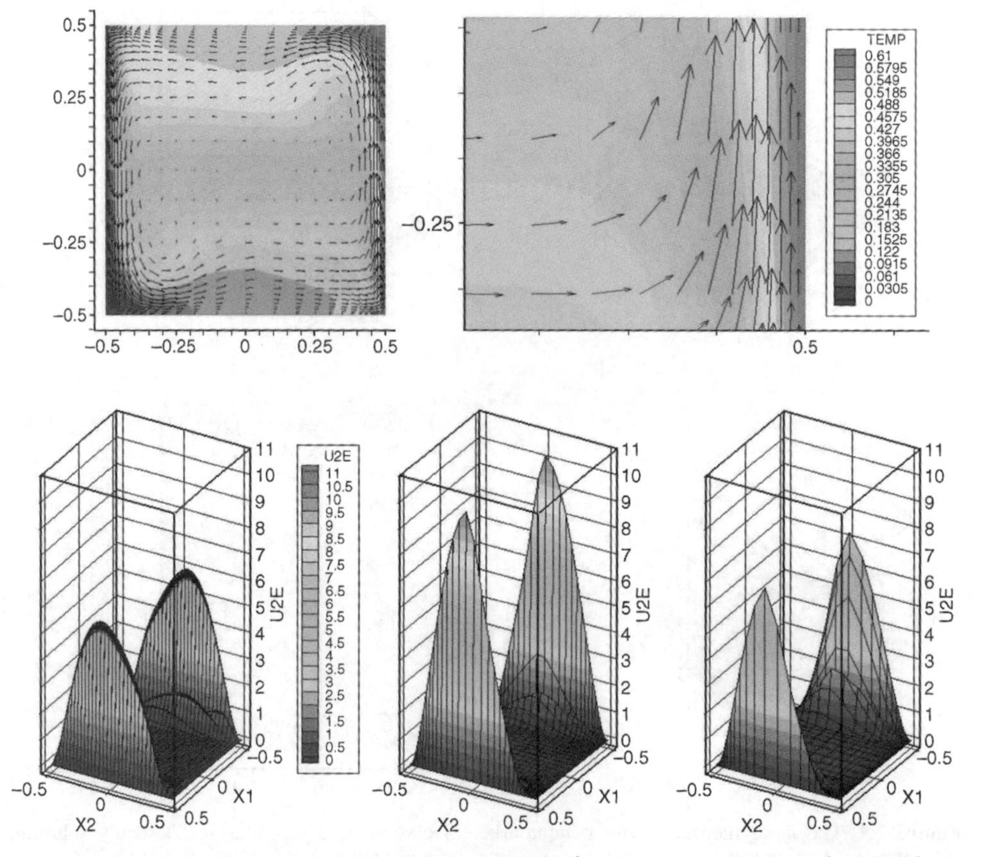

Figure 7.2 The $n = 2$ thermal cavity NS benchmark, GWS^h steady solution, Ra = 10^6, $M = 32^2$ TP $k = 1$ basis, $1.0 \le \text{pr} \le 0.8$; top left, velocity vector distribution; top right, wall jet u_2 velocity distribution: bottom row, pr = (1.0, 0.9, 0.8) distributions of $\left\|u_2^h(t_{converged})\right\|_{E,\Omega_e}$

confirms the best resolution, further confirming this the *optimal* solution-adapted $M=32^2$ $k=1$ TP basis solution based on global and local norm distributions for the entire NS classic state variable.

The $n=2$ natural convection thermal cavity also supports CCM theory implementation *robustness* assessment. GWS^h algorithm iterative convergence set at $\varepsilon<10^{-4}$ generates $\|\phi^h\|_E \Rightarrow O(10^{-8}) \le \varepsilon^2$. This *quantitative* measure confirms the error $DM^h \neq 0$ negligibly impacts velocity solution significance. The $Ra=10^6$ steady solution post-processed P^h distribution, also the terminal ϕ^h and its accumulation $\Sigma\phi^h$ (detailed shortly), are graphed in Figure 7.3. The *genuine* pressure solution is strictly *monotone*, conversely the CCM functions ϕ^h and $\Sigma\phi^h$ exhibit the "$2\Delta x$" dispersive error mode characteristic of the discrete first spatial derivative. Of fundamental impact, imposition of artificial diffusion to smooth ϕ^h *destroys* CCM algorithm functionality.

The thermal cavity supports testing an approximation to CCM theory implementation. Since no Dirichlet BCs exist for pressure (no inflow/outflow), the kinetic action of pressure

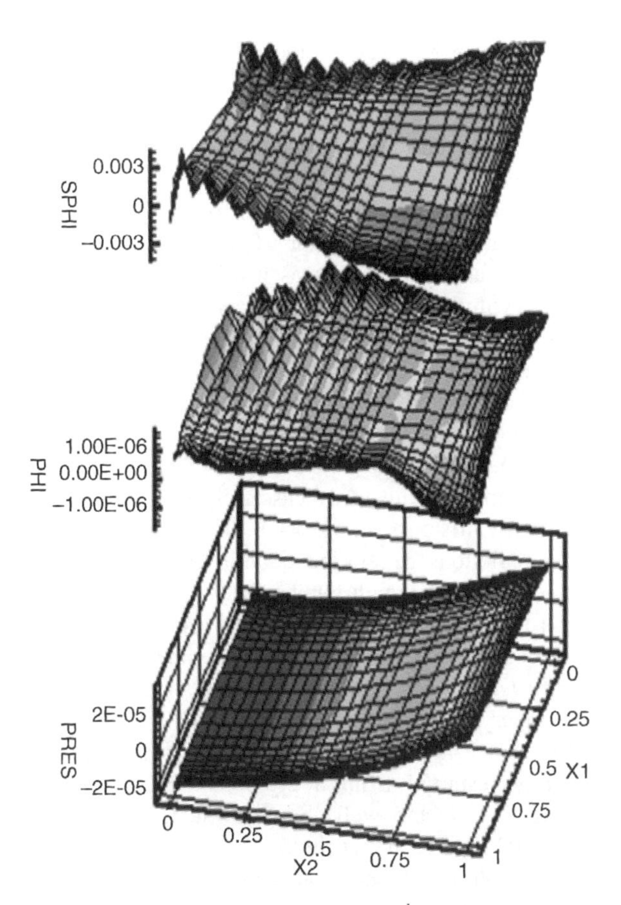

Figure 7.3 The $n=2$ thermal cavity NS benchmark, GWS^h steady solution, $Ra=10^6$, $M=32^2$ TP $k=1$ basis, pr = 0.8; bottom, post-processed genuine pressure P^h; middle, terminal iteration CCM theory function ϕ^h; top, summation of ϕ^h over solution time history

in the GWS^h solution process can be approximated as $P_n^h \approx \sum_{\gamma=1}^{n} \phi^h(t_\gamma)$ in (7.5), which defines the variable $\Sigma\phi^h$ graphed in Figure 7.3. This alters (7.27) to the form

$$\{\text{RES\_U}I(\{P^*|_{n+1}^{p}\})\}_e \Rightarrow [\text{M}20I]_e \left(\sum_{\alpha=1}^{n} \{\text{Q6}(t_\alpha)\}_e + \frac{1}{\theta\Delta t} \sum_{\alpha=1}^{p} \{\delta\text{Q6}(t_{n+1})\}_e^{\alpha} \right) \tag{7.45}$$

Use of (7.45) does not alter cited GWS^h solution characterizations, Figures 7.1–7.2, and $\Sigma\phi^h \equiv \sum_{\alpha=1}^{n} \{\text{Q6}(t_\alpha)\}_e$ constitutes *data*.

Literature citations for square thermal cavity CFD simulation for $\text{Ra} > 10^6$ are rare. Le Quere (1991) documents steady $n = 2$ benchmark-quality solutions for $\text{Ra} = 10^7$ and 10^8, also characterization of the transition to unsteadiness and *chaos* for $\text{Ra} > 10^8$, Le Quere and Behnia (1998). Mitra (2000) reports the requisite GWS^h algorithm uniform mesh refinement study for $\text{Ra} = 10^6$, 10^7, 10^8 for the CCM theory implementation (7.45).

For $\text{Ra} > 10^6$, the insipient separation evident on both horizontal surfaces, Figure 7.2 upper left, transitions to multiple recirculation zones. For the left wall now hot, Figure 7.4 top row graphs the finest mesh $\text{M} = 128^2$ velocity *unit vector* distributions for $\text{Ra} = 10^7$, 10^8 which clearly visualize generated flowfield complexity. The lower graphs present post-processed streamfunction verifying creation of five, then eleven recirculation zones as identified by existence of closed streamfunction contours.

Mitra (2000) also reports a solution-adapted regular mesh refinement study for $\text{Ra} = 10^7$, 10^8 confirming the global energy norm distributions for the NS five-member state variable evidence a common extremum within $0.8 \leq \text{pr} \leq 0.9$ The convergence rate of the matrix iteration algorithm for jacobian (7.40–7.42) was just below quadratic at 1.8$\pm$ independent of time in the unsteady evolution. The excellent *quantitative* agreement at $\text{Ra} = 10^8$ for both the $\text{M} = 64^2$*and* $\text{M} = 128^2$ uniform mesh GWS^h steady solutions with available *point norm* benchmark data is detailed in Table 7.1.

As noted, thermal cavity flowfields for $\text{Ra} > 10^6$ possess increased spectral content, Figure 7.4, eventually transitioning to unsteady leading to chaotic. An air experiment in a vertically elongated 8×1 thermal cavity generates data identifying transition from a steady single recirculation cell to multiple unsteady cascading roll vortices at a critical Rayleigh number, Le Quere (1994). Christon et al. (2002), document a collection of roughly two dozen CFD algorithm predictions generating quantitative agreement with the cavity NS stability theory predicted critical Rayleigh number $\text{Ra}_c \approx 3 \times 10^5$.

The cavity subcritical Ra GWS^h steady temperature distribution is graphed in Figure 7.5 left, and the algorithm determined critical Rayleigh number was $\text{Ra}_c = 4 \times 10^5$, Sahu and Baker (2007). The *a posteriori data* generated by $\text{Ra} > \text{Ra}_c$ GWS^h/*m*GWS^h simulations sought to validate the theory predicting the improved $O(m^3)$ phase accuracy of *m*GWS^h algorithm solution. For an $\text{M} = 40 \times 200$ wall-adapted cartesian mesh, GWS^h/*m*GWS^h solutions were independently restarted from a previous simulation at $\text{Ra} = 3.4 \times 10^6$ and 3.4×10^7, corresponding to just, and well above, critical, and were continued for 1500 time steps at fixed $\Delta t = 0.02$ s for each algorithm.

Having the comparative solution IC identical eliminates the impact of the second term in the theory, (7.8). Over the 1500 time steps the cavity cascading vortex velocity field "turned over" twice, generating a rich multi-scale distribution of thermal entity transport throughout time evolution. Figure 7.5 completion compares the unsteady temperature

Table 7.1 GWS^h algorithm benchmark data, Ra = 10^8

Point	Uniform TP $k=1$ basis mesh			Le Quere
Norm	M = 32 × 32	M = 64 × 64	M = 128 × 128	(1991)
Ψ mid	0.00564	0.00522	0.00536	0.005232
u^h max	0.0306	0.0322	0.0322	0.03219
y coord	0.93	0.926	0.926	0.928
v^h max	0.297	0.221	0.222	0.2222
x coord	0.01	0.01	0.01	0.012
$\lvert\nabla^h \bullet \mathbf{u}^h\rvert$	E−05	2E−07	2E−07	5E−08

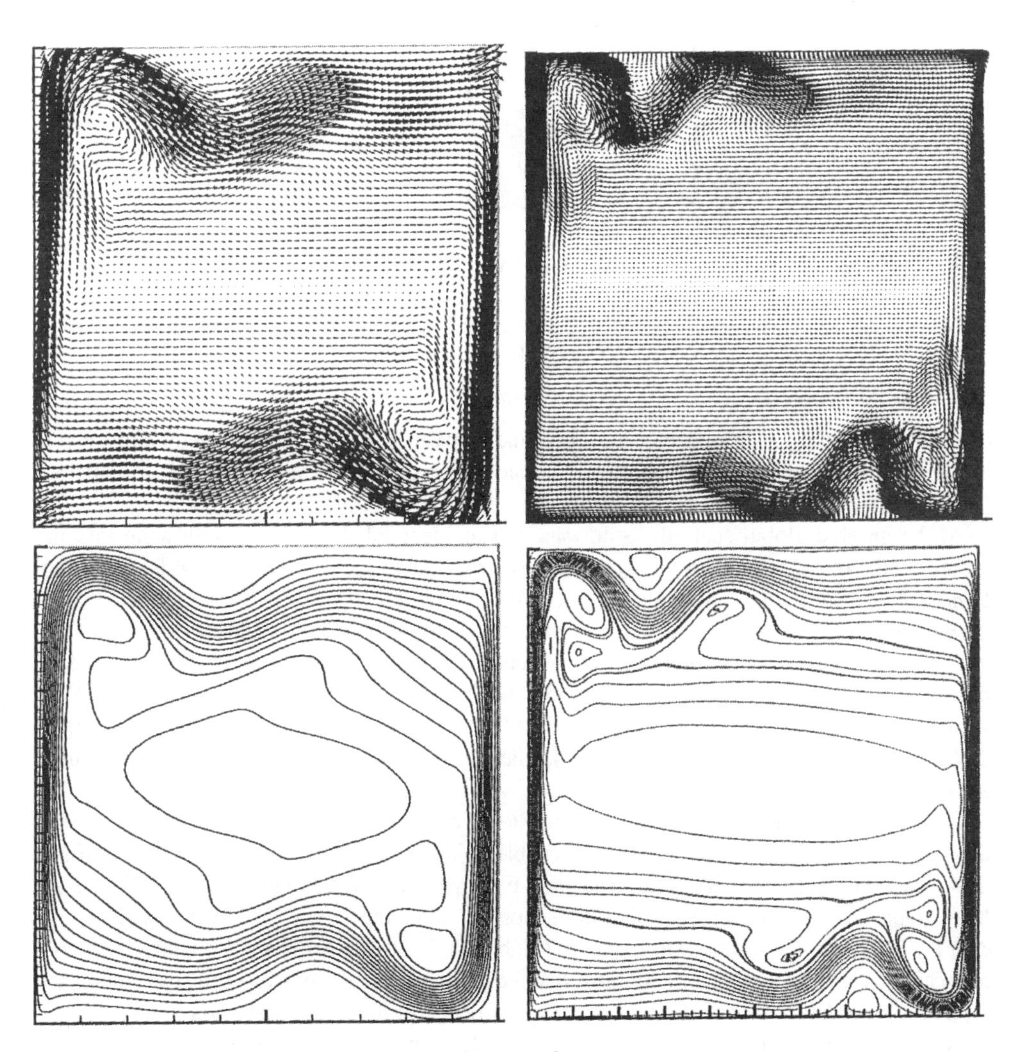

Figure 7.4 The $n=2$ thermal cavity benchmark, GWS^h algorithm steady solutions, uniform TP $k=1$ basis meshes: top left, M = 64^2, Ra = 10^7, top right, M = 128^2, Ra = 10^8 velocity unit vector distributions; bottom row, Ra = 10^7, 10^8 streamfunction distributions, from Mitra (2000)

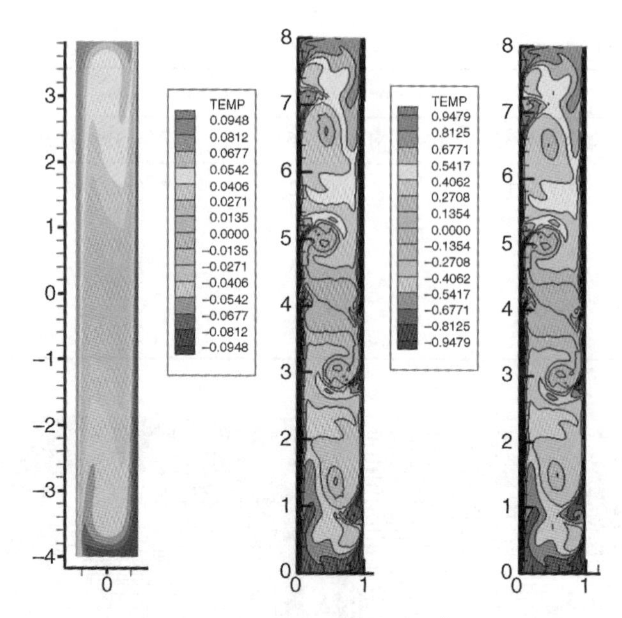

Figure 7.5 The 8×1 thermal cavity validation, GWS^h/mGWS^h algorithm temperature distributions, wall-adapted cartesian $M = 40 \times 200$ TP $k = 1$ basis mesh: left, steady solution at subcritical $Ra = 10^5$; following 1500 time steps at $Ra = 3.4 \times 10^6$ from identical IC, middle, GWS^h solution, right mGWS^h solution. *Source:* Sahu, S. and Baker, A.J. 2007. Reproduced with permission of Wiley

distribution solution pair snapshot at 1500 time steps past IC. The thermal boundary layers are very thin and the shed roll vortices cascading up/down the hot/cold vertical walls are clearly visible at cavity mid-height at this time station.

At the graphic global scale the solutions appear identical. However, windowing into the representative centroidal vortex-encapsulated thermal bubble confirms that GWS^h algorithm $O(m^2)$ phase accuracy non-smoothness of the temperature contours is moderated by the *optimal* $\gamma = -0.5$ mGWS^h algorithm, Figure 7.6. This observation yields to quantification via (7.8) in concert with computing experience verifying thermal cavity solution energy norm convergence is from below, Ericson (2001). At this time station the GWS^h/mGWS^h solution energy norms are $0.663116/3.33277 \times 10^{-4}$ for $Ra = 3.4 \times 10^6$ and $2.086058/2.086188 \times 10^{-7}$ for $Ra = 3.4 \times 10^7$. These data pairs clearly validate mGWS^h optimality, hence linear theory appropriateness.

Concluding optimality confirmation accrues to velocity vector solution evolution *point norm* analyses. Quantitative assessment is scalar-enabled via convergence post-processed streamfunction, since the wall $\psi = 0$ BC guarantees that extrema occur in the cavity interior. Figure 7.7 left graphs a mGWS^h ψ^h solution snapshot computed via a GWS^h written on $\mathcal{D}(\psi) \equiv \nabla^2\psi + \hat{\mathbf{k}} \bullet \nabla \times \mathbf{u} = 0$. In Figure 7.7 right column, the larger span of the tiny(!) ψ^h contour centered at $(x, y) \approx (0.5, 1.5)$ confirms that the mGWS^h solution possesses the extremum, and *a posteriori* nodal data confirms that the extrema occurs throughout time evolution. As *italics* boldface-coded in Table 7.2, the differences are indeed small but in the right direction at *every* time station sampled. In combination, this solution measure range solidly validates the theory predicting mGWS^h algorithm optimality for the classic NS state variable in total.

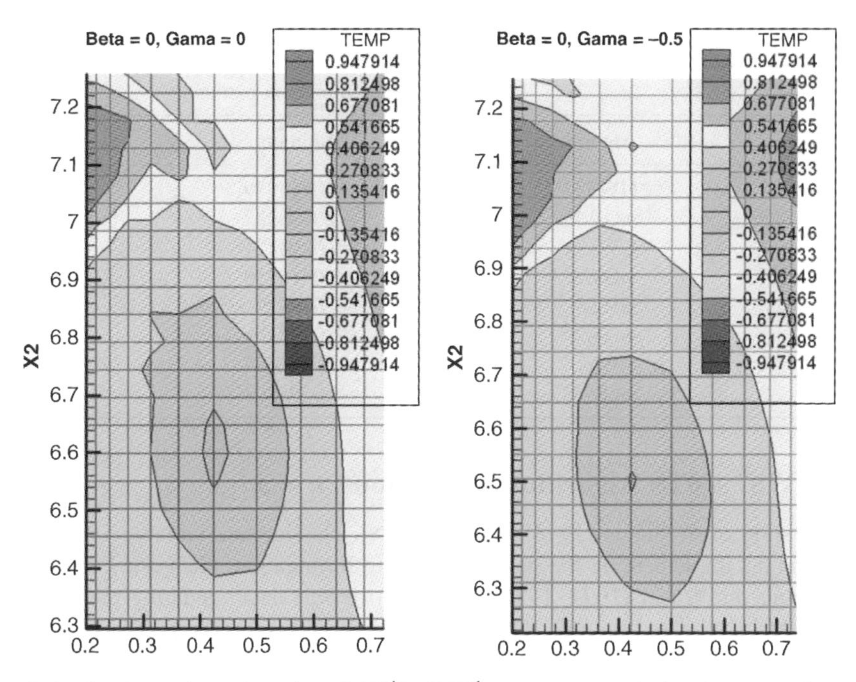

Figure 7.6 The 8×1 thermal cavity, $GWS^h/mGWS^h$ temperature window-in comparison, wall-adapted cartesian M = 40×200 TP $k = 1$ basis mesh, following 1500 time steps at Ra = 3.4×10^6 from identical IC: left, GWS^h solution; right, $mGWS^h$ solution. *Source:* Sahu, S. and Baker, A.J. 2007. Reproduced with permission of Wiley

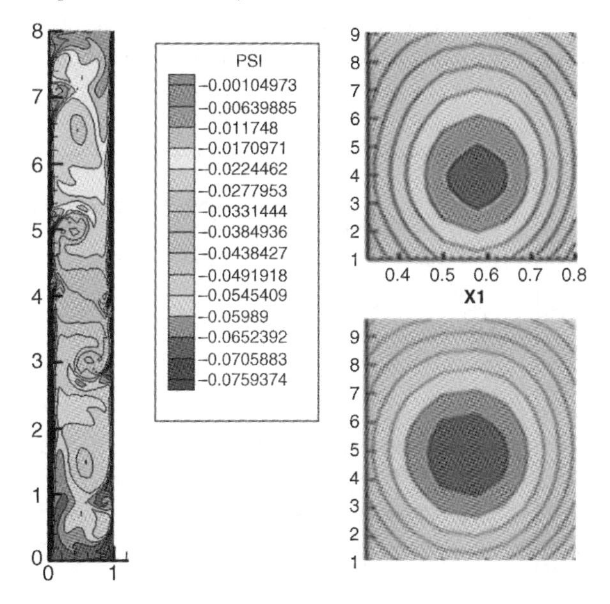

Figure 7.7 The 8×1 thermal cavity, wall-adapted cartesian M = 40×200 TP $k = 1$ basis mesh, Ra = 3.4×10^6: left, $mGWS^h$ streamfunction with color bar; right column, *tiny* streamfunction solution extremum centered at $(x, y) \approx (0.5, 1.5)$; top, GWS^h algorithm, bottom, $mGWS^h$ algorithm

Table 7.2 Streamfunction solution extrema, 8 × 1 thermal cavity, from Sahu and Baker (2007)

		Streamfunction extrema	
Time (s)	Algorithm	Minimum	Maximum
1.40E+02	GWS^h	−0.07959***79***	0.00114***737***
	$mGWS^h$	−0.07959***83***	0.00114***807***
1.49E+02	GWS^h	−0.08***08***421	0.002***8***06
	$mGWS^h$	−0.08***10***310	0.003***3***54
1.50E+02	GWS^h	−0.0795***79***	0.001***14***73
	$mGWS^h$	−0.0797***86***	0.001***85***10
1.63E+02	GWS^h	−0.0***68***7447	0.000***4***458
	$mGWS^h$	−0.07***0***4804	0.00***12***224

7.7 $m\mathbf{GWS}^h + \theta\mathbf{TS}$ Algorithm Validations, $n = 3$

The dissertation of Williams (1993) derives the NS classic state variable $GWS^h/mGWS^h$ $n = 3$ algorithms implementing the genuine pressure formulation, (7.7), for laminar-thermal flow. Williams and Baker (1994, 1996) detail *a posteriori* data validating the GWS^h algorithm for select cavity and through flow geometries. Briefly summarizing these *a posteriori* data analyses.

- Square cross-section duct: validates the Dirichlet/Neumann BC mix defined in (7.6–7.7) for throughflow. Slug profile inflow BC $u(x = 0, y, z)$ generates inflow plane extrema in ϕ^h; moving BC upstream with addition of inviscid inflow section eliminates this inappropriate BC-induced local error. Distance to development of the $u(x, y)$ paraboloidal distribution, the $n = 3$ equivalent of the classic $n = 2$ parabola, agrees qualitatively with data, Figure 7.8.

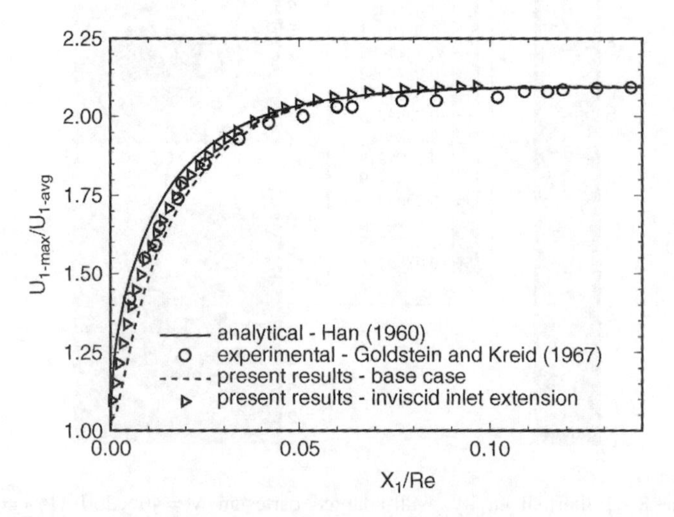

Figure 7.8 GWS^h $n = 3$ throughflow BC validation, developing laminar flow in square cross-section duct, Re = 100, from Williams (1993)

- Window thermal cavity: an $n=3$ thermal cavity with Dirichlet temperature BC sidewalls twice the span of the square adiabatic end walls. Resultant vortex circulation patterns are determined totally dependent on Prandtl number Pr, Mallinson and de Vahl Davis, (1977). GWS^h solution distinct mirror-symmetric vortex circulation patterns are visualized in Figure 7.9 via time evolution tracks of (massless) lagrangian particles inserted near the centroid of each adiabatic end wall for $\mathrm{Ra}=10^4$, $\mathrm{Pr}=0.1$, left, $\mathrm{Pr}=100$, right.
- Two cell thermal enclosure with a door: experiment documents buoyancy-induced circulation pattern distinctions dependent on door/cell width ratio, w/W, Neymark *et al.* (1989). Geometry in Figure 7.10 top left; experiment and GWS^h solution symmetry plane velocity vectors, $\mathrm{Ra}=3.2\times10^{10}$, $w/W=0.2$, compared in bottom graphic pair. Top right, 600 s duration GWS^h steady solution lagrangian particle track tracing out fluid movement between cold and hot cells, particle released at base of door.

The fundamental GWS^h/mGWS^h algorithm validation is prediction of isothermal laminar Re-dependence of the three-dimensional, multi-scale recirculation velocity vector field in the close coupled step wall diffuser experiment of Armaly *et al.* (1983). The characteristic dimensions of the experiment geometry, Figure 7.11, are $h=5.2$ mm and $H=10.1$ mm yielding the expansion ratio $H/h=1.9423$, and $S=4.9$ mm. The channel width $W=180$ mm yields $W/S=36.735$, which ensures that the diffuser symmetry plane velocity vector field is amply distant from the sidewalls. The straight upstream entrance flow channel was 200 mm in length and the exit from the channel was 500 mm downstream from the step face.

The experiment focus was laser-Doppler measurement of the downstream (axial) velocity component recirculation region attachment coordinates on the diffuser symmetry plane intersection with the top and bottom walls as a function of Re. Additionally, for $\mathrm{Re}=397$, 648 axial velocity distributions across the diffuser transverse span at select elevations were recorded, including penetrating the primary recirculation region. The reference definitions for Re are: U, two-thirds measured maximum axial velocity 10 mm upstream of the step face; L, twice the upstream channel height (its hydraulic diameter); ν, the kinematic viscosity of air at room temperature.

These data confirm the flowfield is laminar for $\mathrm{Re}<1200$, thereafter it turns transitional and becomes fully turbulent for $\mathrm{Re}>6600$, Figure 7.12. The primary recirculation bubble extremum penetration downstream is $x_1/S\approx18$ at $\mathrm{Re}=1000$. The upper wall secondary recirculation bubble is initiated at $\mathrm{Re}\approx400$ and coincidentally achieves extremum axial

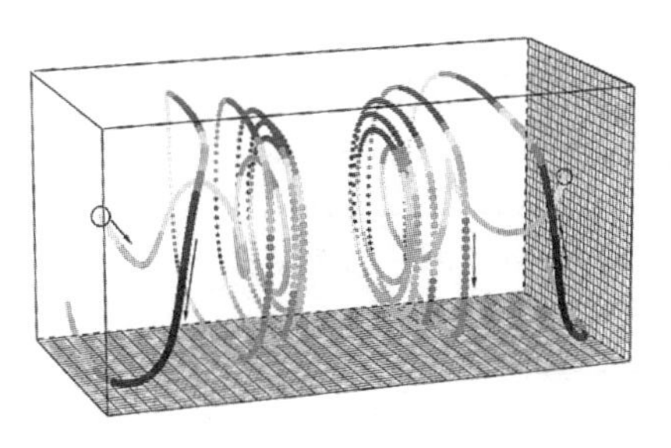
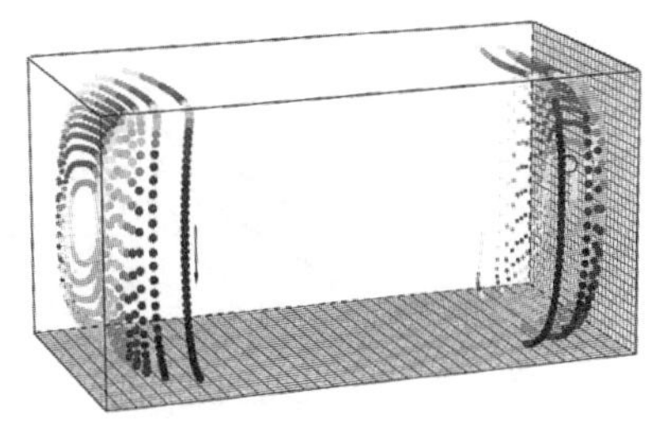

Figure 7.9 GWS^h $n=3$ validation, window thermal cavity: vortex velocity vector lagrangian particle visualizations, $\mathrm{Ra}=10^4$: left, $\mathrm{Pr}=0.1$, right, $\mathrm{Pr}=100$, from Williams (1993)

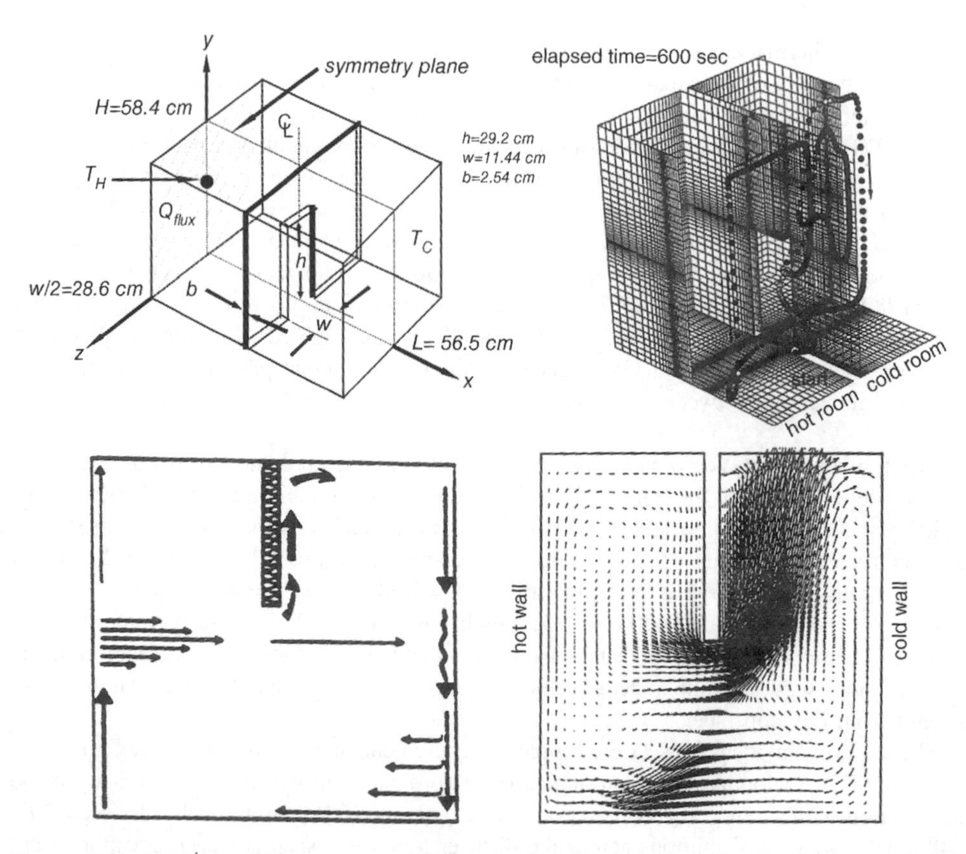

Figure 7.10 GWS^h algorithm $n = 3$ validation, two cell enclosure with door, $\mathrm{Ra} = 3.2 \times 10^{10}$, $\mathrm{Pr} = 6.7$, $w/W = 0.2$: top left, geometry; top right, GWS^h steady velocity lagrangian particle track, 600 s duration: bottom row, velocity vectors on enclosure symmetry plane; left, experiment, right GWS^h steady solution, from Williams and Baker (1994)

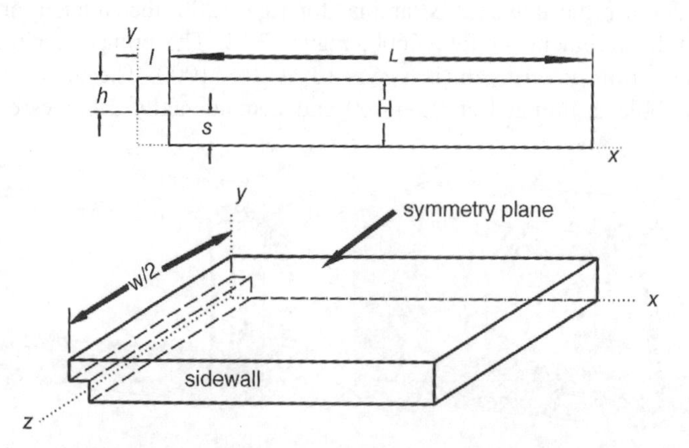

Figure 7.11 Close coupled step wall diffuser geometry of the Armaly *et al.* (1983) experiment. *Source:* Williams, P.T. and Baker A.J. 1997. Reproduced with permission of Wiley

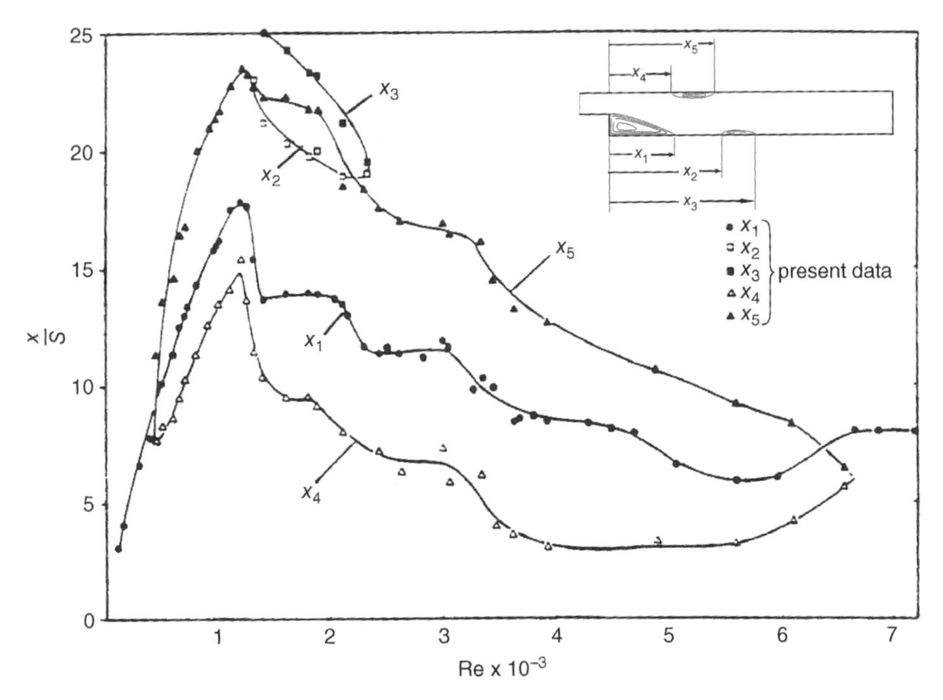

Figure 7.12 Experimental measurement of primary, secondary and tertiary recirculation zone termini on symmetry centerplane, close coupled step wall diffuser geometry of Armaly *et al.* (1983), with permission *J. Fluid Mechanics*. *Source:* Armaly, B.F., Durst, F., Pereira, J.C.F., and Schonung, B. 1983. Reproduced with permission of Cambridge University Press

span $\Delta x_2/S \approx 15$ at Re = 1000. A small tertiary lower wall recirculation region is generated during flow transition to turbulent, and all but the primary recirculation cell disappears when the flow becomes fully turbulent.

As detailed in Williams (1993), extensive compute trials were required to establish a solution domain with well-posed BC-admissable closure segments amenable to early 1990s RISC architecture workstation computing. Inaugural GWS^h algorithm experiments at modest Re evidenced instability ultimately traced to round-off error accumulation, the result of the widely disparate velocity vector rectangular cartesian resolution magnitudes. Rotating the solution domain through dual 45° Euler angles equilibrated the velocity vector resolution energy norms and eliminated this problem.

The GWS^h algorithm encountered a second instability for simulations at Re > 400, ultimately traced to the algorithm $O(m^2)$ dispersive error mode. Its impact was barely discernable in the velocity vector matrix iteration process, but was clearly evident in the genuine pressure solution. The resolution was transition from GWS^h to the β mGWS^h algorithm in concert with NS state variable generalization of the scalar mPDE analysis, Chapter 5, reference Figure 5.8. (Note: this study is published prior to derivation of the optimal coefficient Pah^2/12.)

The TS operation detailed in Section 5.7 involves identification of the flux vector jacobian, in that instance a scalar. The jacobian for the NS PDE system is a matrix for state variable $\{q(\mathbf{x}, t)\} \equiv \{\mathbf{u}, \Theta, Y\}^T$, hence the starting point replacing (5.61) is

$$\mathcal{L}(\{q\}) = \frac{\partial\{q\}}{\partial t} + \frac{\partial\{f_j\}}{\partial x_j} \Rightarrow \{q\}_{,t} + [\mathrm{A}_j]\frac{\partial\{q\}}{\partial x_j} = 0, \ [\mathrm{A}_j] \equiv \frac{\partial\{f_j\}}{\partial\{q\}} \tag{7.46}$$

Retaining only the TS generated β term, (5.66), the resultant NS *m*PDE system replacement for (7.2–7.4) is

$$\mathcal{L}^m(\{q\}) = \mathcal{L}(\{q\}) - \frac{\Delta t}{2}\frac{\partial}{\partial x_j}\left(\beta[A_j][A_k]\frac{\partial\{q\}}{\partial x_k}\right) + O(\Delta t^2) = 0 \tag{7.47}$$

The NS parent PDE system kinetic flux vector is defined in (7.11). Hence for $\{q(\mathbf{x}, t)\} \equiv \{u_j, \Theta, Y\}^T$, $1 \le i \le n$, the jacobian is the diagonal matrix

$$[\mathrm{A}_j] \equiv \frac{\partial\{f_j\}}{\partial\{q\}} = \mathrm{diag}\begin{bmatrix} u_j + u_i\delta_{ij}, 0, 0 & & \\ & , u_j, 0 & \\ & & , u_j \end{bmatrix} \tag{7.48}$$

For *no* summation on i *not italicized* the jacobian matrix product defined in (7.47) is

$$[A_j][A_k] = \mathrm{diag}\begin{bmatrix} u_ju_k + u_i\delta_{ij} + u_ju_\mathrm{i}\delta_{\mathrm{i}k} + u_ku_\mathrm{i}\delta_{\mathrm{i}j} + u_\mathrm{i}u_\mathrm{i}\delta_{\mathrm{i}j}\delta_{\mathrm{i}k}, 0 & , 0 & \\ & , u_ju_k, 0 & \\ & & , u_ju_k \end{bmatrix} \tag{7.49}$$

Proceeding through the Kronecker delta operations generates the final statement for the NS *m*PDE jacobian addition, Noronha (1989)

$$\frac{\beta\Delta t}{2}\left([A_j][A_k]\frac{\partial\{q\}}{\partial x_k}\right) = \frac{\beta\Delta t}{2}\begin{bmatrix} u_ju_k\dfrac{\partial u_i}{\partial x_k} + u_iu_k\dfrac{\partial u_j}{\partial x_k}, 0 & , 0 & \\ & , u_ju_k\dfrac{\partial\Theta}{\partial x_k}, 0 & \\ & & , u_ju_k\dfrac{\partial Y}{\partial x_k} \end{bmatrix} \tag{7.50}$$

Implementing this *m*GWS^h algorithm involves replacing the *m*PDE multiplier Pah^2/12 in (7.12) with $\beta\Delta t/2$ and appending the second velocity tensor product term. The dimensionality of coefficient $\beta\Delta t(u_ju_k)$ remains $D(\mathrm{L}^2\tau^{-1})$, hence the operator is diffusive *iff* the velocity tensor products are non-negative. This *m*GWS^h algorithm resolved the stability problem for all $\mathrm{Re} \le 800$ for $\beta = 0.1$, a very modest coefficient specification, recall Figure 5.8. Its efficacy at $\mathrm{Re} = 648$ is quantified in pressure DOF isobar perspective comparisons on select vertical planes, Figure 7.13, GWS^h solution on top, *m*GWS^h below.

Simulations on the inflow uniform cross-section rectangular duct generated the associated fully developed laminar axial velocity distribution $u^h(x \to \infty, y, z)$. Further experiments proved this velocity as inflow BC could be imposed on a plane as close as 5 mm upstream from the step face. The diffuser symmetric geometry admits simulation on a

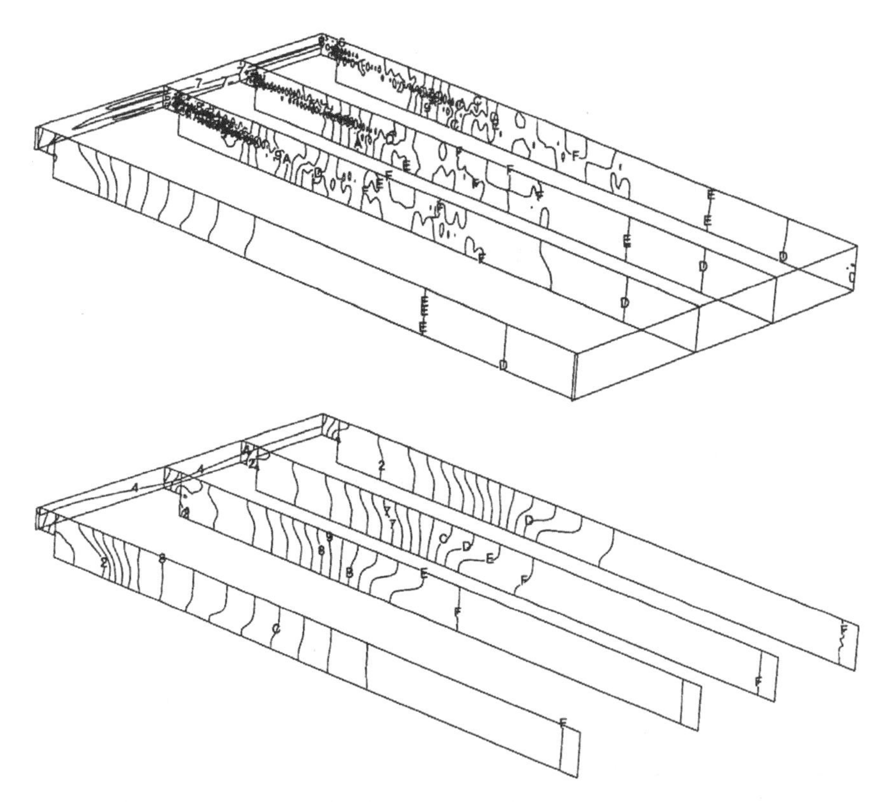

Figure 7.13 $GWS^h/mGWS^h$ algorithm predictions of genuine pressure DOF isobar distribution, Re $= 648$, planar perspective graphics on select transverse vertical planes: top, GWS^h solution; bottom, $\beta = 0.1$ $mGWS^h$ solution, from Williams (1993)

half-domain; the utilized solution-adapted, wall pr-resolved non-uniform $k = 1$ TP basis $M = 42{,}816$ mesh is graphed in Figure 7.14. The NS classic state variable well-posed BCs are:

- inflow plane: fully developed axial $u^h(x = -5, y, z)$, homogeneous Neumann for ϕ^h, zero Dirichlet for v^h, w^h and P^h
- outflow plane: vanishing Neumann for $\mathbf{u}^h$, zero Dirichlet for ϕ^h, non-homogenous Robin for P^h, (7.7)
- symmetry plane: vanishing Neumann except homogeneous Dirichlet for w^h
- no-slip walls: homogenous Dirichlet for $\mathbf{u}^h$, homogeneous Neumann for ϕ^h and P^h.

For completeness, the symmetry plane $n = 2$ GWS^h simulation, Figure 7.15 left (open symbols) is compared with various reported *a posteriori* data for primary recirculation bubble lower wall symmetry plane intercept. Recalling Figure 6.8, all $n = 2$ predictions progressively fail quantitative agreement with the experimental data for $Re > 400$. Conversely, excellent *quantitative agreement* for the $mGWS^h$ steady solution $n = 3$ prediction for $100 \le Re \le 800$, Figure 7.15 right (open symbols) solidly *validates* the algorithm.

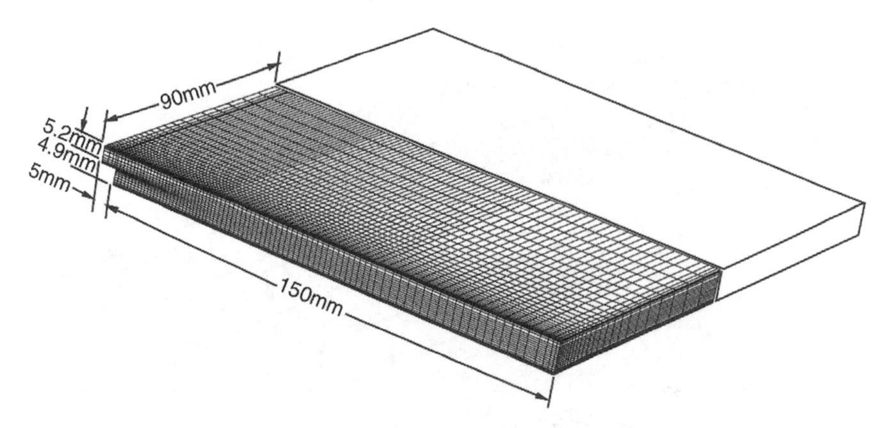

Figure 7.14 Solution-adapted non-uniform M = 42,816 $k = 1$ TP basis discretization supporting $GWS^h/mGWS^h$ algorithm symmetric half-domain predictions of three-dimensional flow in the close coupled step wall diffuser geometry of Armaly *et al.* (1983). *Source:* Williams, P.T. and Baker A.J. 1997. Reproduced with permission of Wiley

These data are clearly superior to the limited Re data of Ku *et al.* (1989), and fully verify that three-dimensionality dominates close-coupled step wall diffuser velocity distributions for all Re > 400, as postulated.

For solution domain scale reference, the diffuser symmetry centerplane $mGWS^h$ $n = 3$ axial velocity DOF distribution for Re = 800 is graphed in Figure 7.16 top. Beneath left and right are window-in graphs of dual recirculation region detachment/reattachment coordinates for Re = 648 and 800. The larger vertical arrows mark the $mGWS^h$ solution; the primary bubble experimental data are denoted x_1/S with the secondary data (spread) via smaller arrow pairs x_2/S. Excellent quantitative agreement is evident for primary bubble reattachment and secondary bubble detachment.

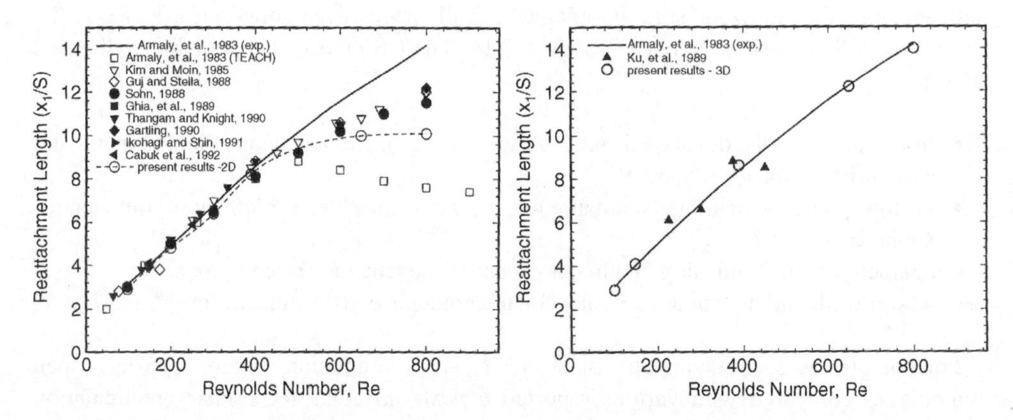

Figure 7.15 GWS^h and comparative algorithm predictions of symmetry centerplane primary recirculation zone reattachment, close coupled step wall diffuser geometry of Armaly *et al.* (1983): left, $n = 2$; right, $mGWS^h$ algorithm (open symbols), $n = 3$. *Source:* Williams, P.T. and Baker A.J. 1997. Reproduced with permission of Wiley

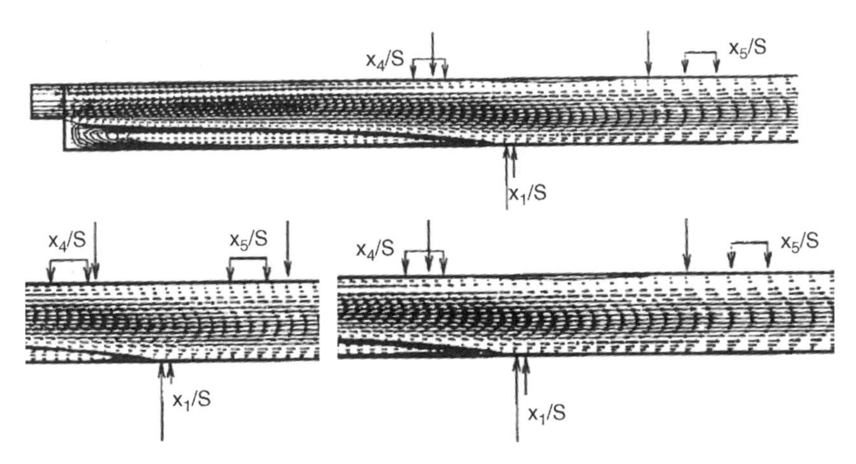

Figure 7.16 $m\text{GWS}^h$ step wall diffuser symmetry centerplane velocity vector distributions: top, Re = 800 full scale; bottom window-in graphs: left, Re = 648; right, Re = 800. *Source:* Williams, P.T. and Baker A.J. 1997. Reproduced with permission of Wiley

The poorer agreement for the secondary bubble reattachment coordinate may result from its *exceptional shallowness*, of order 1 mm, which is *the* significant challenge to experimental accuracy as well as resolution via the $m\text{GWS}^h$ solution M = 42,816 mesh. These $m\text{GWS}^h$ *a posteriori* data do provide, for the first time, quantitative fully three-dimensional resolution of the secondary recirculation bubble extent.

For Re = 500, 648, 800, the numerous recirculation regions, outlined by negative axial velocity DOF interpolation isovels, are graphed in visual perspective, Figure 7.17. The diffuser domain top plane is rotated 180° to expose the undersurface data, and no-slip bounding surface velocity isovel data for first off-wall DOF planes are plotted. The Re = 500 solution is not steady state; both the primary and secondary recirculation bubbles lack cohesive structure and the latter spans the entire half-width. Both develop into well-coordinated steady recirculation regions at Re = 648 and 800, and of prime significance, the secondary recirculation region transitions to an isolated upper surface entity not extending to the sidewalls. The sidewall separation region is always well organized with downstream reach increasing with Re.

As noted, the experiment measured axial velocity distributions across the diffuser transverse span at elevations y = 2.5 mm and 7.5 mm above the floor plane at defined locations downstream from the step. Figure 7.18 top details these specifications pertinent to the flowfield for Re = 648. Measurements at the y = 7.5 mm elevation were also taken for Re = 397 with goal to verify the essential two-dimensionality of the velocity field away from sidewalls. Excellent agreement exists between the $m\text{GWS}^h$ solution and data, graphic below, both of which confirm that flowfield two-dimensionality extends to within $z/Z \sim 0.12$ of the wall.

For Re = 648 the diffuser flowfield is totally dominated by three-dimensional effects. The comparison between $m\text{GWS}^h$ prediction and data at both elevations at axial stations penetrating recirculation flow regions, Figure 7.19, is very revealing regarding near-wall experiment resolution limitation. The solution adapted M = 42,816 non-uniform mesh near-wall resolution exceeds the experiment, limited to $z/Z \sim 0.1$ except at the far downstream

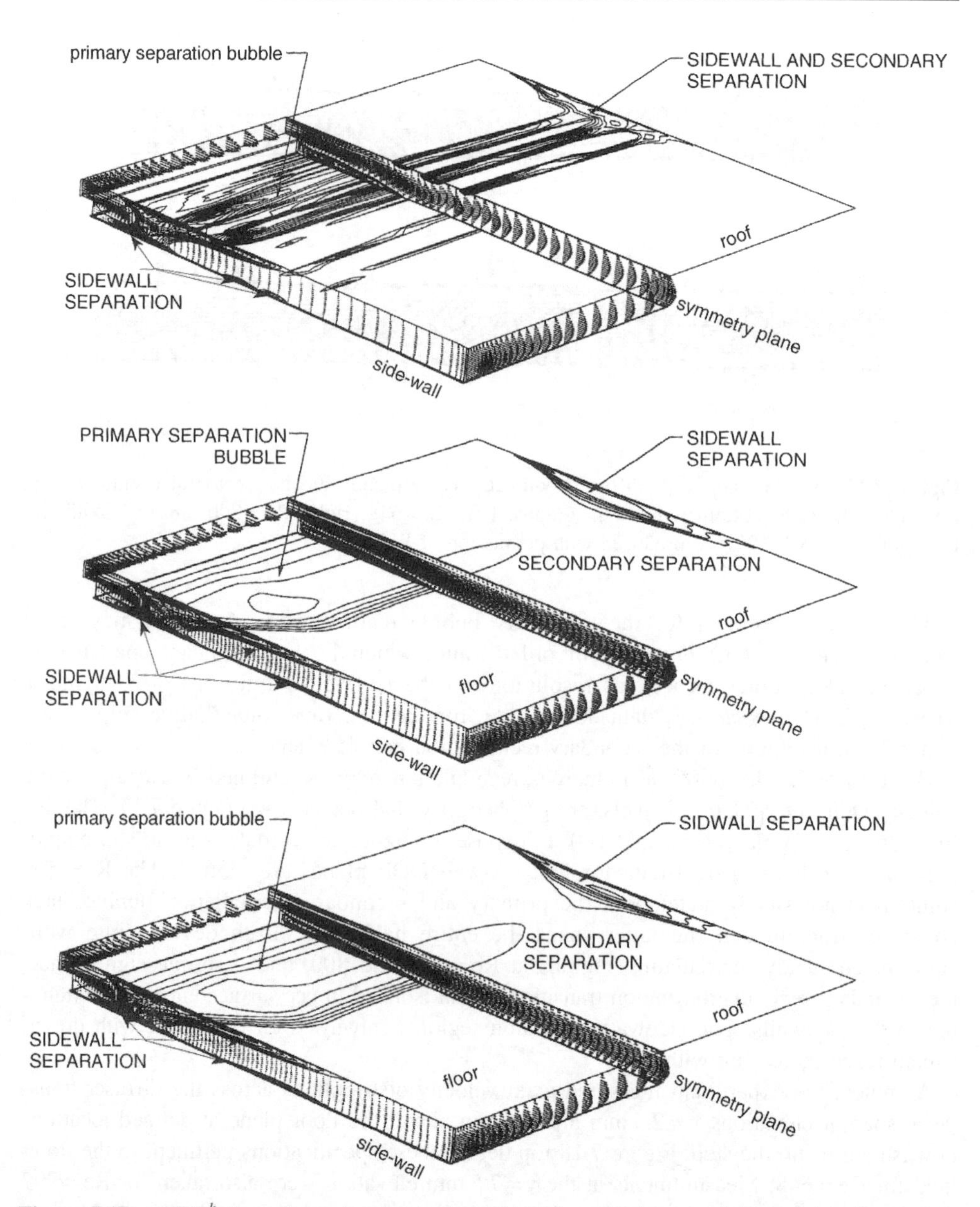

Figure 7.17 $m\text{GWS}^h$ three-dimensional velocity field with multiple recirculation region distribution identification, step wall diffuser, top surface rotated open to expose undersurface data: Re = 500 (not steady state), Re = 648, Re = 800 top to bottom. *Source:* Williams, P.T. and Baker A.J. 1997. Reproduced with permission of Wiley

station $x/S = 12.3$, $y = 2.35$ mm, where one additional data point is recorded at $z/Z \sim 0.02$. Excellent *quantitative agreement* exists in the core flow region throughout. The experiment captures neither the CFD predicted near-wall local extrema nor the reverse flow distributions directly wall-adjacent.

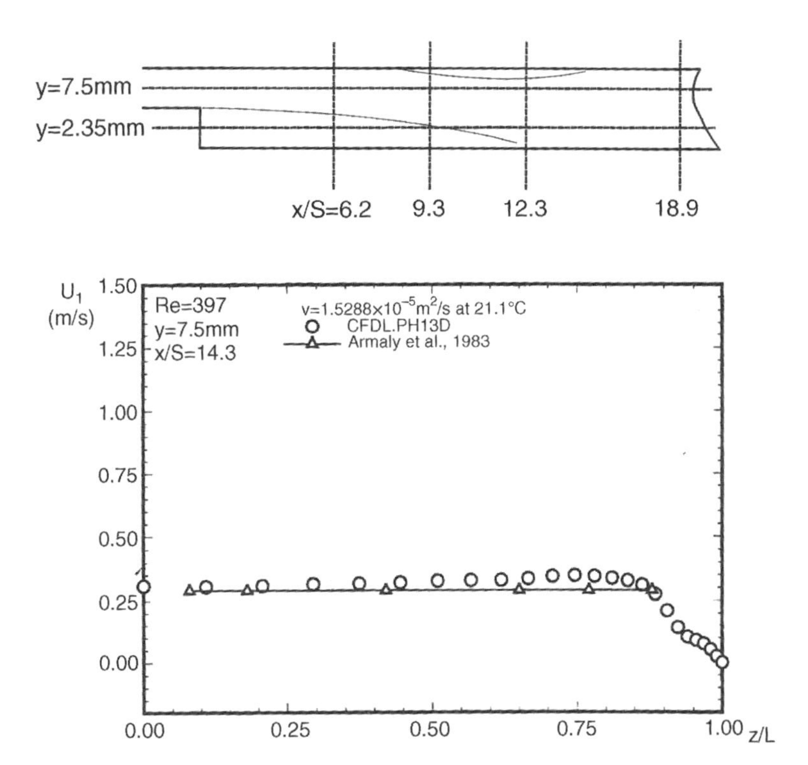

Figure 7.18 Step wall diffuser spanwise distributions of axial velocity, Re = 397, $y = 7.5$ mm, $x/S = 14.3$: $m\text{GWS}^h$ solution (open symbols), experimental data (triangles with line), from Williams (1993)

For the close-coupled step wall diffuser, CFD community simulation focus has been primary and secondary recirculation region intercepts comparison with the experimental data. Viewing Figures 7.17 and 7.19, these CFD results as well as the experiment totally missed the persistent presence of the extensive sidewall separated flow region, predicted to occur at Re = 396 as well, Williams (1993).

Added interpretation of this complex recirculation region results for creation of "*oil flow streaklines*," a legacy experimental surface flow visualization technique, from the $m\text{GWS}^h$ solution on DOF planes one row removed from the diffuser floor and roof. For Re = 800, the solid lines with arrows, Figure 7.20, are streakline distributions in the primary and secondary recirculation bubbles. The solid lines without arrows are the contours of negative axial velocity DOF plotted in Figure 7.17. The prominent vertical-axis vortex near the roof is very shallow.

The inflow channel fully developed axial velocity BC upstream of the step is responsible for this sidewall event. Velocity vector distribution enlargements near the sidewall at Re = 800 are projected onto transverse planes at $x/S = 7.72$, 18.37 downstream from the step face, Figure 7.21. These graphics clearly visualize strong three-dimensionality at the channel step expansion, creating the wall-adjacent very complex vortex structures extending well downstream. The perspectives underneath each are added to fix orientation.

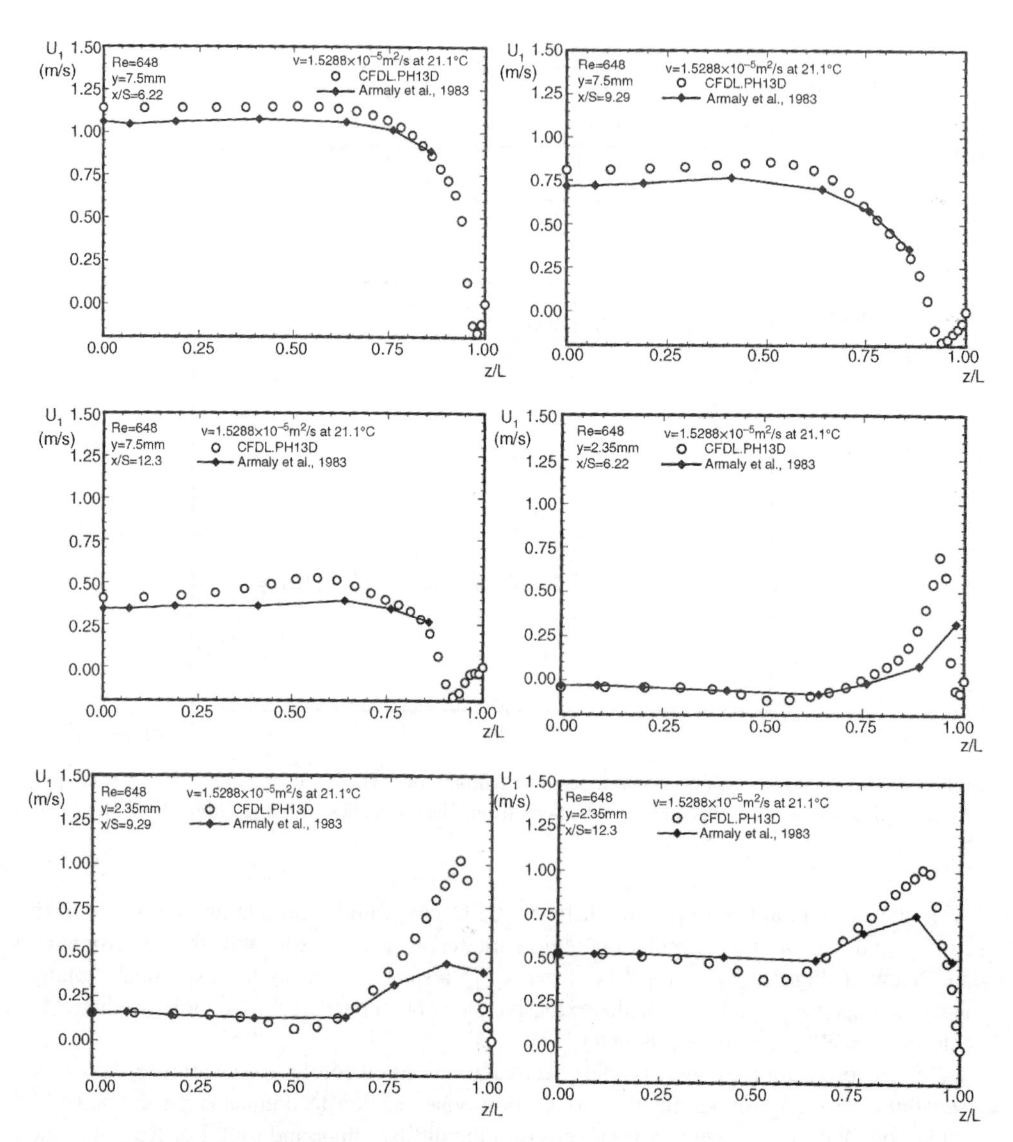

Figure 7.19 Step wall diffuser spanwise distributions of axial velocity, Re = 648, $m\mathrm{GWS}^h$ solution (open symbols), experimental data (triangles with line), left-right, top to bottom: $y = 7.5$ mm, $x/S = 6.22$, 9.29, 12.3; $y = 2.35$ mm, $x/S = 6.22$, 9.29, 12.3, from Williams (1993)

Figure 7.21 top indicates a very significant wall jet is generated on the diffuser floor in the transverse direction by the upstream sidewall momentum deficit modification. Factually, the wall jet is responsible for creating a transverse direction vortex velocity field within the primary recirculation region. This is visualized via a lagrangian particle inserted into the sidewall flow at the step face. The resultant particle tracks are graphed in Figure 7.22 for Re = 389, 648 and 800. The diameter of each sphere is linearly dependent

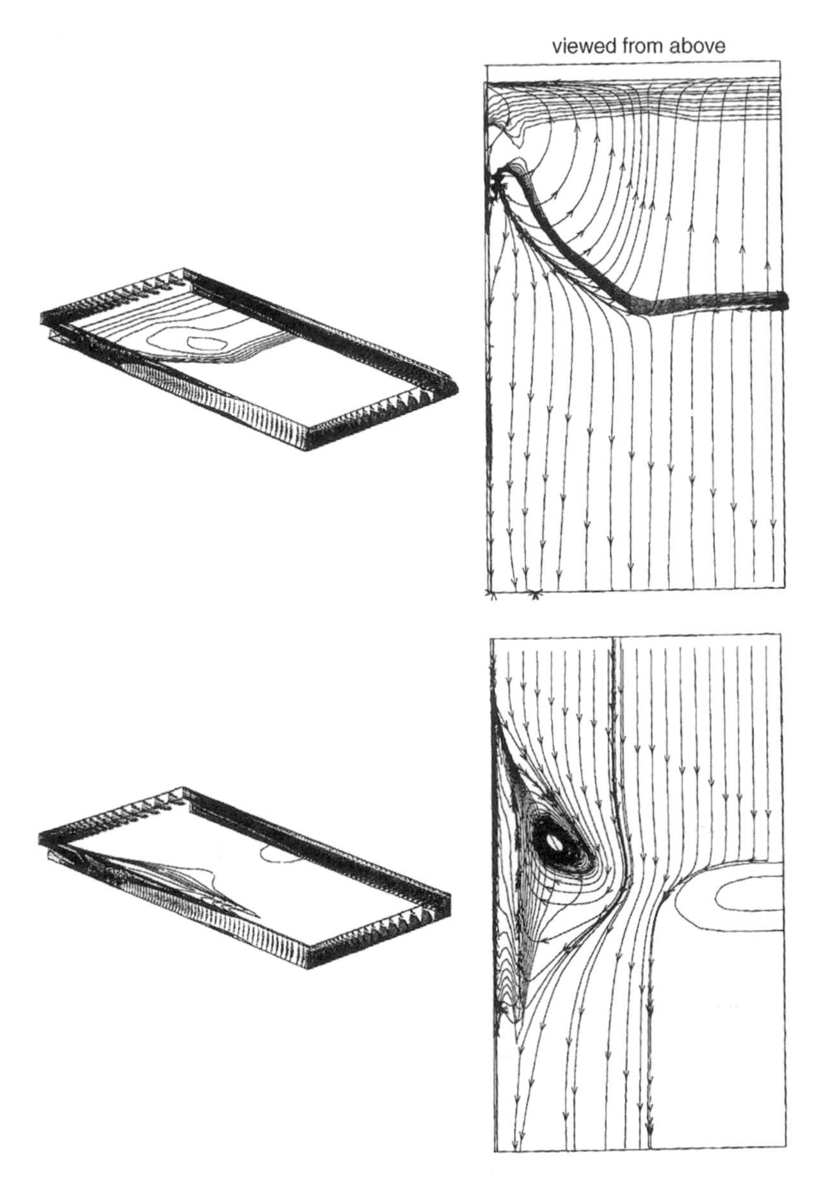

Figure 7.20 Step wall diffuser surface oil flow streakline distributions, from $m\text{GWS}^h$ solution, Re = 800: top, near floor; bottom, near roof. *Source:* Williams, P.T. and Baker A.J. 1997. Reproduced with permission of Wiley

on its elevation above the diffuser floor and particle spacing is at a constant elapsed time interval.

Momentum transport within the primary recirculation region vortex velocity field is clearly Re-dependent. The particle tracks confirm that fluid is totally trapped within this vortex until completing its traverse to the diffuser symmetry plane. The bottom graphic in Figure 7.22 is the lagrangian particle track for insertion near the boundary of the

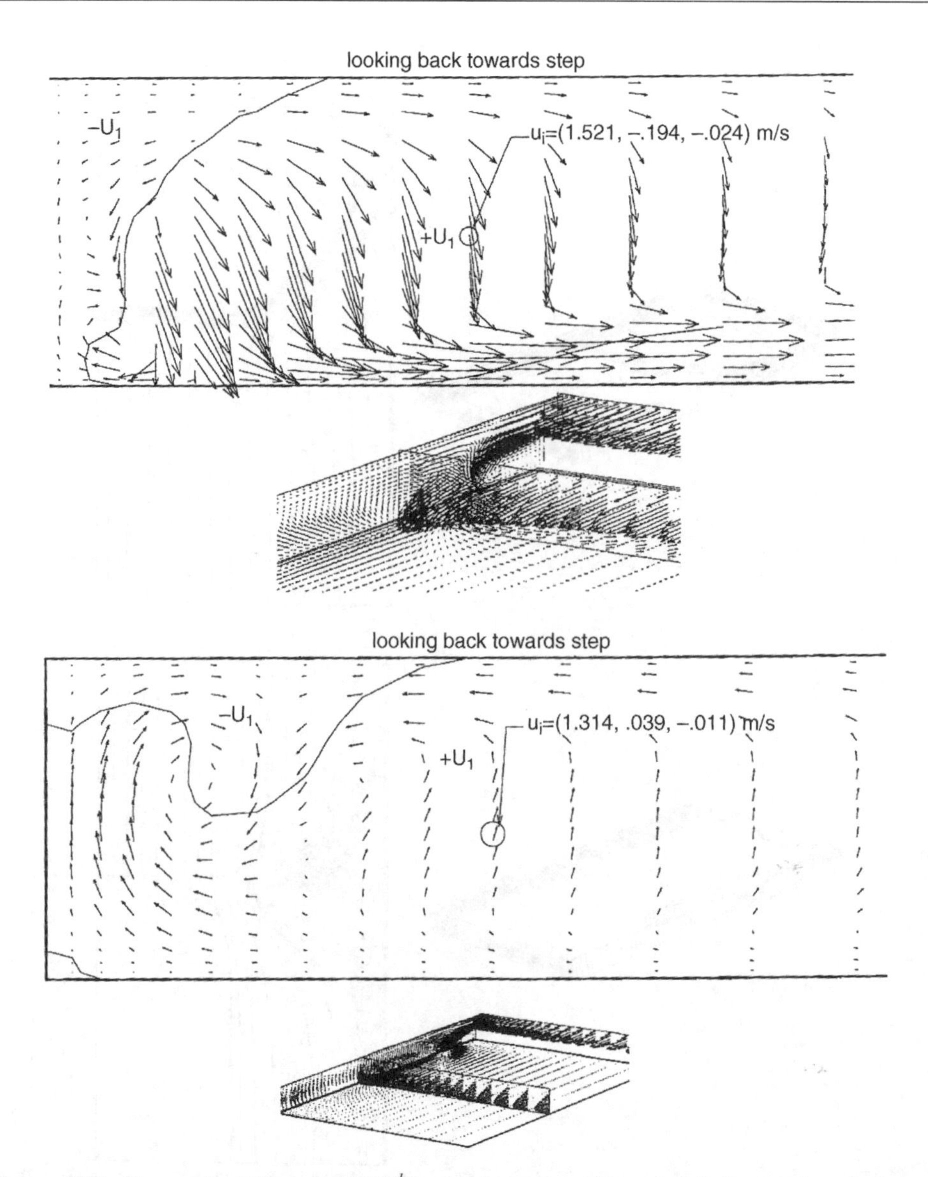

Figure 7.21 Step wall diffuser, $m\text{GWS}^h$ solution perspective window-in graphs of channel expansion-induced dominant sidewall recirculation region, Re = 800, viewed looking towards step: top, $x/S = 7.72$; bottom, $x/S = 18.37$. *Source:* Williams, P.T. and Baker A.J. 1997. Reproduced with permission of Wiley

downstream vertical axis vortex, confirmed in Figure 7.20 bottom. The bubble diameters being nearly uniform indicate a significant vertical velocity component is absent.

In total, this $m\text{GWS}^h$ theory *a posteriori* Re-sequenced CFD data set thoroughly elucidates the fully three-dimensional laminar fluid mechanics characteristic of the classic close-coupled step wall diffuser.

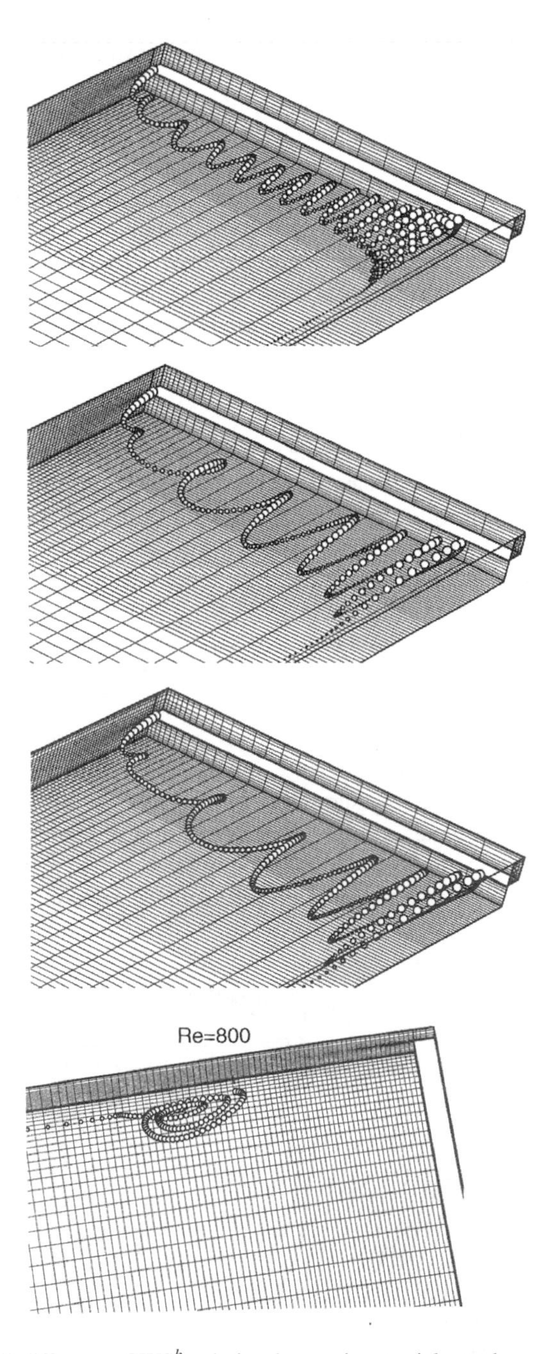

Figure 7.22 Step wall diffuser, $m\text{GWS}^h$ solution lagrangian particle tracks, particle injected at side-wall step lip, transverse wall jet motion in perspective view from symmetry plane, top to bottom: Re = 389, Re = 648, Re = 800; top view for particle injected into the near sidewall vertical axis vortex, Re = 800. *Source:* Williams, P.T. and Baker A.J. 1997. Reproduced with permission of Wiley

7.8 Flow Bifurcation, Multiple Outflow Pressure BCs

Verification, validation *a posteriori* data for cavities and single exit throughflow geometries are complete. Practical applications often involve multiple outflow boundary closure segments for which flow partitioning is *a priori* unknown. Since the admissable NS simulation outflow BC is a pressure specification, this occurrence generates a key issue to *validate*.

Via rigorously derived Robin BC in PDE + BCs system (7.7), accurate flow bifurcation prediction accrues to the CCM pressure projection theory. The human bronchial tree which undergoes order two dozen bifurcations between the trachea terminus and the alveoli is illustrative. Figure 7.23 left graphs the *non-coplanar* geometry for the first six bifurcations. Wong *et al.* (2004) report *a posteriori* data using an $M \sim 1.6E{+}06$ NC $k = 1$ basis mesh to generate a steady $n = 3$, Re = 500 velocity vector field. Using this prediction and for mass fraction IC a rotated gaussian at trachea inflow, the $GWS^h + \theta TS$ predicted mass fraction DOF distribution interpolation on the trachea mid-z plane at time $n\Delta t$ for mass exiting the most distant bronchial branch exit plane is graphed in Figure 7.23 right.

Figure 7.23 data confirm *qualitatively* the multiple outflow pressure BC requirement. A channel with internal splitter plate dividing the flow into distinct channel cross-sectional areas enables genuine pressure BC requirement *quantification*. Using the SIMPLE theory, specifically Dirichlet outflow BC $\phi = 0$ and genuine pressure approximation $\Sigma\phi$, (7.45), the $GWS^h + \theta TS$ Re = 1000 steady solution pressure gradient distributions in each channel generate identical outflow pressure level, Figure 7.24 left. Conversely, for CCM theory genuine pressure Robin BC (7.7), along with $\phi = 0$ on both channel outflow planes, the much reduced pressure gradient in larger cross-section channel leads to the requisite wide channel outflow larger pressure level, Figure 7.24 right.

Finally, a classic fluid dynamics issue warrants exposure. Axial flow redirection at each bifurcation, Figure 7.23, generates a vortex velocity vector distribution in the plane transverse to dominant flow direction. For an $n = 3$ *coplanar* bronchial tree laboratory model, Figure 7.25 top left, GWS^h steady solution velocity vector *a posterior* data for Re = 500 are

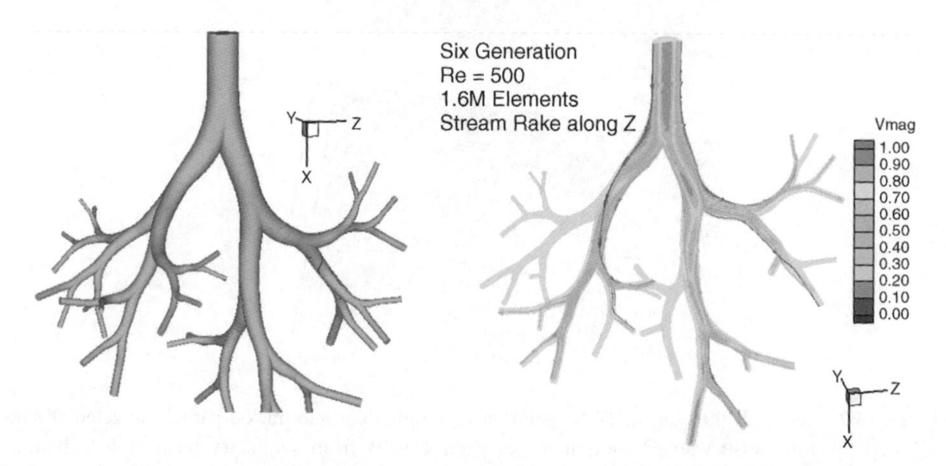

Figure 7.23 Bronchial tree mass transport simulation: left, 6 bifurcation *non-coplanar* $n = 3$ geometry; right, Re = 500, Sc = 1, $M \sim 1.6E{+}06$ NC $k = 1$ basis $GWS^h + \theta TS$ $n = 3$ mass fraction DOF $\{Y(n\Delta t)\}$ interpolation at time mass fraction exits the most distant outflow boundary

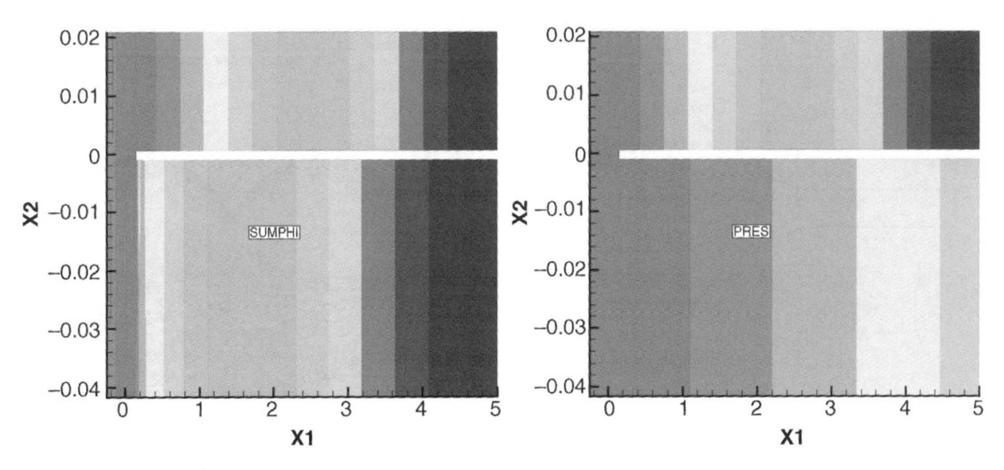

Figure 7.24 $GWS^h + \theta TS$ solution, channel geometry with internal splitter plate, Re = 1000, unitized scale graph of steady pressure distribution, $\phi = 0$ Dirichlet BC on both outflow boundaries; left, SIMPLE-theory $\Sigma\phi$ pressure approximation; right, CCM theory genuine pressure solution for rigorously derived outflow Robin BC

reported, Wong *et al.* (2004). Axial velocity inflow plane BC is fully developed laminar flow paraboloid. For defined cutting planes, mirror symmetric axial velocity color floods are graphed above transverse plane velocity vector DOF mirror symmetric distributions.

7.9 Convection/Radiation BCs in $GWS^h + \theta TS$

It is rare in genuine Navier–Stokes fluid-thermal applications for *a priori* known Dirichlet BCs to exist for D*E*. In fact, the sole occurrence is when the heat exchange medium undergoes phase change, for example boiling/freezing water. That the weak form process precisely embeds all pertinent *convection – radiation* flux boundary statements as the *natural* Robin BC warrants detailed exposition.

The weak form D*E* conduction algorithm development is in Section 2.2. Generalizing (2.1–2.2) to include unsteadiness, fluid convection and radiation, and dividing through by $\rho_0 c_p$ the continuum *Galerkin weak statement* for the *n*-D D*E* PDE + BCs + IC system is

$$
\begin{aligned}
GWS^N &= \int_\Omega \Psi_\beta(\mathbf{x})\left[\frac{\partial T^N}{\partial t} + \mathbf{u}\bullet\nabla T^N - \nabla\cdot\kappa\nabla T^N - s/\rho_0 c_p\right]d\tau \\
&= \int_\Omega \nabla\Psi_\beta\bullet\kappa\nabla T^N d\tau + \int_\Omega \Psi_\beta\left(\frac{\partial T^N}{\partial t} + \mathbf{u}\bullet\nabla T^N - (s/\rho_0 c_p)\right)d\tau \\
&\quad + \int_{\partial\Omega_R\cap\partial\Omega} \Psi_\beta\frac{k}{\kappa}\left[\begin{aligned} &h(T^N - T_{con}) \\ &+\sigma f(\mathrm{vf},\varepsilon)\left((T^N)^4 - T_{rad}{}^4\right) + \mathbf{f}(\mathbf{x},t)\bullet\widehat{\mathbf{n}}\end{aligned}\right]d\sigma \\
&\quad - \int_{\partial\Omega_D\cap\partial\Omega} \Psi_\beta\kappa\nabla T^N\bullet\widehat{\mathbf{n}}d\sigma = \{0\}, for\ 1\le\beta\le N
\end{aligned}
\tag{7.51}
$$

Recall the *data* definitions: ρ_0 is (essentially) constant density, c_p is specific heat, $\kappa \equiv k/\rho_0 c_p$ is thermal diffusivity for *k* thermal conductivity.

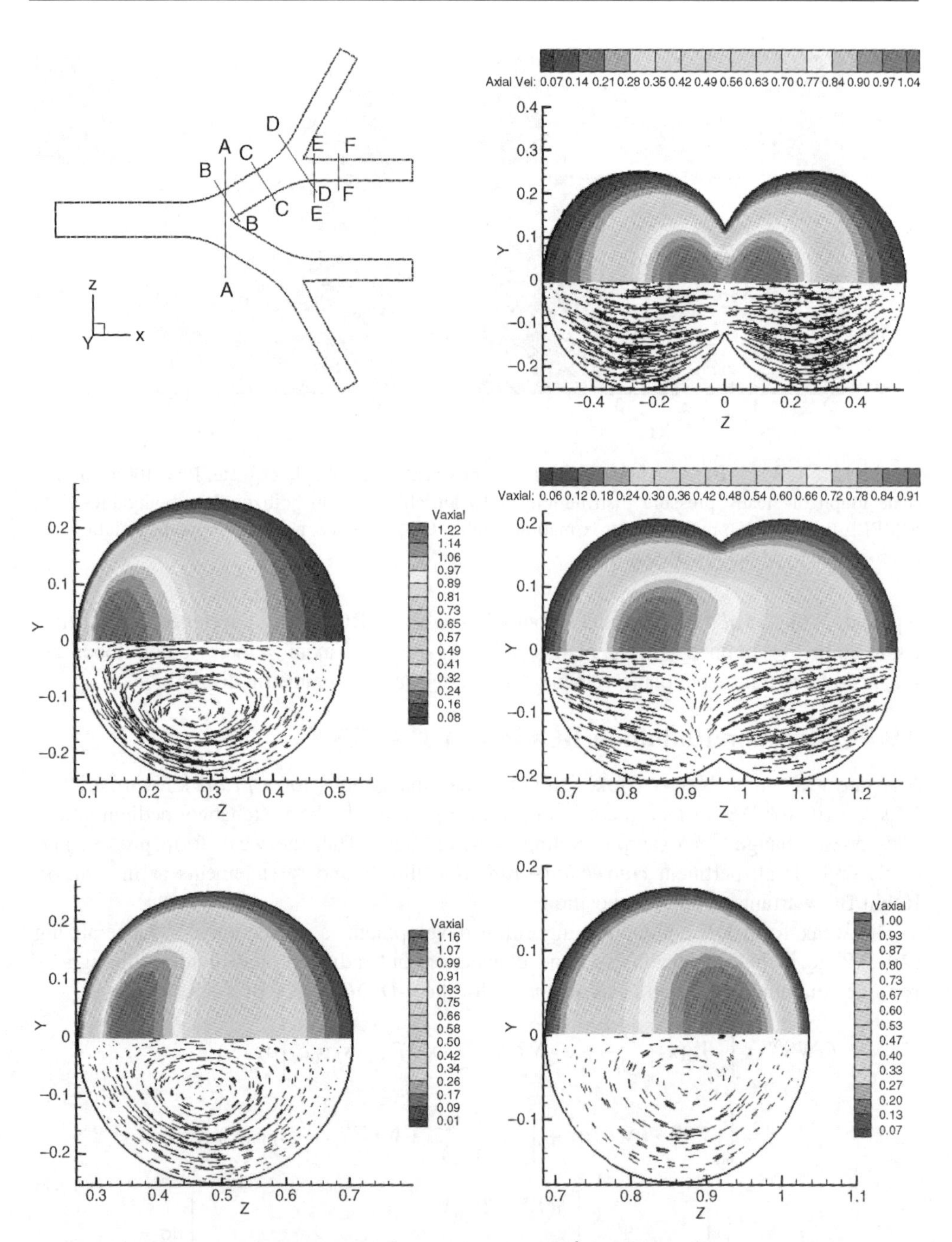

Figure 7.25 Coplanar bronchial tree laboratory model, $GWS^h + \theta TS$ steady velocity solution, $n = 3$, $Re = 500$: top left, geometry with DOF slice locations; clockwise from top right, slice plane AA, DD, FF, CC, BB mirror symmetric axial velocity DOF flood above transverse plane velocity vector DOF mirror symmetric distribution

In the weak from *R*obin BC h is convective heat transfer coefficient for a velocity field with exchange temperature T_{con}. For radiation exchange $\sigma = 5.67\,\mathrm{E}-08\,\mathrm{w/m^2K^4}$ is the Stefan-Boltzmann constant with $f(\mathrm{vf},\varepsilon)$ the *view factor* distribution for exchange entities with emissivity ε and temperature distribution T_{rad}. Finally, $\mathbf{f}(\mathbf{x}, t)$ is any imposed heat flux vector including omnidirectional diffuse sources. The last integral in (7.51) predicts efflux at Dirichlet temperature BC constraints, a decoupled post-solution evaluation hereafter discarded.

Focusing on the *R*obin BC, the $\partial\Omega_R \cap \partial\Omega$ surface integral in (7.51), the algorithm semi-discrete finite element basis implementation for assembly is

$$\int_{\partial\Omega_R\cap\partial\Omega} \Psi_\beta \frac{k}{\kappa}\left[h(T^N - T_{con}) + \sigma f(\mathrm{vf},\varepsilon)\left((T^N)^4 - T_{rad}{}^4\right) + \mathbf{f}(t)\bullet\widehat{\mathbf{n}}\right]\mathrm{d}\sigma$$

$$\Rightarrow \quad (\rho c_p)^{-1}\int_{\partial\Omega_e}\{N\}\begin{pmatrix} h_e\{N\}^T\left(\{Q\}_e - \{\mathrm{Q}C\}_e\right) \\ +\ \sigma f(\mathrm{vf},\varepsilon)_e\{N\}^T\left(\{\mathrm{Q}\}_e^4 - \{\mathrm{Q}R\}_e^4\right) \\ +\ \{N\}^T\{\mathbf{FLX}\bullet\widehat{\mathbf{n}}\}_e \end{pmatrix}\mathrm{d}\sigma \tag{7.52}$$

for $\{\mathrm{Q}C\}_e$ the convection heat transfer exchange temperature DOF and $\{\mathrm{Q}R\}_e$ the radiation exchange temperature DOF. A hypermatrix implementation for fourth power $\{\mathrm{Q}R\}_e$ is possible but discarded as an unnecessary cost. For $n = 3$ and assuming *for now* the BC data h_e and $f(\mathrm{vf},\varepsilon)_e$ are element constants, the element matrix statement companion to (7.52) is

$$\int_{\partial\Omega_R\cap\partial\Omega} \Psi_\beta[\bullet]\mathrm{d}\sigma \Rightarrow$$

$$\frac{\det_e}{\rho c_p}[\mathrm{B200}]\begin{pmatrix} h_e\left(\{Q\}_e - \{\mathrm{Q}C\}_e\right) + \{\mathbf{FLX}\bullet\widehat{\mathbf{n}}\}_e \\ +\sigma f(\mathrm{vf},\varepsilon)_e\left(\{Q\}_e^4 - \{\mathrm{Q}R\}_e^4\right) \end{pmatrix} \tag{7.53}$$

Convective heat transfer comes in two basic flavors, forced and natural, with mixed convection a smooth transition between them. The former is of boundary layer character, Figure 7.26 left, while the latter is of wall jet character, Figure 7.26 right. Incropera and DeWitt (1985) document heat transfer convection coefficient *correlations* in the

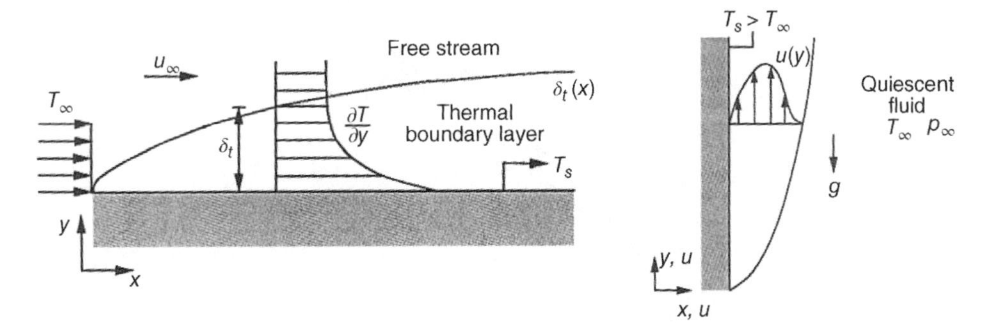

Figure 7.26 Illustration of forced and natural convection character

functional form $h \equiv h(k, \mathrm{Re}, \mathrm{Gr}, \mathrm{Pr}, x)$, where x is distance downstream from onset flow freestream impingement. The correlations are

$$\begin{aligned}
&\textit{forced}\text{ convection} : h = \mathrm{C}_1 k \mathrm{Pr}^{1/3}\mathrm{Re}^{1/2}/x, \mathrm{C}_1 = 0.664 \\
&\textit{natural}\text{ convection} : h = \mathrm{C}_2 k(\mathrm{Gr\,Pr})^{\mathrm{n}}, \\
&\qquad 0.15 \le \mathrm{C}_2 \le 0.54, \text{as} f(\text{surface } \widehat{\mathbf{n}}, \Delta T) \\
&\qquad \mathrm{Gr} = \rho\,\beta\Delta\,T x^3/\mu^2 \\
&\qquad \mathrm{n} = (0.25, 0.3) \text{for } \Delta T < 0, \Delta T > 0 \\
&\textit{mixed}\text{ convection} : \text{smooth transition between the correlations}
\end{aligned} \tag{7.54}$$

Radiation environments involve surface facets possessing emissivity ε typically less than unity, which defines a *grey body*. The energetic balance at the generic surface facet is

$$\text{absorbed} + \text{reflected} + \text{transmitted} \equiv \varepsilon + \rho + \delta = 1 \tag{7.55}$$

For non-transparent facets, a common situation, $\delta = 0$ and the portion reflected is $\rho = 1 - \varepsilon$.

Radiation exchange involves facet geometrical orientation which requires line-of-sight determination called a *viewfactor*. Lambert's cosine law defines viewfactor as the radiation received at surface facet with differential area $\mathrm{d}A_i$ from differential area $\mathrm{d}A_k$ in terms of respective orientation and separation distance r as

$$F_{k\to i} \equiv \frac{1}{A_k}\int_{A_k}\int_{A_i} \frac{\cos\,\phi_k \cos\,\phi_i}{\pi r^2_{k\to i}} \mathrm{d}A_k\, \mathrm{d}A_i \tag{7.56}$$

The geometry and nomenclature are illustrated in Figure 7.27. For a $k=1$ NC/TP basis planar facet implementation the kernel in (7.56) possesses the symmetry property

$$K_{k\to i} = \frac{\cos\,\phi_k \cos\,\phi_i}{\pi r^2_{k\,i}} = K_{i\to k} \equiv K_{i\,k} \tag{7.57}$$

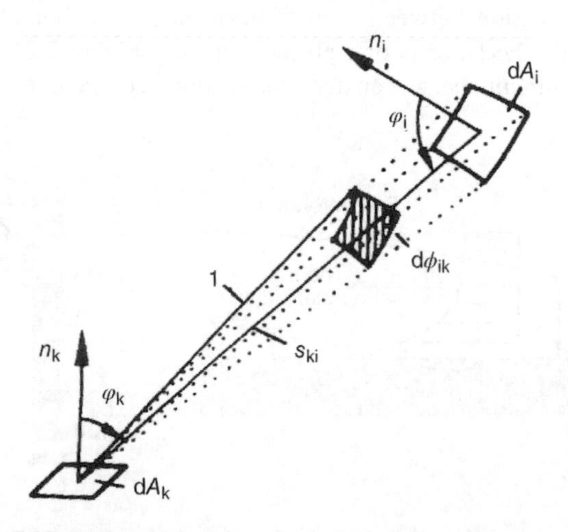

Figure 7.27 Differential surface facets in radiation energy exchange

This leads to the identities $F_{k\to i} = A_i K_{ik}$ and $F_{i\to k} = A_k K_{ik}$ valid only for differential facets. Thus, a viewfactor is a distribution for finite size surface facets, rigorously pointing to a DOF formulation requirement.

This is not common practice, however, as formulation theories intrinsically lead to isothermal facets with constant viewfactors, Siegel and Howell (2001, Table 7.1). For this assumption, the viewfactor becomes an element constant entering the $\mathrm{GWS}^h + \theta\mathrm{TS}$ algorithm as *element data*. For BC (7.52), the radiation exchange surface integral in the discrete implementation takes the form

$$
\begin{aligned}
&\int_{\partial\Omega_{rad}\cap\partial\Omega} \Psi_\beta \frac{k}{\kappa}\sigma f(\mathrm{vf},\varepsilon)\left[(T^N)^4 - T_{rad}^4\right]\mathrm{d}\sigma \\
&\Rightarrow \frac{\sigma}{\rho c_p}\int_{\partial\Omega_e\cap\partial\Omega_{rad}} \{N\}\{N\}^T\left(\varepsilon_{e=i}F_{i\to k}\{Q^4\}_{e=i} - \varepsilon_{e=k}F_{k\to i}\{Q^4\}_{e=k}\right)\mathrm{d}\sigma
\end{aligned}
\tag{7.58}
$$

Element subscript notation in (7.58) enforces that the integrals are performed on distinct elements.

For a *black body* $\varepsilon = 1$ and $A_e = (2, 4)\det_e$ for NC/TP $k = 1$ basis elements. Since K_{ki} is constant and using the relationships following (7.57), the $\mathrm{GWS}^h + \theta\mathrm{TS}$ algorithm BC statement for black body radiation exchange between a pair of surface elements is

$$
\begin{aligned}
&\frac{\sigma\varepsilon}{\rho c_p}\int_{\partial\Omega_e\cap\partial\Omega_{rad}} \{N\}K_{ki}\{N\}^T\left(A_k\{Q^4\}_{e=i} - A_i\{Q^4\}_{e=k}\right)\mathrm{d}\sigma \\
&= \frac{(2,4)\sigma}{\rho c_p}K_{k\,i}(\det_k \det_i)[\mathrm{B200}]\left(\{Q^4\}_i - \{Q^4\}_k\right) \\
&= \frac{(2,4)\sigma}{\rho c_p}[\mathrm{B200}]\left(\det_i F_{i\to k}\{Q^4\}_i - \det_k F_{k\to i}\{Q^4\}_k\right)
\end{aligned}
\tag{7.59}
$$

The alteration to (7.59) for grey body radiation interaction is expressed via Gebhart factors $G_{k\to i}$, Miller *et al.* (2010). The fraction of radiant energy received at a finite surface with element $\mathrm{d}A_i$ from a surface with element $\mathrm{d}A_k$ is

$$
G_{k\to i} \equiv \varepsilon_i F_{k\to i} + \sum_{j=1}^{n} \rho_j F_{k\to j} G_{j\to i}
\tag{7.60}
$$

which reverts to black body radiation in the absence of reflection. For n interacting surfaces (7.60) generates the order n-square full matrix statement in Gebhart factor coupling

$$
\begin{bmatrix}
(F_{11}\rho_1 - 1), \rho_2 F_{12}, & \dots & , \rho_n F_{1n} \\
\rho_1 F_{21}, (F_{22}\rho_2 - 1), & \dots & , \rho_n F_{2n} \\
 & \cdot & \\
\rho_1 F_{n1}, \quad \dots & & , (F_{nn}\rho_n - 1)
\end{bmatrix}
\begin{Bmatrix} G_{1i} \\ G_{2i} \\ \cdot \\ G_{ni} \end{Bmatrix}
= \varepsilon_i \begin{Bmatrix} -F_{1\to i} \\ -F_{2\to i} \\ \cdot \\ -F_{n\to i} \end{Bmatrix}
\tag{7.61}
$$

All self viewfactors F_{kk} are zero hence each matrix diagonal contains -1. There exist as many equations (7.61) as radiation interaction surface facets. Note for efficiency only a single LU decomposition is required for solving n right hand sides.

For this grey body formulation, the radiation contribution to the BC statement (7.52) transformed to the discrete implementation is

$$
\begin{aligned}
&\int_{\partial\Omega_{rad}\cap\partial\Omega} \Psi_\beta \frac{\sigma\varepsilon}{\rho c_p}\left[(T^N)^4 - T_{rad}^4\right] d\sigma \\
&\Rightarrow \frac{\sigma}{\rho c_p}\int_{\partial\Omega_e\cap\partial\Omega_{rad}} \{N\}\{N\}^T \left(\varepsilon_i\{Q^4\}_{e=i} - \sum_{k=1}^{n}\varepsilon_k G_{k\to i}\{Q^4\}_{e=k}\right) d\sigma \\
&= \frac{\sigma}{\rho c_p}[\mathrm{B200}]\left(\varepsilon_i\,\mathrm{det}_i\{Q^4\}_i - \sum_{k=1}^{n}\varepsilon_k\,\mathrm{det}_k G_{k\to i}\{Q^4\}_k\right)
\end{aligned}
\tag{7.62}
$$

The BC net contribution to the $\mathrm{GWS}^h + \theta\mathrm{TS}$ algorithm jacobian (7.21) is

$$
\begin{aligned}
[\mathrm{JAC\_}\{Q\}]_e &\equiv \frac{\partial}{\partial\{Q\}_e}\left[\frac{\sigma}{\rho c_p}[\mathrm{B200}]\left(\varepsilon_i\,\mathrm{det}_i\{Q^4\}_i - \sum_{k=1}^{n}\varepsilon_k\,\mathrm{det}_k G_{k\to i}\{Q^4\}_k\right)\right] \\
&= \frac{4\sigma}{\rho c_p}[\mathrm{B200}]\left(\varepsilon_i\,\mathrm{det}_i\,\mathrm{diag}\left[Q^3\right]_i - \sum_{k=1}^{n}\varepsilon_k\,\mathrm{det}_k G_{k\to i}\,\mathrm{diag}\left[Q^3\right]_k\right)
\end{aligned}
\tag{7.63}
$$

a *very nonlinear* contribution to $[\mathrm{JAC}]_e$.

The Gebhart factor formulation is valid for black body radiation by replacing distinct emissivities with $\varepsilon = 1$ and $G(k\to i)$ with $F(k\to i)$. The Σ_k [. . .] argument in (7.62–7.63) is a *compute-intensive* operation for hundreds of radiation exchange surfaces. Of similar impact, the contribution (7.63) to $[\mathrm{JAC}]_e$ *greatly expands* the bandwidth of the assembled global [JAC]. This can be moderated by zeroing out Gebhart factors $G(k\to i)$, equivalently viewfactors $F(k\to i)$, with magnitude less than some threshold.

7.10 Convection BCs Validation

Data for diurnal radiation-mixed convection heat transfer experiments are reported, Sanders and Rinald (2010). The fully instrumented field experiment, Figure 7.28, comprises elevated horizontally disposed thin and thick aluminum flat plates, each ~0.6 m (2 ft) square and painted flat black ($\varepsilon = 0.97$), exposed to the natural environment for 24 hour data recording periods. This field experiment supports $\mathrm{GWS}^h + \theta\mathrm{TS}$ convection BC validation as dynamic radiation loading is viewfactor-independent via given measured flux data.

These data support validation of $\mathrm{GWS}^h + \theta\mathrm{TS}$ convection heat transfer implementation (7.53–7.54). The thermal diffusivity of aluminum ensures that the temperature variation through the plate thickness is negligible, confirmed by 3-D simulation data. This admits a dimensionally reduced GWS^h formulation highly useful for radiation dominated environments. Restricting (7.23) to DOF $QI = Q4$ only, replacing $[\mathrm{M}]_e$ with $[\mathrm{m}]_e$ and for Δz the

Figure 7.28 Bobcat Salmon diurnal natural environment exposure experiments. *Source:* Sanders, J.S. and Rinald, D.J. 2010. Reproduced with permission of Dr. J. S. Sanders

uniform plate thickness, $[m]_e = \det_e[\text{M200}] \Rightarrow \Delta z \det_e[\text{B200}]$ in (7.24) while the (7.28) alteration is $(\kappa/\det_e)[\text{M2}KK]_e \Rightarrow (\kappa\Delta z/\det_e)[\text{B2}KK]_e$.

Experimentally recorded radiation omnidirectional fluxes that always act in the direction opposing a surface normal $\widehat{\mathbf{n}}$ include sky diffuse long wave (CG3*up*, CG3*dn*) and diffuse solar (Pyro), listed in Figure 7.28. Conversely, direct shortwave from the sun (CM3*up*, CM3*dn*) is the magnitude of the vector **CM3**. Figure 7.29 illustrates these radiation flux data recorded at 5 minute intervals. The 12 April 2006 flux record, graphed at 10 minute time intervals (to avoid symbol clutter), plus onset wind speed record is presented in Figure 7.30.

Data assimilation is detailed in the instrumentation user guide, Kipp and Zonen (2004). As the recorded data are time-dependent single numbers, Figure 7.30, their imposition is uniform on every finite element of plate discretizations. Hence in (7.52) flux interpolation DOF must transition to element scalars as $\{\mathbf{FLX} \bullet \widehat{\mathbf{n}}\}_e \Rightarrow \mathbf{FLX}(t) \bullet \widehat{\mathbf{n}}_e$ leading to the matrix implementation form in (7.53)

$$\begin{aligned}\{\mathbf{FLX} \bullet \widehat{\mathbf{n}}\}_e \Rightarrow \mathbf{FLX}(t) \bullet \widehat{\mathbf{n}}_e &= (\mathbf{CM3}) \bullet \widehat{\mathbf{n}}_{\partial\Omega} - (\text{CG3}up - \text{CG3}dn) - \text{Pyro} \\ &= (\text{CM3}up - \text{CM3}dn)\left(\widehat{\nu}(t) \bullet \hat{\mathbf{j}}\right) - (\text{CG3}up - \text{CG3}dn) - \text{Pyro}\end{aligned} \tag{7.64}$$

with all data time dependent. The dot product of sun unit vector $\widehat{\nu}(t)$ with plate upper surface normal $\hat{\mathbf{j}}$ in (7.64) is a time dependent cosine approaching negative unity at midday at the Alabama test site.

Two additional diffuse radiation exchange mechanisms exist, both described by the Stephan–Boltzmann term in (7.53). Plate upper surfaces are in radiation exchange with the sky semi-infinite half dome with temperature $T_{sky} = (\text{CG3}up)\sigma^{-1/4}$. Plate lower surface radiation exchange with the ground occurs at recorded temperature history T_{grd}.

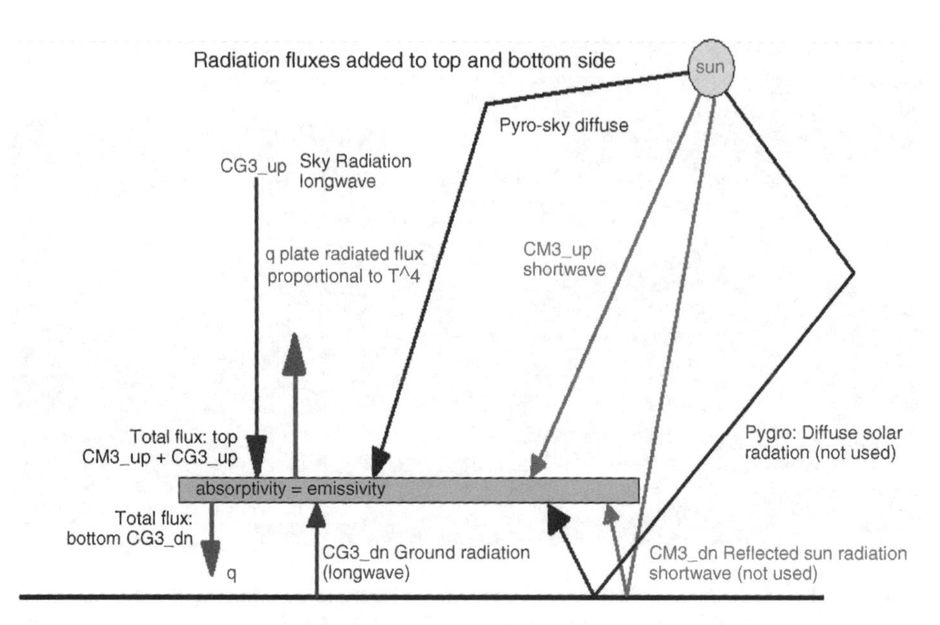

Figure 7.29 Bobcat Salmon experiment recorded radiation processes. *Source:* Sanders, J.S. and Rinald, D.J. 2010. Reproduced with permission of Dr. J. S. Sanders

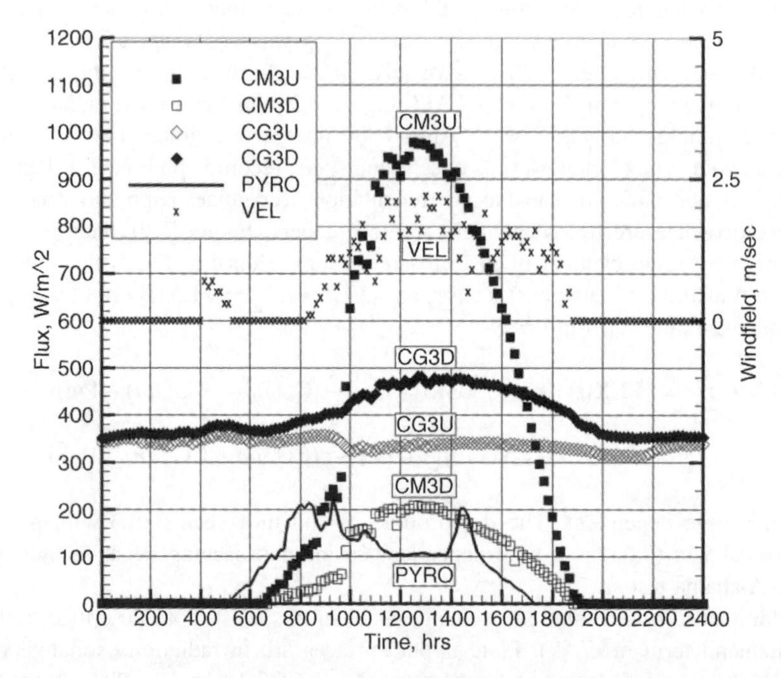

Figure 7.30 Bobcat Salmon radiation flux and onset wind speed 10 minute data record, 12 April. *Source:* Sanders, J.S. and Rinald, D.J. 2010. Reproduced with permission of Dr. J. S. Sanders

With e subscript denoting element-dependent, and defining the unit matrix {ONE} to handle (7.64), $\text{GWS}^N \Rightarrow \text{GWS}^h + \theta\text{TS}$ completion for (7.51) generates the temperature DOF $\{Q\}$ element matrix statement for $\{\text{RES\_}Q]_e$

$$\begin{aligned}\{\text{RES\_}Q\}_e &= \frac{\kappa\Delta z}{\det_e}[\text{B2}KK]_e\{Q\}_e + \frac{h_e\det_e}{\rho c_p}[\text{B200}](\{Q\}_e - T_{air}\{\text{ONE}\}) \\ &+ \frac{\sigma\,\varepsilon\det_e}{\rho c_p}[\text{B200}]\Big(\{Q\}_e^4 - T_{sky}^4\{\text{ONE}\} + \{Q\}_e^4 - T_{grd}^4\{\text{ONE}\}\Big) \\ &+ \frac{\varepsilon\det_e}{\rho c_p}\{\text{B10}\}\begin{pmatrix}(\text{CM3}up - \text{CM3}dn)\big(\widehat{\nu}(t)\bullet\hat{\mathbf{j}}\big) \\ -(\text{CG3}up - \text{CG3}dn) - \text{Pyro}\end{pmatrix}\end{aligned} \tag{7.65}$$

The experiment admits the full range of convection heat transfer processes. The forced convection model requires defining L for Re and distance x. Each plate built-back thermocouple (TC) is located at the upper surface centroid, hence $x = \text{L}/2$ with $\text{L} \approx 0.6\,\text{m}$ (2 ft). Onset flow in the open field setup is definitely boundary layer (BL) for any wind direction, hence the appropriate length scale for Re is x leading to $\text{Re}^{1/2}/x \Rightarrow (\text{U}/\nu x)^{1/2}$ in (7.54). Plate elevations above the ground plane are only $\sim 0.3\,\text{m}$ and the fairly tall grass likely impedes onset flow penetration under each plate. Thereby natural convection logically appears to be the sole underside heat transfer mechanism, subject to verification, with $\Delta T < 0$ in (7.54) indicating plate heating by the surroundings with plate cooling corresponding to $\Delta T > 0$.

For these decisions, and variations thereon, 24 hour simulation *a posteriori* data are reported for several data record days, Freeman *et al.* (2014). Simulation time step is uniformly $\Delta t = 5\,\text{min}$, identical with the data acquisition rate, using $\theta = 0.5$. *A posteriori* DOF data are compared with the complete experimental data, graphed as lines to emphasize time variation and coalescing the five radiation data channels into a net flux (RADnet) for clarity. Also graphed are distributions of convection coefficient on the plate top (ConvT) and bottom (ConvB) surfaces.

For the 12 April 2006 1-inch plate, the $\text{GWS}^h + \theta\text{TS}$ predicted centroid temperature DOF comparison with the co-located TC data confirms firm *quantitative* agreement for the entire 24 hour test, Figure 7.31. Plate heating due to a predawn windfield event, 0400–0600 hrs in Figure 7.30, is accurately captured via ConvT. Plate lower surface natural convection (ConvB) is a minimal contribution during daylight hours, but actually exceeds upper surface forced + natural convection (ConvT) during hours of darkness.

The extremum disparity between predicted and experimental temperatures is $\leq 2\,\text{K}$ throughout. Variations on convection coefficient specifications confirms doubling the Re length scale to $2x$ generates $\sim 8\,\text{K}$ overcooling spread over 1100–1700 h. Allowing forced convection on the plate underside while retaining x in Re produces $\sim 12\text{K}$ midday overcooling, Freeman *et al.* (2014).

For the 12 April 2006 $\frac{1}{8}$-inch plate, the $\text{GWS}^h + \theta\text{TS}$ predicted centroid temperature DOF comparison with the TC data confirms firm *quantitative* agreement, Figure 7.32, with exceptions. The predawn temperature excursion is under-predicted by $\sim 2\,\text{K}$, 0500–0600 h, but otherwise agreement is excellent to 1600 h. The reduced thermal mass of the $\frac{1}{8}$-inch plate admits *significant* oscillations in the TC data record (compare to 1-inch plate data,

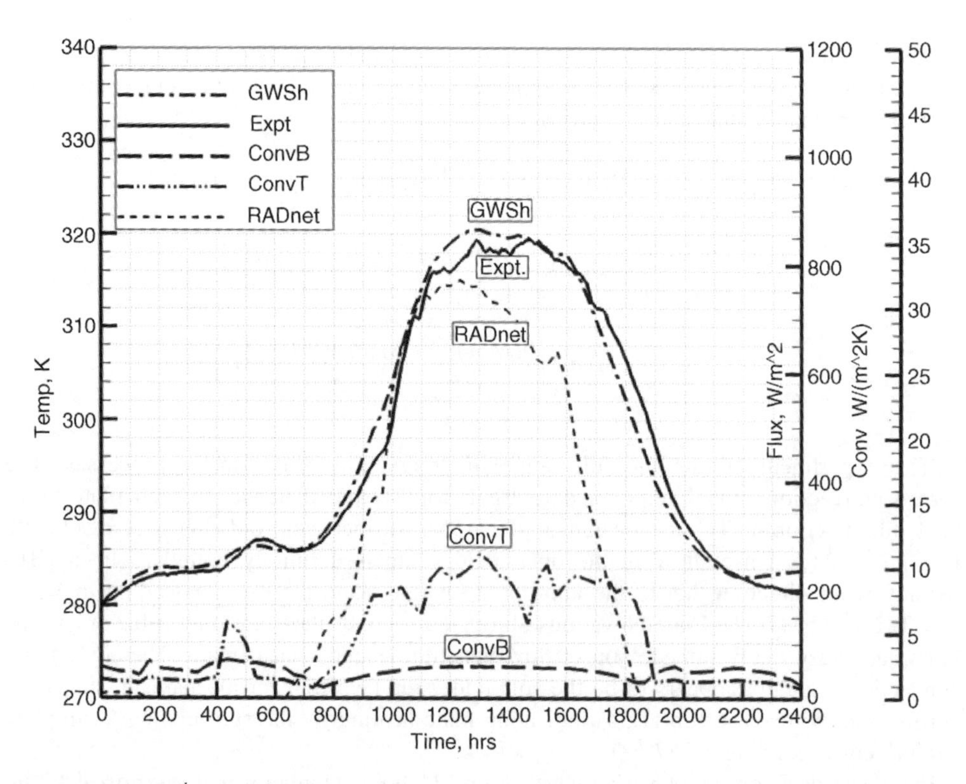

Figure 7.31 GWS^h + θTS 2.5D algorithm prediction of Bobcat Salmon 12 April 2006 experiment, 1-inch plate, from Freeman *et al.* (2014)

Figure 7.31). Plate cooling rate quantitative agreement is excellent during 1600–2000 hours; however, the simulation DOF precedes experiment by ~30 minutes throughout.

The RADnet, ConvT and ConvB graphs for the $\frac{1}{8}$- and 1-inch plates are, of course, identical. Doubling the forced convection correlation length in Re to $2x = \mathrm{L}$ generates plate overcooling by ~8 K during 1200–1600 h. Adding forced convection to the plate underside while retaining x in Re produces ~12 K midday overcooling, Freeman *et al.* (2014).

The Bobcat Salmon data record of 18 April 2006 contains an extended period of reduced radiation, probably due to cloud cover, Figure 7.33. The windfield magnitude is ~30% lower overall than for 12 April 2006, leading to appropriately reduced forced convection levels The darkness wind event now occurs during 2000–2200 h.

For the 18 April 2006 $\frac{1}{8}$-inch plate, the GWS^h + θTS predicted centroid temperature DOF comparison with the TC data confirms excellent *quantitative* agreement, Figure 7.34. The graph of RADnet fully illustrates the highly oscillatory character of the radiation flux field during 1400–1800 hrs during which the TC data record responds dynamically due to plate minimal thermal mass. The quality agreement between prediction and data therein is particularly noteworthy.

The 1-inch plate added thermal capacity significantly moderates the $\frac{1}{8}$ plate highly oscillatory TC data record, Figure 7.35. The RADnet, ConvT and ConvB graphs for the $\frac{1}{8}$- and 1-inch plates are again identical. The significant comparative distinction is the simulation

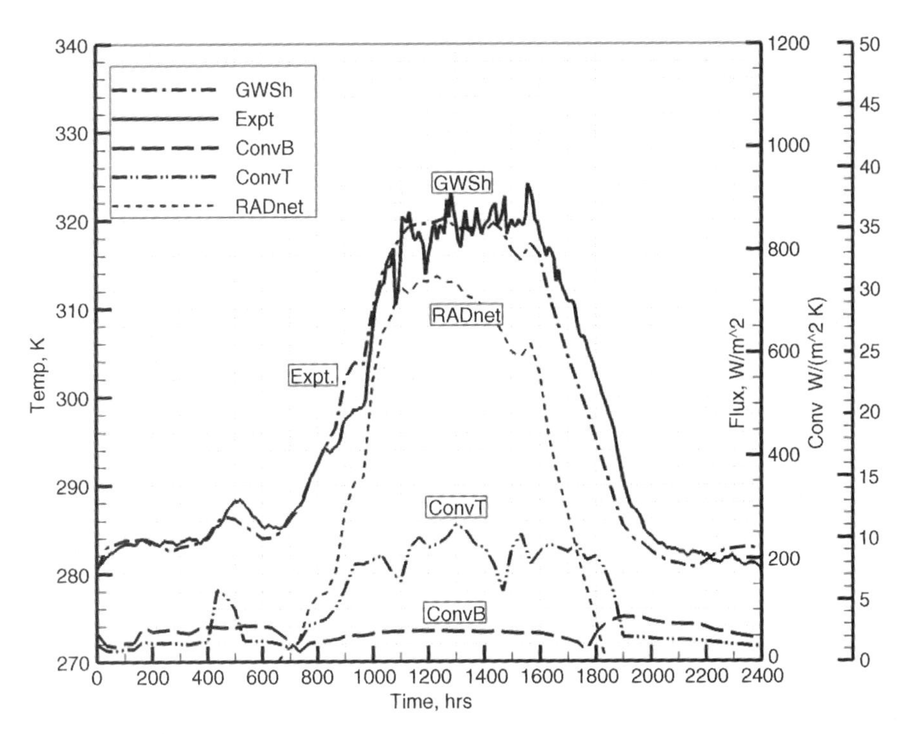

Figure 7.32 GWS^h + θTS 2.5D algorithm prediction of Bobcat Salmon 12 April 2006 experiment, $\frac{1}{8}$-inch plate, from Freeman *et al.* (2014)

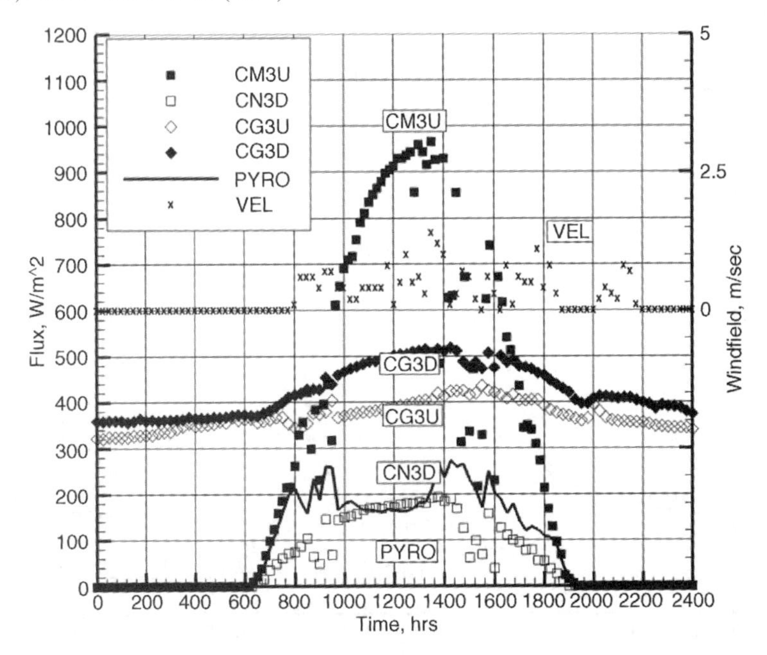

Figure 7.33 Bobcat Salmon radiation flux and onset wind speed 10 minute data record, 18 April. *Source:* Sanders, J.S. and Rinald, D.J. 2010. Reproduced with permission of Dr. J. S. Sanders

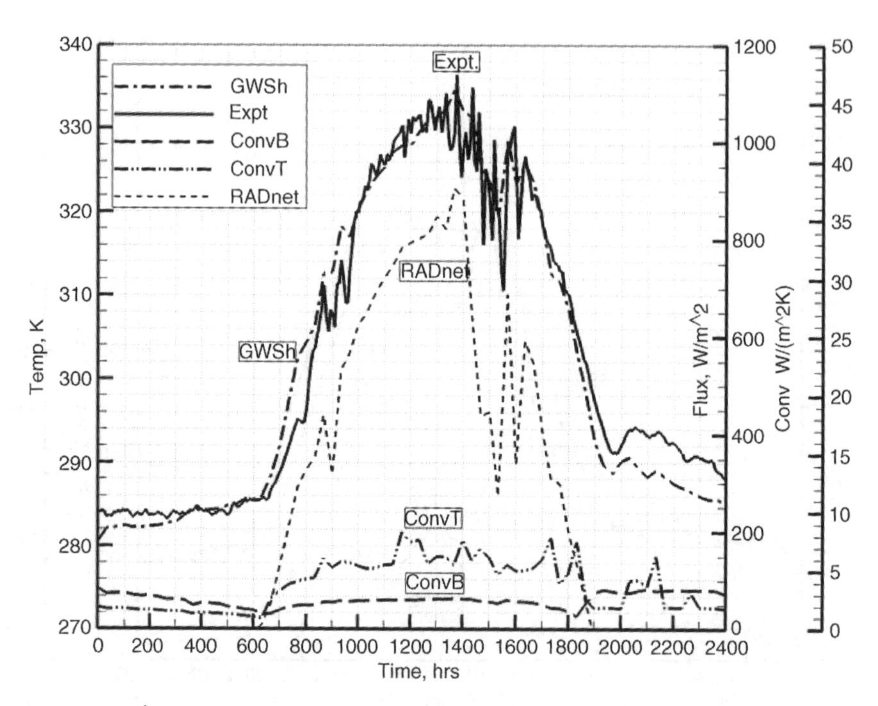

Figure 7.34 $GWS^h + \theta TS$ 2.5D algorithm prediction of Bobcat Salmon 18 April 2006 experiment, $\frac{1}{8}$-inch plate, from Freeman *et al.* (2014)

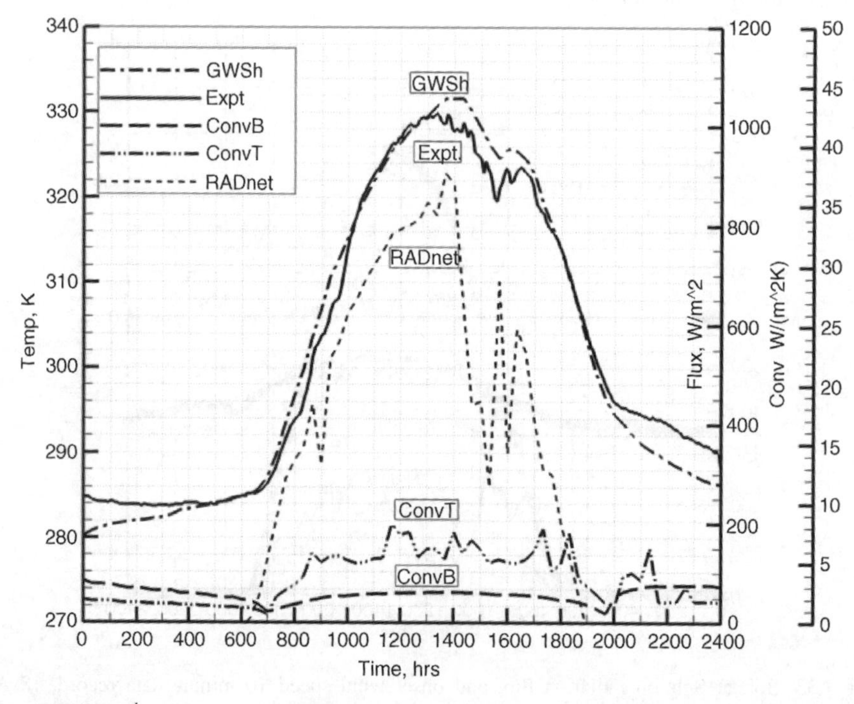

Figure 7.35 $GWS^h + \theta TS$ 2.5D algorithm prediction of Bobcat Salmon 18 April 2006 experiment, 1-inch plate, from Freeman *et al.* (2014)

much smoother DOF distribution during the limited radiation period, 1400–1800 h, leading to prediction exceeding the TC temperature data by $\sim 2 \pm \mathrm{K}$.

In summary, this extensively detailed quality experimental data base enables firm quantitative assessment of the unsteady mixed forced-natural convection correlation BC embedding in the $\mathrm{GWS}^h + \theta\mathrm{TS}$ 2.5D algorithm. Freeman (2012) reports considerable expansion on the range of validation exercises.

7.11 Radiosity, GWS^h Algorithm

The alternative to Gebhart factor grey body radiation implementation employs *radiosity theory*, which exhibits superior compute features for n body surface simulation. Recalling Figure 7.27, radiation $R_k \mathrm{d}A_k$ emanating into the semi-infinite half space from surface differential element $\mathrm{d}A_k$ impinges on a surface differential element $\mathrm{d}A_i$. The portion of this radiation directed towards $\mathrm{d}A_i$ is

$$\mathrm{d}R_{k \to i} \mathrm{d}A_k = R_k \mathrm{d}A_k \cos \phi_k / \pi \tag{7.66}$$

Of this emission, the amount of radiation actually impinging on surface $\mathrm{d}A_i$ located a distance r from $\mathrm{d}A_k$ is

$$\mathrm{d}R_{k \to i} = R_k \, \mathrm{d}A_k \cos \phi_k \, \mathrm{d}\phi_{i\,k}, \mathrm{d}\phi_{i\,k} = \frac{\mathrm{d}A_i \cos \phi_i}{\pi r_{k \to i}^2} \tag{7.67}$$

Direct substitution of the solid angle definition, (7.57), generates the *differential* form of *Lambert's cosine law*

$$\mathrm{d}R_{k \to i} = R_k \, \mathrm{d}A_k \frac{\cos \phi_k \cos \phi_i}{\pi r_{k \to i}^2} \mathrm{d}A_i \tag{7.68}$$

From (7.68) with energy balance (7.66), the radiation emanating from a surface $\mathrm{d}A_i$ per unit time is the sum of self-emission, according to Stefan–Boltzmann, and that reflected from the generic surface element $\mathrm{d}A_k$ as

$$R_i \, \mathrm{d}A_i = \varepsilon_i \sigma T_i^4 \, \mathrm{d}A_i + \rho_i \, \mathrm{d}R_{k \to i} \tag{7.69}$$

Generalizing to n grey body surface facets and substituting (7.68) produces a *Fredholm integral equation* of the second kind

$$R_i = \varepsilon_i \sigma T_i^4 + \rho_i \left[\sum_{k=1}^{n} \int_{A_k} R_k \frac{\cos \phi_k \cos \phi_i}{\pi r_{k \to i}^2} \mathrm{d}A_k \right] \tag{7.70}$$

where R_i is radiation energy per unit time per unit surface area.

The energy efflux q_i at the surface finite area A_i is the difference between the emanated radiation energy R_i and the total incident radiation from all exchange surfaces reaching A_i

$$q_i = R_i - \sum_{k=1}^{n} \int_{A_k} R_k\, K_{i\,k}\, \mathrm{d}A_k \tag{7.71}$$

In combination with (7.70) this defines the net energy efflux from a grey body surface A_i in terms of its *radiosity* R_i as

$$q_i = \frac{\varepsilon_i}{\rho_i}\left[\sigma T_i^4 - R_i\right] \tag{7.72}$$

Note that (7.72) is not valid for black body radiation since $\rho_i = 0$ is not admissable. Insertion of (7.72) into the $\mathrm{GWS}^h + \theta\mathrm{TS}$ algorithm BC statement (7.53) generates the n grey body *much less* compute-intensive alternative to (7.62)

$$\begin{aligned}
&\int_{\partial\Omega_{rad}\cap\partial\Omega} \Psi_\beta \frac{k}{\kappa}\left[(T^N)^4 - T_{rad}^4\right]\mathrm{d}\sigma \\
&\Rightarrow \frac{1}{\rho c_p}\frac{\varepsilon_i}{\rho_i}\int_{\partial\Omega_e\cap\partial\Omega_{rad}} \{N\}\{N\}^T\left(\sigma\{Q^4\}_{e=i} - \{R\}_{e=i}\right)\mathrm{d}\sigma \\
&= \frac{\det_e}{\rho c_p}\frac{\varepsilon_e}{\rho_e}[\mathrm{B200}]\left(\sigma\{Q^4\}_e - \{R\}_e\right)
\end{aligned} \tag{7.73}$$

Note in (7.73) that radiosity is by definition a *distribution* on Ω_e with nodal DOF $\{\mathrm{R}\}_e$, and subscript $e = i$ has been deleted as redundant.

Compared to (7.63) the radiosity formulation BC contribution to $[\mathrm{JAC}]_e$ is similarly the *much less* compute-intensive statement

$$\begin{aligned}
\left[\mathrm{JAC_}\{Q\}\right]_e &\equiv \frac{\partial}{\partial\{Q\}_e}\left[\frac{\det_e}{\rho c_p}\frac{\varepsilon_e}{\rho_\varepsilon}[\mathrm{B200}]\left(\sigma\{Q^4\}_e - \{\mathrm{R}\}_e\right)\right] \\
&= \frac{\det_e}{\rho c_p}\frac{\varepsilon_e}{\rho_\varepsilon}[\mathrm{B200}]\left(4\sigma\,\mathrm{diag}\left[Q^3\right]_e - \frac{\partial\{\mathrm{R}\}_e}{\partial\{Q\}_e}\right) \\
&= \frac{4\det_e\sigma(1-\varepsilon_e)\varepsilon_e}{\rho c_p}[\mathrm{B200}]\mathrm{diag}\left[Q^3\right]_e
\end{aligned} \tag{7.74}$$

as *analytically* derived via (7.55) with $\delta \equiv 0$ and using (7.70).

Obviously the *optimal* algorithm for generation of DOF $\{\mathrm{R}\}_e$ is via $\mathrm{GWS}^N \Rightarrow \mathrm{GWS}^h$ for (7.70), which renders the approximation error orthogonal to the trial space. Following Argyris and Szimmat (1992), the element discrete approximation is $R(\mathbf{x}, t)_e = \{N(\mathbf{x}(\boldsymbol{\eta}))\}^T\{\mathrm{R}(t)\}_e$ for any NC/TP basis spanning the generic domain surface $\Omega_{e=i}$. The GWS^h statement is

$$\{\mathrm{GWS}\}_{e=i} \equiv \int_{\Omega_{e=i}} \{N\}\left[\begin{array}{c}\left(\{N\}^T\{R\}_i - \varepsilon_i\sigma\{N\}^T\{Q^4\}_i\right)\mathrm{d}A_i \\ -\rho_i\left[\sum_{k=1}^{n}\int_{\Omega_{e=k}} \dfrac{\cos\phi_k\cos\phi_i}{\pi r_{k\to i}^2}\{N\}^T\{R\}_k \mathrm{d}A_k\right]\mathrm{d}A_i\end{array}\right] \equiv \{0\} \tag{7.75}$$

Dividing through by ρ_i, inserting the kernel definition (7.57) and completing the defined integrals produces the matrix statements constituting (7.75) as

$$\begin{aligned}\int_{\Omega_i}\frac{1}{\rho_i}\{N\}\{N\}^T\{R\}_i\mathrm{d}A_i &= \frac{1}{\rho_i}\det_i[\mathrm{B200}]\{\mathrm{R}\}_i\\ \int_{\Omega_i}\frac{\varepsilon_i\sigma}{\rho_i}\{N\}\{N\}^T\{Q^4\}_i\mathrm{d}A_i &= \frac{\varepsilon_i\sigma}{\rho_i}\det_i[\mathrm{B200}]\{Q^4\}_i\\ \int_{\Omega_i}\{N\}\left[\int_{\Omega_k}K_{ik}\{N\}^T\{\mathrm{R}\}_k\,\mathrm{d}A_k\right]\mathrm{d}A_i &= K_{ik}\det_k\det_i\{\mathrm{B10}\}\{\mathrm{B10}\}^T\{\mathrm{R}\}_k\end{aligned}\tag{7.76}$$

For the bilinear TP basis, the matrix product $\{\mathrm{B10}\}\{\mathrm{B10}\}^T$ in (7.76) produces a square matrix, all elements of which are unity, and adapting convention is labeled [B1010]. The radiosity GWS^h algorithm direct solve $[\mathrm{A}]\{x\}=\{\mathrm{b}\}$ matrix statement for (7.75–7.76) is

$$\frac{1}{\rho_i}\det_i[\mathrm{B200}]\{\mathrm{R}\}_i-\sum_{k=1}^{n}K_{ik}\det_k\det_i[\mathrm{B1010}]\{\mathrm{R}\}_k=\frac{\varepsilon_i\sigma}{\rho_i}\det_i[\mathrm{B200}]\{Q^4\}_i\tag{7.77}$$

In distinction to the practice of assembly over the elements of Ω^h, (7.77) is *not* assembled but written independently for each element. Assume that Figure 7.27 represents a finite element pair, each of which possesses four DOF $\{\mathrm{R}\}_e$, one at each vertex. Naming these as elements $i=1, 2$ the coupled matrix statements generated by (7.77) are

$$\begin{aligned}i=1:\rho_1^{-1}\det_1[\mathrm{B200}]\{\mathrm{R}\}_1-K_{12}\det_2\det_1[\mathrm{B1010}]\{\mathrm{R}\}_2 &= (\varepsilon_1\sigma/\rho_1)\det_1[\mathrm{B200}]\{Q^4\}_1\\ i=2:\rho_2^{-1}\det_2[\mathrm{B200}]\{\mathrm{R}\}_2-K_{12}\det_1\det_2[\mathrm{B1010}]\{\mathrm{R}\}_1 &= (\varepsilon_2\sigma/\rho_2)\det_2[\mathrm{B200}]\{Q^4\}_2\end{aligned}\tag{7.78}$$

Radiosity DOF coupling is obvious, and the algorithm *hypermatrix* computable statement is

$$\begin{bmatrix}\rho_1^{-1}\det_1[\mathrm{B200}]-K_{12}\det_2\det_1[\mathrm{B1010}]\\ -K_{12}\det_1\det_2[\mathrm{B1010}]+\rho_2^{-1}\det_2[\mathrm{B200}]\end{bmatrix}_{8\times8}\begin{Bmatrix}\{\mathrm{R}\}_1\\ \{\mathrm{R}\}_2\end{Bmatrix}_{8\times1}=\sigma\begin{Bmatrix}\varepsilon_1\rho_1^{-1}\det_1[\mathrm{B200}]\{Q^4\}_1\\ \varepsilon_2\rho_2^{-1}\det_2[\mathrm{B200}]\{Q^4\}_2\end{Bmatrix}_{8\times1}\tag{7.79}$$

For clarity, the global order of each matrix is subscripted in (7.79).

That the GWS^h algorithm generates $\{\mathrm{R}\}_e$ DOF that are nodally independent is confirmed by refining the grid to two elements sharing vertex nodes at the upper facet in Figure 7.27. Their temperature nodal DOF $\{Q\}$ (assembled) are common but the co-located radiosity DOF in $\{\mathrm{R}\}_2$ and $\{\mathrm{R}\}_3$ *are not* common. Preserving an essential matrix notation, (7.77) expanded clearly confirms this algorithm aspect

$$\begin{bmatrix}..[\mathrm{B200}]-K_{12}..[\mathrm{B1010}]-K_{13}..[\mathrm{B1010}]\\ -K_{12}..[\mathrm{B1010}]+..[\mathrm{B200}]-K_{23}..[\mathrm{B1010}]\\ -K_{13}..[\mathrm{B1010}]-K_{23}..[\mathrm{B1010}]+..[\mathrm{B200}]\end{bmatrix}_{12\times12}\begin{Bmatrix}\{\mathrm{R}\}_1\\ \{\mathrm{R}\}_2\\ \{\mathrm{R}\}_3\end{Bmatrix}_{12\times1}=\sigma\begin{Bmatrix}..[\mathrm{B200}]\{Q^4\}_1\\ ..[\mathrm{B200}]\{Q^4\}_2\\ ..[\mathrm{B200}]\{Q^4\}_3\end{Bmatrix}_{12\times1}\tag{7.80}$$

Viewing (7.79–7.80), the GWS^h algorithm generates a fully populated symmetric positive definite matrix amenable to standard solution algorithms. Since viewfactors rather than kernels are typically available as data, the corresponding GWS^h (7.80) restatement is

$$\begin{bmatrix} ..[\mathrm{B200}] - 4F_{1\to2}..[\mathrm{B1010}] - 4F_{1\to3}..[\mathrm{B1010}] \\ -4F_{2\to1}..[\mathrm{B1010}] + ..[\mathrm{B200}] - 4F_{2\to3}..[\mathrm{B1010}] \\ -4F_{3\to1}..[\mathrm{B1010}] - 4F_{3\to2}..[\mathrm{B1010}] + ..[\mathrm{B200}] \end{bmatrix} \begin{Bmatrix} \{\mathrm{R}\}_1 \\ \{\mathrm{R}\}_2 \\ \{\mathrm{R}\}_3 \end{Bmatrix} = \sigma \begin{Bmatrix} ..[\mathrm{B200}]\{Q^4\}_1 \\ ..[\mathrm{B200}]\{Q^4\}_2 \\ ..[\mathrm{B200}]\{Q^4\}_3 \end{Bmatrix} \tag{7.81}$$

The order of each matrix statement is unchanged and the multiplier 4 stems from the $n=2$ bilinear basis relation $A_e = 4\ \mathrm{det}_e$; the NC basis factor is a 2.

7.12 Radiosity BC, Accuracy, Convergence, Validation

Recalling weak form theory, Section 3.4, the GWS^h radiosity algorithm is *optimal* in its class as the approximation error is orthogonal to the trial space. Viewing (7.70), since no *smoothness* requirement exists, convergence is measureable *only* in H^0, equivalently the square root of the L2 norm. The asymptotic error estimate in H^0 and the norm computation is

$$\begin{aligned} \left\|e^h\right\|_{H^0} &\le C\, h^{k+1} \|\mathrm{data}\|_{\mathrm{L}2,\partial\Omega} \\ \left\|R^h\right\|_{H^0} &\equiv \frac{1}{2}\left[\sum_{e=1}^{M} \mathrm{det}_e \{\mathrm{R}\}_e^T [\mathrm{B200}d]\{\mathrm{R}\}_e\right]^{1/2} \end{aligned} \tag{7.82}$$

A semi-analytical solution for an isolated plane facet confirms the radiosity distribution is a *paraboloidal surface* with centroid extremum, Gould (2000). For any facet discretization the GWS^h solution DOF for (7.82) at common geometric nodes will *not* be identical. Specifically, the radiosity distribution will be *piecewise discontinuous* due to the distribution of viewfactors, recall (7.81). Figure 7.36 confirms this for an isothermal plate uniformly discretized into $\mathrm{M} = 8 \times 8$ $k=1$ TP basis elements. The vertical surfaces therein are a graphics package anomaly; the horizontal facets are clearly a piecewise discontinuous interpolation of a paraboloid with centroidal extremum.

A posteriori data supports $k=1$ TP basis implementation algorithm accuracy assessment and adherence to the theory for the facing parallel plate geometry, Figure 7.27. For square plate lateral dimension L, emissivity $\varepsilon = 0.9$, uniform ICs 1000 °C and 20 °C and separation distance L/4, the GWS^h hypermatrix algorithm $\mathrm{M} = 16 \times 16$ temperature solution is in excellent quantitative agreement with the (graphically extracted) benchmark data (symbols), Argyris and Szimmat (1992), Figure 7.37.

For facing plate separation distance L/4, L and 2L, regular mesh refinement *a posteriori* data for $k=1$ TP basis $\mathrm{M} = 1 \times 1,\ 2 \times 2,\ \ldots,\ 22 \times 22$ solutions, the GWS^h radiosity algorithm is confirmed to exhibit quadratic convergence in H^0, Figure 7.38. For the three separation distances and ICs the plate temperatures remain essentially uniform during evolution. However, adequate mesh resolution is *critical* to accuracy, the result of plate viewfactor *distribution* impact on radiosity, as clearly illustrated in Figure 7.36.

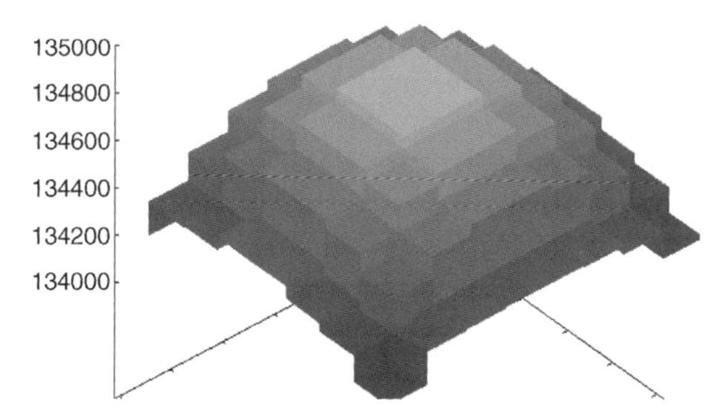

Figure 7.36 GWS^h algorithm $k=1$ TP basis radiosity DOF distribution, isothermal flat plate, M = 8×8 uniform mesh

The selected measure of solution accuracy is evolution time for the cool IC plate to reach its extremum temperature, Figure 7.37. Figure 7.39 graphs hot/cold plate temperatures at predicted time of cold plate extremum for the M mesh sequence. Coarser mesh solution times to the extrema are inaccurate by several minutes; additionally both plate temperatures at this time are low by ~30K. The overlap of the M = 16^2, 22^2 (denoted M = 256, 512) data confirm mesh resolution less than uniform M = 16^2 severely compromises prediction accuracy.

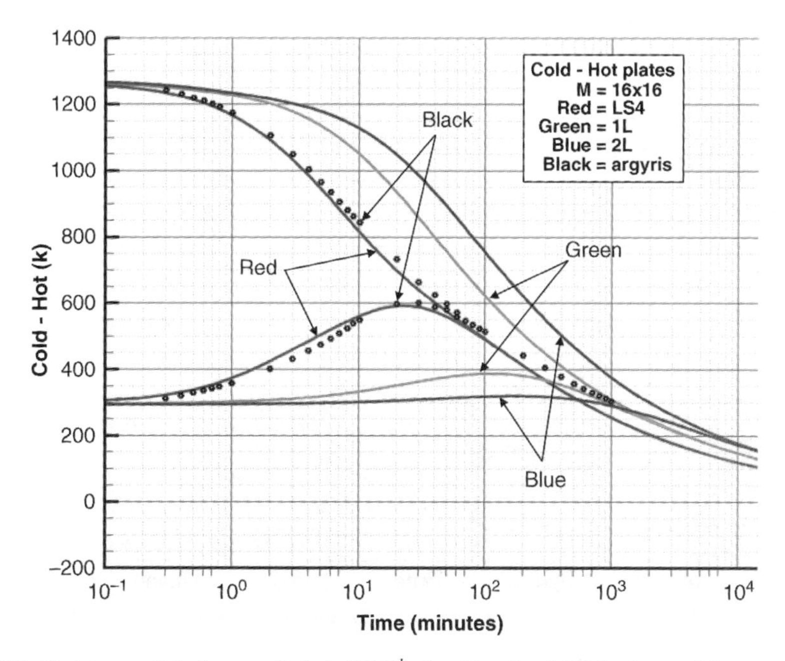

Figure 7.37 Facing parallel plates radiation, GWS^h algorithm $k=1$ TP basis, $\varepsilon=0.9$, separation distance L/4 temperature evolution comparison with benchmark data (symbols) of Argyris and Szimmat (1992), also temperature evolution for separation distances L and 2L

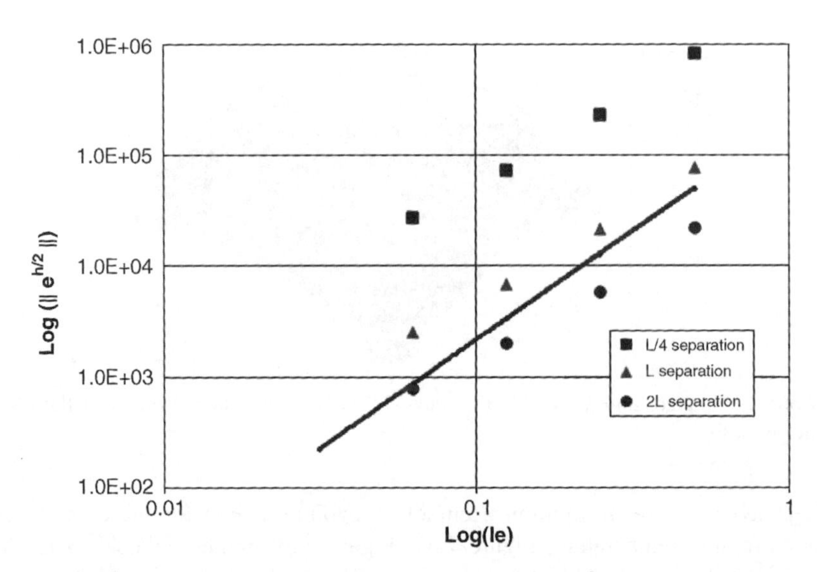

Figure 7.38 GWS^h radiosity algorithm asymptotic convergence in $\boldsymbol{H}^0$, $\boldsymbol{k} = 1$ TP basis, facing parallel plates, $\varepsilon = 0.9$, separation distances L/4, L, 2L at solution time $\boldsymbol{n}\Delta \boldsymbol{t} = 10^1$ min in Figure 7.34, from Freeman *et al.* (2014)

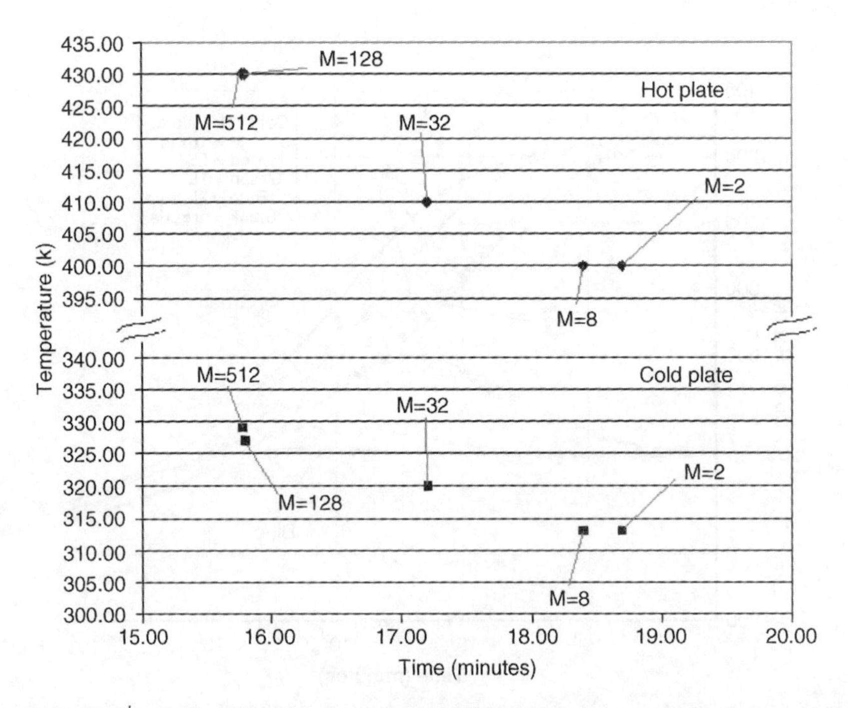

Figure 7.39 GWS^h radiosity algorithm accuracy dependence on facet mesh resolution, temperature vs evolution time to cold plate extremum, facing parallel plates, L/4 separation, $\varepsilon = 0.9$, from Freeman *et al.* (2014)

In Figure 7.37, observe the published benchmark data record stops at the time $\sim 10^3$ minutes for both plate temperatures reaching the cold plate IC. Conversely, for current *a posteriori* data this uniform *equal* plate temperature condition, carried past this time, predicts *isothermal* cooling below the cold IC. The benchmark data non-zero slope at simulation termination confirms this would have occurred had their experiment continued. These observations collectively predict that a non-conservative isothermal heat transfer process exists for this grey body geometry.

The absent enclosure specification for the parallel plate simulation is the medium separating the plates extends to infinity in the lateral directions which neither absorbs nor reflects radiation, hence is perfectly transmissive, recall the energetic balance (7.55). The plate viewfactor distributions not summing to unity is also noted, Freeman (2012), as observation quantification.

For facing plates possessing identical temperatures, the radiosity theory (7.70–7.72) for surface facet i simplifies to

$$R_i \propto \varepsilon_i \sigma T_i^4 \tag{7.83}$$

for $\sigma = 5.67\,\mathrm{E}-08$ w/m^{2}K^4 the Stefan–Boltzmann constant and ε_i plate emissivity. Observe that (7.83) is a proportionality, *not* an equality, since for any uniform temperature facet the radiosity is a *distribution* with centroid extremum, *not* a constant. Finally, recall that radiosity theory is restricted to grey body applications, since $\rho_i = (1 - \varepsilon_i)$ resides in the denominator of (7.72).

Time evolution of the difference between arguments in (7.83), parameterized by plate emissivity ε_i, are graphed in Figure 7.40. Clearly for all but $\varepsilon_i = 0.9999$, compute-admissable

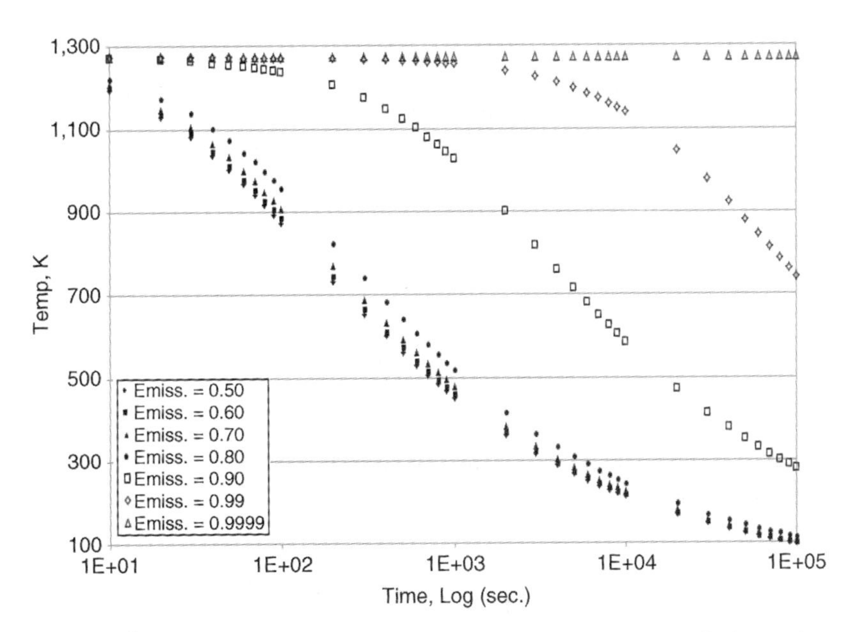

Figure 7.40 GWS^h radiosity algorithm, isothermal facing parallel plates in radiation exchange, time evolution of uniform plate temperature, $0.5 \leq \varepsilon_i \leq 0.9999$, from Freeman *et al.* (2014)

essentially black body definition, imperfect energy absorption with reduced reflection generates the observed energy loss mechanism. The larger the plate surface departure from black body the more rapid is isothermal energy loss.

The *transparent* unspecified separation medium of unbounded lateral extent must constitute the lost energy receiver. It is informative to note that the classic transport reference text, Bird, Stewart and Lightfoot (1960, Figure 14.5-2), graphically sketches the essence of the *physics* process, solidly quantified by these computed *a posteriori* data.

7.13 ALE Thermo-Solid-Fluid-Mass Transport Algorithm

A laser intense thermal processing protocol proved functional for infusing a high-wear-resistant metal surface into a steel substrate. Quality control variability in the UT Space Institute patented LISI (Laser Induced Substrate Infusion) catalyzed a project to formulate a CFD theory to quantify cause/effect relationships. The intrinsic physics processes span thermo-solid interaction, solid melting into a fluid pool enabling convective mass transport, followed by solidification of a potentially geometrically distorted melt pool mass fraction distribution.

The laser processing rig with illustration of operational features is shown in Figure 7.41. The laser provides the melt cover gas with axis angled to avoid workpiece backscatter to the laser. The workpiece, typically a stainless steel, is "painted" with a thin layer ($\sim$50$\pm$ micron) of a precursor saturated with sub-micron particles of the wear agent mixture such as titanium or chromium. Workpiece translation speed beneath the laser is an operational variable, with each pass laterally indexed by beam nominal width. Workpiece translation is plane parallel for marine vessel watertight door latch striker plate manufacture, conversely horizontal axis rotational for an 18-wheeler tractor-trailer rig engagement pin.

The GWS^h + θTS algorithm, Sections 7.3–7.5 absent all $O(h^4, m^3, \Delta t^4)$ *m*PDE augmentations, addresses the LISI statement. The base conservation principle PDE is D*E*, (7.3), augmented with the entire NS PDE system (7.2–7.7), once the solution process detects a melt pool. Accounting for melt pool velocity distribution, hence mass transport, requires implementing an *arbitrary-lagrangian-eulerian* (ALE) algorithm to handle melt pool free surface motion.

The ALE mesh movement algorithm, Hirt *et al.* (1974), is expressed in control volume (CV) context, recall Figure 1.1. The alteration to Reynolds transport theorem (1.6) is

$$\mathrm{d}() \Rightarrow \mathrm{D}() \equiv \frac{\partial}{\partial t}\int_{cv} (\cdot)\mathrm{d}\tau + \oint_{cs} (\cdot)(\mathbf{V}-\mathbf{U})\cdot\hat{\mathbf{n}}\mathrm{d}\sigma \tag{7.84}$$

for **U** the velocity (distribution) of the control surface (CS). For incompressible NS, the ALE algorithm alteration to D*M* (1.10) is

$$\oint_{\mathrm{cs}} (\mathbf{V}-\mathbf{U})\cdot\hat{\mathbf{n}}\,\mathrm{d}\sigma = \int_{\mathrm{cv}} \nabla\cdot(\mathbf{V}-\mathbf{U})\mathrm{d}\tau \Rightarrow \nabla\cdot(\mathbf{V}-\mathbf{U}) = 0 \tag{7.85}$$

with similar alterations inserted into D**P**, D*E* and D*Y*.

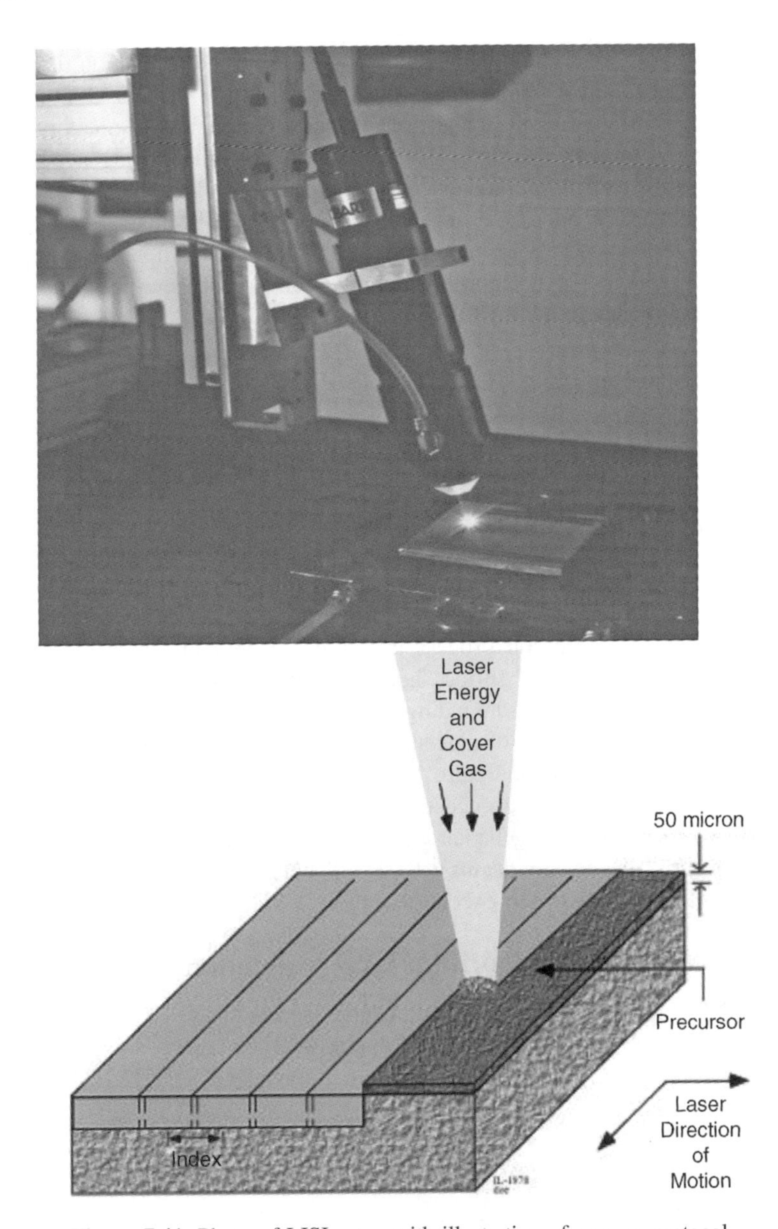

Figure 7.41 Photo of LISI setup with illustration of process protocol

As a LISI melt pool is very shallow, $\sim 100\pm$ microns deep, and transitory the *n*-D ALE algorithm accounts for melt pool surface displacement in planes transverse to workpiece translation. Hence in (7.2–7.4), (7.6), the ALE velocity vector definition is

$$u_i(\mathbf{x}, t) \Rightarrow u_i(\mathbf{x}, t) - U_i(\mathbf{x}, t) \tag{7.86}$$

The workpiece D*E* PDE algorithm operates in isolation up to the simulation time when one or more surface temperature DOF indicates precursor melting. The NS CFD algorithm is then initiated with BCs appropriate for an n-D velocity vector field in a cavity with movable lid. Specifically, precursor surface DM^h constraint potential homogeneous Neumann BC for classic cavity is altered to homogeneous Dirichlet. This admits but *does not* require(!) generation of a melt surface velocity vector in all (any of) $n=3$ directions.

Should surface velocity DOF be predicted during this integration time step Δt, then following NS CFD $GWS^h + \theta TS$ algorithm iterative convergence the position vectors $\mathbf{r}$ to *all* nodes in vertical planes in the melt pool are recomputed via

$$\mathbf{r}_{node}(\mathbf{x}, t_{n+1}) = \mathbf{r}_{node}(\mathbf{x}, t_n) + \Delta t\, \mathbf{u}^h_{node}(\mathbf{x}, t_n) \tag{7.87}$$

The originally rectangular cartesian mesh in the melt pool is thus altered via (7.87) to non-cartesian in vertical planes.

Prior to $GWS^h + \theta TS$ algorithm advanced time $t_{n+1} = t_n + \Delta t$ initiation, all ALE affected mesh plane node coordinates are recomputed via (7.87). The ICs for the melt pool vertical velocity DOF on all planes including the surface are thus nil, as required, which become appropriately altered during algorithm convergence at time t_{n+1}. Additionally, once melting occurs the D*Y* PDE (7.4) solution process is initialized with precursor IC $\{Y\} = \{1.0\}$, substrate IC $\{Y\} = \{0\}$, melt pool surface homogeneous Neumann BC and Dirichlet BCs $\{Y\} = \{0\}$ on all melt pool domain bounding surfaces.

The melt pool volume, surface displacement distribution and pool constituent mixture thus evolve until all surface temperature distribution DOF fall below the pool mixture melting temperature. At this occurrence the melt pool surface distribution is frozen by DM^h potential BC return to homogeneous Neumann. The complete CFD PDE system algorithm continues until all pool temperature DOF fall below that for mixture solidification, whence the algorithm returns to D*E* execution only.

7.14 ALE $GWS^h + \theta TS$ Algorithm LISI Validation

The $GWS^h + \theta TS$ algorithm addressing fundamental aspects of the LISI thermo-solid-fluid-mass transport process is established. LISI operational variations sought characterized include:

- laser-workpiece standoff, $O(\text{mm} \times 100)$
- laser focus plane at or above workpiece, $O(\text{mm})$
- laser energy flux, $O(\text{kW/mm}^2)$
- laser flux direct foot print, $O(\text{mm}^2)$
- footprint exposure laser heating, $O(\text{sec})$
- workpiece translation speed, $O(\text{mm/sec})$
- precursor thickness, $O(\mu\text{m} \times 10)$
- melt pool solidification, $O(\text{sec} \times 100)$

The first three operational variables are interrelated with one constant being laser axis angle $\sim 10°$ off parallel to the workpiece surface normal. Laser-workpiece standoff distance ranges 126–135 mm and the laser focus plane is adjustable. Experimental data confirm laser focus at the precursor plane generates a nominally uniform radiation flux footprint with direct rectangular footprint ~0.6×4 mm with far lateral peaks and rounded corners, Figure 7.42 top left. Elevating the laser focus plane to 6 mm above precursor surface generates an oval ring flux distribution of broader ~1.2×4 mm footprint and diminished lateral peaks, Figure 7.42 top right.

The $\mathrm{GWS}^h + \theta\mathrm{TS}$ simulation radiation flux model yields quality geometric fidelity for in-plane focus. The above-plane focus model geometry is defined of identical span to support direct comparisons, hence is accordingly curvilinear geometry compromised. The flux $\mathbf{f}(\mathbf{x})$ distributions, graphed in perspective to quantify intensity level, are located beneath each data measurement, Figure 7.42. Radiation flux distribution input is via $\mathbf{f}(\mathbf{x}, t)$ in BC (7.51), for fixed spatial dependence and time variation defined by workpiece translation speed.

LISI simulation time integration is initialized with the $\mathrm{GWS}^h + \theta\mathrm{TS}$ algorithm for DE only. The laser flux model distribution is translated at speed over a stationary symmetric-half workpiece. The translation direction (z) solution domain length of 4.8 mm is ~8 flux model footprint widths, hence is sufficient to isolate IC perturbations and finite domain termination. The half domain of width $1 + 2$ mm is discretized by a $k = 1$ TP basis highly solution adapted $\mathrm{M} = 37 \times 39 \times 41$ cartesian mesh. This domain is perspective graphed in Figure 7.43 with out of plane focus flux model distribution superimposed at ~6 footprint

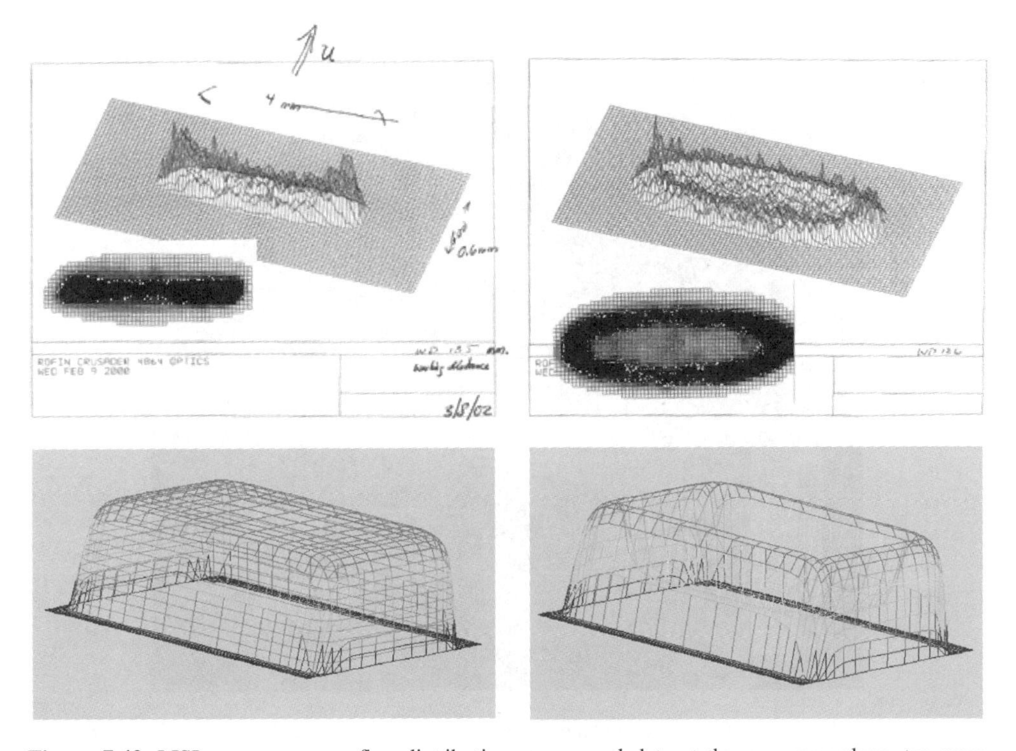

Figure 7.42 LISI process energy flux distributions: measured data at the precursor plane, top row; left, plane focus, right, 6 mm above-plane focus; $\mathrm{GWS}^h + \theta\mathrm{TS}$ algorithm flux model $\mathbf{f}(\mathbf{x})$ distributions, bottom row: left, plane focused, right, identical span, 6 mm above-plane focus

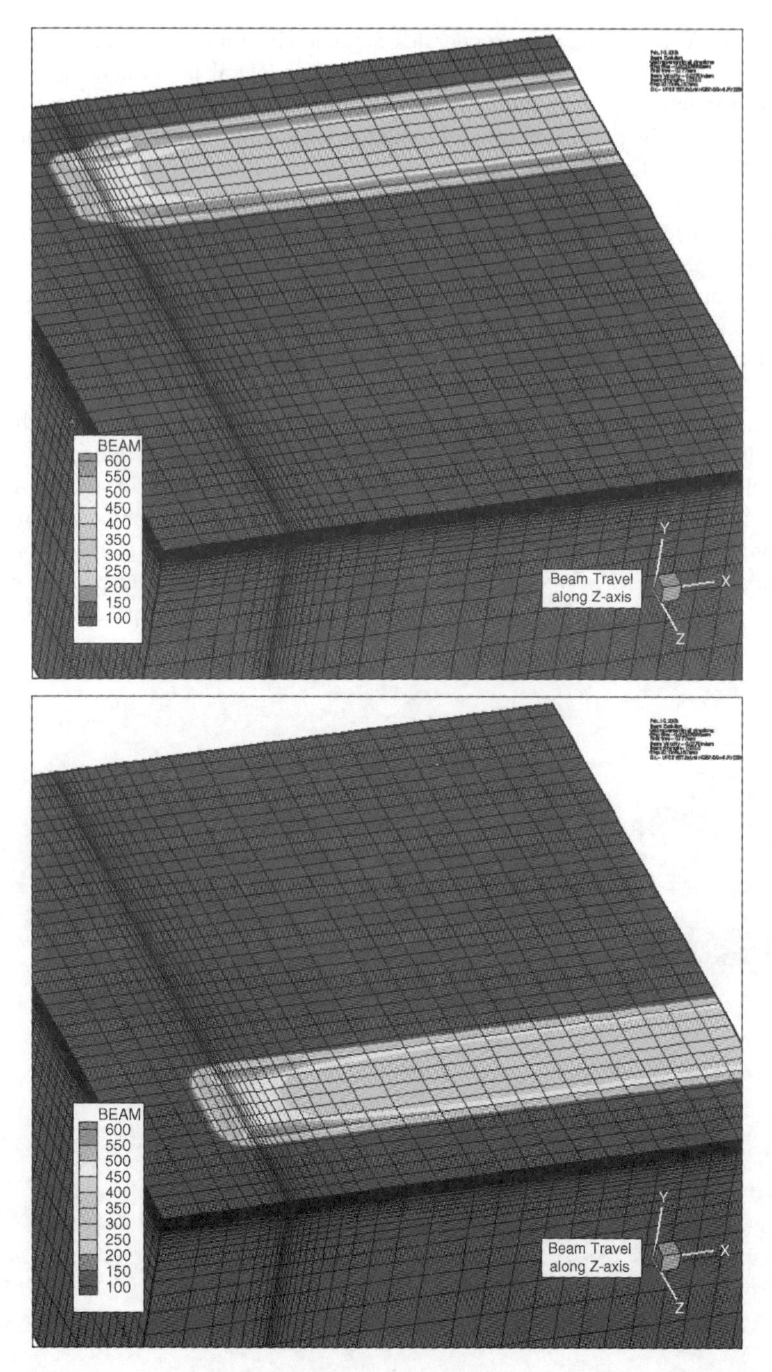

Figure 7.43 LISI process 6 mm above-plane focus flux superposition on D*E* symmetric half domain solution adapted $M = 37 \times 39 \times 41$ cartesian mesh, 126 mm standoff: top, after initiation at t_0; bottom, after $\sim$6 flux footprint width translation

width translation, also at an earlier time confirming flux intensity symmetry. The surface plane node density is just sufficient to resolve the oval ring flux model, conversely is fully adequate for the uniform footprint model.

The D*E* solution domain BCs are everywhere adiabatic. For translation speed 1.7 mm/sec, at time $t_{n+1} = t_0 + n\Delta t$ of the ~6 flux model width translation, Figure 7.43 bottom, the precursor surface temperature DOF distribution is graphed in flood perspective in Figure 7.44. This flux level produces a ~2500 °F surface temperature extrema, just above precursor melt temperature, and the melt pool cools hence solidifies rapidly after passage, Figure 7.44 bottom.

The surface geometry of the solidified melt pool is dependent on radiation flux model energy level. For the 126 mm standoff, completely solidified melt pool mass fraction DOF distribution perspective floods on ALE-algorithm generated surface geometry are compared in Figure 7.45. The 6 mm above-plane focus model flux specifications generate surface extrema of ~2500 °F and ~6900 °F, top and bottom graphs, respectively. The solidified melt pool mass fraction for the former is homogeneous, except for a very narrow region at the end of the laser flux footprint. In comparison, that for the latter specification is not homogeneous and the pool solidified surface is clearly non-planar.

Exploded views of mass fraction perspective with coincident vertical plane mesh distributions are compared in Figure 7.46 for these data. Mesh resolution within the ~100 μm deep melt pool is clearly adequate. The ALE algorithm surface geometry displacement prediction of induced lateral and vertical mesh movement alteration for the higher flux specification is clearly visible.

Qualitative validation of these predictions is afforded by comparison with photomicrographs of LISI test specimen slices, Figure 7.47. The top graph visualizes the sought frozen melt pool homogeneous composition, laser footprint edge transition to workpiece composition and nominally planar solidified surface, compare to Figure 7.46 left. Conversely, the location of the surface depression and elevation in the bottom graph are in excellent qualitative agreement with the higher input flux prediction, Figure 7.46 right, and the melt pool solidified surface is clearly very non-planar.

The solidified melt pool composition and surface geometry distribution are influenced by melting-admitted velocity vector distributions as a function of laser model input flux, focus plane and laser standoff. For the on-plane focus flux model, Figure 7.42 left bottom, and power level generating ~6900 °F surface temperature extrema, Figure 7.48 top/bottom, compares predicted melt pool velocity vector distributions in perspective on select vertical plane node columns with temperature flood for laser standoff distance of 135 mm and 126 mm.

These data are at simulation elapsed time centering the flux footprint traverse at ~6 footprint widths from t_0. For direct comparison vector lengths are identically scaled; note each vector distribution possesses a zero gravity direction component confirming ALE algorithm implementation. For both standoffs the melt pool is solidified at the fourth plane away from t_0, as confirmed by velocity vector absence. For the 135 mm standoff the velocity distributions on the next four planes are of nil magnitude. Conversely, for the 126 mm standoff these planes contain very significant velocity vector distributions, especially near the laser lateral terminus. Directly underneath and behind this laser flux footprint the predicted velocity vector distributions confirm existence of significant stirring in the melt pool.

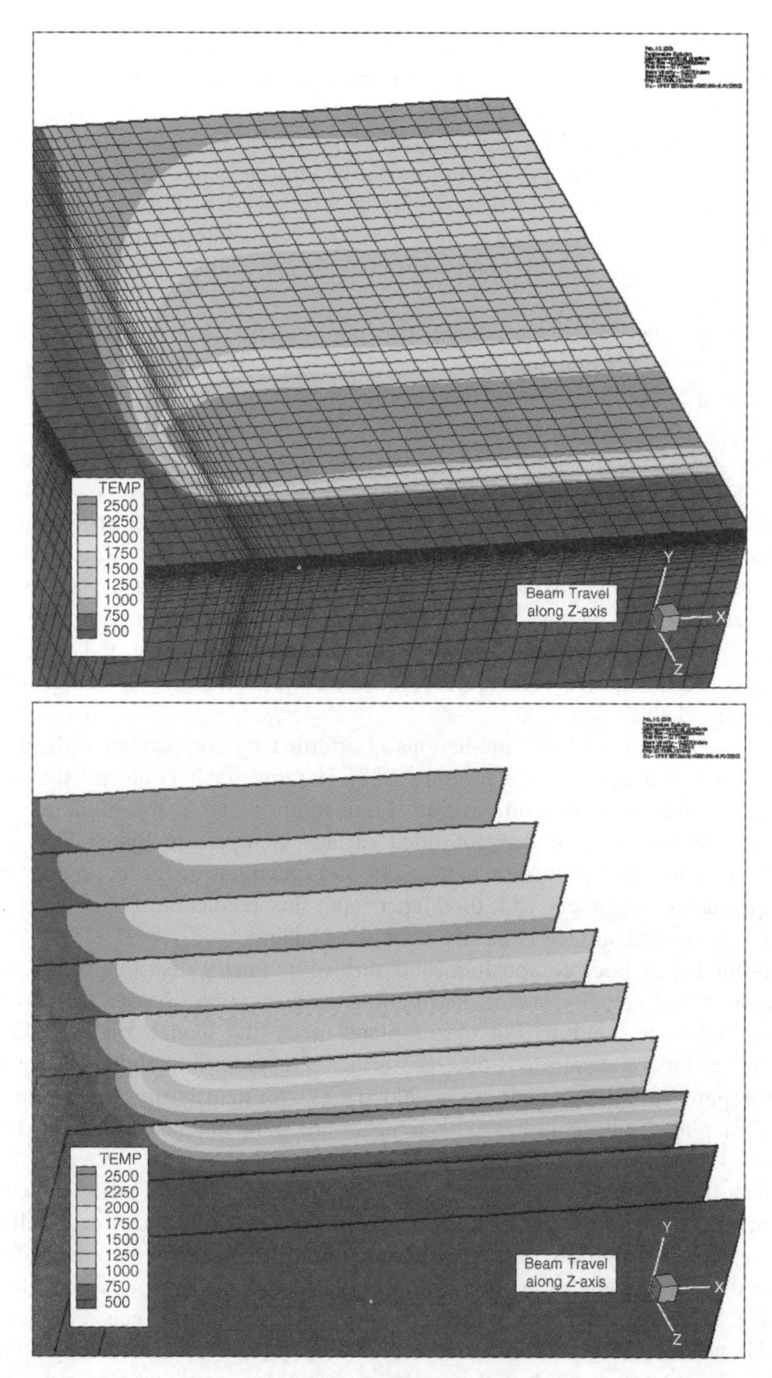

Figure 7.44 $\mathrm{GWS}^h + \theta\mathrm{TS}$ D*E* algorithm temperature DOF distributions, 6 mm off-plane focus flux model, 126 mm standoff, at time $t_{n+1} = t_0 + n\Delta t$ after ~6 flux width translations from t_0: top, precursor surface; bottom, depth perspectives on ~¾ flux width interval vertical planes

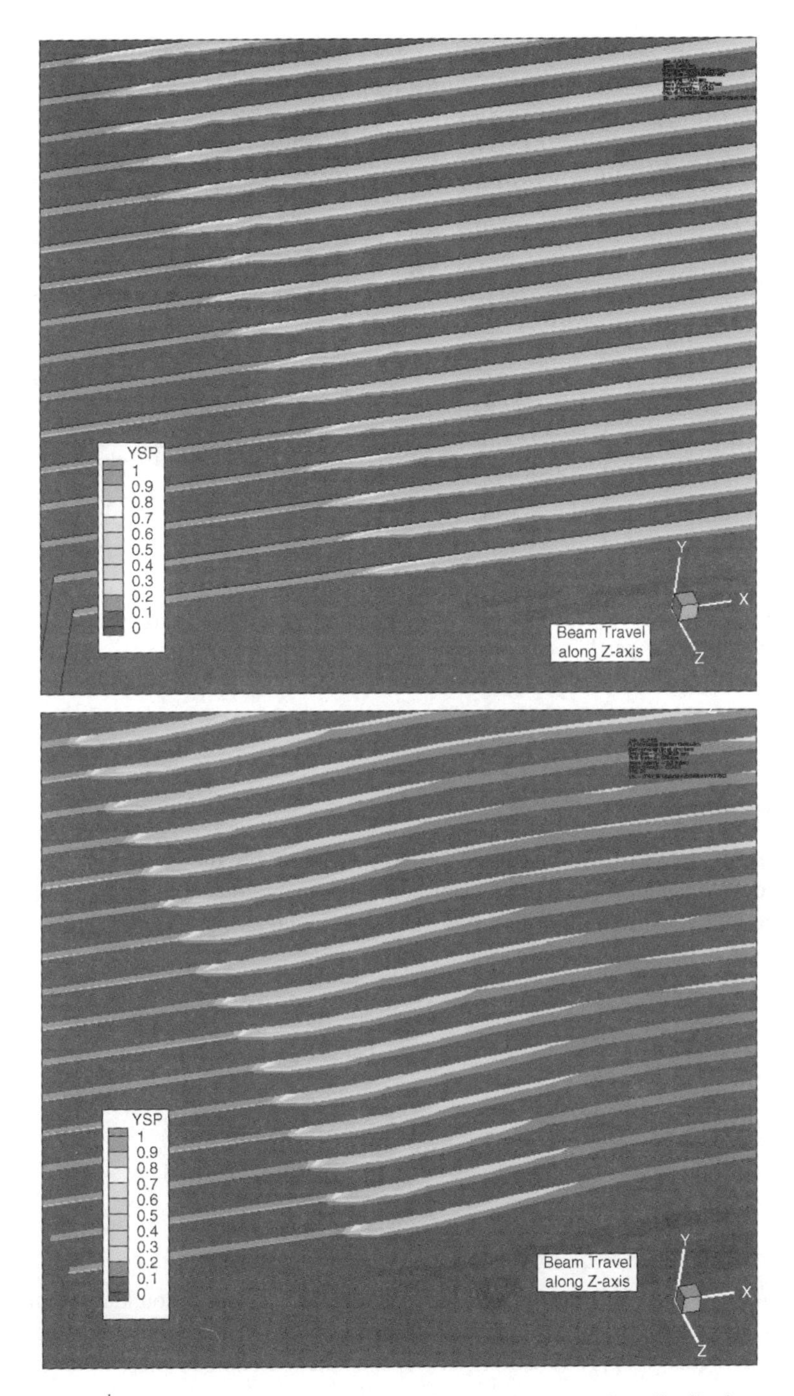

Figure 7.45 $\text{GWS}^h + \theta\text{TS}$ DY algorithm mass fraction DOF perspective distributions on vertical planes, 6 mm off-plane focus flux model, 126 mm standoff, completely solidified melt pool: top, temperature extremum ~2500 °F; bottom, temperature extremum ~6900 °F

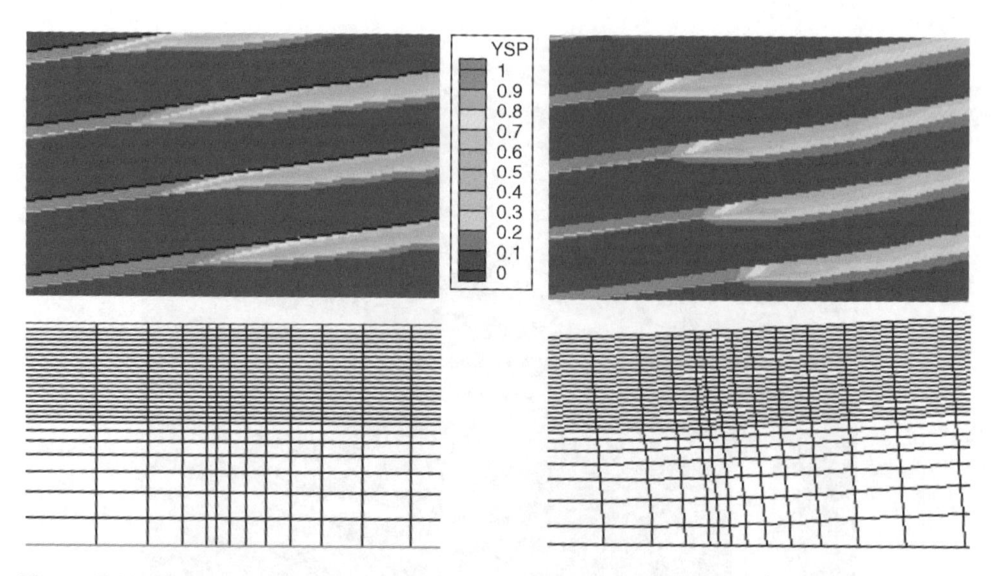

Figure 7.46 Close-in graphics centered at melt pool lateral terminus, $\text{GWS}^h + \theta\text{TS}$ D*Y* algorithm mass fraction DOF perspective on vertical planes, 6 mm off-plane focus flux model, 126 mm stand-off, fully solidified surface: left column, temperature extremum ~2500 °F, snippet of essentially unaltered mesh; right column, temperature extremum ~6900 °F with ALE algorithm displaced mesh snippet

7.15 Summary

Chapter content derives $\text{GWS}^h/m\text{GWS}^h + \theta\text{TS}$ algorithms for the NS classic state variable statement, develops the $k = 1$ TP/NC trial space basis implementation and details validations for the D*M*-constraint algorithm NS PDE system alteration (7.1–7.4)

$$
\begin{aligned}
&\mathrm{D}M: && \nabla \cdot \mathbf{u} = 0 \\
&\mathrm{D}\mathbf{P}: && \frac{\partial \mathbf{u}}{\partial t} + \nabla \cdot \mathbf{u}\mathbf{u} + \frac{1}{\rho_0}\nabla p - \frac{1}{\mathrm{Re}}\nabla \cdot \nabla \mathbf{u} + \frac{\mathrm{Gr}}{\mathrm{Re}^2}\Theta\,\widehat{\mathbf{g}} = 0 \\
&\mathrm{D}E: && \frac{\partial \Theta}{\partial t} + \nabla \cdot \mathbf{u}\Theta - \frac{1}{\mathrm{RePr}}\nabla \cdot \nabla \Theta - s_\Theta = 0 \\
&\mathrm{D}Y: && \frac{\partial Y}{\partial t} + \nabla \cdot \mathbf{u}Y - \frac{1}{\mathrm{ReSc}}\nabla \cdot \nabla Y - s_Y = 0
\end{aligned}
$$

Rearrangement of (7.1–7.4) into a *well-posed* initial value-elliptic boundary value (I-EBV) PDE system is accomplished via *pressure projection* theory, accounting for the constraint of D*M* via the iterative approximation (7.5) to the *kinetic action* of pressure at time t_{n+1}

$$
P^*|_{n+1}^{p} = P_n + \frac{1}{\theta \Delta t}\sum_{\alpha=1}^{p} \delta\phi|_{n+1}^{\alpha}
$$

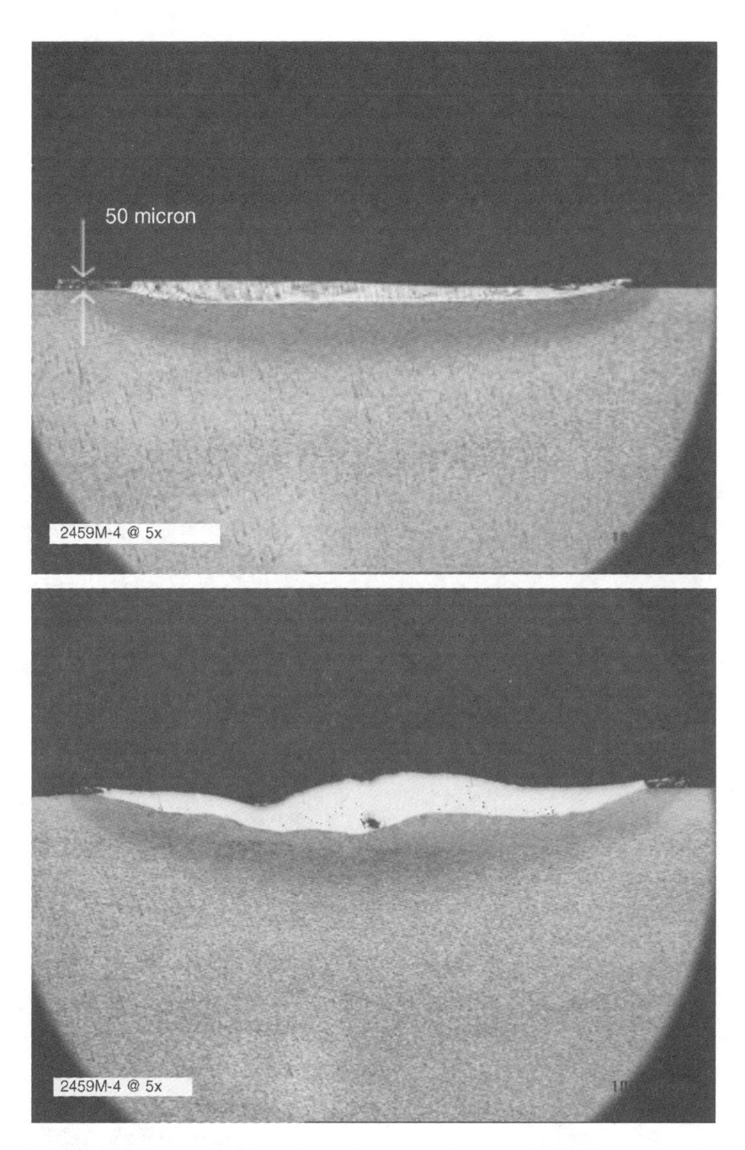

Figure 7.47 LISI process sample slice photomicrograph: top, solidified melt pool planar surface; bottom, solidified melt pool non-planar surface, 50 micron depth denoted

The velocity *potential function* ϕ, at *any* iteration p *and* for *any* approximation $u_i^h(x_i, t_{n+1})$ to a *divergence-free* velocity vector u_i is the solution to the *well-posed* EBV PDE + BCs system (7.6)

$$\mathcal{L}(\phi) = -\nabla^2\phi + \frac{\partial u_i^h}{\partial x_i} = 0$$

$$\ell(\phi) = -\nabla\phi \cdot \hat{\mathbf{n}} - \left(u_i - u_i^h\right)\hat{n}_i = 0$$

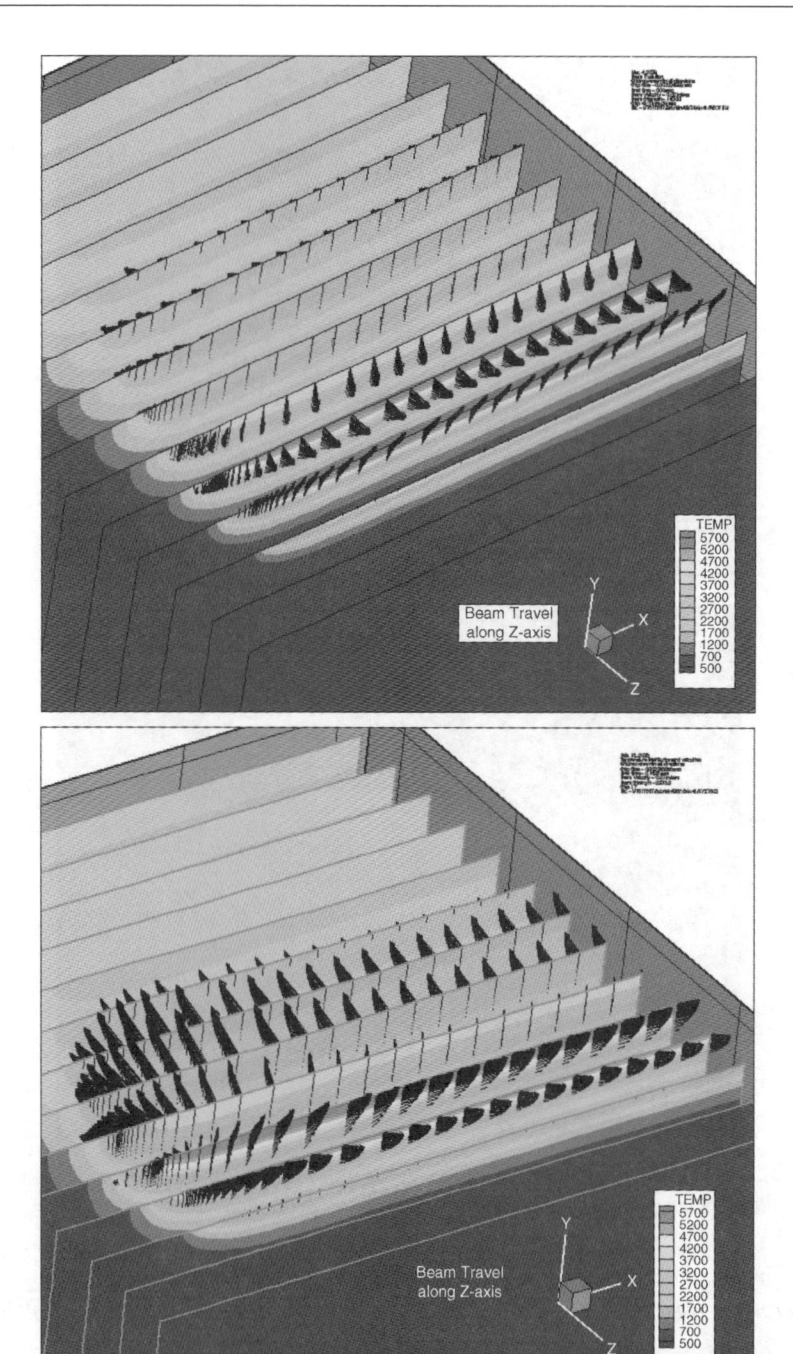

Figure 7.48 $\mathrm{GWS}^h + \theta\mathrm{TS}$ NS algorithm velocity vector DOF perspective distributions on vertical planes, in-focus laser flux model, footprint centered over plane pairs with surface temperature extrema ~6900 °F: top, 135 mm standoff; bottom, 126 mm standoff

The post iterative convergence determination of genuine pressure is via the *well-posed* EBV PDE + BCs system (7.7)

$$\mathcal{L}(P) = -\nabla^2 P - \frac{\partial}{\partial x_j}\left[u_i \frac{\partial u_j}{\partial x_i} + \frac{\mathrm{Gr}}{\mathrm{Re}^2}\Theta \widehat{g}_j\right] = 0$$

$$\ell(P) = \nabla P \cdot \hat{\mathbf{n}} - \left[\frac{\partial \mathbf{u}}{\partial t} - \frac{1}{\mathrm{Re}}\nabla^2 \mathbf{u}\right] \cdot \hat{\mathbf{n}} = 0$$

$m\mathrm{GWS}^h + \theta\mathrm{TS}$ algorithm derivation in classic state variable form alters the I-EBV/EBV PDE system (7.2–7.7) to the mPDE system derived from TS theories detailed in Chapter 5. The resultant *error annihilation* mechanisms for state variable discrete approximation $\{q^h(\mathbf{x}, t)\} = \{u_i^h, \Theta^h, Y^h, \phi^h\}^T$ lead to the asymptotic error estimate (7.8)

$$\left\|\{e^h(n\Delta t)\}\right\|_E \le Ch^{2\gamma}\left[\|\mathrm{data}\|_{L2,\Omega} + \|\mathrm{data}\|_{L2,\partial\Omega}\right] + C_2 \Delta t^{f(\Theta)} \left\|\{q^h(t_0)\}\right\|^2_{H^1(\Omega)}$$
$$\text{for}: \gamma \equiv min(k, (k+1), r-1),\ f(\theta) = (2, 3, 4)$$

for C and C_2 constants. For *smooth data* (7.8) predicts $k = 1$ basis convergence is $O(h^2)$ for $\beta \equiv \Delta t/2$, and improves to $O(h^4)$ for $\beta \equiv h^2\mathrm{Re}/12$ and *sufficiently large* Re. Not explicitly identified in (7.8), the mPDE $k = 1$ basis *optimal* $\gamma \equiv -0.5$ definition improves phase accuracy to $O(m^3)$ for $m \equiv \kappa h$ the non-D wave number.

For I-EBV system (7.2–7.4), upon passage through the Newton iteration algorithm (5.119), generates the NS mPDE system divergence form terminal statement (7.9)

$$\mathcal{L}^m(\{q\}) = [\mathrm{M}]\frac{\partial\{q\}}{\partial t} + \frac{\partial}{\partial x_j}\left[\{f_j\} - \{f_j^d\}\right] - \{s\} = \{0\}$$

with [M] the diagonal mPDE-*augmented* mass matrix, $\{f_j\}$ the NS PDE *kinetic flux vector*, $\{f_j^d\}$ the mPDE-modified *dissipative flux vector* and $\{s\}$ the mPDE-altered source term. The matrix definitions for (7.9) are (7.10–7.13) with $O(h^4)$ mPDE for (7.6–7.7) given in (7.14).

As always $m\mathrm{GWS}^h + \theta\mathrm{TS}$ algorithm derivation starts in the *continuum* with Galerkin criterion *extremization* of the approximation error $\{e^N(\mathbf{x}, t)\} \equiv \{q(\mathbf{x}, t)\} - \{q^N(\mathbf{x}, t)\}$. The ensuing weak statement is (7.16) leading to the matrix iteration solution algorithm (7.17–7.18) coupled with pure EBV weak statement (7.19). The resultant spatially discrete *trial space basis* implementation definition is (7.20)

$$\{q^N(\mathbf{x}, t)\} \equiv \mathrm{diag}\left[\{\Psi(\mathbf{x})\}^T\right]\{Q(t)\} \Rightarrow \{q^h(\mathbf{x}, t)\} \equiv \cup_e\{q(\mathbf{x}, t)\}_e$$
$$\{q(\mathbf{x}, t)\}_e \equiv \mathrm{diag}\left[\{N_k(\boldsymbol{\eta}(\mathbf{x}))\}^T\right]\{Q(t)\}_e$$

which leads to the computable matrix statement (7.21)–(7.22)

$$[\mathrm{JAC}^h\_\{Q\}] \Rightarrow S_e[\mathrm{JAC}\_\{Q\}]_e \equiv S_e\left(\frac{\partial\{\mathrm{F}\_Q\}_e}{\partial\{Q\}_e}\right)$$
$$\{\mathrm{F}\_Q^h\}^p_{n+1} \Rightarrow S_e\{\mathrm{F}\_Q\}^p_{e,n+1}$$
$$\{\mathrm{F}\_QA^h\} \Rightarrow S_e\left([\mathrm{DIFF}]_e\{QA\}_e - \{\mathrm{b}(\{q^h\})\}_e\right)$$

for state variable $\{q^h(\mathbf{x}, t)\} = \{u_i^h, \Theta^h, Y^h, \phi^h; P^h\}^T$

Completion for all matrix contributions to (7.21–7.22) is detailed in Section 7.4. Surprisingly perhaps, *only* six distinct *element independent* matrices constitute the algorithm, [M200] and the five in (7.34) which involve global-local coordinate transformations

$$
\begin{aligned}
[\mathrm{M20}J]_e &= \left(\frac{\partial \eta_k}{\partial x_j}\right)_e [\mathrm{M20}k], [\mathrm{M2}J0]_e = \left(\frac{\partial \eta_k}{\partial x_j}\right)_e [\mathrm{M2}k0] \\
[\mathrm{M300}J]_e &= \left(\frac{\partial \eta_k}{\partial x_j}\right)_e [\mathrm{M300}k] \\
[\mathrm{M2}KK]_e &= \det_e^{-1}\left(\frac{\partial \eta_l}{\partial x_k}\frac{\partial \eta_m}{\partial x_k}\right)_e [\mathrm{M2}lm] \\
[\mathrm{M30}JK]_e &= \det_e^{-1}\left(\frac{\partial \eta_l}{\partial x_j}\frac{\partial \eta_m}{\partial x_k}\right)_e [\mathrm{M30}lm]
\end{aligned}
$$

Accounting for the element-dependent *m*PDE theory multiplier $h^2\mathrm{Re}/12$ alters the scalar multipliers for select $k=1$ TP basis $n=3$ matrices to (7.35)

$$
\begin{aligned}
h^2[\mathrm{M2}KK]_e &= \frac{(2)^{2n/3}}{\det_e^{1/3}}\left(\frac{\partial \eta_l}{\partial x_k}\frac{\partial \eta_m}{\partial x_k}\right)_e [\mathrm{M2}lm] \\
h^2[\mathrm{M300}J]_e &= (2)^{2n/3}\det_e^{2/3}\left(\frac{\partial \eta_k}{\partial x_j}\right)_e [\mathrm{M300}k] \\
\mathrm{Pa}\, h^2[\mathrm{M30}JK]_e &= \frac{(2)^{2n/3}\mathrm{Pa}}{\det_e^{1/3}}\left(\frac{\partial \eta_l}{\partial x_j}\frac{\partial \eta_m}{\partial x_k}\right)_e [\mathrm{M30}lm]
\end{aligned}
$$

with $2^{2n/3}/\det_e{}^{1/3} \Rightarrow 1/(n!)^{2/3}\ \det_c{}^{1/3}$ and $2^{2n/3}\det_e{}^{2/3} \Rightarrow \det_c{}^{2/3}/(n!)^{2/3}$ for the $k=1$ NC basis on $n=3$. Algorithm jacobian derivation, detailed in (7.41–7.44), adds $[\mathrm{M3}J00]_e = \left(\frac{\partial \eta_k}{\partial x_j}\right)_e [\mathrm{M3}k00]$ to (7.34).

The reported range of $n=2$ benchmark problem *a posteriori* data quantitatively *validate* the theory assertion that *equi-distribution* of state variable member energy norms is an accurate predictor of the *optimal* solution adapted mesh for fixed DOF. Performance of the pressure projection algorithm is quantified *robust* in enforcing the measure of $\mathrm{D}M^h \leq O(\varepsilon^2)$ for ε the matrix iteration algorithm convergence requirement. Elongated thermal cavity *a posteriori* data solidly quantify *optimal* γ *m*PDE $O(m^3)$ performance superiority compared to GWS^h.

Data from a variety of $n=3$ benchmark problems complement the $n=2$ results prediction on *linear* theory pertinence. The key assessment is quantified *validation* for the close-coupled step wall diffuser flowfield via detailed comparison with experimental data. For the classic state variable PDE system the (7.50) *m*PDE β term is

$$\frac{\beta\Delta t}{2}\left([A_j][A_k]\frac{\partial\{q\}}{\partial x_k}\right)=\frac{\beta\Delta t}{2}\begin{bmatrix} u_j u_k\dfrac{\partial u_i}{\partial x_k}+u_i u_k\dfrac{\partial u_j}{\partial x_k}, & 0 & , 0 \\ & , u_j u_k\dfrac{\partial\Theta}{\partial x_k} & , 0 \\ & & , u_j u_k\dfrac{\partial Y}{\partial x_k} \end{bmatrix}$$

with phase selective dissipative proven central to algorithm stabilization for all $\mathrm{Re}\leq 800$. Quantitative prediction of primary recirculation region lower surface intercept is excellent, and the upper surface recirculation bubble is predicted to not span the duct cross-section. Excellent *quantitative* agreement with experiment axial velocity transverse plane distributions, with interpretation, is achieved.

The issue of flow simulation in solution domains with multiple efflux surfaces is addressed. Generated *a posteriori* data firmly validate the importance of the genuine pressure Poisson PDE with Robin BCs algorithm. Algorithm prediction in bifurcating flow channels quantifies creation of counter-rotating vortex pairs due to elementary flow turning as observed in experiment.

With solid mathematical validations established, chapter topics turn to assessing weak form theory intrinsic embedding of Robin BCs into the terminal matrix iteration statement. The associated Green–Gauss divergence theorem operation is key to precise algorithm implementation statements (7.52) for convection and radiation heat transfer prediction

$$\int_{\partial\Omega_R\cap\partial\Omega}\Psi_\beta\frac{k}{\kappa}\left[h(T^N-T_{con})+\sigma f(\mathrm{vf},\varepsilon)\left((T^N)^4-T_{rad}{}^4\right)+\mathbf{f}(\mathbf{x},t)\bullet\widehat{\mathbf{n}}\right]\mathrm{d}\sigma$$

$$\Rightarrow(\rho c_p)^{-1}\int_{\partial\Omega_e}\{N\}\begin{pmatrix} h_e\{N\}^T\left(\{Q\}_e-\{QC\}_e\right) \\ +\sigma f(\mathrm{vf},\varepsilon)_e\{N\}^T\left(\{Q\}_e^4-\{QR\}_e^4\right) \\ +\quad \{N\}^T\{\mathbf{FLX}\bullet\widehat{\mathbf{n}}\}_e \end{pmatrix}\mathrm{d}\sigma$$

for $\{QC\}_e$ the convection heat transfer exchange temperature DOF and $\{QR\}_e$ the radiation exchange temperature DOF. For a range of convection heat transfer modes, the literature correlations (7.54) are

$$\begin{aligned}
&\mathit{forced}\ \text{convection}: h=\mathrm{C}_1 k\mathrm{Pr}^{1/3}\mathrm{Re}^{1/2}/x, \mathrm{C}_1=0.664\\
&\mathit{natural}\ \text{convection}: h=\mathrm{C}_2 k(\mathrm{Gr\,Pr})^{\mathrm{n}},\\
&\qquad\qquad 0.15\leq\mathrm{C}_2\leq 0.54, \text{as}\, f(\text{surface}\ \widehat{\mathbf{n}},\Delta T)\\
&\qquad \mathrm{Gr}=\rho\beta\Delta T x^3/\mu^2\\
&\qquad \mathrm{n}=(0.25,0.3)\text{for}\ \Delta T<0,\Delta T>0\\
&\mathit{mixed}\ \text{convection}: \text{smooth transition between the correlations}
\end{aligned}$$

Firm *validation* of the $\mathrm{GWS}^h+\theta\mathrm{TS}$ $k=1$ basis implementation results via excellent *quantitative* agreement between prediction and a range of detailed *unsteady* 24 hour duration experimental data sets.

For radiation heat transfer, the Gebhart factor formulation for viewfactors for multiple grey bodies in radiation exchange is based on *Lambert's cosine law*. The principal detraction to weak form implementation as the algorithm BC is the compute-intensive FOR loop which additionally leads to destruction of the banded character of the algorithm jacobian matrix.

The computationally preferable *radiosity* theory alternative is based on rigorously addressing the *differential form* of Lambert's cosine law. The theory generates the *Fredholm integral equation* of the second kind (7.70)

$$R_i = \varepsilon_i \sigma T_i^4 + \rho_i \left[\sum_{k=1}^{n} \int_{A_k} R_k \frac{\cos\phi_k \cos\phi_i}{\pi r_{k\to i}^2} \mathrm{d}A_k \right]$$

where R_i is radiation energy per unit time per unit surface area. The net energy efflux from a grey body surface A_i replaces the Stefan-Boltzmann definition in terms of surface *radiosity* R_i (7.72)

$$q_i = \frac{\varepsilon_i}{\rho_i}\left[\sigma T_i^4 - R_i\right]$$

A continuum Galerkin weak statement written on (7.70) renders the error orthogonal to the trial space. This *optimal* theory trial space basis implementation is (7.75)

$$\{\mathrm{GWS}\}_{e=i} \equiv \int_{\Omega_{e=i}} \{N\} \left[\begin{array}{c} \left(\{N\}^T\{R\}_i - \varepsilon_i\sigma\{N\}^T\{Q^4\}_i\right)\mathrm{d}A_i \\ -\rho_i\left[\sum_{k=1}^{n}\int_{\Omega_{e=k}} \frac{\cos\phi_k\cos\phi_i}{\pi r_{k\to i}^2}\{N\}^T\{R\}_k \mathrm{d}A_k\right]\mathrm{d}A_i \end{array} \right] \equiv \{0\}$$

which generates an algebraic *hypermatrix* statement and produces *piecewise discontinuous* solutions with asymptotic error estimate and norm definitions (7.82)

$$\begin{aligned} \left\|e^h\right\|_{H^0} &\le C\, h^{k+1}\|\mathrm{data}\|_{L2,\partial\Omega} \\ \left\|R^h\right\|_{H^0} &\equiv \frac{1}{2}\left[\sum_{e=1}^{M}\det{}_e\{R\}_e^T[\mathrm{B200}d]\{\mathrm{R}\}_e\right]^{1/2} \end{aligned}$$

Replacement of Stefan–Boltzmann with (7.72) in the BC eliminates the Gebhart factor detraction of matrix bandwidth destruction at the price of solving the *linear* hypermatrix equation system (7.72). Asymptotic convergence in H^0 is validated via excellent quantitative agreement for the classic facing parallel plate geometry. Isothermal facet TP basis mesh M required to accurately predict the *distribution of radiosity* is quantified.

The chapter final topic details a $\mathrm{GWS}^h + \theta\mathrm{TS}$ algorithm for an intense laser radiation environment seeking quantitative prediction of *thermo-solid-fluid-mass transport* processes. Algorithm design requires implementing an *arbitrary-lagrangian-eulerian* (ALE) mesh movement algorithm to account for fluid thermal-convection generated melt pool free surface displacement. The DM^h pressure projection state variable member BC precisely enables this requirement. Comparison with limited experimental data yields *qualitative validation* of cause–effect characterization for select laser system operational variables.

Exercises

7.2.1 Verify (7.11–7.12) as the proper flux vector representation of (5.2–5.4) upon extraction of all TS theory-generated *m*PDE differential calculus terms in [M].

7.2.2 Proceed through the $(\pm)(\theta)\{\text{these } m\text{PDE terms}\}_n$ operations in $\{\text{RES}\_(\bullet)\}_n$ in (5.119), Section 5.12, hence confirm the validity of (7.10) and (7.13).

7.4.1 Validate the *m*PDE-pressure projection theory unsteady term matrix $[\text{M}]_e$, (7.25).

7.4.2 In (7.33) verify imposing the outflow plane BC Green–Gauss generated term $[\text{A}20S]_e\{\text{U}J\bullet\widehat{n}_j\}_e\big|_\text{w}$ amounts to eliminating the contribution from $[\text{B}2KK]_e\{\text{U}J\bullet\widehat{n}_j\}_e$ at *each* wall node.

7.4.3 Validate one or more of the matrix definitions in (7.34).

7.4.4 Verify that the element matrix multiplier is $h^2/\text{det}_e = 2^{2n/3}/\text{det}_e^{\ 1/3}$ for the $k=1$ TP basis for $n=3$.

7.4.5 Verify the element matrix multiplier $h^2/\text{det}_e = 1/(n!)^{2/3}\ \text{det}_c^{\ 1/3}$ for the $k=1$ NC basis for $n=3$.

7.5.1 Verify the accuracy of the pressure projection theory terms in (7.38) for $1\leq I\leq 3$.

7.5.2 Verify the self-coupling jacobian matrix block (7.41).

7.5.3 Re-express (7.44) by replacing element-dependent matrices and h^2 multipliers with the matrix library, (7.34–7.35).

7.7.1 Verify the NS PDE system jacobian (7.48).

7.7.2 Proceed through the details in forming (7.49), hence validate the jacobian matrix tensor product (7.50).

7.9.1 Confirm insertion of viewfactor into the radiation BC (7.59).

7.9.2 Confirm the Gebhart factor BC element statement (7.62) and the companion element jacobian (7.63).

7.11.1 Confirm the radiosity formulation derivation (7.70).

7.11.2 Proceed through the continuum weak statement details leading to the discrete GWS^h statement (7.73–7.74).

7.11.3 Confirm the element matrix statements (7.76). For the bilinear TP basis, confirm that all elements of the square matrix $\{\text{B}10\}\{\text{B}10\}^T=[\text{B}1010]$ are unity.

7.11.4 Validate the radiosity algorithm assertion that the solution to (7.80) indeed generates multiple nodal DOF at the shared nodes of the discretization.

7.11.5 Verify the radiosity GWS^h algorithm solution restatement for viewfactor, (7.81).

References

Argyris, J. and Szimmat, J. (1992). "An analysis of temperature radiation interchange problems," *J. Computer Methods Applied Mechanics and Engineering*, V. 94.

Armaly, B.F., Durst, F., Pereira, J.C.F. and Schonung, B. (1983). "Experimental and theoretical investigation of backward-facing step flow," *J. Fluid Mechanics*, V. 127.

Bird, R.B., Stewart, W.E. and Lightfoot, E.N. (1960). *Transport Phenomena*, John Wiley, New York.

Christon, M., Gresho, P.M. and Sutton, S.B. (2002). "Computational predictability of time-dependent natural convection flows in enclosures," *J. Numerical Methods Fluids*, V. 40.

de Vahl Davis, G. (1983). "Natural convection of air in a square cavity; a benchmark numerical solution," *J. Numerical Methods Fluids*, V. 3.

Ericson, S.C./Knoxville (2001). "A CFD Laboratory archive supporting the academic process," MSc thesis, University of Tennessee.

Freeman, E.L., (2012). "Validation of Galerkin weak form algorithms for convection-radiation energy exchange prediction," MSc thesis, University of Tennessee.

Freeman, E.L., Baker, A.J., Orzechowski, J.A., Sanders, J.S. and Wong, K.L. (2014). "Validation of a time-accurate weak form algorithm for convection - grey body radiation dominated unsteady heat transfer," *J. Heat and Mass Transfer*, V. XX, in press.

Gould, N. (2000). "Radiation heat transfer between diffuse-grey surfaces using higher order finite elements." Technical Report AIAA:2000-2371.

Hirt, C.W., Amsden, A.A. and Cook, J.L. (1974). "An Arbitrary lagrangian-eulerian Computing Method for All Flow Speeds," *J. Computational Physics*, V. 14 pp. 227–253.

Incropera, F.P. and DeWitt, D.P. (1985). *Fundamentals of Heat and Mass Transfer*, Wiley, New Jersey.

Kipp and Zonen, Inc., (2004). CNR 1 Net Radiometer, Instruction Manual 0340-300, New York USA

Ku, H.C., Hirsh, R.S., Taylor, T.D. and Rosenburg, A.P. (1989). "A pseudo-spectral matrix element method for solution of the three-dimensional incompressible flows and its parallel implementation," *J. Computational Physics*, V. 83.

Le Quere, P. (1991). "Accurate solutions to the square thermally driven cavity at high Rayleigh number," *Computers & Fluids*, V. 21.

Le Quere, P. (1994). "Onset of unsteadiness, routes to chaos and simulations of chaotic flows in cavities heated from the side: a review of present status," *Proc. 10th Int. Heat Transfer Conference*, Brighton, UK, pp. 281–296.

Le Quere, P. and Behnia, M. (1998). "From onset of unsteadiness to chaos in a differentially heated square cavity," *J. Fluid Mechanics*, V. 359.

Mallinson, G.D. and deVahl Davis, G. (1977). "Three dimensional natural convection in a box, a numerical study," *J. Fluid Mechanics*, V. 83, Pt.1.

Miller, F.P., Vandome, A.F. and Brewster, J.M. (2010). *Gebhart Factors*, VDM Publishing House Ltd, England.

Mitra, A.J., (2000). "On performance of a pressure projection algorithm for incompressible Navier–Stokes," MSc thesis, University of Tennessee.

Neymark, J., Boardman, C.R. and Kirkpatrick, A. (1989). "High Rayleigh number natural convection in partially divided air and water filled enclosures," *J. Heat Mass Transfer*, V. 32.

Noronha, W.P. (1989). "Accuracy, convergence and stability of finite element algorithms for incompressible fluid flow," PhD dissertation, University of Tennessee.

Sanders, J.S. and Rinald, D.J. (2010). "Predictive Infrared Signature Model Validation Results for Multiple Threat Ground Systems," *Proceedings, Parallel Meeting of the Military Sensing Symposium, Battlefield Survivability and Discrimination.*

Sahu, S. and Baker, A.J. (2007). "A modified conservation principles theory leading to an optimal Galerkin CFD algorithm," *J. Numerical Methods Fluids*, V. 55.

Siegel, R. and Howell, J.R. (2001). *Thermal Radiation Heat Transfer*, 4th Ed., Taylor and Francis, New York.

Williams, P.T., (1993). "A 3-dimensional time-accurate incompressible Navier–Stokes finite element CFD algorithm," PhD dissertation, University of Tennessee.

Williams, P.T. and Baker, A.J. (1994). "CFD characterization of 3-D natural convection in a two-cell enclosure with a door," *ASHRAE Transactions*, V. 100, Pt.2.

Williams, P.T. and Baker, A.J. (1996). "Incompressible computational fluid dynamics and the continuity constraint method for the 3-D Navier–Stokes equations," *Numerical Heat Transfer, Part B, Fundamentals*, V. 29.

Williams, P.T. and Baker, A.J. (1997). "Numerical simulations of laminar flow over a 3D backward-facing step," *J. Numerical Methods Fluids*, V. 24.

Wong, K.L., Ericson, S.C. and Baker, A.J. (2004). "Flow patterns in a multi-generation bifurcation bronchial tree," poster at SIAM Conference, Parallel Processing for Scientific Computing

8

Time Averaged Navier–Stokes: mGWS^h + θTS algorithm for RaNS, Reynolds stress tensor closure models

8.1 Classic State Variable RaNS PDE System

Text content to this juncture addresses in thoroughness *claimed* optimal continuous GWS/mGWS + θTS CFD algorithm performance for laminar-thermal Navier–Stokes (NS) PDE/mPDE + BCs systems. Comparison of precisely generated *a posteriori* DOF data with *linear* weak form theory prediction of approximation error, asymptotic convergence and optimal mesh solutions *fully verify* theory pertinence to NS *nonlinear* discrete approximate solution characterization. Algorithm prediction comparisons with quality experimental data *validate* solution accuracy, and in the process fully quantify classic flow specification fluid dynamics. Direct DOF-matched comparisons with data generated by prominent alternative CFD theories quantitatively *verify* the optimality claim.

Extension of optimal modified continuous mGWS^h + θTS CFD algorithm to the *time averaged* incompressible NS mPDE + BCs system for turbulent flow simulation is this chapter topic. *Reynolds-averaged Navier–Stokes* (RaNS) is the name attached to this manipulation resultant. Time averaging, as detailed in Chapter 4, injects the *a priori* unknown *Reynolds stress tensor* $\overline{u_i'u_j'}$ into D**P**, along with unknown *Reynolds vectors* $\overline{u_i'q'}$ into RaNS state variable scalar member PDEs.

Deleting overbar/prime notation except where required for clarity, the textbook-appearance of *non-dimensional* RaNS PDE system replacing (7.1–7.4) in visually informative vector-tensor non-rigorous notation is

$$\mathrm{D}M: \quad \nabla \cdot \mathbf{u} = 0 \tag{8.1}$$

$$\mathrm{D}\mathbf{P}: \quad \frac{\partial \mathbf{u}}{\partial t} + \nabla \cdot \left(\mathbf{u}\mathbf{u} + \overline{u_i'u_j'} \right) + \frac{1}{\rho_0}\nabla p - \frac{1}{\mathrm{Re}}\nabla^2 \mathbf{u} + \frac{\mathrm{Gr}}{\mathrm{Re}^2}\Theta\widehat{\mathbf{g}} = 0 \tag{8.2}$$

Optimal MODIFIED CONTINUOUS Galerkin CFD, First Edition. A. J. Baker.

Companion Website: www.wiley.com/go/baker/GalerkinCFD

$$\mathrm{D}E: \quad \frac{\partial \Theta}{\partial t} + \nabla \cdot \left(\mathbf{u}\Theta + \overline{u_j'\Theta'} \right) - \frac{1}{\mathrm{RePr}} \nabla^2 \Theta - s_\Theta = 0 \tag{8.3}$$

$$\mathrm{D}Y: \quad \frac{\partial Y}{\partial t} + \nabla \cdot \left(\mathbf{u}Y + \overline{u_j'Y'} \right) - \frac{1}{\mathrm{ReSc}} \nabla^2 Y - s_Y = 0 \tag{8.4}$$

The state variable $\{q(\mathbf{x}, t)\} \equiv \{\mathbf{u}, P, \Theta, Y\}^T$ for (8.1–8.4) involves *time averaged* velocity, kinematic pressure, temperature and mass fraction. The Reynolds, Grashoff, Prandtl, Schmidt and Eckert number definitions remain $\mathrm{Re} \equiv \mathrm{UL}/\nu$, $\mathrm{Gr} \equiv \mathrm{g}\beta\Delta\mathrm{TL}^3/\nu^2$, $\mathrm{Pr} \equiv \rho_0 \nu c_p/k$, $\mathrm{Sc} = \mathrm{D}/\nu$ and $\mathrm{Ec} \equiv \mathrm{U}^2/c_p\Delta\mathrm{T}$. In the Eckert number and Boussinesq buoyancy model $\Delta\mathrm{T} \equiv (T_{\max} - T_{\min})$ with $\beta \equiv 1/\mathrm{T}_{\mathrm{abs}}$, and $P \equiv p/\rho_0$ is *kinematic* pressure for density ρ_0 otherwise constant. For non-periodic flows $\mathrm{St} \equiv \tau\mathrm{U/L} \equiv 1$ is the usual definition, hence $\tau \equiv \mathrm{L/U}$, and in DE the Peclet number $\mathrm{Pe} = \mathrm{RePr}$ commonly appears.

Time averaging does not alter the mathematical character of (8.2–8.4) as a *constrained* initial-elliptic boundary value (I-EBV) PDE system, with (8.1) requiring *admissable* RaNS solutions $\{q(\mathbf{x}, t)\}$ to possess a *divergence-free* (time averaged) velocity vector. As with the NS algorithm pressure projection theory, the *kinetic action* of pressure at time t_{n+1} may be approximated as

$$P^*\big|_{n+1}^p = P_n + \frac{1}{\theta\Delta t} \sum_{\alpha=1}^{p} \delta\phi\big|_{n+1}^{\alpha} \tag{8.5}$$

for P_n (time averaged) kinematic pressure at time t_n, θ integration implicitness factor and Δt the time step size.

For velocity *potential* function ϕ the EBV PDE + BCs statement at *any* iteration p and for *any* velocity vector approximation $u_i^h(x_i, t_{n+1})$ to u_i remains

$$\begin{aligned} \mathcal{L}(\phi) &= -\nabla^2\phi + \frac{\partial u_i^h}{\partial x_i} = 0 \\ \ell(\phi) &= -\nabla\phi \cdot \widehat{\mathbf{n}} - \left(u_i - u_i^h\right)\widehat{n}_i = 0 \end{aligned} \tag{8.6}$$

In (8.5), $\delta\phi^\alpha$ is the sequence of potential solution iterates generated via (8.6) during CCM algorithm convergent at time t^{n+1}.

As the typical inflow time averaged velocity vector BC is Dirichlet, the ϕ BC (8.6) thereon is homogeneous Neumann, the BC definition for all no-slip walls as well. On all outflow boundary segments the DM-satisfying velocity vector is unknown, hence also its difference with *any* velocity approximation. Therefore, the sole admissible ϕ outflow BC must be Dirichlet.

As with NS practice the measure of *error* in DM, that is, DM^h, is the ϕ *energy norm* $\|\nabla^h \cdot \mathbf{u}^h\| \Rightarrow \|\phi^h\|_E$ with theory requirement being *minimization*. Hence for all nodes on outflow boundaries the Dirichlet BC is $\phi = 0$. Finally, as in Chapter 7 determination of pressure $P(x_i, t_{n+1})$ accrues to the RaNS algorithm convergence *post-process* solution of the *well-posed* EBV PDE + BCs system

$$\begin{aligned} \mathcal{L}(P) &= -\nabla^2 P - \frac{\partial}{\partial x_j}\left[u_i \frac{\partial u_j}{\partial x_i} + \frac{\partial \overline{u_i'u_j'}}{\partial x_i} + \frac{\mathrm{Gr}}{\mathrm{Re}^2}\Theta\widehat{g}_j \right] = 0 \\ \ell(P) &= \nabla P \cdot \widehat{\mathbf{n}} - \left[\frac{\partial \mathbf{u}}{\partial t} - \frac{1}{\mathrm{Re}}\nabla^2\mathbf{u} \right] \cdot \widehat{\mathbf{n}} = 0 \end{aligned} \tag{8.7}$$

In conclusion, the RaNS mixed I-EBV/EBV PDE + BCs system (8.2–8.7) is well-posed for theorizing $m\text{GWS}^h + \theta\text{TS}$ weak form algorithms to generate discrete approximate solutions to the classic state variable time averaged NS conservation principle mPDE system.

8.2 RaNS PDE System Turbulence Closure

A wide range of theoretical frameworks exist generating closure models for $\overline{u_i'u_j'}$ in (8.2). The pervasive basis is variations on mixing length theory leading to $D(\text{L}^2\tau^{-1})$ *eddy viscosity* hypotheses. Algebraic mixing length essentials are detailed in Section 4.7, with transport generalizations leading to BL k-ε and k-ω PDE + BCs systems closure models. Extension to a full Reynolds stress tensor closure algebraic model for 3-D PNS is detailed in Section 4.10.

The historical n-D extension of algebraic mixing length closure to RaNS is Baldwin–Lomax (1978). Baldwin–Barth (1990) developed a single PDE closure model for eddy viscosity; the widely used eddy viscosity transport PDE successor is Spalart–Allmaras (1992). The union of k-ε and/or k-ω dual-PDE n-D extensions with Spalart–Allmaras leads to *detached eddy simulation* (DES) formulations, Spalart (2009). Closure models for the full Reynolds stress tensor exist based on a Reynolds stress tensor transport PDE system augmented with the isotropic dissipation function (ε) PDE, Launder *et al.* (1975).

Chapter focus is to assess $m\text{GWS}^h + \theta\text{TS}$ theory consequences for the RaNS mPDE system rather than performance characterization of the incredible range of closure options/models detailed in the CFD literature. The assumption pervading RaNS closure models is the existence of a *turbulent eddy viscosity* ν^t multiplying the Stokes strain rate tensor written on time averaged velocity

$$\overline{u_i'u_j'} \equiv \tau_{ij} \propto -\nu^t\overline{S}_{ij}, \quad \overline{S}_{ij} \equiv \frac{1}{2}\left(\frac{\partial \overline{u}_i}{\partial x_j} + \frac{\partial \overline{u}_j}{\partial x_i}\right) \tag{8.8}$$

Realizing the trace of τ_{ij} defines turbulent kinetic energy k to within a constant, and extracting the trace of $\overline{S}_{ij}$ via a *turbulent* Lamé parameter λ^t, (8.8) transitions to the equality

$$\tau_{ij} \equiv \frac{2}{3}k\delta_{ij} - \nu^t\left[\frac{\partial \overline{u}_i}{\partial x_j} + \frac{\partial \overline{u}_j}{\partial x_i}\right] - \frac{\lambda^t}{3}\frac{\partial \overline{u}_k}{\partial x_k}\delta_{ij} \tag{8.9}$$

The universal modeling assumption is that k in (8.9) ends up augmenting pressure P. Then since time averaged velocity is divergence-free, the RaNS closure modeling requirement simplifies to the *deviatoric tensor*

$$\tau_{ij}^{\text{D}} \equiv -\nu^t\left[\frac{\partial \overline{u}_i}{\partial x_j} + \frac{\partial \overline{u}_j}{\partial x_i}\right], i \neq j \tag{8.10}$$

Modeling rationale essence and specifics pertinent to a turbulent BL are discussed starting in Section 4.7. Eddy viscosity non-dimensionalized by kinematic viscosity defines the *turbulent Reynolds number* Re^t, recall (1.26)

$$\mathrm{Re}^t \equiv \nu^t/\nu \tag{8.11}$$

The resultant non-dimensional form of (8.8), an exercise, is

$$\overline{u_i'u_j'} \Rightarrow \tau_{ij}^{\mathrm{D}} \equiv -\frac{\mathrm{Re}^t}{\mathrm{Re}}\left[\frac{\partial \bar{u}_i}{\partial x_j}+\frac{\partial \bar{u}_j}{\partial x_i}\right], i \neq j \tag{8.12}$$

Replacing $\nabla^2\mathbf{u}$ in (8.2) with the Stokes tensor in (8.8) symbolized as dyadic $\overline{\mathbf{S}}$, and inserting (8.12) and leads to non-D RaNS D**P**

$$\mathrm{D}\mathbf{P}\!: \frac{\partial \bar{\mathbf{u}}}{\partial t}+\nabla\cdot\bar{\mathbf{u}}\bar{\mathbf{u}}+\nabla\bar{P}-\frac{2}{\mathrm{Re}}\nabla\cdot\left[(1+\mathrm{Re}^t)\overline{\mathbf{S}}\right]+\frac{\mathrm{Gr}}{\mathrm{Re}^2}\overline{\Theta}=0 \tag{8.13}$$

As Re^t easily reaches O(E+03), recall Chapter 4, transparent in (8.13) is eddy viscosity closure model significant augmentation of the *diffusion* level in RaNS D**P** compared to NS, the "1" in $(1+\mathrm{Re}^t)$. The computational benefit to RaNS algorithm performance is substantial diffusive moderation of the *dispersion error* mode intrinsic to discrete CFD algorithms addressing NS at large Re.

The eddy viscosity closure hypothesis is extended to RaNS state variable scalar members by introducing "*turbulent*" Prandtl and Schmidt numbers. An exercise suggests verifying the generated non-D PDE replacements for (8.3 and (8.4)) are

$$\mathrm{D}E\!: \quad \frac{\partial \overline{\Theta}}{\partial t}+\nabla\cdot\bar{\mathbf{u}}\overline{\Theta}-\frac{1}{\mathrm{Re}}\nabla\cdot\left[\left(\frac{1}{\mathrm{Pr}}+\frac{\mathrm{Re}^t}{\mathrm{Pr}^t}\right)\nabla\overline{\Theta}\right]-\bar{s}_\Theta=0 \tag{8.14}$$

$$\mathrm{D}Y\!: \quad \frac{\partial \overline{Y}}{\partial t}+\nabla\cdot\bar{\mathbf{u}}\overline{Y}-\frac{1}{\mathrm{Re}}\nabla\cdot\left[\left(\frac{1}{\mathrm{Sc}}+\frac{\mathrm{Re}^t}{\mathrm{Sc}^t}\right)\nabla\overline{Y}\right]-\bar{s}_Y=0 \tag{8.15}$$

and most RaNS practices assume $\mathrm{Pr}^t \approx \mathrm{Pr}$ and $\mathrm{Sc}^t \approx \mathrm{Sc}$.

The kinematic eddy viscosity definitions for BL dual-PDE closure models, specifically k-ε and k-ω, with k turbulence kinetic energy, ε isotropic dissipation function and $\omega \equiv \varepsilon/k$, are detailed in (4.61–4.63). The k-ε closure model serves chapter requirements. The sought PDE system is the unsteady generalization of (4.73–4.74); hence for $1 \le (i,\ j,\ k) \le n$ and including the low Re^t corrections, (4.76), the chapter subject k-ε closure *non-dimensional* PDE system is

$$\begin{aligned}\mathrm{D}k\!: \frac{\partial k}{\partial t}+\nabla\cdot\bar{\mathbf{u}}k-\frac{1}{\mathrm{Pe}}\nabla\cdot\left[\left(1+\frac{\mathrm{Re}^t}{\mathrm{C}_k}\right)\nabla k\right]-\overline{u_i'u_k'}\frac{\partial \bar{u}_i}{\partial x_k}\\ +\varepsilon+\frac{\mathrm{Gr}}{\mathrm{Re}^2}\frac{\varepsilon}{k}\frac{\partial \Theta}{\partial x_i}\delta_{\mathbf{g}\,i}=0\end{aligned} \tag{8.16}$$

$$\begin{aligned}\mathrm{D}\varepsilon:\ &\frac{\partial \varepsilon}{\partial t}+\nabla\cdot\bar{\mathbf{u}}\varepsilon-\frac{1}{\mathrm{Pe}}\nabla\cdot\left[\left(1+\frac{\mathrm{Re}^t}{\mathrm{C}_\varepsilon}\right)\nabla\varepsilon\right]-\mathrm{C}_\varepsilon^1 f^1\overline{u_i'u_k'}\frac{\varepsilon}{k}\frac{\partial u_i}{\partial x_k}\\ &+\mathrm{C}_\varepsilon^2 f^2\frac{\varepsilon^2}{k}+\mathrm{C}_\varepsilon^3\frac{\mathrm{GrRe}^t}{\mathrm{Re}^3}\frac{\varepsilon}{k}\frac{\partial\Theta}{\partial x_i}\delta_{\mathrm{g}\,i}=0\end{aligned}\tag{8.17}$$

Following Wilcox (2006), the NS viscous "1" has been inserted into (8.17), and (8.12) is the definition for Re^t in (8.16–8.17). The Lam-Bremhorst (1981) low Re^t model definitions are

$$\begin{aligned}\mathrm{Re}^t &= f_\nu\frac{\mathrm{C}_\nu k^2}{\varepsilon}\\ f_\nu &= \left(1-\exp(-0.0165\mathrm{R}_y)\right)^2(1+20.5/\mathrm{Re}^t)\\ f^1 &= 1+(0.05/f_\nu)^3,\quad \mathrm{R}_y=k^{1/2}y/\nu\\ f^2 &= 1-\exp(-\mathrm{Re}^t)^2\end{aligned}\tag{8.18}$$

wherein y is the coordinate spanning a no-slip surface normal. The model constants are $\mathrm{C}_k=1.0$, $\mathrm{C}_\varepsilon=1.3$, $\mathrm{C}_\varepsilon^1=1.44$, $\mathrm{C}_\varepsilon^2=1.92$, $\mathrm{C}_\varepsilon^3=1.0$ and $\mathrm{C}_\nu=0.09$, and in (8.16–8.17) $\delta_{\mathrm{g}\,i}$ aligns the gravity direction.

The no-slip wall BCs are $k=0$ and $\varepsilon=\nu\partial^2k/\partial y^2$; Lam–Bremhorst recommend replacing the latter with $\partial\varepsilon/\partial y=0$, the practice in Chapter 4 and herein. The option to not resolve the low Re^t wall layers remains via turbulent BL *similarity*, recall Figure 4.8. Assuming near wall balance of k production with dissipation, the similarity solution (4.80) enables k-ε model state variable off wall-local DOF determinations

$$\nu^t=\kappa y u_\tau/\mathrm{C}_\nu, k=u_\tau^2/\mathrm{C}_\nu^{1/2}, \tau_w=u_\tau^2/\mathrm{C}_\nu, \varepsilon=\left|u_\tau^3\right|/\kappa y\tag{8.19}$$

Retention of the time derivatives in the eddy viscosity k-ε closed non-D RaNS PDE system (8.13–8.20) admits unsteady solution generation. Such solutions are valid *only* for the timescale of flowfield unsteadiness *much larger* than the span T of time averaging, (4.52).

8.3 RaNS State Variable *m*PDE System

The NS *m*PDE theory is basically unaltered in transition to RaNS. For the dissipative flux vector *m*PDE coefficient $\beta\Delta t/2$, with β a to-be-determined constant of $O(10^{-1})$, $k=1$ basis implementation asymptotic convergence remains $O(h^2)$ for *smooth data*. For the NS *optimal* $O(h^4)$ *m*GWS^h theory, altered to theoretically handle distributed $\varepsilon(q, x)$ in (5.85), the RaNS D**P** *m*PDE coefficient is $h^2\mathrm{Re}/12(1+\mathrm{Re}^t)$. There exists no arbitrary coefficient and the theory constraint of *sufficiently large* Re is certainly academic.

Either *m*PDE theory coefficient is the multiplier on the time averaged velocity tensor product $u_j\, u_k$, with the product defining a *non-positive definite* "dissipation coefficient" in the RaNS *m*PDE system, (5.117). Selecting $h^2\mathrm{Re}/12(1+\mathrm{Re}^t)$ is preferable to $\beta\Delta t/2$, as induced action will be greatly *moderated* in regions of intense turbulence. Thus are identified the *key* sought algorithm performance assessments of accuracy, monotonicity and stabilization via the *m*PDE *theory* dissipative flux vector augmentation, the weak form theory alternative to addition of an ad hoc numerical diffusion operator.

Since $\mathrm{Re}^t(\mathbf{x},\ t)$ is a solution-dependent *distribution*, the $m\mathrm{GWS}^h$ implementation for RaNS *is not* a simple algebraic coefficient modification. The $m\mathrm{GWS}^h + \theta\mathrm{TS}$ algorithm statement will involve the hypermatrix multiplier DOF array $\{1/(1+\mathrm{RET}(t))\}_e$. Recall *m*PDE theory operands address significant order discrete approximation error mechanisms, Figures 5.7–5.8. Additionally, for an extremum Re NS simulation the *m*PDE dissipative operand is verified to phase selectively *annihilate* $O(2\Delta x)$ dispersion error, Figure 6.3. The Table 6.1 data confirm optimal $m\mathrm{GWS}^h + \theta\mathrm{TS}$ NS performance upon replacement of β with $h^2\mathrm{Re}/12$.

The NS-derived *optimal* $m\mathrm{GWS}^h + \theta\mathrm{TS}$ theory is selected for RaNS. To keep development tractable the *m*PDE system omits the NS algorithm $O(h^4)$ Pa-multiplied terms in (5.113). These terms are included and detailed in the chapter summary. For RaNS *continuum* state variable $\{q(\mathbf{x},\ t)\} = \{u_i,\ \Theta,\ Y,\ k,\ \varepsilon,\ \phi,\ \tau^{\mathrm{D}}_{ij};\ P\}$ the *m*PDE flux vector statement is

$$\mathcal{L}^m(\{q\}) = [m]\frac{\partial\{q\}}{\partial t} + \frac{\partial}{\partial x_j}\left[\{f_j\} - \{f_j^d\}\right] - \{s\} = \{0\} \tag{8.20}$$

with $[m]$ the γ *m*PDE (only) altered *mass* matrix, $\{f_j\}$ the unaltered *kinetic flux vector*, $\{f_j^d\}$ the $h^2\mathrm{Re}/12(1+\mathrm{Re}^t)$ *m*PDE augmented *dissipative flux vecto*r and $\{s\}$ the source absent *m*PDE augmentation (an exercise). For $1 \le (i,\ j,\ k) \le n$ the non-D functional forms are

$$[m] \equiv \left[1 - \frac{\gamma\,\Delta t^2}{6}\frac{\partial}{\partial x_j}\left(u_j u_k \frac{\partial}{\partial x_k}\right)\right] \tag{8.21}$$

$$\{f_j(\{q\})\} = \begin{Bmatrix} u_j u_i + P^*|^p_{n+1}\delta_{ij} \\ u_j\Theta \\ u_j Y \\ u_j\,k \\ u_j\,\varepsilon \end{Bmatrix} \tag{8.22}$$

$$\left\{f_j^d(\{q\})\right\} = \begin{Bmatrix} \dfrac{1}{\mathrm{Re}}(1+\mathrm{Re}^t)\left(\dfrac{\partial u_i}{\partial x_j} + \dfrac{\partial u_j}{\partial x_i}\right) + \dfrac{h^2\mathrm{Re}}{6(1+\mathrm{Re}^t)}\left(u_j u_k \dfrac{\partial u_i}{\partial x_k} + u_i u_k \dfrac{\partial u_j}{\partial x_k}\right) \\ \dfrac{1}{\mathrm{Re}}\left(\dfrac{1}{\mathrm{Pr}} + \dfrac{\mathrm{Re}^t}{\mathrm{Pr}^t}\right)\dfrac{\partial\Theta}{\partial x_j} + \dfrac{h^2\mathrm{RePr}}{12(1+\mathrm{Re}^t)(\mathrm{Pr}/\mathrm{Pr}^t)}u_j\,u_k\dfrac{\partial\Theta}{\partial x_k} \\ \dfrac{1}{\mathrm{Re}}\left(\dfrac{1}{\mathrm{Sc}} + \dfrac{\mathrm{Re}^t}{\mathrm{Sc}^t}\right)\dfrac{\partial Y}{\partial x_j} + \dfrac{h^2\mathrm{ReSc}}{12(1+\mathrm{Re}^t)/(\mathrm{Sc}/\mathrm{Sc}^t)}u_j\,u_k\dfrac{\partial Y}{\partial x_k} \\ \dfrac{1}{\mathrm{Pe}}\left(1 + \dfrac{\mathrm{Re}^t}{\mathrm{C}_k}\right)\dfrac{\partial k}{\partial x_j} + \dfrac{h^2\mathrm{Pe}}{12(1+\mathrm{Re}^t/\mathrm{C}_k)}u_j\,u_k\dfrac{\partial k}{\partial x_k} \\ \dfrac{1}{\mathrm{Pe}}\left(1 + \dfrac{\mathrm{Re}^t}{\mathrm{C}_\varepsilon}\right)\dfrac{\partial\varepsilon}{\partial x_j} + \dfrac{h^2\mathrm{Pe}}{12(1+\mathrm{Re}^t/\mathrm{C}_\varepsilon)}u_j\,u_k\dfrac{\partial\varepsilon}{\partial x_k} \end{Bmatrix} \tag{8.23}$$

$$
\{s\} = \left\{ \begin{array}{l} -\dfrac{\mathrm{Gr}}{\mathrm{Re}^2}\Theta \widehat{g}_i \\ \dfrac{\mathrm{Ec}}{\mathrm{Re}}\dfrac{\partial}{\partial x_j}\left(u_i\dfrac{\partial u_j}{\partial x_i}\right) - \dfrac{\partial^2 \tau_{ij}^{\mathrm{D}}}{\partial x_i \partial x_j} \\ s(Y) \\ \tau_{jk}^{\mathrm{D}}\dfrac{\partial u_j}{\partial x_k} - \varepsilon + \dfrac{\mathrm{Gr}}{\mathrm{Re}^2}\dfrac{\varepsilon}{k}\dfrac{\partial \Theta}{\partial x_i}\delta_{gi} \\ \mathrm{C}_\varepsilon^1 f^1 \tau_{jk}^{\mathrm{D}}\dfrac{\varepsilon}{k}\dfrac{\partial u_j}{\partial x_k} - \mathrm{C}_\varepsilon^2 f^2 \dfrac{\varepsilon^2}{k} + \mathrm{C}_\varepsilon^3 \dfrac{\mathrm{Gr}\mathrm{Re}^t}{\mathrm{Re}^3}\dfrac{\varepsilon}{k}\dfrac{\partial \Theta}{\partial x_i}\delta_{gi} \end{array} \right\} \tag{8.24}
$$

In the first row of (8.23), 1/12 is replaced with 1/6 due to the dual tensor product distinction with theory, (5.96). In the next two rows the *m*PDE term coefficients are manipulated to isolate laminar/turbulent non-D group ratios. The obvious simplification results in (8.23) when turbulent Prandtl/Schmidt numbers are defined identical with NS Pr and Sc.

An exercise suggests an algebraic alteration in (8.23) replacing $h^2\mathrm{Re}/12(1+\mathrm{Re}^t)$ with $\beta\Delta t/2$. The algorithm PDE statements for RaNS state variable members $\{\phi;\, P\}$ are (8.6–8.7) with (8.5) defining P^*. As detailed in Chapter 7, while not required for solution accuracy, (5.114) defines the *m*PDE replacement for (8.6 and (8.7)).

The RaNS $m\mathrm{GWS}^N + \theta\mathrm{TS}$ algorithm statement is functionally identical to that for NS, (7.17), with terminal nonlinear algebraic matrix definition (7.18). The discrete implementation, (7.20), again generates the computational form (7.21–7.22).

8.4 RaNS $m\mathrm{GWS}^h + \theta\mathrm{TS}$ Algorithm Matrix Statement

The RaNS modifications to the NS $m\mathrm{GWS}^h + \theta\mathrm{TS}$ partitioned algorithm, Section 7.4–7.5, are modest but detailed. For RaNS algorithm DOF $\{\mathrm{Q}I(t_{n+1})\}$, $1 \le I \le n+11+1$, ordered in the sequence $\{q^h(\mathbf{x},t)\} = \left\{u_i^h, \Theta^h, Y^h, k^h, \varepsilon^h, \phi^h, \tau_{ij}^{Dh}; P^h\right\}^T$, then $I = n+12 = 15$ for $n = 3$ is the DOF for P^h with solution the post-convergence operation defined for NS, (7.33). The discrete algorithm terminal matrix statement (7.23) is

$$
\begin{aligned} \{\mathrm{F\_Q}I\}_e &= [m]\{\mathrm{Q}I_{n+1} - \mathrm{Q}I_n\}_e \\ &\quad + \Delta t\left[\theta\{\mathrm{RES\_Q}I(\{Q\})\}_e\big|_{n+1} + (1-\theta)\{\mathrm{RES\_Q}I(\{Q\})\}_e\big|_n\right] \end{aligned} \tag{8.25}
$$

with $[m]_e$ matrix definition (7.24) unaltered. Since $\mathrm{D}M^h$ is satisfied for RaNS, the NS algorithm kinetic flux vector matrix (7.26) extends directly to RaNS $\{\mathrm{RES\_Q}I\}_e$, $1 \le I \le n+4$

$$
\{\mathrm{RES\_Q}I(\{f_j\})\}_e = \{\mathrm{U}J\}_e^T[\mathrm{M300}J]_e\{\mathrm{Q}I\}_e \tag{8.26}
$$

Further, (7.27) remains valid to augment $\{\mathrm{RES\_U}I(\{f_j\})\}_e$, $1 \le I \le n$, in (8.26) for the pressure projection algorithm

$$
\{\mathrm{RES\_U}I(\{f_j(P^*)\})\}_e = [\mathrm{M20}I]_e\{\mathrm{Q15}(t_n)\}_e + \frac{1}{\theta\Delta t}[\mathrm{M20}I]_e\sum_{\alpha=1}^{p}\{\delta\mathrm{Q8}\}_e^\alpha \tag{8.27}
$$

The dissipative flux vector contribution to the $\{\mathrm{U}I\}_e$ residual, $1 \le I \le n$, following Green–Gauss manipulation alters (7.28) to

$$
\begin{aligned}
&\{\text{RES\_U}I(\{f_j^d\})\}_e \\
&\equiv \int_{\Omega_e} \{N_1\} \frac{\partial}{\partial x_j}\left(\frac{2}{\text{Re}}(1+\text{Re}^t)S_{ij} + \frac{h^2\text{Re}}{6(1+\text{Re}^t)}\left(u_j u_k \frac{\partial u_i}{\partial x_k} + u_i u_k \frac{\partial u_j}{\partial x_k}\right)\right) \text{d}\tau \\
&= -\int_{\Omega_e} \{N_1\} \frac{\partial}{\partial x_j} \begin{pmatrix} \frac{1}{\text{Re}}\{\text{RET}\}_e^T\{N_1\}\left(\frac{\partial\{N_1\}^T}{\partial x_j}\{\text{U}I\}_e + \frac{\partial\{N_1\}^T}{\partial x_i}\{\text{U}J\}_e\right) \\ +\frac{h^2\text{Re}}{6}\begin{pmatrix} \frac{\{\text{UJUK}\}_e^T\{N_1\}}{\{\text{RET}\}_e^T\{N_1\}}\frac{\partial\{N_1\}^T}{\partial x_k}\{\text{U}I\}_e \\ +\frac{\{\text{UIUK}\}_e^T\{N_1\}}{\{\text{RET}\}_e^T\{N_1\}}\frac{\partial\{N_1\}^T}{\partial x_k}\{\text{U}J\}_e \end{pmatrix} \end{pmatrix} \text{d}\tau \\
&\equiv \frac{1}{\text{Re}}\{\text{RET}\}_e^T\left([\text{M30}JJ]_e\{\text{U}I\}_e + [\text{M30}IJ]_e\{\text{U}J\}_e\right) \\
&\quad + \frac{h^2\text{Re}}{6}\begin{pmatrix} \{\text{UJUK/RET}\}_e^T[\text{M30}JK]_e\{\text{Q}I\}_e \\ +\{\text{UIUK/RET}\}_e^T[\text{M30}JK]_e\{\text{Q}J\}_e \end{pmatrix}, \quad 1 \le I \le 3
\end{aligned}
\tag{8.28}
$$

The decision in (8.28) is to implement the *explicitly nonlinear* rational functional $u_j u_k/(1+\text{Re}^t)$ via interpolation of the element nodal DOF ratio. Separately evaluating the $\{\text{RET}\}_e^T\{N_1\}$ integral generates an element scalar (the average) which compromises impact of the distribution of Re^t.

The constants/parameters in (8.23) can be coalesced into the state variable dependent parameter set $\{\text{Pa}(q^h)\}$ and the turbulent counterpart $\{\text{Pa}^t(q^h)\}$. The direct extension on the NS PDE system is $\{\text{Pa}(q^h)\} = \{\text{Re, Re, Re, RePr, ReSc, Pe, Pe}\}^T$. The turbulent parameter *distributions* in (8.23) are $\{1\} + \text{Re}^t\{1, \text{Pr/Pr}^t, \text{Sc/Sc}^t, 1/C_k, 1/C_\varepsilon\}^T$. For definition $\{\text{Pa}^t(q^h)\} \equiv \{1, 1, 1, \text{Pr/Pr}^t, \text{Sc/Sc}^t, 1/C_k, 1/C_\varepsilon\}^T$, which are typically numbers of $O(10^0)$, and then absorbing the NS "{1}" into the Re^t interpolation on Ω_e enables labeling the range of turbulence model closure parameters/functions as $\text{Re}^t\{\text{Pa}^t\}$ throughout (8.23).

Transitioning from array to index notation, $\{\text{Pa}(q^h)\} \Rightarrow \text{Pa(I)}$ and $\{\text{Pa}^t(q^h)\} \Rightarrow \text{Pa}^t\text{(I)}$, with (I) the *indicator* associated one to one with DOF index I. This alteration in (8.28) replaces Re by Pa(I), with I coinciding with $1 \le I \le 3$. Continuing, the dissipative flux vector residual term $\{\text{RES\_Q}I\}_e$ for $n+1 \le I \le n+4$ is

$$
\begin{aligned}
&\{\text{RES\_Q}I(\{f_j^d\})\}_e \\
&= -\int_{\Omega_e} \{N_1\} \frac{\partial}{\partial x_j} \begin{pmatrix} \frac{1}{\text{Pa(I)}}(1+\text{Re}^t\text{Pa}^t(\text{I}))\frac{\partial}{\partial x_j} \\ +\frac{h^2\text{Pa(I)}}{12(1+\text{Re}^t\,\text{Pa}^t(\text{I}))} u_j^h u_k^h \frac{\partial}{\partial x_k} \end{pmatrix} \{N_1\}^T \text{d}\tau \{\text{Q}I\}_e \\
&= \frac{\text{Pa}^t(\text{I})}{\text{Pa(I)}}\{\text{RET}\}_e^T[\text{M30}JJ]_e\{\text{Q}I\}_e \\
&\quad + \frac{h^2\text{Pa(I)}}{12\text{Pa}^t(\text{I})}\{\text{UJUK/RET}\}_e^T[\text{M30}JK]_e\{\text{Q}I\}_e
\end{aligned}
\tag{8.29}
$$

The source matrix for the RaNS state variable partition DOF $\{QI\}_e$, $1 \le I \le n+2$ is the modest alteration on the NS statement (7.29)

$$\{s\}_e = \left\{ \begin{array}{l} \dfrac{-\mathrm{Gr}\,\widehat{g}_I}{\mathrm{Re}^2} \det_e[\mathrm{M200}]\{\mathrm{Q4}\}_e \\ \dfrac{-\mathrm{Ec}}{\mathrm{Re}}\{\mathrm{U}J\}_e^T[\mathrm{M30}JK]_e\{\mathrm{U}K\}_e + [\mathrm{M2}JK]_e\{\mathrm{TAU}JK\}_{e,J \neq K} \\ \det_e[\mathrm{M200}]\{\mathrm{SRC}Y\}_e \end{array} \right\} \tag{8.30}$$

For DOF $\{QI\}_e$, $n+3 \le I \le n+4$, the $k=1$ TP/NC basis algorithm matrix statement for the RaNS additional rows in (8.24) is

$$\begin{aligned} \{s\}_e &= \int_{\Omega_e} \{N_1\} \left\{ \begin{array}{l} \tau_{jk}^{\mathrm{D}} \dfrac{\partial u_j}{\partial x_k} - \varepsilon + \dfrac{\mathrm{Gr}}{\mathrm{Re}^2} \dfrac{\varepsilon}{k} \dfrac{\partial \Theta}{\partial x_i} \delta_{\mathrm{g}\,i} \\ \mathrm{C}_\varepsilon^1 f^1 \tau_{jk}^{\mathrm{D}} \dfrac{\varepsilon}{k} \dfrac{\partial u_j}{\partial x_k} - \mathrm{C}_\varepsilon^2 f^2 \dfrac{\varepsilon^2}{k} + \mathrm{C}_\varepsilon^3 \dfrac{\mathrm{GrRe}^t}{\mathrm{Re}^3} \dfrac{\varepsilon}{k} \dfrac{\partial \Theta}{\partial x_i} \delta_{\mathrm{g}i} \end{array} \right\}^h d\tau \\ &= \left\{ \begin{array}{l} \{\mathrm{TAU}JK\}_e^T[\mathrm{M300}K]_e\{\mathrm{U}J\}_e - \det_e[\mathrm{M200}]\{\mathrm{Q7}\}_e \\ \quad + \dfrac{\mathrm{GrRe}^t}{\mathrm{Re}^2}\{\mathrm{Q7/Q6}\}_e^T[\mathrm{M300}I]_e\{\mathrm{Q4}\}_e\delta_{\mathrm{g}i} \\ \mathrm{C}_\varepsilon^1 f^1 \{\mathrm{TAU}JK*(\mathrm{Q7/Q6})\}_e^T[\mathrm{M300}K]_e\{\mathrm{U}J\}_e \\ \quad - \det_e \mathrm{C}_\varepsilon^2 f^2 \{\mathrm{Q7/Q6}\}_e^T[\mathrm{M3000}]\{\mathrm{EPS}\}_e \\ \quad + \mathrm{C}_\varepsilon^3 \dfrac{\mathrm{Gr}}{\mathrm{Re}^3}\{\mathrm{RET*Q7/Q6}\}_e^T[\mathrm{M300}I]_e\{\mathrm{Q4}\}_e\delta_{\mathrm{g}i} \end{array} \right\} \end{aligned} \tag{8.31}$$

Viewing (8.28–8.31), the weak form implementation choice for RaNS state variable member rational function *nonlinearities*, for example $u_j u_k / (1 + \mathrm{Re}^t)$, (ε/k), $\tau_{jk}^{\mathrm{D}}(\varepsilon / k)$, $\mathrm{Re}^t(\varepsilon/k)$, is nodal DOF ratios interpolated using $\{N_1\}$ as the hypermatrix pre-multiplier. Once decided, the analytical calculus operations precisely generate the required nonlinear matrix structures. The last term Kronecker delta post-multipliers in (8.31) align the gravity unit vector.

RaNS state variable members QI, $n+8 \le I \le n+15$, are described by strictly EBV PDEs and a tensor differential definition. The matrix statement for ϕ^h is unchanged from (7.32)

$$\begin{aligned} \{\mathrm{F\_Q9}\}_e &= \int_{\Omega_e} \{N_1\} \left(-\nabla^2 \phi^h + \frac{\partial u_i^h}{\partial x_i} \right) d\tau \\ &= [\mathrm{M2}JJ]_e\{\mathrm{PHI}\}_e + [\mathrm{M20}I]_e\{\mathrm{U}I\}_e \end{aligned} \tag{8.32}$$

That for P^h differs from (7.33) only by addition of the Reynolds stress tensor in (8.7) leading to

$$\begin{aligned} \{\mathrm{F\_Q15}\}_e &= [\mathrm{M2}KK]_e\{\mathrm{Q15}\}_e + \{\mathrm{U}K\}_e^T[\mathrm{M30}JK]_e\{\mathrm{U}J\}_e \\ &\quad + \frac{\mathrm{Gr}}{\mathrm{Re}^2}[\mathrm{M2}J0]_e\{\mathrm{Q4}\}_e \widehat{g}_j + [\mathrm{N200}]_e \frac{d}{dt}\{\mathrm{U}J \bullet n_j\}_e \\ &\quad + \frac{1}{\mathrm{Re}}\left([\mathrm{N2}KK]_e\{\mathrm{U}J \bullet \widehat{n}_j\}_e - [\mathrm{A20}S]_e\{\mathrm{U}J \bullet \widehat{n}_j\}_e\Big|_{\mathrm{w}} \right) \\ &\quad + [\mathrm{M2}JK]_e\{\mathrm{TAU}JK\}_e \end{aligned} \tag{8.33}$$

Recall that implementing the BC integral $[\mathrm{A20}S]_e\{\mathrm{U}J \bullet \widehat{n}_j\}_e\Big|_{\mathrm{w}}$ amounts to zeroing out corresponding matrix entries from $[\mathrm{N2}KK]_e\{\mathrm{U}J \bullet \widehat{n}_j\}_e$.

The tensor differential definition for the deviatoric Reynolds stress tensor model is homogeneous (8.12). The $\mathrm{GWS}^N \Rightarrow \mathrm{GWS}^h$ process generates the DOF $\{\mathrm{Q}I\}$, $n+9 \leq I \leq n+14$, statements

$$\begin{aligned}\{\mathrm{F\_TAU}JK\}_e &= \int_{\Omega_e}\{N_1\}\left(\tau^{\mathrm{D}}_{jk} + \frac{\mathrm{Re}^t}{\mathrm{Re}}\left[\frac{\partial \bar{u}_j}{\partial x_k} + \frac{\partial \bar{u}_k}{\partial x_j}\right]\right)^h \mathrm{d}\tau, j \neq k \\ &= \det_e[\mathrm{M200}]_e\{\mathrm{TAU}JK\}_e \\ &\quad + \frac{1}{\mathrm{Re}}\{\mathrm{RET}\}_e^T\left([\mathrm{M300}K]_e\{\mathrm{U}J\}_e + [\mathrm{M300}J]_e\{\mathrm{U}K\}_e\right)\end{aligned} \tag{8.34}$$

This completes the $m\mathrm{GWS}^h + \theta\mathrm{TS}$ algorithm for RaNS *m*PDE statements with defined state variable and turbulence closure model. No new algorithm matrices have been generated and the RaNS formulation is replete with hypermatrix statements. All matrices with e subscript require coordinate transformation embedding, recall (7.34). The multiplier h^2 on respective *m*PDE terms must be related to element $\det_e$, as detailed in (7.35) and the paragraph following for $k=1$ TP/NC basis implementations. Replacing *m*PDE theory coefficient $h^2\mathrm{Re}/12(1+\mathrm{Re}^t)$ with $\beta\Delta t/2$ simplifies the terminal lines in dissipative flux vector residual statements (8.28–8.29), a suggested exercise.

Algorithm completion requires jacobian formation for the matrix iterative statement (7.21–7.22). A *quasi*-Newton construction enables direct use of the NS algorithm jacobian, Section 7.5. The RaNS state variable partition $\{u_i, \Theta, Y, \phi\}^h$ is segregated from that for the TKE state variable member partition $\{k, \varepsilon, \tau_{jk}\}$. Maintaining quality iterative convergence for the RaNS partition accrues to retarding by one iteration the $\{\mathrm{RET}\}_e$ DOF update. Similarly the $\{\mathrm{U}J\}_e$ DOF are retarded in the TKE partition jacobian. If the steady solution is sought then DOF retarding is moved from iteration loop to the time step.

The diagonal of the block jacobians for the RaNS DOF partition $\{\mathrm{Q}I\}$, $1 \leq I \leq n+2$, also the TKE partition $\{\mathrm{Q}I\}$, $n+3 \leq I \leq n+4$, is a modest alteration on (7.41). Again neglecting the *m*PDE hypermatrix multiplier nonlinearity $\{\mathrm{U}J\mathrm{U}K\}_e^T$, also $\{\mathrm{U}J\mathrm{U}K/\mathrm{RET}\}_e^T$ and the term in $\{s\}_e$ with multiplier Ec/Re, the RaNS and TKE *self-coupling* partition diagonal blocks replace (7.41) with

$$\begin{aligned}[\mathrm{JAC\_Q}I]_e &\equiv \frac{\partial\{\mathrm{F\_Q}I\}_e}{\partial\{\mathrm{Q}I\}_e}, 1 \leq I \leq n+4 \\ &= \frac{\partial}{\partial\{\mathrm{Q}I\}_e}\left([m]_e\{\mathrm{Q}I_{n+1}\}_e + \theta\Delta t\{\mathrm{RES\_Q}I(\{Q\})\}_e\Big|_{n+1}\right) \\ &\approx \det_e[\mathrm{M200}] + \frac{\gamma\,\Delta t^2}{6}\{\mathrm{U}J\mathrm{U}K\}_e^T[\mathrm{M30}JK]_e \\ &\quad + \theta\Delta t\begin{pmatrix}\{\mathrm{U}J\}_e^T[\mathrm{M300}J]_e + \{\mathrm{U}I\}_e^T[\mathrm{M3}J\mathrm{00}]_e\delta_{IJ} \\ + \dfrac{\mathrm{Pa}^t(\mathrm{I})}{\mathrm{Pa}(\mathrm{I})}\{\mathrm{RET}\}_e^T[\mathrm{M30}JJ]_e \\ + \dfrac{h^2\mathrm{Pa}(\mathrm{I})}{12\mathrm{Pa}^t(\mathrm{I})}\{\mathrm{U}J\mathrm{U}K/\mathrm{RET}\}_e^T[\mathrm{M30}JK]_e\end{pmatrix}\end{aligned} \tag{8.35}$$

The alteration to (8.35) for implementing $\beta\Delta t/2$ is trivial.

The jacobian off-diagonal key stability coupling for the RaNS partition remains (7.42) with DOF index alteration

$$\frac{\partial\{\text{F\_Q}I\}_e}{\partial\{\text{Q8}\}_e} \Rightarrow \frac{\partial\{\text{F\_U}I\}_e}{\partial\{\delta\text{Q8}\}_e} = [\text{M20}I]_e, 1 \le I \le n \tag{8.36}$$

and the $\theta\Delta t$ multiplier again cancels. The RaNS partition jacobian off-diagonal block jacobian completion is

$$\begin{aligned}
\frac{\partial\{\text{F_Q}I\}_e}{\partial\{\text{Q4}\}_e} &= -\theta\Delta t\frac{\text{Gr}\,\widehat{g}_I}{\text{Re}^2}\det{}_e[\text{M200}], 1 \le I \le n \\
\frac{\partial\{\text{F_Q4}\}_e}{\partial\{\text{Q}I\}_e} &= \theta\Delta t\{\text{U}I\}_e^T[\text{M300}J]_e\delta_{IJ}, 1 \le I \le n \\
\frac{\partial\{\text{F_Q5}\}_e}{\partial\{\text{Q}I\}_e} &= \theta\Delta t\{\text{U}I\}_e^T[\text{M300}J]_e\delta_{IJ}, 1 \le I \le n
\end{aligned} \tag{8.37}$$

In the TKE partition jacobian the DOF $\{\text{U}J\}_e$, $\{\text{U}I\}_e$ and $\{\text{U}J\text{U}K\}_e$ in (8.35) are retarded by one iteration. The $\{\text{RET}\}_e$ nonlinearity permeates this partition; the required derivatives are readily established from the continuum definitions

$$\begin{aligned}
\frac{\partial\text{Re}^t}{\partial k} &= \frac{\partial(f_\nu k^2/\varepsilon\text{Re})}{\partial k} = \frac{2f_\nu}{\text{Re}}\frac{k}{\varepsilon} = 2\text{Re}^t/k \\
\frac{\partial\text{Re}^t}{\partial\varepsilon} &= \frac{\partial(f_\nu k^2/\varepsilon\text{Re})}{\partial\varepsilon} = \frac{-f_\nu}{\text{Re}}\frac{k^2}{\varepsilon^2} = -\text{Re}^t/\varepsilon
\end{aligned} \tag{8.38}$$

Using (8.38) an exercise suggests verifying that the TKE partition self-coupling additions to jacobian (8.35) for $\{\text{Q}I\}$, $n+3 \le I \le n+4$ are

$$\begin{aligned}
[\text{JAC_Q6}]_e &\approx \theta\Delta t\begin{pmatrix} \dfrac{2}{\text{ReC}_k}\{\text{TKE}\}_e^T[\text{M3}JJ0]_e\text{diag}\left[\text{RET}/\text{TKE}\right]_e \\ -\text{C}_\varepsilon^3\dfrac{\text{Gr}}{\text{Re}^3}\{\text{RET, Q4}\}_e^T[\text{M3}J00]_e\text{diag}\left[\text{EPS}/\text{TKE}^2\right]_e\delta_{gj} \end{pmatrix} \\
[\text{JAC_Q7}]_e &\approx \theta\Delta t\begin{pmatrix} \dfrac{-1}{\text{ReC}_\varepsilon}\{\text{EPS}\}_e^T[\text{M3}JJ0]_e\text{diag}\left[\text{RET}/\text{EPS}\right]_e \\ +\text{C}_\varepsilon^3\dfrac{\text{Gr}}{\text{Re}^3}\{\text{RET, Q4}\}_e^T[\text{M3}J00]_e\text{diag}\left[1/\text{TKE}\right]_e\delta_{gj} \end{pmatrix}
\end{aligned} \tag{8.39}$$

wherein diag [ ] denotes a diagonal square matrix populated with specified TKE partition DOF ratios. For $\{\text{Q}I\}$, $n+5 \le I \le n+10$, from (8.34) the self-coupling jacobians are

$$[\text{JAC\_TAU}JK]_e \equiv \det{}_e[\text{M200}], 1 \le (J \ne K) \le 3 \tag{8.40}$$

The TKE jacobian partition cross-coupling matrices stemming from the source and dissipative flux vector, using closure variable DOF notation rather than $\{QI\}$ for clarity, are

$$\begin{aligned}\frac{\partial\{\mathrm{F\_TKE}\}_e}{\partial\{\mathrm{EPS}\}_e} &= \theta\Delta t\,\det_e[\mathrm{M200}] \\ &\quad -\frac{\theta\Delta t}{\mathrm{ReC}_k}\{\mathrm{TKE}\}_e^T[\mathrm{M3}JK0]_e\mathrm{diag}\left[\mathrm{TAU}JK*\mathrm{TKE}/\mathrm{EPS}^2\right]_e\end{aligned} \tag{8.41}$$

$$\begin{aligned}\frac{\partial\{\mathrm{F\_EPS}\}_e}{\partial\{\mathrm{TKE}\}_e} &= -\theta\Delta t\begin{pmatrix} \mathrm{C}_\varepsilon^1\{\mathrm{U}J\}_e^T[\mathrm{M3}K00]_e\mathrm{diag}\left[\mathrm{TAU}JK*\mathrm{EPS}/\mathrm{TKE}^2\right]_e \\ +\mathrm{C}_\varepsilon^2\det_e\{\mathrm{EPS}\}_e^T[\mathrm{M3000}]\mathrm{diag}\left[\mathrm{EPS}/\mathrm{TKE}^2\right]_e \\ +\mathrm{C}_\varepsilon^3\dfrac{\mathrm{Gr}}{\mathrm{Re}^3}\{\mathrm{RET},\mathrm{Q4}\}_e^T[\mathrm{M3}J00]_e\mathrm{diag}\left[\mathrm{EPS}/\mathrm{TKE}^2\right]_e\delta_{gj}\end{pmatrix} \\ &\quad +\frac{\theta\Delta t}{\mathrm{ReC}_\varepsilon}\{\mathrm{EPS}\}_e^T[\mathrm{M3}JK0]_e\mathrm{diag}\left[\mathrm{TAU}JK/\mathrm{EPS}\right]_e\end{aligned} \tag{8.42}$$

The cross-coupling jacobians for Reynolds stress tensor DOF are

$$\begin{aligned}\frac{\partial\{\mathrm{F\_TKE}\}_e}{\partial\{\mathrm{TAU}JK\}_e} &= \theta\Delta t\begin{Bmatrix}\{\mathrm{U}J\}_e^T[\mathrm{M3}K00]_e+ \\ \dfrac{1}{\mathrm{ReC}_k}\{\mathrm{TKE}\}_e^T[\mathrm{M3}JK0]_e\mathrm{diag}\left[\mathrm{TKE}/\mathrm{EPS}\right]_e\end{Bmatrix} \\ \frac{\partial\{\mathrm{F\_EPS}\}_e}{\partial\{\mathrm{TAU}JK\}_e} &= \theta\Delta t\begin{Bmatrix}\mathrm{C}_\varepsilon^1\{\mathrm{U}J\}_e^T[\mathrm{M3}K00]_e\mathrm{diag}\left[\mathrm{EPS}/\mathrm{TKE}\right]_e+ \\ \dfrac{1}{\mathrm{ReC}_\varepsilon}\{\mathrm{EPS}\}_e^T[\mathrm{M3}JK0]_e\mathrm{diag}\left[\mathrm{TKE}/\mathrm{EPS}\right]_e\end{Bmatrix}\end{aligned} \tag{8.43}$$

The jacobian matrices accounting for cross-coupling of TKE closure DOF with Reynolds stress tensor DOF are

$$\begin{aligned}\frac{\partial\{\mathrm{F\_TAU}IJ\}_e}{\partial\{\mathrm{TKE}\}_e} &= \theta\Delta t\begin{pmatrix}\dfrac{-2}{3}\det_e[\mathrm{M200}]\delta_{IJ}+ \\ \dfrac{1}{2\mathrm{Re}}\begin{pmatrix}\{\mathrm{U}I\}_e^T[\mathrm{M3}J00]_e \\ +\{\mathrm{U}J\}_e^T[\mathrm{M3}I00]_e\end{pmatrix}\mathrm{diag}\left[\mathrm{RET}/\mathrm{TKE}\right]_e\end{pmatrix} \\ \frac{\partial\{\mathrm{F\_TAU}IJ\}_e}{\partial\{\mathrm{EPS}\}_e} &= \frac{-\theta\Delta t}{2\mathrm{Re}}\begin{pmatrix}\{\mathrm{U}I\}_e^T[\mathrm{M3}J00]_e \\ +\{\mathrm{U}J\}_e^T[\mathrm{M3}I00]_e\end{pmatrix}\mathrm{diag}\left[\mathrm{RET}/\mathrm{EPS}\right]_e\end{aligned} \tag{8.44}$$

Viewing (8.25–8.44), the RaNS $m\mathrm{GWS}^h + \theta\mathrm{TS}$ algorithm is *amply populated* with turbulent closure model DOF *ratios*, due to TKE closure source terms plus Re^t nonlinearity. Selecting the Lam–Bremhorst low turbulence Re^t closure, the wall BC $k=0$ renders not-a-number DOF ratios $\{\mathrm{RET/TKE}\}_e$ and $\mathrm{diag[EPS/TKE]}_e$. The RaNS partition $\{QI\}$ DOF at wall nodes are typically Dirichlet BC constrained hence code practice is to simply not evaluate the offending terms.

Typical RaNS code k-ε closure model practice is to not resolve low Re^t wall layer regions unless absolutely necessary. Hence, via BL similarity theory, the first off-wall

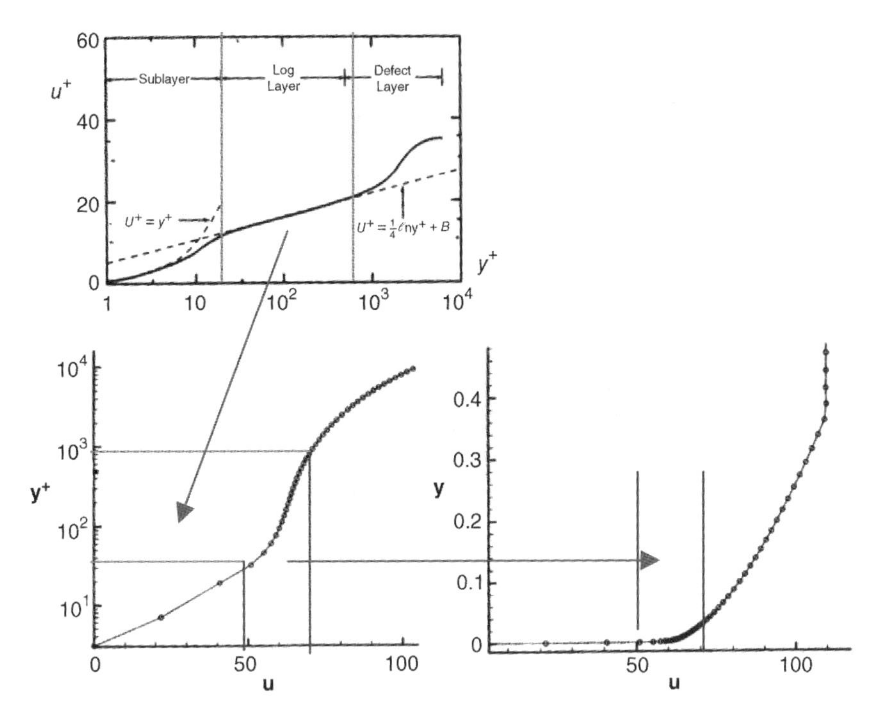

Figure 8.1 Illustration of turbulent BL similarity solution BC validity range for Reynolds stress k-ε closure model DOF, Bradshaw IDENT 2400 data, from Kline *et al.* (1969)

geometric node must be located in the log-law layer, $30 < y^+ < 800$, a solution adaptive remeshing requirement. At such nodes the TKE partition state variables DOF are computed from (8.19) while the RaNS partition DOF are computed via (8.25). For the Bradshaw IDENT 2400 turbulent BL solution summarized in Figure 4.9, Figure 8.1 illustrates the log-law BC validity range in both similarity and traditional BL coordinates.

8.5 RaNS $m\text{GWS}^h$ + θTS Algorithm, Stability, Accuracy

Turbulent flow in a square cross-section duct is the classic validation statement for a RaNS CFD algorithm. However, the deviatoric Reynolds stress tensor model (8.12) with either k-ε or k-ω closure is incapable of generating the experimentally confirmed steady 3-D velocity vector field. The PNS algorithm with *anisotropic* Reynolds stress closure model, Section 4.10, did generate the requisite transverse plane vortex velocity field, including deflection of the near wall axial velocity isovel distribution. In (coarse mesh!) quantitative agreement with experiment, it is totally distinct from the *deviatoric* Re^t model RaNS steady velocity solution, Figure 8.2.

The $m\text{GWS}^h + \theta\text{TS}$ RaNS algorithm with full anisotropic Reynolds stress tensor closure follows this section completion. The requirements for this section reduce to RaNS $m\text{GWS}^h +$ θTS algorithm $k = 1$ basis assessment for solution stability, accuracy and especially *monotonicity*. This requirement is precisely served for a rectangular channel flow specification with homogeneous Neumann BCs applied at farfield lateral domain boundaries.

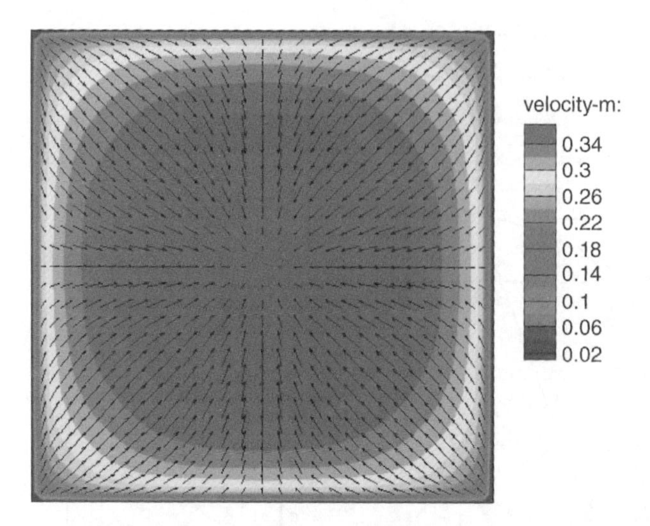

Figure 8.2 RaNS algorithm with deviatoric Reynolds stress k-ε closure, steady velocity vector field, 0.3 m square cross-section duct: transverse plane velocity vector DOF overlay on axial velocity DOF isovel flood

A regular solution adapted mesh exercise determines the $k=1$ TP basis mesh required to accurately resolve a fully developed turbulent BL in a channel absent low Re^t resolution. A half-span transverse solution adapted $\text{M}=16$, $\text{pr}=1.15$ mesh yields acceptable engineering accuracy, the 17-node DOF data (circular symbols) in Figure 8.3. The regular solution adapted $\text{M}=32$, $\text{pr}=1.08$ mesh solution supports excellent *quantitative* agreement with the

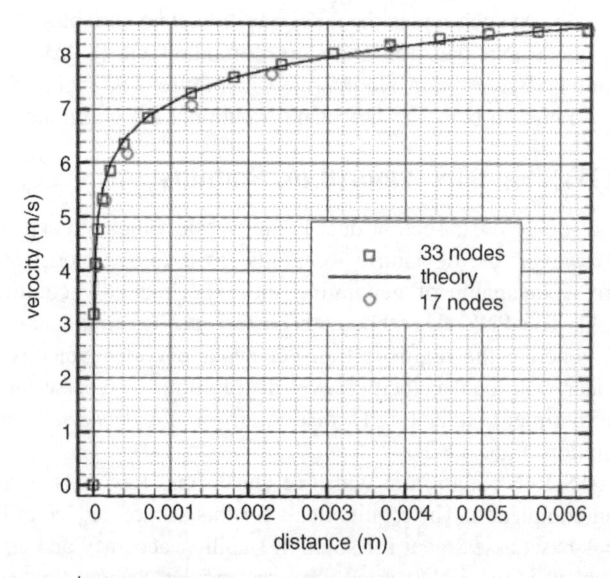

Figure 8.3 RaNS $m\text{GWS}^h+\theta\text{TS}$ algorithm, fully turbulent half-channel symmetry plane M-dependent steady {U1} DOF distribution comparison with BL similarity solution

BL similarity solution, the 33-node DOF data (square symbols) in Figure 8.3. Recall the $\mathrm{M} \approx 50$ solution adapted mesh requirement when resolving the low Re^t BL sublayer, Figure 4.9, extended into the freestream to admit homogeneous Neumann BCs. Summarily, note the first off-wall velocity magnitude DOF lies within the turbulent BL similarity log-law region, Figure 8.1, as required.

The *m*PDE contribution to $m\mathrm{GWS}^h + \theta\mathrm{TS}$ algorithm stability is quantified by inserting a velocity BC *singularity* to generate large local gradients in both the RaNS and TKE state variable partition solution DOF *a posteriori* data. A freestream onset speed $u = 400$ ft/s ($\sim$122 m/s) approaches a narrow channel of unbounded lateral extent and transverse span 0.025 m. The unit Reynolds number is Re/L = 0.4 E+07, equivalently Re = 0.9 E+05 based on channel transverse span. Following a short unbounded "inviscid" flow entrance length without leakage, the geometry BC transitions from tangent to no-slip at axial node plane 18. Figure 8.4 is a half span axial velocity illustration of this BC transition. The channel length is sufficient to admit development of a fully turbulent time averaged axial velocity profile.

The slug onset inflow (non-D) Dirichlet BCs are $\{\mathrm{U1}\} = \{1.0\}$, $\{\mathrm{U2}\} = \{0\} = \{\mathrm{U3}\}$ and $\{\mathrm{RET}\} = \{1\}$. Inviscid flow entrance region transverse boundary $\{\mathrm{U}I\}$ BCs are appropriately homogeneous Neumann and (no leakage) Dirichlet $\{\mathrm{U}L\} = \{0\}$. The $\{\mathrm{U}I\}$ BC switches to no-slip Dirichlet at the channel lip, and thereafter no-slip log-law for the TKE partition variables and homogeneous Neumann for pressure and ϕ^h. BCs for state variable DOF at channel outflow plane are homogeneous Neumann except $\phi^h = 0$ and Robin BC, (8.7), for P^h.

Following experimentation, algorithm *a posteriori* data are generated on a highly solution adapted $\mathrm{M} = 52 \times 32 \times 3$ mesh using progression ratios smoothly transitioning to nearly square *very* small measure elements at entrance channel BC singularity vicinity. The resultant DOF resolution in the downstream fully turbulent flow reach corresponds to the M = 16 data, Figure 8.3. For the *m*PDE system (8.20–8.24), and implementing the projection theory genuine pressure PDE system (8.5–8.7), on this mesh the $k = 1$ TP basis $m\mathrm{GWS}^h + \theta\mathrm{TS}$ algorithm is stable and generates RaNS plus TKE state variable partition unsteady DOF evolution to a nominal periodic solution.

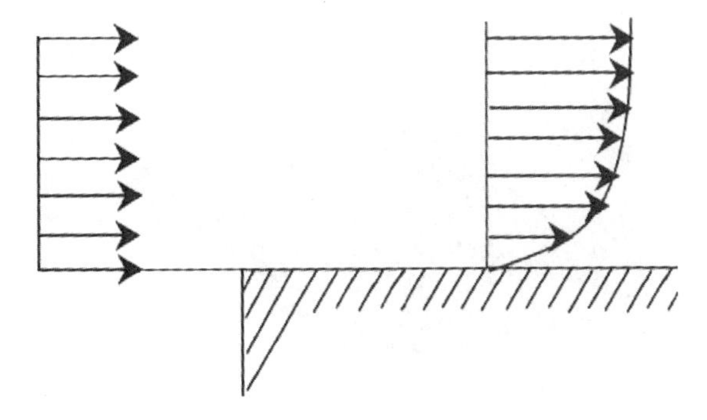

Figure 8.4 RaNS $m\mathrm{GWS}^h + \theta\mathrm{TS}$ algorithm stability assessment, half-span illustration of inviscid inflow encountering a no-slip wall BC singularity at entrance to an unbounded lateral span rectangular channel

The mGWS^h + θTS algorithm *monotonically* alters the inflow plane {U1} slug IC in the BC singularity near region, leading to $35 < y^+ \leq 55$ at first off-wall nodes from ~50 channel heights downstream to the exit plane. Figure 8.5 graphs the RaNS central plane {U1} DOF distribution interpolation snapshot in unitized scale perspective with mesh overlay viewing downstream and upstream. The transition to turbulent flow is smooth with outflow distribution in qualitative agreement with the Figure 8.3 data.

Not confirmable on this graph scale, but quantitatively verified in digital output files, the {U1} and {U2} DOF distributions are *monotone* to $O(\varepsilon)$, for $\varepsilon \leq$ E−04 the matrix iterative convergence requirement. This occurs throughout, critically so in the no-slip BC singularity

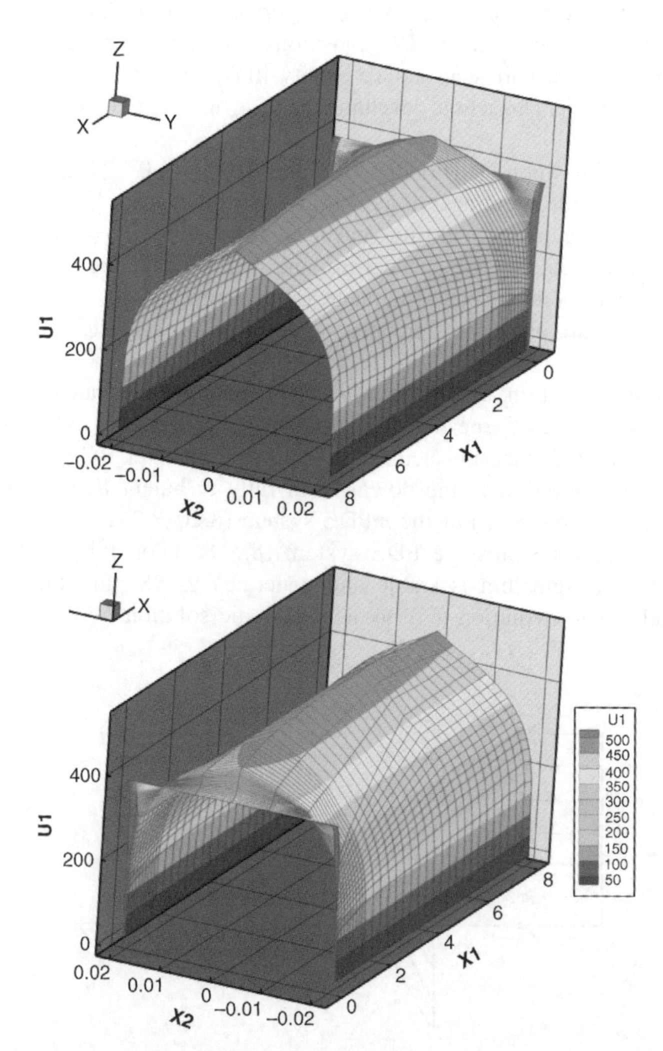

Figure 8.5 RaNS mGWS^h + θTS algorithm terminal {U1} DOF distribution snapshot, center plane unitized scale perspective with mesh overlay, entrance channel, Re = 0.92 E+05: vantage point downstream (top), upstream (bottom)

immediate vicinity. These DOF data further *validate* that the *m*PDE theory-derived dissipative flux vector operand phase selectively *annihilates* the $O(h^2)$ GWS^h significant order error, (7.8), intrinsic to RaNS **DP** discrete implementation, (8.23). These **DP** *m*PDE theory $O(h^2)$ GWS^h error DOF distributions are RaNS solution post-processed via a weak statement on their differential definitions, $\mathcal{D}(m\text{PDE}(\text{Re}^t, u_i)) = 0$ from (8.23), with jacobian [M200].

*m*PDE theory-predicted GWS^h $O(h^2)$ error DOF interpolations, for the Figure 8.5 solution snapshot and on identical scale, are graphed in Figure 8.6. Essentially not visible are dual *very* sharp peaks in {U1} *m*PDE operand at the BC singularity, top. A factor of five smaller but clearly visible are {U2} *m*PDE operand distributions, bottom. Note these $O(h^2)$ error predictions are *null* everywhere except in the immediate vicinity of the BC singularity.

Interpolation close-ups (peaks chopped) confirm that the *m*PDE theory mirror prediction of GWS^h formulation $O(h^2)$ discretization error for {U1}/{U2} is indeed *modulo* $2h$ oscillation, Figure 8.7, top/bottom. Clearly *m*PDE theory precisely predicts $O(h^2)$ dispersion error

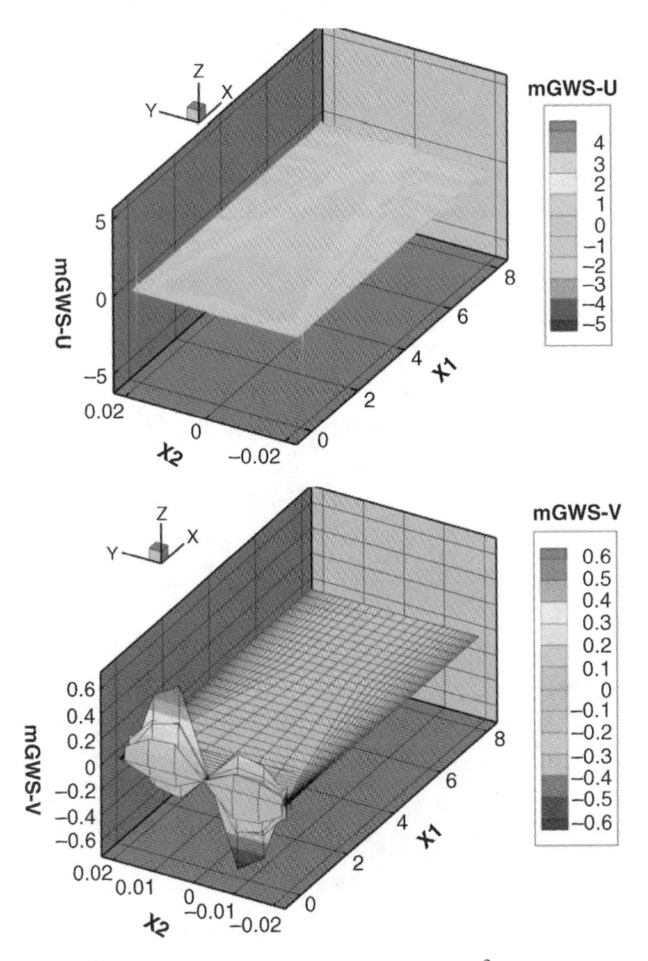

Figure 8.6 RaNS *m*GWS^h + θTS algorithm **DP** *m*PDE theory $O(h^2)$ error annihilator terms in dissipative flux vector, non-D DOF distributions in unitized scale perspective, entrance channel, Re = 0.92 E+05: for {U1} top, for {U2} bottom

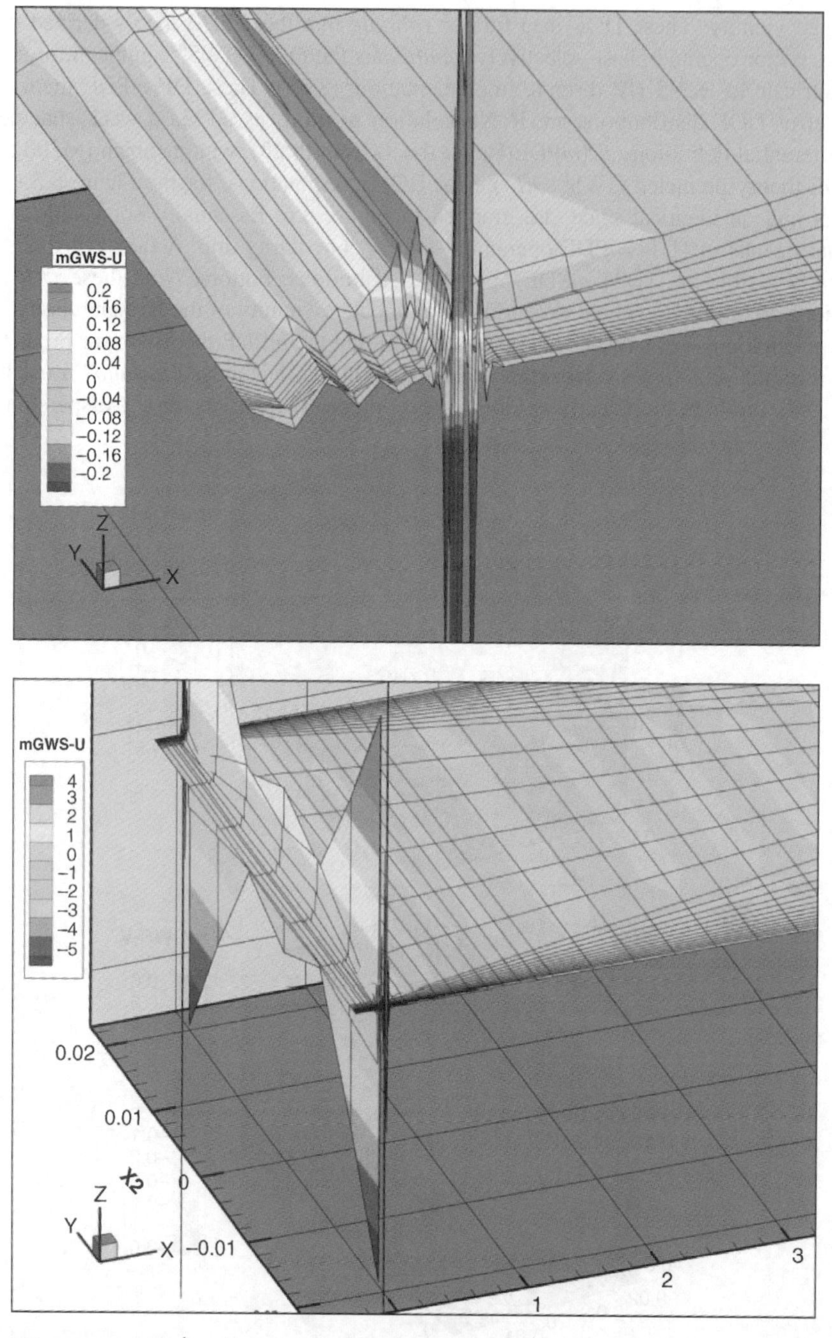

Figure 8.7 RaNS $m\text{GWS}^h + \theta\text{TS}$ algorithm *m*PDE theory $O(h^2)$ error annihilator non-D DOF distribution interpolations, close-up unitized scale perspective with mesh overlay, entrance channel, Re = 0.92 E+05: top, {U1}; bottom {U2}

distribution mirror images that otherwise pollute the GWS^h theory solution. The efficiency of sign alteration *distributions* in tensor products $u_j u_k$ of the *m*PDE operands is antithetical to artificial diffusion operators for dispersion error control, by definition smooth.

TKE partition variable DOF *monotone* distributions are similarly engendered by their *m*PDE theory terms. Very sharp spikes in {EPS} occur at the BC singularities, Figure 8.8 top, with close up, middle, confirming this sharply varying DOF distribution is monotone. The spikes in {TKE} are less pronounced; in concert a smoothly progressive monotone {RET} distribution results, Figure 8.8 bottom.

The terminal $m\mathrm{GWS}^h + \theta\mathrm{TS}$ algorithm assessments are quasi-Newton algorithm iteration performance and efficacy of pressure projection theory continuity constraint $\mathrm{D}M^h$. These graphed RaNS DOF distributions occur on a periodic basis following ~2000 time steps from the slug {U1} IC. The BC singularity engenders a time varying periodic solution which remains stable for over 5000 time steps. This corresponds to an elapsed time of ~4 seconds which for {U1} onset speed BC and ~2.7 m channel length corresponds to flow clearing the entrance channel ~100 times.

The algorithm averages 3 iterations/time step for convergence requirement $\varepsilon \leq \mathrm{E}{-}04$ following the ~1000 time steps required to clear the {U1} slug IC. Iteration convergence rate was super linear and at each pressure post-process the measure of $\mathrm{D}M^h$ enforcement robustness was $\| \{\phi^h(n\Delta t)\} \|_E \leq \mathrm{E} - 06$. The duct pressure solution DOF distribution is linear as expected, Figure 8.9 top, and the close-up at the BC singularity confirms it locally monotone as well, Figure 8.9 bottom.

In distinction to the NS thermal cavity experience, Figure 7.3, at time step convergence the {PHI} DOF distribution is a null level throughout. The close-in graphic at the BC singularity, Figure 8.10, is confirmation. Therefore, the {SUMPHI} component of the projection theory carries the requirement for $\| \{\phi^h(n\Delta t)\} \|_E$ minimization, again via *modulo* mesh scale $2h$ oscillations. This is confirmed in the DOF data in BC region singularity zoom in, Figure 8.10 bottom. Note {SUMPHI} magnitude (dimensional) is $O(10^5)$ smaller than pressure, Figure 8.9.

8.6 RaNS Algorithm BCs for Conjugate Heat Transfer

As stated, in thermal CFD simulations one rarely has available Dirichlet BCs for temperature. Factually, NS and RaNS flow systems invariably interact thermally with the enclosure. In the literature this phenomenon is termed *conjugate heat transfer* to signify directly coupled fluid-structure heat transfer.

Implementing conjugate heat transfer in the $m\mathrm{GWS}^\mathrm{h} + \theta\mathrm{TS}$ algorithm for NS/RaNS involves no more than specifying a unique BC surface for Θ^h in the NS/RaNS state variable partition, since temperature DOF located on no-slip boundaries are no longer BC-eligible. The BCs for all other NS/RaNS and TKE partition DOF are unaltered. The geometric essence is illustrated in Figure 8.11 wherein Ω_{PPNS} signifies the domain for the RaNS/TKE state variable, excepting Θ^h, which spans that denoted ${\Omega_{\mathrm{DE}}}^h$.

The $n = 2$ geometry serves to illustrate. A radiation flux loads the downstream half of one wall of an aluminum conduit enclosing a turbulent flowfield. The flux loaded wall, cross-hatched in Figure 8.11, is discretized into three Θ^h DOF rows. The Ω_{PPNS} domain discretization is $\mathrm{M} = 26$ adapted to both walls for application of log-law TKE partition BCs. The inflow Dirichlet BC for {U1} is a similarity solution wall-adapted profile,

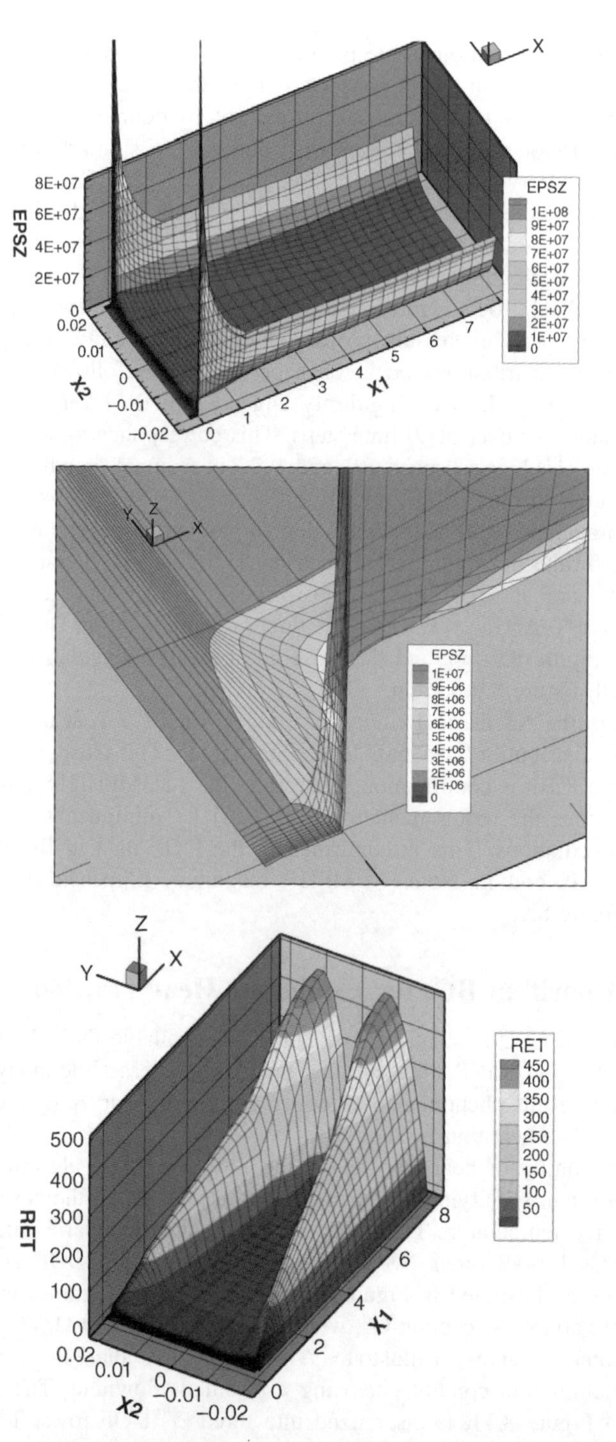

Figure 8.8 RaNS $m\text{GWS}^h + \theta\text{TS}$ algorithm TKE partition DOF distribution interpolation, unitized scale perspective with mesh overlay, entrance channel, Re = 0.92 E+05: top, {EPS}; middle, {EPS} close-up at BC singularity; bottom {RET}

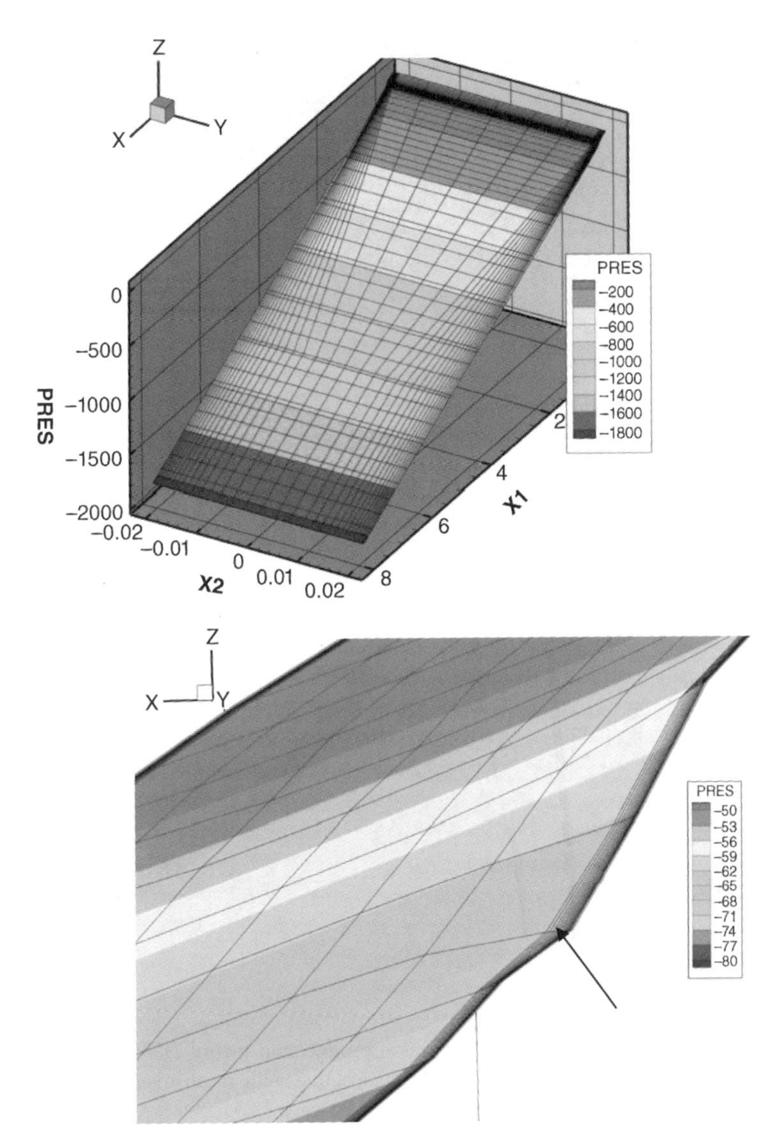

Figure 8.9 RaNS $m\text{GWS}^h + \theta\text{TS}$ algorithm pressure DOF distribution interpolations, unitized scale perspective with mesh overlay, entrance channel, Re = 0.92 E+05: top, global graph viewing upstream; bottom, zoom in at BC singularity (at arrow)

enabling rapid transition to fully turbulent. The temperature BCs are homogeneous Neumann except at the flow entrance wall DOF (only) which is fixed cold by a Dirichlet BC. The opposite channel wall is adiabatic and outflow BCs are homogeneous Neumann except Dirichlet for ϕ^h and Robin BC for P^h.

The RaNS $k = 1$ basis $m\text{GWS}^h + \theta\text{TS}$ algorithm steady temperature DOF distribution in unitized scale perspective is graphed in Figure 8.12. The view looks upstream with the radiation flux wall segment located upper left wherein the temperature DOF distribution is

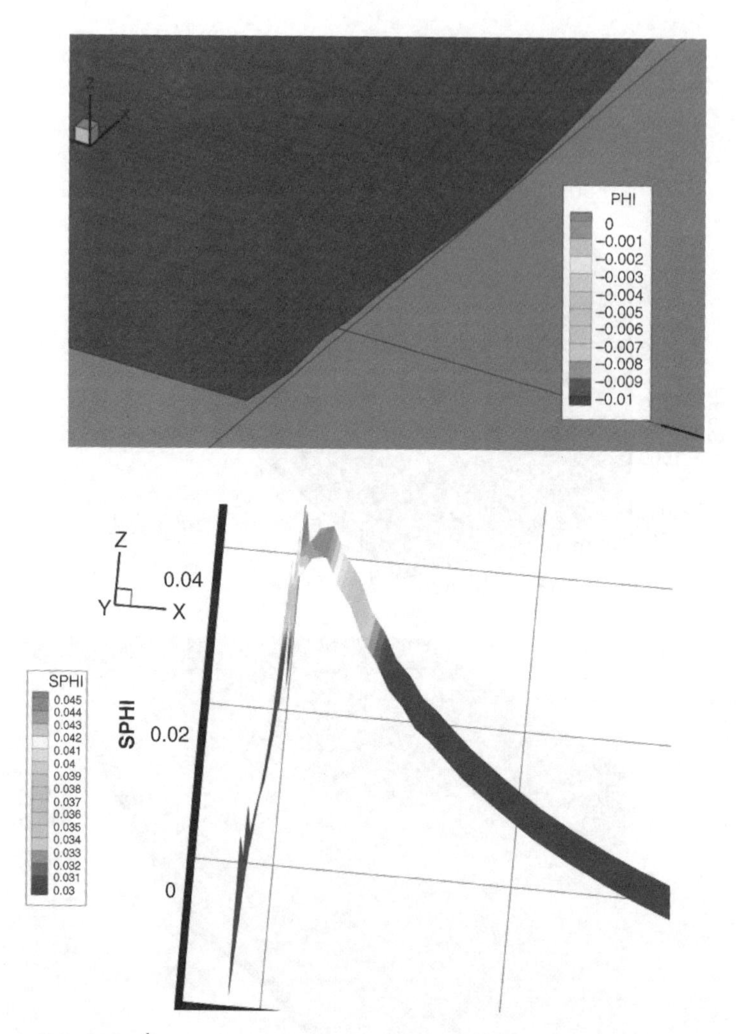

Figure 8.10 RaNS mGWS^h + θTS algorithm pressure projection theory DOF distribution interpolations, zoom in at BC singularity, unitized scale perspective with mesh overlay, entrance channel, Re = 0.92 E+05: top, {PHI}; bottom, {SUMPHI} edge-on view

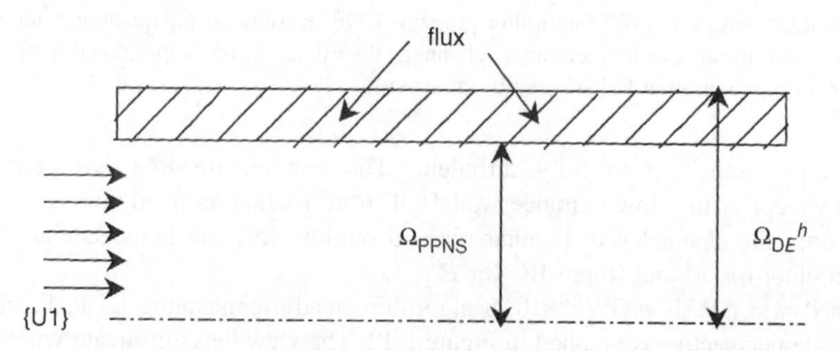

Figure 8.11 Domain definitions for conjugate heat transfer simulation

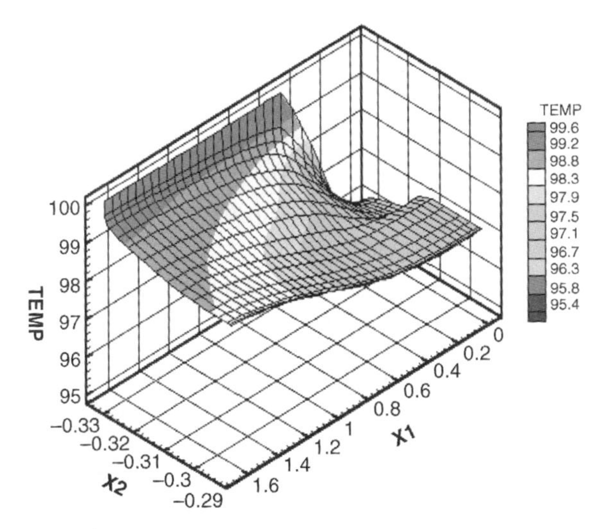

Figure 8.12 RaNS mGWS^h + θTS algorithm conjugate heat transfer simulation, steady temperature DOF distribution, unitized scale perspective with mesh overlay

computed axially uniform. Metallic high thermal conductivity heats the adiabatic wall from flux initiation upstream to the inflow Dirichlet BC. In concert with the adiabatic opposite wall, the net adiabatic BC effect is to elevate the temperature of the incoming fluid above its inflow IC, with distribution responsive to the single wall node Dirichlet BC constrained.

8.7 RaNS Full Reynolds Stress Closure PDE System

The alternative to the RaNS eddy viscosity dual-PDE closure is a complete Reynolds stress transport PDE system. The key theoretical improvement is eliminating the deviatoric stress tensor assumption, (8.8–8.10), enabling prediction of a fully *anisotropic* Reynolds stress tensor. For incompressible flow with overbar for time average and superscript prime for the variation, the *dimensional* transport PDE system for kinematic Reynolds stress tensor $\overline{u_i'u_j'}$ is, Marvin (1983)

$$\begin{aligned}
\mathcal{L}\left(\overline{u_i'u_j'}\right) &= \frac{\partial(\overline{u_i'u_j'})}{\partial t} + \frac{\partial \bar{u}_k(\overline{u_i'u_j'})}{\partial x_k} - \frac{\partial}{\partial x_k}\left[\bar{\nu}\frac{\partial(\overline{u_i'u_j'})}{\partial x_k}\right] \\
&+ \left(\overline{u_j'u_k'}\frac{\partial \bar{u}_i}{\partial x_k} + \overline{u_i'u_k'}\frac{\partial \bar{u}_j}{\partial x_k}\right) + 2\bar{\nu}\overline{\frac{\partial u_i'}{\partial x_k}\frac{\partial u_j'}{\partial x_k}} \\
&- \overline{\frac{p'}{\rho}\left(\frac{\partial u_i'}{\partial x_j} + \frac{\partial u_j'}{\partial x_i}\right)} + \beta\left(g_i\overline{u_j'T'} + g_j\overline{u_i'T'}\right) \\
&+ \frac{\partial}{\partial x_k}\left[\overline{u_i'u_j'u_k'} + \frac{p'}{\rho}\left(\delta_{j\,k}u_i' + \delta_{i\,k}u_j'\right)\right] = 0
\end{aligned} \tag{8.45}$$

The first three terms in (8.45) are the familiar time, convection and molecular diffusion differential operators. The fourth is Reynolds stress production coupling to the time averaged velocity strain rate. The fifth term is the time averaged expression for viscous dissipation of $\overline{u_i'u_j'}$. The definition of scalar *isotropic dissipation* function accrues to its index contraction

$$\frac{2}{3}\delta_{ij}\varepsilon \equiv 2\overline{\nu}\overline{\frac{\partial u_i'}{\partial x_k}\frac{\partial u_j'}{\partial x_k}} \tag{8.46}$$

The next two source terms in (8.45) are pressure strain and buoyancy effects while the last, when modeled, transitions to turbulent diffusion.

Models for the last four terms in (8.45) are required to achieve closure. The resultant form for (8.45) is termed the *Reynolds Stress Model* (RSM) PDE. For the last term, via a generalized gradient-diffusion model approach, Daly–Harlow (1975) derived

$$\left[\overline{u_i'u_j'u_k'} + \overline{\frac{p'}{\rho}\left(\delta_{jk}u_i' + \delta_{ik}u_j'\right)}\right] \cong -C_s\frac{k}{\varepsilon}\overline{u_k'u_l'}\frac{\partial\overline{u_i'u_j'}}{\partial x_l} \tag{8.47}$$

for C_s a model constant. The similar gradient-diffusion model approach for the buoyancy term, upon introduction of a turbulent Prandtl number, generates

$$\beta\left(g_i\overline{u_j'T'} + g_j\overline{u_i'T'}\right) \cong -\frac{\beta\nu^t}{\mathrm{Pr}^t}\left(g_i\frac{\partial T}{\partial x_j} + g_j\frac{\partial T}{\partial x_i}\right) \tag{8.48}$$

where ν^t is turbulent eddy viscosity.

A wide range of models for the pressure strain term have been derived. The basic linear model, Gibson and Launder (1978), Fu *et al.* (1987), Launder (1989), matures into three contributions

$$\overline{\frac{p'}{\rho}\left(\frac{\partial u_i'}{\partial x_j} + \frac{\partial u_j'}{\partial x_i}\right)} \cong \Phi_{ij}^1 + \Phi_{ij}^2 + \Phi_{ij}^3 \tag{8.49}$$

called in sequence *slow* pressure-strain (return to isotropy) term, *rapid* pressure-strain term and the *wall reflection* term. The slow term model is the deviatoric tensor

$$\Phi_{ij}^1 \cong -C_1\frac{\varepsilon}{k}\left[\overline{u_i'u_j'} - \frac{2}{3}k\delta_{ij}\right] \tag{8.50}$$

for k turbulent kinetic energy and $C_1 = 1.8$.

The rapid term closure model employs the deviator of the linear combination of convection, production and buoyancy terms in (8.45)

$$RS_{ij} = \left(\overline{u_j'u_k'}\frac{\partial\overline{u}_i}{\partial x_k} + \overline{u_i'u_k'}\frac{\partial\overline{u}_j}{\partial x_k}\right) - \frac{5\beta}{6}\left(g_i\overline{u_j'T'} + g_j\overline{u_i'T'}\right) - \frac{\partial\overline{u}_k(\overline{u_i'u_j'})}{\partial x_k} \tag{8.51}$$

which leads to the linear rapid pressure strain deviator

$$\Phi_{ij}^2 \cong -C_2\left[RS_{ij} - \frac{2}{3}RS_{kk}\delta_{ij}\right] \tag{8.52}$$

with $C_2 = 0.60$.

The wall reflection term in (8.49) accounts for redistribution of Reynolds normal stresses near a no-slip surface, damping the normal component while enhancing the wall-parallel components. Denoting wall surface normal as $\mathbf{n}_j, \mathbf{n}_k$ the closure model stress deviator is

$$\begin{aligned}\Phi_{ij}^3 \cong & -C_3\frac{\varepsilon}{k}\left[\overline{u_i'u_k'}\,\mathbf{n}_j\mathbf{n}_k + \overline{u_j'u_k'}\,\mathbf{n}_i\mathbf{n}_k - \frac{2}{3}\overline{u_k'u_m'}\mathbf{n}_m\mathbf{n}_k\,\delta_{ij}\right]\frac{C_5}{d}\frac{k^{3/2}}{\varepsilon} \\ & -C_4\frac{\varepsilon}{k}\left[\Phi_{ik}^2\,\mathbf{n}_j\mathbf{n}_k + \Phi_{jk}^2\,\mathbf{n}_i\mathbf{n}_k - \frac{2}{3}\Phi_{km}^2\,\mathbf{n}_m\mathbf{n}_k\,\delta_{ij}\right]\frac{C_5}{d}\frac{k^{3/2}}{\varepsilon}\end{aligned} \tag{8.53}$$

The coefficients are $C_3 = 0.33$, $C_4 = 0.20$, $C_5 = (C_\nu)^{3/4}/\kappa$ where $\kappa = 0.4187$ is the von Karman constant with $C_\nu = 0.09$ the coefficient in the eddy viscosity definition

$$\nu^t \equiv C_\nu\frac{k^2}{\varepsilon} \tag{8.54}$$

The geometric parameter d is the perpendicular distance from the wall to the first off-wall DOF.

Viewing (8.46–8.54), the modeled Reynolds stress tensor transport PDE closure requires addition of the k-ε PDE system along with turbulent eddy viscosity definition. The PDE for k is generated (an exercise) by contraction over the tensor index pair in (8.45) producing

$$\begin{aligned}\mathcal{L}(k) = & \frac{\partial k}{\partial t} + \bar{u}_j\frac{\partial k}{\partial x_j} + \frac{\partial}{\partial x_k}\left[\left(\mathrm{C}_k^{-1}\frac{k}{\varepsilon}\overline{u_l'u_k'} - \frac{\delta_{lk}}{\mathrm{Pr}}\right)\frac{\partial k}{\partial x_l}\right] \\ & - \overline{u_i'u_k'}\frac{\partial \bar{u}_j}{\partial x_k} + \varepsilon + \frac{2\beta\nu^t}{\mathrm{Pr}^t}\frac{\partial T}{\partial x_i}\delta_{\mathrm{g}\,i} = 0\end{aligned} \tag{8.55}$$

The ε transport PDE *model*, Tennekes and Lumley (1974), augmented with molecular diffusion and thermal source terms, is

$$\begin{aligned}\mathcal{L}(\varepsilon) = & \frac{\partial \varepsilon}{\partial t} + \bar{u}_j\frac{\partial \varepsilon}{\partial x_j} + \frac{\partial}{\partial x_k}\left[\left(\mathrm{C}_\varepsilon^{-1}\frac{k}{\varepsilon}\overline{u_l'u_k'} - \frac{\delta_{lk}}{\mathrm{Pr}}\right)\frac{\partial \varepsilon}{\partial x_l}\right] \\ & -\mathrm{C}_\varepsilon^1\overline{u_i'u_k'}\frac{\varepsilon}{k}\frac{\partial \bar{u}_i}{\partial x_k} + \mathrm{C}_\varepsilon^2\frac{\varepsilon^2}{k} + \mathrm{C}_\varepsilon^3\,\beta\nu^t\frac{\varepsilon}{k}\frac{\partial T}{\partial x_i}\delta_{\mathrm{g}\,i} = 0\end{aligned} \tag{8.56}$$

Mathematically non-fundamental alterations to this development occur upon replacing the k-ε with the k-ω dual-PDE closure, but the modeling specifics are truly distinct. The requirement remains to close the dissipation term in (8.45–8.46); the k-ω closure model defines

$$\varepsilon_{ij} \equiv \frac{2}{3}\beta^{*}_{RSM} k\,\omega\,\delta_{ij} \tag{8.57}$$

where β^{*}_{RSM} is the RSM alteration to the β^{*} function in (4.63).

Higher order pressure-strain closure models are available which add implementation complexity but do not fundamentally alter the mathematical character of (8.45–8.57) as I-EBV. BCs are thus required *a priori* known and specified on the solution domain closure entirety. For the low Re^t wall layer resolved, as detailed in Section 8.2, the BC for k is Dirichlet zero hence also the normal stress components $\overline{u_i'u_i'} = 0, i = 1, 2, 3$, no summation. Homogeneous Neumann, proposed for ε, is also appropriate for the shear stresses, $\partial\overline{u_i'u_k'}/\partial x_j = 0, i \neq k$.

Resolution of low Re^t wall layer reach requires modifications to the detailed RSM closure, as identified for the k-ε model. This alteration, with its extensive DOF count, renders this RSM model rather compute intensive. The alternative BC formulation employs BL similarity theory, (8.19), adapted to n dimensions. Wilcox (2006, Figure 2.5) graphs the distributions of measured Reynolds stress normal components in the wall-dominated reach of a flat plate turbulent BL. For x_j spanning the dominant (BL) flow, wall normal and wall transverse directions, $1 \leq j \leq 3$, a rough ordering of Reynolds normal stress near-wall magnitudes is 4:2:3.

In terms of friction velocity u_τ, (4.81), suggested Dirichlet BC specifications are, ANSYS (2011)

$$\frac{\overline{u_1'u_1'}}{u_\tau^2} = 5.1, \frac{\overline{u_2'u_2'}}{u_\tau^2} = 1.0, \frac{\overline{u_3'u_3'}}{u_\tau^2} = 2.3, -\frac{\overline{u_1'u_2'}}{u_\tau^2} = 1.0 \tag{8.58}$$

with homogeneous Neumann BCs for the other two Reynolds shear stress components.

Viewing the Reynolds stress tensor transport PDE closure (8.47–8.56) confirms the dominant presence of the rational function k/ε and to a lesser extent ε/k. As discussed, ε is a rather ill-behaved variable away from regions of strong shear which tends to destabilize discrete CFD algorithm computations. This has prompted replacement of k/ε and its product with $\overline{u_l'u_k'}$ with gradient-diffusion models. For example, (8.46) is replaced with

$$\left[\overline{u_i'u_j'u_k'} + \frac{\overline{p'}}{\rho}\left(\delta_{jk}u_i' + \delta_{ik}u_j'\right)\right] \cong C_s \frac{k}{\varepsilon}\overline{u_k'u_l'}\frac{\partial\overline{u_i'u_j'}}{\partial x_l} \approx -\frac{\nu^t}{C_k^a}\frac{\partial\overline{u_i'u_j'}}{\partial x_k} \tag{8.59}$$

where $C_k^a = 0.6$ is distinct from the k-ε closure standard specification $C_k = 1.0$ in (8.16). Similarly derived replacement models for (8.55–8.56) are

$$\left(C_k^{-1}\frac{k}{\varepsilon}\overline{u_l'u_k'} - \frac{\delta_{lk}}{\mathrm{Pe}}\right)\frac{\partial k}{\partial x_l} \approx -\left(\nu + \frac{\nu^t}{C_k}\right)\frac{\partial k}{\partial x_k} \tag{8.60}$$

$$C_\varepsilon^{-1}\frac{k}{\varepsilon}\overline{u_l'u_k'}\frac{\partial\varepsilon}{\partial x_l} \approx -\left(\nu + \frac{\nu^t}{C_\varepsilon}\right)\frac{\partial\varepsilon}{\partial x_k} \tag{8.61}$$

These closure models become identical to the turbulent eddy viscosity forms, (8.16–8.17), upon non-dimensionalizing.

8.8 RSM Closure $m\text{GWS}^h + \theta\text{TS}$ Algorithm

The RSM *m*PDE system is a minor variation on the RaNS flux vector statement (8.20). The RaNS time averaged non-D *continuum* state variable remains $\{q(\mathbf{x}, t)\} = \{u_i, \Theta, Y, k, \varepsilon, \phi, \tau_{ij}; P\}$. Defining $\tau_{ij} \equiv \overline{u_i' u_j'}$ as an *m*PDE theory addressable transport variable the time averaged RaNS *m*PDE system remains

$$\mathcal{L}^m(\{q\}) = [m]\frac{\partial\{q\}}{\partial t} + \frac{\partial}{\partial x_j}\left[\{f_j\} - \{f_j^d\}\right] - \{s\} = \{0\} \tag{8.20}$$

The *mass* matrix $[m]$ now inherits significance regarding $O(m^3)$ phase accurate scalar transport as the RaNS RSM closure is indeed *unsteady*. The matrix is unaltered from (8.21)

$$[m] \equiv \left[1 - \frac{\gamma\Delta t^2}{6}\frac{\partial}{\partial x_j}\left(u_j u_k \frac{\partial}{\partial x_k}\right)\right] \tag{8.21}$$

with $k = 1$ basis optimal parameter $\gamma = -0.5$. The RSM closure *kinetic flux vector* $\{f_j\}$ becomes

$$\{f_j(\{q\})\} = \begin{Bmatrix} u_j u_i + P^*\big|_{n+1}^{p}\delta_{ij} \\ u_j\Theta \\ u_j Y \\ u_j\, k \\ u_j\, \varepsilon \\ u_j\, \tau_{ik} \end{Bmatrix} \tag{8.62}$$

with Reynolds stress tensor label altered to τ_{ik} for index consistency.

The dissipative flux vector is similarly altered for τ_{ik}. Inserting the simplified eddy viscosity closure models (8.59–8.61) generates

$$\left\{f_j^d(\{q\})\right\} = \begin{Bmatrix} \frac{2}{\text{Re}}(1+\text{Re}^t)S_{ij} + \frac{h^2\text{Re}}{6(1+\text{Re}^t)}\left(u_j u_k \frac{\partial u_i}{\partial x_k} + u_i u_k \frac{\partial u_j}{\partial x_k}\right) \\ \frac{1}{\text{Re}}\left(\frac{1}{\text{Pr}} + \frac{\text{Re}^t}{\text{Pr}^t}\right)\frac{\partial\Theta}{\partial x_j} + \frac{h^2\text{RePr}}{12(1+\text{Re}^t)(\text{Pr}/\text{Pr}^t)}u_j\, u_k \frac{\partial\Theta}{\partial x_k} \\ \frac{1}{\text{Re}}\left(\frac{1}{\text{Sc}} + \frac{\text{Re}^t}{\text{Sc}^t}\right)\frac{\partial Y}{\partial x_j} + \frac{h^2\text{ReSc}}{12(1+\text{Re}^t)/(\text{Sc}/\text{Sc}^t)}u_j\, u_k \frac{\partial Y}{\partial x_k} \\ \frac{1}{\text{Pe}}\left(1 + \frac{\text{Re}^t}{\text{C}_k}\right)\frac{\partial k}{\partial x_j} + \frac{h^2\text{Pe}}{12(1+\text{Re}^t/\text{C}_k)}u_j\, u_k \frac{\partial k}{\partial x_k} \\ \frac{1}{\text{Pe}}\left(1 + \frac{\text{Re}^t}{\text{C}_\varepsilon}\right)\frac{\partial \varepsilon}{\partial x_j} + \frac{h^2\text{Pe}}{12(1+\text{Re}^t/\text{C}_\varepsilon)}u_j\, u_k \frac{\partial \varepsilon}{\partial x_k} \\ \frac{1}{\text{Re}}(1+\text{Re}^t)\frac{\partial\tau_{ik}}{\partial x_j} + \frac{h^2\text{Re}}{12(1+\text{Re}^t)}u_j\, u_l \frac{\partial\tau_{ik}}{\partial x_l} \end{Bmatrix} \tag{8.63}$$

The pervasive repetition of term structure for RSM closure in the RaNS dissipative flux vector is clearly evident. The source matrix $\{s\}$ absent the $O(h^4)$ mPDE terms (a suggested exercise) is

$$\{s\} = \left\{ \begin{array}{l} -\dfrac{\mathrm{Gr}}{\mathrm{Re}^2}\Theta\widehat{g}_i \\ \dfrac{\mathrm{Ec}}{\mathrm{Re}}\dfrac{\partial}{\partial x_j}\left(u_i\dfrac{\partial u_j}{\partial x_i}\right) + \dfrac{\partial^2 \tau_{ij}}{\partial x_i \partial x_j} \\ s(Y) \\ \tau_{jk}\dfrac{\partial u_j}{\partial x_k} - \varepsilon - \dfrac{\mathrm{Gr}}{\mathrm{Re}^2}\dfrac{\varepsilon}{k}\dfrac{\partial \Theta}{\partial x_i}\delta_{\mathbf{g}\,i} \\ \mathrm{C}^1_\varepsilon f^1 \tau_{jk}\dfrac{\varepsilon}{k}\dfrac{\partial u_j}{\partial x_k} - \mathrm{C}^2_\varepsilon f^2 \dfrac{\varepsilon^2}{k} - \mathrm{C}^3_\varepsilon \dfrac{\mathrm{GrRe}^t}{\mathrm{Re}^3}\dfrac{\varepsilon}{k}\dfrac{\partial \Theta}{\partial x_i}\delta_{\mathbf{g}\,i} \\ \Phi^1_{ij} + \Phi^2_{ij} + \Phi^3_{ij} - \dfrac{2}{3}\varepsilon\,\delta_{ij} - \dfrac{\mathrm{GrRe}^t}{\mathrm{Re}^2\mathrm{Pr}^t}\left(g_i\dfrac{\partial \Theta}{\partial x_j} + g_j\dfrac{\partial \Theta}{\partial x_i}\right) \end{array} \right\} \tag{8.64}$$

The RSM closure (8.62–8.64) alteration to the RaNS mPDE flux vectors and source, (8.22–8.24), are certainly detailed but not fundamental. Thereby, the Section 8.4 k-ε closure of the RaNS $m\mathrm{GWS}^h + \theta\mathrm{TS}$ algorithm remains the essential description for theory implementation via the $k = 1$ TP/NC basis.

The predominance of the rational function ε/k and its product with τ_{jk} generates additional nonlinearities in $\{s\}$. These are discretely implemented with precision via element DOF ratio multipliers on hypermatrix statements, recall (8.31). This again facilitates construction of accurate jacobian contributions via calculus. As turbulent eddy viscosity remains a nonlinear parameter in the RSM closure, the differential relations (8.38) contribute to jacobian formation precision.

In closing, the rigorously derived RSM closure significantly alters (8.63) due to tensor embedding. Reverting to the mixed vector/tensor notation introduced in (8.1–8.4) for visual clarity, and substituting $\beta\Delta t/2$ for $h^2\mathrm{Re}/12(1+\mathrm{Re}^t)$ for the exposure, the replacement for (8.63) is

$$\left\{f^d_j(\{q\})\right\} = \left\{ \begin{array}{l} \left(\dfrac{2}{\mathrm{Re}}\delta_{jk} - \overline{u'_j u'_k}\right)S_{ik} + \dfrac{\beta\Delta t}{2}\left(u_j u_k\dfrac{\partial u_i}{\partial x_k} + u_i u_k\dfrac{\partial u_j}{\partial x_k}\right) \\ \left(\dfrac{1}{\mathrm{RePr}}\delta_{jk} - \overline{u'_k\Theta'}\right)\dfrac{\partial \Theta}{\partial x_k} + \dfrac{\beta\Delta t}{2}u_j u_k\dfrac{\partial \Theta}{\partial x_k} \\ \left(\dfrac{1}{\mathrm{ReSc}}\delta_{jk} - \overline{u'_k Y'}\right)\dfrac{\partial Y}{\partial x_k} + \dfrac{\beta\Delta t}{2}u_j u_k\dfrac{\partial Y}{\partial x_k} \\ \left(\dfrac{1}{\mathrm{RePr}}\delta_{jk} - \dfrac{1}{\mathrm{C}_k}\dfrac{k}{\varepsilon}\overline{u'_j u'_k}\right)\dfrac{\partial k}{\partial x_k} + \dfrac{\beta\Delta t}{2}u_j u_k\dfrac{\partial k}{\partial x_k} \\ \left(\dfrac{1}{\mathrm{RePr}}\delta_{jk} - \dfrac{1}{\mathrm{C}_\varepsilon}\dfrac{k}{\varepsilon}\overline{u'_j u'_k}\right)\dfrac{\partial \varepsilon}{\partial x_k} + \dfrac{\beta\Delta t}{2}u_j u_k\dfrac{\partial \varepsilon}{\partial x_k} \\ \left(-C_s\dfrac{k}{\varepsilon}\overline{u'_j u'_l}\right)\dfrac{\partial \tau_{ik}}{\partial x_l} + \dfrac{\beta\Delta t}{2}u_j u_l\dfrac{\partial \tau_{ik}}{\partial x_l} \end{array} \right\} \tag{8.65}$$

Note that (8.65) lacks closure due to the presence of $\overline{u_k'\Theta'}$ and $\overline{u_k'Y'}$. The typical resolution is definition of turbulent Prandtl and Schmidt numbers, whence these two entries in (8.65) recover the (8.63) form upon the identification $\overline{u_j'u_k'} \equiv \tau_{j\,k}$.

8.9 RSM Closure Model Validation

As no surprise the classic validation for the RSM closure (8.62–8.64) is steady turbulent flow in a straight rectangular cross-section duct. As detailed in Section 4.13, the historical RaNS PNS algorithm closed with anisotropic *algebraic* Reynolds stress model generated a transverse plane counter-rotating vortex pair symmetric about the bisector of a square cross-section duct quadrant. These historical CFD *a posteriori* data are in firm (compute resource limited) *quantitative* agreement with the experimental data, Melling and Whitelaw (1976), Figure 4.18.

The presented RaNS $m\text{GWS}^h + \theta\text{TS}$ algorithm has not been implemented for RSM closure. However, a commercial CFD code, (ANSYS, 2011), provides the opportunity to validate the RSM closure via transverse plane outflow DOF distribution reflection as the inflow plane Dirichlet BC. This enables a very modest meshing in the direction of predominant flow, and at algorithm convergence generates a fully developed RaNS turbulent flowfield DOF distribution on all transverse planes.

For modestly wall-attracted M = 33x33 transverse plane mesh, with log-law BCs and Re = 0.42 E+05 matching the experiment, the BC manipulation converges to a steady turbulent velocity field containing a quadruple of transverse plane counter-rotating vortices symmetrically disposed about each quadrant bisector, Figure 8.13 top. The extremum is ~1% of onset speed U, in excellent quantitative agreement with the experiment. The vortex magnitude is sufficient to alter axial velocity isovels from being wall parallel, Figure 8.13 bottom.

The anisotropic RSM closure indeed captures the square duct turbulent flowfield kinetics with *quantitative* validation resting on comparison with experiment. The Melling and Whitelaw (1976) validation data is restricted to the duct lower right quadrant. Figure 8.14 compares this transverse plane velocity vector distribution with the CFD solution, appropriately cropped. Identical velocity vector scaling therein documents excellent *quantitative* agreement, except for the experiment not exhibiting the CFD solution precise mirror symmetry.

Close examination of the non-dimensional axial velocity isovel contour distributions, Figure 8.15, confirms that the CFD solution core velocity vector field is considerably higher speed than that of the experiment. In particular, the CFD solution transverse plane area inside the $u^h/\text{U} = 0.975$ isovel is a factor of ~4 larger than the experiment. Notwithstanding, quantitative agreement on shape and distribution of the isovel contours is excellent.

It is well established that Reynolds stress tensor normal component anisotropy is the vortex causal mechanism. Comparison of the CFD solution $\overline{u_2'u_2'}$ non-D distribution with experiment, Figure 8.16, confirms really quite excellent quantitative agreement, realizing how difficult this measurement must have been experimentally. The CFD data confirm the experiment was unable to resolve these normal stress tensor minimal magnitudes.

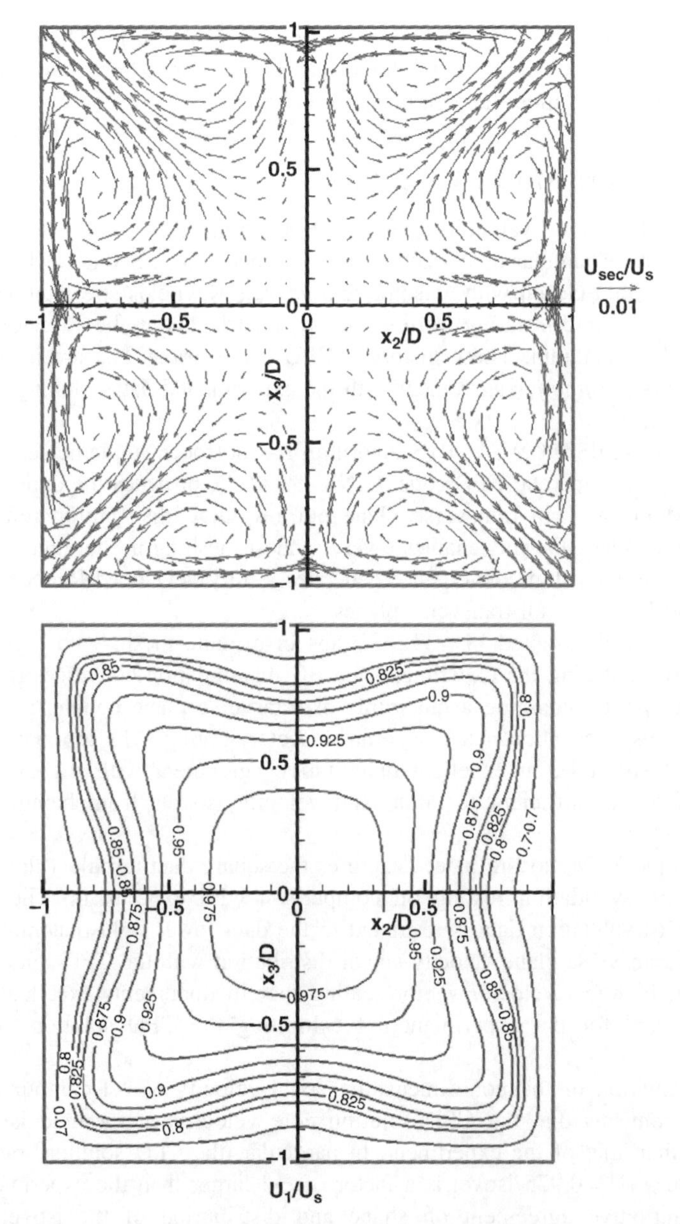

Figure 8.13 RSM closed RaNS algorithm steady solution, turbulent flow in square cross-section duct, Re = 0.42 E+05, M = 33×33 wall-attracted transverse plane mesh, log-law BCs

8.10 Geologic Borehole Conjugate Heat Transfer

Thermal resource extraction typically involves relatively small diameter holes drilled deeply into the earth mantle. A green example is the compact vertical borehole heat exchanger (BHE) operational in a ground-source heat pump (GHP) installation. The BHE device possesses

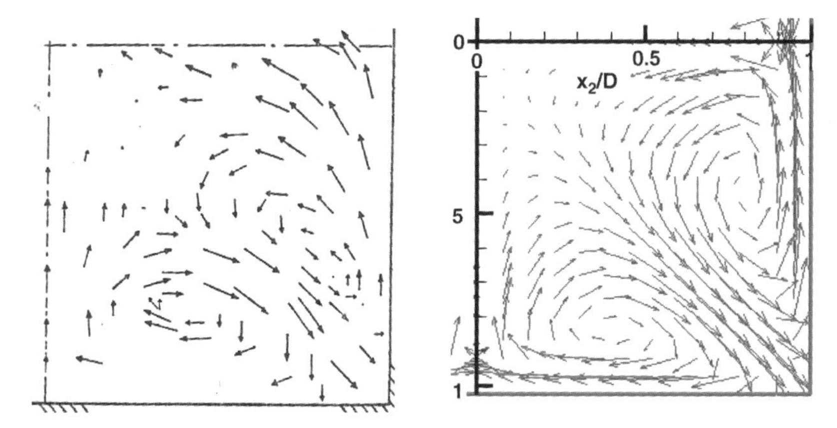

Figure 8.14 Turbulent flow in a square cross-section duct, Re = 0.42 E+05, transverse plane velocity vector distribution: left, experiment of Melling and Whitelaw (1976), with permission, *J. Fluid Mechanics*; right, RSM closed RaNS steady solution, wall-attracted M = 33x33 transverse plane mesh, log-law BCs. *Source:* Melling, A. and Whitelaw, J.H. 1976. Reproduced with permission of Cambridge University Press

exceptionally disparate characterizing length scales ranging from bore diameter $O(\mathrm{m}^{-2})$ to borehole depth $O(\mathrm{m}^{+2})$. Thereby mesh resolution sufficient to support accurate three-dimensional fully coupled conjugate heat transfer CFD simulation leads to an impractical mesh number M or mesh aspect ratios acknowledged to generate inaccurate results.

An iterative algorithm coupling local RaNS predictions with a *parabolic* RaNS (PRaNS) algorithm for D*E* is derived and validated for steady three-dimensional fully coupled BHE conjugate heat transfer prediction, Baker *et al.* (2011). Since BHE conduit diameters of

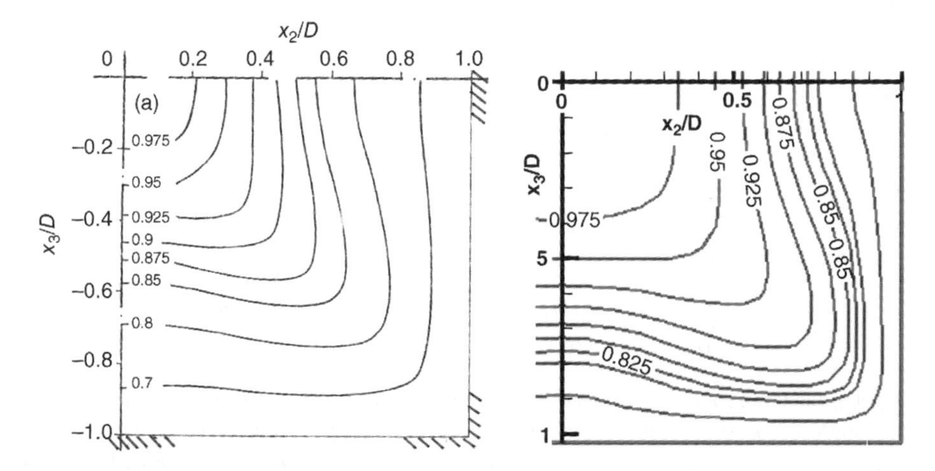

Figure 8.15 Turbulent flow in a square cross-section duct, Re = 0.42 E+05, transverse plane axial velocity distribution: left, experiment of Melling and Whitelaw (1976), reprinted with permission, *J. Fluid Mechanics*: right, RSM RaNS steady solution, wall-attracted M = 33x33 transverse plane mesh, log-law BCs. *Source:* Melling, A. and Whitelaw, J.H. 1976. Reproduced with permission of Cambridge University Press

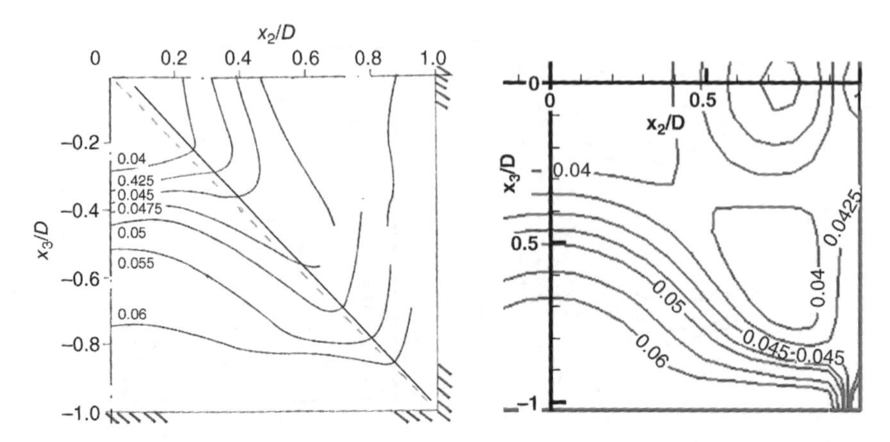

Figure 8.16 Turbulent flow in a square cross-section duct, Re = 0.42 E+05, transverse plane normal stress $\overline{u_2'u_2'}$ distribution: left, experiment of Melling and Whitelaw (1976), reprinted with permission, *J. Fluid Mechanics*: right, RSM RaNS steady solution, wall-attracted M = 33×33 transverse plane mesh, log-law BCs. *Source:* Melling, A. and Whitelaw, J.H. 1976. Reproduced with permission of Cambridge University Press

$O(10^{-2})$ m are coupled with fluid travel distances of $O(10^2)$ m, the conduit axial velocity distribution rapidly becomes fully developed and unidirectional with transverse plane velocity magnitude insignificant.

Hence it is not required to actually solve full RaNS *or* complete PRaNS systems. The exception to this simplification occurs during flow direction reversal from return to supply conduit at borehole depth in the BHE end cap. The developed algorithm addresses this exception. Finally, the naming of *return* and *supply* conduits appears counter-intuitive but does adhere to HVAC industry nomenclature.

In the parabolic simplification the BHE conduit axial velocity vector $w(x, y, z) \Rightarrow w(x, y)$ distribution becomes input *data* computed once only via the reflected BC RaNS algorithm procedure. The PRaNS PDE requirement thus reduces to addressing only D*E* for conjugate heat transfer coupling among the BHE return and supply velocity fields with the conduit piping, grout and surrounding geologic formation.

The parabolic simplification to non-D RaNS I-EBV D*E*, (8.3), recall (4.109), for steady flow within a BHE conduit is

$$\mathcal{L}(\Theta) = w\frac{\partial\Theta}{\partial z} - \frac{1}{\text{RePr}}\left[\frac{\partial}{\partial x}\left((1+\kappa^t)\frac{\partial\Theta}{\partial x}\right) + \frac{\partial}{\partial y}\left((1+\kappa^t)\frac{\partial\Theta}{\partial y}\right)\right] = 0 \tag{8.66}$$

for Θ potential temperature. The z coordinate is parallel to w hence normal to x, y planes, κ^t is turbulent *thermal diffusivity* with “1” denoting the non-D molecular continuum value.

Since $w(x, y)$ is non-zero *only* in a conduit the steady EBV D*E* PDE pertinent to conduit piping, grout and geologic formation is

$$\mathcal{L}(\Theta) = -\kappa\left[\frac{\partial^2\Theta}{\partial x^2} + \frac{\partial^2\Theta}{\partial y^2}\right] = 0 \tag{8.67}$$

where $\kappa = \rho c_p/k$ is the placeholder for *piecewise discontinuous* non-fluid BHE system constituent thermal diffusivities.

The requirement is to join (8.66) and (8.67) into a unique D*E* PDE spanning the entire $n = 3$ solution domain. Dividing (8.67) by Re then multiplying through by Pr identifies a non-D parameter α, the ratio of non-fluid to fluid thermal diffusivity in (8.67). This enables D*E* PDE coalescence into an I-EBV PDE valid throughout the entire solution domain

$$\mathcal{L}(\Theta) = w\frac{\partial\Theta}{\partial z} - \frac{1}{\mathrm{RePr}}\left[\frac{\partial}{\partial x}\left((\alpha + \kappa^t)\frac{\partial\Theta}{\partial x}\right) + \frac{\partial}{\partial y}\left((\alpha + \kappa^t)\frac{\partial\Theta}{\partial y}\right)\right] = 0 \qquad (8.68)$$

In (8.68) w and κ^t are non-zero *only* in the fluid circuits wherein $\alpha \equiv 1$. Elsewhere α is constituent thermal diffusivity normalized by that of the fluid. The characteristic length and velocity scales appropriate for Re are the BHE conduit hydraulic diameter and area-averaged volume flow rate and Pr is for the working fluid.

Since (8.68) is an I-EBV PDE with z spanning the initial value direction an initial condition (IC) is required. Farfield encompassing BCs are also required; the solution adapted mesh domain extending to $\sim 10^3$ BHE diameters admits farfield homogeneous Neumann BC imposition. The iterative $\mathrm{GWS}^h + \theta\mathrm{TS}$ PRaNS algorithm process for (8.68) is:

- via a RaNS code compute the conduit fully developed laminar or turbulent axial velocity distribution $w(x, y)$ and turbulent thermal diffusivity distribution $\kappa^t\ (x, y)$ from $\mathrm{Re}^t(x, y)$ and a Pr^t assumption
- define an IC $\Theta(x, y, z)$ as a linear variation in z from ground plane to the uniform subterranean temperature over a vertical distance Δz, then constant at the subterranean level to the BHE end cap; define α for all non-fluid constituents of the installation
- starting at the ground plane $z = 0$ and for the return conduit fluid IC $\Theta(x, y, 0)$, integrate (8.68) *down* the return conduit terminating at the BHE end cap
- store this temperature distribution estimate $\Theta^h(x, y, z)$ DOF, then map the return conduit endcap plane solution $\Theta^h(x, y, z_{\mathrm{cap}})$ onto the supply conduit plane as the DOF distribution IC for the upward integration sweep
- integrate (8.68) *up* the supply conduit over $z_{\mathrm{cap}} \le z \le 0$, imposing as *data* the just determined fluid temperature $\Theta^h(x, y, z)$ DOF distribution inside the return conduit *only*
- store the generated 3-D $\Theta^h(x, y, z)$ iterate distribution, then starting with IC $\Theta(x, y, 0)$ integrate (8.68) *down* the return conduit terminating at the BHE endcap while imposing as fixed distributed *data* the previously computed $\Theta^h(x, y, z)$ DOF distribution inside the supply conduit *only*.

Repeat this supply/return/... conduit space marching process for (8.68) until the 3-D temperature distribution converges to a unique solution $\Theta^h(x, y, z)$ inside both BHE fluid conduits *and* throughout the piping, grout and geologic formation domain.

BHE operates via thermal conduction to/from the geologic formation through the grout and conduit wall to the conduit fluid, hence advection to the GHP. The performance optimization opportunity is to replace the conventional BHE U-tube totally encased in low conductivity grout, Figure 8.17 left. A tested design alteration is the bisected (BiSec) BHE, Baker *et al.* (2011), which minimizes the surrounding low diffusivity grout thickness while significantly

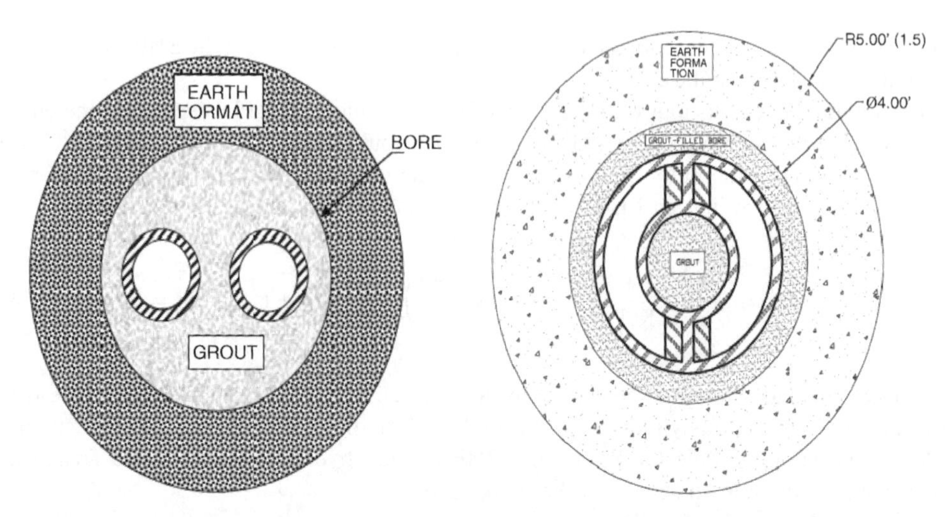

Figure 8.17 Vertical borehole heat exchanger (BHE) transverse plane cross-section: left, traditional U-tube; right, bisected (BiSec) design

increasing conduit heat conduction surface area for transfer to the fluid, Figure 8.17 right. Note the BiSec grouting-in process is via the hollow central core; the grout solidifies therein, further contributing to thermal isolation between supply and return conduits.

A BiSec BHE field installation with long time operational temperature performance data has a borehole depth of 90 m. The cross-section average flowrate is $U = 0.39$ m/s with hydraulic diameter $D = 0.024$ m. The conduit fluid is water, $Pr = 6.58$, and for these data $Re_D = 10{,}900$ a *transitional* flow Reynolds number. Indeed, using BL similarity BCs, (8.19), the RaNS k-ε closure model prediction of transverse plane velocity distribution $w(x, y)$ failed the y^+ test, recall Figure 4.8. Wall adjacent mesh refinement enabled specification and use of the low Re^t model, (8.18). Figure 8.18 summarizes this solution; note the BiSec mid-cross-section velocity $w^h(s)$ DOF distribution, lower left, is at best only approximately turbulent.

The second RaNS CFD component is predicting the non-PRaNS state variable distribution in the BiSec endcap turnaround at 90 m depth. This 3-D RaNS solution domain extends 50 BiSec diameters in all directions, discretized into an $M \approx 2\times10^6$ boundary fitted mesh. The endcap entrance/exit plane RaNS solutions are compared left/right in Figure 8.19. Top to bottom, flow reversal generates a transverse plane vortex pair which totally alters the temperature distribution and nominally quadruples the turbulent diffusivity level extremum.

The velocity vector field exiting the end cap is fully 3-D. But from entrance length considerations this non-PNS distribution will wash out to a $w^h(x, y)$ distribution within $\sim 1\%$ of supply conduit travel to the surface. The PRaNS D*E* algorithm accommodation is to homogenize the return conduit endcap plane temperature distribution as the $\Theta^h(x, y, z_{cap})$ DOF IC for each supply conduit sweep. Figure 8.20 summarizes, top to bottom, the *nearfield* converged BiSec solution for (8.68) in terms of solution adapted mesh (dashed line is grout outer boundary), end cap plane $\Theta^h(x, y, z_{cap})$ distributions showing homogenized supply IC, and ground plane converged $\Theta^h(x, y, z = 0)$ distribution.

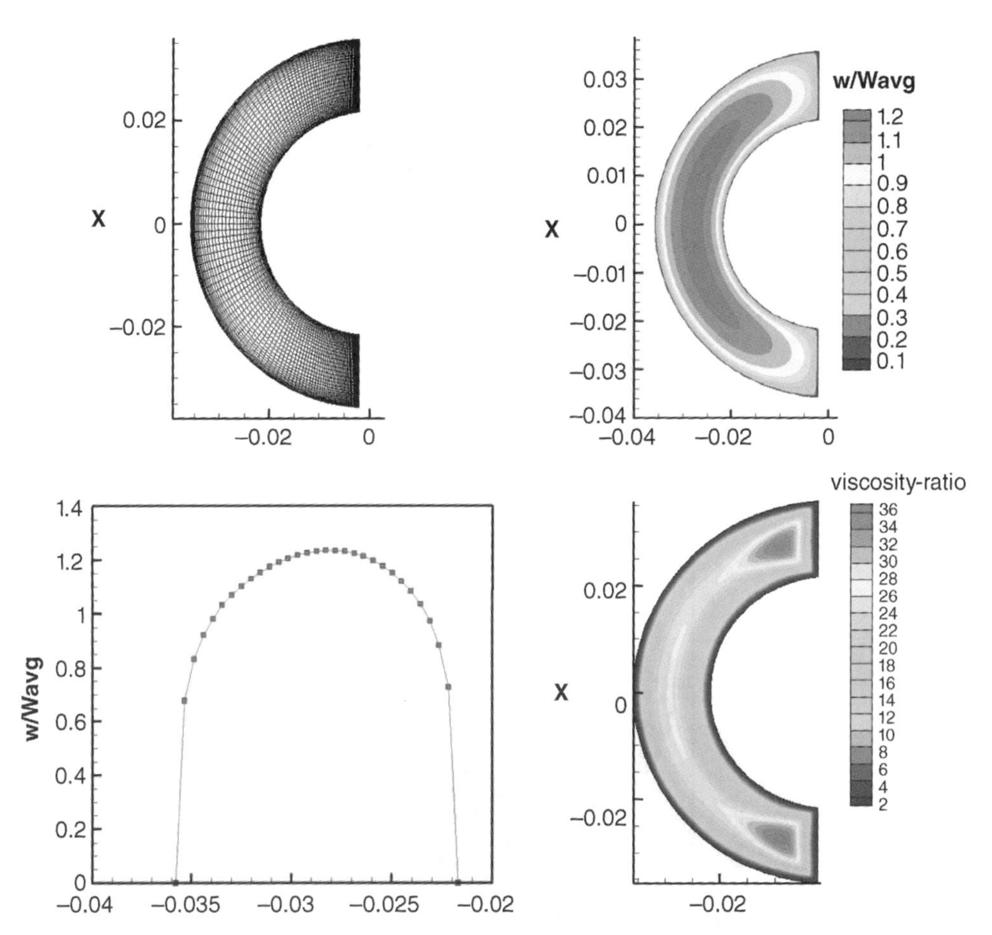

Figure 8.18 ANSYS CFD code steady solution, k-ε closure, turbulent flow assumption, BiSec BHE geometry, $\text{Re}_\text{D} = 10{,}600$, clockwise: transverse plane mesh for low Re^t BCs; non-D axial CFD velocity distribution $w^h(x, y)$; non-D turbulent thermal diffusivity distribution $\kappa^T(x, y)$, $\text{Pr}^t \equiv \text{Pr}$; BiSec cross-section velocity $w^h(s)$ DOF distribution

The PRaNS algorithm iteration sequence converges to a steady three-dimensional temperature distribution $\Theta^h(x, y, z)$ in five return/supply conduit sweeps. The return/supply conduit flow average dimensional temperature $\overline{T}(z)$ distribution with depth is graphed in Figure 8.21, left/right. The BiSec end cap location is at mid-abscissa. In the absence of first iteration data the return conduit temperature prediction is reflected into the supply conduit during the first downward sweep, labeled 1 in the right graph. This poor approximation rapidly washes out, and both conduit temperature distributions thereafter exhibit uniform monotone convergence to the steady solution.

In summer air conditioning mode the BiSec field installation fluid average temperature recorded change was $\Delta\overline{T} = 5.6 \pm 0.1$ K. Data measurements were taken at 10 minute intervals over a 14 hour period, Baker *et al.* (2011, Figure 9). Table 8.1 lists the parametric

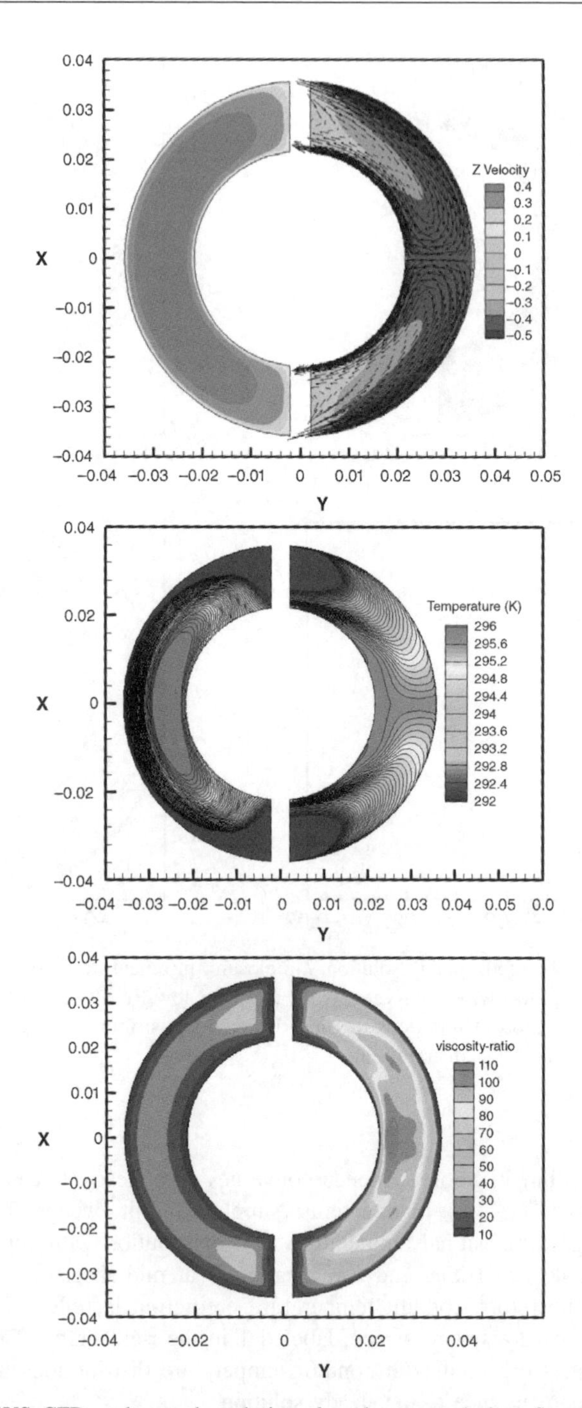

Figure 8.19 ANSYS CFD code steady solution, *k*-ε closure, turbulent flow, BiSec BHE endcap return/supply flow reversal, $Re_D = 10{,}600$, left/right in each graphic, top to bottom: axial velocity floods with transverse plane vector overlay; temperature distributions; non-D turbulent diffusivity distributions

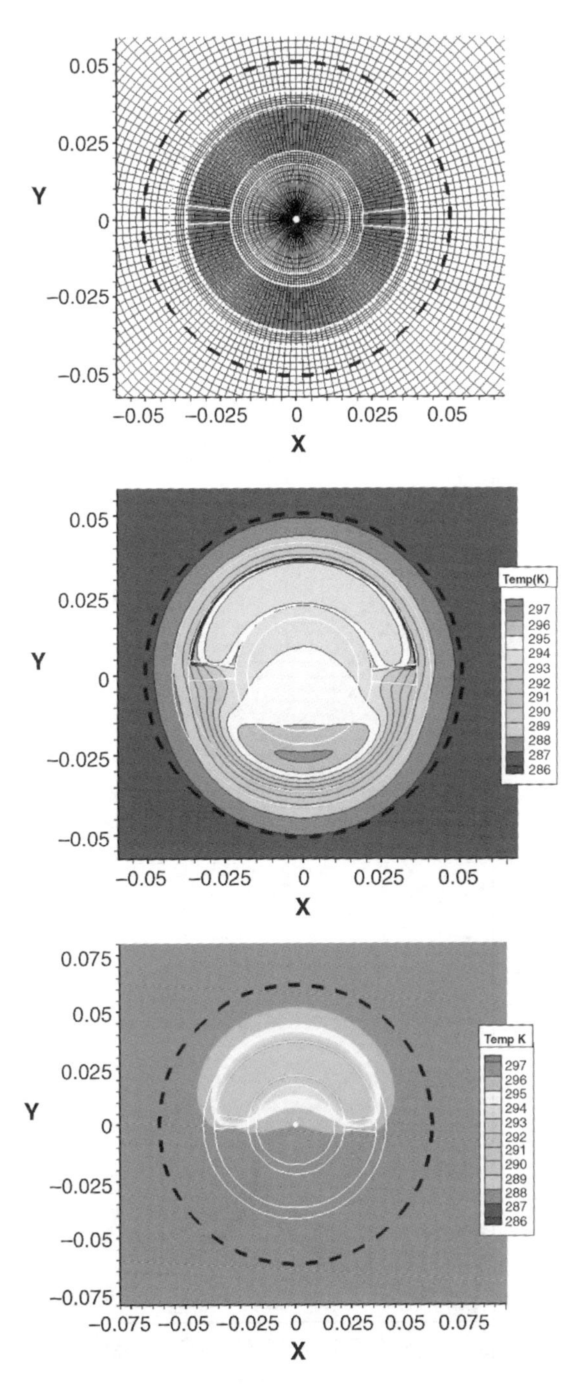

Figure 8.20 BiSec PRaNS D*E* algorithm near field solution, $Re_D = 10{,}600$, top to bottom: solution adapted mesh, endcap plane $\Theta^h(x, y, z_{cap})$ distribution showing homogenized supply conduit IC (at top), ground plane converged solution $\Theta^h(x, y, z = 0)$, from Baker *et al.* (2011)

Table 8.1 BiSec PRaNS validation test parametric data

BiSec component	non-D thermal diffusivity	diffusivity ratio α
piping	0.2176	1.435
grout	0.4663	3.070
formation	0.9735	6.562
fluid	0.1520	1.000

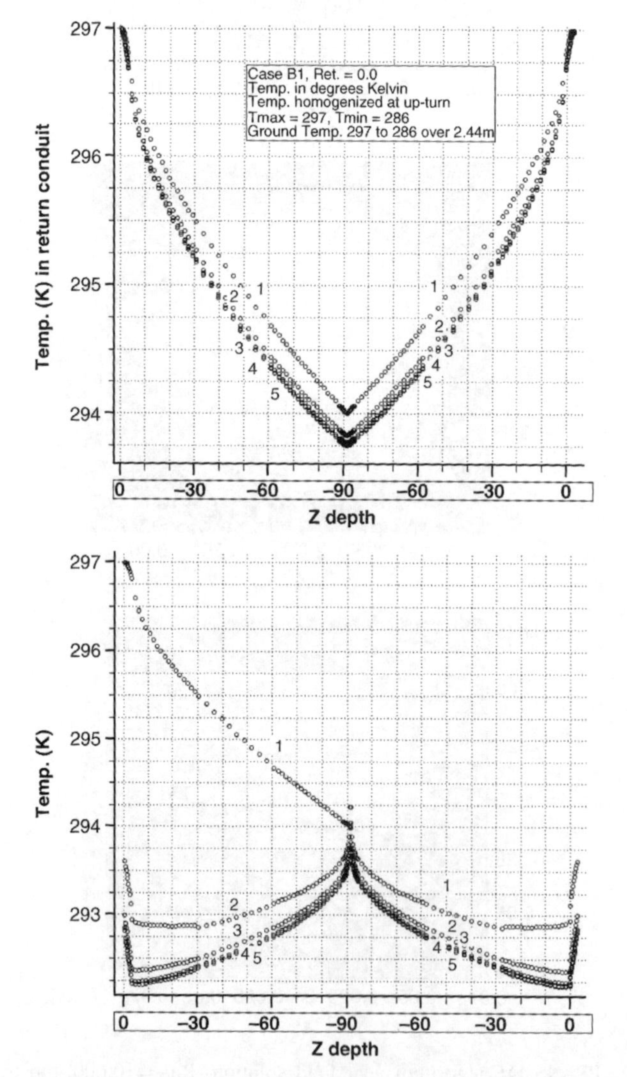

Figure 8.21 BiSec BHE PRaNS algorithm $\overline{T}(z)$ convergence, $\mathrm{Re_D} = 10{,}600$, $\mathrm{Pr} = 6.58$, geologic $T(x, y, z)$ interpolated from ground plane to subterranean temperature over $\Delta z = 2.4$ m, supply conduit homogenized $\Theta^h(x, y, z_{cap})$ DOF IC: top, return conduit; bottom, supply conduit

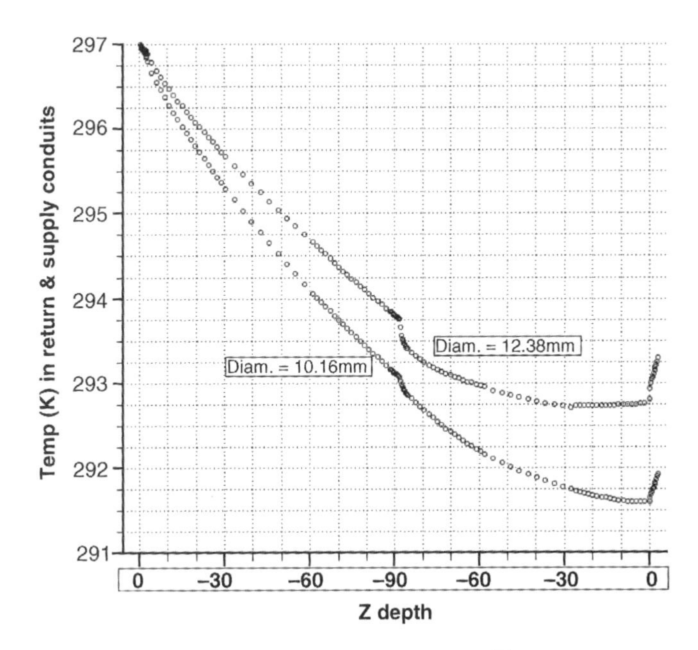

Figure 8.22 BiSec BHE PRaNS algorithm converged solutions $\overline{T}(z)$, $\text{Re}_\text{D} = 10{,}600$, $\text{Pr} = 6.58$, standard and doubled grout thickness

input data for (8.68) for this installation. For laminar flow the PRaNS algorithm prediction is $\Delta\overline{T} = 3.3$ K. The transitional turbulent flow prediction is $\Delta\overline{T} = 5.1$ K, which agrees with measurements within ~10%. This agreement level constitutes firm validation of the PRaNS algorithm.

In addition to conduit increased heat conduction surface area, a design feature of BiSec is minimization of surrounding grout thickness, the necessary low diffusivity separation between conduit and formation. Maintaining all data specifications constant the PRaNS algorithm performance prediction for doubling the grout thickness to ~12 mm is $\Delta\overline{T} = 3.7$ K, Figure 8.22, a ~30% degradation of performance.

The prime *observable* in PRaNS algorithm converged solution $\overline{T}(z)$ distributions, Figure 8.21 bottom, also Figure 8.22, is that the *net* BHE performance contribution from the supply conduit flow is *only* $\Delta\overline{T} \approx 1$ K. The algorithm predicts that this poor performance is due to supply conduit fluid heating in traversing the terminal ~2.4 m of the formation at summer ground temperature. These data clearly predict that system performance improvement would result for BiSec flow partition redesign to minimize supply conduit fluid residence time.

Retaining installation diameter and grout thickness, the BiSec return conduit cross-section is increased 30% with a corresponding decrease in supply cross-section. The return conduit hydraulic diameter alteration is compensated by decreased volume average flow rate which preserves $\text{Re}_\text{D} = 10{,}600$. Retaining identical data specifications and IC definitions, the PRaNS algorithm predicts modified BiSec thermal performance is $\Delta\overline{T} = 7.5$ K, a 47%(!) improvement, only a small portion of which results from supply conduit loss reduction to $\Delta\overline{T} = 0.20$ K.

These PRaNS predictions clearly illustrate the role of computational "*what if*" experiments providing quantitative insight for design optimization. Of key pertinence, once set up, the computing effort of this solution strategy is absolutely minimal, involving only minutes of execution time on a laptop PC. Baker *et al.* (2011) cite options for algorithm integration into a global multi-wellbore BHE field prediction algorithm for long-term (multi-year) formation installation performance estimation.

8.11 Summary

Chapter content extends the Navier–Stokes (NS) classic state variable pressure projection theory *optimal modified continuous* Galerkin $m\text{GWS}^h + \theta\text{TS}$ algorithm to the *time averaged* incompressible NS *m*PDE system for thermal-turbulent flow simulation. Time averaging generates the *Reynolds-averaged Navier–Stokes* (RaNS) PDE system which requires closure for the *a priori* unknown *Reynolds stress tensor* $\overline{u_i'u_j'}$ and companion unknown *Reynolds vectors* $\overline{u_j'q'}$.

The *non-dimensional* RaNS PDE system in visually informative vector-tensor mixed notation is presented as (8.1–8.4)

$$\begin{aligned}
&\mathrm{D}M: && \nabla \cdot \mathbf{u} = 0 \\
&\mathbf{DP}: && \frac{\partial \mathbf{u}}{\partial t} + \nabla \cdot \left(\mathbf{uu} + \overline{u_i'u_j'}\right) + \frac{1}{\rho_0}\nabla p - \frac{1}{\mathrm{Re}}\nabla^2 \mathbf{u} + \frac{\mathrm{Gr}}{\mathrm{Re}^2}\Theta\,\widehat{\mathbf{g}} = 0 \\
&\mathrm{D}E: && \frac{\partial \Theta}{\partial t} + \nabla \cdot \left(\mathbf{u}\Theta + \overline{u_j'\Theta'}\right) - \frac{1}{\mathrm{RePr}}\nabla^2 \Theta - s_\Theta = 0 \\
&\mathrm{D}Y: && \frac{\partial Y}{\partial t} + \nabla \cdot \left(\mathbf{u}Y + \overline{u_j'Y'}\right) - \frac{1}{\mathrm{ReSc}}\nabla^2 Y - s_Y = 0
\end{aligned}$$

with state variable *time averaged* velocity, kinematic pressure, temperature and mass fraction, $\{q(\mathbf{x}, t)\} \equiv \{\mathbf{u}, P, \Theta, Y\}^T$.

The pervading RaNS closure model is a *turbulent eddy viscosity* ν^t scalar multiplier on the Stokes strain rate tensor written on time averaged velocity, (8.8)

$$\overline{u_i'u_j'} \equiv \tau_{ij} \propto -\nu^t\,\overline{S}_{ij},\ \overline{S}_{ij} \equiv \frac{1}{2}\left(\frac{\partial \overline{u}_i}{\partial x_j} + \frac{\partial \overline{u}_j}{\partial x_i}\right)$$

Closure (8.8) is ultimately simplified to a *deviatoric tensor* model (8.10)

$$\tau_{ij}^{\mathrm{D}} \equiv -\nu^t\left[\frac{\partial \overline{u}_i}{\partial x_j} + \frac{\partial \overline{u}_j}{\partial x_i}\right], i \neq j$$

with ν^t determined in concert with k-ε and/or k-ω dual-PDE systems.

Non-dimensionalizing leads to the *turbulent Reynolds number* definition (8.11)

$$\mathrm{Re}^t \equiv \nu^t/\nu$$

leading to closure model statement (8.12)

$$\overline{u_i'u_j'} \Rightarrow \tau_{ij}^{\mathrm{D}} \equiv -\frac{\mathrm{Re}^t}{\mathrm{Re}}\left[\frac{\partial \overline{u}_i}{\partial x_j} + \frac{\partial \overline{u}_j}{\partial x_i}\right], i \neq j$$

whence (8.2) becomes closed in the form (8.13)

$$\mathbf{DP}: \quad \frac{\partial \bar{\mathbf{u}}}{\partial t} + \nabla \cdot \overline{\mathbf{u}\mathbf{u}} + \nabla \bar{P} - \frac{2}{\text{Re}} \nabla \cdot \left[(1 + \text{Re}^t) \bar{\mathbf{S}} \right] + \frac{\text{Gr}}{\text{Re}^2} \bar{\Theta} = 0$$

Since $\text{Re}^t \leq 10^3$ is typical, transparent in (8.13) is that the eddy viscosity closure model significantly augments *diffusion* levels in the RaNS PDE compared to NS, the "1" in $(1 + \text{Re}^t)$. The benefit to RaNS algorithm performance is enhanced diffusive moderation of the *dispersion error* mode intrinsic to discrete CFD algorithms addressing NS at large Re.

The remainder of Section 8.2 details extension of Chapter 4 BL closure methodology to RaNS. Altering the NS *optimal* $O(h^4)$ $m\text{GWS}^h$ theory to handle $\varepsilon(q, x)$ in (5.85), the RaNS **DP** mPDE coefficient is predicted as $h^2\text{Re}/12(1 + \text{Re}^t)$, wherein $\text{Re}^t(\mathbf{x}, t)$ is an explicit *non-linear distribution*. The earlier derived mPDE scalar coefficient $\beta\Delta t/2$ is admissible but discarded as non-optimal. Completion of the RaNS optimal mPDE system results upon insertion of the $O(m^3)$ differential term with $k = 1$ TP/NC basis predicted optimal coefficient $\gamma = -0.5$.

For these decisions and *arbitrary* data, the available (linear) asymptotic convergence theory *error estimate* (7.8) remains pertinent to RaNS $m\text{GWS}^h + \theta\text{TS}$ NS state variable discrete approximate solutions

$$\left\| \{e^h(n\Delta t)\} \right\|_E \leq Ch^{2\gamma} \left[\|\text{data}\|_{L2,\Omega} + \|\text{data}\|_{L2,\partial\Omega} \right] + C_2 \Delta t^{f(\theta)} \left\| \{q^h(t_0)\} \right\|^2_{H^m(\Omega)}$$
$$\text{for}: \gamma \equiv min(k, (k+1), r-1), f(\theta) = (2, 3, 4)$$

for C and C_2 constants. The mPDE caveat is $(k + 1) \Rightarrow (1 + 1) = 2$ in (7.8) for $m\text{GWS}^h + \theta\text{TS}$ $k = 1$ TP/NC basis solutions at *sufficiently large* Re, an academic requirement for RaNS. For *smooth data* this *optimal modified continuous* Galerkin $k = 1$ basis algorithm is predicted to generate RaNS solutions exhibiting the potential for $O(h^4, m^3, \Delta t^4)$ asymptotic convergence for $\theta = 0.5$.

For *continuum* non-D state variable $\{q(\mathbf{x}, t)\} = \{u_i, \Theta, Y, k, \varepsilon, \phi, \tau_{ij}; P\}$ the RaNS mPDE flux vector system is defined in (8.20)

$$\mathcal{L}^m(\{q\}) = [m]\frac{\partial\{q\}}{\partial t} + \frac{\partial}{\partial x_j}\left[\{f_j\} - \{f_j^d\}\right] - \{s\} = \{0\}$$

with $[m]$ the optimal γ mPDE (only) altered *mass* matrix, $\{f_j\}$ the unaltered *kinetic flux vector*, $\{f_j^d\}$ the mPDE augmented *dissipative flux vecto*r and $\{s\}$ the source. For $1 \leq (i, j, k) \leq n$ the non-D functional forms, *absent* the NS mPDE theory derived $O(h^4)$ contributions, are detailed in (8.21) – (8.24)

$$[m] \equiv \left[1 - \frac{\gamma\,\Delta t^2}{6}\frac{\partial}{\partial x_j}\left(u_j u_k \frac{\partial}{\partial x_k}\right)\right]$$

$$
\{f_j(\{q\})\} = \begin{Bmatrix} u_j u_i + P^*|_{n+1}^{p} \delta_{ij} \\ u_j \Theta \\ u_j Y \\ u_j k \\ u_j \varepsilon \end{Bmatrix}
$$

$$
\{f^d(\{q\})\} = \begin{Bmatrix} \frac{1}{\text{Re}}(1+\text{Re}^t)\left(\frac{\partial u_i}{\partial x_j} + \frac{\partial u_j}{\partial x_i}\right) + \frac{h^2 \text{Re}}{6(1+\text{Re}^t)}\left(u_j u_k \frac{\partial u_i}{\partial x_k} + u_i u_k \frac{\partial u_j}{\partial x_k}\right) \\ \frac{1}{\text{Re}}\left(\frac{1}{\text{Pr}} + \frac{\text{Re}^t}{\text{Pr}^t}\right)\frac{\partial \Theta}{\partial x_j} + \frac{h^2 \text{RePr}}{12(1+\text{Re}^t)(\text{Pr}/\text{Pr}^t)} u_j u_k \frac{\partial \Theta}{\partial x_k} \\ \frac{1}{\text{Re}}\left(\frac{1}{\text{Sc}} + \frac{\text{Re}^t}{\text{Sc}^t}\right)\frac{\partial Y}{\partial x_j} + \frac{h^2 \text{Re}Sc}{12(1+\text{Re}^t)/(\text{Sc}/\text{Sc}^t)} u_j u_k \frac{\partial Y}{\partial x_k} \\ \frac{1}{\text{Pe}}\left(1 + \frac{\text{Re}^t}{\text{C}_k}\right)\frac{\partial k}{\partial x_j} + \frac{h^2 \text{Pe}}{12(1+\text{Re}^t/\text{C}_k)} u_j u_k \frac{\partial k}{\partial x_k} \\ \frac{1}{\text{Pe}}\left(1 + \frac{\text{Re}^t}{\text{C}_\varepsilon}\right)\frac{\partial \varepsilon}{\partial x_j} + \frac{h^2 \text{Pe}}{12(1+\text{Re}^t/\text{C}_\varepsilon)} u_j u_k \frac{\partial \varepsilon}{\partial x_k} \end{Bmatrix}
$$

$$
\{s\} = \begin{Bmatrix} -\frac{\text{Gr}}{\text{Re}^2}\Theta \widehat{g}_i \\ \frac{\text{Ec}}{\text{Re}}\frac{\partial}{\partial x_j}\left(u_i \frac{\partial u_j}{\partial x_i}\right) - \frac{\partial^2 \tau_{ij}^{\text{D}}}{\partial x_i \partial x_j} \\ s(Y) \\ \tau_{jk}^{\text{D}} \frac{\partial u_j}{\partial x_k} - \varepsilon + \frac{\text{Gr}}{\text{Re}^2}\frac{\varepsilon}{k}\frac{\partial \Theta}{\partial x_i}\delta_{\mathbf{g}\,i} \\ \text{C}_\varepsilon^1 f^1 \tau_{jk}^{\text{D}} \frac{\varepsilon}{k}\frac{\partial u_j}{\partial x_k} - \text{C}_\varepsilon^2 f^2 \frac{\varepsilon^2}{k} + \text{C}_\varepsilon^3 \frac{\text{GrRe}^t}{\text{Re}^3}\frac{\varepsilon}{k}\frac{\partial \Theta}{\partial x_i}\delta_{\mathbf{g}\,i} \end{Bmatrix}
$$

Inserting the absent *m*PDE theory $O(h^4)$ contributions leads to appropriately revised definitions. Recalling (5.122–5.123), $[m]$ in (8.20–8.21) is replaced with [M], while (8.24) picks up data evaluations at the previous time step n, recall (5.123). These revised RaNS algorithm matrices are

$$
[\text{M}] = \text{diag}\begin{bmatrix} 1 - \frac{\gamma \Delta t^2}{6}\frac{\partial}{\partial x_j}\left(u_j u_k \frac{\partial}{\partial x_k}\right) + \frac{\theta h^2}{6}\left(\frac{u_j \text{Re}}{(1+\text{Re}^t)} - \frac{\partial}{\partial x_j}\right)\frac{\partial}{\partial x_j} \\ 1 - \frac{\gamma \Delta t^2}{6}\frac{\partial}{\partial x_j}\left(u_j u_k \frac{\partial}{\partial x_k}\right) + \frac{\theta h^2}{6}\left(\frac{u_j \text{RePr}}{(1+\text{Re}^t)(\text{Pr}/\text{Pr}^t)} - \frac{\partial}{\partial x_j}\right)\frac{\partial}{\partial x_j} \\ 1 - \frac{\gamma \Delta t^2}{6}\frac{\partial}{\partial x_j}\left(u_j u_k \frac{\partial}{\partial x_k}\right) + \frac{\theta h^2}{6}\left(\frac{u_j \text{ReSc}}{(1+\text{Re}^t)/(\text{Sc}/\text{Sc}^t)} - \frac{\partial}{\partial x_j}\right)\frac{\partial}{\partial x_j} \\ 1 - \frac{\gamma \Delta t^2}{6}\frac{\partial}{\partial x_j}\left(u_j u_k \frac{\partial}{\partial x_k}\right) + \frac{\theta h^2}{6}\left(\frac{u_j \text{Pe}}{(1+\text{Re}^t/\text{C}_k)} - \frac{\partial}{\partial x_j}\right)\frac{\partial}{\partial x_j} \\ 1 - \frac{\gamma \Delta t^2}{6}\frac{\partial}{\partial x_j}\left(u_j u_k \frac{\partial}{\partial x_k}\right) + \frac{\theta h^2}{6}\left(\frac{u_j \text{Pe}}{(1+\text{Re}^t/\text{C}_\varepsilon)} - \frac{\partial}{\partial x_j}\right)\frac{\partial}{\partial x_j} \end{bmatrix}
$$

$$\{s\} = \left\{ \begin{array}{l} -\dfrac{\mathrm{Gr}}{\mathrm{Re}^2}\Theta \widehat{g}_i - \dfrac{h^2}{6}\left[\dfrac{u_j\mathrm{Re}}{(1+\mathrm{Re}^t)}\dfrac{\partial u_i}{\partial x_j} - \dfrac{\partial^2 u_i}{\partial x_j^2}\right]_n \\ \dfrac{\mathrm{Ec}}{\mathrm{Re}}\dfrac{\partial}{\partial x_j}\left(u_i\dfrac{\partial u_j}{\partial x_i}\right) + \dfrac{\partial^2 \tau_{ij}^{\mathrm{D}}}{\partial x_i \partial x_j} \\ \qquad -\dfrac{h^2}{6}\left[\dfrac{u_j\mathrm{RePr}}{(1+\mathrm{Re}^t)(\mathrm{Pr}/\mathrm{Pr}^t)}\dfrac{\partial \Theta}{\partial x_j} - \dfrac{\partial^2 \Theta}{\partial x_j^2}\right]_n \\ s(Y) - \dfrac{h^2}{6}\left[\dfrac{u_j\mathrm{ReSc}}{(1+\mathrm{Re}^t)/(\mathrm{Sc}/\mathrm{Sc}^t)}\dfrac{\partial Y}{\partial x_j} - \dfrac{\partial^2 Y}{\partial x_j^2}\right]_n \\ \tau_{jk}^{\mathrm{D}}\dfrac{\partial u_j}{\partial x_k} - \varepsilon + \dfrac{\mathrm{Gr}}{\mathrm{Re}^2}\dfrac{\varepsilon}{k}\dfrac{\partial \Theta}{\partial x_i}\delta_{\mathbf{g}\,i} - \dfrac{h^2}{6}\left[\dfrac{u_j\mathrm{Pe}}{(1+\mathrm{Re}^t/\mathrm{C}_k)}\dfrac{\partial k}{\partial x_j} - \dfrac{\partial^2 k}{\partial x_j^2}\right]_n \\ \mathrm{C}_\varepsilon^1 f^1 \tau_{jk}^{\mathrm{D}}\dfrac{\varepsilon}{k}\dfrac{\partial u_j}{\partial x_k} - \mathrm{C}_\varepsilon^2 f^2\dfrac{\varepsilon^2}{k} + \mathrm{C}_\varepsilon^3\dfrac{\mathrm{GrRe}^t}{\mathrm{Re}^3}\dfrac{\varepsilon}{k}\dfrac{\partial \Theta}{\partial x_i}\delta_{\mathbf{g}\,i} \\ \qquad -\dfrac{h^2}{6}\left[\dfrac{u_j\mathrm{Pe}}{(1+\mathrm{Re}^t/\mathrm{C}_k)}\dfrac{\partial \varepsilon}{\partial x_j} - \dfrac{\partial^2 \varepsilon}{\partial x_j^2}\right]_n \end{array} \right\}$$

Modifications to the NS $m\mathrm{GWS}^h + \theta\mathrm{TS}$ partitioned algorithm are modest but detailed. For RaNS DOF $\{\mathrm{Q}I(t_{n+1})\}$, $1 \le I \le n + 11 + 1$, ordered in the sequence $\left\{q^h(\mathbf{x}, t)\right\} = \left\{u_i^h, \Theta^h, Y^h, k^h, \varepsilon^h, \phi^h, \tau_{ij}^{\mathrm{D}\,h}; P^h\right\}^T$, the $k=1$ TP/NC trial space basis implemented algorithm terminal matrix statement is (8.25)

$$\begin{aligned} \{\mathrm{F\_Q}I\}_e &= [m]\{\mathrm{Q}I_{n+1} - \mathrm{Q}I_n\}_e \\ &\quad + \Delta t\left[\theta\{\mathrm{RES\_Q}I(\{Q\})\}_e\big|_{n+1} + (1-\theta)\{\mathrm{RES\_Q}I(\{Q\})\}_e\big|_n\right] \end{aligned}$$

The kinetic flux vector matrix (7.26) extends directly to RaNS $\{\mathrm{RES\_Q}I\}_e$, $1 \le I \le n+4$, also the NS pressure projection theory matrix statement (7.27) for $1 \le I \le n$.

A decision is required for implementing the *explicitly nonlinear* mPDE term rational functional $u_j u_k/(1+\mathrm{Re}^t)$. Interpolation of the nodal DOF $\{\mathrm{U}J\mathrm{U}K/\mathrm{RET}\}_e^T$ embeds distributional selectivity. Hence in (8.25) the $\{\mathrm{U}I\}_e$ residual, $1 \le I \le n$, compact form for (8.28) is

$$\begin{aligned} \{\mathrm{RES\_U}I(\{f^d\})\}_e &\equiv \frac{1}{\mathrm{Re}}\{\mathrm{RET}\}_e^T\left([\mathrm{M30}JJ]_e\{\mathrm{U}I\}_e + [\mathrm{M30}I\,J]_e\{\mathrm{U}J\}_e\right) \\ &\quad + \frac{h^2\mathrm{Re}}{6}\begin{pmatrix} \{\mathrm{U}J\mathrm{U}K/\mathrm{RET}\}_e^T[\mathrm{M30}JK]_e\{\mathrm{Q}I\}_e \\ +\{\mathrm{U}I\mathrm{U}K/\mathrm{RET}\}_e^T[\mathrm{M30}JK]_e\{\mathrm{Q}J\}_e \end{pmatrix}, 1 \le I \le 3 \end{aligned}$$

The constants/parameters in the remaining dissipative flux vector entries (8.23) are coalesced into $\{\mathrm{Pa}(q^h)\}$ and its turbulent counterpart $\{\mathrm{Pa}^t(q^h)\}$. $\{\mathrm{Pa}(q^h)\} = \{$Re, Re, Re, RePr, ReSc, Pe, Pe$\}^T$ is obvious for the lead terms. Absorbing the NS "{1}" into the Re^t interpolation, the turbulent parameter multiplier array for (8.23) becomes $\{\mathrm{Pa}^t(q^h)\} \equiv \{1, 1, 1, \mathrm{Pr}/\mathrm{Pr}^t, \mathrm{Sc}/\mathrm{Sc}^t,$

$1/C_k$, $1/C_\varepsilon\}^T$. This enables RaNS turbulence model closure *distributions* to be expressed as $\mathrm{Re}^t\{\mathrm{Pa}^t\}$ on Ω_e for all entries in (8.23). Transitioning from array to index notation the dissipative flux vector residual $\{\mathrm{RES\_Q}I\}_e$ for $n+1 \le I \le n+4$, compact (8.29) is

$$\{\mathrm{RES\_Q}I(\{f^d\})\}_e = \frac{\mathrm{Pa}^t(\mathrm{I})}{\mathrm{Pa(I)}}\{\mathrm{RET}\}_e^T[\mathrm{M30}JJ]_e\{\mathrm{Q}I\}_e + \frac{h^2\mathrm{Pa}(I)}{12\mathrm{Pa}^t(I)}\{\mathrm{U}J\mathrm{U}K/\mathrm{RET}\}_e^T[\mathrm{M30}JK]_e\{\mathrm{Q}I\}_e$$

The source term matrix for $1 \le I \le n+4$ is detailed in (8.30–8.31) and the pressure projection matrix statement is (8.32–8.33). The $\mathrm{GWS}^N \Rightarrow \mathrm{GWS}^h$ process for the deviatoric Reynolds stress tensor differential definition replacement for (8.12) generates algorithm statements for DOF $\{\mathrm{Q}I\}$, $n+9 \le I \le n+14$, as compact (8.34)

$$\{\mathrm{F\_TAU}JK\}_e = \det{}_e[\mathrm{M200}]_e\{\mathrm{TAU}JK\}_e + \frac{1}{\mathrm{Re}}\{\mathrm{RET}\}_e^T\left([\mathrm{M300}K]_e\{\mathrm{U}J\}_e + [\mathrm{M300}J]_e\{\mathrm{U}K\}_e\right), j \ne k$$

The *quasi*-Newton jacobian segregates the RaNS and TKE state variable partitions $\{u_i, \Theta, Y, \phi\}^h$ and $\{k, \varepsilon, \tau_{jk}\}$. The diagonal of the block jacobians for RaNS DOF $\{\mathrm{Q}I\}$, $1 \le I \le n+2$, and TKE DOF $\{\mathrm{Q}I\}$, $n+3 \le I \le n+4$, neglecting select *m*PDE nonlinearities is compact (8.35)

$$[\mathrm{JAC\_Q}I]_e \approx \det{}_e[\mathrm{M200}] + \frac{\gamma\,\Delta t^2}{6}\{\mathrm{U}J\mathrm{U}K\}_e^T[\mathrm{M30}JK]_e + \theta\Delta t\begin{pmatrix} \{\mathrm{U}J\}_e^T[\mathrm{M300}J]_e + \{\mathrm{U}I\}_e^T[\mathrm{M3}J\mathrm{00}]_e\delta_{IJ} \\ + \dfrac{\mathrm{Pa}^t(\mathrm{I})}{\mathrm{Pa(I)}}\{\mathrm{RET}\}_e^T[\mathrm{M30}JJ]_e \\ + \dfrac{h^2\mathrm{Pa(I)}}{12\mathrm{Pa}^t(\mathrm{I})}\{\mathrm{U}J\mathrm{U}K/\mathrm{RET}\}_e^T[\mathrm{M30}JK]_e \end{pmatrix}$$

The RaNS $m\mathrm{GWS}^h + \theta\mathrm{TS}$ algorithm matrix iteration statement is *amply populated* with turbulent closure model DOF *ratios* due to TKE closure source terms plus Re^t nonlinearity. Accurate jacobian computational forms are enabled via hypermatrix statements. Handling $\{\mathrm{RET}\}_e$ nonlinearity is supported by the continuum derivatives (8.38)

$$\frac{\partial \mathrm{Re}^t}{\partial k} = \frac{\partial(f_\nu k^2/\varepsilon\mathrm{Re})}{\partial k} = \frac{2f_\nu}{\mathrm{Re}}\frac{k}{\varepsilon} = 2\mathrm{Re}^t/k$$

$$\frac{\partial \mathrm{Re}^t}{\partial \varepsilon} = \frac{\partial(f_\nu k^2/\varepsilon\mathrm{Re})}{\partial \varepsilon} = \frac{-f_\nu}{\mathrm{Re}}\frac{k^2}{\varepsilon^2} = -\mathrm{Re}^t/\varepsilon$$

The chapter is completed with *a posteriori* data discussion assessing RaNS $m\mathrm{GWS}^h + \theta\mathrm{TS}$ algorithm performance regarding accuracy, convergence and solution monotonicity.

A regular solution-adapted non-uniform mesh study confirms convergence to the turbulent BL similarity solution. For confined flows a transverse span M = 32, pr = 1.15 dual wall-attracted mesh supports engineering accuracy using log-law BCs for the TKE partition.

The truly critical assessment is $m\text{GWS}^h + \theta\text{TS}$ RaNS algorithm solution *monotonicity*, quantitatively confirmed for an entrance channel flow with a no-slip wall BC singularity. The induced highly localized RaNS and TKE state variable partition solution extremum distributions are universally *monotone*. This is the direct result of GWS^h algorithm $O(h^2)$ approximation error *annihilation* by the dissipative flux vector nonlinear *m*PDE term $u_j u_k/(1 + \text{Re}^t)$ discretely implemented via DOF $\{\text{UJU}K/\text{RET}\}_e^T$. The *a posteriori* data clearly confirm *m*PDE theory precisely predicts the mirror of the $O(h^2)$ dispersion error distributions that pollute the GWS^h theory solution.

This leads to the $m\text{GWS}^h + \theta\text{TS}$ algorithm with a full Reynolds stress tensor closure model *m*PDE system. The validation experiment is turbulent flow in a square duct with *a posteriori* data clearly confirming the fundamental limitation of the industry standard *deviatoric* Reynolds stress tensor closure modeling.

Chapter content concludes with a coupling of RaNS and PRaNS formulations generating an iterative algorithm strategy for addressing fully coupled conjugate heat transfer in systems with largely disparate length scales. Algorithm validation is supported by experimental data for a specific borehole heat exchanger (BHE) installation. The algorithm strategy is clearly applicable to any resource extraction borehole system typified by characteristic length scales disparity.

Exercises

8.2.1 Verify the steps leading to the non-D eddy viscosity deviatoric tensor closure model (8.12).

8.2.2 Confirm the k-ε closure non-D PDE system (8.16–8.17).

8.3.1 Confirm correctness of the *m*PDE terms defined in (8.23) with respect to laminar/turbulent non-D group ratios.

8.3.2 Derive the replacement for (8.24) for insertion of *m*PDE theory augmentation via the Section 5.11 defined procedure.

8.3.3 Re-express the *m*PDE terms in (8.23) replacing $h^2\text{Re}/12(1 + \text{Re}^t)$ with $\beta\Delta t/2$.

8.4.1 Validate accuracy of the algorithm RaNS partition flux vector residual combination (8.26–8.27) and (8.28).

8.4.2 Verify the algorithm statement (8.31) for the k-ε closure model source term contributions.

8.4.3 Assess correctness of the RaNS state variable partition algorithm jacobian (8.35), in particular validity of the coefficient 1/12 for $1 \leq I \leq n$.

8.4.4 Using definitions (8.38) verify the TKE partition algorithm jacobian statement (8.39) coupling for $\{\text{RET}\}_e$.

8.4.5 Verify the algorithm TKE partition closure model cross coupling jacobians (8.41–8.44).

8.7.1 Contract the tensor index pair in (8.45), hence verify (8.55) for turbulence kinetic energy

8.7.2 Non-dimensionalize (8.60–8.61), hence verify the assertion they are identical to (8.16–8.17).

References

ANSYS (2011). *ANSYS FLUENT Theory Guide*, ANSYS, Inc., Canonsburg, PA.

Baker, A.J., Lin, C.-X., Orzechowski, J.A. and Gordon, C. (2011). "Fully Coupled 3-D Conjugate Heat Transfer Algorithm for Vertical Geothermal Heat Exchanger Performance Prediction," *J. Numerical Heat Transfer, Part B*: V. 60 pp. 47–167.

Baldwin, B.S. and Lomax, H. (1978). "Thin-layer approximation and algebraic model for separated turbulent flows," Technical Paper AIAA:78-257.

Baldwin, B.S. and Barth, T.J. (1990). "A one-equation turbulence transport model for high Reynolds number wall-bounded flows," Technical Report NASA TM-102847, also Technical Paper AIAA:91-610.

Daly, B.J. and Harlow, F.H. (1970). "Transport Equations in Turbulence," *Physics Fluids*, V. 13 pp. 2634–2649.

Fu, S., Launder, B.E. and Leschziner, M.A. (1987). "Modeling Swirling Recirculating Jet Flow with Reynolds Stress Transport Closures," *Proc. 6th Symposium on Turbulent Shear Flows*, Toulouse, France.

Gibson, M.M. and Launder, B.E. (1978). "Ground Effects on Pressure Fluctuations in the Atmospheric Boundary Layer," *J. Fluid Mechanics*, V. 86, pp. 491–511.

Kline, S.J., Morkovin, M.V., Sovran, G. and Cockrell, D.J. (1969). "*Computation of Turbulent Boundary Layers* – Volume I," 1968 AFOSR-IFP-Stanford Conference, Thermosciences Division, Stanford University, CA.

Lam, C.K.C. and Bremhorst, K. (1981). "A modified form of *k*-ε model for predicting wall turbulence," *ASME Transactions*, V. 103.

Launder, B.E. (1989). "Second-moment closure and its use in modeling turbulent industrial flows," *J. Numerical Methods in Fluids*, V. 9, pp. 963–985.

Launder, M.M., Reece, G.J. and Rodi, W. (1975). "Progress in the Development of a Reynolds Stress Turbulence Closure," *J. Fluid Mechanics*, V. 68 (3), pp. 537–566.

Marvin, J.G. (1983). "Turbulence Modeling for Computational Aerodynamics," *AIAA Journal*, V. 21 pp. 941–955.

Melling, A. and Whitelaw, J.H. (1976). "Turbulent flow in a rectangular duct," *J. Fluid Mechanics*, V. 78.

Spalart, P.R. (2009). "Detached Eddy Simulation," in *Annual Review of Fluid Mechanics*, V. 41, pp. 181–202.

Spalart, P.R. and Allmaras, S.R. (1992). "A one-equation turbulence model for aerodynamic flows," Technical Paper AIAA:92-439.

Tennekes, H. and Lumley, J.L. (1974). *A First Course in Turbulence*, MIT Press, Cambridge MA.

Wilcox, D.W., (2006). *Turbulence Modeling for* CFD, 3rd Edition, DCW Industries, La Canada, CA.

9

Space Filtered Navier–Stokes: $GWS^h/mGWS^h + \theta TS$ for space filtered Navier–Stokes, modeled, analytical closures

9.1 Classic State Variable LES PDE System

The classic state variable $mGWS^h + \theta TS$ algorithm for *time averaged* incompressible RaNS *m*PDE system is detailed in Chapter 8. *Convolution* with a filter defined on the NS PDE system spatial domain is the alternative manipulation leading to a turbulent flow simulation capability via a discretely implemented CFD algorithm. The literature terms this the *large eddy simulation* (LES) PDE system.

Derivation and performance assessment of Galerkin and *optimal modified continuous* Galerkin $GWS^h/mGWS^h + \theta TS$ algorithms for LES PDE systems is the chapter subject. The starting point remains the incompressible NS PDE system

$$\mathrm{D}M: \quad \nabla \cdot \mathbf{u} = 0 \tag{9.1}$$

$$\mathrm{D}\mathbf{P}: \quad \mathrm{St}\frac{\partial \mathbf{u}}{\partial t} + \nabla \cdot \mathbf{uu} + \nabla P - \frac{1}{\mathrm{Re}}\nabla \cdot \nabla \mathbf{u} + \frac{\mathrm{Gr}}{\mathrm{Re}^2}\Theta\,\widehat{\mathbf{g}} = 0 \tag{9.2}$$

$$\mathrm{D}E: \quad \mathrm{St}\frac{\partial \Theta}{\partial t} + \nabla \cdot \mathbf{u}\Theta - \frac{1}{\mathrm{RePr}}\nabla \cdot \nabla \Theta - s_\Theta = 0 \tag{9.3}$$

$$\mathrm{D}Y: \quad \mathrm{St}\frac{\partial Y}{\partial t} + \nabla \cdot \mathbf{u}Y - \frac{1}{\mathrm{ReSc}}\nabla \cdot \nabla Y - s_Y = 0 \tag{9.4}$$

with state variable $\{q(\mathbf{x}, t)\} = \{\mathbf{u}, p, \Theta, Y\}^T$, the *non-dimensional* (non-D) velocity vector, pressure, temperature and mass fraction Y. The Stanton, Reynolds, Grashoff, Prandtl and Schmidt number definitions are $\mathrm{St} \equiv \tau \mathrm{U/L}$, $\mathrm{Re} \equiv \mathrm{UL}/\nu$, $\mathrm{Gr} \equiv \mathrm{g}\beta\Delta\mathrm{TL}^3/\nu^2$, $\mathrm{Pr} \equiv \rho_0 \nu c_p/k$ and $\mathrm{Sc} = \mathrm{D}/\nu$, for D the binary diffusion coefficient. For the Boussinesq buoyancy approximation $\Delta\mathrm{T} \equiv (T_{max} - T_{min})$ with $\beta \equiv 1/\mathrm{T}_{abs}$, and $P \equiv p/\rho_0$ is kinematic pressure. St is typically unity, hence $\tau \equiv \mathrm{L/U}$, and in D*E* the Peclet number Pe commonly replaces RePr.

Optimal MODIFIED CONTINUOUS Galerkin CFD, First Edition. A. J. Baker.

Companion Website: www.wiley.com/go/baker/GalerkinCFD

The fundamental length scale L for (9.1–9.4) resides in the Reynolds number Re. NS flowfields are smooth, stable and typically single scale for modest Re, whence L pertains to the container, for example hydraulic diameter. As Re increases past a threshold the flowfield transitions to unsteady, then *chaotic*, whereupon characteristic length scales proliferate ranging from container dimension to molecular mean free path. Of note, the goal of a *direct numerical simulation* (DNS) formulation is to resolve the entire length scale spectrum.

The intent of LES alteration is to resolve the *significant* scales, stopping well short of mean free path. Convolution with a filter of *measure* δ explicitly recognizes that state variable quantification is limited to the *resolved* scale, with everything of scale smaller than that *unresolved*. Filter measure δ is correlated with mesh measure h and that spectral content of $O(2h)$ is not resolvable is the well established guideline.

Velocity vector is the usual LES exposition vehicle. Denoting *resolved scale* by overbar, *unresolved scale* by superscript prime, the LES fundamental statement of velocity scale resolution is

$$u_j(\mathbf{x},t) \equiv \overline{u}_j(\mathbf{x},t) + u_j'(\mathbf{x},t) \tag{9.5}$$

As always, NS convection nonlinearities in (9.2–9.4) generate the closure requirement. In tensor index notation and deferring details to latter sections, convolution of **DP** (9.2) with a filter of measure δ replaces $u_j u_i$ with $\overline{u_j u_i}$. Multiplying (9.5) by the identical resolution for u_i leads to

$$\overline{u_j u_i} \equiv \overline{(\overline{u}_j + u_j')(\overline{u}_j + u_j')} = \overline{\overline{u}_j \overline{u}_i} + \overline{\overline{u}_j u_i'} + \overline{u_j' \overline{u}_i} + \overline{u_j' u_i'} \tag{9.6}$$

For q denoting *scalar* state variable members Y, Θ, the space filtered NS PDE convection term nonlinearities lead to

$$\overline{u_j q} \equiv \overline{(\overline{u}_j + u_j')(\overline{q} + q')} = \overline{\overline{u}_j \overline{q}} + \overline{\overline{u}_j q'} + \overline{u_j' \overline{q}} + \overline{u_j' q'} \tag{9.7}$$

No terms in *quadruples* (9.6–9.7) are directly related to LES state variable members $\overline{u}_i$ and $\overline{q}$ of the space filtered PDE system replacing (9.1–9.4). Consequential approximations to rigorous handling of (9.6) lead to LES legacy *subgrid scale* (SGS) tensor *modeling* which sets the chapter tone. The continuing content details first principal developments addressing rigorous accounting of (9.6–9.7) including space filtering on *bounded domains*.

This *key* theoretical topic requires augmenting the convolved NS PDE system with *boundary commutation error* (BCE) integrals, also NS state variable extension exterior to the defined domain in the sense of *distributions*. Associated therewith is the requirement for resolved scale state variable non-homogeneous Dirichlet boundary condition (BC) identification. Theory resolutions detailed in closing this chapter lead to the prospect of an *m*PDE *essentially analytical* closure without any modeling component.

9.2 Space Filtered NS PDE System

Space filtering involves the formal mathematical operation of *convolution* of the NS PDE system (9.1–9.4) with a filter. The $n=1$ statement is

$$\overline{u}(x,t) \equiv \int_{-\infty}^{\infty} g_\delta(y)\, u(x-y,t)\mathrm{d}y \tag{9.8}$$

for $g_\delta(y)$ the filter of span δ and coordinate system y doubly infinite.

The mathematical requirements for a filter are linearity

$$\overline{u_i + \lambda v_i} = \overline{u}_i + \lambda \overline{v}_i \tag{9.9}$$

and that differentiation and filtering *commute*

$$\overline{\left(\frac{\partial u_i}{\partial x_j}\right)} = \left(\frac{\partial \overline{u}_i}{\partial x_j}\right), \quad \overline{\left(\frac{\partial u_i}{\partial t}\right)} = \left(\frac{\partial \overline{u}_i}{\partial t}\right), \; 1 \le (i,j) \le n \tag{9.10}$$

Convolution in n dimensions is symbolized by *

$$\overline{u}_i(x_j, t) \equiv g_\delta {}^* u_i(x_j, t) \tag{9.11}$$

with δ the *measure* (diameter) of the filter. Filter candidates include a box, the original cartesian FD cell, a box in transform space (called sharp spectral) and a gaussian. Definition of a specific filter does not impact the result of convolution. The NS PDE system (9.1–9.4) convolved with any filter is

$$\mathcal{L}(\rho_0) = \frac{\partial \overline{u}_j}{\partial x_j} = 0 \tag{9.12}$$

$$\mathcal{L}(\overline{u}_i) = \mathrm{St}\frac{\partial \overline{u}_i}{\partial t} + \frac{\partial}{\partial x_j}\left[\overline{u_j u_i} + \frac{\overline{p}}{\rho_0}\delta_{ij} - \frac{1}{\mathrm{Re}}\left(\frac{\partial \overline{u}_i}{\partial x_j} + \frac{\partial \overline{u}_j}{\partial x_i}\right)\right] + \frac{\mathrm{Gr}}{\mathrm{Re}^2}\overline{\Theta}\hat{\mathbf{g}}_{\mathbf{i}} = 0 \tag{9.13}$$

$$\mathcal{L}(\overline{\Theta}) = \mathrm{St}\frac{\partial \overline{\Theta}}{\partial t} + \frac{\partial}{\partial x_j}\left[\overline{u_j \Theta} - \frac{1}{\mathrm{RePr}}\frac{\partial \overline{\Theta}}{\partial x_j}\right] - \frac{\mathrm{Ec}}{\mathrm{Re}}\frac{\partial}{\partial x_j}\left(\frac{\partial \overline{u_j u_i}}{\partial x_i}\right) = 0 \tag{9.14}$$

$$\mathcal{L}(\overline{Y}) = \mathrm{St}\frac{\partial \overline{Y}}{\partial t} + \frac{\partial}{\partial x_j}\left[\overline{u_j Y} - \frac{1}{\mathrm{ReSc}}\frac{\partial \overline{Y}}{\partial x_j}\right] - s\overline{Y} = 0 \tag{9.15}$$

As essentially universally presented in the LES literature, (9.13–9.15) omit the boundary commutation error (BCE) surface integrals that result from convolution on a bounded domain, for example a no-slip surface segment. Addressing this LES formality is deferred, pending completion of closure modeling for which BCE integral omission is of no consequence.

Neither spatial filtering nor BCE integral existence alters the character of LES PDE system (9.12–9.15), with (9.6–9.7), as *constrained* initial-elliptic boundary value (I-EBV). As always, (9.12) requires that *admissable* LES state variable solutions $\{q(\mathbf{x}, t)\}$ possess a *divergence-free* resolved scale velocity vector. The second term in the 1/Re bracket in (9.13) is thus identically zero. However, as in the LES closure modeling literature, this complete term is expressed as resolved scale velocity strain rate (Stokes) tensor.

Suppressing overbar notation except where required for clarity, the LES PDE system *resolved scale* state variable remains $\{q(\mathbf{x}, t)\} = \{u_i, \Theta, Y, \phi; P\}$. As with well-posed NS and RaNS $m\mathrm{GWS}^h + \theta\mathrm{TS}$ algorithms, projection theory implements the *kinetic action* of pressure at time t_{n+1} as

$$P^*|_{n+1}^{p} = P_n + \frac{1}{\theta\Delta t}\sum_{\alpha=1}^{p}\delta\phi|_{n+1}^{\alpha} \tag{9.16}$$

for P_n (space filtered) kinematic pressure at time t_n with Δt the time step and θ integration implicitness factor.

The EBV PDE + BCs statement for velocity *potential* function ϕ, at iteration p and for *any* resolved scale velocity vector approximation $u_i^h(x_i, t_{n+1})$ to u_i, remains

$$\begin{aligned}\mathcal{L}(\phi) &= -\nabla^2\phi + \frac{\partial u_i^h}{\partial x_i} = 0 \\ \ell(\phi) &= -\nabla\phi\cdot\hat{\mathbf{n}} - \left(u_i - u_i^h\right)\hat{n}_i = 0\end{aligned} \tag{9.17}$$

In (9.16), $\delta\phi^\alpha$ is the sequence of potential solution iterates generated via (9.17) during algorithm convergence at time t^{n+1}.

The typical inflow resolved scale velocity vector BC is Dirichlet, hence thereon and on no-slip walls (9.17) confirms that the ϕ BC is homogeneous Neumann. On all outflow boundary segments the D*M*-satisfying velocity vector is unknown, hence also its difference with *any* velocity approximation. Therefore, the sole admissible ϕ outflow BC is homogeneous Dirichlet. As with NS the *energy norm* $\|\nabla^h\cdot\mathbf{u}^h\| \Rightarrow \|\phi^h\|_E$ is the measure of DM^h*error* with theory requirement *minimization*.

Finally, as thoroughly developed, determination of resolved scale pressure $P(x_i, t_{n+1})$ following LES algorithm convergence is the post-process solution of the *well-posed* EBV PDE + BCs system

$$\begin{aligned}\mathcal{L}(P) &= -\nabla^2 P - \frac{\partial}{\partial x_j}\left[\frac{\partial\overline{u_j u_i}}{\partial x_i} + \frac{\text{Gr}}{\text{Re}^2}\Theta\widehat{g}_j\right] = 0 \\ \ell(P) &= \nabla P\cdot\hat{\mathbf{n}} - \left[\frac{\partial\mathbf{u}}{\partial t} - \frac{1}{\text{Re}}\nabla^2\mathbf{u}\right]\cdot\hat{\mathbf{n}} = 0\end{aligned} \tag{9.18}$$

Thereby the LES mixed I-EBV plus EBV PDE + BCs system (9.13–9.18) is well-posed for theorizing *m*GWS^h + θTS weak form algorithms generating discrete approximate solutions to the space filtered NS classic state variable *m*PDE system. Note all members therein are explicitly time dependent, in distinction to the RaNS time averaged state variable wherein solution unsteadiness is limited to timescales much larger than the integration interval T, (4.52).

9.3 SGS Tensor Closure Modeling for LES

The pioneering closure model for a spatially filtered form of isothermal **DP** is that of Smagorinsky (1963). On a rectangular cartesian finite difference (FD) mesh the hypothesized *box-filtered* NS **DP** is mathematically *identical* with isothermal RaNS **DP**, (8.1), in possessing a single Reynolds stress tensor to be modeled. The Smagorinsky model is of mixing length theory (MLT) *eddy viscosity* origin closed with a single *constant* C_S. It is distinguished from the RaNS MLT closure *only* by C_S replacing the van Driest wall proximity length scale damping function, (4.60).

The LES closure modeling literature spans five decades, equally voluminous with that for RaNS. This requirement results from the fact the convection nonlinearity $\overline{u_j u_i}$ resident in convolved **DP**, (9.13), is *not* directly related to PDE system resolved scale velocity state variable $\bar{u}_j$. Topical discussions on mathematical hypotheses, assumptions and simplifications leading to *subgrid scale* (SGS) tensor closure models for isothermal **DP** (9.13) manipulations are detailed in Berselli *et al.* (2006), Sagaut (2006).

In summary, two distinct arguments lead to $\overline{u_j u_i}$ resolution both of which *significantly* alter rigorous **DP**, (9.13), following tensor quadruple (9.6) insertion. These assumptions end up replacing convolved convection nonlinearity $\overline{u_j u_i}$ with the resolved scale velocity tensor product $\bar{u}_j \bar{u}_i$ and augment (9.13) with a *single* (Reynolds) stress tensor τ_{ij} to be modeled.

The *double decomposition* argument, Sagaut (2006, Chapter 3), leads to Reynolds stress tensor definition

$$\begin{aligned} \tau_{ij} &\equiv \overline{u_j u_i} - \overline{\bar{u}_j \bar{u}_i} \\ &= (\overline{\bar{u}_j u_i'} + \overline{u_j' \bar{u}_i}) + \overline{u_j' u_i'} \equiv C_{ij} + R_{ij} \end{aligned} \tag{9.19}$$

The second line in (9.19) accrues to (9.6) direct substitution, and the LES literature labels these the resolved-unresolved scale interaction *cross-stress tensor* pair C_{ij} and strictly unresolved scale *Reynolds subgrid scale* tensor R_{ij}.

These rigorous theory dependencies are intrinsically embedded in *any* SGS tensor model for τ_{ij} without defining detail. The alternate *triple decomposition* theory assumes all terms in (9.6) are expressible in terms of resolved scale velocity, hence forming $\overline{\bar{u}_j \bar{u}_i}$ requires a second filter application. As a remedy Leonard (1974) proposed

$$\begin{aligned} \overline{\bar{u}_j \bar{u}_i} &= (\overline{\bar{u}_j \bar{u}_i} - \bar{u}_j \bar{u}_i) + \bar{u}_j \bar{u}_i \\ &\equiv L_{ij} + \bar{u}_j \bar{u}_i \end{aligned} \tag{9.20}$$

which defines the *Leonard stress* tensor L_{ij}.

Substituting (9.6) the functionality now embedded in Reynolds stress tensor τ_{ij} becomes

$$\begin{aligned} \tau_{ij} &\equiv \overline{u_j u_i} - \bar{u}_j \bar{u}_i \\ &= L_{ij} + C_{ij} + R_{ij} \end{aligned} \tag{9.21}$$

adding the Leonard stress to cross-stress pair *and* Reynolds subgrid tensors. As with (9.19) subsequent SGS tensor models for τ_{ij} are formed without detailing convolution generated resolved-unresolved scale interactions.

The net impact of these approximating manipulations, LES literature pervasive, is replacing rigorously convolved **DP**, (9.13) with (9.6) insertion, with the truly fundamental *simplification*

$$\mathcal{L}(\bar{u}_i) = \frac{\partial \bar{u}_i}{\partial t} + \frac{\partial}{\partial x_j}\left[\bar{u}_j \bar{u}_i + \bar{P}\delta_{ij} - \frac{2}{\text{Re}}\bar{S}_{ij} + \tau_{ij}\right] + \frac{\text{Gr}}{\text{Re}^2}\overline{\Theta}\widehat{g}_i = 0 \tag{9.22}$$

That (9.22) is the legacy LES D**P** PDE, closed via τ_{ij} modeling, is truly *transformative* as it is *identical*(!) with time averaged RaNS D**P**, (8.2) with (8.8) insertion, and defining the Stokes strain rate tensor

$$\overline{S}_{ij} \equiv \frac{1}{2}\left(\frac{\partial \overline{u}_i}{\partial x_j} + \frac{\partial \overline{u}_j}{\partial x_i}\right) \tag{9.23}$$

Thereby, it is not surprising that SGS tensor models exhibit mathematical similarity with RaNS closures, typically an eddy viscosity multiplier on a *deviatoric* strain rate tensor. As with RaNS, $\tau_{ij} \Rightarrow \tau_{ij}^{\mathrm{D}}$ upon assuming the trace $\tau_{kk}\delta_{ij}/3$ absorbed into resolved scale pressure. The functional form for such SGS tensor models is

$$\tau_{ij}^{\mathrm{D}} \cong f\big(\mathrm{C}_S(\mathbf{x},t),\ \delta^2,\ \big|\overline{S}_{ij}\big|\big)\overline{S}_{ij},\ i \neq j \tag{9.24}$$

for $\mathrm{C}_S(\mathbf{x}, t)$ the Smagorinsky "constant" replaced with a computed *dynamic* distribution and $\left|\overline{S}_{ij}\right|$ a norm of (9.23).

$\mathrm{C}_S(\mathbf{x}, t), \delta^2, \left|\overline{S_{ij}}\right|$ defines the eddy viscosity with selection of filter, of measure δ, a *post* (9.22) identification decision. Adequate energetic capture requires $\delta \geq 2h$, for h the mesh measure, Pope (2000). Hence, (9.24) inserted into the (9.22) dissipative flux vector constitutes an $O(h^2)$ diffusion operator moderated by $\mathrm{C}_S(\mathbf{x}, t)$ distributions.

Term-by-term comparison of (9.24) with time averaged BL D**P** MLT closure (4.60) is informative. Filter measure δ replaces Prandtl hypothesis mixing length l_m with both squared. Norm $\left|\overline{S}_{ij}\right|$ exists in MLT $n = 3$ closures, for example Baldwin and Lomax (1978), as generalization of $n = 2$ BL strain rate $|\partial\overline{u}/\partial y|$. Hence, $\mathrm{C}_S(\mathbf{x}, t)$ is the replacement of van Driest length scale damping function ω. Closure model (9.24) dimensionality is $D(\mathrm{L}^2\tau^{-1})$, in accord with pioneering theorization on momentum exchange in an incompressible shear layer, Prandtl (1904).

The theoretical consequence of arguments leading to (9.22–9.24) is that embedded resolved-unresolved scale interaction plus strictly unresolved scale functionalities

$$\tau_{ij}^{\mathrm{D}} \approx f\big(C_{ij} + R_{ij}; L_{ij} + C_{ij} + R_{ij}\big) \tag{9.25}$$

are assumed adequately represented by models involving *only* resolved scale velocity. That quality *a posteriori* data century old exist for SGS tensor models of form (9.24) certainly attests to select pertinence of the century old Prandtl theorization.

The ultimately simple Smagorinsky constant C_S proved totally inadequate prompting a sequence of alterations catalyzed by Leonard stress identification (9.20). The ensuing principally successful formulations, Germano at al, (1991), Lilly (1992), are termed *dynamic-subgrid scale* models with Smagorinsky C_S replaced with *distributions* $\mathrm{C}_S(\mathbf{x}, t)$ computed during solution adaptive mesh refinement processes.

In addition to those cited, SGS tensor models of form (9.24) include dynamic Smagorinsky–Lilly, structure function, kinetic energy of the subgrid scale, Yoshizawa and Horiuti (1985), scale similar and mixed scale closures, refer to Sagaut (2006). That such SGS tensor models can contribute excess $O(h^2)$ diffusion has been addressed via numerics alteration to locally monotone FD schemes, Boris (1990), Boris *et al.* (1992). Monotone integrated LES (MILES), the original implicit LES (ILES) algorithm, Shah and Ferziger (1995), further alters the SGS model resolved scale velocity prediction via *approximate deconvolution* (AD) onto a finer mesh. AD augmented with a subgrid-scale estimation (SGSE) model is reported, Domaradzki and Saiki (1997), Domaradzki and Yee, (2000).

AD utilization in the fully turbulent regime has generated *a posteriori* data in quantitative agreement with DNS data, cf. Stolz and Adams (1999), Stolz *et al.* (2001). An alternate ILES theorization addresses diffusive deficiencies via *high resolution* flux vector differencing schemes, Grinstein and Fureby (2004), Margolin *et al.* (2006), Grinstein *et al.* (2007). Numerics of this genre include flux-corrected transport (FCT), piecewise parabolic method (PPM) and total variation diminishing (TVD), cf. Harten (1983).

SGS tensor model alterations seeking improved transitional Re applicability include insertion of van Driest wall-damping and a Klebanoff-type intermittency correction, Piomelli *et al.* (1990), dynamic SGS model alterations, Germano *et al.* (1991), Lilly (1992), filtered structure function, Metais and Lesieur (1992), low Re corrections, Voke and Yang (1995), high-pass filtered eddy viscosity models, Stolz *et al.* (2003). Schlatter *et al.* (2004) detail *a posteriori* data validation for a transitional Re on rather coarse meshes using AD.

Despite the fundamental theoretical simplifications leading to D**P** (9.22), hence SGS tensor models (9.25), the LES literature documents considerable success for a range of Re. Via the assumptions generating (9.22) the omission of BCE integral in (9.22), also identification of resolved scale velocity *genuine* no-slip wall Dirichlet BCs, is of no consequence. As with RaNS closure modeling, extension of (9.22–9.25) to state variable scalar members, if considered, involves *turbulent* Prandtl/Schmidt number hypotheses.

With (9.22–9.25), mGWS$^h + \theta$TS algorithm implementation of an SGS tensor model closed mPDE alteration to (9.22) presents no fundamental challenge. Factually, if an alternative to SGS tensor closure is not available no need exists for this chapter. Fortunately rigorous mathematical operations on (9.6–9.18), including BCE integrals, lead to derivation of an *essentially analytical* LES closure theory alternative to (9.22–9.24), valid for all Re.

9.4 Rational LES Theory Predictions

The *rational* LES (RLES) theory, Galdi and Layton (2000), is based on Padé rational polynomial interpolations of gaussian filter Fourier transform. This enables formal completion of convolution, Fourier transformation and deconvolution mathematical operations necessary to rigorously predict mathematical expressions for the NS convolution identified tensors (9.6).

Polynomial interpolations of gaussian filter Fourier transform are compared in Figure 9.1, gaussian the solid curves, interpolations the dashed curves. The second-order Taylor series, which leads to the LES legacy SGS tensor *gradient model*, constitutes a truly poor approximation of the transform distribution becoming non-positive at roughly measure half-span. Second- and fourth-order rational Padé polynomial interpolations generate significant fidelity improvements.

The polynomial definitions are

$$\text{second-order Taylor:}\quad e^{ax_i} = 1 + ax_i + O\left(a^2 x_i^2\right) \tag{9.26}$$

$$\text{second-order Padé:}\quad e^{ax_i} = \frac{1}{1 + ax_i} + O\left(a^2 x_i^2\right) \tag{9.27}$$

$$\text{fourth order Padé:}\quad e^{ax_i} = \frac{1}{1 + ax_i + a^2 x_i^2} + O\left(a^4 x_i^4\right) \tag{9.28}$$

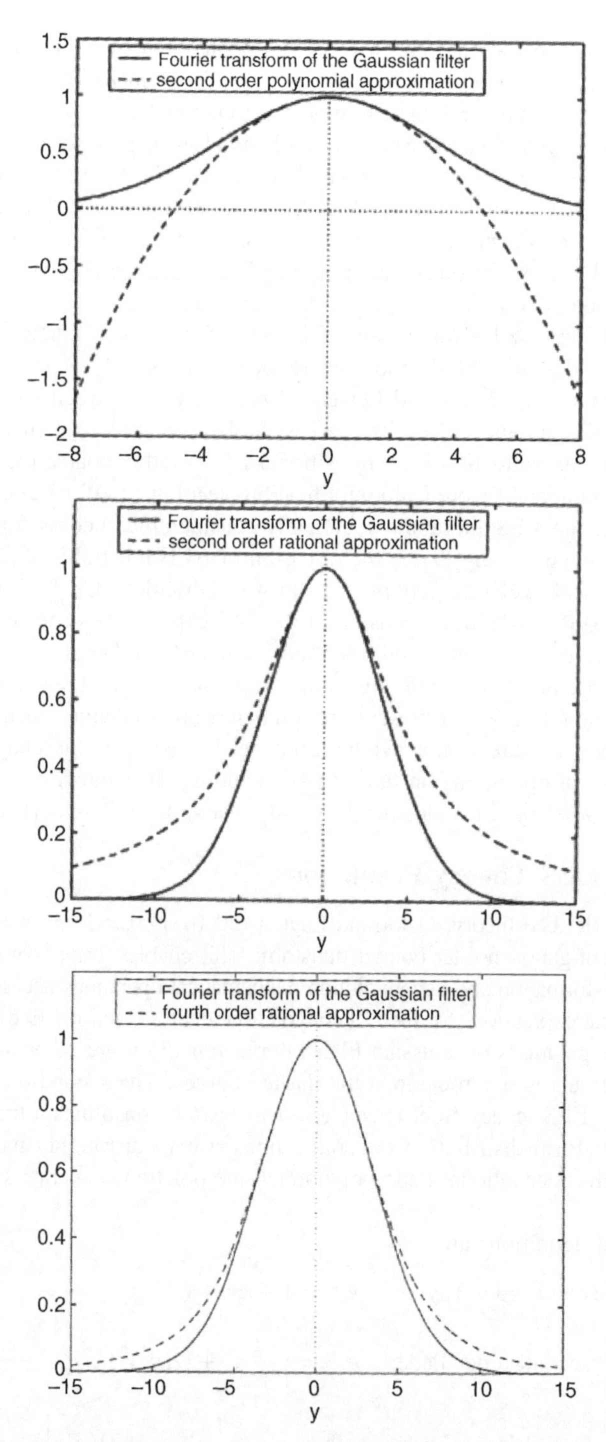

Figure 9.1 Polynomial interpolation of gaussian filter Fourier transform: top, second-order Taylor series; middle, second-order Padé; bottom, fourth-order Padé. *Source:* John, V. 2004. Reproduced with permission from Springer

For reference, the gaussian filter definition

$$g_\delta(x_i) = \left(\frac{\gamma}{\pi\delta^2}\right)\exp\left(-\frac{\gamma}{\delta^2}x_i^2\right) \tag{9.29}$$

contains as *data* the measure δ and shape factor γ. Symbolizing Fourier transformation as $F(\bullet)$, by tensors in (9.6

$$\begin{aligned} F\left(\overline{\bar{u}_j\bar{u}_i}\right) &= F(g_\delta)F\left(\bar{u}_j\bar{u}_i\right) \\ F\left(\overline{\bar{u}_j u_i'}\right) &= F(g_\delta)\left(F\left(\bar{u}_j\right)^* F(u_i')\right) \\ F\left(\overline{u_j'\bar{u}_i}\right) &= F(g_\delta)\left(F\left(u_j'\right)^* F(\bar{u}_i)\right) \end{aligned} \tag{9.30}$$

Since $F(g_\delta) \neq 0$, the Fourier transforms of u_i and u_i' are

$$F(u_i) = \frac{F(g_\delta)F(u_i)}{F(g_\delta)} = \frac{F(\bar{u}_i)}{F(g_\delta)} \tag{9.31}$$

$$F(u_i') = \left(\frac{1}{F(g_\delta)} - 1\right)F(\bar{u}_i) \tag{9.32}$$

and inserting (9.32) into (9.30) generates

$$\begin{aligned} F\left(\overline{\bar{u}_j\bar{u}_i}\right) &= F(g_\delta)F\left(\bar{u}_j\bar{u}_i\right) \\ F\left(\overline{\bar{u}_j u_i'}\right) &= F(g_\delta)\left(F\left(\bar{u}_j\right)^* \left(\frac{1}{F(g_\delta)} - 1\right)F(\bar{u}_i)\right) \\ F\left(\overline{u_j'\bar{u}_i}\right) &= F(g_\delta)\left(\left(\frac{1}{F(g_\delta)} - 1\right)F\left(\bar{u}_j\right)^* F(\bar{u}_i)\right) \end{aligned} \tag{9.33}$$

Since the Fourier transform of resolved scale velocity vector is available, an approximation is required for $F(g_\delta)$ and $1/F(g_\delta)$ to complete evaluation of (9.33). From LES literature the second-order TS interpolation of the gaussian of measure δ and shape factor γ generates

$$\begin{aligned} F\left(g_\delta(\delta, y_j)\right) &= 1 - \frac{\left\|y_j\right\|_2^2}{4\gamma}\delta^2 + O\left(\delta^4\right) \\ \frac{1}{F\left(g_\delta(\delta, y_j)\right)} &= 1 + \frac{\left\|y_j\right\|_2^2}{4\gamma}\delta^2 + O\left(\delta^4\right) \end{aligned} \tag{9.34}$$

Inserting (9.34) into (9.33) and using well-known properties of Fourier transformations leads to

$$\begin{aligned} F\left(\overline{\bar{u}_j\bar{u}_i}\right) &= F\left(\bar{u}_j\bar{u}_i\right) + \frac{\delta^2}{4\gamma}F\left(\frac{\partial^2\left(\bar{u}_j\bar{u}_i\right)}{\partial x_k \partial x_k}\right) + O\left(\delta^4\right) \\ F\left(\overline{\bar{u}_j u_i'}\right) &= -\frac{\delta^2}{4\gamma}F\left(\bar{u}_j\frac{\partial^2 \bar{u}_i}{\partial x_k \partial x_k}\right) + O\left(\delta^4\right) \\ F\left(\overline{u_j'\bar{u}_i}\right) &= -\frac{\delta^2}{4\gamma}F\left(\frac{\partial^2 \bar{u}_j}{\partial x_k \partial x_k}u_i\right) + O\left(\delta^4\right) \end{aligned} \tag{9.35}$$

Applying the inverse Fourier transform, predictions for the lead three tensors in (9.6) in physical space are

$$\begin{aligned}\overline{\bar{u}_j\bar{u}_i} &= \bar{u}_j\bar{u}_i + \frac{\delta^2}{4\gamma}\frac{\partial^2(\bar{u}_j\bar{u}_i)}{\partial x_k \partial x_k} + O(\delta^4)\\ \overline{\bar{u}_j u_i'} &= -\frac{\delta^2}{4\gamma}\bar{u}_j\frac{\partial^2\bar{u}_i}{\partial x_k \partial x_k} + O(\delta^4)\\ \overline{u_j'\bar{u}_i} &= -\frac{\delta^2}{4\gamma}\frac{\partial^2\bar{u}_j}{\partial x_k \partial x_k}\bar{u}_i + O(\delta^4)\end{aligned} \tag{9.36}$$

Then using well-known kinematic relations, and for series truncation at $O(\delta^4)$, (9.36) leads to *analytical* ordered prediction of LES theory tensor lead triple

$$\overline{\bar{u}_j\bar{u}_i} + \overline{\bar{u}_j u_i'} + \overline{u_j'\bar{u}_i} = \bar{u}_j\bar{u}_i + \frac{\delta^2}{2\gamma}\frac{\partial \bar{u}_j}{\partial x_k}\frac{\partial \bar{u}_i}{\partial x_k} + O(\delta^4) \tag{9.37}$$

For isothermal (Gr = 0) flow and omitting the Re^{-1} term in (9.13), (9.37) inserted therein generates the SGS tensor gradient model LES literature reported. In application to isotropic decaying turbulence, comparison with experiment showed quality correlations, Winckelmans *et al.* (2001). Conversely, numerous reported *a posteriori* data confirm that the gradient model does not dissipate sufficient energy, yielding an unstable CFD algorithm algebraic solution process, Vreman (1995). Stabilizing the algorithm via Smagorinsky model alteration generates excessive diffusion, Clark *et al.* (1979).

RLES theory resolution is to replace TS interpolation with compact *rational* Padé polynomials, a process possessing ready extension to higher-order compact approximations which are monotone and *non-negative*, Figure 9.1. The second-order Padé interpolation (9.27) of the gaussian filter Fourier transform leads to (9.34) replacements

$$\begin{aligned}F\big(g_\delta(\delta, y_j)\big) &= \frac{\delta^2}{1 + \dfrac{\|y_j\|_2^2}{4\gamma}} + O(\delta^4)\\ \frac{1}{F\big(g_\delta(\delta, y_j)\big)} &= 1 + \frac{\|y_j\|_2^2}{4\gamma}\delta^2 + O(\delta^4)\end{aligned} \tag{9.38}$$

Inserting (9.38) into (9.32) and taking the inverse Fourier transform produces to $O(\delta^4)$

$$\begin{aligned}\overline{\bar{u}_j\bar{u}_i} &= \left[\delta_{ij} - \frac{\delta^2}{4\gamma}\nabla^2\right]^{-1}(\bar{u}_j\bar{u}_i) + O(\delta^4)\\ \overline{\bar{u}_j u_i'} &= -\frac{\delta^2}{4\gamma}\left[\delta_{ij} - \frac{\delta^2}{4\gamma}\nabla^2\right]^{-1}\left(\bar{u}_j\frac{\partial^2\bar{u}_i}{\partial x_k \partial x_k}\right) + O(\delta^4)\\ \overline{u_j'\bar{u}_i} &= -\frac{\delta^2}{4\gamma}\left[\delta_{ij} - \frac{\delta^2}{4\gamma}\nabla^2\right]^{-1}\left(\frac{\partial^2\bar{u}_j}{\partial x_k \partial x_k}\bar{u}_i\right) + O(\delta^4)\end{aligned} \tag{9.39}$$

whereupon (9.37) is replaced with

$$\overline{\overline{u}_j\overline{u}_i} + \overline{\overline{u}_j u_i'} + \overline{u_j'\overline{u}_i} = \overline{u}_j\overline{u}_i + \frac{\delta^2}{2\gamma}\left[I - \frac{\delta^2}{4\gamma}\nabla^2\right]^{-1}\left(\frac{\partial\overline{u}_j}{\partial x_k}\frac{\partial\overline{u}_i}{\partial x_k}\right) + O(\delta^4) \tag{9.40}$$

The second-order Padé prediction (9.40) for the lead three tensors in (9.6) identifies the *matrix inverse differential operator*

$$[\mathrm{A}] \equiv \left[I - \frac{\delta^2}{4\gamma}\nabla^2\right]^{-1} \tag{9.41}$$

for I the identity matrix. The laplacian in [A] is a second-order EBV differential operator on resolved scale velocity strain rate tensor product, (9.40). The RLES literature cites (9.41) as defining "the *auxiliary problem*."

Viewing Figure 9.1, fourth-order rational Padé gaussian Fourier transform interpolation generates a truly accurate approximation. For the appropriate revision to (9.38) and proceeding through the operational sequence (9.39–9.40), the suggested exercise will confirm the analytical $O(\delta^6)$ prediction

$$\begin{aligned}\overline{\overline{u}_j\overline{u}_i} + \overline{\overline{u}_j u_i'} + \overline{u_j'\overline{u}_i} = \overline{u}_j\overline{u}_i + \frac{\delta^2}{2\gamma}\left[I - \frac{\delta^2}{4\gamma}\nabla^2 + \frac{\delta^4}{32\gamma^2}\nabla^4\right]^{-1} \times \\ \left[\begin{array}{l}\dfrac{\partial\overline{u}_i}{\partial x_k}\dfrac{\partial\overline{u}_j}{\partial x_k} - \dfrac{\delta^4}{16\gamma^2}\left(\nabla^2\overline{u}_j\nabla^2\overline{u}_i + \dfrac{\partial\nabla^2\overline{u}_i}{\partial x_k}\dfrac{\partial\overline{u}_j}{\partial x_k}\right) \\ \quad - \dfrac{\delta^4}{16\gamma^2}\left(\dfrac{\partial\overline{u}_i}{\partial x_k}\dfrac{\partial\nabla^2\overline{u}_j}{\partial x_k} + \nabla^2\left(\dfrac{\partial\overline{u}_i}{\partial x_k}\dfrac{\partial\overline{u}_j}{\partial x_k}\right)\right)\end{array}\right] + O(\delta^6)\end{aligned} \tag{9.42}$$

This defines the altered matrix inverse differential operator

$$[\mathrm{A}] = \frac{\delta^2}{2\gamma}\left[I - \frac{\delta^2}{4\gamma}\nabla^2 + \frac{\delta^4}{32\gamma^2}\nabla^4\right]^{-1} \tag{9.43}$$

and I remains the identity matrix. [A] is now a *biharmonic* EBV statement via ∇^4 operating on a truly detailed array of higher-order differential operations on resolved velocity strain rate tensor products with laplacian operations. Thus (9.43) corresponds to a mathematically more detailed *auxiliary problem*.

LES theory predicted fourth tensor in (9.6) is the *dissipation* mechanism at the *unresolved scale threshold* replacing unfiltered NS viscous dissipation at molecular scale. Via the identified operators and operations, the predictions for second-order Taylor and second-order Padé interpolations are, respectively

$$\overline{u_j'u_i'} = \frac{\delta^4}{16\gamma^2}\left[\nabla^2\overline{u}_j\nabla^2\overline{u}_i\right] + O(\delta^6) \tag{9.44}$$

$$\overline{u_j'u_i'} = \frac{\delta^4}{16\gamma^2}\left[I - \frac{\delta^2}{4\gamma}\nabla^2\right]^{-1}\left[\nabla^2\overline{u}_j\nabla^2\overline{u}_i\right] + O(\delta^6) \tag{9.45}$$

Both are formally negligible since the lead term multiplier δ^4 is the truncation order in (9.37) and (9.40). Prediction (9.45) is not trivial however, as RLES theory has *analytically* identified $O(\delta^4) < O(\overline{u_j' u_i'}) < O(\delta^2)$. The fourth-order Padé $\overline{u_j' u_i'}$ determination is not published, but the significant term coefficient is certainly not larger than δ^4.

Thereby, RLES theory predicts that SGS tensor models of functionality (9.25), universally of $O(\delta^2)$, are *too* significant to represent LES theory dissipation at the *unresolved scale threshold*. This identification adds theoretical substance to SGS tensor closure practice observation that $O(\delta^2) \sim O(h^2)$ models tend to excess diffusion.

In summary, to $O(\delta^4)$, RLES theory *analytically* derives (9.40) and (9.42), respectively, for the first three convolved NS **D**P PDE generated tensors (9.6). The theory fails to identify a mathematically significant fourth tensor but does predict the bound $O(\delta^4) < O(\overline{u_j' u_i'}) < O(\delta^2)$. Padé interpolations generate EBV *auxiliary problem* matrix inverse differential operators [A], (9.41) and (9.43). These represent algorithm implementation challenges as Green–Gauss divergence theorem differential projection generates an encompassing boundary condition (BC) requirement, *a priori* unknown.

Of truly substantial fluid dynamics theory significance, RLES predictions are completely devoid of the word *turbulent*! All manipulations are rigorous mathematical operations defined for/by NS PDE system convolution. Consequently, (9.13) with (9.6) implemented via (9.41) and/or (9.43) are Reynolds number Re *unconstrained*, hence theoretically appropriate for laminar, transitional and/or turbulent resolved scale velocity vector distribution prediction.

9.5 RLES Unresolved Scale SFS Tensor Models

Upon invoking the word *turbulent*, the RLES literature reports eddy viscosity genre closure models for the *subfilter scale* (SFS) tensor $\overline{u_j' u_i'}$ meeting $O(\delta^3)$ bound. These models are distinguished from *subgrid scale* (SGS) tensor models, (9.25), RLES theory predicted inappropriate at $O(\delta^2)$.

Reasoning that energetic dissipation at unresolved scale threshold depends on kinetic energy of the smallest resolved eddies, and following theorization credited to Kolmogorov, Iliescu *et al.* (1998) proposed the eddy viscosity model

$$\nu^t \propto l_m \sqrt{\frac{1}{2}\rho \|u_i'\|_2^2} \equiv C_S \frac{\delta^3}{\gamma} \left\|\nabla^2 \overline{u}_i\right\|_{L2} + O(\delta^4) \tag{9.46}$$

$\left\|\nabla^2 \overline{u}_i\right\|_{L2}$ is the L2 norm of the resolved scale velocity vector with C_S a "Smagorinsky" constant required determined. Defining the mixing length as $l_m \equiv \delta$ generates the requisite $O(\delta^3)$.

Alternatively, assuming dissipation at the unresolved scale threshold depends on the average resolved scale velocity, a sequence of mathematical formalities leads to an alternative $O(\delta^3)$ eddy viscosity model, John (2004, 2005)

$$\nu^t \equiv C_S \frac{\delta}{\gamma} \left\|\overline{u}_j - [A(\overline{u}_i)]\right\|_{L2} + O(\delta^4) \tag{9.47}$$

Therein [A] is the $O(\delta^2)$ matrix inverse operator (9.41) with C_S a model constant to be determined. Unresolved scale SFS tensor closures then result upon inserting eddy viscosity *models* (9.46) and/or (9.47) as the diffusion coefficient on resolved scale velocity strain rate tensor (9.23).

Performance *a posteriori* data for these SFS tensor closure models, including comparison with order alterations to the Smagorinsky SGS tensor model (9.25), are reported, John (2004, Ch. 11). The analysis statement with *benchmark* solution is an unbounded $n=2$ channel mixing layer with characteristic length scale the resolved scale velocity initial condition (IC) shear thickness σ_0, Figure 9.2.

The weak form *natural* homogeneous Neumann, BC is precisely farfield appropriate for unfiltered and filtered NS D**P** on $n=2$. Additionally, it is *assumed* appropriate as EBV auxiliary problem BC. The resolved scale velocity vector IC is

$$\mathbf{u}_0(y) = \begin{Bmatrix} U_\infty \tanh(2y/\sigma_0) \\ 0 \end{Bmatrix} + c_{\text{noise}} U_\infty \{\nabla \times \psi \bullet \mathbf{k}\}$$
$$\psi = \exp\left(-(2y/\sigma_0)^2\right)(\cos(8\pi x) + \cos(20\pi x)) \tag{9.48}$$

for c_{noise} defining perturbation streamfunction ψ level, recall Section 6.1. The Reynolds number characteristic length scale L is σ_0. Algorithm DOF on outflow boundary segments are reflected to corresponding DOF on inflow boundary segments as time dependent Dirichlet BCs. This strategy admits a domain of x axis finite span.

The ψ velocity perturbation (9.48) spectral content leads to *a priori* known vortex number and translation frequency. For $\sigma_0 = 1/14$, four vortices become initialized on axial span $x \in (-1, 1)$. For $U_\infty \equiv 1.0$, $\text{Re} = U_\infty \sigma_0 / 0.14\ \text{E}{-06} = 10{,}000$. The resultant characteristic timescale is $t^* = \sigma_0 / U_\infty = 14$. Time integration employs a θ implicit fractional step algorithm for fixed $\Delta t = 0.1\ t^* = (140)^{-1}$ s.

Solution final time definition is $n\Delta t \equiv 200\ t^* \sim 14.285$ s. Algorithm *a posteriori* data characterization employs vorticity thickness distribution $\sigma(t)$ evolution. Recalling $\mathbf{\Omega} \bullet \mathbf{k} \equiv \omega \equiv \nabla \times \mathbf{u} \bullet \mathbf{k}$

$$\sigma(t) \equiv 2U_\infty / sup_{y \in [-1,1]} |\langle\omega\rangle(t, y)| \tag{9.49}$$

for vorticity integral measure normalization

$$\langle\omega\rangle(t, y) = \int_{-1}^{1} \omega(t, y, x) \mathrm{d}x \Big/ \int_{-1}^{1} \mathrm{d}x = \frac{1}{2} \int_{-1}^{1} \omega(t, y, x) \mathrm{d}x \tag{9.50}$$

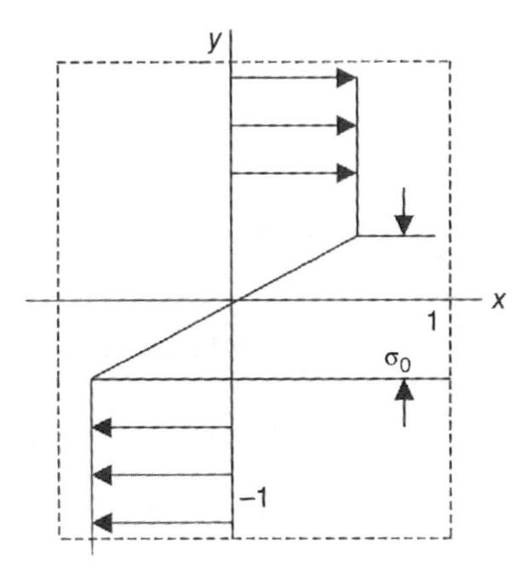

Figure 9.2 Initial condition for channel mixing layer prediction

Data presentation is $\sigma(t)/\sigma_0$ which confirms the IC $\sigma(t=0)=\sigma_0$. Solution generated $\sigma(t)$ are computed on all grid lines parallel to x from which the *supremum* is extracted for (9.49).

The Galerkin weak form $n=2$ discrete implementation employs the Q_2/P_1^{disc} tensor product (TP) basis. Each finite element is spanned by a Lagrange complete biquadratic (nine-node) for velocity and a three-node discontinuous linear Lagrange for pressure. This basis is verified to satisfy the LBB *inf sup* condition (5.15), Gresho and Sani (1998). Defined reflection BCs are required specified *only* for velocity DOF coincident with inflow/outflow element surface nodes, as pressure DOF lie at nodes interior to element boundaries.

This algorithm for unfiltered isothermal $n=2$ classic state variable NS PDE system (7.1–7.2) generates the benchmark solution via a *multi-grid* strategy cycled from $M=4\times4$ through eight *uniform* mesh refinements to $M=512\times512$. The terminal mesh contains 2+ million velocity and ~0.8 million pressure DOF. Linear algebra iterative convergence to euclidean norm E−12 is stated mandatory.

The statistically converged $M=512\times512$ mesh NS *unsteady* evolution solution $\mathbf{u}^h(\mathbf{x},t)$ is filtered to $\overline{\mathbf{u}^h}(\mathbf{x},t)$ via convolution with a gaussian filter of measure $\delta\equiv\sqrt{2}h=\sqrt{2}/32$. For the level 5 $M=64\times64$ mesh with h the velocity DOF disposition measure, $\delta=\sqrt{2}/32$ meets the adequate energetic capture requirement $\delta\geq 2h$, Pope (2000). This benchmark solution post-processed to planar vorticity $\overline{\omega^h}(\mathbf{x},t)\equiv \text{curl}\,\overline{\mathbf{u}^h}\bullet\mathbf{k}$ clearly visualizes mixing layer evolution at decile time stations on $0\leq n\Delta t\leq 200\,\text{s}$, Figure 9.3 left to right, top to bottom.

On the scale of Figure 9.3, vorticity distributions predicted for $\overline{\mathbf{u}^h}(\mathbf{x},t)$ on $M=64\times64$ from $\mathbf{u}^h(\mathbf{x},t)$ on $M=512\times512$ are visually indistinguishable throughout $0<n\Delta t\leq 140\,\text{s}$, John (2004, Chapter 11). The last three output time station distributions for $\mathbf{u}^h(\mathbf{x},t)$ are graphed in Figure 9.4. Comparing to Figure 9.3, convolution slightly blurs contour details around the dominant vortex.

John (2004) defines two strategies for handling the auxiliary problem inverse differential operator [A], (9.40), with universal imposition of homogeneous Neumann BCs. From properties of Fourier transformation

$$g_\delta * \overline{u_i} = \left[I - \frac{\delta^2}{4\gamma}\nabla^2\right]^{-1}\overline{u_i} \tag{9.51}$$

the operator [A] is thus interpreted to define convolution leading to the replacement of (9.40) with

$$\overline{\overline{u}_j\overline{u}_i} + \overline{\overline{u}_j u_i'} + \overline{u_j'\overline{u}_i} \approx \overline{u}_j\overline{u}_i + \frac{\delta^2}{2\gamma} g_\delta * \left(\frac{\partial \overline{u}_j}{\partial x_k}\frac{\partial \overline{u}_i}{\partial x_k}\right) \tag{9.52}$$

The alternative strategy involves interpretation of [A] as a backwards Euler time integration algorithm for I-EBV PDE + BCs statement

$$\begin{aligned} &\frac{\partial q}{\partial t} - \nabla^2 q = \frac{4\gamma}{\delta^2} f, \quad \text{in } [0,T]\times\Omega, T\equiv\delta^2/4\gamma \\ &\left(\hat{n}\bullet\nabla\right)q \equiv 0, \quad\quad \text{in } [0,T]\times\partial\Omega, \quad (q(0,t_0)) = 0 \text{ in } \Omega \end{aligned} \tag{9.53}$$

For time step definition $\Delta t_1 \equiv \delta^2/4\gamma$ this corresponds to EBV PDE

$$\left[I - \frac{\delta^2}{4\gamma}\nabla^2\right] q(t_1) = f(t_1) \text{ in } \Omega \tag{9.54}$$

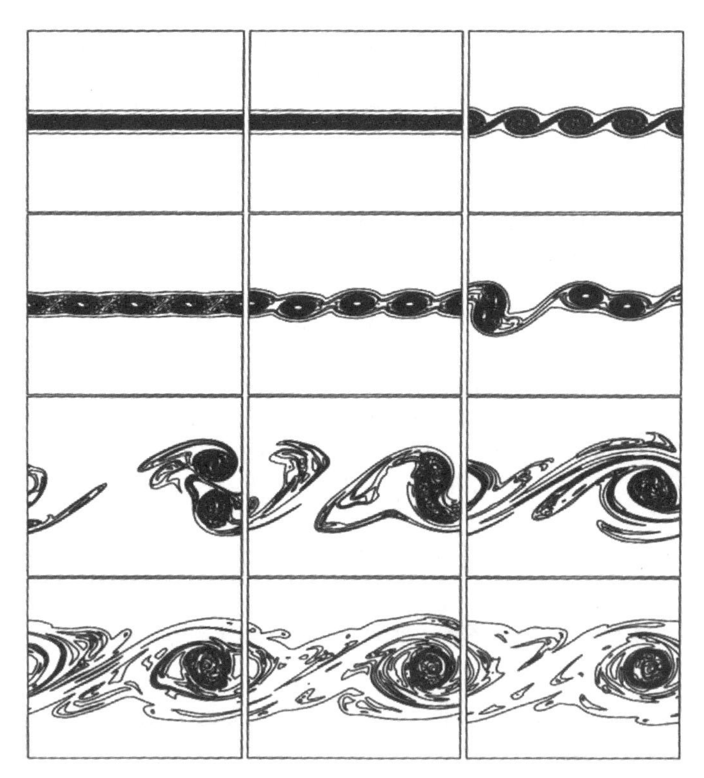

Figure 9.3 $\mathrm{GWS}^h + \theta\mathrm{TS}$ NS vorticity evolution, level 8 mesh solution $\mathbf{u}^h(\mathbf{x}, t)$ convolved to $\overline{\mathbf{u}^h}(\mathbf{x}, t)$ on level 5 mesh, $\delta \equiv \sqrt{2}/32$, at $n\Delta t = 0$, 20, 30, 50, 70, 80, 100, 120, 140, 160, 180, 200 s, left to right, top to bottom. *Source:* John, V. 2004. Reproduced with permission from Springer

For auxiliary problem handled via convolution (9.52), argument functionality was extended periodically exterior to $x \in (-1.1)$ and by zero in the transverse direction. For direct solving as (9.54), reflective BCs were applied on inflow/outflow boundary DOF with homogeneous Neumann BCs, (9.53), applied in the transverse direction.

Absent a model for SFS tensor $\overline{u_j' u_i'}$, the $\overline{\mathbf{u}^h}(\mathbf{x}, t)$ $\mathrm{GWS}^h + \theta\mathrm{TS}$ algorithm Q_2/P_1^{disc} implementing second-order Padé cross-stress tensor pair prediction is unstable, exhibiting algebraic divergence for $n\Delta t > 10$ on the level 5 mesh. All tested SFS and/or modified Smagorinsky SGS tensor model insertions netted a stable solution process.

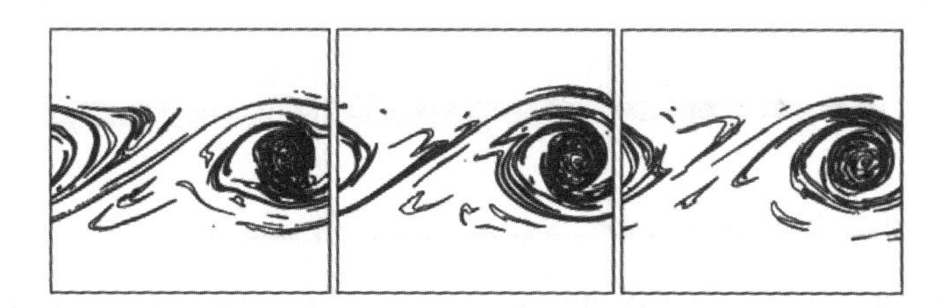

Figure 9.4 $\mathrm{GWS}^h + \theta\mathrm{TS}$ NS vorticity evolution on level 8 mesh, from solution $\mathbf{u}^h(\mathbf{x}, t)$, $n\Delta t = 160$, 180, 200 s, left to right. *Source:* John, V. 2004. Reproduced with permission from Springer

Figure 9.5 directly compares vorticity DOF interpolation *a posteriori* data from $M=64 \times 64$ post-processed $\overline{\mathbf{u}^h}(\mathbf{x},t)$ for SFS and altered SGS tensor model implementations at $n\Delta t=0$, 20, 30 s. From top to bottom by rows, the Q_2/P_1^{disc} $\text{GWS}^h+\theta\text{TS}$ algorithm SFS/SGS tensor models are:

- none, level 8 benchmark convolved to $\overline{\mathbf{u}^h}(\mathbf{x},t)$, $\delta \equiv \sqrt{2}/32$
- Smagorinsky SGS tensor model (9.25), $C_S=0.01$
- Iliescu *et al.* SFS tensor model (9.47), $C_S=0.5$, augmented with Smagorinsky SGS tensor model (9.25), $C_S=0.01$
- Iliescu *et al.* SFS tensor model (9.47), $C_S=0.5$

Each modified Smagorinsky implementation fails to generate the four vortex row solution at $n\Delta t=30$ s, no doubt due to excess diffusion. Deleting Smagorinsky augmentation, the $O(\delta^3)$ SFS tensor model (9.47) gives modest indication of vortex generation but exhibits no quantitative agreement with the filtered benchmark.

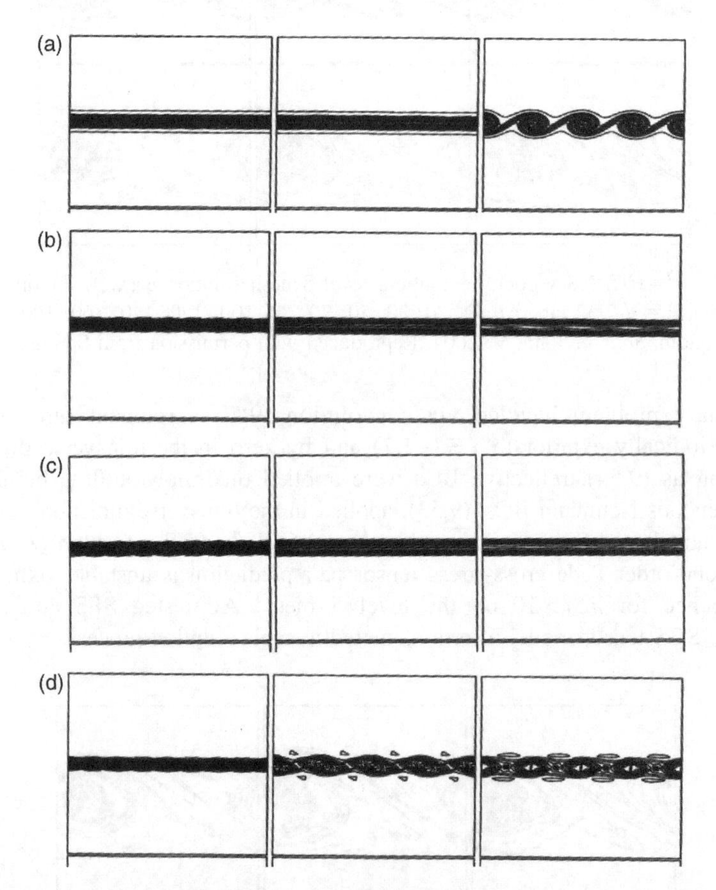

Figure 9.5 $\text{GWS}^h+\theta\text{TS}$ Q_2/P_1^{disc} vorticity distributions post-processed from $M=64\times64$ solutions $\overline{\mathbf{u}^h}(\mathbf{x},t)$, $n\Delta t=0$, 20, 30 s left to right: for text detailed SFS, altered SGS tensor model closures, top to bottom. *Source:* John, V. 2004. Reproduced with permission from Springer

For the $O(\delta^3)$ SFS tensor model (9.47) without Smagorinsky augmentations, vorticity distributions post-processed from $M = 64 \times 64$ solutions $\overline{\mathbf{u}^h}(\mathbf{x}, t)$ generated by cited *auxiliary problem* solution strategies are compared with the filtered benchmark at $n\Delta t = 50$, 70, 80 s in Figure 9.6. By rows:

- none, level 8 benchmark convolved to $\overline{\mathbf{u}^h}(\mathbf{x}, t)$, $\delta \equiv \sqrt{2}/32$
- Iliescu *et al.* SFS tensor (9.47), $C_S = 0.5$, via auxiliary problem solution (9.54) with homogeneous Neumann BCs
- Iliescu *et al.* SFS tensor (9.47), $C_S = 0.5$, via convolution (9.52), axial data extrapolation periodic, transverse extrapolation nil

No $\overline{\mathbf{u}^h}(\mathbf{x}, t)$ solution post-processed vorticity distribution on $M = 64 \times 64$ for selected SFS and/or altered SGS tensor closure model combinations, Figures 9.5–9.6, exhibits quantitative or even *qualitative* agreement with the benchmark. The solution generated via auxiliary problem [A] convolution solution strategy for homogeneous Neumann BCs is clearly furthest from acceptable.

9.6 Analytical SFS Tensor/Vector Closures

The convolved NS D**P** PDE theory role for SFS tensor, (9.6), hence also convolution-generated SFS *vectors* in (9.7), is dissipation at the *unresolved scale threshold*. Literature

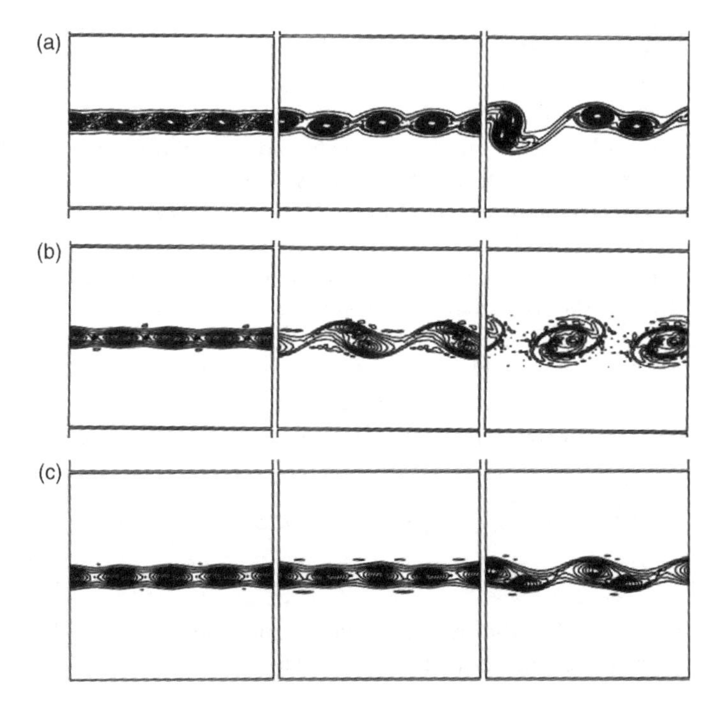

Figure 9.6 $\text{GWS}^h + \theta\text{TS}$ Q_2/P_1^{disc} vorticity distributions post-processed from $M = 64 \times 64$ solutions $\overline{\mathbf{u}^h}(\mathbf{x}, t)$, $n\Delta t = 50$, 70, 80 s, top to bottom: $\overline{\mathbf{u}^h}(\mathbf{x}, t)$ convolved benchmark; [A] via direct solution; [A] via convolution. *Source:* John, V. 2004. Reproduced with permission from Springer

reported SFS tensor closure models employ resolved scale velocity dynamics *hypotheses* to approximate the functionality defined in (9.25), for example (9.46), (9.47). The geometrically elementary but precisely designed mixing layer benchmark *a posteriori* data confirm that cited models generate substandard performance.

From the vantage point of RaNS CFD, as amply *a posteriori* data validated to this juncture, nonlinear cascading of CFD algorithm discrete implementation induced $O(2h)$ phase *dispersion error* is *the*(!) dominant energetic mechanism at the unresolved scale threshold. Full attestation of this observation is generated by universal appending of *artificial diffusion* operators to stabilize RaNS FD/FV discrete implementation CFD codes.

From the perspective of dispersion error annihilation, Figures 6.3, 7.13 and those in Section 8.5, the *m*PDE theory-generated nonlinear GWS^h theory $k=1$ basis $O(h^2)$ error annihilation augmentation of the NS PDE dissipative flux vector (5.117) constitute *analytically derived* SFS tensor-vector candidates. As generalized in Section 8.3, and inserting a coefficient C_S to be determined, the explicitly nonlinear SFS tensor analytical closure involving only resolved scale velocity vector is

$$-\overline{u'_j u'_i} \equiv \frac{C_S(\delta)h^2\mathrm{Re}}{6}\left(\overline{u}_j\overline{u}_k\frac{\partial \overline{u}_i}{\partial x_k} + \overline{u}_i\overline{u}_k\frac{\partial \overline{u}_j}{\partial x_k}\right) \tag{9.55}$$

Interpretation of (9.55) as of eddy viscosity genre with $D(\mathrm{L}^2\tau^{-1})$ "viscosity coefficient" $h^2\mathrm{Re}/6\,\overline{u}_j\overline{u}_i$ is inaccurate, as the resolved scale velocity tensor products $\overline{u}_j\overline{u}_k$ and $\overline{u}_i\overline{u}_k$ therein are *non-positive definite*. The *modus operandi* of this *m*PDE theory $k=1$ basis augmentation is fully documented *dissipative/anti-dissipative modulo* $O(2h)$ mesh scale oscillations.

Chapters 5–8 $m\mathrm{GWS}^h + \theta\mathrm{TS}$ algorithm *a posteriori* data fully validate that NS/RaNS *m*PDE theory definition $C_S \equiv 1.0$ in (9.55) generates $O(2h)$ dispersion error *annihilation* for NS/RaNS *m*PDE theory $k=1$ NC/TP basis implementations. These data further confirm that the modality is *without* infusion of measureable artificial diffusion for $n>1$, LES theory cited a requirement. Finally, that (9.55) as an SFS tensor can promote anti-dissipation admits the LES theory requirement that closure enables *backscatter*, the energetic cascade from smaller to larger scales, identically the action of phase dispersion error.

Adequate energetic capture requires $\delta \geq 2h$ and defining a non-constant δ induces commutation error, detailed shortly. Altering the capture constraint to an equality confirms that the scalar multiplier on (9.55) tensor products is $O(\delta^2)$. Therefore, for this analytical SFS tensor to be RLES theory admissible inserted C_S cannot exceed $O(\delta)$.

Because (9.40) with (9.55) is an analytical determination, no need exists for "turbulent" Prandtl/Schmidt number model hypotheses. Expanding (9.7) for DE and DY state variable scalar members

$$\overline{u_j\Theta} \equiv \overline{(\overline{u}_j + u'_j)(\overline{\Theta} + \Theta')} = \overline{\overline{u}_j\overline{\Theta}} + \overline{\overline{u}_j\Theta'} + \overline{u'_j\overline{\Theta}} + \overline{u'_j\Theta'} \tag{9.56}$$

$$\overline{u_jY} \equiv \overline{(\overline{u}_j + u'_j)(\overline{Y} + Y')} = \overline{\overline{u}_j\overline{Y}} + \overline{\overline{u}_jY'} + \overline{u'_j\overline{Y}} + \overline{u'_jY'} \tag{9.57}$$

Direct extension on the second-order Padé interpolation processes (9.38–9.40) leads to *analytical* prediction of the lead *vector* triples in (9.56–9.57)

$$\overline{\overline{u}_j\overline{\Theta}} + \overline{\overline{u}_j\Theta'} + \overline{u'_j\overline{\Theta}} = \overline{u}_j\overline{\Theta} + \left[I - \frac{\delta^2}{4\gamma}\nabla^2\right]^{-1}\frac{\delta^2}{2\gamma}\frac{\partial \overline{u}_j}{\partial x_k}\frac{\partial \overline{\Theta}}{\partial x_k} + O(\delta^4) \tag{9.58}$$

$$\overline{\bar{u}_j\bar{Y}} + \overline{\bar{u}_j Y'} + \overline{u_j'\bar{Y}} = \bar{u}_j\bar{Y} + \frac{\delta^2}{2\gamma}\left[I - \frac{\delta^2}{4\gamma}\nabla^2\right]^{-1}\frac{\partial\bar{u}_j}{\partial x_k}\frac{\partial\bar{Y}}{\partial x_k} + O(\delta^4) \tag{9.59}$$

Similarly, direct extension on *m*PDE theory derivation (9.55) generates *analytical* SFS *vector* closures for the terminal vectors in (9.56–9.57), Grubert (2006)

$$-\overline{u_j'\Theta'} \equiv C_S(\delta)\frac{h^2\text{RePr}}{12}\left[\bar{u}_j\bar{u}_k\frac{\partial\bar{\Theta}}{\partial x_j}\right] \tag{9.60}$$

$$-\overline{u_j'Y'} \equiv C_S(\delta)\frac{h^2\text{ReSc}}{12}\left[\bar{u}_j\bar{u}_k\frac{\partial\bar{Y}}{\partial x_j}\right] \tag{9.61}$$

The summary accomplishment of RLES + *m*PDE theory is derivation of an *essentially analytical* closure for LES rigorous theory-generated tensor/vector quadruples (9.6–9.7) without inserting the word *turbulent*. *Only* the RLES theory constraint $C_S \leq O(\delta)$ for (9.55), (9.60–9.61) requires the caveat "essentially". Meeting this RLES theory requirement renders no longer applicable the $O(h^4)$ *smooth data* convergence rate predicted for NS *optimal* $m\text{GWS}^N \Rightarrow m\text{GWS}^h k = 1$ TP/NC basis algorithm (5.120). This appears an *academic* loss as regular solution adapted meshes M sufficiently refined to verify the theory prediction in *genuine* LES compute practice are likely impractical.

9.7 Auxiliary Problem Resolution Via Perturbation Theory

RLES theory Padé polynomial interpolations generate the *auxiliary problem*, the name attached to EBV matrix inverse differential operators [A], (9.41), (9.43). Insertion of second-order Padé closure (9.40) into (9.13) yields

$$\begin{aligned}\mathcal{L}(\bar{u}_i) = \text{St}\frac{\partial\bar{u}_i}{\partial t} + \frac{\partial}{\partial x_j}\left[\bar{u}_j\bar{u}_i + \bar{P}\,\delta_{ij} - \frac{2}{\text{Re}}\bar{S}_{ij}\right] + \frac{\text{Gr}}{\text{Re}^2}\bar{\Theta}\hat{g}_i \\ + \frac{\partial}{\partial x_j}\left(\frac{\delta^2}{2\gamma}\left[I - \frac{\delta^2}{4\gamma}\nabla^2\right]^{-1}\left(\frac{\partial\bar{u}_j}{\partial x_k}\frac{\partial\bar{u}_i}{\partial x_k}\right)\right) + O(\delta^3) = 0\end{aligned} \tag{9.62}$$

substituting resolved scale velocity strain rate tensor $\bar{S}_{ij}$, (9.23). The truncation level is $O(\delta^3)$ as (9.62) is absent an SFS tensor.

LES identification of the *unresolved scale* leads to RLES auxiliary problem theoretical resolution. From classical mechanics *perturbation theo*ry, "Most useful approximations (in fluid mechanics theory) are valid when one or more of the parameters or variables in the problem is small (or large).", van Dyke (1975, Ch. 1.2). Convolution filter measure δ exponentiations resident in RLES predictions (9.40) and (9.42) correlate precisely with this observation.

LES non-D resolved scale velocity vector $\bar{u}_i$ is by definition $O(1)$, as is convection tensor product second filtering $\overline{\bar{u}_j\bar{u}_i}$. In context of NS CFD algorithm discrete implementation, spectral content is not resolvable on a mesh of span $2h$ hence unresolved scale velocity vector u_i' cannot exceed $O(h)$. Via the adequate energetic capture constraint $\delta \geq 2h$, u_i' is thus determined $O(\delta)$.

RLES theory predicts $\overline{u_j'u_i'}$ is $O(\delta^3)$, two orders smaller than u_i' order. As $O(\delta)$ and $O(\delta^3)$ thus bound the cross-stress tensor pair $\overline{\bar{u}_j u_i'} + \overline{u_j'\bar{u}_i}$ their order cannot be other than $O(\delta^2)$. Therefore, via fluid mechanics perturbation theory (9.40) and (9.42) *each* predict

$$\overline{\bar{u}_j\bar{u}_i} = \bar{u}_j\bar{u}_i + O(\delta^2) \tag{9.60}$$

Note that (9.60) lends theoretical fluid dynamics substantiation to LES SGS tensor closure practice of replacement of $\overline{\bar{u}_j\bar{u}_i}$ with $\bar{u}_j\bar{u}_i$ plus an $O(\delta^2)$ function, regularly selected an SGS stress tensor model, (9.22–9.24).

Extracting (9.60) from second-order Padé prediction (9.40) yields

$$\overline{\bar{u}_j u_i'} + \overline{u_j'\bar{u}_i} = \frac{\delta^2}{2\gamma}\left[I - \frac{\delta^2}{4\gamma}\nabla^2\right]^{-1}\left(\frac{\partial\bar{u}_j}{\partial x_k}\frac{\partial\bar{u}_i}{\partial x_k}\right) + O(\delta^4) \tag{9.61}$$

and for fourth-order Padé prediction (9.42)

$$\begin{aligned}\overline{\bar{u}_j u_i'} + \overline{u_j'\bar{u}_i} = \frac{\delta^2}{2\gamma}&\left[I - \frac{\delta^2}{4\gamma}\nabla^2 + \frac{\delta^4}{32\gamma^2}\nabla^4\right]^{-1} \times \\ &\left[\begin{array}{l}\dfrac{\partial\bar{u}_i}{\partial x_k}\dfrac{\partial\bar{u}_j}{\partial x_k} - \dfrac{\delta^4}{16\gamma^2}\left(\nabla^2\bar{u}_j\nabla^2\bar{u}_i + \dfrac{\partial\nabla^2\bar{u}_i}{\partial x_k}\dfrac{\partial\bar{u}_j}{\partial x_k}\right) \\ \quad - \dfrac{\delta^4}{16\gamma^2}\left(\dfrac{\partial\bar{u}_i}{\partial x_k}\dfrac{\partial\nabla^2\bar{u}_j}{\partial x_k} + \nabla^2\left(\dfrac{\partial\bar{u}_i}{\partial x_k}\dfrac{\partial\bar{u}_j}{\partial x_k}\right)\right)\end{array}\right] + O(\delta^6)\end{aligned} \tag{9.62}$$

So *indeed* the cross-stress tensor pair is RLES theory confirmed $O(\delta^2)$.

Classic fluid mechanics perturbation theory thereby generates theoretical resolution of the RLES theory generated auxiliary problem. For second-order Padé, multiplying (9.61) through by [A] yields

$$\left[I - \frac{\delta^2}{4\gamma}\nabla^2\right]\left(\overline{\bar{u}_j u_i'} + \overline{u_j'\bar{u}_i}\right) = \frac{\delta^2}{2\gamma}\left(\frac{\partial\bar{u}_j}{\partial x_k}\frac{\partial\bar{u}_i}{\partial x_k}\right) + O(\delta^4) \tag{9.63}$$

Labeling the cross-stress tensor pair $c_{ij}(\mathbf{x}, t) \equiv \overline{\bar{u}_j u_i'} + \overline{u_j'\bar{u}_i}$, coalescing coefficients in (9.63) generates the *harmonic* EBV statement

$$\mathcal{L}(c_{ij}) = -\nabla^2 c_{ij} + \frac{4\gamma}{\delta^2}c_{ij} - 2\frac{\partial\bar{u}_i}{\partial x_k}\frac{\partial\bar{u}_j}{\partial x_k} = 0 \tag{9.64}$$

characterization of LES rigorous theory identified *resolved-unresolved scale interaction* velocity tensor product pair (9.6).

The auxiliary problem strategy (9.54) bears functional resemblance to (9.63) upon deletion of backwards Euler unsteadiness and time t_1. Of *theoretical* fluid mechanics pertinence, since (9.64) is harmonic with constant coefficients, generated solutions $c_{ij}(\mathbf{x}, t)$ are *symmetric*, *realizable* and translation and Galilean *invariant*, thereby meeting *all* fundamental mathematical requirements for a fluid mechanics tensor field, Pope (2000). Deriving encompassing BCs is required for EBV PDE system (9.64) to be well-posed, addressed later in this chapter.

For completeness, fourth-order Padé prediction (9.42) with coefficient rearrangement produces constant coefficient harmonic EBV

$$\mathcal{L}(c_{ij}) = -\nabla^4 c_{ij} + \frac{8\gamma}{\delta^2}\nabla^2 c_{ij} - \frac{32\gamma^2}{\delta^4} c_{ij} - \left[\begin{array}{c} \dfrac{16\gamma}{\delta^2}\dfrac{\partial \overline{u}_i}{\partial x_k}\dfrac{\partial \overline{u}_j}{\partial x_k} - \dfrac{\delta^2}{\gamma}\left(\nabla^2\overline{u}_j\nabla^2\overline{u}_i + \dfrac{\partial \nabla^2\overline{u}_i}{\partial x_k}\dfrac{\partial \overline{u}_j}{\partial x_k}\right) \\ -\dfrac{\delta^2}{\gamma}\left(\dfrac{\partial \overline{u}_i}{\partial x_k}\dfrac{\partial \nabla^2\overline{u}_j}{\partial x_k} + \nabla^2\left(\dfrac{\partial \overline{u}_i}{\partial x_k}\dfrac{\partial \overline{u}_j}{\partial x_k}\right)\right) \end{array} \right] = 0 \tag{9.65}$$

The practical utility of GWS^N discrete implementation of (9.65) is doubtful. Dual Green–Gauss divergence theorem operations required to symmetrize biharmonic (∇^4) differentiability generates gradient BC identification requirements. Embedded high-order ∇^2 operations on gradient operator products places impractical demands on weak form trial space completeness degree requirements.

For scalar state variable members in LES theory PDE system (9.13–9.15), perturbation theory resolution of the RLES second-order Pade auxiliary problem is the direct extension

$$\overline{\overline{u}_j\overline{\Theta}} = \overline{u}_j\overline{\Theta} + O(\delta^2), \overline{\overline{u}_j\overline{Y}} = \overline{u}_j\overline{Y} + O(\delta^2) \tag{9.66}$$

and substitution into (9.57) and (9.60) generates

$$\overline{\overline{u}_j\Theta'} + \overline{u_j'\overline{\Theta}} = \frac{\delta^2}{2\gamma}\left[I - \frac{\delta^2}{4\gamma}\nabla^2\right]^{-1}\frac{\partial \overline{u}_j}{\partial x_k}\frac{\partial \overline{\Theta}}{\partial x_k} + O(\delta^4) \tag{9.67}$$

$$\overline{\overline{u}_jY'} + \overline{u_j'\overline{Y}} = \frac{\delta^2}{2\gamma}\left[I - \frac{\delta^2}{4\gamma}\nabla^2\right]^{-1}\frac{\partial \overline{u}_j}{\partial x_k}\frac{\partial \overline{Y}}{\partial x_k} + O(\delta^4) \tag{9.68}$$

Labeling scalar resolved-unresolved scale interaction *vector* pairs

$$v_j^{\Theta} \equiv \overline{\overline{u}_j\Theta'} + \overline{u_j'\overline{\Theta}}, \quad v_j^{Y} \equiv \overline{\overline{u}_jY'} + \overline{u_j'\overline{Y}} \tag{9.69}$$

the resultant *harmonic* EBV statements are

$$\mathcal{L}\left(v_j^{\Theta}\right) = -\nabla^2 v_j^{\Theta} + \frac{4\gamma}{\delta^2}v_j^{\Theta} - 2\frac{\partial \overline{u}_j}{\partial x_k}\frac{\partial \overline{\Theta}}{\partial x_k} = 0 \tag{9.70}$$

$$\mathcal{L}\left(v_j^{Y}\right) = -\nabla^2 v_j^{Y} + \frac{4\gamma}{\delta^2}v_j^{Y} - 2\frac{\partial \overline{u}_j}{\partial x_k}\frac{\partial \overline{Y}_{\alpha}}{\partial x_k} = 0 \tag{9.71}$$

As with (9.64), encompassing BCs are required identified for (9.70–9.71) to be well-posed following weak form Green–Gauss divergence theorem enforcement.

9.8 LES Analytical Closure (*ar*LES) Theory

The *essentially analytical* closure for LES theory rigorously convolved PDE system (9.13–9.18) is derived, facilitated by coupling RLES second-order Padé prediction with *m*PDE error annihilation theory interpreted within context of classic fluid mechanics perturbation theory. Referenced hereon as *ar*LES theory, only order requirement $C_S(\delta)$ compromises *ar*LES theory closure from being totally analytical. Meeting this requirement likely degrades Chapter 5 $m\mathrm{GWS}^h+\theta\mathrm{TS}$ $k=1$ TP/NC basis algorithm *smooth data* uniform mesh TS estimate $O(h^4, m^3, \Delta t^4)$ to $O(h^3, m^3, \Delta t^2)$ as *m*PDE theory GWS^h $k=1$ basis $O(h^2)$ error annihilation remains functional.

The *m*PDE system for *ar*LES theory is a direct extension on Chapter 5 *optimal* NS *m*PDE theory for the space filtered I-EBV and EBV PDE systems. Suppressing overbar and prime notation except where required for clarity, the *ar*LES I-EBV + EBV *m*PDE system *continuum* state variable, semicolon partitioned according to (eventual) jacobian coupling, is

$$\{q(x_j,t)\} = \left\{u_i, \Theta, Y, \phi; c_{ij}, v_j^{\Theta}, v_j^{Y}; P\right\}^T \tag{9.72}$$

For members in (9.72) satisfying the identified I-EBV PDE system, the now familiar *m*PDE flux vector statement is

$$\mathcal{L}^m(\{q\}) = [m]\frac{\partial\{q\}}{\partial t} + \frac{\partial}{\partial x_j}\left[\{f_j\} - \{f_j^d\}\right] - \{s\} = \{0\} \tag{9.73}$$

for $[m]$ the *optimal* γ diagonal mass matrix alteration, $\{f_j\}$ the *ar*LES theory *kinetic flux vector*, $\{f_j^d\}$ the theory *dissipative flux vector* and $\{s\}$ the source. For $1 \le (i, j, k) \le n$ the non-D matrix statements are

$$[m] \equiv \mathrm{diag}\left[1 - \frac{\gamma\,\Delta t^2}{6}\frac{\partial}{\partial x_j}\left(u_j u_k \frac{\partial}{\partial x_k}\right)\right] \tag{9.74}$$

$$\{f_j(\{q\})\} = \begin{Bmatrix} u_j u_i + c_{ij} + P*|_{n+1}^{P}\delta_{ij} \\ u_j\Theta + v_j^{\Theta} \\ u_j Y + v_j^{Y} \end{Bmatrix} \tag{9.75}$$

$$\{f^d(\{q\})\} = \begin{Bmatrix} \dfrac{2}{\mathrm{Re}}S_{ij} + \dfrac{C_S(\delta)h^2\mathrm{Re}}{6}\left(u_j u_k\dfrac{\partial u_i}{\partial x_k} + u_i u_k\dfrac{\partial u_j}{\partial x_k}\right) \\ \dfrac{1}{\mathrm{RePr}}\dfrac{\partial\Theta}{\partial x_j} + \dfrac{C_S(\delta)h^2\mathrm{RePr}}{12}u_j\,u_k\dfrac{\partial\Theta}{\partial x_k} \\ \dfrac{1}{\mathrm{ReSc}}\dfrac{\partial Y_\alpha}{\partial x_j} + \dfrac{C_S(\delta)h^2\mathrm{ReSc}}{12}u_j\,u_k\dfrac{\partial Y_\alpha}{\partial x_k} \end{Bmatrix} \tag{9.76}$$

$$\{s\} = \begin{Bmatrix} -\dfrac{\mathrm{Gr}}{\mathrm{Re}^2}\Theta\widehat{g}_i \\ \dfrac{\mathrm{Ec}}{\mathrm{Re}}\begin{bmatrix} \dfrac{\partial}{\partial x_j}\left(u_i\dfrac{\partial u_j}{\partial x_i}\right) + \dfrac{\partial^2 c_{ij}}{\partial x_j\partial x_i} \\ -\dfrac{\partial^2}{\partial x_j\partial x_i}\left(\dfrac{C_S(\delta)h^2\mathrm{Re}}{6}\left(u_j u_k\dfrac{\partial u_i}{\partial x_k} + u_i u_k\dfrac{\partial u_j}{\partial x_k}\right)\right) \end{bmatrix} \\ s(Y) \end{Bmatrix} \tag{9.77}$$

PDE statements for *ar*LES state variable scalar members $\{\phi; P\}$ remain (9.17–9.18) with (9.16) defining P^*. The linearly *decoupled* harmonic EBV PDE systems for $O(\delta^2)$ resolved-unresolved scale interaction state variable members are (9.64), (9.71–9.72). It may be advantageous to alter these to $O(h^4)$ *m*PDE form (5.114).

There exists no *a priori* constraint on Reynolds number Re and the *ar*LES theory $m\mathrm{GWS}^N + \theta\mathrm{TS}$ algorithm statement is identically that for NS, (7.17), with terminal nonlinear iterative matrix statement (7.18). The discrete implementation (7.20) generates computable forms (7.21–7.22). For $k=1$ TP/NC trial space basis implementation, the *asymptotic error estimate* for $m\mathrm{GWS}^h + \theta\mathrm{TS}$ algorithm solutions to I-EBV PDE system (9.73) is

$$\left\|\{e^h(n\Delta t)\}\right\|_E \leq Ch^{2\gamma}\left[\|\mathrm{data}\|_{L2,\Omega} + \|\mathrm{data}\|_{L2,\partial\Omega}\right] + C_2\Delta t^{f(\theta)}\left\|\{q^h(t_0)\}\right\|^2_{H^m(\Omega)}$$
$$\text{for: } \gamma \equiv \min(k(k+1/2), r-1), f(\theta) = (2,3) \tag{9.78}$$

with $(k+1/2)$ anticipated for *m*PDE theory $C_S(\delta)$ alteration. The alteration of (9.78) for GWS^h algorithm solutions for *ar*LES theory harmonic EBV PDE systems is

$$\left\|\{e^h(n\Delta t)\}\right\|_E \leq Ch^{2\gamma}\left[\|\mathrm{data}\|_{L2,\Omega} + \|\mathrm{data}\|_{L2,\partial\Omega}\right]$$
$$\text{for: } \gamma \equiv \min(k, (k+1), r-1) \tag{9.79}$$

with $(k+1)$ the prediction for EBV *m*PDE definition (5.114).

9.9 *ar*LES Theory $m\mathrm{GWS}^h$ + θTS Algorithm

*ar*LES theory $m\mathrm{GWS}^h + \theta\mathrm{TS}$ $k=1$ TP/NC basis algorithm matrix statement bears closest resemblance to the NS algorithm, Section 7.4, for now omitting the $O(\Delta t^4)$ term augmentations. The I-EBV partition $O(1)$ discrete state variable $\{q^h(\mathbf{x},t)\} = \{u_i^h, \Theta^h, Y^h, \phi^h; P^h\}^T$ has DOF $\{\mathrm{Q}I(t_{n+1})\}$, $1 \leq I \leq n+4$. DOF $\{\mathrm{Q}I(\bullet)\}$, $1 \leq I \leq 3$, 6, are selectively denoted $\{\mathrm{U}I(\bullet)\}$ and {PHI $(\bullet)\}$ for clarity. The solution of (9.18) for P^h DOF $\{\mathrm{Q}I(t_{n+1})\}$, $I = n+4$, remains the linear decoupled post-convergence operation.

Avoiding repetitive detail, the $m\mathrm{GWS}^h + \theta\mathrm{TS}$ matrix statement on generic finite element domain Ω_e

$$\begin{aligned}\{\mathrm{F\_Q}I\}_e &= \left[m(\gamma, \{\mathrm{U}J\}_e)\right]_e \{\mathrm{Q}I_{n+1} - \mathrm{Q}I_n\}_e \\ &+ \Delta t\left[\theta\{\mathrm{RES\_Q}I(\{Q\})\}_e\Big|_{n+1} + (1-\theta)\{\mathrm{RES\_Q}I(\{Q\})\}_e\Big|_n\right]\end{aligned} \tag{9.80}$$

is assembled into the global matrix statement. For $1 \leq I \leq n+2$, following Green–Gauss divergence theorem and discarding generated surface integrals (due to typical BCs) the element matrix $[m]_e$ is unaltered from (7.24)

$$[m]_e = \det_e[\mathrm{M200}] + \frac{\gamma\,\Delta t^2}{6}\{\mathrm{U}J\mathrm{U}K\}_e^T[\mathrm{M30}JK]_e \tag{9.81}$$

The kinetic flux vector contribution in $\{\mathrm{RES\_Q}I\}_e$, $1 \leq I \leq n+2$, includes $O(\delta^2)$ state variable partition $\{q^h(\mathbf{x},t)\} = \left\{c_{ij}^h, v_j^{\Theta h}, v_j^{Yh}\right\}^T$ with DOF $\{\mathrm{Q}I(t_{n+1})\}$, $n+5 \leq I \leq n+17$. Denoting DOF as $\{\mathrm{C}IJ\}$, VΘJ, VY$J\}$ for clarity, for $1 \leq I \leq n$ and recalling (7.26–7.27), the kinetic flux vector element matrix in (9.80) is

$$
\begin{aligned}
\{\text{RES\_Q}I(\{f_j\})\}_e &= \int_{\Omega_e} \{N_1\} \frac{\partial}{\partial x_j} u_j^h \{N_1\}^T \, d\tau \{\text{Q}I\}_e + \int_{\Omega_e} \{N_1\} \frac{\partial \{N_1\}^T}{\partial x_j} d\tau \{\text{C}IJ\}_e \\
&\quad + \int_{\Omega_e} \{N_1\} \frac{\partial}{\partial x_i} \left\{ P_n^h + \frac{1}{\theta \Delta t} \sum_{\alpha=1}^{p} \delta \phi^h |_{n+1}^{\alpha} \right\} d\tau \\
&= \{\text{U}J\}_e^T [\text{M300}J]_e \{\text{Q}I\}_e + [\text{M20}J]_e \{\text{C}IJ\}_e \\
&\quad + [\text{M20}I]_e \{\text{Q7}(t_n)\}_e + \frac{1}{\theta \Delta t} [\text{M20}I]_e \sum_{\alpha=1}^{p} \{\delta \text{Q6}\}_e^{\alpha}
\end{aligned}
\tag{9.82}
$$

The kinetic flux vector element matrix statement for $n+1 \leq I \leq n+2$ in (9.80) is

$$
\begin{aligned}
\{\text{RES\_Q}I(\{f_j\})\}_e &= \int_{\Omega_e} \{N_1\} \frac{\partial}{\partial x_j} u_j^h \{N_1\}^T \, d\tau \{\text{Q}I\}_e + \int_{\Omega_e} \{N_1\} \frac{\partial \{N_1\}^T}{\partial x_j} d\tau \begin{Bmatrix} \text{V}\Theta J \\ \text{VY}J \end{Bmatrix}_e \\
&= \{\text{U}J\}_e^T [\text{M300}J]_e \{\text{Q}I\}_e + [\text{M20}J]_e \begin{Bmatrix} \text{V}\Theta J \\ \text{VY}J \end{Bmatrix}_e
\end{aligned}
\tag{9.83}
$$

Following Green–Gauss divergence theorem enforcement an exercise suggests verifying the (9.76) dissipative flux vector element matrix $\{\text{RES\_U}I\}_e$, $1 \leq I \leq n$, is the modest alteration to (8.28)

$$
\begin{aligned}
\{\text{RES\_U}I(\{f_j^d\})\}_e &= \frac{2}{\text{Re}} [\text{M20}J]_e \{\text{S}IJ\}_e \\
&\quad + \frac{\text{C}_\text{S}(\delta) h^2 \text{Re}}{6} \begin{pmatrix} \{\text{U}J\text{U}K\}_e^T [\text{M30}JK]_e \{\text{U}I\}_e \\ + \{\text{U}I\text{U}K\}_e^T [\text{M30}JK]_e \{\text{U}J\}_e \end{pmatrix}
\end{aligned}
\tag{9.84}
$$

with $\{SIJ\}_e$ the differential definition weak form algorithm DOF for resolved scale velocity strain rate tensor.

Recalling RaNS algorithm non-D group notation Pa(I), where I (*not* italics) is $O(1)$ state variable *indicator* index, the dissipative flux vector (9.76) element matrix $\{\text{RES\_Q}I\}_e$, $n+1 \leq I \leq n+2$, is

$$
\begin{aligned}
\{\text{RES\_Q}I(\{f_j^d\})\}_e &= \frac{1}{\text{Pa(I)}} [\text{M2}KK]_e \{\text{Q}I\}_e \\
&\quad + \frac{\text{C}_\text{S}(\delta) h^2 \text{Pa(I)}}{12} \{\text{U}J\text{U}K\}_e^T [\text{M30}JK]_e \{\text{Q}I\}_e
\end{aligned}
\tag{9.85}
$$

The element matrix statement for source term (9.77) is a modest alteration to (7.29–7.31). An exercise suggests verifying

$$
\{s\}_e = \left\{ \begin{array}{l}
\dfrac{-\text{Gr} \, \widehat{g}_I}{\text{Re}^2} \det_e [\text{M200}] \{\text{Q4}\}_e \\
\dfrac{\text{Ec}}{\text{Re}} \left(\{\text{U}I\}_e^T [\text{M30}JI]_e \{\text{U}J\}_e + [\text{M2}JI]_e \left(\{\text{C}IJ\}_e - \{\text{SFS}IJ\}_e \right) \right) \\
\det_e [\text{M200}] \{\text{SRC}Y\}_e
\end{array} \right\}
\tag{9.86}
$$

for $\{\mathrm{SFS}IJ\}_e$ the SFS stress tensor differential definition weak form solution DOF.

GWS^h statement for DM^h constraint remains (7.32)

$$\{\mathrm{F\_Q6}\}_e = [\mathrm{M2}KK]_e\{\mathrm{PHI}\}_e + [\mathrm{M20}I]_e\{\mathrm{U}I\}_e \tag{9.87}$$

while that for P^h is the alteration to (7.33)

$$\begin{aligned}\{\mathrm{F\_Q7}\}_e =& [\mathrm{M2}KK]_e\{\mathrm{Q7}\}_e + \{\mathrm{U}I\}_e^T[\mathrm{M30}JI]_e\{\mathrm{U}J\}_e \\ &+ [\mathrm{M2}JI]_e\left(\{\mathrm{C}IJ\}_e - \{\mathrm{SFS}IJ\}_e\right) \\ &+ \frac{\mathrm{Gr}}{\mathrm{Re}^2}[\mathrm{M2}J0]_e\{\mathrm{Q4}\}_e\widehat{g}_j + [\mathrm{N200}]_e\frac{\mathrm{d}}{\mathrm{d}t}\{\mathrm{U}J \bullet n_j\}_e \\ &+ \frac{1}{\mathrm{Re}}\left([\mathrm{N2}KK]_e\{\mathrm{U}J \bullet \widehat{n}_j\}_e - [\mathrm{A20}S]_e\{\mathrm{U}J \bullet \widehat{n}_j\}_e\Big|_w\right)\end{aligned} \tag{9.88}$$

The GWS^h algorithm for algebraically decoupled *resolved-unresolved scale interaction* tensor/vector harmonic EBV PDEs for DOF $\{\mathrm{Q}I(t_{n+1})\}$, $n+5 \le I \le n+17$, is classical. Using $O(\delta^2)$ state variable tensor index notation for clarity, the element matrix statements for (9.64), (9.70), (9.71) are

$$\begin{aligned}\{\mathrm{F\_C}IJ\}_e =& [\mathrm{M2}KK]_e\{\mathrm{C}IJ\}_e + \frac{4\gamma}{\delta^2}\det_e[\mathrm{M200}]\{\mathrm{C}IJ\}_e \\ &- 2\{\mathrm{U}I\}_e^T[\mathrm{M3}K0K]_e\{\mathrm{U}J\}_e - [\mathrm{N20}K]_e\{\mathrm{C}IJ\}_e\hat{n}_K\end{aligned} \tag{9.89}$$

$$\begin{aligned}\{\mathrm{F\_V\Theta}J\}_e =& [\mathrm{M2}KK]_e\{\mathrm{V\Theta}J\}_e + \frac{4\gamma}{\delta^2}\det_e[\mathrm{M200}]\{\mathrm{V\Theta}J\}_e \\ &- 2\{\mathrm{Q4}\}_e^T[\mathrm{M3}K0K]_e\{\mathrm{U}J\}_e - [\mathrm{N20}K]_e\{\mathrm{V\Theta}J\}_e\hat{n}_K\end{aligned} \tag{9.90}$$

$$\begin{aligned}\{\mathrm{F\_VY}J\}_e =& [\mathrm{M2}KK]_e\{\mathrm{VY}J\}_e + \frac{4\gamma}{\delta^2}\det_e[\mathrm{M200}]\{\mathrm{VY}J\}_e \\ &- 2\{\mathrm{Q5}\}_e^T[\mathrm{M3}K0K]_e\{\mathrm{U}J\}_e - [\mathrm{N20}K]_e\{\mathrm{VY}J\}_e\hat{n}_K\end{aligned} \tag{9.91}$$

The matrices $[\mathrm{N20}K]_e\{\mathrm{Q}I\}_e\hat{n}_K$ in (9.89–9.91) are Green–Gauss divergence theorem generated Neumann BCs.

Two parametric operations complete the *ar*LES theory *m*GWS^h + θTS algorithm. GWS$^N \Rightarrow$ GWS^h on *differential definition* form of (9.23) determines resolved scale velocity strain rate tensor DOF $\{\mathrm{S}IJ\}_e$ for (9.84) via element matrix statement

$$\begin{aligned}\mathrm{GWS}^N\left(\overline{S}_{ij}\right) &= \int_\Omega \Psi_\beta(x)\left(\overline{S}_{ij} - \frac{1}{2}\left(\frac{\partial \overline{u}_i}{\partial x_j} + \frac{\partial \overline{u}_j}{\partial x_i}\right)\right)^N \mathrm{d}\tau \equiv 0, \forall\beta \\ &\equiv \mathrm{S}_e\int_{\Omega_e}\{N_1\}\left(\overline{S}_{ij} - \frac{1}{2}\left(\frac{\partial \overline{u}_i}{\partial x_j} + \frac{\partial \overline{u}_j}{\partial x_i}\right)\right)^h \mathrm{d}\tau \\ &\Rightarrow \{\mathrm{F\_S}IJ\}_e = \det_e[\mathrm{M200}]\{\mathrm{S}IJ\}_e - \frac{1}{2}\begin{pmatrix}[\mathrm{M20}J]_e\{\mathrm{U}I\}_e \\ +[\mathrm{M20}I]_e\{\mathrm{U}J\}_e\end{pmatrix}\end{aligned} \tag{9.92}$$

As with all differential definition GWS^h statements, the *sole* admissable constraint prior to solution is *a priori* known $\{\mathrm{S}IJ\}_e = \{0\}$ on domain boundary segments $\partial\Omega$.

The identical $\text{GWS}^N \Rightarrow \text{GWS}^h$ process for SFS tensor differential definition rearrangement of (9.55) generates

$$\text{GWS}^N\left(\overline{u_j'u_i'}\right) = \int_\Omega \Psi_\beta(x)\left(\overline{u_j'u_i'} + \frac{C_S(\delta)h^2\text{Re}}{6}\left(u_ju_k\frac{\partial u_i}{\partial x_k} + u_iu_k\frac{\partial u_j}{\partial x_k}\right)\right)^N d\tau \equiv 0, \forall\beta$$
$$\Rightarrow \{\text{F\_SFS}IJ\}_e = \det_e[\text{M200}]\{\text{SFS}IJ\}_e + \frac{C_S(\delta)h^2\text{Re}}{6}\begin{pmatrix}\{\text{U}J\text{U}K\}_e^T[\text{M300}K]_e\{\text{U}I\}_e \\ +\{\text{U}I\text{U}K\}_e^T[\text{M300}K]_e\{\text{U}J\}_e\end{pmatrix} \quad (9.93)$$

Except for [M200] all matrices in (9.81–9.93) are subscripted e indicating Ω_e dependence. Replacement with the element-independent library matrices is via coordinate transformation imposition, Section 7.4. For the *ar*LES theory algorithm

$$[\text{M20}J]_e = \left(\frac{\partial\eta_k}{\partial x_j}\right)_e[\text{M20}k], [\text{M2}J0]_e = \left(\frac{\partial\eta_k}{\partial x_j}\right)_e[\text{M2}k0]$$
$$[\text{M300}J]_e = \left(\frac{\partial\eta_k}{\partial x_j}\right)_e[\text{M300}k], [\text{M300}K]_e = \left(\frac{\partial\eta_j}{\partial x_k}\right)_e[\text{M300}j]$$
$$[\ \text{M2}KK]_e = \det_e^{-1}\left(\frac{\partial\eta_l}{\partial x_j}\frac{\partial\eta_m}{\partial x_k}\right)_e[\text{M2}lm] \quad (9.94)$$
$$[\text{M30}JK]_e = \det_e^{-1}\left(\frac{\partial\eta_l}{\partial x_j}\frac{\partial\eta_m}{\partial x_k}\right)_e[\text{M30}lm]$$
$$[\text{M3}K0K]_e = \det_e^{-1}\left(\frac{\partial\eta_l}{\partial x_k}\frac{\partial\eta_m}{\partial x_k}\right)_e[\text{M3}l0m]$$

Matrices for *ar*LES theory *m*PDE SFS tensor/vector closure contain scalar factor h^2 which, recall Section 7.4 following (7.43), is proportional to measure $\det_e$ of $k=1$ TP/NC basis elements Ω_e. For TP basis $h^3 \approx V_e = 2^n\det_e$ and replacing proportionality with equality leads to $h^2/\det_e = 2^{2n/3}/\det_e{}^{1/3}$. The (9.94) alterations are

$$h^2[\text{M2}KK]_e = \frac{(2)^{2n/3}}{\det_e^{1/3}}\left(\frac{\partial\eta_l}{\partial x_j}\frac{\partial\eta_m}{\partial x_k}\right)_e[\text{M2}lm]$$
$$h^2[\text{M300}J]_e = (2)^{2n/3}\det_e^{2/3}\left(\frac{\partial\eta_k}{\partial x_j}\right)_e[\text{M300}k] \quad (9.95)$$
$$\text{Pa}\,h^2[\text{M30}JK]_e = \frac{(2)^{2n/3}\text{Pa}}{\det_e^{1/3}}\left(\frac{\partial\eta_l}{\partial x_j}\frac{\partial\eta_m}{\partial x_k}\right)_e[\text{M30}lm]$$

For $k=1$ NC basis, scalar multiplier ratio alterations in (9.95) are $2^{2n/3}/\det_e{}^{1/3} \Rightarrow 1/(n!)^{2/3}\det_c{}^{1/3}$ and $2^{2n/3}\det_e{}^{2/3} \Rightarrow \det_c{}^{2/3}/(n!)^{2/3}$.

9.10 *ar*LES Theory $m\text{GWS}^h + \theta\text{TS}$ Completion

The linear algebra jacobian completes the *ar*LES theory algorithm via calculus operations on $m\text{GWS}^h + \theta\text{TS}$ element matrix statements. The quasi-Newton NS jacobian, Section 7.4, is appropriate for I-EBV state variable partition $\left\{q^h(\mathbf{x},t)\right\} = \left\{u_i^h, \Theta^h, Y^h, \phi^h\right\}^T$ upon select term deletions. From (7.21), via (9.81–9.85), neglecting mPDE matrix multiplier $\{\text{U}J\text{U}K\}_e^T$ nonlinearities, conversely accounting for these term post-multipliers $\{\text{U}J\}_e$, leads to I-EBV self-coupled state variable partition DOF $\{QI\}$, $1 \le I \le n+2$, diagonal block jacobian

$$
\begin{aligned}
[\text{JAC\_Q}I]_e &\equiv \frac{\partial\{\text{F\_Q}I\}_e}{\partial\{\text{Q}I\}_e}, 1 \le I \le n+2 \\
&= \frac{\partial}{\partial\{\text{Q}I\}_e}\left([m]_e\{\text{Q}I_{n+1}\}_e + \theta\Delta t\{\text{RES\_Q}I(\{Q\})\}_e\big|_{n+1}\right) \\
&\approx \det{}_e[\text{M200}] + \frac{\gamma\,\Delta t^2}{6}\{\text{U}J\text{U}K\}_e^T[\text{M30}JK]_e \\
&+\theta\Delta t\left(\begin{array}{l} \{\text{U}J\}_e^T[\text{M300}J]_e + \{\text{U}I\}_e^T[\text{M3}J\text{00}]_e\ \delta_{IJ} + \dfrac{1}{\text{Pa}}[\text{M2}KK]_e \\ + \dfrac{\text{C}_\text{S}\text{Pa}\,h^2}{6}\left(\{\text{U}J\text{U}K\}_e^T[\text{M30}JK]_e + \{\text{U}J\}_e^T[\text{M3}KJ0]_e\text{diag}[\text{U}K]_e\right)\end{array}\right)
\end{aligned}
\tag{9.96}
$$

As for NS algorithm, iterative stability requires direct coupling of $\text{D}M^h$ potential solution. From (9.82), the off-diagonal block jacobian element matrices for $1 \le I \le n$ are

$$
\frac{\partial\{\text{F\_Q}I\}_e}{\partial\{\text{Q6}\}_e} \Rightarrow \frac{\partial\{\text{F\_U}I\}_e}{\partial\{\delta\text{Q6}\}_e} = [\text{M20}I]_e \tag{9.97}
$$

with cancellation of the $\theta\Delta t$ multiplier. The off-diagonal block jacobian element matrix completions for $1 \le I \le n$ are

$$
\begin{aligned}
\frac{\partial\{\text{F\_Q}I\}_e}{\partial\{\text{Q4}\}_e} &= -\theta\Delta t\frac{\text{Gr}\,\widehat{g}_I}{\text{Re}^2}\det{}_e[\text{M200}] \\
\frac{\partial\{\text{F\_Q4}\}_e}{\partial\{\text{Q}I\}_e} &= \theta\Delta t\left(\begin{array}{l}\{\text{U}I\}_e^T[\text{M300}J]_e\delta_{IJ} \\ + \dfrac{\text{Ec}}{\text{Re}}\left(\begin{array}{l}\{\text{U}J\}_e^T[\text{M3}I0J]_e \\ +\{\text{U}I\}_e^T[\text{M30}JJ]_e\end{array}\right)\end{array}\right) \\
\frac{\partial\{\text{F\_Q5}\}_e}{\partial\{\text{Q}I\}_e} &= \theta\Delta t\{\text{U}I\}_e^T[\text{M300}J]_e\delta_{IJ}
\end{aligned}
\tag{9.98}
$$

The linear decoupled jacobian for *ar*LES theory $O(\delta^2)$ state variable members $\left\{q^h(\mathbf{x},t)\right\} = \left\{c_{ij}^h, v_j^{\Theta h}, v_j^{Yh}\right\}^T$ is identical for all DOF $\{\text{Q}I(t_{n+1})\}$, $n+5 \le I \le n+17$. Differentiating (9.89–9.91)

$$
\frac{\partial\{\text{F\_Q}I\}_e}{\partial\{\text{Q}I\}_e} = [\text{M2}KK]_e + \frac{4\gamma}{\delta^2}\det{}_e[\text{M200}] \tag{9.99}
$$

Except for [M200] all matrices in (9.96–9.99) are subscripted e indicating Ω_e dependence. Those with an h^2 multiplier require $\det_e$ exponent modifications, (9.95).

9.11 *ar*LES Theory Implementation Diagnostics

While LES *essentially analytical* (*ar*LES) theory can be implemented via *any* discrete CFD procedure, text content fully substantiates that the $m\text{GWS}^h + \theta\text{TS}$ $k = 1$ basis implementation is optimal in the discrete peer group. Specifically, this decision leads to precise theory translation into a stable, efficient, artificial diffusion-free and convergent compute platform for rigorously convolved LES PDE system on a mesh of measure h with *uniform* gaussian measure δ. Such solutions are *diagnostic* via generation of $O(1, \delta^2, \delta^3)$ resolved scale, resolved-unresolved scale interaction and strictly unresolved scale DOF distributions directly *responsive* to *classic* fluid dynamics theory non-D groups Re, Gr, Pr, Sc. In summary:

- $O(\delta^2)$ resolved-unresolved scale interaction state variables explicitly impact $O(1)$ resolved scale state variables, kinetic flux vector (9.75)
- *a priori* decision of gaussian filter of uniform measure δ directly embedded in rigorously formed LES PDE system, (9.75) with (9.64), (9.70–9.71)
- perturbation theory organized $O(\delta^2)$ cross-stress tensor guaranteed symmetric, realizable and translation and Galilean invariant, meeting all NS tensor necessary requirements
- *m*PDE theory predicted $O(\delta^3)$ subfilter scale (SFS) tensors/vectors validated phase selective *dissipative/anti-dissipative* at *unresolved scale threshold*, (9.76)
- SFS tensor *non-positive definite*, a necessary requirement to enable prediction of *backscatter*
- *optimal* $k = 1$ TP/NC basis NS algorithm $O(m^3)$ phase accuracy retained, *m*PDE mass matrix (9.75)
- *ar*LES theory closure addresses *complete* LES state variable absent turbulent Prandtl/Schmidt number hypotheses
- theory derivation mathematically rigorous, devoid of *turbulent* arguments, potentially applicable $\forall$ Re

Laminar-thermal NS and *ar*LES theory directly comparative $m\text{GWS}^h + \theta\text{TS}$ diagnostic data are reported on the $n = 2$ symmetry plane of an $8 \times 1 \times$ transversely wide thermal cavity, Grubert (2006). This geometry has been identified a virtual fluid dynamics laboratory as, " . . . the spatial structure of the flow is made of vertical and horizontal boundary layers, corner structures, and a stratified core . . . to give rise to very complex time behaviours (sic) resulting from several instability mechanisms . . . ", Le Quere and Behnia (1998). Via classic NS PDE system jacobian eigenvalue spectrum analysis, Xin and Le Quere (2001), theory predicts the 8×1 thermal cavity exhibits a *critical* Rayleigh number, $\text{Ra}_{\text{CR}} \approx 3.1$ E+05, above which a steady thermal flowfield cannot persist. Recalling $\text{Ra} \equiv \text{GrPr}$ and for air, $\text{Pr} = 0.71$, *a posteriori* data generated by some three dozen algorithms in a "CFD shoot-out" confirm the $n = 2$ 8×1 cavity solution is unsteady at $\text{Ra} = 3.4$ E+05, Christon et al. (2002).

Thermal cavity characteristic U is undefined; the buoyant flow characterization $\text{Gr}/\text{Re}^2 \equiv 1$ generates $\text{U} \equiv (\text{gL}\beta\Delta\text{T})^{1/2}$, Gebhart *et al.* (1988, Ch. 10). Rayleigh number $\text{Ra} \equiv \text{GrPr}$ definitions ranging E+04 $\leq$ Ra $\leq$ E+08 for standard atmosphere air, $\text{Pr} = 0.71$, thus correspond to E+02 < Re < 2.0 E+04 bracketing laminar through transitional Re.

*ar*LES theory $m\text{GWS}^h + \theta\text{TS}$ and NS $\text{GWS}^h + \theta\text{TS}$ *a posteriori* data are generated on identical $k = 1$ TP basis discretizations of the $n = 2$ 8×1 thermal cavity. Since bounded domain LES resolved scale state variable BCs are *a priori* unknown, in concert with omission of BCE integrals in (9.12–9.14), the only *ar*LES theory implementation yielding directly comparable *a posteriori* data is a detached eddy simulation (DES).

The cavity $n=2$ plane is discretized into $\mathrm{M}=326\times46$ $k=1$ TP basis elements with $\mathrm{M}=318\times38$ *square* elements of *uniform* measure $h=0.025$ spanning the interior, the *ar*LES *partition*. Gaussian filter *uniform* measure admits theory *commutation* requirement, (9.10), and $\delta\equiv2h=0.05$ meets the energetic capture requirement. The remaining mesh populates the enclosing DES partition pr-graded to each cavity wall. DES domain wall-normal span is equal to *ar*LES partition element measure $h=0.025$ and is discretized into $\mathrm{M}=4$ elements. The wall-normal span of wall-adjacent element is $\sim h/10$. Figure 9.7 graphs the cavity upper right corner mesh partition with nodal velocity vector overlay on vorticity flood, arrow denotes partition shared $\partial\Omega$ vertex node.

No-slip BCs are specified for velocity DOF on DES partition cavity walls. Temperature BCs are Dirichlet on vertical surfaces and homogeneous Neumann on horizontal surfaces. Green–Gauss theorem imposition on $O(\delta^2)$ variable harmonic PDEs (9.64), (9.70), (9.71), generate encompassing BC requirements *a priori* unknown. Resolution is to apply homogeneous Neumann BCs on all DES-*ar*LES common boundary segments $\partial\Omega$, hence $O(\delta^2)$ state variables are *not present* in the DES partition.

The DOF IC for NS and *ar*LES-DES simulations is $\{\mathrm{Q}I(t_0)\}=\{0\}$ except for temperature DOF linear interpolation between hot and cold wall Dirichlet BCs. As for all cavity simulations, homogeneous Neumann is the encompassing BC for $\mathrm{D}M^h$ potential function and post-processed pressure.

Key sought NS and *ar*LES-DES *a posteriori* data comparison is the impact of $O(\delta^2)$ state variable distributions on resolved scale predictions as *f*(Re). Laminar flow results for all Ra < E+07 hence an eddy viscosity model is not appropriate in the DES partition. This

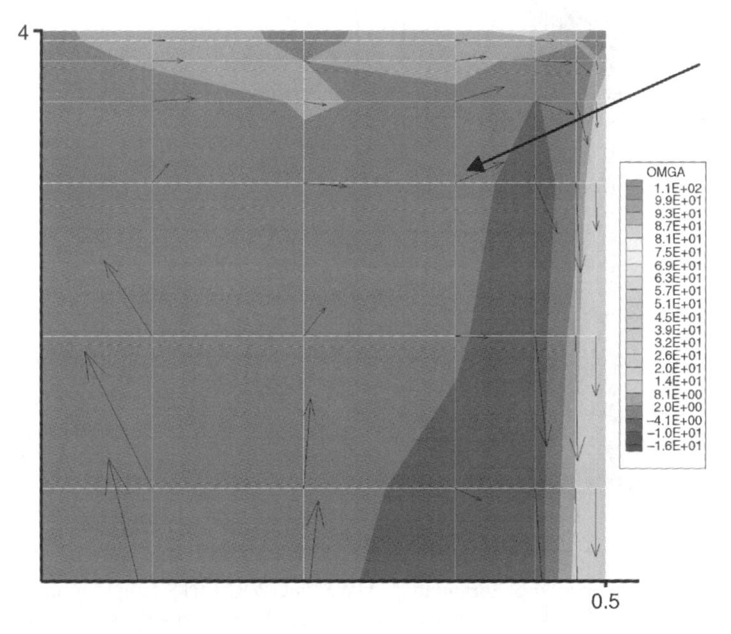

Figure 9.7 *ar*LES-DES partitioned domain mesh, upper right corner of 8 × 1 thermal cavity, vorticity DOF flood with velocity vector overlay, arrow locates partition shared $\partial\Omega$ vertex node

is also the decision for Ra = E+08, Re = 11,876, to maintain uniform the Re progression DOF distribution comparisons. The NS laminar $GWS^h + \theta TS$ algorithm proved stable over the Ra range hence no *m*PDE operands are inserted. *ar*LES-DES $mGWS^h + \theta TS$ algorithm is stable for SFS tensor/vector specification $C_S(\delta) \equiv C\delta$, C = 0.25, (9.55), (9.60), Grubert (2006), in agreement with RLES theory order bound (9.45).

For Ra = E+04, Re = 119, NS and *ar*LES-DES formulations generate *indistinguishable* steady DOF distributions with single cavity spanning vortex. For Ra = E+05, Re = 375, each formulation generates *quasi-steady* solutions with domain filling single recirculation cell, Figure 9.8 left. The NS and *ar*LES-DES solution snapshot temperature DOF isotherm interpolations *exactly* overlay, Figure 9.8 right, except near the mirror symmetric pair of vertically oriented "isotherm finger" prominences denoted by the arrows.

Continued time accurate ($\theta = 0.5$) execution of each formulation confirms solution isotherm features undergo *bounded* oscillation about the locations graphed in Figure 9.8. For execution continuations to Ra = 4.0 E+05, just larger than $Ra_{CR} \approx 3.1$ E+05, isotherm

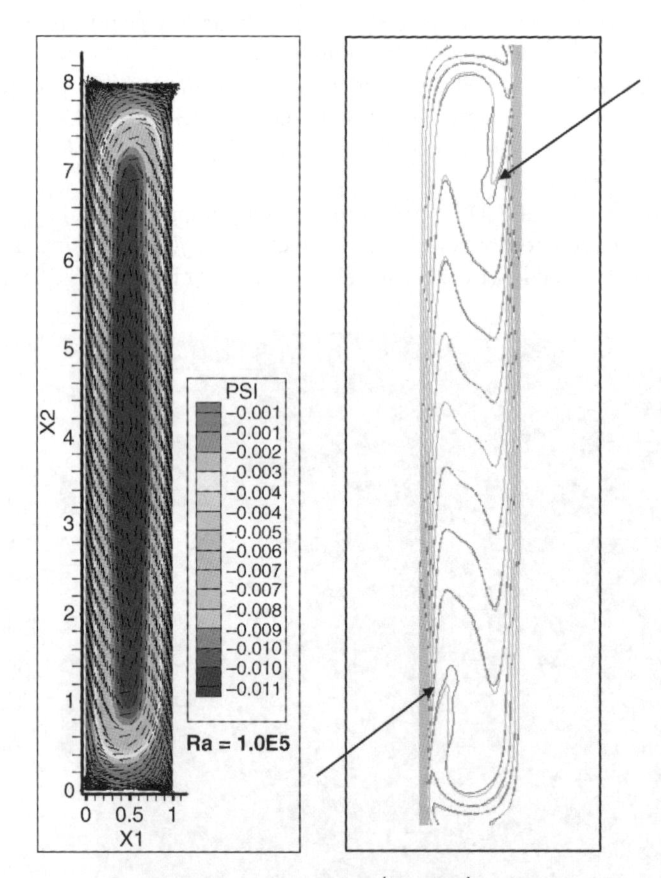

Figure 9.8 NS and *ar*LES-DES formulation $GWS^h/mGWS^h + \theta TS$ solution comparison, 8 × 1 thermal cavity: left, Ra = E+04 steady streamfunction flood with velocity vector (only every fifth) overlay; right, Ra = E+05 quasi-steady isotherm distribution overlay, NS red, *ar*LES-DES green, arrows denote mirror symmetric "isotherm finger" prominences

prominence oscillations turn *unbounded* inducing transition to unsteady solutions for both formulations. This constitutes $\text{GWS}^h + \theta\text{TS}$ algorithm quantitative *validation* by direct comparison with NS stability theory and benchmark data.

For cold start IC Ra = E+06, Re = 1191, time evolution dynamic graphics of (every fifth) velocity vector overlay on streamfunction flood confirm NS and *ar*LES-DES formulation unsteady DOF distributions remain visually *indistinguishable*. After IC (not a solution) washout, each generates a precisely mirror symmetric *periodic* unsteady solution constituted of *multi-scale* wall-attached translating vortices interacting with cavity central vortex structures, globally convected in an overall clockwise circulation pattern. Generated multi-scale content is very sensitive to Ra, compare *ar*LES-DES formulation snapshots for Ra = E+06 and Ra = 3.4 E+06, Figure 9.9 left/right. The latter specification is nominally one order larger than Ra_{CR} and yields Re = 1585.

Published square cross-section thermal cavity NS algorithm steady *a posteriori* data without exception exhibit *precise mirror symmetry* for all Ra ≤ E+08, Chapters 6, 7, see also Le Quere (1983). The *a posteriori* data snapshots, Figure 9.9, follow long time $t - t_0 \approx 100\,\text{s}$ execution of the unsteady $\text{GWS}^h + \theta\text{TS}$ algorithm, therefore the IC is thoroughly flushed. Both NS and *ar*LES-DES formulation time-accurate unsteady, multi-scale DOF evolutions remain *precisely* mirror symmetric throughout which constitutes an additional *validation*.

A scalar measure of 8 × 1 cavity volumetric transport accrues to GWS^h post-process of streamfunction, (6.12). The wall BC is uniformly $\{\text{PSI}(t)\} = \{0\}$, hence ψ^h DOF extrema exist in the interior. Computed extremum $\Delta\psi^h$ cavity volumetric transport for Ra = E+05, E+06, 3.4E+06 are $\Delta\psi^h = 0.011, 0.026, 0.054$, confirming cavity flowfield vigor essentially doubles with each Ra increment.

The corresponding Reynolds numbers are Re = 375, 1190, 1585 confirming all flowfields are indeed laminar. In summary, NS and *ar*LES-DES prediction *a posteriori* data comparison observables are:

- Ra = E+05, Re = 375, *ar*LES-DES and NS *single scale* quasi-steady isotherm distributions precisely overlay except for *small scale* bounded oscillation "isotherm fingers"
- $\text{Ra} > \text{Ra}_{\text{CR}} \approx 3.1\text{E+05}$, *ar*LES-DES and NS *steady* single scale solutions both transition to *unsteady* periodic in agreement with theoretical prediction of critical Rayleigh number existence
- Ra = 3.4 E+06, Re = 1585, *ar*LES-DES and NS unsteady *multi-scale* solution evolution *dynamics* are visually *indistinguishable*

These exact comparison *a posteriori* data *validate* *ar*LES theory closure $O(\delta^2)$ state variables, and $O(\delta^3)$ SFS tensor/vector, do not impact unfiltered NS prediction of steady single scale and unsteady multi-scale flowfields for laminar Re specification.

Increasing cold start IC to Ra = E+07, Re = 3750, is sufficient for spectral filtering inherent in *ar*LES $O(\delta^2)$ closure to moderate NS fine scale prediction. Upon IC washout the NS solution unsteady periodic velocity vector distribution contains a range of translating roll vortex structures with measures spanning cavity width (L/8) to a fraction thereof, clearly visible in post-processed ψ^h solution snapshot, Figure 9.10 left. The ψ^h distribution contours confirm that this unsteady, multi-scale flowfield (snapshot) is precisely mirror symmetric.

The NS Ra = E+07 snapshot volumetric transport scalar measure is $\Delta\psi^h = 0.366$, larger than that for Ra = E+06 by an order of magnitude plus. Visual verification that mesh M

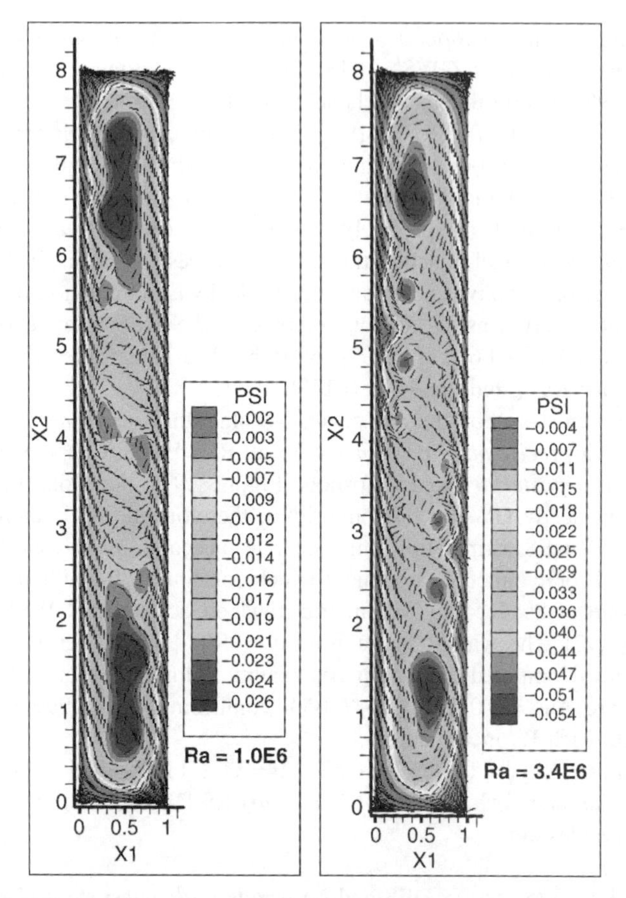

Figure 9.9 *ar*LES-DES GWS^h + θTS algorithm periodic unsteady solution snapshots, 8 × 1 thermal cavity, velocity vector (every fifth) overlay on streamfunction flood: left, Ra = E+06; right, Ra = 3.4 E+06

supports satisfactory spectral resolution of roll vortex structure scale cascade accrues to unitized scale ψ^h flood with velocity vector (all) overlay, Figure 9.10 right.

Following very dynamic transition from identical cold start IC, both NS and *ar*LES-DES formulations generate precisely mirror symmetric periodic multi-scale, multi-vortex solution evolution. Both are characterized by local small scale roll vortex structures traversing up/down the hot/cold walls, interacting with cavity interior vortices in an overall global clockwise rotation.

Directly comparative T^h DOF flood snapshots at $t - t_0 = 48$ s, Figure 9.11, confirm NS solution fine spectral content is theory $O(\delta^2)$ state variable moderated in the *ar*LES-DES solution. Filtering is most evident in the cavity central reach $-1.0 < y < 1.0$ wherein the *ar*LES solution lacks the NS "backwards S" local T^h isotherm feature. NS solution mirror symmetric T^h DOF "circular pairs" at cavity top and bottom are also removed by $O(\delta^2)$ variable impact. Both solutions predict cavity central ~50% reach is isothermal except for *very* large wall-normal gradients at Dirichlet BC walls, not discernable on scale of Figure 9.11.

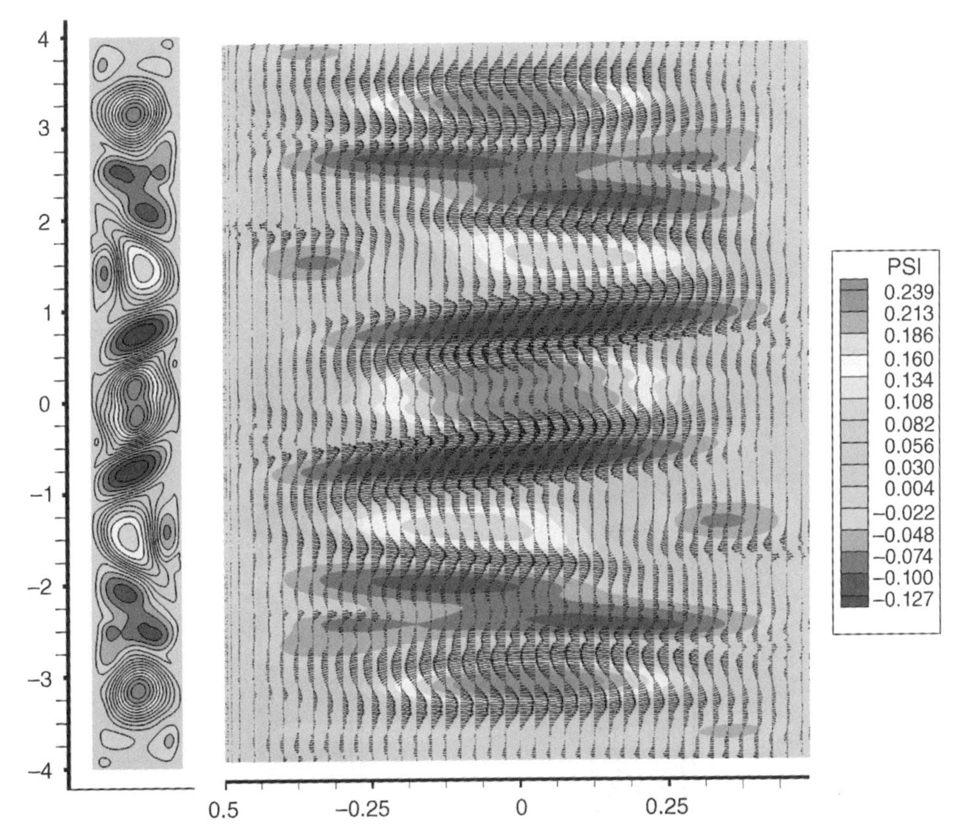

Figure 9.10 $GWS^h + \theta TS$ algorithm NS unsteady solution snapshot, 8×1 thermal cavity, Ra = E+07, Re = 3750, $t - t_0 = 12$ s: left, streamfunction flood with contours; right, unitized scale streamfunction flood with velocity vector (all) overlay

Streamfunction DOF flood with contours post-processed from $\mathbf{u}^h$ at $t - t_0 = 48$ s, are directly compared, Figure 9.12. The cavity central core NS ω^h distribution confirms the T^h feature correlates with $\mathbf{u}^h$, completely absent in the *ar*LES-DES prediction. DOF extrema for ω^h are not visible, as with T^h, existing strictly adjacent to walls in the DES partition.

These *ar*LES-DES Ra = E+07 *a posteriori* data generate *quantitative diagnostics* on theory $O(1, \delta^2, \delta^3)$ state variable relative significance. The selected solution feature is wall-attached roll vortex growing/collapsing up the hot wall, located at the left wall y ~ 1.6, Figure 9.10 unitized scale graphic. Windowed-in dimensional state variable DOF floods with velocity vector overlay snapshots, Figure 9.13, lead to theory pertinent diagnostics:

- top left: $O(1)$ $\overline{T}^h$ extremum is $O(10^1)$, very steep wall-normal gradient terminus aligned with *ar*LES-DES common $\partial\Omega$; local roll vortex-encapsulated elevated temperature bubble lies entirely in *ar*LES partition
- top right: $O(\delta^2)$ interaction vector $v_j^{T^h} \equiv \left(\overline{\overline{u}_j T'} + \overline{u_j' \overline{T}}\right)^h$, $j = 2$, extremum $O(10^{-1})$ in *ar*LES partition at vortex wall separation of $\overline{T}^h$; DES partition zero definition merges smoothly with *ar*LES-computed nil level in cavity interior

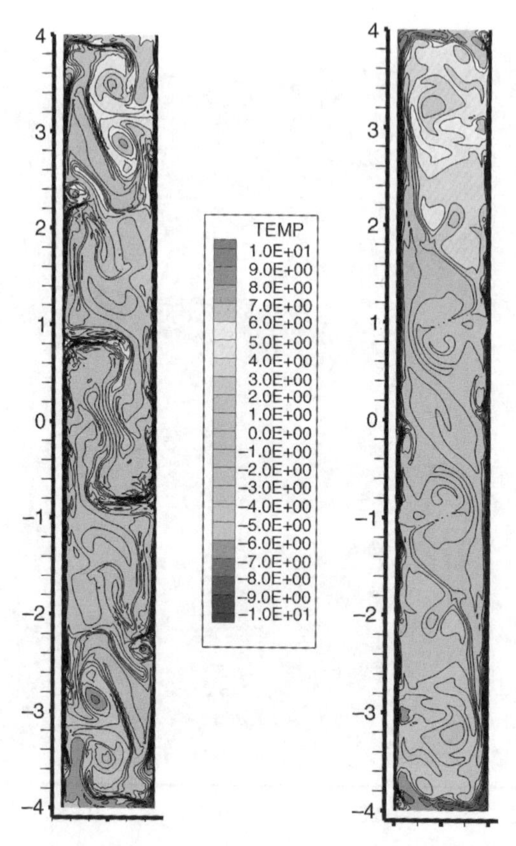

Figure 9.11 $\mathrm{GWS}^h/m\mathrm{GWS}^h + \theta\mathrm{TS}$ algorithm periodic unsteady solution snapshots, 8×1 thermal cavity, Ra = E+07, Re = 3750, $T^h(t - t_0 = 48\ \mathrm{s})$ distribution: left, NS laminar; right, *ar*LES theory

- middle left: $O(1)$ convection nonlinearity $\overline{u}_i^h \overline{u}_j^h$ shear, $i = 1$, $j = 2$, extremum is $O(10^{-1})$, overlies vortex footprint spanning full range of extrema, significance constrained to *ar*LES partition
- middle right: $O(\delta^2)$ interaction tensor $c_{ij}^h \equiv \left(\overline{u}_i u_j' + u_i' \overline{u}_j\right)^h$ shear, $i = 1$, $j = 2$, extrema $O(10^{-3})$ is two orders smaller than $\overline{u}_i^h \overline{u}_j^h$, aligned with roll vortex boundaries; DES partition zero definition merges smoothly with *ar*LES-computed nil level
- bottom left: $O(\mathrm{Re}^{-1})$ Stokes tensor $\overline{S}_{ij}{}^h$, $i = 2$, $j = 2$, extrema $O(10^{-3})$ matches $c_{12}{}^h$, opposite sign extrema coincide with vortex coalescence at wall
- bottom right: $O(\delta^3)$ SFS tensor $\overline{u_i' u_j'}^h$ shear, $i = 1$, $j = 2$, extrema $O(10^{-3})$ matches $c_{12}{}^h$; broadly *computed* nil in *ar*LES-DES domain, co-located extrema paralleling roll vortex velocity maximum confirms robust $O(h^2)$ dispersion error action

*ar*LES theory diagnostic data *quantitatively* predict feature adequate mesh resolution, the alternative to *eyeball norm* assessment of resolved scale data, recall Figure 9.10. RLES theory state variable prediction $O(1, \delta^2)$ is DOF data confirmed. Specifically, $\overline{T}^h$ is $O(10^1)$ while $v_2^T \equiv \overline{\overline{u}_2 T' + u_2' \overline{T}}$ is $O(10^{-1})$, similarly $\overline{u}_1\, \overline{u}_2$ is $O(10^{-1})$ and $c_{1\,2} = \overline{\overline{u}_1 u_2' + u_1' \overline{u}_2}$ is

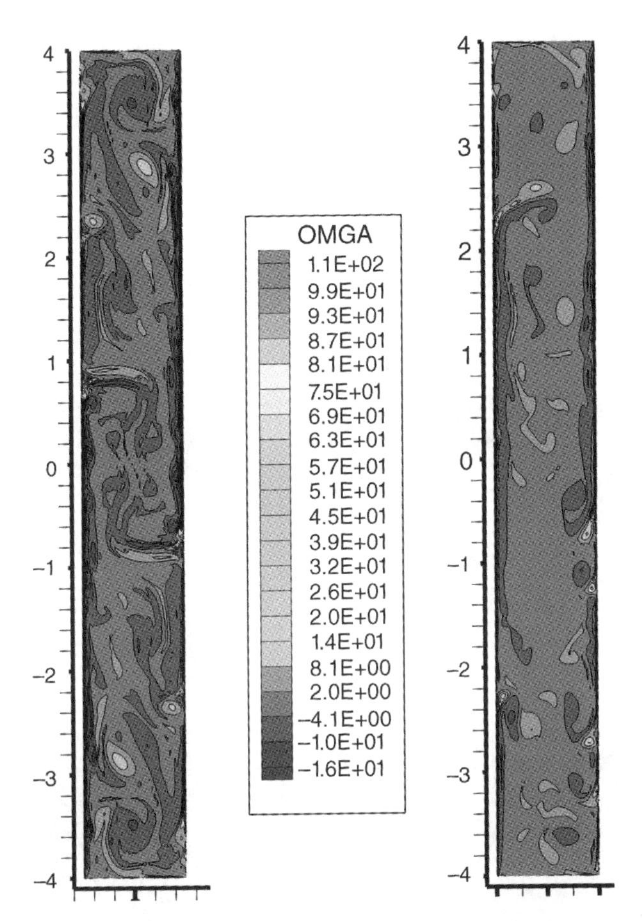

Figure 9.12 $\mathrm{GWS}^h/m\mathrm{GWS}^h + \theta\mathrm{TS}$ algorithm periodic unsteady solution snapshots, 8×1 thermal cavity, Ra = E+07, $\omega^h(t - t_0 = 48\,\mathrm{s})$ distributions: left, NS laminar; right, *ar*LES theory

$O(10^{-3})$. These $O(10^{-2})$ distinctions agree precisely with theory. Thereby the mesh M is quantified *adequate* for resolution of LES resolved scale state variable DOF local features.

Ra ≡ E+08, Re = 11,900, completes the *ar*LES-DES 8×1 cavity *a posteriori* data set. Re > E+04 is anticipated sufficient for laminar-turbulent transition, hence these data are likely compromised by eddy viscosity model absence in the DES partition. Identifying and implementing a validated transition model is stated beyond study scope, Grubert (2006). These data are thus pertinent only for $O(1, \delta^2, \delta^3)$ state variable relative significance comparison with Ra = E+07 data.

The *ar*LES-DES $m\mathrm{GWS}^h + \theta\mathrm{TS}$ algorithm undergoes an exceptionally dynamic transition from interpolated wall temperature BC data IC, leading to characteristic unsteady, periodic, multi-scale solution by $t - t_0 = 14\,\mathrm{s}$. Continued time accurate ($\theta = 0.5$) execution confirms unsteady solution periodicity persists exhibiting precise mirror symmetric $O(1, \delta^2)$ state variable DOF distributions.

The Ra = E+08 T^h and ω^h DOF snapshots, Figure 9.14, confirm further spectral content diminution from Ra = E+07, Figures 9.11–9.12. The cavity central three-quarters is now *isothermal*

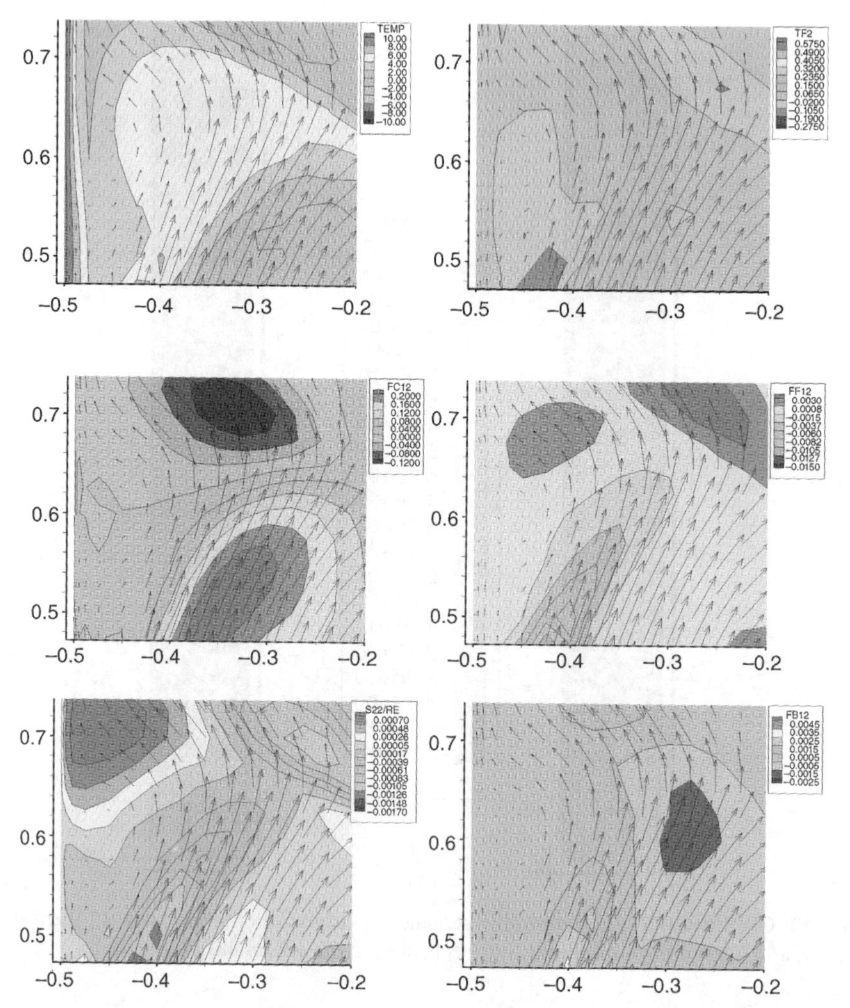

Figure 9.13 *ar*LES-DES $O(1, \delta^2, \delta^3)$ predictions, periodic unsteady state variable floods with velocity vector overlay, hot wall roll vortex region close-up, 8×1 thermal cavity, Ra = E+07, solution snapshot at $t - t_0 = 48$ s: left to right, top to bottom: $\overline{T}$, $\overline{\overline{u}_2 T'} + \overline{u_2' \overline{T}}$, $\overline{\overline{u_1}\,\overline{u_2}}$, $\overline{\overline{u}_1 u_2'} + \overline{u_1' \overline{u}_2}$, $\overline{S}_{2\,2}$, $\overline{u_1' u_2'}$

except directly wall adjacent where Dirichlet BCs are the extrema. Predicted Ra = E+07 spectral content is thoroughly removed by $O(\delta^2)$ state variable distribution. The one order larger ω^h extremum occurs in the wall layer, while the cavity central range is only $\omega^h = \pm 50\,\mathrm{s}^{-1}$.

Ra = E+08 resolved scale velocity vector prediction is attached wall jets without embedded vortices, a total alteration from Ra = E+07, Figure 9.14. The solution temperature feature is now translating *thermal fingers* protruding wall normal into the cavity, clearly illustrated in unitized scale flood with velocity vector (all) overlay, Figure 9.15 right. The number of cavity central vortices is about double with overall intensity substantially diminished in comparison to Ra = E+07. Post-processed $\Delta\psi^h = 0.66$ confirms volumetric transport is about doubled.

Ra = E+08 *ar*LES-DES *a posteriori* data quantify the RLES theory $O(1, \delta^2, \delta^3)$ state variable relative significance. Selected solution feature is wall-normal thermal finger at $y \sim -0.5$ in

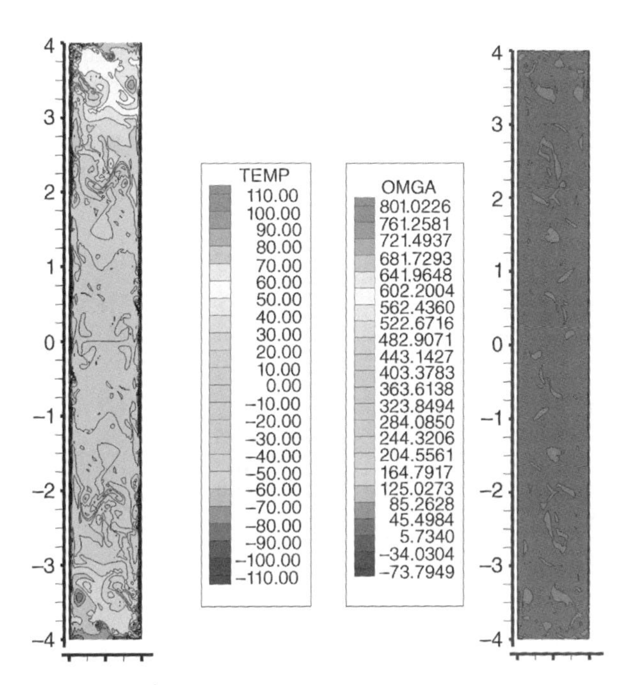

Figure 9.14 *ar*LES-DES *m*GWS^h + θTS algorithm periodic unsteady solution snapshots, 8 × 1 thermal cavity, Ra = E+08, $t - t_0$ = 14 s: left, temperature flood with contours; right, vorticity flood with contours

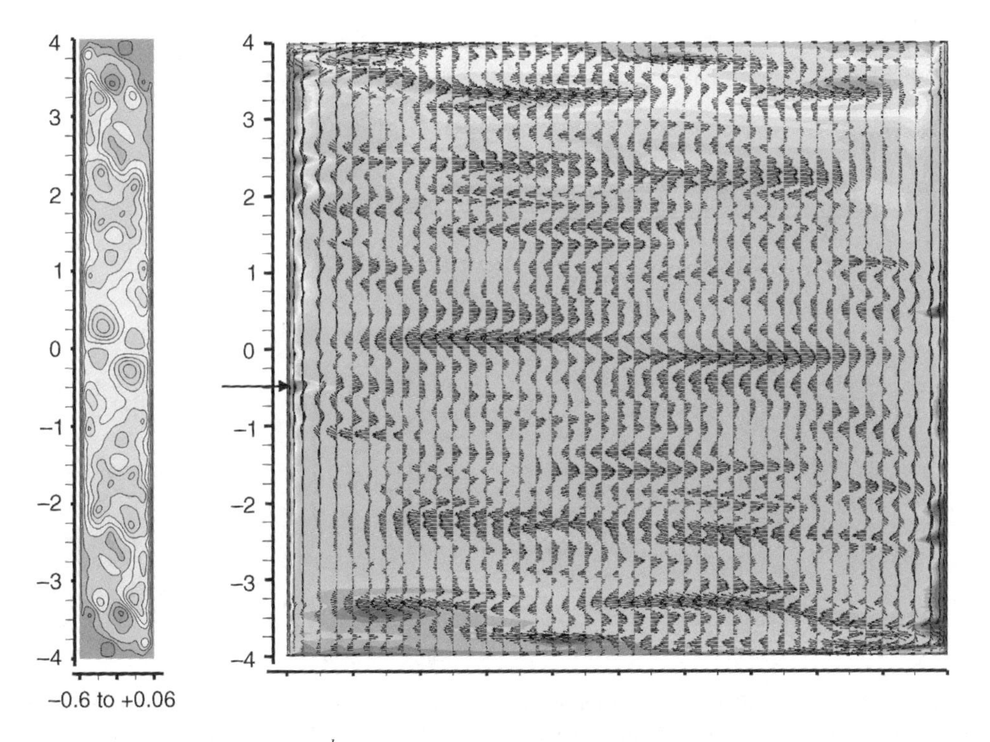

Figure 9.15 *ar*LES-DES *m*GWS^h + θTS algorithm unsteady solution snapshot, 8 × 1 thermal cavity, Ra = E+08, $t - t_0$ = 14 s: left, streamfunction flood with contours; right, unitized scale temperature flood with velocity vector (all) overlay

unitized scale graph Figure 9.15 right. Diagnostic data are summarized as state variable DOF floods with velocity vector (all) overlay, Figure 9.16, with distinguishing features:

- top left: theory $O(1)$ $\overline{T}^h$ extrema $O(10^2)$, wall-normal steep gradient thermal finger just penetrates to *ar*LES partition; wall jet velocity extrema occur one node into *ar*LES partition shared $\partial\Omega$

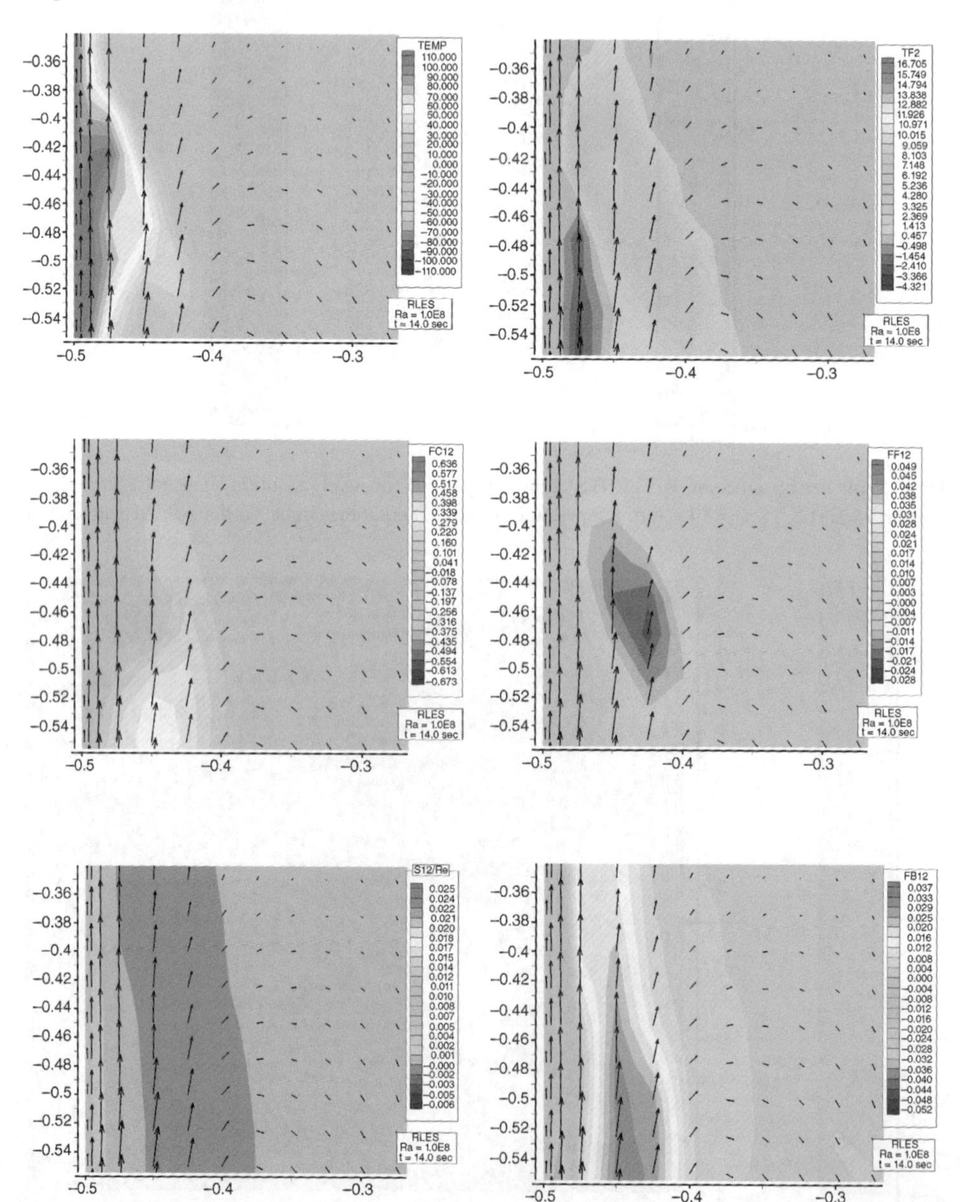

Figure 9.16 *ar*LES-DES *m*GWS^h + θTS algorithm periodic unsteady wall jet thermal finger flood with velocity vector overlay snapshot, 8 × 1 thermal cavity, Ra = E+08, t = 14 s: left to right, top to bottom: $\overline{T}$, $\overline{\overline{u}_2 T'} + \overline{u_2' \overline{T}}$, $\overline{\overline{u}_1}\,\overline{\overline{u}_2}$, $\overline{\overline{u}_1 u_2'} + \overline{u_1' \overline{u}_2}$, $\overline{S}_{2\,2}$, $\overline{u_1' u_2'}$

- top right: $O(\delta^2)$ thermal vector $v_j^{Th} \equiv \left(\overline{\bar{u}_j T'} + \overline{u_j' \bar{T}}\right)^h j=2$, is *two orders* more significant at $O(10^1)$, extremum in *ar*LES partition; DES partition contains next significance, nil definition therein merges smoothly with *ar*LES computed nil level
- middle left: theory $O(1)$ convection nonlinearity $\bar{u}_i^h \bar{u}_j^h$ shear, $i=1$, $j=2$, one order larger at $O(10^0)$, exists only at cavity top/bottom turnarounds; thermal finger level is five times smaller
- middle right: $O(\delta^2)$ tensor $c_{ij}{}^h \equiv \left(\overline{\bar{u}_i u_j'} + \overline{u_i \bar{u}_j}\right)^h$ shear, $i=1$, $j=2$, $O(10^{-2})$ extremum remains two orders smaller than $\bar{u}_i^h \bar{u}_j^h$. DES partition absence merges with *ar*LES computed nil level
- bottom left: $O(\mathrm{Re}^{-1})$ Stokes tensor $\bar{S}_{ij}{}^h$ normal, $i=2$, $j=2$, extremum at $O(10^{-2})$ is one order larger, computed nil in wall jet
- bottom right: $O(\delta^3)$ SFS tensor $\overline{u_i' u_j'}^h$ shear, $i=1$, $j=2$, extremum one order larger at $O(10^{-2})$, now surrounds wall jet $\bar{\mathbf{u}}^h$ maximum in *ar*LES partition

RLES theory $O(1, \delta^2, \delta^3)$ state variable relative significance is not confirmed in these Ra = E+08 data snapshots. Specifically, $\bar{T}^h$ is BC definition one order larger at $O(10^2)$ while $v_2^T \equiv \overline{\bar{u}_2 T'} + \overline{u_2' \bar{T}}$ has increased two orders to $O(10^1)$, one order larger than theory prediction. In distinction, $\overline{u_1}\,\overline{u_2}$ at $O(10^0)$ and $c_{12} = \overline{\bar{u}_1 u_2'} + \overline{u_1' \bar{u}_2}$ at $O(10^{-2})$ are in agreement with $O(\delta^2) \approx$ E-02.

These Ra = E+08 diagnostic data comparisons confirm the mesh adequate for resolved scale velocity *only*. Specifically, data order in δ *quantify* the mesh *inadequate* for resolved scale temperature. This is eyeball norm confirmable in the number of wall-normal DOF supporting $\bar{T}^h$ thermal finger, Figure 9.16 top left. Additionally $v_2^T \equiv \overline{\bar{u}_2 T'} + \overline{u_2' \bar{T}}$ and $c_{1\,2} = \overline{\bar{u}_1 u_2'} + \overline{u_1' \bar{u}_2}$ DOF distributions *and* extrema, top two right side graphics, exhibit *no* similarity. This distinction also exists in the Ra = E+07 diagnostics, top two right side graphics in Figure 9.13, confirming *ar*LES theory prediction of resolved-unresolved scale interaction thermal dynamics truly distinct from flow dynamics.

9.12 RLES Theory Turbulent BL Validation

Rigorous BC identification for space filtered NS resolved scale state variables on *wall bounded* domains is a requirement. Filter penetrating the wall, Figure 9.17, appends *boundary commutation error* (BCE) integrals to the LES PDE system (9.12–9.15). *ar*LES-DES

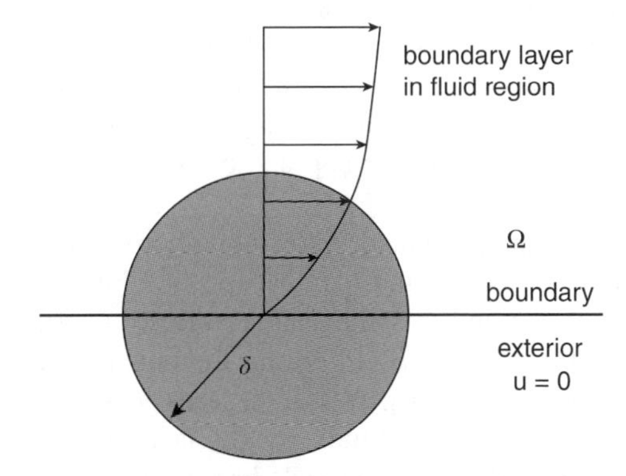

Figure 9.17 Bounded domain filter penetration issue

domain partitioning by design circumvented this issue, as do legacy SGS tensor closure models via assumptions (9.22–9.24) replacing theory-generated tensor/vector quadruples (9.6–9.7).

A second formulation admitting omission of BCE integrals defines a distributed filter $\delta(\mathbf{x})$ vanishing at the wall. Termed *near wall resolution* (NWR), this definition violates rigorous theory since filtering and differentiation, recall (9.10), no longer commute on $\Omega \cup \partial\Omega$, Dunca *et al.* (2003)

$$\overline{\left(\frac{\partial \mathbf{u}}{\partial x_i}\right)} \neq \left(\frac{\partial \bar{\mathbf{u}}}{\partial x_i}\right), 1 \leq i \leq n \tag{9.100}$$

Theoretical characterization of commutation error distribution via TS analyses for very smooth functions exists, Ghosal and Moin (1995), Fureby and Tabor (1997), Vasilyev *et al.* (1998). Using special filter kernels the theory predicts commutation error is $O(h^2)$ for h the mesh measure. Accepting this order error, the advantage of NWR is admissibility of unfiltered NS no-slip Dirichlet BC for the resolved scale velocity. The significant disadvantage is computational cost scaling as $O(\mathrm{Re}^{2.5})$, very close to that of DNS, Chapman (1979).

RLES theory *a posteriori* data are reported for an NWR $\delta(\mathbf{x})$ specification addressing isothermal fully turbulent boundary layer (BL) flow in a channel, Iliescu and Fischer (2003). A pseudo-spectral Galerkin algorithm implements the RLES gradient closure (9.37), the second-order Padé prediction (9.40) and a Smagorinsky SGS tensor model (9.25) with length scale $\delta(y)$ diminution via van Driest damping, (4.60). All closures tested are lacking an SFS tensor model and, in accord with legacy RaNS algorithms, also omit the Re^{-1} viscous term in (9.13).

The solution domain and mesh are rectangular cartesian with periodic BCs applied on domain axial (x, flow direction) and spanwise (z) bounding planes. Recalling (4.136) for parabolic NS, an axial pressure gradient is adjusted dynamically to engender mass flow conservation $\mathrm{D}M^h$ to within an unreported bound.

The filter spatial distribution is $\delta(\mathbf{x}) \equiv [\Delta x \Delta z \Delta y(y)]^{0.5}$ for Δx, Δz the extremum distance separating Gauss–Lobatto–Legendre (GLL) quadrature coordinates in the x, z directions, and $\Delta y(y)$ an interpolation from *zero* at the wall to twice the mesh span Δy at channel centerline. For utilized mesh the similarity coordinate of the first off-wall node is reported $y^+ = 1.7$.

Homogeneous Neumann BCs are stated admissable for Green–Gauss divergence theorem impacted *auxiliary problem* operator [A], (9.41), when implementing $\delta(\mathbf{x})$. In omitting the SFS tensor model the authors implement "a filter which removes 2%–5% of the highest velocity mode to stabilize the Galerkin formulation . . . which does not compromise spectral accuracy." Without defining detail this appears to constitute *artificial diffusion* operator augmentation.

Reported *a posteriori* data are generated following a thoroughly rigorous compute protocol. Mathematical manipulations enabling comparison of LES closure generated resolved scale velocity, Reynolds stress and rms fluctuating velocity distribution *a posteriori* data with the fine DNS data of Moser *et al.* (1999) are thoroughly detailed. Solved as an initial-value problem the IC specification perturbation leads to statistically steady isothermal fully turbulent channel flow in finite time. Sought is *a posteriori* data for similarity Reynolds numbers $\mathrm{Re}_\tau = 180$, 395 based on wall shear velocity u_τ, (4.80).

Graphed in BL similarity coordinates, axial resolved scale velocity DOF distributions for altered SGS tensor model, gradient model and RLES second-order theory prediction are *indistinguishable* for both Re_τ, Figure 9.18. All *a posteriori* data exhibit excellent *quantitative* agreement with DNS validation for $\mathrm{Re}_\tau = 180$. For $\mathrm{Re}_\tau = 395$ each exhibits modest departure in the viscous sublayer-lower log layer BL reach, $10 < y^+ < 100$. Solution post predictions of Re_τ are smaller than design by $\sim 1\%$, which attests to design protocol precision.

Reynolds shear stress τ_{xy} distributions for each Re_τ are direct output from altered SGS tensor model but must be reconstructed from gradient closure and RLES prediction data. For abscissa scaled $1 - |y|$ with y normalized on channel half-width, τ_{xy} predictions are compared with fine DNS data in Figure 9.19. The Smagorinsky SGS tensor model with van Driest damping generates distributions in *total disagreement* with DNS data. The gradient closure and second-order Padé theory data distributions are indistinguishable with $\sim 30\%$ error extremum centered at $\sim 10\pm\%$ displacement from the channel wall.

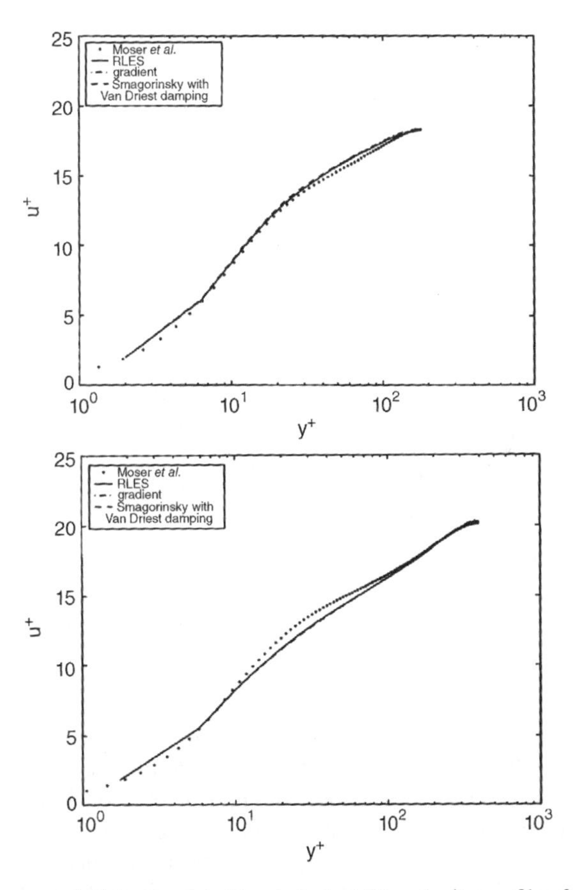

Figure 9.18 Pseudo-spectral Galerkin algorithm turbulent BL velocity profiles for altered SGS tensor model, gradient closure and RLES second-order Padé prediction compared with fine DNS data: top, $\mathrm{Re}_\tau = 180$; bottom, $\mathrm{Re}_\tau = 395$. *Source:* Iliescu, T. and Fisher, P.F. 2003. Reproduced with permission of American Institute of Physics

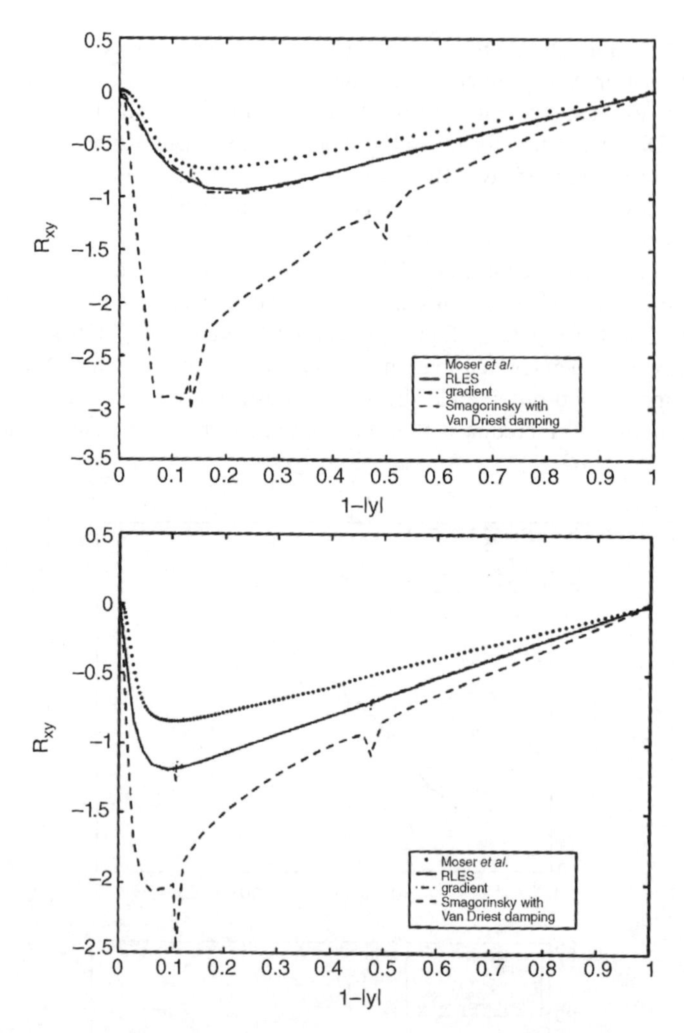

Figure 9.19 Pseudo-spectral Galerkin algorithm Reynolds shear stress τ_{xy} profiles for SGS tensor model, gradient closure and RLES second-order Padé prediction compared with fine DNS data: top, $Re_\tau = 180$; bottom, $Re_\tau = 395$. *Source:* Iliescu, T. and Fisher, P.F. 2003. Reproduced with permission of American Institute of Physics

The comparative disagreement with DNS data is reversed for Reynolds normal (rms) stress distributions predicted/reconstructed from *a posteriori* data. For $Re_\tau = 180$, altered Smagorinsky SGS tensor model prediction exhibits excellent qualitative agreement for all three rms correlations except ±10% overprediction of near wall extrema, Figure 9.20. The data for gradient closure and RLES second-order Padé theory *reconstructions* are indistinguishable except directly wall adjacent where they exceed the DNS data by ~100%.

For $Re_\tau = 395$, the Smagorinsky SGS tensor model data exhibit excellent quantitative agreement for $\sqrt{u'u'}$ and $\sqrt{v'v'}$, but $\sqrt{w'w'}$ is lower than DNS data by ~30% adjacent to the wall. The gradient closure and RLES theory reconstructions are again indistinguishable, Figure 9.21.

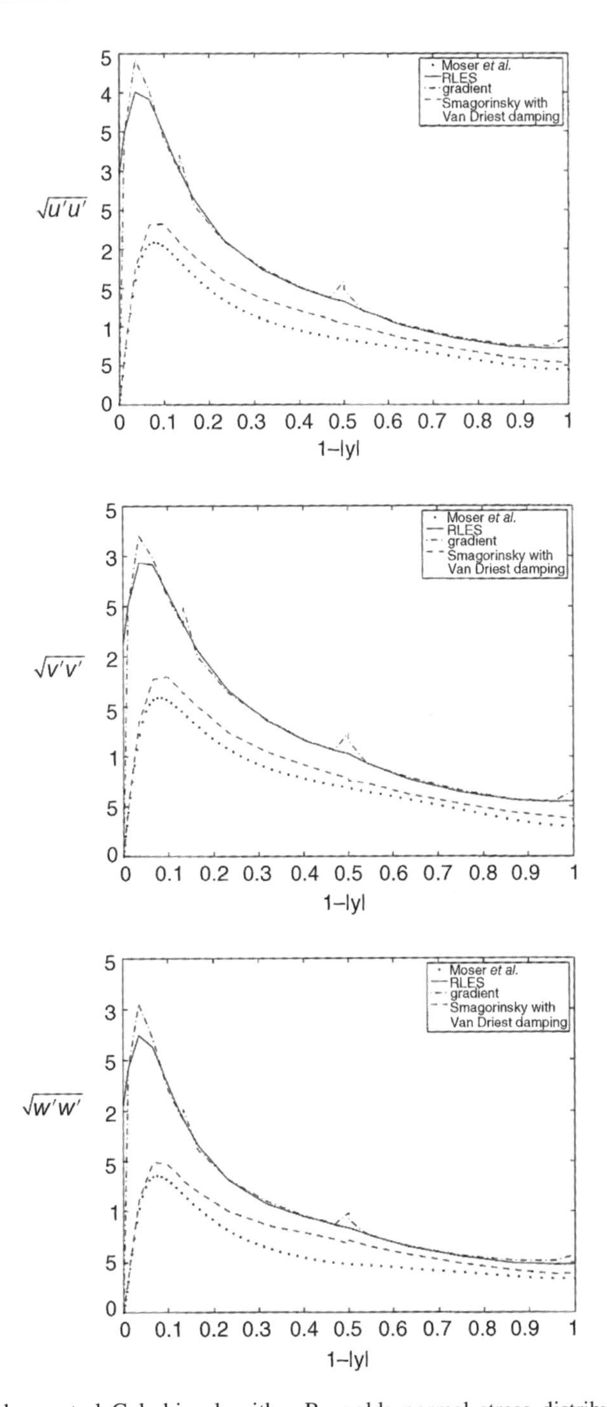

Figure 9.20 Pseudo-spectral Galerkin algorithm Reynolds normal stress distributions for an SGS tensor model, gradient closure and RLES second-order Padé prediction compared with fine DNS data, $Re_\tau = 180$: top to bottom, $\sqrt{u'u'}$, $\sqrt{v'v'}$, $\sqrt{w'w'}$. *Source:* Iliescu, T. and Fisher, P.F. 2003. Reproduced with permission of American Institute of Physics

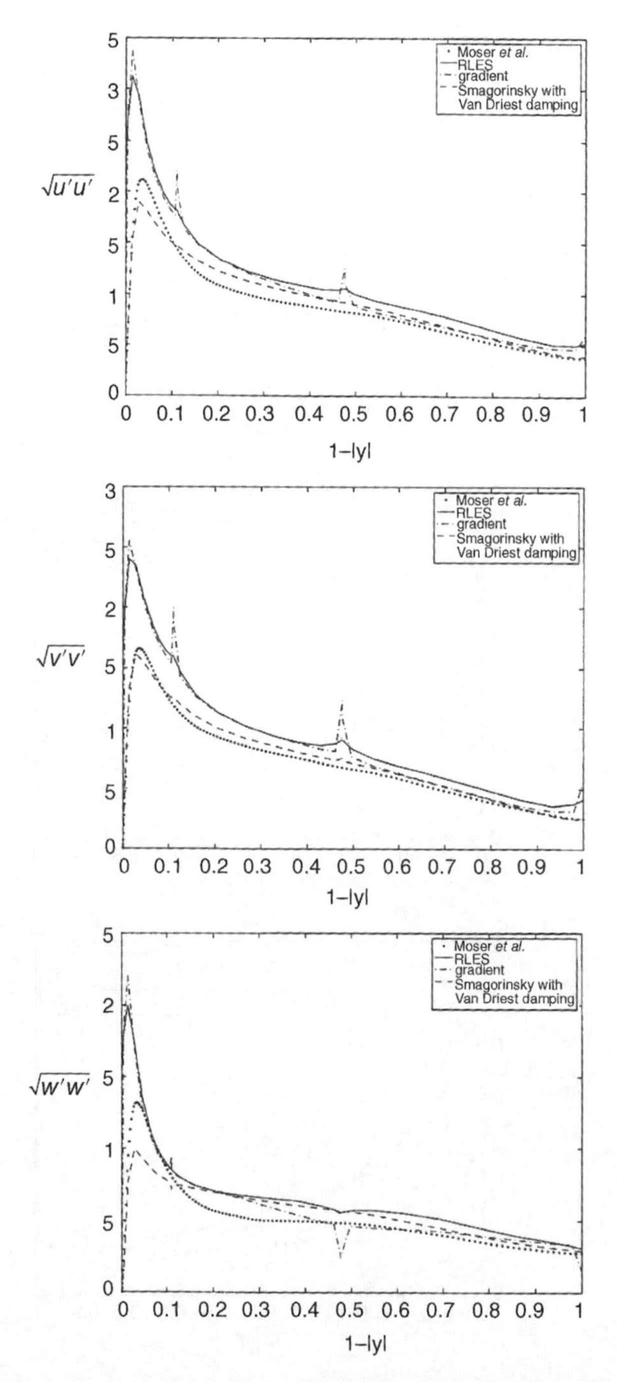

Figure 9.21 Pseudo-spectral Galerkin algorithm Reynolds normal stress distributions for an SGS tensor model, gradient closure and RLES second-order Padé prediction compared with fine DNS data, $\mathrm{Re}_\tau = 395$: top to bottom, $\sqrt{u'u'}$, $\sqrt{v'v'}$, $\sqrt{w'w'}$, from Iliescu and Fischer (2003), reprinted with permission *Physics of Fluids*. *Source:* Iliescu, T. and Fisher, P.F. 2003. Reproduced with permission of American Institute of Physics

Relative to $\text{Re}_\tau = 180$ data they exhibit significantly improved quantitative agreement except in the near wall reach wherein solution reconstructions exceed DNS data by ~60%.

Although not discernible in Figures 9.20, 9.21 scale, an author cited detraction to RLES second-order theory solution reconstructions is non-zero Reynolds normal (rms) stress computed at the wall. This is attributed to use of homogeneous Neumann BCs for operator [A] following Green–Gauss divergence theorem imposition. Not stated but relevant to poor Reynolds shear stress reconstruction comparisons, the RLES second-order Padé theory generates a prediction for *resolved-unresolved scale interaction* tensor pair (9.8). It is *not* a Reynolds stress tensor model.

To summarize, *a posteriori* data confirm pseudo-spectral Galerkin weak form CFD algorithm implementation of gradient and/or RLES second-order Padé prediction with distributed filter $\delta(\mathbf{x})$ generates resolved scale velocity turbulent BL DOF distributions in excellent *quantitative* agreement with DNS data. Predicted $O(h^2)$ commutation error is not assessed via a regular mesh refinement study. LES and DNS Reynolds stress and rms distribution disagreement is focused in the laminar sublayer – low log layer reach. This is precisely the BL region most sensitive to use of unfiltered NS no-slip BCs for resolved scale velocity, also operator [A] homogeneous Neumann BCs.

9.13 Space Filtered NS PDE System on Bounded Domains

Isothermal space filtered NS **DP** PDE system (9.22) is absent the *boundary commutation error* (BCE) integral, which results from filtering on a *bounded domain*, Fureby and Tabor (1997), recall Figure 9.17. Theory approximation (9.22), hence SGS tensor model (9.24) approximations to (9.25), enables this omission. In distinction RLES theory Padé predictions preserve resolved scale **DP** PDE rigor, substitute (9.40) or (9.42) into (9.13) with (9.6), hence this subject.

As convolution and differentiation *do not commute* on a bounded domain, (9.100), unfiltered NS state variables must first be extended exterior to PDE domain Ω. Recalling Section 3.4 and Table 3.1, in vector notation NS **DP** extended first-order weak derivatives $\partial\mathbf{u}/\partial t, \nabla\mathbf{u}, \nabla\cdot\mathbf{u}, \nabla\cdot\mathbf{uu}$ are well defined on $\Omega \in \mathbb{R}^n$. Hence these functions possess the following regularities, Dunca *et al.* (2003)

$$\begin{aligned}&\mathbf{u} \in \left(H_0^1(\mathbb{R}^n)\right)^n,\ P \in \mathrm{L2}(\mathbb{R}^n),\ for\, t \in [0, T]\\&\mathbf{u} \in \left(H^1(0, t)\right)^n,\ for\ \mathbf{x} \in \mathbb{R}^n\end{aligned} \tag{9.101}$$

As identified in (9.101), $\mathbf{u} \notin \left(H^2(\mathbb{R}^n)\right)^n$ and $P \notin H^1(\mathbb{R}^n)^n$ do not meet NS **DP** PDE (9.2) differentiation requirements. Hence, the NS **DP** *full* stress tensor $\mathbf{D}(\mathbf{u}, P) \equiv -\nabla P + (2/\text{Re})\nabla\cdot\mathbf{S}(\mathbf{u})$ must be defined in the sense of *distributions*, Kolmogorov and Fomin (1975), Schwartz (1966). It then follows that extended NS state variable $\{\mathbf{u}, P\}$ does fulfill isothermal **DP** PDE (9.2) altered to

$$\frac{\partial\mathbf{u}}{\partial t} + \nabla\cdot\mathbf{uu} + \nabla P - \frac{2}{\text{Re}}\nabla\cdot\mathbf{S}(\mathbf{u}) = \mathbf{f} + \int_{\partial\Omega}\mathbf{D}(\mathbf{u}, P)(s)\mathbf{n}(s)\phi(s)\text{d}s \tag{9.102}$$

for all piecewise continuous functions $\phi(s) \in C_0^\infty(\Omega) \subset H_0$, for coordinate system s spanning $\partial\Omega$ characterized by outward pointing unit normal $\mathbf{n}(s)$ and for *data* $\mathbf{f}$.

For the NS state variable defined in distributional sense, convolution and differentiation do commute on a wall-bounded domain. The convolution operation then generates space filtered thermal NS D**P** (9.13) replacement on bounded domains in the form

$$\begin{aligned}\mathcal{L}(\overline{u}_i) = \mathrm{St}\frac{\partial \overline{u}_i}{\partial t} + \frac{\partial}{\partial x_j}\left[\overline{u_j u_i} + \overline{P}\,\delta_{ij} - \frac{1}{\mathrm{Re}}\left(\frac{\partial \overline{u}_i}{\partial x_j} + \frac{\partial \overline{u}_j}{\partial x_i}\right)\right] \\ + \frac{\mathrm{Gr}}{\mathrm{Re}^2}\overline{\Theta}\hat{g}_i - \mathrm{A}_\delta\left(D_{ij}(u_i, P)\right) = 0\end{aligned} \tag{9.103}$$

with

$$\mathrm{A}_\delta\left(D_{ij}(u_i, P)(x_k, t)\right) \equiv \int_{\partial\Omega} g_\delta(x_k - s_k)\left[-P\delta_{ij} + \frac{1}{\mathrm{Re}}\left(\frac{\partial u_i}{\partial x_j} + \frac{\partial u_j}{\partial x_i}\right)\right]\hat{n}_j \mathrm{d}s \tag{9.104}$$

defining the filtered D**P** *boundary commutation error* (BCE) integral.

The BCE integrand involves *unfiltered* NS state variable members. Convolution with filter $g_\delta(x_k - s_k)$ transforms the evaluated integral to space filtered state variable members. Bounded domain space filtered NS PDE system completion for (9.14–9.15) results for BCE integral additions

$$\begin{aligned}\mathcal{L}(\overline{\Theta}) = \mathrm{St}\frac{\partial \overline{\Theta}}{\partial t} + \frac{\partial}{\partial x_j}\left[\overline{u_j \Theta} - \frac{1}{\mathrm{RePr}}\frac{\partial \overline{\Theta}}{\partial x_j}\right] - \frac{\mathrm{Ec}}{\mathrm{Re}}\frac{\partial}{\partial x_j}\left(\frac{\partial \overline{u_j u_i}}{\partial x_i}\right) \\ - \int_{\partial\Omega} g_\delta(x_k - s_k)\left[\frac{1}{\mathrm{RePr}}\frac{\partial \Theta}{\partial x_j}\right]\hat{n}_j \mathrm{d}s = 0\end{aligned} \tag{9.105}$$

$$\begin{aligned}\mathcal{L}(\overline{Y}) = \mathrm{St}\frac{\partial \overline{Y}}{\partial t} + \frac{\partial}{\partial x_j}\left[\overline{u_j Y} - \frac{1}{\mathrm{ReSc}}\frac{\partial \overline{Y}}{\partial x_j}\right] - s_{\overline{Y}} \\ - \int_{\partial\Omega} g_\delta(x_k - s_k)\left[\frac{1}{\mathrm{ReSc}}\frac{\partial Y}{\partial x_j}\right]\hat{n}_j \mathrm{d}s = 0\end{aligned} \tag{9.106}$$

Pressure projection theory handles $\mathrm{D}M^h$ as prescribed for NS PDE system redefinition (9.103–9.106). The distributional sense requirement appends BCE integrals to the theory linear Poisson PDE + BCs (9.17–9.18)

$$\begin{aligned}\mathcal{L}(\overline{\phi}) = -\nabla^2\overline{\phi} + \frac{\partial \overline{u}_i^h}{\partial x_i} - \int_{\partial\Omega} g_\delta(x_k - s_k)\left[\frac{\partial \phi}{\partial x_j}\right]\hat{n}_j \mathrm{d}s = 0 \\ \ell(\overline{\phi}) = -\nabla\overline{\phi}\cdot\hat{\mathbf{n}} - \left(\overline{u}_i - \overline{u}_i^h\right)\hat{n}_i = 0\end{aligned} \tag{9.107}$$

$$\begin{aligned}\mathcal{L}(\overline{P}) = -\nabla^2\overline{P} - \frac{\partial}{\partial x_j}\left[\frac{\partial \overline{u_j u_i}}{\partial x_i} + \frac{\mathrm{Gr}}{\mathrm{Re}^2}\overline{\Theta}\overline{g}_j\right] - \int_{\partial\Omega} g_\delta(x_k - s_k)\left[\frac{\partial P}{\partial x_j}\right]\hat{n}_j \mathrm{d}s = 0 \\ \ell(\overline{P}) = \nabla\overline{P}\cdot\hat{\mathbf{n}} - \left[\frac{\partial \overline{\mathbf{u}}}{\partial t} - \frac{1}{\mathrm{Re}}\nabla^2\overline{\mathbf{u}}\right]\cdot\hat{\mathbf{n}} = 0\end{aligned} \tag{9.108}$$

9.14 Space Filtered NS Bounded Domain BCs

The LES literature references the NWR alternative *near wall modeling* (NWM), wherein the no-slip wall Dirichlet BC is replaced with a non-homogeneous Neumann BC *model.*

This eliminates the complicating issue of D**P** no-slip BC satisfied *only* by unfiltered NS velocity. Neumann and (nonlinear) Robin BCs are reported, Piomelli and Balaras (2002), John *et al.* (2004), John and Liakos (2006). A wall model BC compendium is detailed in Sagaut (2006, Ch. 9.2.2), slip-with-drag wall BC modeling is summarized, Berselli *et al.* (2006, Ch. 10), BC-pertinent rigorous *approximate deconvolution* (AD) mathematical implications are presented in Layton and Rebholz (2012).

Experimental data pertinent to near wall modeling are reported, Marusic *et al.* (2001). Piomelli (2008) summarizes wall BC DES-LES hybrid models with DES partition closure based on a similarity log-law assumption, and/or implementing the turbulent BL or RaNS PDE system therein with weak coupling to LES partition state variable.

Aside from P appearing in (9.102), the BCE integrals in (9.102–9.108) are of identical appearance with Green–Gauss divergence theorem generated $\mathrm{GWS}^h/m\mathrm{GWS}^h+\theta\mathrm{TS}$ non-homogeneous Neumann BC on bounding surfaces. This duality provides resolution strategy insight for NS rigorously filtered PDE system BCE integrals, *also* derivation of *ar*LES theory state variable BCs *a priori* known *only* for unfiltered NS state variable.

Recalling Section 2.2, (2.10–2.12), and retaining D**P** (9.102–9.104) for exposition, for any solution approximation supported by trial space $\Psi_\alpha(\mathbf{x})$, $1\leq\alpha\leq N$, the *continuum* Galerkin weak form is

$$
\begin{aligned}
\mathrm{GWS}^N &= \int_\Omega \Psi_\beta(\mathbf{x})\mathcal{L}(\overline{u}_i)\mathrm{d}\tau \equiv 0, \quad \forall\beta=1,N\\
&= \int_\Omega \Psi_\beta(\mathbf{x})\left[\begin{array}{l}\mathrm{St}\dfrac{\partial\overline{u}_i}{\partial t}+\dfrac{\partial}{\partial x_j}\left[\overline{u_j u_i}+\overline{P}\,\delta_{ij}-\dfrac{1}{\mathrm{Re}}\left(\dfrac{\partial\overline{u}_i}{\partial x_j}+\dfrac{\partial\overline{u}_j}{\partial x_i}\right)\right]\\ \qquad +\dfrac{\mathrm{Gr}}{\mathrm{Re}^2}\overline{\Theta}\hat{g}_i-\mathrm{A}_\delta\left(\mathrm{D}_{ij}(u_i,P)\right)\end{array}\right]\mathrm{d}\tau\\
&= \int_\Omega \Psi_\beta(\mathbf{x})\left[\mathrm{St}\frac{\partial\overline{u}_i}{\partial t}+\frac{\partial}{\partial x_j}\left[\overline{u_j u_i}+\overline{P}\,\delta_{ij}\right]+\frac{\mathrm{Gr}}{\mathrm{Re}^2}\overline{\Theta}\hat{g}_i\right]\mathrm{d}\tau\\
&\quad +\int_\Omega \frac{\partial\Psi_\beta(\mathbf{x})}{\partial x_j}\left[\frac{2}{\mathrm{Re}}\overline{S}_{ij}\right]\mathrm{d}\tau-\int_{\partial\Omega}\Psi_\beta(\mathbf{x})\left[\frac{1}{\mathrm{Re}}\left(\frac{\partial\overline{u}_i}{\partial x_j}+\frac{\partial\overline{u}_j}{\partial x_i}\right)\right]\hat{n}_j\mathrm{d}\sigma\\
&\quad -\int_\Omega \Psi_\beta(\mathbf{x})\left[\int_{\partial\Omega} g_\delta(x_k-s_k)\left[-P\delta_{ij}+\frac{1}{\mathrm{Re}}\left(\frac{\partial u_i}{\partial x_j}+\frac{\partial u_j}{\partial x_i}\right)\right]\hat{n}_j\,\mathrm{d}s\right]\mathrm{d}\sigma
\end{aligned}
\tag{9.109}
$$

The first surface integral is the Neumann BC for $\overline{u}_i$; on impervious bounding surface segments the second term dot product vanishes identically. This Green–Gauss theorem generated integrand thus exhibits functionally similar to BCE integrand, the second integral in (9.109), the second term in which also vanishes on impervious bounding surface segments.

Hence, non-homogeneous Neumann BC and weak form BCE integral evaluation involve similar calculus operations. One approach to spatially filtered NS state variable BCE integral evaluation is the *approximate deconvolution boundary condition* (ADBC) algorithm. The $n=1$ formulation for filtered DE, Borggaard and Iliescu (2006), is extended to $n=2$ domains Ω with non-homogeneous data specified on a bounding surface $\partial\Omega$, Sekachev and Baker (2014).

Altering the reference state variable to resolved scale velocity, at any time t ADBC algorithm determination of $\overline{\mathbf{u}}(\mathbf{x}_b)$ at impervious bounding surface nodes $\mathbf{x}_b\in\partial\Omega$ involves unfiltered $\mathbf{u}(\mathbf{x}_b, t)$ with *a priori* known homogeneous Dirichlet no-slip BC. The calculus operation is *convolution* with the gaussian filter

$$\overline{\mathbf{u}}(\mathbf{x}_b) = (g_\delta{}^*\mathbf{u}(\mathbf{x}_b)) = \int_\Omega g_\delta(\mathbf{x}_b - \mathbf{y})\mathbf{u}(\mathbf{y})d\mathbf{y},\ \mathbf{x}_b \in \partial\Omega \tag{9.110}$$

The integral in (9.110) is defined over a domain Ω encompassing $\partial\Omega$, hence $\mathbf{u}(\mathbf{x}, t)$ must be known *inside* Ω as well as on $\partial\Omega$. Meeting this requirement in practice is fully moderated by gaussian filter g_δ rapid decay away from $\mathbf{x}_b$. Thereby (9.110) is required evaluated *only* within the $\delta \approx 2h$ radius *hemisphere* in the near vicinity of each $\mathbf{x}_b \in \partial\Omega$, the shaded region in planar graph Figure 9.22.

The reference implements numerical *quadrature* for evaluating the integrals in (9.110), recall Section 3.9. DOF located at unfilled nodes *only* in Figure 9.22 enter the quadrature rule for $\overline{\mathbf{u}}(\mathbf{x}_b)$. The required additional unfiltered $\mathbf{u}(\mathbf{x}, t)$ data in Ω are generated via *approximate deconvolution* (AD) for gaussian with measures δ and γ

$$\mathbf{u}(\mathbf{x}) \cong \overline{\mathbf{u}}(\mathbf{x}) - \frac{\delta^2}{4\gamma}\nabla^2\overline{\mathbf{u}}(\mathbf{x}) + O(\delta^4) \tag{9.111}$$

The linear laplacian therein is discretely approximated using an appropriate order TS or completeness degree k trial space basis, Section 3.11. The resultant solution statement for (9.110) with (9.111) is

$$\begin{aligned}\overline{\mathbf{u}}(\mathbf{x}_b) \cong &\sum_{i \in I} w_i g_\delta(\mathbf{x}_b - \mathbf{x}_i)\left(\overline{\mathbf{u}}(\mathbf{x}) - \frac{\delta^2}{4\gamma}\nabla^2\overline{\mathbf{u}}(\mathbf{x})\right) \\ &+ \sum_{i \in B} w_i g_\delta(\mathbf{x}_b - \mathbf{x}_i)(\mathbf{u}(\mathbf{x}_b))\end{aligned} \tag{9.112}$$

The index sets I and B correspond with nodes on domain *I*nterior and *B*oundary, respectively, with $\mathbf{x}_i$ denoting quadrature coordinates for w_i the respective quadrature rule weights.

Evaluation of (9.112) generates $\overline{\mathbf{u}}(\mathbf{x}_b)$ DOF at nodes $\mathbf{x}_b \in \partial\Omega$ on domain bounding $\partial\Omega$. This enables Neumann BC evaluation in (9.109), alternatively *a priori* known DOF *non-homogeneous* Dirichlet BCs at all nodes $\mathbf{x}_b \in \partial\Omega$.

9.15 ADBC Algorithm Validation, Space Filtered D*E*

ADBC algorithm accuracy/convergence *a posteriori* data are reported for a non-D space filtered D*E* PDE + BCs + IC $n = 2$ *verification* statement, Sekachev (2013). Without fluid

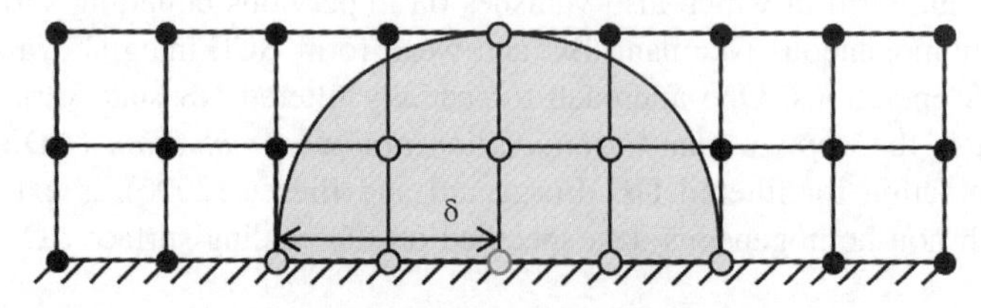

Figure 9.22 Planar projection of hemispherical domain for convolution (9.110) with node identifications

convection, for $\mathrm{RePr} \equiv 1$ the subject unfiltered D*E* with non-homogeneous Dirichlet BC and IC is

$$\text{D}E\text{:} \quad \frac{\partial \Theta}{\partial t} - \nabla^2 \Theta - s_\Theta = 0, \text{ on } \Omega \subset \mathbb{R}^2 \times [0, T) \tag{9.113}$$

$$\text{BCs:} \quad \Theta(\mathbf{x}_b, t) = \Theta_b, \text{ on } \partial\Omega \subset \mathbb{R}^1 \times [0, T) \tag{9.114}$$

$$\text{IC:} \quad \Theta(\mathbf{x}, t = t_0) = \Theta_0, \text{ on } \Omega \subset \mathbb{R}^2 \tag{9.115}$$

for $\mathbb{R}^n$ n-D euclidean space. The domain for (9.113) is the unit square $\Omega = \{-1 \le x \le 1; -1 \le y \le 1\}$ and *data* specifications for (9.114–9.115) are

$$\begin{aligned} s_\Theta &= 2\pi^2 \sin(\pi x)\sin(\pi y) \\ \Theta_b &= \begin{cases} \sin(\pi x)/2\pi, & \text{on}\{x, y = 0\} \\ 0, & \text{elsewhere on } \partial\Omega \end{cases} \\ \Theta_0 &= \frac{\sin(\pi x)\sinh(\pi - \pi y)}{2\pi\sinh(\pi)} + \frac{1}{2\pi^2}\sin(\pi x)\sin(\pi y) \end{aligned} \tag{9.116}$$

Classic SOV complementary-particular EBV PDE methodology, Section 2.4, generates the *analytical* unfiltered solution

$$\begin{aligned} \Theta(\mathbf{x}, t) = &\frac{\sin(\pi x)\sinh(\pi - \pi y)}{2\pi\sinh(\pi)} + \frac{1}{2\pi^2} e^{-2\pi^2 t}\sin(\pi x)\sin(\pi y) \\ &+ \left(1 - e^{-2\pi^2 t}\right)\sin(\pi x)\sin(\pi y) \end{aligned} \tag{9.117}$$

which contains spectral content fully supporting ADBC algorithm accuracy/convergence quantification. Formed therefrom are space filtered *verification* solutions $\overline{\Theta}(\delta, \mathbf{x}, t)$ generated by convolution with gaussian filter of measure δ. As detailed in Section 9.13, unfiltered Θ must first be extended exterior to PDE domain Ω. This is identically the operation required for *auxiliary problem* operator [A] resolution via convolution, Figure 9.6 bottom.

Denoting extended domain as $\Omega^+ \subset \mathbb{R}^2$, the D*E* statement *data* extensions are

$$\begin{aligned} &\Theta_b(x, y) && \Rightarrow \sin(\pi x)/2\pi, \quad \text{for}\{0 \le x \le 1; y \le 0\} \\ &\left.\begin{aligned} &\Theta^{ext}(x, y, t) \\ &\Theta_0^{ext}(x, y, t_0) \end{aligned}\right\} && \Rightarrow \{0, \quad \text{for } x, y \notin \Omega \cup \{0 \le x \le 1; y \le 0\} \\ &\qquad s_\Theta^{ext} = 0, \quad \text{for } x, y \notin \Omega \end{aligned} \tag{9.118}$$

D*E* (9.113) is then convolved with gaussian filter

$$g_\delta(x, y) \equiv \left(\frac{6}{\pi\delta^2}\right)\exp\left(\frac{6\,(x^2 + y^2)}{\delta^2}\right) \tag{9.119}$$

which generates *filtered* state variable D*E* containing BCE integral

$$\frac{\partial\overline{\Theta}}{\partial t} - \nabla^2\overline{\Theta} - \overline{s}_\Theta - \int_{\partial\Omega} g_\delta(x_k - s_k)\frac{\partial\Theta}{\partial x_j}\hat{n}_j\mathrm{d}s = 0, \ \text{ in } \Omega^+ \subset \mathbb{R}^2 \times [0, t) \tag{9.120}$$

The base $n=2$ cartesian $\mathrm{M}=20\times20$ uniform $k=1$ TP basis mesh measure is $\Delta x=0.05=\Delta y$. For gaussian filter measure $\delta=0.2$, adequate energetic capture is ensured for element measure $h=\delta/4$. $\mathrm{GWS}^h+\theta\mathrm{TS}$ for filtered PDE (9.120) employs constant time step $\Delta t=0.001$ with final time definition $n\Delta t\equiv T=0.05$.

For any mesh M, hence δ, the verification solution $\overline{\Theta}(\delta, x, y, t)$ results by filtering analytical solution (9.114) from which the *exact* filtered Dirichlet BC data are extracted. BCE integral in (9.120) can be computed exactly via analytical solution Θ, (9.117). Space filtered source $\overline{s}_\Theta$ is exactly calculable via convolution with (9.119). Exact $\overline{\Theta}(\delta, x, y, t)$, data and BCE integral are generated using *Mathematica*, Wolfram Research Inc., (2008). $\mathrm{GWS}^h+\theta\mathrm{TS}$ algorithm *a posteriori* DOF data are generated using *FEmPSE*, a Matlab toolbox-enable finite element problem solving environment (PSE), Baker (2012), for $k=1$ TP basis option.

For D*E* scalar variable restatement of ADBC algorithm (9.111–9.112), validation protocol independently assesses domain extension, BCE integral omission, its discrete approximation and filtered Dirichlet BC data error mechanisms. Error measure is the difference between filtered exact solution $\overline{\Theta}(\delta, x, y, t)$ and mesh of measure h computed filtered solution $\overline{\Theta}^h(\delta, x, y, t)$.

Computing $\overline{\Theta}^h$ using exact boundary data and exact BCE integral evaluation but without the $\Omega^+ \subset \mathbb{R}^2$ Dirichlet BC data extension, (9.118), compromises BCE integral in (9.120) evaluation accuracy. The $T=0.05$ error measure $\overline{\Theta}_{exact} - \overline{\Theta}^h_{\Omega^+\subset\mathbb{R}^2}$, Figure 9.23, confirms the error is confined directly adjacent to Dirichlet BC boundary $\partial\Omega(x,\ y=0)$ with extremum magnitude 0.050.

The mechanism dominating Figure 9.23 error is data *non-smoothness* in the vicinity of non-homogeneous Dirichlet BC data. The gaussian filter of measure δ extends well outside

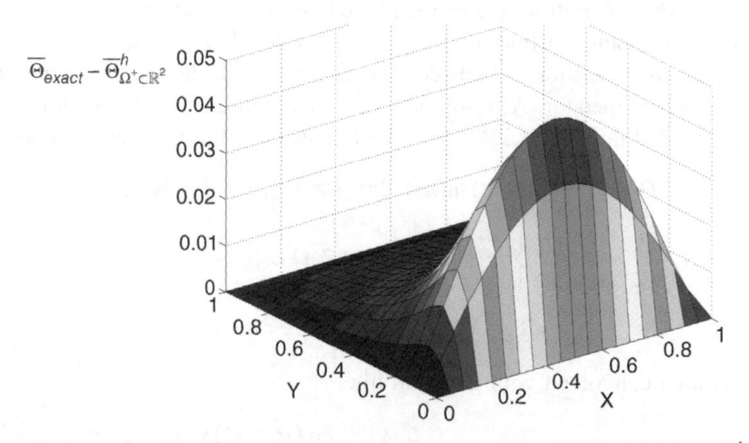

Figure 9.23 ADBC algorithm filtered solution max norm error interpolation $\overline{\Theta}_{exact} - \overline{\Theta}^h_{\Omega^+\subset\mathbb{R}^2}$, exact boundary data, exact BCE integral, Dirichlet BC data not extended to $\Omega^+ \subset \mathbb{R}^2$, $T=0.05$

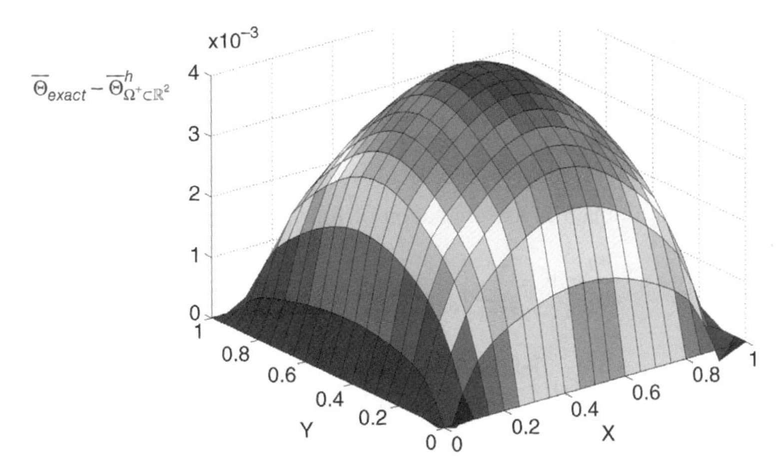

Figure 9.24 ADBC algorithm solution discrete approximation max norm error interpolation $\overline{\Theta}_{exact} - \overline{\Theta}^h_{\Omega^+\subset\mathbb{R}^2}$, Dirichlet BC data domain extension, exact boundary data, exact BCE integral, $T=0.05$

the solution domain boundary $\partial\Omega$, recall Figure 9.10. For exact boundary data and exact BCE integral, with Dirichlet BC data now extended to $\Omega^+ \subset \mathbb{R}^2$, error $\overline{\Theta}_{exact} - \overline{\Theta}^h_{\Omega^+\subset\mathbb{R}^2}$ is nil adjacent to Dirichlet BC $\partial\Omega$, Figure 9.24. The extremum, located at domain center, is reduced over one order to 0.0043.

For data in Figures 9.23, 9.24, the BCE integral contribution to (9.120) is calculated exactly using analytical solution (9.117) on all four boundary segments $\partial\Omega$. Retaining data extension $\Omega^+ \subset \mathbb{R}^2$ and exact boundary data but removing the BCE integral from (9.120) generates error extrema directly adjacent to all four $\partial\Omega$, Figure 9.25. The error sign is reversed from Figures 9.23, 9.24 with extremum increased by a factor of seven to $|0.027|$.

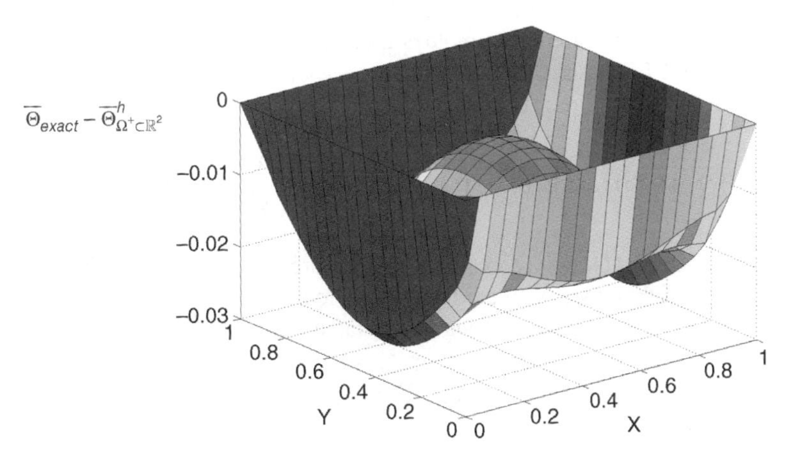

Figure 9.25 ADBC algorithm solution discrete approximation error max norm $\overline{\Theta}_{exact} - \overline{\Theta}^h_{\Omega^+\subset\mathbb{R}^2}$, extended data, exact boundary data, BCE integral omitted, $T=0.05$

These *a posteriori* data firmly quantify the requirement for data extension to $\Omega^+ \subset \mathbb{R}^2$ *and* to not omit the BCE integral. The reference implements convolution of D*E* BCE integral

$$\mathrm{A}_\delta\left(\mathrm{D}_j(\Theta)(x_k,t)\right) \equiv \int_{\partial\Omega} g_\delta(x_k - s_k)\frac{\partial\Theta}{\partial x_j}\hat{n}_j\mathrm{d}s \tag{9.121}$$

on the $n=2$ cartesian domain Ω as summation of four line integrals

$$\begin{aligned}
\int_{\partial\Omega} g_\delta(x_k - s_k)\frac{\partial\Theta}{\partial x_j}\hat{n}_j\mathrm{d}s &= \sum_{i=1}^{4}\int_{\partial\Omega_i} g_\delta(x_k - s_i)\frac{\partial\Theta}{\partial x_j}\hat{n}_j\mathrm{d}s_i \\
&= -\int_0^1 g_\delta(x-0, y-s_2)\frac{\partial\Theta(0,s_2)}{\partial x}\mathrm{d}s_2 + \int_0^1 g_\delta(x-1, y-s_2)\frac{\partial\Theta(1,s_2)}{\partial x}\mathrm{d}s_2 \\
&\quad - \int_0^1 g_\delta(x-s_1, y-0)\frac{\partial\Theta(s_1,0)}{\partial y}\mathrm{d}s_1 + \int_0^1 g_\delta(x-s_1, y-1)\frac{\partial\Theta(s_1,1)}{\partial y}\mathrm{d}s_1
\end{aligned} \tag{9.122}$$

Newton–Cotes quadrature formulae are recommended for (9.122), as high degree interpolation polynomial error mechanisms are well understood, Runge (1901). Termed the *composite rule*, each integral in (9.122) is subdivided into sequential five node subsets. Quadrature for each subset is via the five-point closed Newton–Cotes "Boole's Rule," Boole and Moulton, (1960)

$$\begin{aligned}
\int_a^b f(x_i)\mathrm{d}x &= \frac{b-a}{90}\left(7f(x_0) + 32f(x_1) + 12f(x_2) + 32f(x_3) + 7f(x_4)\right) \\
&\quad - \frac{(b-a)^7}{1935360}\frac{d^6 f(\zeta)}{\mathrm{d}x^6}, \quad a < \zeta < b, x_i = \frac{i(b-a)}{n(=5)}
\end{aligned} \tag{9.123}$$

The ordinary derivative of unfiltered variable Θ in each line integral in (9.122) is discretely approximated. The reference employs a uniform mesh second order one-sided $n=1$ TS. The $n=1$, $k=2$ FE basis $O(h^2)$ comparative *stencils* for line element Ω_e with left and right node DOF indexed Q_j are, Baker (2012, Ch. 5.5)

$$[\mathrm{A201}]_e\{Q\}_e \Rightarrow \frac{1}{2h}\begin{pmatrix} -Q_{j+2} + 4Q_{j+1} - 3Q_j \\ 3Q_j \quad - 4Q_{j-1} + Q_{j-2} \end{pmatrix} + O(h^2) \tag{9.124}$$

The unfiltered DOF $\{Q\} \to Q_j$ residing *on* $\partial\Omega$ are *a priori* given Dirichlet BC data, (9.116), while the other two unfiltered DOF are located in Ω and are unknown. These unfiltered DOF are recovered from $\overline{\Theta}$ via D*E* restated AD algorithm (9.111)

$$\Theta(\mathbf{x}) \cong \overline{\Theta}(\mathbf{x}) - \frac{\delta^2}{24}\nabla^2\overline{\Theta}(\mathbf{x}) + O(\delta^4) \tag{9.125}$$

The reference employs square mesh $O(h^2)$ FD stencil for the laplacian in (9.125), identically the uniform mesh centered node $k=1$ NC basis laplacian assembly (3.105).

Laplacian discrete approximation is *the* dominant data error mechanism in AD algorithm (9.125). Using exact filtered solution $\overline{\Theta}^{exact}_{\Omega^+\subset\mathbb{R}^2}$ domain extended, with exact boundary data and exact BCE integral, time evolution of the M = 20 × 20 uniform mesh *unfiltered* solution $\Theta^h(\mathbf{x}, t)$ DOF *error* extremum

$$\left|\{e^h(t)\}\right| = 100 \times \left|\left(\{Q(t)\}_{exact} - \{Q(t)\}_{approx}\right)/\{Q(t)\}_{exact}\right| \tag{9.126}$$

is graphed in Figure 9.26. The quite large initial ~12% error asymptotes to ~2% by $T = 0.05$.

ADBC algorithm final validation replaces exact BCE integral, then exact Dirichlet BC data with numerical quadrature. The AD algorithm is thus applied to both (9.125) and

$$\overline{\Theta}(\mathbf{x}_b) = (g_\delta * \Theta)(\mathbf{x}_b) = \int_{\Omega\subset\mathbb{R}^2} g_\delta(\mathbf{x}_b - \mathbf{y})\Theta(\mathbf{y})d\mathbf{y}, \ \ \mathbf{x}_b \in \partial\Omega \tag{9.127}$$

The domain extension $\Omega^+ \subset \mathbb{R}^2$ supporting accurate quadrature replaces the DOF block enclosed hemisphere, Figure 9.24, with encompassing sphere, planar projection the shaded circle in Figure 9.27. This extension fully accounts for gaussian filter penetration of domain Dirichlet BC boundary, hence is appropriate for BCE integral (9.121) quadrature as well. The mesh block added sphere-external DOF are of no computational consequence due to filter measure δ cutoff.

Efficient numerical quadrature on $\Omega^+ \subset \mathbb{R}^2$ accrues to tensor products of Boole's rule (9.123). Computation of $\overline{\Theta}(\mathbf{x}_b)$ Dirichlet BC DOF data, (9.127), is thus accomplished via quadrature rule

$$\overline{\Theta}(\mathbf{x}_b) \cong \sum_{i=1}^{N_x}\sum_{j=1}^{N_y} w_i w_j g_\delta\left(x_b - x_i, y_b - y_j\right)\Theta(\mathbf{x}) \tag{9.128}$$

for N_x and N_y quadrature degree and w_i and w_j the respective weights. Coding (9.128) is elementary as is theory/practice extension to $n = 3$. For handling DOF at nodes $\mathbf{x}_b$ less than

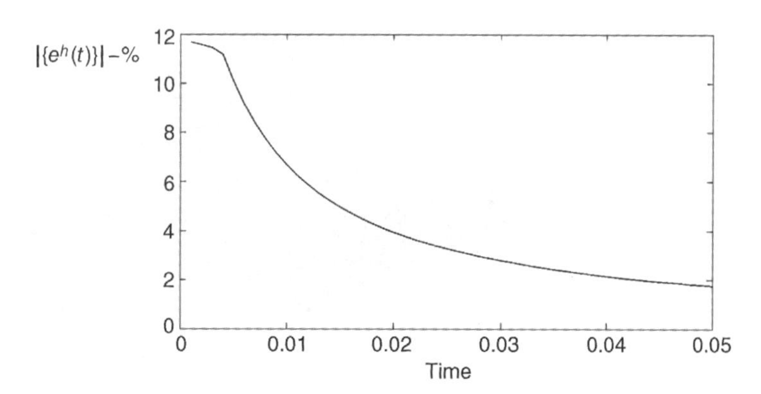

Figure 9.26 AD algorithm $\Theta^h(\mathbf{x}, t)$ extremum DOF error evolution in $\left|\{e^h(t)\}\right|$, extended domain, exact boundary data, exact BCE integral

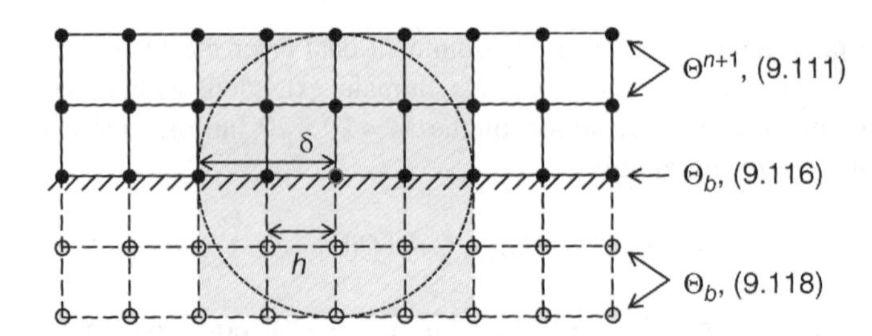

Figure 9.27 Extended domain $\Omega^+ \subset \mathbb{R}^2$ for convolution (9.111) and AD algorithms (9.125), (9.127), with node identifications, designations for Dirichlet BC data

δ distant from a domain corner, Boole's rule is replaced with reduced span Newton–Cotes formulae, specifically $O(h^2)$ two-node trapezoidal rule, $O(h^4)$ Simpson's three-node and Simpson's ⅓ four-node rules.

The shaded square region enclosing the circle of radius δ, Figure 9.27, contains only 12 DOF. However, to retain composite quadrature rule advantages each such region is designated to contain 5×5 DOF. On $\partial\Omega$ with non-homogeneous Dirichlet BCs, evaluating (9.128) employs Boole's rule (9.123) in each direction. No data extensions occur along $\partial\Omega$ segments with homogeneous Dirichlet BCs.

For $\Omega^+ \subset \mathbb{R}^2$ and exact Dirichlet BC data, the ADBC algorithm discrete *filtered* solution $\overline{\Theta}^h$ *error* distribution at $T = 0.05$ for BCE integral (9.121) via quadrature (9.123) is graphed in Figure 9.28. It is visually indistinguishable from that for $\overline{\Theta}^h$ generated for exact BCE integral, Figure 9.24, with identical extrema. The exceptions are local error of minor magnitude at each endpoint node on the non-homogeneous Dirichlet BC $\partial\Omega$, likely due to the $O(h^2)$ quadrature rule.

Implementing identified quadrature protocols for *both* non-homogeneous Dirichlet BCs *and* BCE integral, generated $\overline{\Theta}^h$ solution error distribution at $T = 0.05$ is graphed in

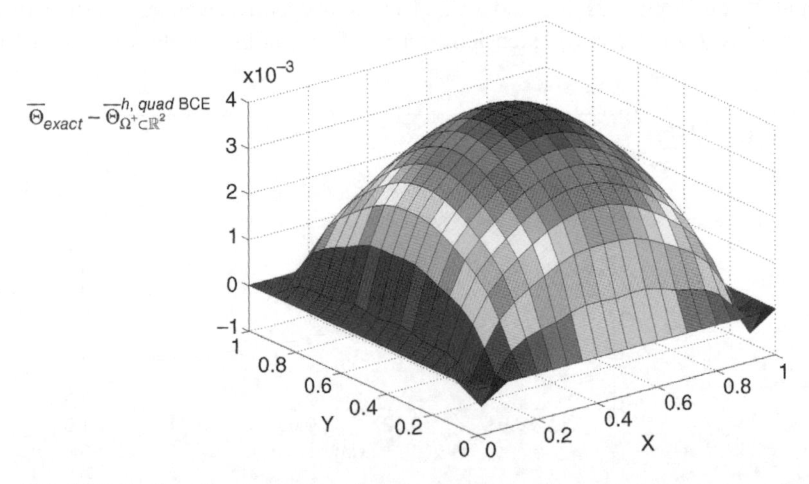

Figure 9.28 ADBC algorithm solution discrete approximation error max norm distribution $\overline{\Theta}_{exact} - \overline{\Theta}^{h,\,quad\ BCE}_{\Omega^+ \subset \mathbb{R}^2}$, extended data, exact Dirichlet BC data, BCE integral via quadrature, $T = 0.05$

Figure 9.29. The extremum is unaltered from that for exact Dirichlet BC and BCE integral. The BCE quadrature corner node local error in Figure 9.28 is now distributed along the entire non-homogeneous Dirichlet BC $\partial\Omega$ with absolute magnitude about half that of the extremum.

The discrete numerics generating these *a posteriori* data are second-order accurate, equivalent to $k=1$ NC/TP basis implementation. Asymptotic error estimate (9.78) is pertinent to assess convergence under regular mesh refinement. This estimate is directly quantifiable, as detailed in (2.105–2.106), *only* for $\left\|\{q^h(t_0)\}\right\|^2_{H^m(\Omega)}$ *and* $\left[\|\mathrm{data}\|_{L2,\Omega} + \|\mathrm{data}\|_{L2,\partial\Omega}\right]$ in (9.78) mesh M independent. Neither requirement is met for the solution data set since $\overline{\Theta}^h(\delta, \mathbf{x}, t)$ is δ dependent by definition $h=\delta/4$. Acknowledging that this compromises slope prediction accuracy, $T=0.05$ solution *a posteriori* data do confirm *monotone* convergence under uniform mesh refinement, Table 9.1.

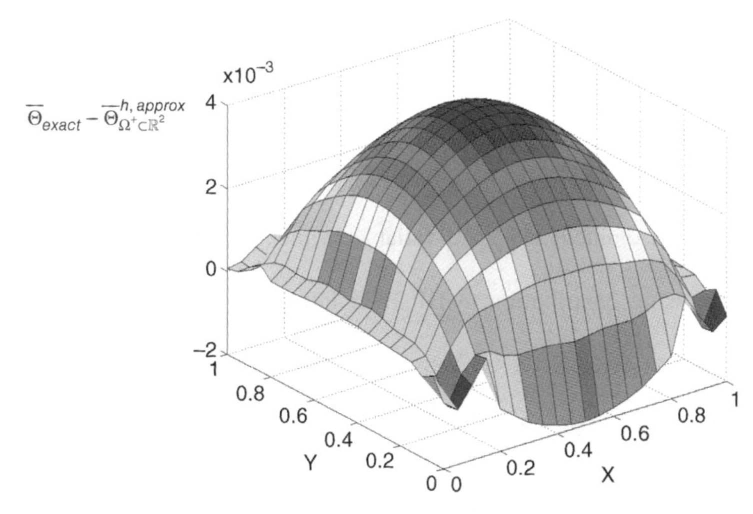

Figure 9.29 ADBC algorithm solution discrete approximation error max norm distribution $\overline{\Theta}_{exact} - \overline{\Theta}^{h,\,quad\ \mathrm{BCE}}_{\Omega^+\subset\mathbb{R}^2}$, extended data, Dirichlet BC data via quadrature, BCE integral via quadrature, time $T=0.05$

Table 9.1 ADBC GWS^h + θTS algorithm $k=1$ basis space filtered solution $\overline{\Theta}^h$ asymptotic convergence rate estimate in D*E* energy norm $\left\|e^h(t=T)\right\|_E$

Mesh M	$\left\|\overline{\Theta}^h\right\|_E$	$\Delta\left\|\overline{\Theta}^{\Delta h/2}\right\|_E$	$\left\|e^{\Delta h/2}\right\|_E$	Slope
8×8	0.4359	—	—	—
16×16	0.7558	0.3199	0.1066	—
32×32	0.9174	0.1615	0.0538	0.99
64×64	0.9875	0.0702	0.0234	1.20

9.16 *ar*LES Theory Resolved Scale BCE Integrals

The $m\text{GWS}^h + \theta\text{TS}$ algorithm for *ar*LES theory *m*PDE system (9.13–9.18) including BCE integrals (9.103–9.106) is derived. Insertion into *m*PDE system (9.73) alters *only* the source matrix (9.77)

$$\{s\} = \left\{ \begin{array}{c} -\dfrac{\text{Gr}}{\text{Re}^2}\Theta\widehat{g}_i + \displaystyle\int_{\partial\Omega} g_\delta(x_k - s_k)\left[-P\delta_{ij} + \dfrac{1}{\text{Re}}\left(\dfrac{\partial u_i}{\partial x_j} + \dfrac{\partial u_j}{\partial x_i}\right)\right]\hat{n}_j \text{d}s \\ \dfrac{\text{Ec}}{\text{Re}}\left[\begin{array}{l} \dfrac{\partial}{\partial x_j}\left(u_i\dfrac{\partial u_j}{\partial x_i}\right) + \dfrac{\partial^2 c_{ij}}{\partial x_j \partial x_i} \\ -\dfrac{\partial^2}{\partial x_j\partial x_i}\left(\dfrac{\text{C}_\text{S}(\delta)h^2\text{Re}}{6}\left(u_j u_k\dfrac{\partial u_i}{\partial x_k} + u_i u_k\dfrac{\partial u_j}{\partial x_k}\right)\right) \end{array}\right] \\ + \displaystyle\int_{\partial\Omega} g_\delta(x_k - s_k)\left[\dfrac{1}{\text{RePr}}\dfrac{\partial\Theta}{\partial x_j}\right]\hat{n}_j \text{d}s \\ s(Y) + \displaystyle\int_{\partial\Omega} g_\delta(x_k - s_k)\left[\dfrac{1}{\text{ReSc}}\dfrac{\partial Y}{\partial x_j}\right]\hat{n}_j \text{d}s \end{array} \right\} \tag{9.129}$$

Overbar notation is omitted; NS state variable members in BCE integrands are *always unfiltered*!

Kinematic pressure contribution to D**P** BCE integral (9.104), located in (9.129) first row, is

$$\text{A}_\delta\left(\text{D}_{ij}(P\delta_{ij}\,\hat{n}_j)(x_k, t)\right) \equiv \int_{\partial\Omega} g_\delta(x_k - s_k)P\delta_{ij}\,\hat{n}_j \text{d}s \tag{9.130}$$

The Kronecker delta restricts contributions to boundary segments with normal parallel to unfiltered u_i. For practical Reynolds number Re, the homogeneous reduction of Neumann BC (9.18) applies to $P(\mathbf{x}, t)$ on all bounding wall segments $\partial\Omega \cap \partial\Omega_{\hat{n}_i}$. Non-homogeneous Neumann BC in (9.18) applies to outflow segments $\partial\Omega$. Hence only inflow surface nodes $\mathbf{x}_b \in \partial\Omega_{\hat{n}_i}$ can contain non-homogeneous Dirichlet BC data.

The $m\text{GWS}^h + \theta\text{TS}$ pressure projection algorithm post-processes $\overline{P}^{n+1}$ via (9.108), upon iterative convergence at t_{n+1}. For state variable alteration in AD algorithm (9.111), P^{n+1} on $\partial\Omega$ is predictable via

$$P^{n+1}(\mathbf{x}) \cong \overline{P}^{n+1}(\mathbf{x}) - \frac{\delta^2}{4\gamma}\nabla^2\overline{P}^{n+1}(\mathbf{x}) + O(\delta^4) \tag{9.131}$$

with $\overline{P}^{n+1}$ *a priori* known everywhere on $\Omega \cup \partial\Omega$. Accurate laplacian evaluation therein requires domain *and* $\overline{P}^{n+1}$ data extension to $\Omega^+ \subset \mathbb{R}^n$ encompassing $\partial\Omega$, Figure 9.27.

For D*E* on $n=2$ uniform meshes, Sekachev (2013) solves (9.131) approximating the laplacian with (3.105). The fully general alternative to $\Omega^+ \subset \mathbb{R}^3$ extension is via GWS^N on *differential definition* replacement of (9.131)

$$\mathcal{D}(P^{n+1}) \equiv P^{n+1} - \overline{P}^{n+1} + \frac{\delta^2}{4\gamma}\nabla^2\overline{P}^{n+1} = 0 + O(\delta^4) \tag{9.132}$$

Deleting $O(\delta^4)$, in precise DOF notation $\mathrm{GWS}^N \Rightarrow \mathrm{GWS}^h$ transitions (9.132) to the element matrix statement

$$\begin{aligned}
\{\mathrm{F_Q}P^{n+1}\}_e &= \det_e[\mathrm{M200}d]\{\mathrm{Q_}P^{n+1}\}_e - \det_e[\mathrm{M200}d]\left\{\mathrm{Q_}\overline{P}^{n+1}\right\}_e \\
&\quad - \frac{\delta^2}{4\,\gamma\det_e}[\mathrm{M2}KKd]_e\left\{\mathrm{Q_}\overline{P}^{n+1}\right\}_e \\
&\quad + \frac{\delta^2}{4\gamma}[\mathrm{N20}Kd]_e\left\{\mathrm{Q_}\overline{P}^{n+1}\right\}_e \hat{n}_K \\
\left[\mathrm{JAC_Q}P^{n+1}\right]_e &= \det_e[\mathrm{M200}d]
\end{aligned} \tag{9.133}$$

Matrix suffix d emphasizes that linear (9.133) may be implemented for *any* $k \geq 1$ trial space basis. Further, mPDE laplacian modification (5.114) is appropriate for (9.132). Asymptotic error estimate for assembled (9.133) solution is (9.79) with $k = 1$ exponent $(k+1) = 2$ for the mPDE alteration. Gaussian filter *uniform* measure δ and shape factor γ in (9.133) are *data*.

The Green–Gauss divergence theorem surface integral in (9.133) is deleted as DOF $\left\{\mathrm{Q\_}\overline{P}^{n+1}\right\}_e$ are *a priori* known. The sole DOF constraint admissible in solving assembled (9.133) is $\{\mathrm{Q\_}P^{n+1}\}_e = \{0\}$, a typical BC for inflow $\partial\Omega$ nodes, which generates DOF distribution $\{\mathrm{Q\_}P^{n+1}\} \equiv \{\mathrm{Q\_}P(t_{n+1})\}$ on $\Omega^+ \subset \mathbb{R}^n$. Recalling assembly operator S_e the GWS^h unfiltered pressure distribution at $m\mathrm{GWS}^h + \theta\mathrm{TS}$ algorithm converged solution time level t_{n+1} is

$$P^{h,n+1} \equiv P^h(\mathbf{x}, t_{n+1}) = \mathrm{S}_e\{N_k(\mathbf{x})\}^T\{\mathrm{Q\_}P(t_{n+1})\}_e \tag{9.134}$$

Substituting (9.134) into (9.130) enables BCE integral completion

$$\overline{P}^{n+1}(\mathbf{x}_b \in \partial\Omega_{\hat{n}_i}) = \int_\Omega g_\delta(x_k - x_i)\mathrm{S}_e\left(\{N_k(\mathbf{x})\}^T\{\mathrm{Q\_}P(t_{n+1})\}_e\right)\delta_{ij}\,\hat{n}_j \mathrm{d}\tau, \mathbf{x}_b \in \partial\Omega_{\hat{n}_i} \tag{9.135}$$

Quadrature of (9.135) on $\Omega^+ \subset \mathbb{R}^n$ is via n-D generalization and state variable alteration of (9.128). Taking care with index notation the resultant *filtered* pressure DOF at *each* pertinent node $\mathbf{x}_b \in \partial\Omega_{e,\hat{n}_j}$ is

$$\mathrm{Q\_}\overline{P}(t_{n+1})\Big|_{\mathbf{x}_b \in \partial\Omega_{e,\hat{n}_j}} = \sum_{l=1}^{N_x}\sum_{m=1}^{N_y}\sum_{n=1}^{N_z}\begin{pmatrix} w_l w_m w_n g_\delta(x_b - x_l, y_b - y_m, z_b - z_n) \\ \times \quad \mathrm{S}_e\left(\{N_k(\mathbf{x})\}^T\{\mathrm{Q\_}P(t_{n+1})\}_e\right)\end{pmatrix}\delta_{ij}\,\hat{n}_j \tag{9.136}$$

Retaining composite rule, N_x, N_y, N_z are Newton–Cotes formula degree with w_i, w_j, w_k the respective weights, altered as detailed for nodes $\mathbf{x}_b \in \partial\Omega_{\hat{n}_i}$ less than δ distant from domain corners. For the $\theta = 0.5$ $m\text{GWS}^h + \theta\text{TS}$ algorithm (9.136) data are organized into DOF $\{\text{Q\_}\overline{P}(t_{n+1})\}$ and constitute time t_n *data* in forming (9.16) at time t_{n+1}.

Returning to GWS^N (9.109), and the discussion thereafter, the second terms in both surface integral integrands vanish identically as the resolved scale velocity vector is divergence free. Deleting this term the velocity strain rate tensor term in D**P** BCE integrand (9.129) reduces to

$$\text{A}_\delta\left(D_{ij}(u_i, \hat{n}_j)(x_k, t)\right) \equiv \int_{\partial\Omega} g_\delta(x_k - s_k)\frac{\partial u_i}{\partial x_j}\hat{n}_j \text{d}s \tag{9.137}$$

For domain extension $\Omega^+ \subset \mathbb{R}^n$ domain bounding surface BCs for *unfiltered* velocity u_i therein are non-homogeneous Dirichlet at inflow (typically) and homogeneous Dirichlet on no-slip walls. Well-posed $m\text{GWS}^h + \theta\text{TS}$ requires homogeneous Neumann BCs on outflow segments $\partial\Omega$. Therefore, the integrand in (9.137) is *non-vanishing* everywhere except on outflow segments $\partial\Omega$.

The AD differential definition for (9.137) integrand is

$$\mathcal{D}\left(\frac{\partial u_i}{\partial x_j}\hat{n}_j\right) \equiv \frac{\partial u_i}{\partial x_j}\hat{n}_j - \frac{\partial \overline{u}_i}{\partial x_j}\hat{n}_j + \frac{\delta^2}{4\gamma}\nabla^2\left(\frac{\partial \overline{u}_i}{\partial x_j}\hat{n}_j\right) = 0 + O(\delta^4) \tag{9.138}$$

Deleting $O(\delta^4)$, noting that $\hat{n}_j$ is a common multiplier, labeling element DOF $\{\text{Q\_}\nabla_j\text{U}I\}_e$, $1 \leq I,\ j \leq n$, and proceeding through $\text{GWS}^N \Rightarrow \text{GWS}^h$ transitions (9.138) to the element matrix statement

$$\begin{aligned}
\{\text{F_Q}\nabla_j\text{U}I\}_e &= \text{det}_e[\text{M}200d]\{\text{Q_}\nabla_j\text{U}I\}_e - \text{det}_e[\text{M}200d]\{\text{Q_}\nabla_j\overline{\text{U}}I\}_e \\
&\quad - \frac{\delta^2}{4\gamma\,\text{det}_e}[\text{M}2KKd]_e\{\text{Q_}\nabla_j\overline{\text{U}}I\}_e \\
&\quad + \frac{\delta^2}{4\gamma}[\text{N}20Kd]_e\{\text{Q_}\nabla_j\overline{\text{U}}I\}_e\hat{n}_K,\ 1 \leq I \leq n \\
[\text{JAC_Q}\nabla_j\text{U}I]_e &= \text{det}_e[\text{M}200d]
\end{aligned} \tag{9.139}$$

Matrix suffix d again denotes (9.139) implementation options detailed after (9.133).

Green–Gauss theorem generated surface integral in (9.139) is deleted as sole non-Dirichlet BC is homogeneous Neumann. In solving assembled (9.139) the only applicable DOF constraint is at outflow $\partial\Omega$ nodes where DOF $\left\{\text{Q\_}\nabla_j\text{U}I\widehat{n}_j\right\}_e$ *a priori* vanish. The ensuing solution generates for any time t *unfiltered* velocity strain rate DOF distributions $\left\{\text{Q\_}\nabla_j\text{U}I\widehat{n}_j\right\}$.

GWS^h solution unfiltered velocity vector strain rate tensor dot product with outwards pointing normal is

$$\left(\frac{\partial u_i^h}{\partial x_j}\hat{n}_j\right)(\mathbf{x}, t) = \text{S}_e\left(\{N_k(\mathbf{x})\}^T\{\text{Q\_}\nabla_j\text{U}I(t)\}_e\right)\hat{n}_j \tag{9.140}$$

Substituting (9.140) into (9.137) completes BCE integrals

$$\left(\frac{\partial \bar{u}_i}{\partial x_j}\hat{n}_j\right)(\mathbf{x}_b \in \partial\Omega_{\hat{n}_j}) = \int_\Omega g_\delta(x_k - x_i)S_e\left(\{N_k(\mathbf{x})\}^T\{\text{Q\_}\nabla_j \text{U}I(t)\hat{n}_j\}\right)\text{d}\tau, \mathbf{x}_b \in \partial\Omega_{\hat{n}_j} \quad (9.141)$$

For $\Omega^+ \subset \mathbb{R}^n$, taking care with index notation and recalling $\mathbf{x}_b$ exist on all segments $\partial\Omega$ except outflow, convolution (9.141) quadrature generates the DOF *data* at each pertinent node $\mathbf{x}_b \in \partial\Omega_{e,\hat{n}_j}$

$$\text{Q\_}\nabla_j\overline{\text{U}}I(t)\hat{n}_j\Big|_{x_b \in \partial\Omega_{e,\hat{n}_j}} \cong \sum_{l=1}^{N_x}\sum_{m=1}^{N_y}\sum_{n=1}^{N_z}\begin{pmatrix} w_l w_m w_n\, g_\delta(x_b - x_l, y_b - y_m, z_b - z_n) \\ \times\, S_e\left(\{N_k(\mathbf{x})\}^T\{\text{Q\_}\nabla_j\text{U}I(t)\hat{n}_j\}_e\right)\end{pmatrix} \quad (9.142)$$

for N_x, N_y, N_z Newton–Cotes formula degree with weights w_i.

Thus, via convolution and GWS^h differential definition, the D**P** BCE integral is quantified to generate resolved scale velocity vector DOF distributions $\{\text{Q\_}\nabla_j\overline{\text{U}}I(t)\hat{n}_j\}$ for $1 \leq (I,\ j) \leq n$ on $\mathbf{x}_b \in \partial\Omega_{\hat{n}_i}$ At any time t these DOF multiplied by 1/Re are appropriately inserted into *ar*LES theory mGWS^h + θTS algorithm as data.

Upon alteration of state variable, sequence (9.138–9.142) generates quadrature evaluation of the BCE integrals in PDEs (9.105) and (9.106). Thereby, at appropriate nodes $\mathbf{x}_b \in \partial\Omega_{\hat{n}_i}$ and any solution time t, resolved scale temperature and mass fraction BCE integrals are DOF quantified

$$\left\{\text{Q\_}\nabla_j\overline{\Theta}(t)\hat{n}_j\right\}\Big|_{\mathbf{x}_b \in \partial\Omega_{e,\hat{n}_j}}, \left\{\text{Q\_}\nabla_j\overline{Y}(t)\hat{n}_j\right\}\Big|_{\mathbf{x}_b \in \partial\Omega_{e,\hat{n}_j}}$$

and inserted into mGWS^h + θTS algorithm as *data* with appropriate Re, Pr, Sc multipliers.

The remaining topic is BCE integrals in pressure projection closure PDEs (9.107–9.108). The BC for unfiltered potential is everywhere homogeneous Neumann except at outflow, where the Dirichlet BC is always homogeneous. Therefore, the BCE integral in (9.107) vanishes identically. The sequence (9.138–9.142) leads to quadrature prediction of (9.108) BCE integral involving unfiltered pressure normal gradient at inflow and outflow nodes $\mathbf{x}_b \in \partial\Omega_{\hat{n}_i}$ only. Determined are DOF *data* $\{\text{Q\_}\nabla_j\bar{P}(t_{n+1})\}_{\mathbf{x}_b \in \partial\Omega_{\hat{n}_i}}$ implemented in the mGWS^h + θTS algorithm convergence post-process operation.

9.17 Turbulent Resolved Scale Velocity BC *Optimal* Ω^h-δ

Rigorously space filtered NS PDE system BCE integrands contain *unfiltered* NS state variable members. Detailed BCE integral AD-GWS^h differential definition algorithm strategy, completed with numerical quadrature, confirms that the *ar*LES theory mPDE with BCE integrals system is completely evaluable. Thus *ar*LES theory for filtered D**P** is verified well posed via resolved scale velocity vector *non-homogeneous* Dirichlet BC (9.112).

This has been accomplished *without* stating the word *turbulent* and is devoid of any specific resolved scale velocity near wall hypothesis. The fact that both NS and RaNS velocity vector state variable members satisfy the no-slip BC leads to derivation of a "turbulent" resolved scale velocity vector *non-homogeneous* Dirichlet BC (9.112), Sekachev

(2013, Section 6.2). The assumption is that unfiltered **u,** last line in (9.109), also (9.111), must as well correspond to the *time averaged* turbulent velocity near wall (BL) distribution RaNS predicted for large Re.

The theory retains gaussian filter *uniform* measure δ to eliminate non-uniform δ(**x**) commutation error (9.100). RaNS *validated* comparative data are generated via the "PIPE" software, Wilcox (2006), for embedded low Re^t *k*-ω model (4.63). The PIPE FD algorithm solves RaNS axial **DP** in BL similarity variables. Generated solution for wall adapted non-uniform M = 200 mesh is validated via the DNS data of Mansour *et al.* (1988), Wilcox (2006, Figure 4.30). This solution first off-wall node corresponds to $y^+ \equiv u_\tau y/\nu = 0.1$, hence possesses quality resolution of turbulent BL low Re^t near wall distribution.

Addressed is the symmetry plane of an $n = 3$ channel of half-height H for similarity Reynolds numbers $\mathrm{Re}_\tau \equiv u_\tau H/\nu = 180$, 395, 590, recall friction velocity definition $u_\tau \equiv (\tau_w/\rho)^{1/2}$. Symmetry plane steady isothermal axial **DP** *time averaged* (double overbar) reduces to

$$\mathcal{L}(\overline{\overline{u}}) = \frac{d\overline{\overline{P}}}{dx} - \frac{d}{dy}\left[\tau_{lam} - \overline{\overline{\tau}}_{x\,y}\right] = 0 \tag{9.143}$$

for $\tau_{lam} \equiv \mathrm{Re}^{-1}\left(d\overline{\overline{u}}/dy\right)$ and $\overline{\overline{\tau}}_{x\,y}$ the RaNS closure shear stress tensor model. The SOV analytical solution for (9.143) confirms constant axial pressure gradient balancing the linear (in *y*) difference between laminar and turbulent shear stress distributions, Figure 9.30. Note therein dominance of τ_{lam} in both near wall reaches.

Inserting validated $\overline{\overline{\tau}}_{x\,y}(y)$ as *data*, Chapter 8 RaNS $\mathrm{GWS}^h + \theta\mathrm{TS}$ algorithm simplified to (9.143) exactly reproduces DNS validated $\overline{\overline{u}}^h$ turbulent BL velocity profile, Sekachev (2013, Figure B.17). *ar*LES theory space filtered **DP** (9.103–9.104), simplified to channel

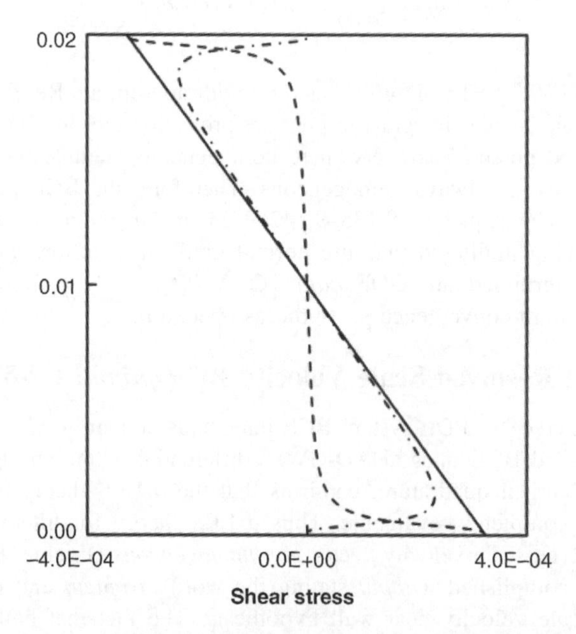

Figure 9.30 Channel symmetry plane RaNS BL shear stress distributions, $\mathrm{Re}_\tau = 180$: τ_{lam} (dashed curve), $\overline{\overline{\tau}}_{x\,y}$ (dot-dash curve), $\tau_{lam} - \overline{\overline{\tau}}_{x\,y}$ (solid line)

symmetry plane steady axial flow emulating (9.143) is

$$\mathcal{L}(\overline{u}) = \frac{\partial\overline{u}}{\partial t} + \frac{\mathrm{d}\overline{P}}{\mathrm{d}x} + \frac{\partial}{\partial y}\left[c_{xy} - \frac{1}{\mathrm{Re}}\frac{\partial\overline{u}}{\partial y} + \overline{u'v'}\right] - \int_{\partial\Omega} g_{\delta}(x-s)\left[-P + \frac{1}{\mathrm{Re}}\frac{\partial u}{\partial y}\right]\mathrm{d}s = 0 \quad (9.144)$$

The time derivative is added to enable $\mathrm{GWS}^h + \theta\mathrm{TS}$ time marching to the steady solution. Since steady solutions are unaffected by approach thereto, $\mathrm{D}M^h$ algorithm and convection nonlinearities are omitted in (9.144) as both vanish in the limit.

For gaussian filter uniform measure δ and *any* solution adapted non-uniform mesh M, the requisite RaNS *validation* is generated by space filtering $\overline{\overline{u}}$, $\overline{\overline{P}}$, $\overline{\overline{\tau}}_{xy}$ DOF interpolations using *Mathematica*, Wolfram Research, (2008). BCE integrals in (9.144) can then be exactly computed using data $\overline{\overline{u}}^h(y)$ and $\overline{\overline{P}}^h(x)$.

The resolved scale velocity non-homogeneous Dirichlet BC AD-GWS^h-BCE algorithm verification sequence is as detailed for D*E*, Section 9.15. $\mathrm{GWS}^h + \theta\mathrm{TS}$ algorithm directly comparative *a posteriori* DOF for (9.144) are generated by replacing c_{xy} with $\overline{\overline{\tau}}_{xy}$, $\overline{P}$ with $\overline{\overline{P}}$ and deletion of the SFS tensor $\overline{u'v'}$. For channel $k = 1$ basis $\mathrm{M} = 150$ uniform transverse mesh, exact data steady $\overline{u}^h$ DOF predictions for uniform $\delta = 0.002$, 0.00053, Figure 9.31 top, bottom, are compared with RaNS $\mathrm{M} = 150$ $\overline{\overline{u}}^h_{\mathrm{filtered}}$, the solid curves. BCE integral omission in (9.144) is again confirmed the dominant error mechanism (diamond symbols) which responds to filter measure δ reduction. For BCE integral inclusion essentially exact agreement between $\mathrm{M} = 150$ $\overline{\overline{u}}^h_{\mathrm{filtered}}$ and $\overline{u}^h$ DOF distribution is insensitive to δ.

With exact data verification established, a solution adapted mesh refinement study identifies AD-GWS^h-BCE algorithm *measure strategy* (h, δ) leading to $\overline{\overline{u}}^h_{\mathrm{filtered}}$ DOF distribution upon BCE integral (9.112) evaluation. AD statement (9.111) differential definition alteration for "unfiltered" velocity u_{AD} is

$$\mathcal{D}(u_{\mathrm{AD}}) \equiv u_{\mathrm{AD}} - \overline{u} + \frac{\delta^2}{4\gamma}\nabla^2\overline{u} = 0 + O(\delta^4) \quad (9.145)$$

Deleting $O(\delta^4)$, $\mathrm{GWS}^N \Rightarrow \mathrm{GWS}^h$ (9.15) transitions (9.145) to the element matrix statement

$$\begin{aligned}\{\mathrm{F\_Q}U\mathrm{AD}\}_e &= \det_e[\mathrm{M200}d]\{\mathrm{Q\_}U\mathrm{AD}\}_e - \det_e[\mathrm{M200}d]\{\mathrm{Q\_}\overline{U}\}_e \\ &\quad - \frac{\delta^2}{4\,\gamma\det_e}[\mathrm{M2}KKd]_e\{\mathrm{Q\_}\overline{U}\}_e + \frac{\delta^2}{4\gamma}[\mathrm{N20}Kd]_e\{\mathrm{Q\_}\overline{U}\}_e\hat{n}_K \\ [\mathrm{JAC\_Q}U\mathrm{AD}]_e &= \det_e[\mathrm{M200}d]\end{aligned} \quad (9.146)$$

Assembling (9.146), reducing jacobian order for DOF $\{\mathrm{Q\_}U\mathrm{AD}\}_e$ corresponding to $u_{\mathrm{AD}}(\mathbf{x}_b) \equiv \overline{\overline{u}}^h(\mathbf{x}_b) = 0$, then solving generates DOF $\{\mathrm{Q\_}U\mathrm{AD}\}$ distribution.

RaNS validation $\overline{\overline{u}}^h_{\mathrm{filtered}}(y, \delta^+)$ DOF are generated for a wall adapted $\mathrm{pr} = \pm 1.0043\,\mathrm{M} = 1000$ mesh. For gaussian filter uniform similarity definition $\delta^+ \equiv \mathrm{u}_\tau\delta/\nu$, AD-$\mathrm{GWS}^h$ algorithm *a posteriori* data are reported for $36 \geq \delta^+ \geq 4.5$. Resultant wall normal u^h_{AD} and $\overline{\overline{u}}^h_{\mathrm{filtered}}$ DOF interpolations are compared in BL similarity coordinates (top) and near wall

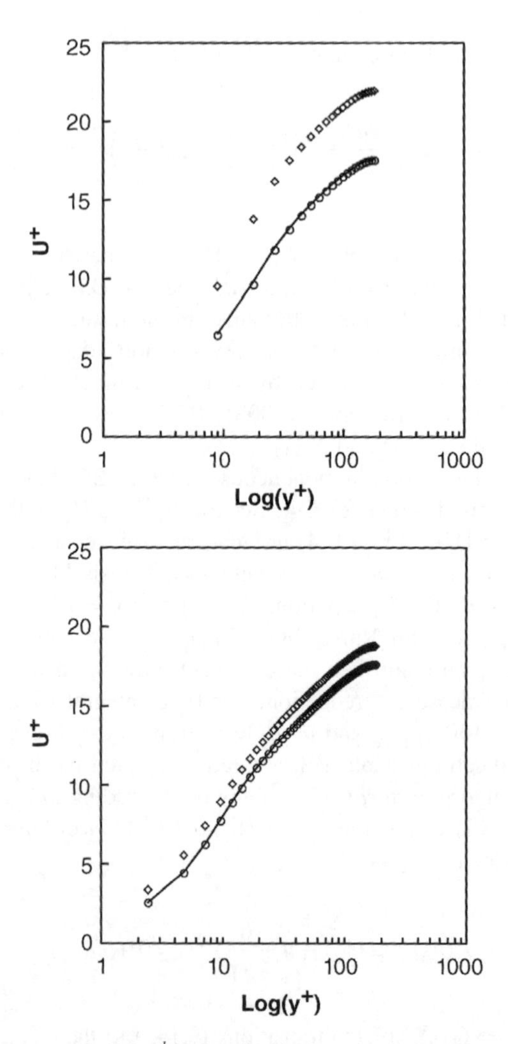

Figure 9.31 *ar*LES theory steady BL $\overline{u}^h$ DOF distributions, turbulent channel flow symmetry plane, validation $\overline{\overline{u}}^h_{\text{filtered}}(y^+)$ (solid curve), from steady solution of (9.144), BCE integral omitted (diamond symbol), BCE integral included (circle symbol): top, $\delta = 0.002$; bottom, $\delta = 0.00053$

region linear in y^+ (bottom), Figure 9.32. The error in u^h_{AD} is restricted to BL wall adjacent reach $y^+ < 0.25\delta^+$ and responds to δ^+ reduction via slope rotation about BL similarity coordinate $y^+ \sim 4\pm$.

Error localization in turbulent BL similarity *linear* sublayer $u^+ = y^+$, Figure 4.8, is genesis for u^h_{AD} wall slope error moderation. Omitted when (9.111) is FD addressed, Section 9.15, $\text{GWS}^N \Rightarrow \text{GWS}^h$ Green–Gauss divergence theorem generates the enclosing *non-homogeneous* Neumann BC surface integral on $\partial\Omega$. By not imposing *a priori* known $u_{\text{AD}}(\mathbf{x}_b) \equiv 0$ no-slip BCs, associated rows in jacobian (9.146) are *not* reduced out prior to solution, hence DOF $\{\text{Q\_}U\text{AD}\}$ on $\mathbf{x}_b \in \partial\Omega_{\hat{n}_i}$ become an output.

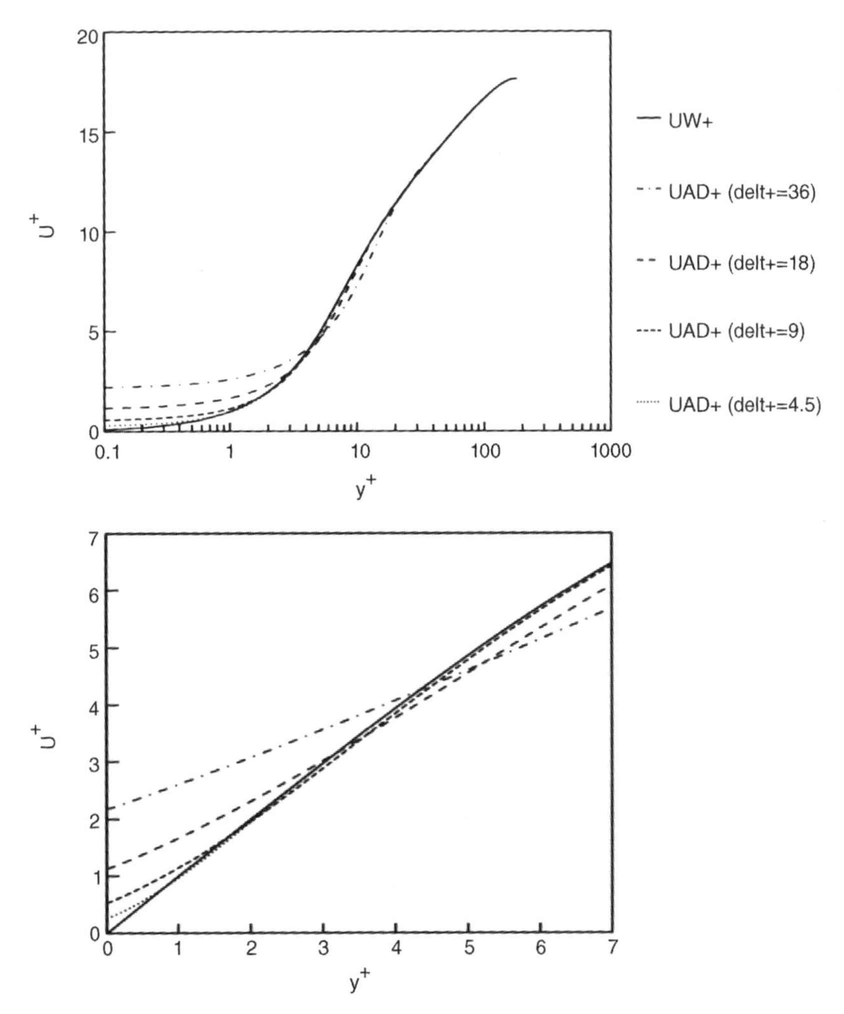

Figure 9.32 AD-GWS^h differential definition algorithm prediction of BL self-similar velocity U+ DOF interpolations, fully turbulent $n=3$ channel flow, validation $\overline{\overline{u}}^h_{\text{filtered}}$ (solid curve), for $36 \geq \delta^+ \geq 4.5$: top, in BL similarity coordinates; bottom, linear in y$^+$

Direct implementation of Neumann BC $[\text{N20}Kd]_e\{\text{Q\_}\overline{U}\}_e \hat{n}_K$ is contraindicated as $\mathbf{x}_b \in \partial\Omega$ DOF distribution $\{\text{Q\_}\overline{U}\}$ is sought output, *not* input data! Resolution replaces $[\text{N20}Kd]_e\{\text{Q\_}\overline{U}\}_e \hat{n}_K$ with $[\text{N200}d]_e\{\text{dQ\_}\overline{U}/\text{d}n\}_e$ with DOF an estimation of $\nabla\overline{u} \bullet \hat{n} \neq 0$ at $\mathbf{x}_b \in \partial\Omega$ where u^h_{AD} should vanish via the no-slip BC.

That $u^h_{\text{AD}}(y^+)$ error is restricted to *linear* similarity reach $\text{u}^+ = \text{y}^+$ enables accurate estimation of DOF $\{\text{dQ\_}\overline{U}/\text{d}n\}_e$. Near wall solution adapted mesh refinement *a posteriori* data, Sekachev (2013, Section 6.2.4), assesses AD-GWS^h algorithm non-homogeneous Dirichlet BC (DBC) *measure constraint* $\delta \equiv \text{n}h$, n an integer, for near wall reach discretization, Figure 9.33. For gaussian *uniform* δ the first four wall adjacent Ω_e are of identical measure $h = \delta/16$. The fifth Ω_e measure is $h = \delta/4$ which locates the sixth node exactly $\delta/2$

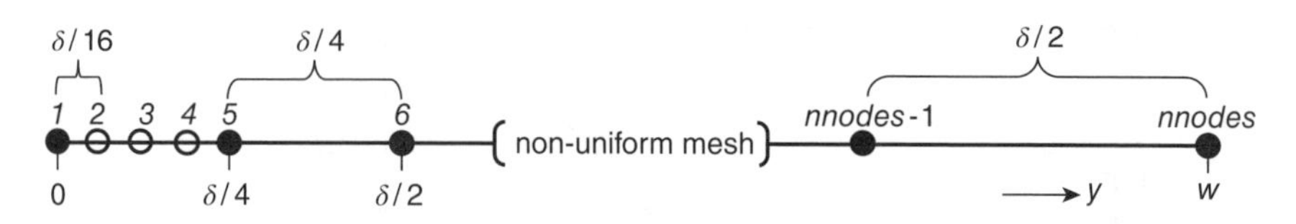

Figure 9.33 AD-GWS^h-DBC algorithm wall region discretization supporting $\{dQ\_\overline{U}/dn\}_e$ estimation at nodes $\mathbf{x}_b \in \partial\Omega$, no-slip wall BC $u_{AD} \equiv 0$ not enforced

distant from a no-slip BC wall node. Thereafter pr-adapted meshing is admissible to wall remote Ω_e of measure $h = \delta/2$, the energetic capture requirement.

For this near wall discretization AD-GWS^h-DBC algorithm employs an $O(h^3)$ end point TS to evaluate DOF $\{dQ\_\overline{U}/dn\}$ at $\mathbf{x}_b \in \partial\Omega$, node 1 in Figure 9.33, using DOF at nodes 2, 3, 4. This wall node DOF determination completes the data necessary to evaluate (9.146), hence determination of DOF $\{Q\_UAD\}_e$ on $\Omega^+ \subset \mathbb{R}^n$. **DP** BCE integrand term $(\partial u_i/\partial x_j)\hat{n}_j$, first line in (9.129), requires DOF $\{dQ\_UAD/dn\}_e$ at all nodes $\mathbf{x}_b \in \partial\Omega$. These data are generated via an $O(h^p)$ wall normal TS using DOF $\{Q\_UAD\}_e$ at nodes 5, 6, 7, . . . order p appropriate, all distant $\geq \delta/4$ from no-slip BC wall, Figure 9.33.

AD-GWS^h-BCE-DBC algorithm strategy accuracy/convergence assessment of *measure constraint* $\delta \equiv nh$ is documented, Sekachev (2013, Tables A.5–A.9). Mesh refinement *a posteriori* data for $20 \leq M \leq 640$ channel spanning wall normal discretizations, uniform except for near wall discretization, Figure 9.33, are reported. These data confirm that du^h_{AD}/dy^+ error at "unfiltered" u^h_{AD} no-slip BC nodes $\mathbf{x}_b \in \partial\Omega$, generated by defined $O(h^3)$ TS DOF prediction $\{dQ\_\overline{U}/dn\}$, responds to constraint $\delta \equiv nh$, $2 \leq n \leq 32$. Error magnitude $|du^h_{AD}/dy|$ distribution for measure constraint $\delta \equiv nh$, $2 \leq n \leq 16$, h uniform measure of $20 \leq M \leq 640$ channel spanning discretizations, is graphed in Figure 9.34.

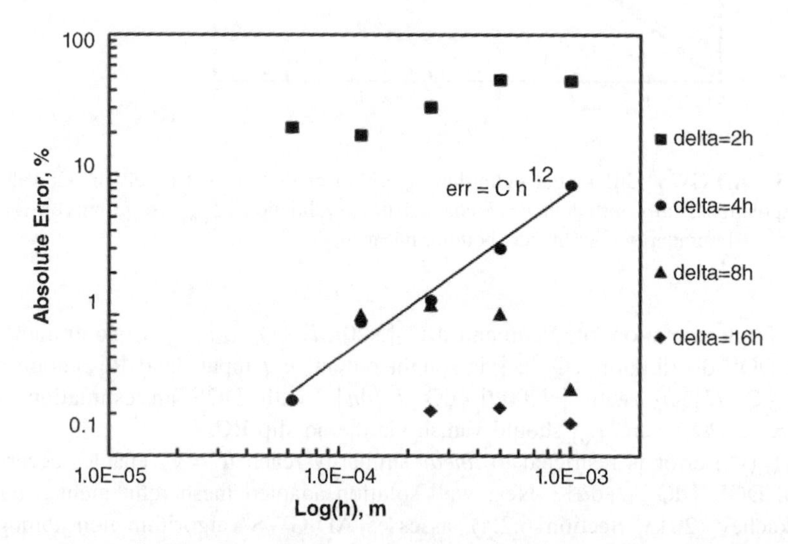

Figure 9.34 AD-GWS^h-BCE-DBC algorithm distributions of $|du^h_{AD}/dy|$ error at "unfiltered" u^h_{AD} no-slip BC nodes $\mathbf{x}_b \in \partial\Omega$, for constraint $\delta \equiv nh$, $2 \leq n \leq 16$, fully turbulent channel flow, $L = 0.01$ m

Surprisingly, energetic capture minimum n = 2 generates non-convergent data with excessive error for all h. For n = 4 the coarsest channel spanning M = 20 uniform discretization (h = 0.1 non-D) error is ~10%. This specification *only* exhibits monotone *convergence* rate ~1.2 under h refinement, the line in Figure 9.34. The n = 4 modest discretization h = 0.012 error ~1% matches that for n = 8, the data trend for which is divergent from h = 0.1 prediction ~0.2%. The n = 8, h = 0.1, also n = 16, $0.1 \leq h \leq 0.012$, data exhibit smallest error ~0.2% which is achieved for n = 4 for h = 0.006.

Unsteady axial **DP** (9.144) validation of AD-GWS^h-BCE-DBC algorithm is for $Re_\tau = 180$, channel spanning uniform M = 144 discretization of measure h = 0.013, n = 16, plus no-slip wall local discretization Figure 9.33. For IC uniform $\bar{u}(x, y) = 0.3$ m/s, filtered RaNS $\bar{\bar{u}}^h_{\text{filtered}}$ direct comparison results for $\bar{\bar{\tau}}_{xy}$ replacing c_{xy} and $\overline{u'v'}$ deletion. In similarity coordinates, lower half channel computed steady {Q_$U(y)$} DOF distribution exhibits excellent quantitative agreement with validation $\bar{\bar{u}}^h_{\text{filtered}}$, the solid curve in Figure 9.35. Omitting BCE integral generates the dot-dash {Q_$U(y)$} DOF interpolation therein again quantifying inclusion importance.

Isolation of BCE integral discrete evaluation error generated by the AD-GWS^h-BCE-DBC algorithm is enabled by inserting the exact data $\bar{\bar{u}}^h_{\text{filtered}}$ therein. Steady solution error $\left|\bar{\bar{u}}^h_{\text{filtered}} - \bar{u}^h\right|$ distribution in near wall physical coordinate y, Figure 9.36, confirms that the AD-GWS^h-BCE-DBC algorithm prediction (dot-dash curve) parallels exact data evaluation (solid curve) to $O(<\text{E}-03)$ throughout.

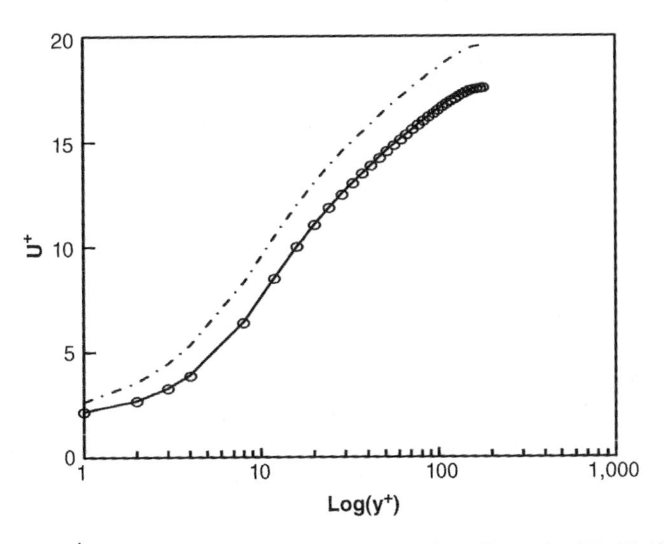

Figure 9.35 AD-GWS^h-BCE-DBC algorithm validation, axial **DP** steady {Q_$U(y)$} DOF distribution (open symbols) compared with RaNS validation $\bar{\bar{u}}^h_{\text{filtered}}$ (solid curve), h = 0.013, $\delta = 16h$, in similarity variables, dot-dash DOF {Q $U(y)$} interpolation omits BCE integral

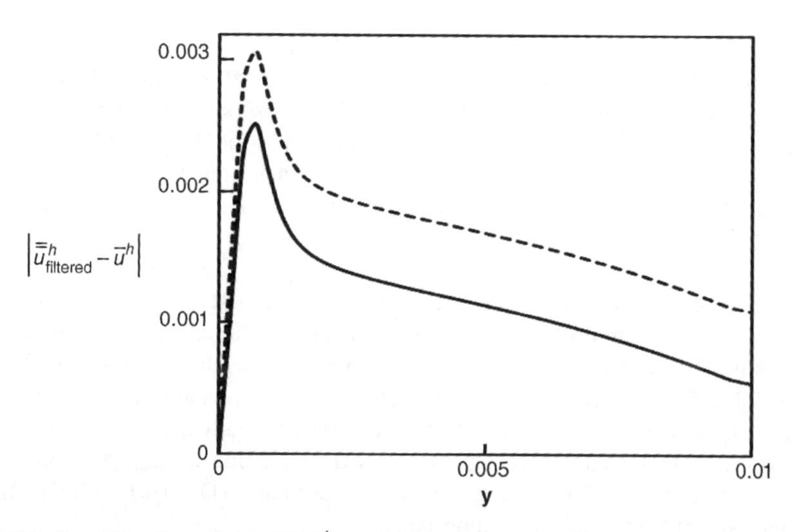

Figure 9.36 Quantification of AD-GWS^h-BCE-DBC algorithm BCE discrete convolution error, steady **DP** channel flow, $\mathrm{Re}_\tau = 180$, $\delta = 16h$, axial resolved scale velocity $\overline{u}^h$, channel spanning uniform $k = 1$ basis $\mathrm{M} = 144$ ($h = 0.013$) plus no-slip wall local discretization

9.18 Resolved Scale Velocity DBC Validation ∀ Re

Turbulent BL profile validated AD-GWS^h-BCE-DBC algorithm measure constraint $\delta \equiv \mathrm{n}h$ pertinence to ∀ Re is assessed. Algorithm *a posteriori* data are generated for *ar*LES theory complete *m*PDE system (9.103–9.108), in distinction to Section 9.17 axial **DP** restriction. Selected geometry is an $n = 3$ semi-infinite flat plate with $k = 1$ basis wall normal span pr-adapted $\mathrm{M} = 68$ discretization. Validation base is the laminar *unfiltered* $u_1(y)$ Blasius series BL solution for no-slip wall BC, Schlichting (1979, Part B).

For E+04 ≤ Re ≤ E+06 and no-slip BC wall near discretization, Figure 9.33, constraint $\delta \equiv 8h$ GWS^h + θTS algorithm steady *a posteriori* data are extracted from the $n = 3$ plate symmetry plane solution, Sekachev (2013, Section 6.2). Omitting *ar*LES theory $O(\delta^2, \delta^3)$ variables and BCE integral (9.104), non-homogeneous Dirichlet BC (9.110–9.112) algorithm generates symmetry plane steady $\overline{u_1^+}^h(y^+)$ DOF interpolation labeled U$^+$(AD BC) in Figure 9.37. Disagreement with Blasius validation, labeled U$^+$(no-slip BC), starts at $y^+ \approx 50$ and continues to the wall. Including BCE integral restricts $\overline{u_1^+}^h(y^+)$ disagreement with unfiltered Blasius to $y^+ \leq 1.0\pm$. This DOF interpolation, labeled U$^+$(AD BC;BCE), again verifies BCE integral omission *the* major error source.

Implementing *ar*LES theory complete $O(1, \delta^2, \delta^3)$ state variable, symmetry plane steady $\overline{u_1^+}^h(y^+)$ DOF interpolation U$^+$(AD BC;BCE;RLES) is visually unaltered from U$^+$(AD BC; BCE), Figure 9.37. Thereby, *ar*LES theory $O(\delta^2, \delta^3)$ state variables are *again* confirmed of negligible influence on NS laminar Re solution prediction, recall Section 9.11. These data quantitatively *validate* AD-GWS^h-BCE-DBC algorithm measure constraint strategy $\delta \equiv \mathrm{n}h$ appropriate ∀ Re.

9.19 *ar*LES $O(\delta^2)$ State Variable Bounded Domain BCs

*ar*LES theory $O(1, \delta^2)$ state variable resolution of RLES auxiliary problem operator [A], (9.41), generates harmonic PDE systems (9.64), (9.70), (9.71). Each linear EBV is derived

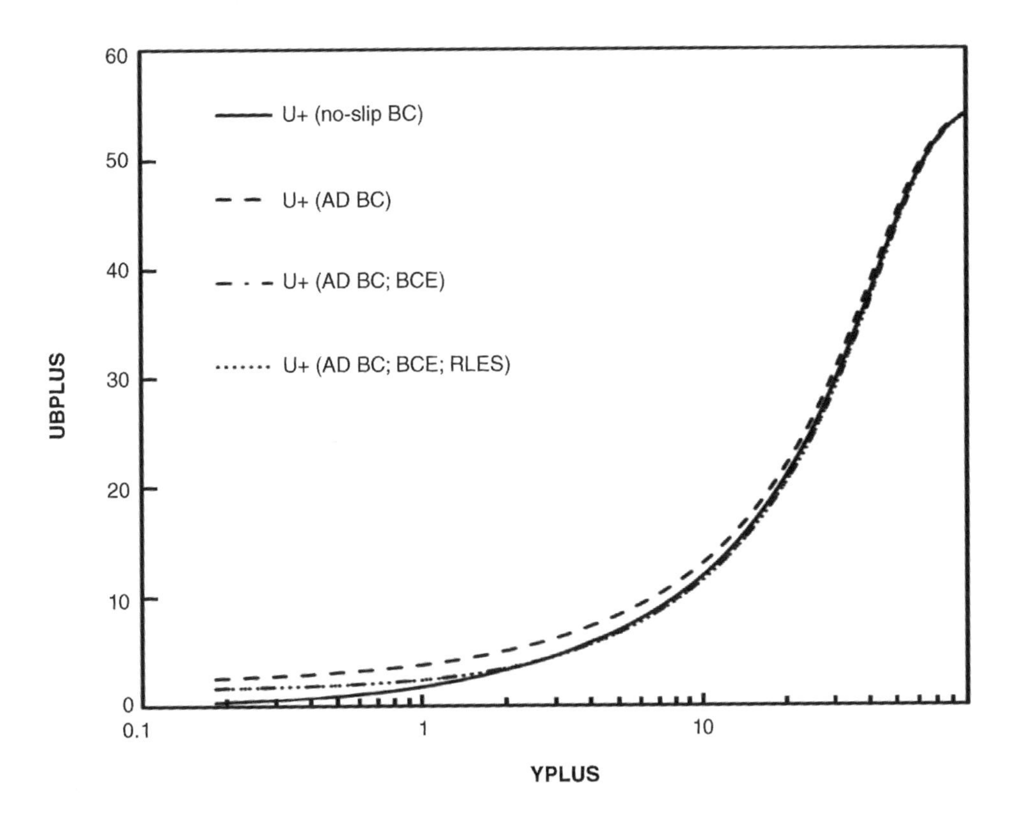

Figure 9.37 AD-GWS^h-BCE-DBC algorithm validation, $n=3$ laminar flat plate flow, Re = E+06, assessment sequence, $\overline{u_1^+}^h(y^+)$ DOF similarity coordinate interpolation compared with unfiltered Blasius series no-slip wall BC solution

for the space filtered state variable hence BCE integrals are not appended. As EBV, encompassing bounded domain BCs must be identified for $O(\delta^2)$ *resolved-unresolved scale interaction* state variables $\overline{c}_{ij}^{n+1}$ and $\overline{v}_j^{n+1}$, overbar notation now added for clarity.

Cross stress tensor pair derivation is initialized by AD definition of *non-physical* unfiltered c_{ij}

$$c_{ij} = \overline{c}_{ij} - \frac{\delta^2}{4\gamma}\nabla^2\overline{c}_{ij} + O(\delta^4) \tag{9.147}$$

Differential definition alteration of (filtered) c_{ij} EBV PDE is

$$\mathcal{D}(\overline{c}_{ij}) \equiv -\nabla^2\overline{c}_{ij} + \frac{4\gamma}{\delta^2}\overline{c}_{ij} - 2\frac{\partial\overline{u}_i}{\partial x_k}\frac{\partial\overline{u}_j}{\partial x_k} = 0 \tag{9.148}$$

Multiplying (9.148) through by $\delta^2/4\gamma$ and subtracting from (9.147) generates at any time t

$$c_{ij} = \frac{\delta^2}{2\gamma}\left(\frac{\partial\overline{u}_i}{\partial x_k}\frac{\partial\overline{u}_j}{\partial x_k}\right) + O(\delta^4) \tag{9.149}$$

Thus, strictly mathematical definition (9.147) is determined to $O(\delta^4)$ to be a constant times resolved scale velocity vector strain rate products involving gaussian filter measure data.

Domain bounding surface non-homogeneous Dirichlet BC DOF prediction on $\Omega^+ \subset \mathbb{R}^n$ is convolution generated

$$\begin{aligned}\bar{c}_{ij}(\mathbf{x}_b \in \partial\Omega_{\hat{n}_k}) &\equiv \left(g_\delta * c_{ij}\right), (\mathbf{x}_b \in \partial\Omega_{\hat{n}_k}) \\ &\equiv \int_\Omega g_\delta(x_m - x_l) c_{ij}(x_l, t) \mathrm{d}\tau, \mathbf{x}_b \in \partial\Omega_{\hat{n}_k} \\ &= \frac{\delta^2}{2\gamma} \int_\Omega g_\delta(x_m - x_l) \left(\frac{\partial \bar{u}_i}{\partial x_k} \frac{\partial \bar{u}_j}{\partial x_k}\right)_{x_l, t} \mathrm{d}\tau, \mathbf{x}_b \in \partial\Omega_{\hat{n}_k}\end{aligned} \tag{9.150}$$

The strain rate products in (9.150) integrand are evaluated via $\mathrm{GWS}^N \Rightarrow \mathrm{GWS}^h$ on its differential definition

$$\mathcal{D}\left(d\bar{u}_i d\bar{u}_j\right) \equiv d\bar{u}_i d\bar{u}_j - \frac{\partial \bar{u}_i}{\partial x_k} \frac{\partial \bar{u}_j}{\partial x_k} = 0 \tag{9.151}$$

which leads to the element matrix statement

$$\begin{aligned}\left\{\mathrm{F\_QD\bar{U}}I\mathrm{D\bar{U}}J\right\}_e &= \det_e[\mathrm{M}200d]\left\{\mathrm{Q\_D\bar{U}}I\mathrm{D\bar{U}}J\right\}_e \\ &\quad - \frac{1}{\det_e}\left\{\bar{\mathrm{U}}I\right\}_e^T [\mathrm{M}3K0Kd]_e \left\{\bar{\mathrm{U}}J\right\}_e, 1 \le (I, J) \le n \\ \left[\mathrm{JAC\_QD\bar{U}}I\mathrm{D\bar{U}}J\right]_e &= \det_e[\mathrm{M}200d]\end{aligned} \tag{9.152}$$

Assembled and solved, constraining only those DOF on $\partial\Omega$ possessing a nil entry, yields the convolution integrand distribution on $\Omega^h \cup \partial\Omega^h$

$$\left(\frac{\partial \bar{u}_i}{\partial x_k} \frac{\partial \bar{u}_j}{\partial x_k}\right)_{x,t}^h = \mathrm{S}_e\left(\{N(\mathbf{x})\}^T \left\{\mathrm{D\bar{U}}I\mathrm{D\bar{U}}J(t)\right\}_e\right) \tag{9.153}$$

Extending these DOF data to $\Omega^+ \subset \mathbb{R}^n$, evaluation via numerical quadrature (9.136) generates the requisite $\bar{c}_{ij}^h$ non-homogeneous Dirichlet BC DOF at all pertinent $\mathbf{x}_b \in \partial\Omega$

$$\begin{aligned}\overline{CIJ}(t)\big|_{x_b \in \partial\Omega_{\hat{n}_k}} &= \frac{\delta^2}{2\gamma} \int_{\Omega^+} g_\delta(x_m - x_l) \left(\frac{\partial \bar{u}_i}{\partial x_k} \frac{\partial \bar{u}_j}{\partial x_k}\right)^h \mathrm{d}\tau, \mathbf{x}_b \in \partial\Omega_{\hat{n}_k} \\ &= \frac{\delta^2}{2\gamma} \sum_{l=1}^{N_x} \sum_{m=1}^{N_y} \sum_{n=1}^{N_z} \begin{pmatrix} w_l w_m w_n g_\delta(x_b - x_l, y_b - y_m, z_b - z_n) \\ \times \mathrm{S}_e\left(\{N(\mathbf{x})\}^T \left\{\mathrm{D\bar{U}}I\mathrm{D\bar{U}}J(t)\right\}_e\right) \end{pmatrix}_{\mathbf{x}_b \in \partial\Omega_{\hat{n}_k}}\end{aligned} \tag{9.154}$$

The sequence (9.147–9.149) also identifies *ar*LES theory resolved-unresolved scale interaction *unfiltered* state variable vectors

$$v_j^\Theta(\mathbf{x}, t) = \frac{\delta^2}{2\gamma} \frac{\partial \bar{u}_j}{\partial x_k} \frac{\partial \bar{\Theta}}{\partial x_k} + O(\delta^4) \tag{9.155}$$

$$v_j^Y(\mathbf{x}, t) = \frac{\delta^2}{2\gamma} \frac{\partial \bar{u}_j}{\partial x_k} \frac{\partial \bar{Y}}{\partial x_k} + O(\delta^4) \tag{9.156}$$

State variable alterations in (9.150–9.154) lead to required non-homogeneous Dirichlet BC DOF data at domain bounding surface nodes $\mathbf{x}_b \in \partial\Omega$

$$\overline{\mathrm{V}^{\Theta}J(t)}\Big|_{\mathbf{x}_b \in \partial\Omega_{\hat{n}_k}} = \frac{\delta^2}{2\gamma}\int_{\Omega^+} g_\delta(x_m - x_l)\left(\frac{\partial\overline{\Theta}}{\partial x_k}\frac{\partial \overline{u}_j}{\partial x_k}\right)^h d\tau, \mathbf{x}_b \in \partial\Omega_{\hat{n}_k}$$
$$= \frac{\delta^2}{2\gamma}\sum_{l=1}^{N_x}\sum_{m=1}^{N_y}\sum_{n=1}^{N_z}\begin{pmatrix} w_l w_m w_n g_\delta(x_b - x_l, y_b - y_m, z_b - z_n) \\ \times \mathrm{S}_e\left(\{N(\mathbf{x})\}^T\{\mathrm{D\overline{\Theta}D\overline{U}}J(t)\}_e\right)\end{pmatrix}_{\mathbf{x}_b \in \partial\Omega_{\hat{n}_k}} \tag{9.157}$$

$$\overline{\mathrm{V}^{Y}J(t)}\Big|_{\mathbf{x}_b \in \partial\Omega_{\hat{n}_k}} = \frac{\delta^2}{2\gamma}\int_{\Omega^+} g_\delta(x_m - x_l)\left(\frac{\partial\overline{Y}}{\partial x_k}\frac{\partial \overline{u}_j}{\partial x_k}\right)^h d\tau, \mathbf{x}_b \in \partial\Omega_{\hat{n}_k}$$
$$= \frac{\delta^2}{2\gamma}\sum_{l=1}^{N_x}\sum_{m=1}^{N_y}\sum_{n=1}^{N_z}\begin{pmatrix} w_l w_m w_n g_\delta(x_b - x_l, y_b - y_m, z_b - z_n) \\ \times \mathrm{S}_e\left(\{N(\mathbf{x})\}^T\{\mathrm{D\overline{Y}D\overline{U}}J(t)\}_e\right)\end{pmatrix}_{\mathbf{x}_b \in \partial\Omega_{\hat{n}_k}} \tag{9.158}$$

With identified AD-GWS^h-BCE-DBC algorithm, *ar*LES theory I-EBV/EBV *m*PDE system (9.103–9.108), (9.16–9.18), (9.64), (9.70–9.71) is confirmed *well-posed* for complete $O(1,\ \delta^2)$ state variable $\forall$ Re for $k=1$ basis $m\mathrm{GWS}^h + \theta\mathrm{TS}$ algorithm, Section 9.9–9.10. QED.

9.20 Well-Posed *ar*LES Theory $n = 3$ Validation

Well-posed *ar*LES theory $m\mathrm{GWS}^h + \theta\mathrm{TS}$ algorithm validation addresses the $8\times 1\times$ transversely wide thermal cavity on $n=3$, Figure 9.38. Implementing validated measure constraint $\delta \equiv \mathrm{n}h$ AD-GWS^h-BCE-DBC algorithm for all $O(1,\ \delta^2)$ state variables enables rigorous replacement of *ar*LES-DES formulation, Section 9.11, $\forall$ Re.

Unfiltered potential temperature Θ hot/cold wall Dirichlet BC specification is uniform on opposing cavity vertical surfaces. Adiabatic Neumann BC is enforced on top, bottom and back walls, the latter plane $\mathrm{z}=0$. Plane $\mathrm{z}=8$ is the symmetry mid-plane of an assumed $8\times 1\times 16$ cavity. Symmetry plane BC specifications are zero through flow, $u_3=0$, with all other state variable member BCs homogeneous Neumann.

Cavity $\mathrm{D}M^h$ potential function BC is everywhere homogeneous Neumann. Reported *a posteriori* data are generated using the simplified constraint option (7.45), hence $\mathcal{L}(\overline{P})$ (9.108) with uniformly homogeneous Neumann BCs is not post-processed. Cold start IC is linear interpolation of unfiltered Θ Dirichlet BCs with all other *ar*LES theory state variable DOF defined zero.

Recalling Section 9.11, for $\mathrm{Ra} = \mathrm{E}{+}05 < \mathrm{Ra}_{\mathrm{CR}}$, hence $\mathrm{Re} < \mathrm{E}{+}03$, *ar*LES-DES prediction confirms theory in generating a single recirculation cell filling the cavity. The DOF interpolation exhibits exact mirror symmetry with vertically oriented "isotherm finger" pair undergoing bounded oscillation, Figure 9.8 right. Transition to unsteady for $\mathrm{Ra} = 4\ \mathrm{E}{+}05 > \mathrm{Ra}_{\mathrm{CR}} \approx 3.1\ \mathrm{E}{+}05$ is confirmed by *ar*LES-DES data.

For $\mathrm{Ra} = 4\mathrm{E}{+}05$, just larger than $\mathrm{Ra}_{\mathrm{CR}}$, $\mathrm{Re} = 756$, $\mathrm{Pr} = 0.71$, well-posed *ar*LES theory $8\times 1\times 8$ cavity symmetry plane non-D $\overline{\Theta}^h(x, y, z = 8, n\Delta t < 170\,\mathrm{s})$ prediction exhibits exact mirror symmetry, Figure 9.39. Realizing that $n=3$ hot wall BC is now on the right, this bounded oscillatory symmetry plane solution DOF interpolation exhibits excellent

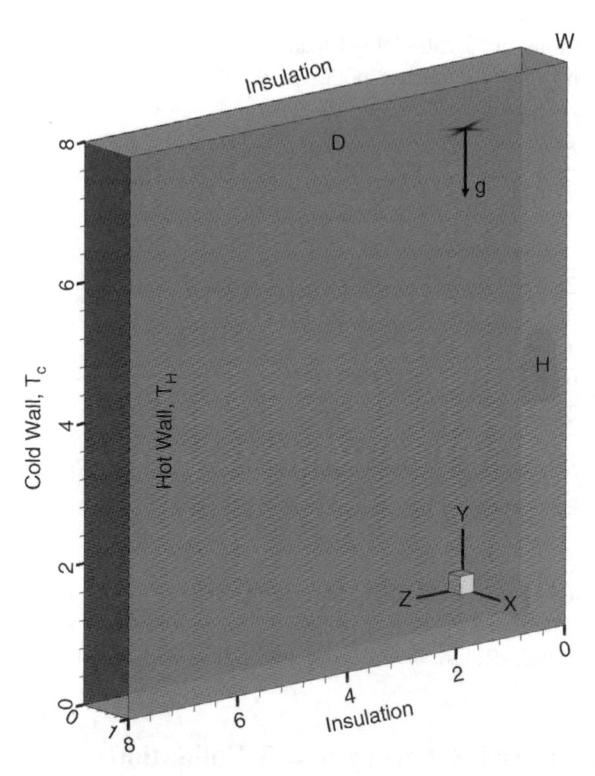

Figure 9.38 $8 \times 1 \times 16$ thermal cavity symmetric half domain, $n = 3$, unfiltered NS potential temperature BCs defined

quantitative agreement with *ar*LES-DES prediction, Figure 9.8 right. Continued $n = 3$ execution to $n\Delta t > 200$ s transitions $\overline{\Theta}^h(\mathbf{x}, n\Delta t)$ bounded oscillation to unsteady periodic, Sekachev (2013, Section 6.4.5), *quantitatively* validating well-posed *ar*LES theory.

Well-posed *ar*LES theory $8 \times 1 \times 8$ thermal cavity specification for a potentially "turbulent" Reynolds number, Re ≥ E+04, results for Ra ≡ E+08. The *a priori* data specifications are ΔT ≡ 111 K (200 °R), standard atmosphere air, Pr = 0.71, and L = 0.218 m (0.68 ft). For these data the Boussinesq buoyancy relation $\mathrm{Gr} \equiv \mathrm{Re}^2$ yields U = 0.92 m/s (2.87 f/s) hence Re = 11,824 > E+04.

The $k = 1$ TP basis $\mathrm{M} = 206 \times 64 \times 40$ ($y \times x \times z$) mesh is z direction (only) pr > 1 adapted to the symmetry plane z = 8. At all walls with no-slip wall BC the discretization is wall adjacent augmented as defined, Figure 9.33. Simulation requires gaussian filter *uniform* measure δ definition. Reported DNS *a posteriori* data for mid-height horizontal plane vertical and horizontal velocity distributions for a 4×1 thermal cavity at Ra = E+10, Trias *et al.* (2010), predict wall remote velocity extrema. For AD-GWS^h-BCE-DBC algorithm DOF node located δ/2 wall distant to not occur inside the DNS prediction, wall normal uniform mesh measure is defined $h = 0.0027$ m (0.0085 ft, non-D 0.0125). For algorithm *best practice* measure constraint $\delta \equiv 4h$, $\delta = 0.011$ m (0.034 ft, non-D 0.05) places AD-GWS^h-BCE-DBC algorithm computation at abscissa coordinate E−02, Figure 9.34, with predicted error level ~0.5%.

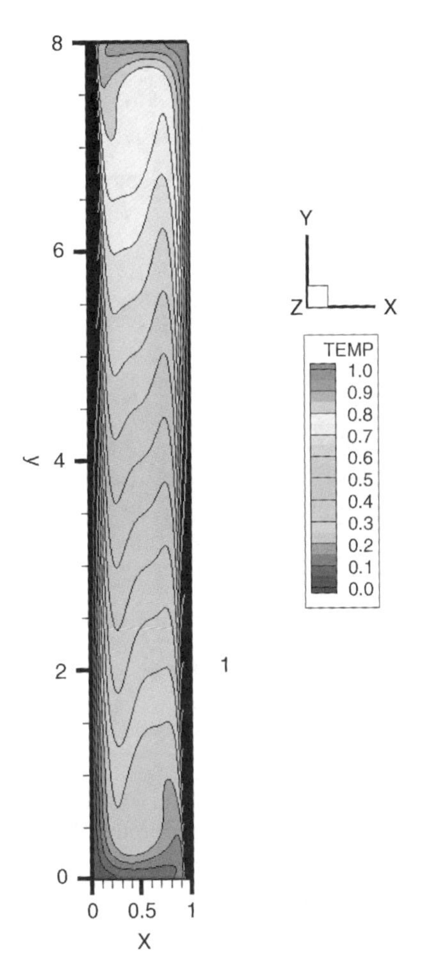

Figure 9.39 Well-posed *ar*LES theory quantitative validation, $8 \times 1 \times 8$ thermal cavity symmetry plane $\overline{\Theta}^h(x, y, z = 8, n\Delta t < 170\,\text{s})$ flood with contours, Ra = 4.0 E+05 > $\text{Ra}_{\text{CR}} \approx 3.1$ E+05, Re = 756, Pr = 0.71

AD-GWS^h-BCE-DBC algorithm non-homogeneous Dirichlet BCs (9.111–9.112) are generated for resolved scale velocity wall tangential DOF *only*. Wall normal velocity component DOF BC is homogeneous Dirichlet. Non-homogeneous Dirichlet BCs for $\overline{\Theta}^h$ DOF are AD-GWS^h-BCE-DBC algorithm generated on cavity surfaces with *unfiltered* ΔT temperature data specification.

Ra-continuation for decile increments from Ra = E+04 leads to Ra = E+08 solution initiation. Following ample time for dynamic evolution to Ra = E+08, well-posed *ar*LES theory unsteady, periodic solution $\overline{\Theta}^h(\mathbf{x}, n\Delta t)$ exhibits very large wall normal gradients bounding an essentially isothermal cavity mid-volume. In unitized scale perspective, $\overline{\Theta}^h(\mathbf{x}, n\Delta t)$ snapshot floods with contours for DOF on planes z = 8, 4, δ/2 are graphed in Figure 9.40. Contour distributions confirm that wall normal "isotherm

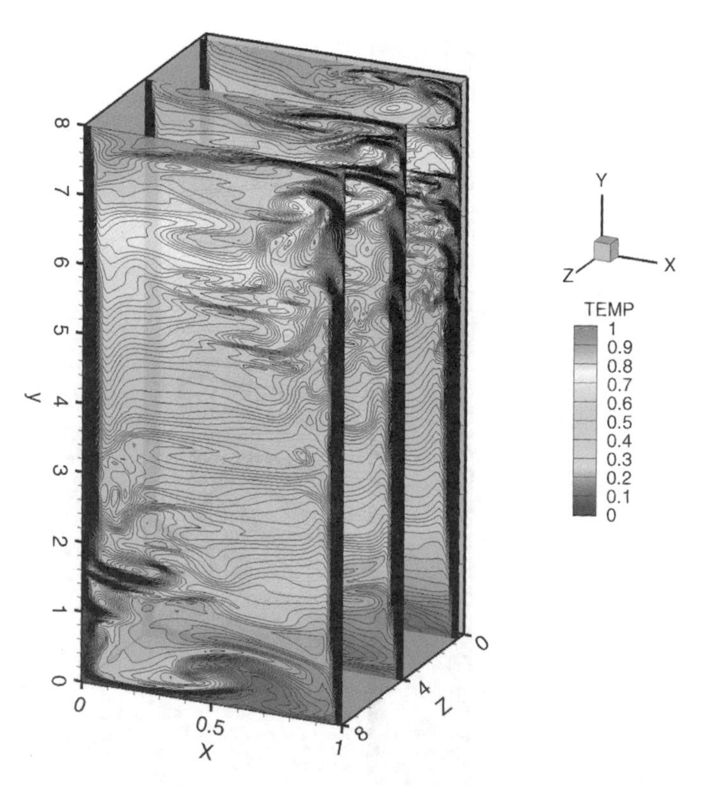

Figure 9.40 $8 \times 1 \times 8$ symmetric half thermal cavity, Ra = E+08, Re = 11,824, Pr = 0.71, well-posed *ar*LES theory $\overline{\Theta}^h(x, y, n\Delta t)$ snapshot unitized scale perspective flood with contours, DOF on planes z = 8, 4, δ/2

fingers" penetrate into the cavity interior, and that $\overline{\Theta}^h(\mathbf{x}, n\Delta t)$ is now only approximately mirror symmetric.

Enhanced visual interpretation of the intense thermal dynamics generated for cavity Ra = E+08 specification is supported by unitized scale perspectives of $\overline{\Theta}^h(\mathbf{x}, n\Delta t)$ distributions on cavity select horizontal planes, Figure 9.41. Graphed data emanate from DOF surfaces displaced from cavity walls by $|\Delta x_i| = \delta/2$, node 5 in Figure 9.35. Viewpoint perspective is towards the cavity end wall. Data presentation as surfaces clearly confirms temperature variation in the z direction is very modest compared to - elevation distinctions. Further, well organized temperature structures confirmed existent at the hot wall bottom, left, also the top of the cold wall, right, lose this organization in traversing vertical walls on which unfiltered temperature specification ΔT is imposed.

*ar*LES theory solution resolved scale velocity vector $\overline{\mathbf{u}}^h(\mathbf{x}, n\Delta t)$ snapshot unitized scale perspective planar flood *magnitude* interpolation on planes z = 8, 4, δ/2, Figure 9.42, clearly delineates wall jets traversing the hot wall upwards, right, and cold wall downwards, left. The wall normal thermal finger features in $\overline{\Theta}^h(\mathbf{x}, n\Delta t)$ coincide with wall jet separations, each of which generates a wall traversing roll vortex. Associated instability is

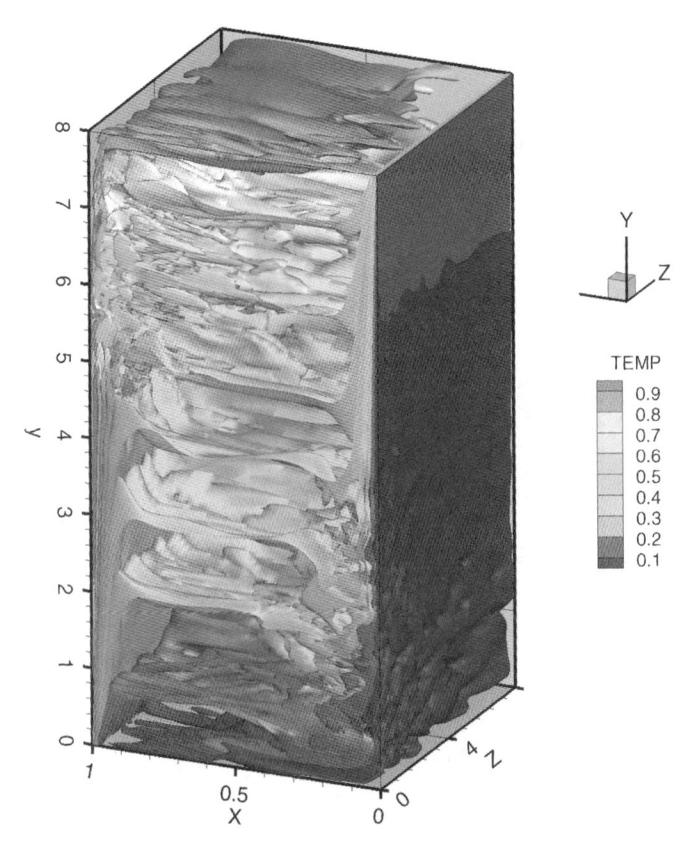

Figure 9.41 $8 \times 1 \times 8$ symmetric half thermal cavity, Ra = E+08, Re = 11,824, Pr = 0.71, well-posed *ar*LES theory select horizontal plane $\overline{\Theta}^h(\mathbf{x}, n\Delta t)$ surface perspectives, from DOF vertical planes displaced from cavity walls by $|\Delta x_i| = \delta/2$

predicted more dynamic on the hot wall. Resultant local cavity corner distinctions in the z-component of velocity vector $\overline{\mathbf{u}}^h(\mathbf{x}, n\Delta t)$ are responsible for the absence of mirror symmetry observed in $\overline{\Theta}^h(\mathbf{x}, n\Delta t)$, Figure 9.40.

From plane surfaces displaced from ΔT specified cavity surfaces by $|\Delta x_i| = \delta/2$, specifically node 5 in Figure 9.33, resolved scale velocity magnitude $|\overline{\mathbf{u}}^h(\mathbf{x}, n\Delta t)|$ distribution perspectives, Figure 9.43, confirm cavity central dynamic penetration at elevations corresponding to wall jet separations. Cavity end wall suppression of wall jet creation, left graphic, is clearly distinguished from symmetry plane data, right image. This data presentation additionally quantifies the more intense instability within the hot wall jet.

Symmetry plane features of well-posed *ar*LES theory Ra = E+08, Re = 11,824 $\{\overline{q}^h(\mathbf{x}, t)\}$ snapshot are clearly distinguished from the *ar*LES-DES prediction, Figure 9.17. The directly comparative unitized scale projection of symmetry plane resolved scale velocity vector on temperature flood is Figure 9.44. Current data clearly quantify presence of more robust cavity top/bottom surface wall jets as well as cavity isothermal mid-region essential absence of the vortex structure distribution evident in Figure 9.15 right. These data additional render clearly visible snapshot observed non-symmetric temperature distribution.

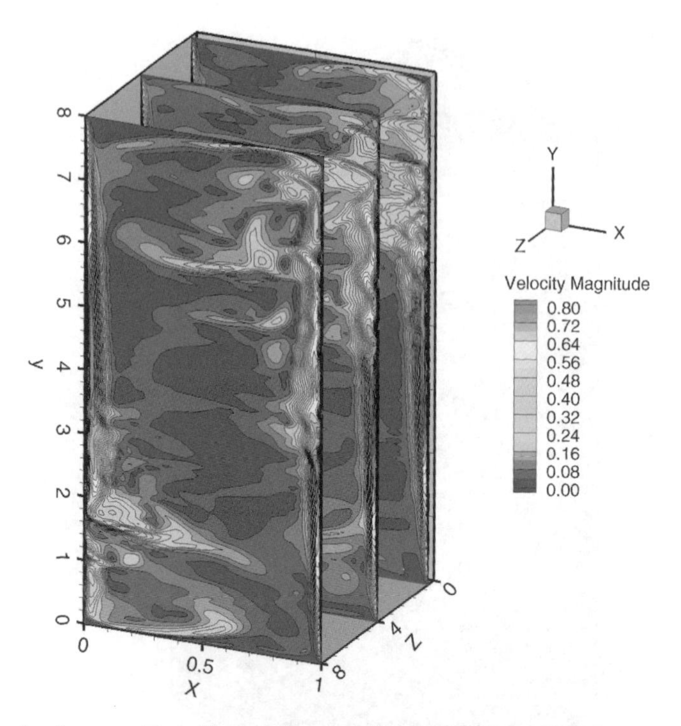

Figure 9.42 $8 \times 1 \times 8$ symmetric half thermal cavity, Ra = E+08, Re = 11,824, Pr = 0.71, well-posed *ar*LES theory $|\bar{\mathbf{u}}^h(x, y, n\Delta t)|$ snapshot flood with contours, DOF on planes z = 8, 4, δ/2

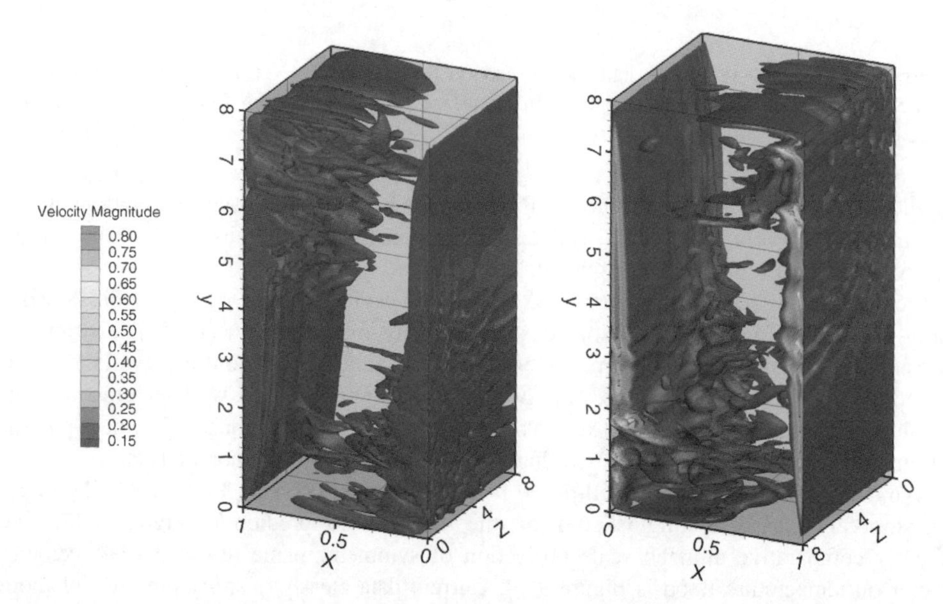

Figure 9.43 $8 \times 1 \times 8$ symmetric half thermal cavity, Ra = E+08, Re = 11,824, Pr = 0.71, well-posed *ar*LES theory, $|\bar{\mathbf{u}}^h(\mathbf{x}, n\Delta t)|$ isothermal surface perspectives on planes displaced by $|\Delta x_i| = \delta/2$ from walls: left, wall plane; right, symmetry plane

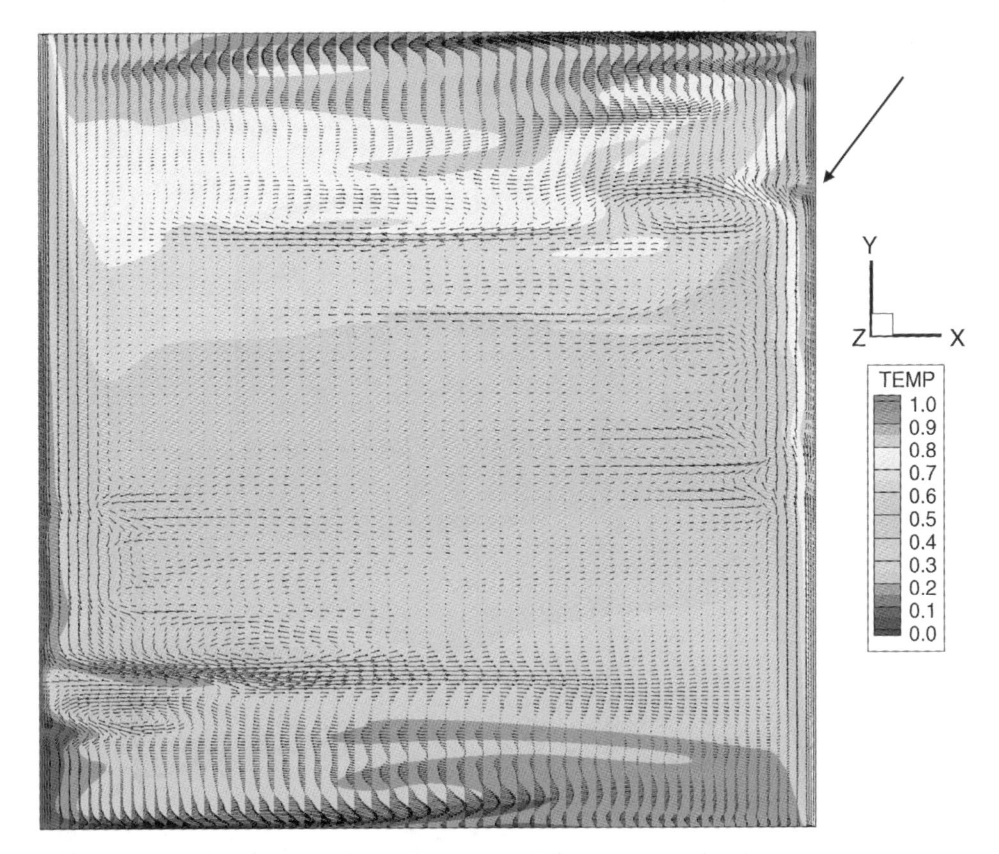

Figure 9.44 $8 \times 1 \times 8$ symmetric half thermal cavity, Ra = E+08, Re = 11,824, Pr = 0.71, well-posed *ar*LES theory, unitized symmetry plane $\overline{\Theta}^h(\mathbf{x}, n\Delta t)$ flood with velocity vector (all) overlay, arrow denotes hot wall jet separation thermal feature

Distinctions summarily result from well-posed theory prediction-implementation of AD-GWS^h-BCE-DBC algorithm for the complete *ar*LES $O(1, \delta^2)$ state variable at bounding surface nodes $\mathbf{x}_b \in \partial\Omega$. Cavity wall surface $\overline{\mathbf{u}}^h(\mathbf{x}, n\Delta t) \bullet \hat{\mathbf{s}}$ perspective, Figure 9.45, for $\hat{\mathbf{s}}$ unit tangent vector, confirms computed non-zero DOF at all nodes $\mathbf{x}_b \in \partial\Omega$. Vector density also visualizes mesh resolution along with cavity corner induced floor-ceiling wall jets *anti-parallel* to those in Figure 9.44, close-up bottom right. The homogeneous Neumann BC symmetry plane perspective $\overline{\mathbf{u}}^h(\mathbf{x}, n\Delta t) \bullet \hat{\mathbf{s}}$ DOF close-up, upper right, scales surface BC magnitude distinctions.

A fundamental axiom when solving an EBV PDE is, "You get out what you put in, *especially* on the boundaries, specifically, the BCs!" As O(E+04) time steps $n\Delta t$ are required to evolve this Ra = E+08 solution from its initiation, accumulation from AD-GWS^h-BCE-DBC algorithm *discrete* BC generation at $\mathbf{x}_b \in \partial\Omega$ may well contribute to observed lack of exact mirror symmetry. This conjecture can be quantified via a *regular* solution adapted mesh refinement study for $\delta = 4h$, *a posteriori* data for which is not available at this writing.

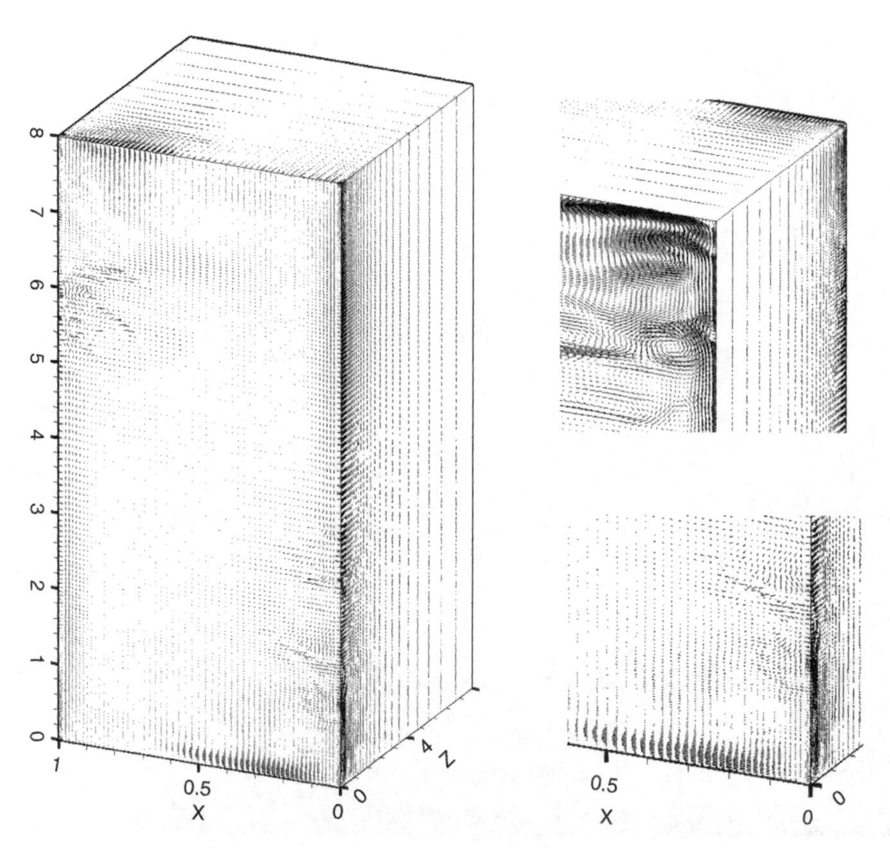

Figure 9.45 $8 \times 1 \times 8$ symmetric half thermal cavity, Ra = E+08, Re = 11,824, Pr = 0.71, well-posed *ar*LES theory, $\overline{\mathbf{u}}^h(\mathbf{x}, n\Delta t) \bullet \hat{\mathbf{s}}$ DOF surface distributions, $\hat{\mathbf{s}}$ surface tangent unit vector

As predicted by cavity comparative DNS data, Trias *et al.* (2010), selected mesh M with constraint $\delta \equiv 4h$ enables adequate resolved scale near wall $\overline{\mathbf{u}}^h \bullet \hat{j}$ distribution resolution. Figure 9.46 graphs $z = 8$ symmetry plane $\overline{\mathbf{u}}^h \bullet \hat{j} = \overline{v}^h(x, y, n\Delta t)$ DOF near hot wall distributions on horizontal planes $y = 1$, 3, 5, 7. The $y = 5$, 7 profiles bound predicted thermal finger ejection occurring at right wall $y \approx 6.5$ wall jet separation, Figure 9.44.

Well-posed *ar*LES theory assertion $\forall$ Re is validated for the Ra = E+08, Re = 11,824 solution snapshot data. Resolved scale velocity hot wall jet $\overline{v^+}^h$ DOF distributions at cited four horizontal planes are BL similarity coordinate log y^+ compared, Figure 9.47 top, to validated RaNS turbulent BL solution v^+ (labeled Wilcox) and turbulent BL sublayer similarity correlation $y^+ = v^+$. Not surprisingly, $y^+ = v^+$ *is* the Blasius series laminar BL solution similarity correlation for $y^+ \leq 10$, Sekachev (2013, Figure B.56).

Well-posed *ar*LES theory Re = 11,824 solution hot wall jet $\overline{v^+}^h(x, y = 1, n\Delta t)$ DOF interpolation distribution, Figure 9.47 bottom, in exactly paralleling correlation $y^+ = v^+$ clearly corresponds to a laminar BL profile in wall adjacent reach. In clear distinction, $\overline{v^+}^h(x, y = 5, n\Delta t)$ DOF interpolation distribution asymptotes to RaNS turbulent BL *validated* solution v^+ (Wilcox) in departing sublayer reach.

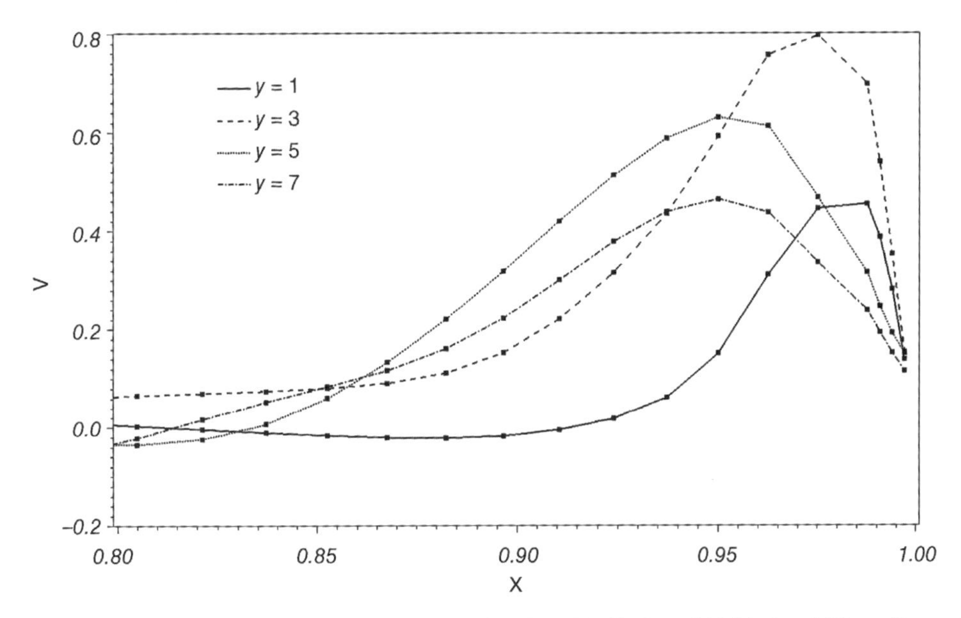

Figure 9.46 $8 \times 1 \times 8$ symmetric half thermal cavity, Ra = E+08, Re = 11,824, Pr = 0.71, well-posed *ar*LES theory hot wall resolved scale velocity wall jet profiles $\bar{\mathbf{u}}^h \bullet \hat{j} = \bar{v}^h(x, y, n\Delta t)$, plane elevations $y = 1, 3, 5, 7$

Thereby, well-posed arLES theory $8 \times 1 \times 16$ thermal cavity symmetry plane Re = 11,824 near wall resolved scale velocity DOF distributions validate this analytical theory-intrinsic prediction of laminar-turbulent *transition*. This is enabled solely by AD-GWS^h-BCE-DBC algorithm prediction of resolved scale tangential velocity *wall node* DOF increase from $v^+(y=1) \sim 1$ to $v^+(y=5) = 3$ at *unfiltered* NS no-slip BC nodes $\mathbf{x}_b \in \partial\Omega$. These data are necessary but not sufficient to fully *validate* analytical theory assertion ∀ Re applicability without any modeling component.

9.21 Well-Posed *ar*LES Theory *n* = 3 Diagnostics

Well-posed *ar*LES theory $O(\delta^2)$ state variable members plus SFS tensor/vector $O(\delta^3)$ closure, for $C_S \equiv O(\delta)$, explicitly impact $O(1)$ state variable solution process at each time step. For cavity solution snapshot *a posteriori* data, $O(\delta^2)$ state variable cartesian resolution $\left\{q^h_{\delta^2}(\mathbf{x}, n\Delta t)\right\} \Rightarrow \left\{c^h_{ij}, v^h_j\right\}$ *extrema* predict $O(1)$ state variable relative impact. In decreasing significance $O(\delta^2)$ non-D data ranges $\{0.0 \Leftrightarrow 0.50\}$ for c^h_{22}, $\{-0.15 \Leftrightarrow 0.15\}$ for c^h_{23} and $\{0.0 \Leftrightarrow 0.05\}$ for v^h_2.

Unitized scale perspective c^h_{22} and c^h_{23} *isosurface* distributions viewed parallel to z axis are compared, Figure 9.48 top. To no surprise c^h_{22} extrema are concentrated in the wall jet distribution. Conversely, those for c^h_{23} are buried in the cavity interior with distribution giving the visual appearance of chaos. Comparable isosurface graphs for cross stress tensor diagonal entries c^h_{11} and c^h_{33}, Figure 9.48 bottom, exhibit greater distinction of finite scale distributions. DOF extrema range $\{0.0 \Leftrightarrow 0.08\}$ and $\{0.0 \Leftrightarrow 0.001\}$ respectively, and in

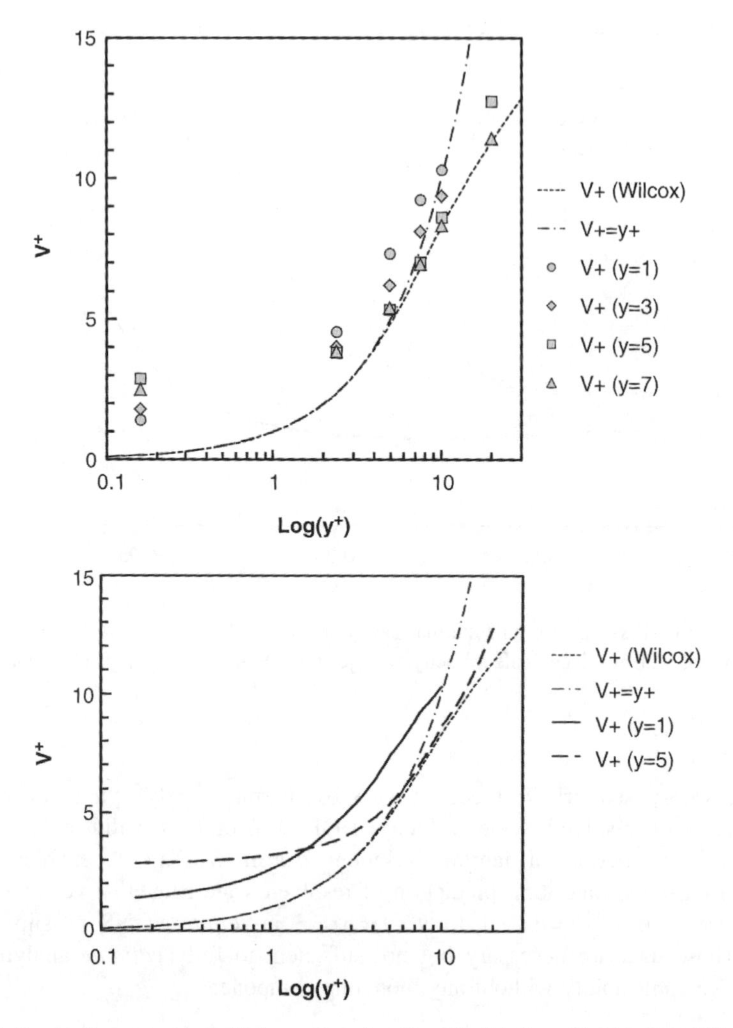

Figure 9.47 $8\times1\times8$ symmetric half thermal cavity, Ra = E+08, Re = 11,824, Pr = 0.71, well-posed *ar*LES theory solution, hot wall $\overline{v}^h(x, y = 1, 3, 5, 7, n\Delta t)$ near wall distributions, similarity coordinates: top, DOF data; bottom, DOF distribution interpolations

accord with the NS tensor field theory *necessary* requirement all diagonal entries $c_{ij}^h(n\Delta t), i = j$ are *ar*LES theory predicted *symmetric positive definite*.

Cross stress tensor pair shear c_{12}^h and c_{13}^h isosurface perspectives are compared, Figure 9.49 top. Extrema ranges $\{-0.08 \Leftrightarrow 0.03,\ -0.03 \Leftrightarrow 0.02\}$ confirm resolved scale state variable impact is secondary to $c_{2\,3}^h$ by a factor of ~5. Theory companion thermal vector v_2^h and v_3^h extrema are $\{0.0 \Leftrightarrow 0.05,\ -0.02 \Leftrightarrow 0.02\}$, perspectives at Figure 9.49 bottom. Thermal vector v_1^h ranges $\{0.0 \Leftrightarrow 0.005\}$, hence v_3^h dominates by ~2.

Well-posed *ar*LES theory solution diagnostics focus is thermal finger at hot wall jet separation, Figure 9.44 arrow, which due to absence of a transitional SGS tensor closure model is inaccurately predicted by *ar*LES-DES formulation, Figure 9.16. Generated *a*

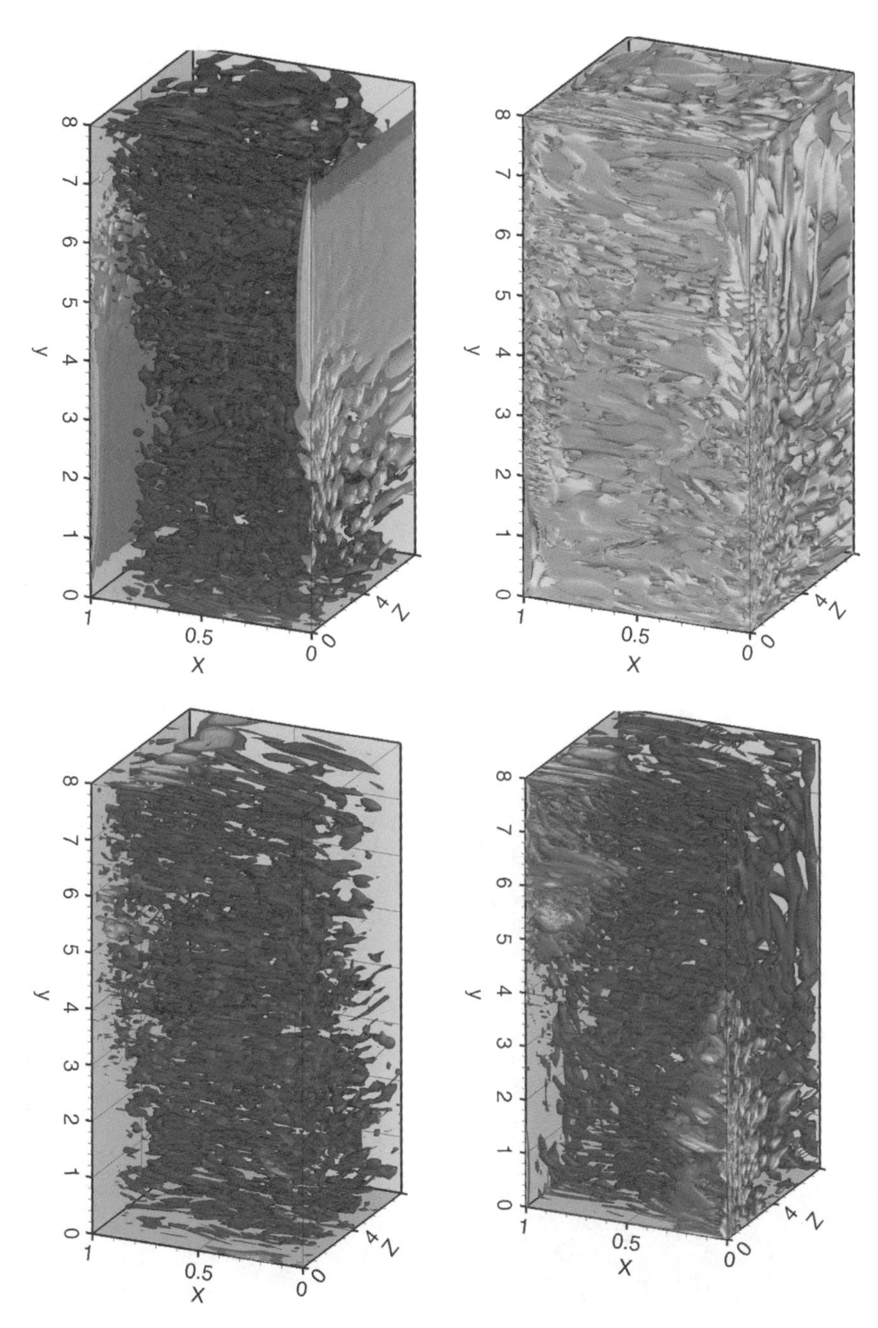

Figure 9.48 $8 \times 1 \times 8$ symmetric half thermal cavity, Ra = E+08, Re = 11,824, Pr = 0.71, well-posed *ar*LES theory $O(\delta^2)$ state variable solution snapshot isosurface perspective: clockwise from top left, c^h_{22}, c^h_{23}, c^h_{33}, c^h_{11}

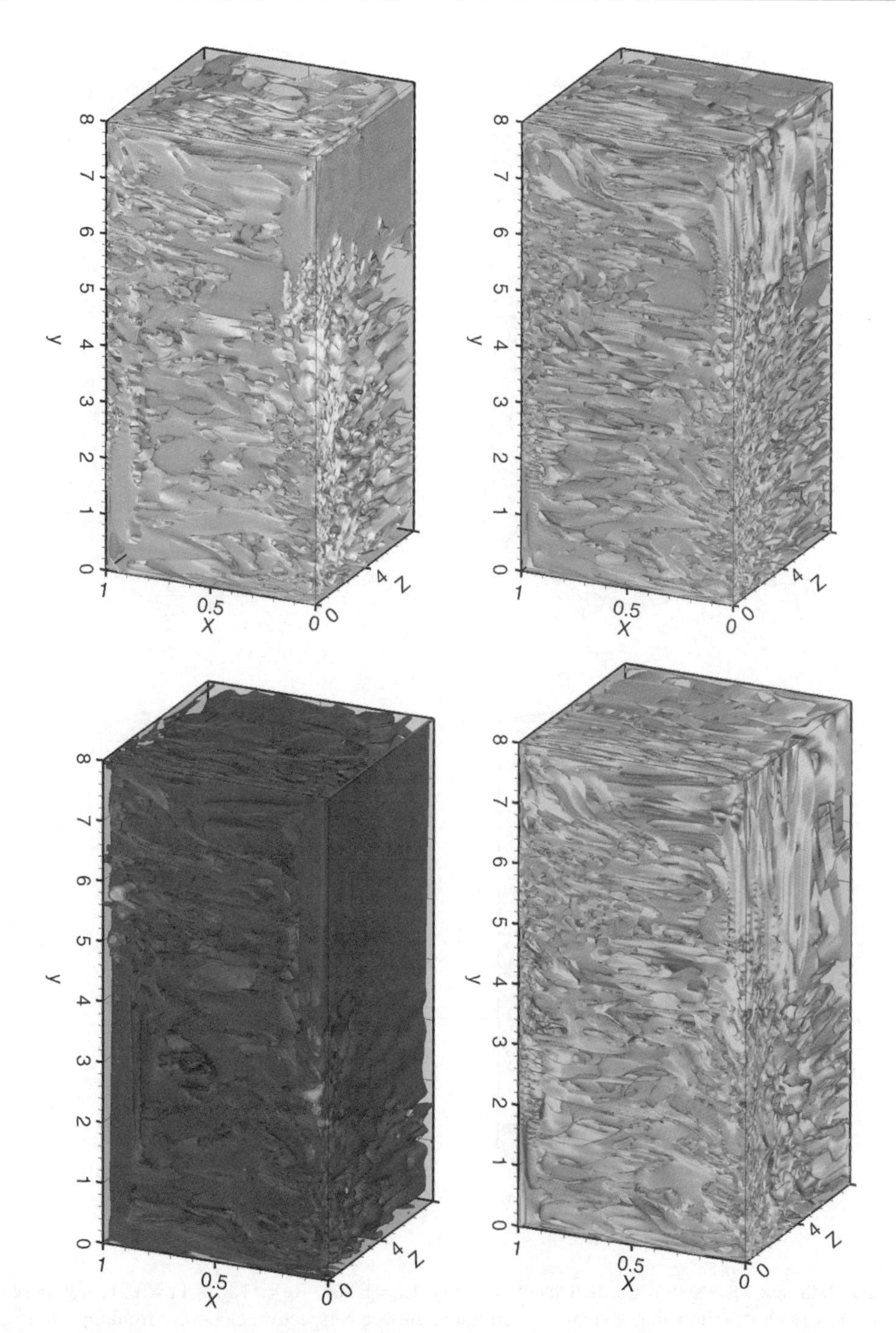

Figure 9.49 $8\times1\times8$ symmetric half thermal cavity, Ra = E+08, Re = 11,824, Pr = 0.71, well-posed *ar*LES theory $O(\delta^2)$ state variable solution snapshot isosurface perspective: clockwise from top left, c_{12}^h, c_{13}^h, v_3^h, v_2^h

posteriori data quantify relative magnitude and distribution of theory $O(1, \delta^2, \delta^3)$ state variable contributions to *kinetic* and *dissipative* flux vectors, (9.75–9.76). DOF interpolation with velocity vector overlay, Figure 9.50, supports quantitative assessments:

- top left: resolved scale temperature non-D $\overline{\Theta}^h$ is $O(10^0)$, dimensional $O(10^2)$, ample mesh resolution is evident

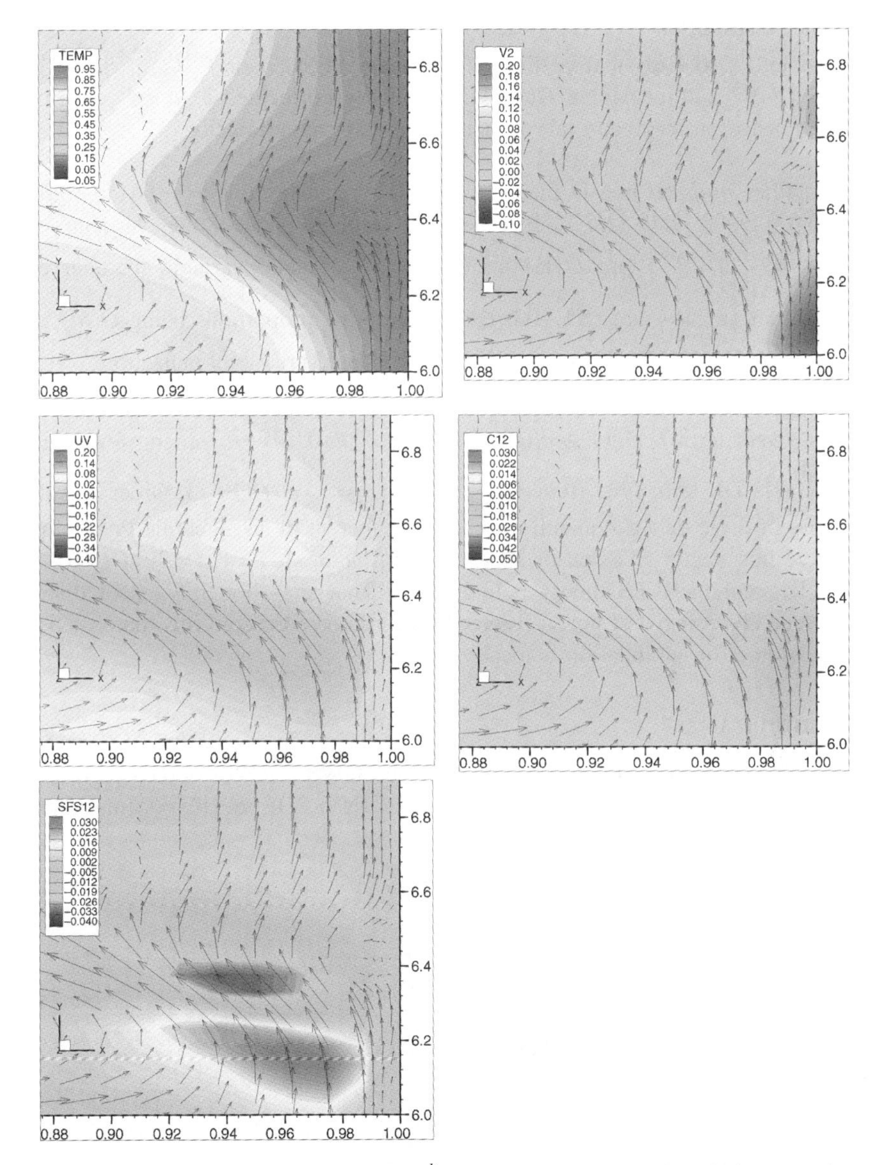

Figure 9.50 Well-posed *ar*LES theory $m\text{GWS}^h + \theta\text{TS}$ periodic unsteady solution snapshot at hot wall thermal event, close-in flood with velocity vector overlay, $8 \times 1 \times 16$ thermal cavity symmetry plane, Ra = E+08: left to right, top to bottom: $\overline{T}$, $\overline{\overline{u}_2 T'} + \overline{u_2' \overline{T}}$, $\overline{\overline{u_1}\;\overline{u_2}}$, $\overline{\overline{u}_1 u_2'} + \overline{u_1' \overline{u}_2}$, $\overline{u_1' u_2'}$

- top right: $O(\delta^2)$ kinetic flux vector thermal vector pair $f_j^h\left(\overline{\bar{u}_j\Theta'} + \overline{u_j'\bar{\Theta}}\right)^h$, $j=2$, non-D $O(10^{-2})$, dimensional $\sim O(10^0)$, ample mesh resolution is apparent
- middle left: $O(1)$ kinetic flux vector convection nonlinearity $\bar{u}_i^h \bar{u}_j^h$ $i=1$, $j=2$, non-D $O(10^{-1})$, dimensional $O(10^{-1})$, velocity vector magnitudes in separation bubble are miniscule
- middle right: $O(\delta^2)$ kinetic flux vector shear cross stress tensor pair $f_j^h \equiv \left(\overline{\bar{u}_i u_j'} + \overline{u_i'\bar{u}_j}\right)^h$, $i=1$, $j=2$, non-D $O(10^{-2})$, dimensional $O(10^{-3})$, exhibits no significance distribution
- bottom left: $O(\delta^3)$ dissipative flux vector shear SFS tensor $f_j^{dh}\left(\overline{u_i' u_j'}^h\right)$, $i=1$, $j=2$, non-D $O(10^{-2})$, dimensional $O(10^{-3})$, essentially co-located extrema confirm dominant influence on resolved scale velocity prediction monotonicity

Gaussian filter *uniform* non-D measure $\delta = 0.05$ translates to $\delta = \sim 0.01$ m (0.0333 ft), hence $O(\delta^2) \sim$ E−03 and $O(\delta^3) >$ E−05. $O(1)$ resolved scale state variable extrema $\{\bar{q}^h(\mathbf{x}, n\Delta t)\} \Rightarrow \left\{\left|\bar{u}_j^h\right|, \bar{T}^h\right\}$ are O\{E+00, E+02\}. $O(\delta^2)$ state variable member extrema $\left\{q_{\delta^2}^h(\mathbf{x}, n\Delta t)\right\} \Rightarrow \left\{c_{ij}^h, v_j^h\right\}$ are O\{E−03, E−01\}. Extrema proportions being $O(\delta^2)$ generate *quantitative* validation of RLES-perturbation theory state variable organization.

Well-posed *ar*LES theory eliminates need for a turbulent Prandtl number hypothesis. D*E* kinetic flux vector term $f_j^h\left(\overline{\bar{u}_j\Theta'} + \overline{u_j'\bar{\Theta}}\right)^h$, $j=2$, at $\sim O(10^0)$ is $\sim$3 orders more significant than comparable D**P** term $f_j^h \equiv \left(\overline{\bar{u}_i u_j'} + \overline{u_i'\bar{u}_j}\right)^h$, $i=1$, $j=2$, at $O(10^{-3})$. Order and distribution disparity of these data do quantify limitations inherent in an algebraic Pr^t assumption.

Dissipative flux vector SFS tensor shear term $f_j^{dh}\left(\overline{u_i' u_j'}^h\right)$, $i=1$, $j=2$, at $O(10^{-3})$ is an order plus larger than RLES theory bound $O(\delta^3)$. That snapshot SFS tensor distribution extrema essentially co-exist at hot wall jet separation trajectory confirms that *m*PDE theory $O(h^2)$ dispersion error annihilation is locally robust.

9.22 Summary

Chapter content derives, implements and validates $\mathrm{GWS}^h/m\mathrm{GWS}^h + \theta\mathrm{TS}$ algorithms for space filtered incompressible-thermal Navier–Stokes (NS) PDE/*m*PDE systems. The departure point remains (9.1–9.4)

$$\mathrm{D}M: \quad \nabla \cdot \mathbf{u} = 0$$

$$\mathrm{D}\mathbf{P}: \quad \mathrm{St}\frac{\partial \mathbf{u}}{\partial t} + \nabla \cdot \mathbf{uu} + \nabla P - \frac{1}{\mathrm{Re}}\nabla \cdot \nabla \mathbf{u} + \frac{\mathrm{Gr}}{\mathrm{Re}^2}\Theta\,\widehat{\mathbf{g}} = 0$$

$$\mathrm{D}E: \quad \mathrm{St}\frac{\partial \Theta}{\partial t} + \nabla \cdot \mathbf{u}\Theta - \frac{1}{\mathrm{RePr}}\nabla \cdot \nabla \Theta - s_\Theta = 0$$

$$\mathrm{D}Y: \quad \mathrm{St}\frac{\partial Y}{\partial t} + \nabla \cdot \mathbf{u}Y - \frac{1}{\mathrm{ReSc}}\nabla \cdot \nabla Y - s_Y = 0$$

for state variable $\{q(\mathbf{x}, t)\} = \{\mathbf{u}, p, \Theta, Y\}^T$ non-D velocity vector, pressure, temperature and mass fraction with Re, Pr, Gr and Sc the well known non-dimensional groups.

Termed the *large eddy simulation* (LES) PDE system, theory intent is to resolve significant scales in generated flowfields stopping well short of mean free path. The LES system results from NS system convolution with a spatial filter of measure δ, which precisely quantifies the resolved scale. Every scale smaller than δ is unresolved leading to the precise statement of velocity vector delineation (9.5)

$$u_j(\mathbf{x},t) \equiv \overline{u}_j(\mathbf{x},t) + u_j'(\mathbf{x},t)$$

As always, the nonlinearity in **DP** (9.2) generates the closure requirement, mathematically quantified in (9.6) upon (9.5) substitution

$$\overline{u_j u_i} \equiv \overline{(\overline{u}_j + u_j')(\overline{u}_j + u_j')} = \overline{\overline{u}_j\overline{u}_i} + \overline{\overline{u}_j u_i'} + \overline{u_j'\overline{u}_i} + \overline{u_j' u_i'}$$

Extending definition (9.5) to LES PDE scalar state variable members $\{q(\mathbf{x},t)\}$ generates (9.7)

$$\overline{u_j q} \equiv \overline{(\overline{u}_j + u_j')(\overline{q} + q')} = \overline{\overline{u}_j\overline{q}} + \overline{\overline{u}_j q'} + \overline{u_j'\overline{q}} + \overline{u_j' q'}$$

and no terms in (9.6) or (9.7) are resolved scale state variable members of derived LES PDE system (9.13–9.15). Also detailed, (9.12) is handled via pressure projection theory (9.16–9.18).

A large LES literature details closure models for space filtered **DP** (9.13), derived for mathematically consequential assumptions on tensor quadruple (9.6). The nominally universal alteration of (9.13) is (9.22)

$$\mathcal{L}(\overline{u}_i) = \frac{\partial \overline{u}_i}{\partial t} + \frac{\partial}{\partial x_j}\left[\overline{u}_j\overline{u}_i + \overline{P}\delta_{ij} - \frac{2}{\text{Re}}\overline{S}_{ij} + \tau_{ij}\right] + \frac{\text{Gr}}{\text{Re}^2}\overline{\Theta}\widehat{g}_i = 0$$

for $\overline{S}_{ij}$ resolved scale velocity strain rate tensor (9.23).

The single Reynolds stress tensor τ_{ij} inserted in (9.22) is a truly consequential simplification of LES theory tensor quadruple (9.6). This renders LES **DP** PDE (9.22) indistinguishable from unsteady time averaged RaNS **DP** PDE, (8.12–8.13). Hence, RaNS $m\text{GWS}^h + \theta\text{TS}$ algorithm, Section 8.4, is readily edited to address (9.22).

Closure for *a priori* unknown τ_{ij} in (9.22) is via a subgrid scale (SGS) tensor model. The representative functional form is (9.24)

$$\tau_{ij}^{\text{D}} \equiv f(\text{C}_\text{S}(\mathbf{x},t), \delta^2, |\overline{S}_{ij}|)\overline{S}_{ij}, i \neq j$$

$\text{C}_\text{S}(\mathbf{x},t)$ denotes a dynamic Smagorinsky constant, a solution-dependent distribution, $(\text{C}_\text{S}(\mathbf{x},t), \delta^2, |\overline{S}_{ij}|)$ corresponds to a $D(\text{L}^2\tau^{-1})$ eddy viscosity coefficient with $|\overline{S}_{ij}|$ a norm of (9.23).

Elimination of assumptions generating (9.22) requires rigorous addressing of LES theory predictions (9.6–9.7), bounded domain PDE system augmentation with boundary commutation error (BCE) integrals and derivation of *a priori* known resolved scale state variable non-homogeneous Dirichlet BCs. A contribution thereto is rational LES (RLES) theory based on rational Padé interpolation of the gaussian filter Fourier transform. Second-order

Padé RLES theory analytically predicts (9.40) for the lead three tensors in (9.6)

$$\overline{\bar{u}_j\bar{u}_i} + \overline{\bar{u}_j u_i'} + \overline{u_j'\bar{u}_i} = \bar{u}_j\bar{u}_i + \frac{\delta^2}{2\gamma}\left[I - \frac{\delta^2}{4\gamma}\nabla^2\right]^{-1}\left(\frac{\partial \bar{u}_j}{\partial x_k}\frac{\partial \bar{u}_i}{\partial x_k}\right) + O(\delta^4)$$

containing the matrix inverse differential operator $[\cdot]^{-1}$, RLES literature referenced the auxiliary problem. Second-order Padé RLES theory also predicts (9.45) for the fourth tensor in (9.6

$$\overline{u_j'u_i'} = \frac{\delta^4}{16\gamma^2}\left[I - \frac{\delta^2}{4\gamma}\nabla^2\right]^{-1}\left[\nabla^2\bar{u}_j\nabla^2\bar{u}_i\right] + O(\delta^6)$$

The significant term coefficient δ^4 identifies this result trivial but does generate LES theory bound $O(\delta^4) < O(\overline{u_j'u_i'}) < O(\delta^2)$.

Fluid mechanics perturbation theory theoretically resolves RLES auxiliary problem implementation via the energetic capture constraint $\delta \geq 2h$. Non-D LES resolved scale velocity $\bar{u}_i$ is $O(1)$, true also for convection tensor product second filtering $\overline{\bar{u}_j\bar{u}_i}$. By definition unresolved scale velocity u_i' is $O(h)$, equivalently $O(\delta)$ via the constraint.

Thus derivation (9.40) exponentiation on small parameter δ admits predictions (9.60–9.61)

$$\begin{aligned}
\overline{\bar{u}_j\bar{u}_i} &= \bar{u}_j\bar{u}_i + O(\delta^2)\\
\overline{\bar{u}_j u_i'} + \overline{u_j'\bar{u}_i} &= \frac{\delta^2}{2\gamma}\left[I - \frac{\delta^2}{4\gamma}\nabla^2\right]^{-1}\left(\frac{\partial \bar{u}_j}{\partial x_k}\frac{\partial \bar{u}_i}{\partial x_k}\right) + O(\delta^4)
\end{aligned}$$

Coalescing coefficients in (9.61), labeling resolved-unresolved scale interaction tensor product pair $c_{ij}(\mathbf{x}, t) \equiv \overline{\bar{u}_j u_i'} + \overline{u_j'\bar{u}_i}$, these terms in (9.6) become identified solutions to *harmonic* EBV PDE system (9.64)

$$\mathcal{L}(c_{ij}) = -\nabla^2 c_{ij} + \frac{4\gamma}{\delta^2}c_{ij} - 2\frac{\partial \bar{u}_i}{\partial x_k}\frac{\partial \bar{u}_j}{\partial x_k} = 0$$

Of classic theoretical fluid mechanics pertinence, *ar*LES theory identified $c_{ij}(\mathbf{x}, t)$ tensor will be symmetric, realizable and translation and Galilean invariant thus meeting all *necessary* mathematical requirements for an NS tensor field. For space filtered D*E* and D*Y* resolved-unresolved scale interaction vector pairs are solutions to (9.70–9.71)

$$\begin{aligned}
\mathcal{L}\left(v_j^{\Theta}\right) &= -\nabla^2 v_j^{\Theta} + \frac{4\gamma}{\delta^2}v_j^{\Theta} - 2\frac{\partial \bar{u}_j}{\partial x_k}\frac{\partial \Theta}{\partial x_k} = 0\\
\mathcal{L}\left(v_j^{Y}\right) &= -\nabla^2 v_j^{Y} + \frac{4\gamma}{\delta^2}v_j^{Y} - 2\frac{\partial \bar{u}_j}{\partial x_k}\frac{\partial Y_\alpha}{\partial x_k} = 0
\end{aligned}$$

LES theory role for $\overline{u_j'u_i'}$, the fourth term in (9.6) named the subfilter scale (SFS) tensor, is dissipation at $O(2h)$ unresolved scale threshold. RLES literature documents substandard performance *for* $O(\delta^3)$ closure *models* hypothesized on classic fluid mechanics principles.

RaNS CFD literati recognize that $O(h)$ discretization generated phase dispersion error is the energetic mechanism required dissipated.

*m*PDE discrete $O(h^2)$ dispersion error annihilation theory generates NS dissipative flux vector (5.117) augmentation, an alternative to numerical diffusion, which analytically identifies SFS tensor (9.55)

$$-\overline{u_j' u_i'} \equiv \frac{C_S(\delta)h^2\mathrm{Re}}{6}\left(\overline{u}_j\overline{u}_k\frac{\partial \overline{u}_i}{\partial x_k} + \overline{u}_i\overline{u}_k\frac{\partial \overline{u}_j}{\partial x_k}\right)$$

Interpretation of $D(\mathrm{L}^2\tau^{-1})$ coefficient $\frac{h^2\mathrm{Re}}{6}\overline{u}_j\overline{u}_k$ as eddy viscosity genre is incorrect as resolved scale velocity tensor products therein are non-positive definite. Detailed *a posteriori* data validate that *m*PDE theory augmentation *modus operandi* is dissipative/ anti-dissipative *modulo* $O(2h)$ oscillations.

Meeting RLES theory bound and energetic constraint $\delta \geq 2h$ is via inserted coefficient $C_S(\delta)$. Identical *m*PDE theory operations analytically derive SFS vectors (9.60–9.61)

$$-\overline{u_j'\Theta'} \equiv C_S(\delta)\frac{h^2\mathrm{RePr}}{12}\left[\overline{u}_j\overline{u}_k\frac{\partial\overline{\Theta}}{\partial x_j}\right]$$

$$-\overline{u_j' Y'} \equiv C_S(\delta)\frac{h^2\mathrm{ReSc}}{12}\left[\overline{u}_j\overline{u}_k\frac{\partial\overline{Y}}{\partial x_j}\right]$$

which eliminate the need for turbulent Prandtl–Schmidt number hypotheses.

Thereby, an *essentially analytical* LES (*ar*LES) theory for space filtered NS PDE system *without* any simplifying assumption is derived. Suppressing filter notation the *ar*LES theory I-EBV + EBV *m*PDE system *continuum* state variable is (9.72)

$$\{q(x_j,t)\} = \left\{u_i, \Theta, Y, \phi; c_{ij}, v_j^{\Theta}, v_j^{Y}; P\right\}^T$$

Theory I-EBV *m*PDE flux vector statement is (9.73

$$\mathcal{L}^m(\{q\}) = [m]\frac{\partial\{q\}}{\partial t} + \frac{\partial}{\partial x_j}\left[\{f_j\} - \{f_j^d\}\right] - \{s\} = \{0\}$$

with definitions (9.74)–(9.77)

$$[m] \equiv \left[1 - \frac{\gamma\,\Delta t^2}{6}\frac{\partial}{\partial x_j}\left(u_j u_k\frac{\partial}{\partial x_k}\right)\right]$$

$$\{f_j(\{q\})\} = \left\{\begin{array}{l} u_j u_i + c_{ij} + P^*|_{n+1}^{p}\delta_{ij} \\ u_j\Theta + v_j^{\Theta} \\ u_j Y + v_j^{Y} \end{array}\right\}$$

$$\left\{f_j^d(\{q\})\right\} = \left\{\begin{array}{l} \dfrac{2}{\text{Re}}S_{ij} + \dfrac{\text{C}_\text{S}(\delta)h^2\text{Re}}{6}\left(u_j u_k \dfrac{\partial u_i}{\partial x_k} + u_i u_k \dfrac{\partial u_j}{\partial x_k}\right) \\ \dfrac{1}{\text{RePr}}\dfrac{\partial \Theta}{\partial x_j} + \dfrac{\text{C}_\text{S}(\delta)h^2\text{RePr}}{12} u_j u_k \dfrac{\partial \Theta}{\partial x_k} \\ \dfrac{1}{\text{ReSc}}\dfrac{\partial Y}{\partial x_j} + \dfrac{\text{C}_\text{S}(\delta)h^2\text{ReSc}}{12} u_j u_k \dfrac{\partial Y}{\partial x_k} \end{array}\right\}$$

$$\{s\} = \left\{\begin{array}{l} -\dfrac{\text{Gr}}{\text{Re}^2}\Theta \widehat{g}_i \\ \dfrac{\text{Ec}}{\text{Re}}\left[\begin{array}{l} \dfrac{\partial}{\partial x_j}\left(u_i \dfrac{\partial u_j}{\partial x_i}\right) + \dfrac{\partial^2 c_{ij}}{\partial x_j \partial x_i} \\ -\dfrac{\partial^2}{\partial x_j \partial x_i}\left(\dfrac{\text{C}_\text{S}(\delta)h^2\text{Re}}{6}\left(u_j u_k \dfrac{\partial u_i}{\partial x_k} + u_i u_k \dfrac{\partial u_j}{\partial x_k}\right)\right) \end{array}\right] \\ s(Y) \end{array}\right\}$$

Theory EBV PDE systems remain (9.16–9.18), (9.64), (9.70–9.71).

*ar*LES theory $m\text{GWS}^N + \theta\text{TS}$ algorithm statement is functionally identical to NS, Section 7.3, nonlinear iterative matrix statement (7.17) with $k = 1$ TP/NC trial space basis form detailed in Sections 9.9–9.10. The asymptotic error estimate for $m\text{GWS}^N + \theta\text{TS}$ algorithm *a posteriori* data for (9.73–9.77) is (9.78)

$$\left\|\{e^h(n\Delta t)\}\right\|_E \leq Ch^{2\gamma}\left[\|\text{data}\|_{\text{L}2,\Omega} + \|\text{data}\|_{\text{L}2,\partial\Omega}\right] + \text{C}_2\Delta t^{f(\theta)}\left\|\{q^h(t_0)\}\right\|^2_{H^1(\Omega)}$$
$$\text{for: } \gamma \equiv \min(k, (k+1/2), r-1), f(\theta) = (2, 3)$$

That for EBV PDE/*m*PDE systems is (9.79)

$$\left\|\{e^h(n\Delta t)\}\right\|_E \leq Ch^{2\gamma}\left[\|\text{data}\|_{\text{L}2,\Omega} + \|\text{data}\|_{\text{L}2,\partial\Omega}\right]$$
$$\text{for: } \gamma \equiv \min(k, (k+1), r-1)$$

for $(k+1) = 2$ accorded for EBV $\text{GWS}^h m\text{PDE}$ alteration (5.114).

Analytically derived *ar*LES theory I-EBV + EBV *m*PDE system (9.73–9.77) plus (9.16–9.18), (9.64), (9.70–9.71) is without the BCE integrals required for rigorously derived LES PDE system pertinence to bounded domains. Rigorous mathematical processes confirm that resolution is unfiltered NS domain extension, hence state variable interpretation in the sense of distributions, Section 9.13. The resultant alteration to convolved D**P** (9.13) is (9.103)

$$\mathcal{L}(\overline{u}_i) = \text{St}\frac{\partial \overline{u}_i}{\partial t} + \frac{\partial}{\partial x_j}\left[\overline{u_j u_i} + \overline{P}\,\delta_{ij} - \frac{1}{\text{Re}}\left(\frac{\partial \overline{u}_i}{\partial x_j} + \frac{\partial \overline{u}_j}{\partial x_i}\right)\right]$$
$$+ \frac{\text{Gr}}{\text{Re}^2}\overline{\Theta}\hat{\text{g}}_i - \text{A}_\delta\left(\text{D}_{ij}(u_i, P)\right) = 0$$

D**P** BCE integral definition is (9.104)

$$\mathrm{A}_\delta\left(\mathrm{D}_{ij}(u_i,P)(x_k,t)\right) \equiv \int_{\partial\Omega} g_\delta(x_k - s_k)\left[-P\delta_{ij} + \frac{1}{\mathrm{Re}}\left(\frac{\partial u_i}{\partial x_j} + \frac{\partial u_j}{\partial x_i}\right)\right]\hat{n}_j \mathrm{d}s$$

Dissipative term existence in space filtered NS PDE system (9.14–9.15), also laplacian operators in (9.17–9.18), lead to (9.105–9.108) for scalar state variable members and DM^h theory variables. BCE integrands contain *unfiltered* NS state variable members, surface values for which are typically *a priori* known, specifically the no slip wall BC.

BCE integral inclusion in *ar*LES theory *m*PDE system (9.73–9.77) alters only the source matrix (9.77) to (9.129)

$$\{s\} = \left\{\begin{array}{c} -\dfrac{\mathrm{Gr}}{\mathrm{Re}^2}\Theta\widehat{g}_i + \displaystyle\int_{\partial\Omega} g_\delta(x_k - s_k)\left[-P\delta_{ij} + \frac{1}{\mathrm{Re}}\left(\frac{\partial u_i}{\partial x_j} + \frac{\partial u_j}{\partial x_i}\right)\right]\hat{n}_j \mathrm{d}s \\ \dfrac{\mathrm{Ec}}{\mathrm{Re}}\left[\begin{array}{c} \dfrac{\partial}{\partial x_j}\left(u_i\dfrac{\partial u_j}{\partial x_i}\right) + \dfrac{\partial^2 c_{ij}}{\partial x_j \partial x_i} \\ -\dfrac{\partial^2}{\partial x_j \partial x_i}\left(\dfrac{\mathrm{C_S}(\delta)h^2\mathrm{Re}}{6}\left(u_j u_k\dfrac{\partial u_i}{\partial x_k} + u_i u_k\dfrac{\partial u_j}{\partial x_k}\right)\right) \end{array}\right] \\ + \displaystyle\int_{\partial\Omega} g_\delta(x_k - s_k)\left[\frac{1}{\mathrm{RePr}}\frac{\partial\Theta}{\partial x_j}\right]\hat{n}_j \mathrm{d}s \\ s(Y) + \displaystyle\int_{\partial\Omega} g_\delta(x_k - s_k)\left[\frac{1}{\mathrm{ReSc}}\frac{\partial Y}{\partial x_j}\right]\hat{n}_j \mathrm{d}s \end{array}\right\}$$

Overbar notation is omitted as NS state variables in BCE integrands are always unfiltered.

Aside from P appearing in (9.104), BCE integrals in (9.102–9.108), hence (9.129), calculus operations appear similar with Green–Gauss divergence theorem generated GWS^h/ mGWS^h + θTS non-homogeneous Neumann BC on bounding surfaces, (9.109). This duality provides insight for design of BCE integral evaluation, *also* state variable BCs *a priori* known *only* for the unfiltered NS state variable.

For resolved scale velocity at domain bounding surface nodes $\mathbf{x}_b \in \partial\Omega$, convolution identity is (9.110)

$$\overline{\mathbf{u}}(\mathbf{x}_b) = (g_\delta{}^* \mathbf{u}(\mathbf{x}_b)) = \int_\Omega g_\delta(\mathbf{x}_b - \mathbf{y})\mathbf{u}(\mathbf{y})\mathrm{d}\mathbf{y},\ \ \mathbf{x}_b \in \partial\Omega$$

Generation of data $\mathbf{u}^{n+1}$ required to evaluate (9.110) is via *approximate deconvolution* (AD), (9.111)

$$\mathbf{u}(\mathbf{x}) \cong \overline{\mathbf{u}}(\mathbf{x}) - \frac{\delta^2}{24}\nabla\overline{\mathbf{u}}(\mathbf{x}) + O(\delta^4)$$

Numerical quadrature (9.112) leads to domain surface node $\bar{\mathbf{u}}(\mathbf{x}_b)$ DOF determination

$$\bar{\mathbf{u}}(\mathbf{x}_b) = \sum_{i \in I} w_i g_\delta(\mathbf{x}_b - \mathbf{x}_i)\left(\bar{\mathbf{u}}(\mathbf{x}) - \frac{\delta^2}{24}\nabla^2\bar{\mathbf{u}}(\mathbf{x})\right)$$
$$+ \sum_{i \in B} w_i g_\delta(\mathbf{x}_b - \mathbf{x}_i)(\mathbf{u}(\mathbf{x}_b))$$

for index sets I and B corresponding to nodes on domain *I*nterior and on the *B*oundary, respectively, with $\mathbf{x}_i$ and w_i quadrature nodes and weights.

Denoting (9.110–9.112) the ADBC algorithm, *a posteriori* data for a *verification* DE statement with non-homogeneous Dirichlet BC confirms variable and data are required extended to $\Omega^+ \subset \mathbb{R}^n$, Figure 9.27. These data further confirm BCE integral omission the dominant error source. ADBC algorithm solutions exhibit monotone convergence with uniform mesh refinement for linear $k=1$ basis discrete numerics equivalence, Table 9.1.

For state variable alteration to $\overline{P}(\mathbf{x}_b \in \partial\Omega_{\hat{n}_i})$ in (9.110–9.112), ADBC algorithm readily generates BCE integral lead term, first row in (9.129). Extension to AD-GWS^h-BCE-DBC algorithm directly addresses unfiltered NS state variable derivative ubiquitous appearance in the remaining BCE integrands. The identity for **DP** BCE integrand second term absent Re^{-1}

$$\mathrm{A}_\delta\left(\mathrm{D}_{ij}(u_i, \hat{n}_j)(x_k, t)\right) \equiv \int_{\partial\Omega} g_\delta(x_k - s_k)\frac{\partial u_i}{\partial x_j}\hat{n}_j \mathrm{d}s$$

is non-vanishing everywhere except on outflow surface segments $\partial\Omega$. The AD differential definition replacement for (9.137) is (9.138)

$$\mathcal{D}\left(\frac{\partial u_i}{\partial x_j}\hat{n}_j\right) \equiv \frac{\partial u_i}{\partial x_j}\hat{n}_j - \frac{\partial \bar{u}_i}{\partial x_j}\hat{n}_j + \frac{\delta^2}{4\gamma}\nabla^2\left(\frac{\partial \bar{u}_i}{\partial x_j}\hat{n}_j\right) = 0 + O(\delta^4)$$

Deleting $O(\delta^4)$, noting $\hat{n}_j$ is a common multiplier, for DOF label $\{\mathrm{Q\_\nabla}_j\mathrm{U}I\}_e$, $1 \leq I, j \leq n$, $\mathrm{GWS}^N \Rightarrow \mathrm{GWS}^h$ transitions (9.138) to element matrix statement (9.139)

$$\{\mathrm{F\_Q}\nabla_j\mathrm{U}I\}_e = \det_e[\mathrm{M}200d]\{\mathrm{Q\_}\nabla_j\mathrm{U}I\}_e - \det_e[\mathrm{M}200d]\{\mathrm{Q\_}\nabla_j\overline{\mathrm{U}}I\}_e$$
$$- \frac{\delta^2}{4\gamma \det_e}[\mathrm{M}2KKd]_e\{\mathrm{Q\_}\nabla_j\overline{\mathrm{U}}I\}_e$$
$$+ \frac{\delta^2}{4\gamma}[\mathrm{N}20Kd]_e\{\mathrm{Q\_}\nabla_j\overline{\mathrm{U}}I\}_e\hat{n}_K, 1 \leq I \leq n$$
$$[\mathrm{JAC\_Q}\nabla_j\mathrm{U}I]_e = \det_e[\mathrm{M}200d]$$

Green–Gauss theorem generated surface integral therein is deleted as sole non-Dirichlet BC is homogeneous Neumann on outflow $\partial\Omega$. Applicable DOF constraint in solving assembled (9.139) is only where DOF $\{\mathrm{Q\_}\nabla_j\mathrm{U}I\}_e \bullet \widehat{n}_j$ vanish which generates *unfiltered* NS velocity strain rate DOF distributions $\{\mathrm{Q\_}\nabla_j\mathrm{U}I\}, i \leq (I,j) \leq n$.

GWS^h solution unfiltered NS velocity vector strain rate tensor dot product substitution fully defines (9.137) integrand functionality (9.141)

$$\left(\frac{\partial \bar{u}_i}{\partial x_j}\hat{n}_j\right)_{\mathbf{x}_b \in \partial\Omega_{\hat{n}_i}} = \int_{\partial\Omega} g_\delta(x_k - x_i)\mathrm{S}_e\left(\{N_k(\mathbf{x})\}^T\{\mathrm{Q\_\nabla}_j\mathrm{U}I(t)\}_e\right)\hat{n}_j \mathrm{d}\tau, \mathbf{x}_b \in \partial\Omega_{\hat{n}_j}$$

Inserting Re^{-1}, for $\Omega^+ \subset \mathbb{R}^n$ at all pertinent nodes $\mathbf{x}_b \in \partial\Omega$, numerical quadrature of (9.141) generates BCE integral DOF $\{\mathrm{Q\_\nabla}_j\mathrm{U}I(t)\hat{n}_j\}$ data contributions (9.142) at any time t for N_x, N_y, N_z Newton–Cotes formula degree with weights w_i, w_j, w_k.

$$\mathrm{Q\_\nabla}_j\overline{\mathrm{U}}I(t)\hat{n}_j\Big|_{\mathbf{x}_b \in \partial\Omega_{e,\hat{n}_j}} \cong \sum_{l=1}^{N_x}\sum_{m=1}^{N_y}\sum_{n=1}^{N_z}\begin{pmatrix} w_i w_m w_n\, g_\delta(x_b - x_l, y_b - y_m, z_b - z_n) \\ \times \mathrm{S}_e\left(\{N_k(\mathbf{x})\}^T\{\mathrm{Q\_\nabla}_j\mathrm{U}I(t)\}\right)\end{pmatrix}\hat{n}_j$$

For appropriate state variable alterations, AD-GWS^h-BCE-DBC algorithm (9.138–9.142) generates BCE integral DOF data $\left\{\mathrm{Q\_\nabla}_j\overline{\Theta}(t)\hat{n}_j\right\}\Big|_{\mathbf{x}_b \in \partial\Omega_{e,\hat{n}_j}}$ and $\left\{\mathrm{Q\_\nabla}_j\overline{Y}(t)\hat{n}_j\right\}\Big|_{\mathbf{x}_b \in \partial\Omega_{e,\hat{n}_j}}$.

Invoking *turbulent* for the first time(!), *ar*LES theory resolved scale velocity vector *non-homogeneous* Dirichlet BC is derived assuming unfiltered $\mathbf{u}$ in (9.111) is RaNS *time averaged* turbulent BL velocity $\bar{\bar{u}} \bullet \widehat{\mathbf{s}}$ for $\widehat{\mathbf{s}}$ unit tangent vector on $\partial\Omega$. AD statement (9.111) is replaced with differential definition (9.145) for u_{AD}

$$\mathcal{D}(u_{\mathrm{AD}}) \equiv u_{\mathrm{AD}} - \bar{u} + \frac{\delta^2}{4\gamma}\nabla^2\bar{u} = 0 + O(\delta^4)$$

Deleting $O(\delta^4)$, $\text{GWS}^N \Rightarrow \text{GWS}^h$ transitions (9.145) to element matrix statement (9.146)

$$\begin{aligned}\{\mathrm{F\_Q}U\mathrm{AD}\}_e &= \det_e[\mathrm{M200}d]\{\mathrm{Q\_}U\mathrm{AD}\}_e - \det_e[\mathrm{M200}d]\{\mathrm{Q\_}\overline{U}\}_e \\ &\quad - \frac{\delta^2}{4\gamma\det_e}[\mathrm{M2}KKd]_e\{\mathrm{Q\_}\overline{U}\}_e + \frac{\delta^2}{4\gamma}[\mathrm{N20}Kd]_e\{\mathrm{Q\_}\overline{U}\}_e\hat{n}_K \\ [\mathrm{JAC\_Q}U\mathrm{AD}]_e &= \det_e[\mathrm{M200}d]\end{aligned}$$

Not reducing jacobian for DOF $\{\mathrm{Q\_}U\mathrm{AD}\}_e$ at nodes $\mathbf{x}_b \in \partial\Omega_{\hat{n}_i}$ *a priori* known no-slip, that is, $u_{\mathrm{AD}}(\mathbf{x}_b) \equiv \bar{\bar{\mathbf{u}}}(\mathbf{x}_b) \bullet \widehat{\mathbf{s}} = 0$, and replacing $\text{GWS}^N \Rightarrow \text{GWS}^h$ Green–Gauss divergence theorem *non-homogeneous* Neumann BC matrix $[\mathrm{N20}Kd]_e\{\mathrm{Q\_}\overline{U}\}_e\hat{n}_K$ with $[\mathrm{N200}d]_e\{\mathrm{dQ\_}\overline{U}/\mathrm{d}n\}_e$ admits prediction of $u^h_{\mathrm{AD}}(\mathbf{x})$ distribution.

Regular mesh refinement *a posteriori* data confirm computed "turbulent" $u^h_{\mathrm{AD}}(\mathbf{x})$ error is restricted to near wall reach region of BL similarity range $y^+ < 0.25\delta^+$. This enables accurate estimation of normal derivative DOF $\{\mathrm{dQ\_\overline{U}/d}n\}$. These data confirm gaussian filter *uniform* measure constraint $\delta \equiv 4h$ *optimal*, for similarity reach normal mesh measure $h = \delta/16$ to $h = \delta/4$, Figure 9.33, exhibiting monotone asymptotic convergence under wall normal uniform measure h refinement, Figure 9.34.

Derived AD-GWS^h-BCE-DBC algorithm generation of resolved scale velocity non-homogeneous Dirichlet BCs (DBC) is validated accurate for large Re laminar BL profiles, Section 9.18. Algorithm strategy is further confirmed fully appropriate for *ar*LES theory $O(\delta^2)$ state variable members. AD definition for non-physical unfiltered cross stress tensor c_{ij} (9.147)

$$c_{ij} = \overline{c}_{ij} - \frac{\delta^2}{4\gamma}\nabla^2\overline{c}_{ij} + O(\delta^4)$$

leads to ordered statement (9.149) valid at any time t

$$c_{ij} = \frac{\delta^2}{2\gamma}\left(\frac{\partial\overline{u}_i}{\partial x_k}\frac{\partial\overline{u}_j}{\partial x_k}\right) + O(\delta^4)$$

Convolution (9.150)

$$\begin{aligned}\overline{c}_{ij}(\mathbf{x}_b \in \partial\Omega_{\hat{n}_k}) &\equiv \left(g_\delta * c_{ij}\right), (\mathbf{x}_b \in \partial\Omega_{\hat{n}_k})\\ &\equiv \int_\Omega g_\delta(x_m - x_l)c_{ij}(x_l, t)\mathrm{d}\tau, \mathbf{x}_b \in \partial\Omega_{\hat{n}_k}\\ &= \frac{\delta^2}{2\gamma}\int_\Omega g_\delta(x_m - x_l)\left(\frac{\partial\overline{u}_i}{\partial x_k}\frac{\partial\overline{u}_j}{\partial x_k}\right)_{x_l,t}\mathrm{d}\tau, \mathbf{x}_b \in \partial\Omega_{\hat{n}_k}\end{aligned}$$

coupled with GWS$^N \Rightarrow$ GWS^h on differential definition (9.151)

$$\mathcal{D}\left(d\overline{u}_i d\overline{u}_j\right) \equiv d\overline{u}_i d\overline{u}_j - \frac{\partial\overline{u}_i}{\partial x_k}\frac{\partial\overline{u}_j}{\partial x_k} = 0$$

leads to element matrix statement (9.152)

$$\begin{aligned}\left\{\mathrm{F\_QD\overline{U}}I\mathrm{D\overline{U}}J\right\}_e &= \det_e[\mathrm{M200}d]\left\{\mathrm{Q\_D\overline{U}}I\mathrm{D\overline{U}}J\right\}_e\\ &\quad - \frac{1}{\det_e}\left\{\mathrm{\overline{U}}I\right\}_e^T[\mathrm{M3}K\mathrm{0}Kd]_e\left\{\mathrm{\overline{U}}J\right\}_e, 1 \le (I,J) \le n\\ \left[\mathrm{JAC\_QD\overline{U}}I\mathrm{D\overline{U}}J\right]_e &= \det_e[\mathrm{M200}d]\end{aligned}$$

Requisite $\overline{c}_{ij}^h$ non-homogeneous Dirichlet BC DOF on $\mathbf{x}_b \in \partial\Omega$ are computed via numerical quadrature (9.154)

$$\begin{aligned}\overline{C}IJ(t)\big|_{\mathbf{x}_b \in \partial\Omega_{\hat{n}_k}} &= \frac{\delta^2}{2\gamma}\int_{\Omega^+} g_\delta(x_m - x_l)\left(\frac{\partial\overline{u}_i}{\partial x_k}\frac{\partial\overline{u}_j}{\partial x_k}\right)^h \mathrm{d}\tau, \mathbf{x}_b \in \partial\Omega_{\hat{n}_k}\\ &= \frac{\delta^2}{2\gamma}\sum_{l=1}^{N_x}\sum_{m=1}^{N_y}\sum_{n=1}^{N_z}\left(\begin{matrix} w_l w_m w_n g_\delta(x_b - x_l, y_b - y_m, z_b - z_n)\\ \times \mathrm{S}_e\left(\{N(\mathbf{x})\}^T\left\{\mathrm{D\overline{U}}I\mathrm{D\overline{U}}J(t)\right\}_e\right)\end{matrix}\right)_{\mathbf{x}_b \in \partial\Omega_{\hat{n}_k}}\end{aligned}$$

For $O(\delta^2)$ *unfiltered* state variable LES vector identifications (9.155–9.156)

$$v_j^{\Theta}(\mathbf{x},t) = \frac{\delta^2}{2\gamma}\frac{\partial \bar{u}_j}{\partial x_k}\frac{\partial \overline{\Theta}}{\partial x_k} + O(\delta^4)$$

$$v_j^{Y}(\mathbf{x},t) = \frac{\delta^2}{2\gamma}\frac{\partial \bar{u}_j}{\partial x_k}\frac{\partial \overline{Y}}{\partial x_k} + O(\delta^4)$$

AD-GWS^h-BCE-DBC algorithm directly extended leads to requisite non-homogeneous Dirichlet BC DOF definitions (9.157–9.158). Thus, *ar*LES theory I-EBV plus EBV state variable *m*PDE system (9.103–9.108), (9.16–9.18), (9.64), (9.70–9.71) with AD-GWS^h-BCE-DBC algorithm is confirmed *well-posed* ∀ Re for identified *m*GWS$^h + \theta$TS algorithm, Sections 9.9–9.10.

Algorithm validation addresses $8\times1\times16$ thermal cavity symmetric half on $n=3$. Ra = E+05 solution thermal finger pair bounded oscillation prediction, Figure 9.39, is in *quantitative* agreement with *ar*LES-DES prediction, Figure 9.8 right. Indexing to Ra 4.0 E+05 > $\mathrm{Ra_{CR}} \approx 3.1$ E+05, well-posed *ar*LES theory Re = 756 solution transitioning to unsteady periodic constitutes well-posed *ar*LES theory *m*GWS$^h + \theta$TS algorithm $k=1$ basis implementation validation.

Well-posed *ar*LES theory Ra = E+08, Re = 11,824 $n=3$ *a posteriori* data validate ∀ Re assertion. Detailed assessment of hot wall jet DOF distributions confirms prediction of laminar-turbulent transition *without* any "turbulent" modeling component, Figure 9.47. Resolved scale velocity non-homogeneous Dirichlet BC prediction is confirmed critical to transition *prediction*. The lack of solution exact mirror symmetry is attributable to fully three dimensional velocity vector dynamics induced in cavity geometry corners with Dirichlet BC temperature walls.

Local diagnostics for well-posed *ar*LES theory Ra = E+08, Re = 11,824 solution snapshot in near vicinity of hot wall jet separation, Figure 9.50, validates RLES-perturbation theory state variable $O(1, \delta^2)$ organization. These *a posteriori* data further quantitatively validate adequate mesh resolution, proper $O(\delta^3)$ SFS tensor/vector closure dissipation at the unresolved scale threshold, hence stable *m*GWS$^h + \theta$TS algorithm performance without infusion of artificial diffusion.

Exercises

9.4.1 Validate the Fourier transformations in (9.30).

9.4.2 Confirm the Fourier transformations (9.33).

9.4.3 Verify the Fourier transformations for the second-order Taylor series approximation (9.35).

9.4.4 Confirm the second-order Taylor series inverse Fourier transforms (9.36), hence validate the closure (9.37).

9.4.5 Repeat 9.4.2–9.4.4 for the second-order Padé approximation hence confirm the RLES closure (9.40).

9.4.6 Repeat 9.4.5 for the 4th order Padé approximation hence validate the RLES closure (9.42).

9.4.7 Validate the TS unresolved scale tensor closure (9.44).

9.4.8 Verify the Padé theory unresolved scale tensor closure (9.45).

9.7.1 Verify the correctness of EBV statement (9.64).

9.7.2 Repeat exercise 9.7.1 for statements (9.65), (9.70) and (9.71).

9.9.1 For *ar*LES theory PDE system kinetic flux vector (9.75), confirm algorithm statements (9.82) and (9.83).

9.9.2 For *ar*LES PDE system dissipative flux vector (9.76) confirm algorithm statements (9.84) and (9.85).

9.9.3 Verify the Stokes stress tensor weak form (9.92).

9.9.4 Confirm correctness of (9.93) for SFS stress tensor.

9.10.1 Verify the last line in jacobian element matrix set (9.96).

9.10.2 Confirm the jacobian element matrix for $\{Q4\}_e$ in (9.98).

9.10.3 Verify that jacobian element matrix (9.99) is valid for the entire *ar*LES theory $O(\delta^2)$ state variable partition.

References

Baker, A.J. (2012). *Finite Elements ⇔ Computational Engineering Sciences*, John Wiley, London.

Baldwin, B.S. and Lomax, H. (1978). "Thin-layer approximation and algebraic model for separated turbulent flows," Technical Paper AIAA:78-257.

Berselli, L.C., Iliescu, T. and Layton, W.J. (2006). *Mathematics of Large Eddy Simulation of Turbulent Flows*, Springer, Heidelberg.

Boole. G. and Moulton, J.F. (1960). *A treatise on the Calculus of Finite Differences*, 2nd Ed. revised, Dover, New York.

Borggaard, J. and Iliescu, T. (2006). "Approximate Deconvolution Boundary Conditions for Large Eddy Simulation," *Appl. Math Letters*, V. 19, pp. 735–740.

Boris, P.J. (1990). "On Large Eddy Simulation Using Subgrid Turbulence Models", in J.L. Lumley, ed., *Whither Turbulence? Turbulence at the Crossroads*, Springer, New York, pp. 344.

Boris, P.J., Grinstein, F.F., Oran, E.S. and Kolbe, R.L. (1992). "New Insights into Large Eddy Simulation", *Fluid Dynamics Research*, V. 10, pp. 199–228.

Chapman, D.R. (1979). Computational aerodynamics development and outlook," *AIAA Journal*, V. 17, pp. 129.

Christon, M.A., Gresho, P.M. and Sutton, S.B. (2002). "Computational predictability of time-dependent natural convection flows in enclosures (including a benchmark solution)," *J. Numerical Methods Fluids*, V. 40, pp. 953–980.

Clark, R.A., Ferziger, J.H. and Reynolds, W.C. (1979). "Evaluation of subgrid scale models using an accurately simulated turbulent flow," *J. Fluid Mechanics*, V. 91, p. 1.

Domaradzki, A.J. and Saiki, E.M. (1997). "A Subgrid-Scale Model Based on the Estimation of Unresolved Scales of Turbulence", *Physics Fluids*, V. 9, p. 2148.

Domaradzki, A.J. and Yee, P.P. (2000). "The Subgrid-Scale Estimation Model for High Reynolds Number Turbulence," *Physics Fluids*, V. 12, pp. 193–196.

Dunca, A., John, V. and Layton, W.J. (2003). "The commutation error of the space averaged Navier-Stokes equations on a bounded domain," *J. Mathematical Fluid Mechanics*, V. 5, pp. 1–27.

Fureby, C. and Tabor, G. (1997). "Mathematical and Physical Constraints on Large-Eddy Simulation," *J. Theoretical Computational Fluid Dynamics*, V. 9, pp. 85–102.

Galdi, G.P. and Layton, W.J. (2000). "Approximation of the larger eddies in fluid motion, II. A model for space filtered flow," *J. Math Modeling Methods Applied Science*, V. 10, pp. 343–350.

Gebhart, B., Jaluria, Y., Mahajan, R.L. and Sammakia, B., (1988). *Buoyancy-Induced Flow and Transport*, Hemisphere, New York.

Germano, M., Piomelli, U., Moin, P. and Cabot, W. (1991). "A Dynamic subgrid-scale eddy viscosity model," *Physics of Fluids* A, V. 3, pp. 1760–1765.

Ghosal, S. and Moin, P. (1995). "The Basic Equations for the Large Eddy Simulation of Turbulent Flows in Complex Geometry, " *J. Computational Phys.*, V. 118, pp. 24–37.

Gresho, P.M. and Sani, R.L. (1998). *Incompressible Flow and the Finite Element Method*, John Wiley, Chichester.

Grinstein, F.F. and Fureby, C. (2004). "From Canonical to Complex Flows: Recent Progress on Monotonically Integrated LES", *Comp. Science Engineering.*, V. 6, pp. 37–49.

Grinstein, F.F., Margolin, L.G. and Rider, W.J. eds., (2007). *Implicit Large Eddy Simulation: Computing Turbulent Fluid Dynamics*, Cambridge University Press, England.

Grubert, M. (2006). "Development of a Potentially Accurate and Efficient LES CFD Algorithm to Predict Heat and Mass Transport in Inhabited Spaces," PhD dissertation, University of Tennessee.

Harten, A. (1983). "High Resolution Schemes for Hyperbolic Conservation Laws", *J. Computational Physics* V. 49, pp. 357–393.

Iliescu, T. and Layton, W.J. (1998). "Approximating the Larger Eddies in Fluid Motion, III. the Boussinesq Model for Turbulent Fluctuations," Analete Stiintifice ale Universitatii, Al. I. Caza, Tomal XLIV, s.I.a, Matematica, V. 44, pp. 245–261.

Iliescu, T. and Fischer, P.F. (2003). "Large eddy simulation of turbulent channel flows by the rational large eddy simulation model," *Physics Fluids*, V. 15, pp. 3036–3047.

John, V., Layton, W. and Sahin, N. (2004). "Derivation and Analysis of Near Wall Models for Channel and Recirculating Flows," *Computational Math Applications*, V. 48, pp. 1135–1151.

John, V. (2004). *Large Eddy Simulation of Turbulent Incompressible Flows, Analytical and Numerical Results for a Class of LES Models*, Lecture Notes in Computational Science and Engineering, V. 34, Springer, Berlin.

John, V. (2005). "An assessment of two models for the subgrid scale tensor in the rational LES model," *J. Comp. Applied Mathematics*, V. 163, pp. 57–80.

John, V. and Liakos, A. (2006). "Time-dependent flow across a step: the slip with friction boundary condition," *J. Numerical Methods Fluids*, V. 50, pp. 713–731.

Kolmogorov, A.N. and Fomin, S.V., (1975). *Introductory Real Analysis*, Dover, New York.

Layton, W.J. and Rebholz, L. (2012). *Approximate Deconvolution Models of Turbulence*, Lecture Notes in Mathematics, V. 2042, Springer, Berlin.

Leonard, A. (1974). "Energy Cascade in Large-eddy Simulations of Turbulent Fluid Flows," *Advances in Geophysics* A, V. 18, pp. 237–248.

Le Quere, P. (1983). "Accurate solutions to the square thermally driven cavity at high Raleigh number," *J. Computers and Fluids*, V. 2, pp. 29–41.

Le Quere, P., and Behnia, M. (1998). "From onset of unsteadiness to chaos in a differentially heated square cavity," *J. Fluid Mechanics*, V. 359, p. 81–107.

Lilly, D.K. (1992). "A proposed modification of the Germano subgrid-scale closure method," *Physics Fluids* A, V. 4, pp. 663–635.

Mansour, N.N., Kim, J. and Moin, P. (1988). "Reynolds stress and dissipation rate budgets in turbulent channel flow," *J. Fluid Mechanics*, V. 194, pp. 15–44.

Margolin, L.G., Rider, W.J. and Grinstein, F.F. (2006). "Modeling Turbulent Flow with Implicit LES," *J. Turbulence*, V. 7, pp. 1–27.

Marusic, I., Kunkel, G.J. and Porte-Agel, F. (2001). "Experimental study of wall boundary conditions for large-eddy simulation," *J. Fluid Mechanics*, V. 446, p. 309.

Metais, O. and Lesieur, M. (1992). "Spectral Large-Eddy Simulation of Isotropic and Stably Stratified Turbulence", *J. Fluid Mechanics*, V. 239, pp. 157–194.

Moser, D.R., Kim, J. and Mansour, N.N. (1999). "Direct numerical simulation of turbulent channel flow up to $Re_\tau = 590$," *Physics of Fluids*, V. 11, pp. 943–945.

Piomelli, U., Zang, T.A., Speziale, C.G. and Hussaini, M.Y. (1990). "On the Large-Eddy Simulation of Transitional Wall-Bounded Flows," *Physics Fluids* A, V. 2, pp. 257–265, 1990.

Piomelli, U. and Balaras, E. (2002). "Wall-Layer Models for Large-Eddy Simulations," *Annual Review Fluid Mechanics*, V. 34, pp. 349–374.

Piomelli, U. (2008). "Wall-Layer Models for Large-Eddy Simulations," *Progress in Aerospace Sciences*, V. 44, pp. 437–446.

Pope, S.B. (2000). *Turbulent Flows*, Cambridge University Press, New York.

Prandtl, L. (1904). "Uber Flussigkeitsbewegung bei sehr kleiner Reibung," Proc. 3rd *Int. Mathematics Congress*, Heidelberg, Germany, pp. 484–491.

Runge, C. (1901). "Uber empirische Funktionen und die Interpolation zwischen aquidistanten Ordinaten," *Zeitschrift fur Mathematik und Physik*, pp. 224–243.

Sagaut, P. (2006). *Large Eddy Simulation for Incompressible Flows*, 3rd ed., Springer, Berlin.

Salenger, A.G., Lehouco, R.B., Pawlowski, R.B. and Shadid, J.N. (2002). "Computational Bifurcation and Stability Studies of the 8:1 Thermal Cavity Problem," *J. Numerical Methods Fluids*, V. 40, pp. 1059–1073.

Schlatter, P., Stolz, S. and Kleiser, L. (2004). "LES of Transitional Flows Using the Approximate Deconvolution Model", *J. Heat and Fluid Flow*, V. 25, pp. 549–558.

Schlichting, H. (1979). *Boundary Layer Theory*, McGraw-Hill, NY.

Schwartz, L. (1966). *Theorie des distributions*, Hermann, Paris.

Sekachev, M. (2013). "Essentially analytical theory closure for thermal- incompressible space filtered Navier-Stokes PDE system on bounded domains," PhD dissertation, University of Tennessee.

Shah, K.B. and Ferziger, J.H. (1995). "A New Non-Eddy Viscosity Subgrid-Scale Model and Its Application to Channel Flow", CTR Annual Research Briefs, Stanford University

Smagorinsky, J. (1963). "General Circulation Experiments with the Primitive Equations, I. The Basic Experiments," *Monthly Weather Revue*, V. 91, pp. 99–164.

Stolz, S. and Adams, N.A. (1999). "An Approximate Deconvolution Procedure for Large-Eddy Simulation", *Physics Fluids*, V. 11, pp. 1699–1701.

Stolz, S., Adams, N.A. and Kleiser, L. (2001). "An Approximate Deconvolution Model for Large-Eddy Simulation with Application to Incompressible Wall-Bounded Flows", *Physics Fluids*, V. 13, pp. 997–1015.

Stolz, S., Schlatter, P., Meyer, D. and Kleiser, L. (2003). "High-Pass Filtered Eddy-Viscosity Models for LES," *Proceedings* DLES-5, Munich, Germany.

Trias, F.X., Verstappen, R.W.C.P, Gorobets, A., Soria, M. and Oliva, A. (2010). "Parameter-free symmetry-preserving regularization of a turbulent differentially heated cavity," *Computers and Fluids*, V. 39, pp. 1815–1831.

Van Dyke, M. (1975). *Perturbation Methods in Fluid Mechanics*, Academic Press, New York.

Vasilyev, O.V., Lund, T.S. and Moin, P. (1998). "A General Class of Commutative Filters for LES in Complex Geometries," *J. Computational Phys.*, V. 146, pp. 82–104.

Voke, P. and Yang, Z. (1995). "Numerical Study of Bypass Transition," *Physics Fluids*, V. 7, pp. 2256–2264.

Vreman, B. (1995). "Direct and large-eddy simulation of the compressible turbulent mixing layer," PhD thesis, University of Twente, Netherlands.

Wilcox, D.C. (2006). *Turbulence Modeling for CFD*, 3rd Edition, DCW Industries, La Canada, CA.

Winckelmans, G.S., Wray, A.A., Vasilyev, G.V. and Jeanmart, H. (2001). "Explicit filtering large-eddy simulations using the tensor-diffusivity model supplemented by a dynamic Smagorinsky term," *Physics of Fluids*, V. 13, p. 1385.

Wolfram Research, Inc. (2008). The *Mathematica* software system.

Xin, S., and Le Quere, P. (2001). "Linear stability analyses of natural convection flowfields in differentially heated square cavity with conducting horizontal walls," *Physics of Fluids*, V. 13, p. 2529–2542.

Xin, S. and Le Quere, P. (2002). "An extended Chebyshev pseudo-spectral benchmark for the 8:1 differentially heated cavity," *J. Numerical Methods Fluids*, V. 40, pp. 981–998.

Yoshizawa, A. and Horiuti, K. (1985). "A Statistically-Derived Subgrid Scale Kinetic Energy Model for the Large-Eddy Simulation of Turbulent Flows", *J. Physical Society Japan*, V. 54, pp. 2834–2839.

10

Summary - VVUQ: verification, validation, uncertainty quantification

10.1 Beyond Colorful Fluid Dynamics

The 1970s and 1980s witnessed burgeoning interest in what is now established as finite element spatial semi-discrete implementation of a weak form CFD algorithm. Hence was prompted a continuum of annual international conferences topically documented in the Wiley Interscience monograph series *Finite Elements in Fluids*, Gallagher *et al.* (1975–1988).

Time period monograph, textbook, conference and archival publications typically detailed *a posteriori* data generated on truly coarse meshes, the consequence of limited accessible compute power. Time frame coincident emerged progressively improved color graphics PC software enabling conference presentation transition away from monochrome. The ensuing practice of *gilding* marginally generated CFD solution data with full color visuals prompted Prof. Habashi (Concordia University, now McGill) to conclude that CFD really stood for *colorful fluid dynamics*.

CFD professional community self correction matured in the mid-1990s leading to an AIAA Fluid Dynamics technical committee-authored guide classifying CFD content/completeness requirements for conference/archival publication submittal, AIAA (1998). Broadly endorsed by US national and military establishment laboratories, aerospace industry and CFD academics, reportable *a posteriori* data was categorized *verification*, *benchmark* or *validation* (VBV) for these definitions:

- *verification*: comparison of CFD data with an exact solution, e.g., the rotating cone
- *benchmark*: prediction comparison with an alternative CFD solution independently assessed accurate, e.g., laminar driven/thermal cavities
- *validation*: comparison of CFD data with quality experimental data, e.g., laminar step wall diffuser, turbulent square duct

Optimal MODIFIED CONTINUOUS Galerkin CFD, First Edition. A. J. Baker.

Companion Website: www.wiley.com/go/baker/GalerkinCFD

In addition, citing CFD data generated on a single mesh was cause for rejection, initiating the practice of solution adapted mesh refinement.

The VBV protocol has well served the CFD community to present in thwarting promulgation of substandard colorful fluid dynamics *a posteriori* data. A measure of solution quality quantification that emerged in VBV practice was reporting the "*mesh independent solution*" without rigorous definition of a mathematical measure of independence. Via weak form theory perspective, such a solution exists if and only if its *degrees of freedom* (DOF) distribution corresponds one to one with knot coefficients of an interpolation of the exact solution. This is a truly *unlikely* event in Navier-Stokes PDE systems CFD practice, which emphasizes the requirement to rigorously formalize *uncertainty quantification* (UQ).

10.2 Observations on Computational Reliability

The US National Academy of Sciences published a comprehensive assessment of complex computational mechanics *reliability*, National Academies Press (2012). Surveying *verification, validation* and *uncertainty quantification* (VVUQ) definitions, and their respective mathematical foundations, the report identifies computational mechanics research topics considered requisite for practice *outcome* improvement.

Directly quoting the report (page 4), "the areas identified for verification research are:

- development of goal-oriented *a posteriori* error-estimation methods that can be applied to mathematical models that are more complicated than linear elliptic partial differential equations (PDEs).
- development of algorithms for goal-oriented *error estimates* that scale well on massively parallel architectures, especially given complicated grids (including adaptive-mesh grids).
- development of methods to estimate *error bounds* when meshes cannot resolve important scales. An example is turbulent fluid flow.
- development of reference solutions, including "manufactured" solutions, for the kinds of complex mathematical models described above.
- for computational models that are composed of simpler components, including hierarchical models: development of methods that use numerical-error estimates from the simpler components, along with information about how the components are coupled, to produce *numerical-error estimates* for the overall model."

These NAS report identifications complement the two decades old VBV categorization, in particular requiring mathematical rigor for the non-quantified issue of "mesh independent solution." The VV partition of VVUQ translated into "plain English" is:

- *verification* – "Are we solving the equations right?"
- *validation* – "Are we solving the right equations?"

The UQ partition needs no translation. It demands *rigorous* formalizing of error *quantification*, *a posteriori* error *estimation*, error *bounding* and, by extension, quantifying mesh resolution *inadequacy* in complex computational practice.

Text detailed VBV *a posteriori* data, Chapters 3–9, solidly validate continuum weak form linear theory error estimation, bounding, quantification, and summarily *optimal* mesh solution identification. That *linear* theory precisely predicts performance of classic/optimal modified continuous Galerkin CFD spatially discrete algorithms for explicitly *nonlinear* incompressible-thermal NS/RaNS/LES PDE/*m*PDE systems is thus VVUQ-topics fully assessed. Thereby, mathematically rigorous weak formalisms are validated to address individually and comprehensively weak form CFD algorithm solution error *quantification* NAS report-cited as required.

For incompressible-thermal NS/RaNS/LES, this closing chapter content summarizes weak form CFD theory attributes specifically pertinent to the UQ partition. While *m*PDE theory is extended to select compressible NS systems, for example potential flow isentropic (weak) shock capturing, Section 3.12, PRaNS hypersonic aerothermodynamics, Appendix B, summary content is *smooth* solution *only* pertinent to compressible NS/Euler weak form CFD. This excludes key *non-smooth* solution features, specifically strong shocks, contact discontinuities, addressed by alternative weak formulations including *discontinuous* Galerkin (DG), Cockburn (1999), *discontinuous* non-Galerkin, Toro (1999), LeVeque (2002), characteristics *m*PDE Galerkin, Iannelli (2006).

10.3 Solving the Equations Right

For NS/RaNS/LES PDE/*m*PDE systems possessing *smooth* solutions, weak form theory addresses "solving the equations right" with precision. The theory *always* identifies the *continuum approximation* to the subject PDE/*m*PDE system state variable $\{q(\mathbf{x},t)\}$

$$\{q(\mathbf{x},t)\} \approx \left\{q^N(\mathbf{x},t)\right\} \equiv \text{diag}\left[\{\Psi(\mathbf{x})\}^T\right]\{Q(t)\} \tag{10.1}$$

wherein diag[•] is a *diagonal* matrix with elements the *trial space* $\{\Psi(\mathbf{x})\}$. This set of functions resides in H^m, the space of *all* functions possessing m derivatives in all combinations that exist and are square integrable. The time dependent coefficient matrix $\{Q(t)\}$ contains the continuum approximation unknown *degrees of freedom* (DOF).

The alternative to (10.1) wherein the trial space includes time is discarded as not competitive. Weak form theory *optimal* Galerkin criterion requires that the *approximation error*

$$\{e^N(\mathbf{x},t)\} \equiv \{q(\mathbf{x},t)\} - \{q^N(\mathbf{x},t)\} \tag{10.2}$$

the *continuum* distribution intrinsic to (10.1), be rendered *orthogonal* to the trial space $\{\Psi(\mathbf{x})\}$. An interpolation followed by an *extremization* pair, Section 2.2, transitions this set of *scalar* expressions into a matrix statement of order *identical* with DOF $\{Q(t)\}$ defined in (10.1)

$$\left\{\text{GWS}^N(t)\right\} \equiv \int_\Omega \text{diag}\left[\{\Psi(\mathbf{x})\}\right]\mathcal{L}^m\left(\left\{q^N\right\}\right)\text{d}\tau \equiv \{0\} \tag{10.3}$$

This *very large* order ordinary differential equation (ODE) system expresses the time derivative of DOF $\{Q(t)\}$. Substitution into any Taylor series (TS) ODE algorithm generates the terminal *continuum* weak form approximation nonlinear algebraic matrix statement with iterative solution strategy

$$\begin{aligned}
\left[\mathrm{JAC_}\{Q\}\right]\{\delta Q\}^{p+1} &= -\{\mathrm{F_}\{Q\}\}_{n+1}^{p} \equiv -\left(\left\{\mathrm{GWS}^N(t)\right\} + \theta\mathrm{TS}\right) \\
\{Q\}_{n+1}^{p+1} &= \{Q\}_{n+1}^{p} + \{\delta Q\}^{p+1} = \{Q\}_n + \sum_{\alpha=0}^{p}\{\delta Q\}^{\alpha+1} \\
\left[\mathrm{JAC_}\{Q\}\right] &\equiv \frac{\partial\{\mathrm{F_}\{Q\}\}}{\partial\{Q\}}
\end{aligned} \tag{10.4}$$

Weak form theorization is *complete*(!) in (10.1–10.4), which clearly identifies any NS/RaNS/LES PDE/*m*PDE system state variable *approximation*. Of VVUQ fundamental pertinence, the approximate solution generated via (10.1–10.4), *if*(!) attainable, is theoretically *optimal* among its peers via the Galerkin criterion approximation error orthogonality *constraint* (10.3).

With reference to Table 3.1, the sole decision remaining is the trial space $\{\Psi(\mathbf{x})\}$. Options with labels are:

- direct numerical simulation (DNS), $\Psi(\mathbf{x}) \in H^m$, of global domain span, for $m \approx \mathrm{DOF}^{1/3}$
- pseudo-spectral, $\Psi(\mathbf{x}) \in H^m$ for relatively large $m \ll \mathrm{DOF}^{1/3}$ on unions of relatively large span subdomains
- spatially semi-discrete, $\Psi\left(\mathbf{x}^h\right) \subset H^h \in H^m$ for $m = k \leq 3$ on unions of very small span subdomains of measure h.

The literature on *direct numerical simulation* (DNS) decision for NS is rich, also pseudo-spectral for select NS/RaNS/LES PDE systems. These options come with versatility limitations, in particular state variable boundary conditions (BCs). This issue alone thwarts further consideration compared to the *spatially semi-discrete* decision.

Thus becomes identified the *trial space basis* $\{N_k(\mathbf{x})\}$, a finite-dimensional subset of H^m constituted of modest *completeness* degree k Lagrange/Hermite polynomials. This selection responds precisely to UQ partition issues regarding *rigorous* error *estimation*, *quantification* and *bounding* theories with unconstrained versatility *especially* regarding BCs.

For any *discretization* of an NS/RaNS/LES PDE/*m*PDE + BCs spatial domain with enclosing boundary, that is, $\Omega \cup \partial\Omega \approx \Omega^h \cup \partial\Omega^h$, superscript h denoting mesh measure, the mathematically *rigorous* transition from weak form continuum approximation (10.1) to spatially semi-discrete is

$$\begin{aligned}
\left\{q^N(\mathbf{x},t)\right\} &\equiv \mathrm{diag}\left[\{\Psi(\mathbf{x})\}^T\right]\{Q(t)\} \Rightarrow \left\{q^h(\mathbf{x},t)\right\} \equiv \cup_e\{q(\mathbf{x},t)\}_e \\
\{q(\mathbf{x},t)\}_e &\equiv \mathrm{diag}\left[\{N_k(\boldsymbol{\eta}(\mathbf{x}))\}^T\right]\{Q(t)\}_e
\end{aligned} \tag{10.5}$$

As fully detailed, Chapters 3–9, approximation statement (10.5) *immediately* renders (10.4) computable *and* solvable. Furthermore, *piecewise continuous* solution (10.5) admit *analytical* differential-integral calculus operations in forming the computable statement.

A stationary variational boundary value theoretical analysis for decision (10.5) with rigor *proves* that the weak form Galerkin criterion trial space basis implementation generates the *optimal* solution, that is, the *best possible*(!) among peer discretization alternatives with identical DOF. Hence alternative FD/FV discretization methodology, shown herein to invariably constitute non-Galerkin weak forms, need not be considered.

For NS/RaNS/LES state variable $\{q^h(\mathbf{x},t)\}$ members adhering to an initial-elliptic boundary value (I-EBV) PDE + BCs, this linear theory rigorously extended to initial-value predicts the *error bound*

$$\begin{aligned}\left\|\{e^h(n\Delta t)\}\right\|_E \leq Ch^{2\gamma}\left(\left\{\|\text{data}\|_{\text{L2},\Omega}\right\} + \left\{\|\text{data}\|_{\text{L2},\partial\Omega}\right\}\right) + \text{C}_2\Delta t^{f(\theta)}\|\{Q(t_0)\}\|^2_{H^1(\Omega)} \\ \text{for}: \gamma \equiv \min(k, r-1), f(\theta) = (2,3)\end{aligned} \tag{10.6}$$

Incompressible NS/RaNS/LES state variables invariably include scalar members described by quasi-linear Poisson EBV PDEs + BCs. Pervasive nonlinear state variable coupling requires the resultant EBV form of (10.6) to be time dependent

$$\begin{aligned}\left\|e^h(n\Delta t)\right\|_E \leq Ch^{2\gamma}\left(\|\text{data}\|_{\text{L2},\Omega} + \|\text{data}\|_{\text{L2},\partial\Omega}\right) \\ \text{for}: \gamma \equiv \min(k, r-1)\end{aligned} \tag{10.7}$$

In linear theory predictions (10.6–10.7), $\left\|\{e^h(n\Delta t)\}\right\|_E$ is the set of global *energy norms* for each state variable member at solution evolution time $n\Delta t$. Index k is trial space basis completeness degree, an implementation decision. Error norm magnitudes are *bounded* by state variable specific L2 norms $\left(\left\{\|\text{data}\|_{\text{L2},\Omega}\right\} + \left\{\|\text{data}\|_{\text{L2},\partial\Omega}\right\}\right)$, the theory predicted *data* measure for *everything defined beforehand* on the PDE domain and its boundary.

Should $r < k+1$ then *non-smoothness* of these *data* is predicted to dominate convergence rate, validated for elementary D*E* examples in Section 2.9. Additionally, linear weak form theory proves that the *optimal* mesh solution is that for which (10.6–10.7) attain their *extrema* for fixed DOF. This is predicted associated with *equi-distribution* of element level state variable specific energy norms $\left\|\{e^h(n\Delta t)\}\right\|_e$ throughout $\Omega^h = \cup\,\Omega_e$.

Variational boundary value theory does not address initial-value discretization compromised $\{q^h(\mathbf{x}, n\Delta t)\}$ *spectral content* propagation error. ODE algorithm selection is the second decision in forming (10.4–10.5), invariably *implicit* to handle NS/RaNS/LES nonlinearity. Taylor series (TS) theoretical modal analyses *analytically* quantify time discretization error mechanisms as $O(m^{\text{p}})$, for integer $\text{p} \geq 1$ in non-D wave number $m \equiv \kappa h$.

TS theory predicts that Galerkin criterion weak form discrete implementation mass matrix extremizes spectral content propagation *accuracy* in the discretization peer group for Lagrange trial space basis degree $1 \leq k \leq 3$. Further, *optimal modified* Galerkin *m*PDE *analytical* theory derived mass matrix augmentation

$$m\text{PDE}(\gamma) \equiv -\frac{\gamma\,\Delta t^2}{6}\frac{\partial}{\partial x_j}\left(u_j u_k \frac{\partial}{\partial x_k}\right) \tag{10.8}$$

is predicted and *a posteriori* data validated *optimal* for $\gamma \equiv -(0.5,\ 0.4,\ 0.33333\ \ldots)$ for $1 \leq k \leq 3$.

Finally, absent in (10.1–10.8) theoretical synopsis is the incredible complication of radiation energy exchange *nonlinear* BC in NS/RaNS/LES PDE systems. Replacement of classic Stefan–Boltzmann with *radiosity* theory leads to a Galerkin weak formulation generating an *optimal piecewise discontinuous* semi-discrete solution with rigorous error *bound*

$$\left\| e^h \right\|_{H^0} \leq \mathrm{C}\, h^{k+1} \|\mathrm{data}\|_{\mathrm{L2},\partial\Omega} \tag{10.9}$$

notably without the $r < k+1$ smoothness caveat in (10.6).

Thus is summarized a thoroughly formal CFD algorithm strategy responsive to "solving the equations right" for NS/RaNS/LES PDE/*m*PDE systems. *m*PDE theory rigorously predicts $k=1$ basis *optimal* performance, among the $k=1$ equivalent peer group, via alteration of (10.3) to

$$\left\{ m\mathrm{GWS}^N(t) \right\} \equiv \int_\Omega \mathrm{diag}\left[\{\Psi(\mathbf{x})\}\right] \mathcal{L}^m\left(\{q^N\}\right) \mathrm{d}\tau \equiv \{0\} \tag{10.10}$$

Detailed comparison of Chapters 3–9 *a posteriori* data with linear weak form theory fully validates that (10.1–10.10) defines the optimal performance CFD algorithm with attributes:

- error *quantification*
- *a posteriori* error *estimation*
- error *bounding*
- spectral content propagation accuracy *extremization*
- phase selective $k = 1$ basis $O(h^2)$ *dispersion* error annihilation
- *monotone* solution generation
- error *extremization* optimal mesh quantification
- mesh resolution *inadequacy* measure
- efficient *radiation* implementation with error *bound*.

This answer to "Are we solving the equations right?" does address in completeness VVUQ, specifically the cited UQ partition requirements.

10.4 Solving the Right Equations

The subsequent decision fundamental to VVUQ is "solving the right equations." This is the "physics" decision among laminar NS, time averaged NS (RaNS) and space filtered NS (LES). Not specifically NAS report cited but precisely pertinent to CFD VVUQ is identification of *fundamental* detractions to each option:

- NS laminar: very limited discrete implementation pertinence
- RaNS: *single deviatoric* Reynolds stress closure via kinematic eddy viscosity *models*, fully turbulent flow with low Re^t additional modeling, phase error stabilization tends to excess *artificial diffusion*, scalar closure via turbulent Prandtl/Schmidt number *models*
- RaNS(RSM): *full* Reynolds stress tensor closure eliminates deviatoric assumption, numerous embedded deviatoric tensor *models*, RSM eddy viscosity *model* can lead to algebraic instability, RaNS state variable size doubled, plus cited RaNS detractions

- LES(SGS): theory stress tensor *quadruple* simplified to *single* Reynolds stress, RaNS-LES **DP** PDEs *identical*(!), resolved scale BCs identical to NS and RaNS, deviatoric SGS tensor $O(h^2)$ *models* tend to excess diffusion, turbulent Prandtl/Schmidt number *models*
- LES(RLES): second-order rational theory analytically predicts LES tensor lead *triple*, generates EBV *auxiliary problem* with *a priori* unknown BCs, SFS tensor trivial but predicts bound $O(\delta^2) > O(\overline{u'_j u'_i}) > O(\delta^4)$, reported SFS tensor models exhibit sub-standard performance, state variable bounded domain BCs issue unresolved
- LES(*ar*LES): fluid mechanics perturbation theory organized $O(1, \delta^2, \delta^3)$ state variable size is large, completely analytical theory $\forall$ Re *without* modeling, validation immature

The laminar NS selection is discarded as only DNS makes sense. RaNS widely used single Reynolds stress tensor closure *models*, Chapters 4 and 8, are of *deviatoric* eddy viscosity mixing length theory (MLT*)* genre

$$\overline{u'_i u'_j} \Rightarrow \tau^D_{ij} \equiv -\nu^t \overline{S}_{ij}, i \neq j; \quad \overline{S}_{ij} \equiv \frac{1}{2}\left[\frac{\partial \overline{u}_i}{\partial x_j} + \frac{\partial \overline{u}_j}{\partial x_i}\right] \tag{10.11}$$

CFD algorithm *a posteriori* data for (10.11) fail the turbulent square duct validation. This is not altered replacing MLT algebraic models with dual transport (TKE model) PDE systems. Typical BCs derived via boundary layer (BL) similarity theory, detailed model modifications required for addressing low Re^t near wall resolution.

The full Reynolds stress model (RSM), a less common RaNS code practice, eliminates RaNS deviatoric tensor assumption. State variable size is doubled by addition of $\mathcal{L}(\overline{u'_i u'_j})$ transport PDE + BCs system (8.45), plus $\mathcal{L}(\varepsilon)$ of RaNS TKE closure model. RSM state variable BCs require turbulent BL and near wall low Re^t interpretations. RSM closure *a posteriori* data pass the turbulent square duct validation.

To summarize, RaNS closure modeling generates solution-dependent eddy viscosity *distributions* with *nonlinear* functionality

$$\nu^t(\mathbf{x}, t) \equiv f\left(\mathrm{C}_\alpha; (\omega l)^2, \left|\overline{S}_{ij}\right|; k^2/\varepsilon, k\omega; -\frac{k}{\varepsilon}\overline{u'_k u'_l}\right) \tag{10.12}$$

via mixing length theory (MLT), dual transport (TKE) PDE systems and/or RSM. C_α represents the numerous embedded model constants. RSM eddy viscosity, the last functional in (10.12), is suggested replaced with TKE model ν^t, (8.59–8.61), should algebraic instability be experienced in practice.

Literature dominant LES subgrid scale (SGS) tensor closure models are also of eddy viscosity genre with solution dependent *nonlinear* functionality

$$\nu^t(\mathbf{x}, t) \equiv f\left(\mathrm{C}_S(\mathbf{x}, t), \delta^2, \left|\overline{S}_{ij}\right|\right) \tag{10.13}$$

Closure model commonality with RaNS, (10.12), is evident and not surprising as legacy LES and RaNS **DP** PDE systems are *identical* prior to model insertion. Prandtl/Schmidt number model closure hypotheses for RaNS/LES heat/mass transport are typically scalar multipliers on (10.12–10.13).

RaNS code *literati* fully appreciate that discrete implementations are nonlinearly *unstable* exhibiting algebraic divergence in finite time or iteration number. Instability "correction" is to *augment* $\mathbf{DP}^h$ dissipative flux vector, (8.20), with an $O(h^p)$, $p \le 3$ typically, artificial diffusion operator. The functional essence of this alteration is

$$\left\{f_j^d(\{q = u_i\})\right\} \Rightarrow \left\{(\nu + \nu^t + \nu^{art})\left(\frac{\partial u_i}{\partial x_j} + \frac{\partial u_j}{\partial x_i}\right)\right\} \tag{10.14}$$

The RaNS CFD FD/FV literature contains a wide variety of ν^{art} operators, cf. Figure 5.8, with *modus operandi* diffusion of discretization-intrinsic $O(h^2)$ dispersion error. Recommended RaNS CFD code practice is to increase p once solution DOF have transitioned sufficiently from the IC which is never a RaNS state variable solution. Algorithm iterative instability moderation is relatively less detailed in the LES(SGS) literature. This perhaps results from SGS tensor models, typically constituting an $O(h^2)$ diffusion operator, generating adequate stabilization.

Weak form *m*PDE significant order error annihilation theory *analytically* derives a dissipative flux vector nonlinear differential *term* augmentation of (10.14) comparative form

$$\left\{f_j^d(\{q = u_i\})\right\} \Rightarrow \left\{\left(\nu + \nu^t + \frac{h^2}{12\nu} u_j\, u_k\right)\left(\frac{\partial u_i}{\partial x_j} + \frac{\partial u_j}{\partial x_i}\right)\right\} \tag{10.15}$$

While yet to be proven rigorously, *a posteriori* data solidly document that the *non-positive definite* $u_j\, u_k$ tensor products therein are validated a dissipative/anti-dissipative *annihilator* of discretization-induced $O(h^2)$ dispersion error. The *modulo* $2h$ functionality is responsible for verified $m\mathrm{GWS}^h + \theta\mathrm{TS}$ algorithm NS/RaNS state variable *monotone* DOF distributions for BC-singular driven cavity and entrance channel specifications, Figures 6.3, 8.6, 8.7.

RaNS and LES(SGS) eddy viscosity closure models (10.12–10.13) are just that: *models* verified to possess successes and failures. Dispersion error stabilization of RaNS algorithms via artificial diffusion infusion is the legacy practice. Too little leads to instability or non-monotone DOF distributions, too much compromises genuine diffusion mechanism prediction. LES(SGS) algorithm implementation of high-order FD/FV numerics is one documented approach to moderation of the excess diffusion inherent in $O(h^2)$ SGS tensor models (10.13).

In distinction *m*PDE nonlinear augmentation (10.15) dissipative/anti-dissipative $O(h^2)$ dispersion error *annihilation* is the consequence of analytical theory, not a model. RaNS and LES(SGS) single Reynolds stress tensor models are not derived as stabilizing mechanisms, although they certainly function as diffusion operators. Since (10.15) is analytical, hence amenable to any discrete implementation, then "Are we solving the right equations?" for any NS/RaNS/LES(SGS) *model* closed spatially semi-discrete algorithm should consider replacing artificial diffusion operators with the *m*PDE *nonlinear* augmentation (10.15) implemented as detailed in Section 8.3–8.4.

10.5 Solving the Right Equations Without Modeling

Time averaging the NS **DP** PDE fails to identify either a resolved scale or the dissipation requirement RaNS code addressed by artificial diffusion. Conversely, convolving the NS

DP PDE with *any* spatial filter clearly identifies both via rigorous derivation of LES theory stress tensor *quadruple*

$$\overline{u_j u_i} \equiv \overline{(\bar{u}_j + u'_j)(\bar{u}_i + u'_i)} = \overline{\bar{u}_j \bar{u}_i} + \overline{\bar{u}_j u'_i} + \overline{u'_j \bar{u}_i} + \overline{u'_j u'_i} \tag{10.16}$$

Spatial filter measure δ identifies the resolved scale, energetic capture requirement correlates δ with mesh measure h, and dissipation requirement at the unresolved scale threshold is identified with the last tensor in (10.16).

For these reasons at least, space filtered NS is theoretically *preferable* to RaNS. In legacy practice, however, LES(SGS) **DP** closure modeling compromises (10.16) rigor by assuming that resolved-unresolved scale interaction plus strictly unresolved scale *functionality*, the last three tensors therein, is adequately represented by a *single* tensor involving resolved scale velocity *only*, specifically,

$$\begin{aligned}\tau_{ij}^{\mathrm{D}} &\approx f\left(\overline{\bar{u}_j u'_i} + \overline{u'_j \bar{u}_i} + \overline{u'_j u'_i}\right) \\ &\equiv f(C_S(\mathbf{x}, t), \delta^2, |\bar{S}_{ij}|)\bar{S}_{ij}\end{aligned} \tag{10.17}$$

Thus LES(SGS) and RaNS **DP** PDEs, prior to model implementation, are rendered *identical*. An analytical alternative to τ_{ij}^{D} closure modeling is via Padé rational polynomial interpolation of gaussian filter Fourier transform. The *ordered* rational LES (RLES) process generates analytical expressions for the lead three tensors in (10.16). That pertinent to a $\mathbf{DP}^h$ CFD algorithm is the second-order RLES prediction

$$\overline{\bar{u}_j \bar{u}_i} + \overline{\bar{u}_j u'_i} + \overline{u'_j \bar{u}_i} = \bar{u}_j \bar{u}_i + \frac{\delta^2}{2\gamma}\left[I - \frac{\delta^2}{4\gamma}\nabla^2\right]^{-1}\left(\frac{\partial \bar{u}_j}{\partial x_k}\frac{\partial \bar{u}_i}{\partial x_k}\right) + O(\delta^4) \tag{10.18}$$

Gaussian filter measures are δ and γ and (10.18) is an expansion in $O(\delta^{\mathrm{p}})$ for p even integers. The matrix inverse differential operator therein embedded is documented an RLES theory implementation challenge.

Enforcing the energetic capture constraint $\delta \geq 2h$, classic fluid dynamics *perturbation theory* resolves this issue in predicting $\overline{\bar{u}_j \bar{u}_i} = \bar{u}_j \bar{u}_i + O(\delta^2)$. Thereby, to $O(\delta^4)$ the resolved-unresolved scale interaction tensor pair in (10.16) are identified solutions to the harmonic EBV PDE system

$$\left[I - \frac{\delta^2}{4\gamma}\nabla^2\right]\left(\overline{\bar{u}_j u'_i} + \overline{u'_j \bar{u}_i}\right) = \frac{\delta^2}{2\gamma}\left(\frac{\partial \bar{u}_j}{\partial x_k}\frac{\partial \bar{u}_i}{\partial x_k}\right) + O(\delta^4) \tag{10.19}$$

As system (10.19) is linear constant coefficient, generated *cross stress* tensors $c_{ij} \equiv \left(\overline{\bar{u}_j u'_i} + \overline{u'_j \bar{u}_i}\right)$ will be *symmetric*, *realizable* and translation and Galilean *invariant*, thereby meeting all theoretical requirements for an NS tensor field.

Space filtering NS PDEs for state variable *scalar* members $q(\mathbf{x}, t)$ generates the *vector quadruples*

$$\overline{u_j q} \equiv \overline{(\bar{u}_j + u'_j)(\bar{q} + q')} = \overline{\bar{u}_j \bar{q}} + \overline{\bar{u}_j q'} + \overline{u'_j \bar{q}} + \overline{u'_j q'} \tag{10.20}$$

Perturbation theory predicts $\overline{\overline{u_j}\overline{q}} = \overline{u_j}\overline{q} + O(\delta^2)$, and to $O(\delta^4)$ scalar *resolved-unresolved scale interaction* state variable members are solutions to harmonic EBV PDE systems, for example for species mass fraction

$$\left[I - \frac{\delta^2}{4\gamma}\nabla^2\right]\left(\overline{\overline{u}_j Y'} + \overline{u'_j \overline{Y}}\right) = \frac{\delta^2}{2\gamma}\left(\frac{\partial \overline{u}_j}{\partial x_k}\frac{\partial \overline{Y}}{\partial x_k}\right) + O(\delta^4) \quad (10.21)$$

Asymptotic error estimate (10.7) precisely predicts GWS^h algorithm accuracy/convergence for EBV PDE systems (10.19) and (10.21).

The second-order RLES theory fails to derive a prediction for **D**P unresolved scale velocity tensor product $\overline{u'_j u'_i}$ because the significant term multiplier is δ^4, the truncation order in (10.19). Theory does predict the bound $O(\delta^4) < O(\overline{u'_j u'_i}) < O(\delta^2)$ for the *subfilter scale* (SFS) tensor. The *m*PDE lead order error annihilation theory analytically derives a **D**P SFS tensor

$$-\overline{u'_j u'_i} \equiv \frac{C_S(\delta)h^2\text{Re}}{6}\left(\overline{u}_j\overline{u}_k\frac{\partial \overline{u}_i}{\partial x_k} + \overline{u}_i\overline{u}_k\frac{\partial \overline{u}_j}{\partial x_k}\right) \quad (10.22)$$

validated *operational* at unresolved scale *threshold* $\delta \approx 2h$, the rigorous LES theory requirement. Defining $C_S \equiv C\delta$, for C an $O(10^0)$ constant, meets the RLES theory $O(\delta^3)$ requirement for (10.22). The *m*PDE significant order error annihilation theory analytically identifies SFS vector closures, specifically for mass fraction

$$-\overline{u'_j Y'} \equiv C_S(\delta)\frac{h^2\text{ReSc}}{12}\left[\overline{u}_j\overline{u}_k\frac{\partial \overline{Y}}{\partial x_j}\right] \quad (10.23)$$

To summarize, a rigorously derived model-free response to the query, "Are we solving the right equations?" appears to be the essentially *analytical* (*ar*LES) LES PDE system closure (10.19–10.23). The *ar*LES theory state variable members are $O(1, \delta^2, \delta^3)$ with gaussian filter uniform measure δ defining unresolved scale threshold. Theory derivation is without the word "turbulent," hence no *a priori* assumption exists regarding admissible Reynolds number Re.

10.6 Solving the Right Equations Well-Posed

The analytical, model-free reply to, "Are we solving the right NS/RaNS/LES equations?" summarizes to:

- LES(*ar*LES theory): union of second-order RLES and *m*PDE error annihilation theories *analytically* derives $O(1, \delta^2, \delta^3)$ state variable system via classic fluid mechanics perturbation theory, unresolved scale threshold defined by gaussian filter of uniform measure δ, in theory applicable ∀ Re.

In rigorous NS PDE system convolution the filter penetrates the boundary ∂Ω of LES PDE system domain Ω, Figure 9.17. The near wall resolution (NWR) approach circumvents this by defining a non-uniform measure filter δ(**x**) vanishing on ∂Ω. As filtering and differentiation thereby no longer *commute*

$$\overline{\left(\frac{\partial \mathbf{u}}{\partial x_i}\right)} \neq \left(\frac{\partial \overline{\mathbf{u}}}{\partial x_i}\right), \quad 1 \leq i \leq n \tag{10.24}$$

this definition violates theory and is documented to induce an $O(h^2)$ error distributed on $\Omega \cup \partial\Omega$.

For filter of uniform measure δ, rigorous theory for $\partial\Omega$ penetration requires NS state variable extension in the sense of *distributions*. Consequently, exact NS convolution generates an LES PDE system containing *boundary commutation error* (BCE) integrals, for example for resolved scale D**P**

$$A_\delta\big(D_{ij}(u_i, P)(x_k, t)\big) \equiv \int_{\partial\Omega} g_\delta(x_k - s_k)\left[-P\delta_{ij} + \frac{1}{\mathrm{Re}}\left(\frac{\partial u_i}{\partial x_j} + \frac{\partial u_j}{\partial x_i}\right)\right]\hat{n}_j ds \tag{10.25}$$

BCE integrands contain *unfiltered* NS state variable members. At geometric nodes $\mathbf{x}_b \in \partial\Omega$ the significant integrand strain rate term in (10.25) can be predicted by *approximate deconvolution* (AD). Altered to *differential definition*

$$\mathcal{D}\left(\frac{\partial u_i}{\partial x_j}\hat{n}_j\right) \equiv \frac{\partial u_i}{\partial x_j}\hat{n}_j - \frac{\partial \overline{u}_i}{\partial x_j}\hat{n}_j + \frac{\delta^2}{4\gamma}\nabla^2\left(\frac{\partial \overline{u}_i}{\partial x_j}\hat{n}_j\right) = 0 + O(\delta^4) \tag{10.26}$$

then deleting $O(\delta^4)$, $\mathrm{GWS}^N \Rightarrow \mathrm{GWS}^h$ generates the D**P** BCE integrand discrete approximation on $\Omega^h \cup \partial\Omega^h$

$$\begin{aligned}&\left(\frac{\partial \overline{u}_i}{\partial x_j}\hat{n}_j\right)(\mathbf{x}_b \in \partial\Omega_{\hat{n}_j})\\ &= \int_{\Omega^+} g_\delta(x_k - x_i)\mathrm{S}_e\big(\{N_k(x_i)\}^T\{\mathrm{Q\_}\nabla_j\mathrm{U}I(t)\}\big)\hat{n}_j d\tau, \mathbf{x}_b \in \partial\Omega_{\hat{n}_j}\end{aligned} \tag{10.27}$$

Upon domain extension $\Omega^+ \subset \mathbb{R}^n$, Figure 9.27, numerical quadrature of (10.27) defined integrals generates BCE integral DOF *data* at $\mathbf{x}_b \in \partial\Omega$

$$\begin{aligned}&\frac{1}{\mathrm{Re}}\{\mathrm{Q\_}\nabla_j\overline{\mathrm{U}}I(t)\}_e\hat{n}_j\big|_{x_b \in \partial\Omega_{e,\hat{n}_j}}\\ &\cong \frac{1}{\mathrm{Re}}\sum_{i=1}^{N_x}\sum_{j=1}^{N_y}\sum_{k=1}^{N_z}\begin{pmatrix} w_i w_j w_k g_\delta\big(x_b - x_i, y_b - y_j, z_b - z_i\big) \\ \times\ \mathrm{S}_e\big(\{N_k(x_i)\}^T\{\mathrm{Q\_}\nabla_j\mathrm{U}I(t)\}\big)\end{pmatrix}\hat{n}_j\end{aligned} \tag{10.28}$$

for N_x, N_y, N_z quadrature formula degree with weights w_i, w_j, w_k. Thereby, all BCE integrals for *ar*LES theory $O(1)$ state variable PDEs are evaluable, generated solutions of which adhere to asymptotic error estimate (10.7).

The *ar*LES theory bounded domain PDE system with BCE integrals is initial value-elliptic boundary value (I-EBV) for resolved scale $O(1)$ state variable members, pure EBV for $O(\delta^2)$ members. For an EBV statement to be *well-posed*, boundary conditions (BCs) are required *a priori* specified on the entirety of $\partial\Omega$. Resolved scale velocity at bounding surface nodes $\mathbf{x}_b \in \partial\Omega$ is convolution defined

$$\bar{\mathbf{u}}(\mathbf{x}_b) = (g_\delta * \mathbf{u}(\mathbf{x}_b)) = \int_\Omega g_\delta(\mathbf{x}_b - \mathbf{y})\mathbf{u}(\mathbf{y})\mathrm{d}\mathbf{y}, \mathbf{x}_b \in \partial\Omega \tag{10.29}$$

Generation of integrand data $\mathbf{u}$ is theoretically predictable via AD

$$\mathbf{u}(\mathbf{x}) \cong \bar{\mathbf{u}}(\mathbf{x}) - \frac{\delta^2}{24}\nabla\bar{\mathbf{u}}(\mathbf{x}) + O(\delta^4) \tag{10.30}$$

The simplifying approximations leading to legacy LES(SGS) D**P** are without BCE integrals, hence utilize Dirichlet BC $\bar{\mathbf{u}}(\mathbf{x}_b) \equiv \mathbf{u}(\mathbf{x}_b)$. This approximation constitutes truncation of (10.30) at $O(\delta^2)$, identically the *significance* order of SGS tensor models (10.13). Mathematical correspondence accrues to filter measure vanishing on $\partial\Omega$, LES(SGS) enforceable via $C_S(\mathbf{x}, t)$, (10.13), RLES theory examined for a $\delta(\mathbf{x})$ definition, Section 9.12.

Rigorous resolution requires derivation of $O(1)$ state variable *non-homogeneous* Dirichlet BCs on $\mathbf{x}_b \in \partial\Omega$. For filtered D**P**, assuming unfiltered $\mathbf{u}$ in (10.29) correlates with large Re NS and/or RaNS fully turbulent BL profiles $\bar{\bar{\mathbf{u}}}(\mathbf{x}) \bullet \widehat{\mathbf{s}} \equiv u_{\mathrm{AD}}(y^+)$ is the enabler, for y^+ the BL similarity coordinate and $\widehat{\mathbf{s}}$ the $\partial\Omega$ unit tangent vector.

AD statement (10.30) is replaced with scalar u_{AD} differential definition

$$\mathcal{D}(u_{\mathrm{AD}}) \equiv u_{\mathrm{AD}} - \bar{u} + \frac{\delta^2}{4\gamma}\nabla^2\bar{u} = 0 + O(\delta^4) \tag{10.31}$$

Deleting $O(\delta^4)$, *not* imposing *a priori* known NS/RaNS no-slip BC $u_{\mathrm{AD}}|_{\mathbf{x}_b} \equiv 0$, $\mathrm{GWS}^N \Rightarrow \mathrm{GWS}^h$ transitions (10.31) to a linear matrix statement containing Green–Gauss divergence theorem generated *non-homogeneous* Neumann BC $\nabla\bar{u}^h \bullet \hat{n}$ at nodes $\mathbf{x}_b \in \partial\Omega$. For specifically designed near wall normal discretization, Figure 9.33, error in $\nabla\bar{u}^h \bullet \hat{n} \Rightarrow \mathrm{DOF}\{\mathrm{d}\bar{U}/\mathrm{d}n\}_e$ is determined restricted to BL similarity *sublayer* reach $y^+ < 0.25\delta^+$.

For LES-GWS^h theory measure *constraint* $\delta \equiv 4h$, encompassing energetic capture requirement $\delta \geq 2h$, generation of non-homogeneous Dirichlet BC (DBC) resolved scale DOF at nodes $\mathbf{x}_b \in \partial\Omega$ involves a modest alteration to validated AD-GWS^h-BCE algorithm. Regular mesh refinement *a posteriori* data confirm monotone convergence $u_{\mathrm{AD}}|_{\mathbf{x}_b} \Rightarrow 0$, also $u_{\mathrm{AD}}(y^+)$ excellent quantitative agreement with RaNS similarity BL profile on $0.1 \leq y^+ \leq 1000$. Added data confirm accurate prediction of laminar BL profiles $u_{\mathrm{AD}}(y^+)$ for $\mathrm{E{+}04} \leq \mathrm{Re} \leq \mathrm{E{+}06}$. Thereby, analytically derived *ar*LES theory assertion $\forall$ Re pertinence is validated for $O(1)$ state variable members.

Non-homogeneous Dirichlet BCs for *ar*LES theory $O(\delta^2)$ state variable members are derived via AD definition of *non-physical* unfiltered variables. With clarifying overbar notation

$$c_{ij} = \bar{c}_{ij} - \frac{\delta^2}{4\gamma}\nabla^2\bar{c}_{ij} + O(\delta^4) \tag{10.32}$$

substituting EBV (10.19) yields

$$c_{ij} = \frac{\delta^2}{2\gamma}\left(\frac{\partial \bar{u}_i}{\partial x_k}\frac{\partial \bar{u}_j}{\partial x_k}\right) + O(\delta^4) \tag{10.33}$$

At nodes $\mathbf{x}_b \in \partial\Omega$, $\bar{c}_{ij} \equiv \left(\overline{\bar{u}_j u'_i} + \overline{u'_j \bar{u}_i}\right)$ convolution generates

$$\begin{aligned}\bar{c}_{ij}(\mathbf{x}_b \in \partial\Omega_{\hat{n}_k}) &\equiv \left(g_\delta * c_{ij}\right), (\mathbf{x}_b \in \partial\Omega_{\hat{n}_k})\\ &= \frac{\delta^2}{2\gamma}\int_\Omega g_\delta(x_m - x_l)\left(\frac{\partial \bar{u}_i}{\partial x_k}\frac{\partial \bar{u}_j}{\partial x_k}\right)_{x_l} \mathrm{d}\tau, \mathbf{x}_b \in \partial\Omega_{\hat{n}_k}\end{aligned} \tag{10.34}$$

$\mathrm{GWS}^N \Rightarrow \mathrm{GWS}^h$ on convolution argument differential definition

$$\mathcal{D}\left(d\bar{u}_i d\bar{u}_j\right) \equiv d\bar{u}_i d\bar{u}_j - \frac{\partial \bar{u}_i}{\partial x_k}\frac{\partial \bar{u}_j}{\partial x_k} = 0 \tag{10.35}$$

leads to a linear matrix statement for DOF $\{\mathrm{Q\_D\bar{U}ID\bar{U}}J\}$. Well-posed requirement for $\bar{c}^h_{ij}$ non-homogeneous Dirichlet BC DOF on $\mathbf{x}_b \in \partial\Omega$ results via numerical quadrature on extended domain Ω^+

$$\begin{aligned}\{\overline{C}IJ(t)\}_{\mathbf{x}_b \in \partial\Omega_{\hat{n}_k}} &= \frac{\delta^2}{2\gamma}\int_{\Omega^+} g_\delta(x_m - x_l)\left(\frac{\partial \bar{u}_i}{\partial x_k}\frac{\partial \bar{u}_j}{\partial x_k}\right)^h \mathrm{d}\tau, \mathbf{x}_b \in \partial\Omega_{\hat{n}_k}\\ &= \frac{\delta^2}{2\gamma}\sum_{l=1}^{N_x}\sum_{m=1}^{N_y}\sum_{n=1}^{N_z}\left(\begin{array}{l} w_l w_m w_n g_\delta(\mathbf{x}_b - x_l, y_b - y_m, z_b - z_n) \\ \times\ \mathrm{S}_e\left(\{N(\mathbf{x})\}^T\{\mathrm{D\bar{U}ID\bar{U}}J(t)\}_e\right)\end{array}\right)_{x_b \in \partial\Omega_{\hat{n}_k}}\end{aligned} \tag{10.36}$$

The sequence (10.32–10.36) completes $O(\delta^2)$ state variable member non-homogeneous Dirichlet BC determination starting with

$$v_j^\Theta(\mathbf{x}, t) = \frac{\delta^2}{2\gamma}\frac{\partial \bar{u}_j}{\partial x_k}\frac{\partial \overline{\Theta}}{\partial x_k} + O(\delta^4) \tag{10.37}$$

$$v_j^Y(\mathbf{x}, t) = \frac{\delta^2}{2\gamma}\frac{\partial \bar{u}_j}{\partial x_k}\frac{\partial \overline{Y}}{\partial x_k} + O(\delta^4) \tag{10.38}$$

10.7 Well-Posed Right Equations Optimal CFD

Well-posed *ar*LES theory filter ordered *continuum* state variable is

$$\{q(\mathbf{x}, t)\} = \left\{u_i, \Theta, Y, c_{ij}, v_j^\Theta, v_j^Y; \phi, P\right\}^T \tag{10.39}$$

The $O(1)$ state variable mPDE I-EBV + BCE flux vector statement is

$$\mathcal{L}^m(\{q\}) = [m]\frac{\partial\{q\}}{\partial t} + \frac{\partial}{\partial x_j}\left[\{f_j\} - \{f_j^d\}\right] - \{s\} = \{0\} \tag{10.40}$$

Non-D matrix definitions for (10.40) are:

$$[m] = \text{diag}\left[1 - \frac{\gamma\,\Delta t^2}{6}\frac{\partial}{\partial x_j}\left(u_j u_k \frac{\partial}{\partial x_k}\right)\right] \tag{10.41}$$

$$\{f_j(\{q\})\} = \begin{Bmatrix} u_j u_i + c_{ij} + P^*|_{n+1}^{p}\delta_{ij} \\ u_j\Theta + v_j^{\Theta} \\ u_j Y + v_j^{Y} \end{Bmatrix} \tag{10.42}$$

$$\left\{f_j^d(\{q\})\right\} = \begin{Bmatrix} \dfrac{2}{\text{Re}}S_{ij} + \dfrac{C_S(\delta)h^2\text{Re}}{6}\left(u_j u_k \dfrac{\partial u_i}{\partial x_k} + u_i u_k \dfrac{\partial u_j}{\partial x_k}\right) \\ \dfrac{1}{\text{RePr}}\dfrac{\partial\Theta}{\partial x_j} + \dfrac{C_S(\delta)h^2\text{RePr}}{12}u_j\,u_k\dfrac{\partial\Theta}{\partial x_k} \\ \dfrac{1}{\text{ReSc}}\dfrac{\partial Y}{\partial x_j} + \dfrac{C_S(\delta)h^2\text{ReSc}}{12}u_j\,u_k\dfrac{\partial Y}{\partial x_k} \end{Bmatrix} \tag{10.43}$$

$$\{s\} = \begin{Bmatrix} -\dfrac{\text{Gr}}{\text{Re}^2}\Theta\widehat{g}_i + \displaystyle\int_{\partial\Omega} g_\delta(x_k - s_k)\left[-P\delta_{ij} + \dfrac{1}{\text{Re}}\left(\dfrac{\partial u_i}{\partial x_j} + \dfrac{\partial u_j}{\partial x_i}\right)\right]\hat{n}_j\text{d}s \\ \dfrac{\text{Ec}}{\text{Re}}\left[\begin{matrix} \dfrac{\partial}{\partial x_j}\left(u_i\dfrac{\partial u_j}{\partial x_i}\right) + \dfrac{\partial^2 c_{ij}}{\partial x_j \partial x_i} \\ -\dfrac{\partial^2}{\partial x_j\partial x_i}\left(\dfrac{C_S(\delta)h^2\text{Re}}{6}\left(u_j u_k\dfrac{\partial u_i}{\partial x_k} + u_i u_k \dfrac{\partial u_j}{\partial x_k}\right)\right)\end{matrix}\right] \\ + \displaystyle\int_{\partial\Omega} g_\delta(x_k - s_k)\left[\dfrac{1}{\text{RePr}}\dfrac{\partial\Theta}{\partial x_j}\right]\hat{n}_j\text{d}s \\ s(Y) + \displaystyle\int_{\partial\Omega} g_\delta(x_k - s_k)\left[\dfrac{1}{\text{ReSc}}\dfrac{\partial Y}{\partial x_j}\right]\hat{n}_j\text{d}s \end{Bmatrix} \tag{10.44}$$

*ar*LES theory $O(\delta^2)$ *resolved-unresolved scale interaction* tensor/vector state variable members reside in the resolved scale *kinetic* flux vector, (10.42). This is *totally distinct* from LES(SGS) and RaNS closure model universal placement in algorithm dissipative flux vectors. BCE integrals reside solely in source term (10.44). The $k=1$ NC/TP trial space basis *optimal* determination in (10.41) is $\gamma \equiv -0.5$ generating $O(m^3)$ spectral content propagation accuracy, $m \equiv \kappa/h$ non-D wave number.

The EBV + BCE + BCs PDE system for pressure projection theory DM^h state variable members is

$$\begin{aligned}\mathcal{L}(\overline{\phi}) &= -\nabla^2\overline{\phi} + \frac{\partial \overline{u}_i^h}{\partial x_i} - \int_{\partial\Omega} g_\delta(x_k - s_k)\left[\frac{\partial \phi}{\partial x_j}\right]\hat{n}_j \mathrm{d}s = 0 \\ \ell(\overline{\phi}) &= -\nabla\overline{\phi}\cdot\hat{\mathbf{n}} - \left(\overline{u}_i - \overline{u}_i^h\right)\hat{n}_i = 0\end{aligned} \tag{10.45}$$

$$\begin{aligned}\mathcal{L}(\overline{P}) &= -\nabla^2\overline{P} - \frac{\partial}{\partial x_j}\left[\frac{\partial \overline{u_j u_i}}{\partial x_i} + \frac{\mathrm{Gr}}{\mathrm{Re}^2}\overline{\Theta}\widehat{g}_j\right] - \int_{\partial\Omega} g_\delta(x_k - s_k)\left[\frac{\partial P}{\partial x_j}\right]\hat{n}_j ds = 0 \\ \ell(\overline{P}) &= \nabla\overline{P}\cdot\hat{\mathbf{n}} - \left[\frac{\partial \overline{\mathbf{u}}}{\partial t} - \frac{1}{\mathrm{Re}}\nabla^2\overline{\mathbf{u}}\right]\cdot\hat{\mathbf{n}} = 0\end{aligned} \tag{10.46}$$

Derived for space filtered variables, EBV PDE systems for *ar*LES theory $O(\delta^2)$ state variable members are without BCE integrals

$$\mathcal{L}(c_{ij}) = -\nabla^2 c_{ij} + \frac{4\gamma}{\delta^2}c_{ij} - 2\frac{\partial \overline{u}_i}{\partial x_k}\frac{\partial \overline{u}_j}{\partial x_k} = 0 \tag{10.47}$$

$$\mathcal{L}\left(v_j^\Theta\right) = -\nabla^2 v_j^\Theta + \frac{4\gamma}{\delta^2}v_j^\Theta - 2\frac{\partial \overline{u}_j}{\partial x_k}\frac{\partial \overline{\Theta}}{\partial x_k} = 0 \tag{10.48}$$

$$\mathcal{L}\left(v_j^Y\right) = -\nabla^2 v_j^Y + \frac{4\gamma}{\delta^2}v_j^Y - 2\frac{\partial \overline{u}_j}{\partial x_k}\frac{\partial \overline{Y}}{\partial x_k} = 0 \tag{10.49}$$

Well-posed *a priori* known non-homogeneous Dirichlet BC predictions, also BCE integral evaluations, are enabled by derived AD-GWS^h-BCE-DBC algorithm. Theory optimal measure constraint $\delta \equiv 4h$ is *a posteriori* validated pertinent $\forall$ Re.

To conclude, the *right* solution for the *right* NS *m*PDE system manipulation for incompressible-thermal NS is the *modified continuous* Galerkin GWS$^N \Rightarrow m$GWS$^h + \theta$TS algorithm for model-free *ar*LES theory. In response to NAS report VVUQ assessment third bullet, *resolution adequacy* is quantified via $O(1, \delta^2, \delta^3)$ *diagnostic* data comparison with RLES/fluid mechanics perturbation theory predictions. Specifically, $O(\delta^2)$ state variable member local *extrema* being commensurate with $O(1)$ data *quantifies* filter measure δ is too large. Theory measure constraint $\delta \equiv 4h$ provides practical guidance on requisite *regular solution adapted* mesh refinement. QED.

10.8 The Right Closing Caveat

The mathematically rigorous response to the query, "Are we solving the right NS/RaNS/LES equations optimally?" summarizes to:

- LES(well-posed *ar*LES theory): the union of second-order RLES and *m*PDE error annihilation theories in concert with fluid mechanics perturbation theory *analytically* derives I-EBV plus EBV PDE system for $O(1, \delta^2)$ state variables, bounded domain BCE integrals rigorously included, resolved scale defined by gaussian filter of uniform measure δ; theoretical development, completed in the *continuum*, is guaranteed well-posed via

AD-GWS^h-BCE-DBC algorithm *a priori* derived non-homogeneous Dirichlet BCs at nodes $\mathbf{x}_b \in \partial\Omega$, theory validated pertinent $\forall$ Re; *optimal* mGWS$^h + \theta$TS $k = 1$ TP basis discrete implementation *a posteriori* data confirm model-free laminar-turbulent BL transition prediction, validation otherwise an immature subject.

The dissertation deriving, implementing and validating well-posed *ar*LES theory GWS$^N \Rightarrow m$GWS$^h + \theta$TS with AD-GWS^h-BCE-DBC algorithm is recently defended, Sekachev (2013). The enthusiasm with which Chapter 10 concludes this text anticipates that continuing *a posteriori* data generation will moderate validation immature status.

References

AIAA (1998). "AIAA Guide for the Verification and Validation of Computational Fluid Dynamics Simulations," American Institute of Aeronautics and Astronautics, report AIAA:G-077-1998e.

Cockburn, B. (1999). "Discontinuous Galerkin methods for convection-dominated problems," in Barth, T. and Deconink, H. (eds), *High Order Methods for Computational Physics, Lecture Notes in Computational Science and Engineering*, V. 9, pp. 69–224, Springer Verlag.

Gallagher, R.H., Oden, J.T., and Zienkiewicz, O.C. (1975–1988), Principal Editors. *Finite Elements in Fluids*, Volumes 1–7, Wiley Interscience, London.

Iannelli, J. (2006). *Characteristics Finite Element Methods in Computational Fluid Dynamics*, Springer, Heidelberg.

LeVeque, R.Q. (2002). *Finite Volume Methods for Hyperbolic Problems*, Cambridge University Press.

National Academies Press (2012). "Assessing the Reliability of Complex Models: Mathematical and Statistical Foundations of Verification, Validation, and Uncertainty Quantification," National Research Council Report, Washington DC.

Sekachev, M. (2013). "Essentially analytical theory closure for thermal-incompressible space filtered Navier–Stokes PDE system on bounded domains," PhD dissertation, University of Tennessee.

Toro, E.F. (1999). *Riemann Solvers and Numerical Methods for Fluid Dynamics*, Springer.

Appendix A

Well-Posed *ar*LES Theory PICMSS Template

Space filtered NS well-posed *ar*LES theory $k=1$ TP basis $m\text{GWS}^h + \theta\text{TS}$ algorithm is detailed, Sections 9.8–9.10, 9.16–9.19, in text element matrix syntax. The $n=3$ unsteady $8\times1\times16$ symmetric half thermal cavity *a posteriori* data detailed in Chapters 9.20–9.21 were generated using the open source PICMSS (*P*arallel *I*nteroperable *C*omputational *M*echanics *S*ystem *S*imulator) platform, a joint development of UT CFD Laboratory and the UT/ORNL Joint Institute for Computational Science (JICS).

Following is a key topics extraction of thc PICMSS template file for well-posed *ar*LES theory $m\text{GWS}^h + \theta\text{TS}$ algorithm. It is available at www.wiley.com/go/baker/GalerkinCFD/arLEStheory. Template snippets correlated with algorithm $O(1,\delta^2,\delta^3)$ state variable member element matrix statements are shortened by columns of (.).

```
PRIMARY_VARIABLES 27
  U V W TEMP PHI
  SPHI
  C11 C12 C13 C22 C23 C33
  T1 T2 T3
  DUBDY DVBDY DWBDY DTBDY
  UAD VAD WAD TAD
  UBCE VBCE WBCE TBCE *

  SECONDARY_VARIABLES 0
  *
  FIXED_VARIABLES 5
  DCONST XCVAL YCVAL ZCVAL TCVAL *

  DERIVED_VARIABLES 6
 UU  2  U U
 UV  2  U V
 UW  2  U W
```

Optimal MODIFIED CONTINUOUS Galerkin CFD, First Edition. A. J. Baker.

Companion Website: www.wiley.com/go/baker/GalerkinCFD

```
 VV 2 V V
 VW 2 V W
 WW 2 W W
 *
 OPERATORS 93
 OP_1    *   c200t  *
 OP_2    *   c20x   *
 OP_3    *   c20y   *
 OP_4    *   c20z   *
 OP_5    U   c300x  *
 OP_6    V   c300y  *
 OP_7    W   c300z  *
 OP_8    *   c2kk   *
 OP_9    U   c3x00  *
 OP_10   U   c3y00  *
 OP_11   U   c3z00  *
.
.
.
 OP_90 UU  c30zzh *
 OP_91 UV  c30zzh *
 OP_92 VV  c30zzh *
 OP_93 *   cn200  *

NUMBER_OF_SETS 7

UVWPHI_EQUATION_SET 0:

OPERATORS 61
  OP_1  OP_2  OP_3  OP_4  OP_5  OP_6  OP_7  OP_8  OP_9  OP_10
  OP_11 OP_12 OP_13 OP_14 OP_15 OP_16 OP_17 OP_18 OP_19 OP_20
  OP_21 OP_22 OP_23 OP_24 OP_25 OP_26 OP_27 OP_28 OP_29 OP_30
  OP_62 OP_63 OP_64 OP_65 OP_66 OP_67 OP_68 OP_69 OP_70 OP_71
  OP_72 OP_73 OP_74 OP_75 OP_76 OP_77 OP_78 OP_79 OP_80 OP_81
  OP_82 OP_83 OP_84 OP_85 OP_86 OP_87 OP_88 OP_89 OP_90 OP_91
  OP_92

EQUATIONS 5

RHS_U 29
  ONE OP_1 U
  ONE OP_5 U
  ONE OP_6 U
  ONE OP_7 U
  ONE OP_2 C11
  ONE OP_3 C12
  ONE OP_4 C13
.
.
```

```
RHS_V 30
  ONE OP_1 V
  ONE OP_5 V
  ONE OP_6 V
  ONE OP_7 V
  ONE OP_2 C12
  ONE OP_3 C22
  ONE OP_4 C23
.
.
RHS_W 29
  ONE OP_1 W
  ONE OP_5 W
  ONE OP_6 W
  ONE OP_7 W
  ONE OP_2 C13
  ONE OP_3 C23
  ONE OP_4 C33
.
.
RHS_TEMP 18
  ONE OP_1 TEMP
  ONE OP_5 TEMP
  ONE OP_6 TEMP
  ONE OP_7 TEMP
  ONE OP_2 T1
  ONE OP_3 T2
  ONE OP_4 T3
.
RHS_PHI 4
  ITDELT OP_2 U
  ITDELT OP_3 V
  ITDELT OP_4 W
  ITDELT OP_8 PHI
.
JAC_U_by_U 18
  MONE OP_1
  MONE OP_5
  MONE OP_6
  MONE OP_7
  MONE OP_9
.
.
JAC_U_by_V 4
.
.
JAC_U_by_W 4
.
.
```

```
JAC_U_by_TEMP 0
.
JAC_U_by_PHI 1
  MITDELT OP_2
.
JAC_V_by_U 4
.
JAC_V_by_V 18
.
JAC_V_by_W 4
  MONE OP_14
.
JAC_V_by_TEMP 1
.
JAC_V_by_PHI 1
.
JAC_W_by_U 4
.
.
.
NO_NEU_BC_TYPE_U 0

NO_NEU_BC_TYPE_V 0

NO_NEU_BC_TYPE_W 0

NO_NEU_BC_TYPE_TEMP 0

NO_NEU_BC_TYPE_PHI 0
```

SPHI_EQUATION_SET 1:

CIJ_EQUATION_SET 2:

```
OPERATORS 11
  OP_8 OP_30 OP_50 OP_51 OP_52 OP_53 OP_54 OP_55 OP_56 OP_57 OP_58

EQUATIONS 6

RHS_C11 5
.
RHS_C12 5
.
RHS_C13 5
.
.
.
JAC_C11_by_C11 2
  MONE OP_8
  TTTT55 OP_30
```

```
JAC_C11_by_C12 0
JAC_C11_by_C13 0
JAC_C11_by_C22 0
JAC_C11_by_C23 0
JAC_C11_by_C33 0
.
JAC_C12_by_C12 2
.
JAC_C13_by_C13 2
.
JAC_C22_by_C22 2
.
JAC_C23_by_C23 2
.
JAC_C33_by_C33 2
.
.
NO_NEU_BC_TYPE_C11 0
.
.
NO_NEU_BC_TYPE_C33 0
```

TJ_EQUATION_SET 3:

```
OPERATORS 5
  OP_8 OP_30 OP_59 OP_60 OP_61

EQUATIONS 3

RHS_T1 5
  ONE     OP_8   T1
  TTTT54  OP_30  T1
  TTTT56  OP_59  U
  TTTT56  OP_60  U
  TTTT56  OP_61  U
.
RHS_T2 5
.
RHS_T3 5
.
JAC_T1_by_T1 2
.
.
JAC_T2_by_T2 2
.
.
JAC_T3_by_T3 2
  MONE   OP_8
  TTTT55 OP_30
```

```
NO_NEU_BC_TYPE_T1 0

NO_NEU_BC_TYPE_T2 0

NO_NEU_BC_TYPE_T3 0
```

DUBDY_EQUATION_SET 4:

```
OPERATORS 0

EQUATIONS 4

RHS_DUBDY DUBDY -222
  U

RHS_DVBDY DVBDY -222
  V

RHS_DWBDY DWBDY -222
  W

RHS_DTBDY DTBDY -222
  TEMP
```

UVWTAD_EQUATION_SET 5:

```
OPERATORS 3
  OP_8 OP_30 OP_93

EQUATIONS 4

RHS_UAD 3
  ONE   OP_30 UAD
  MONE  OP_30 U
  CE1   OP_8  U

RHS_VAD 3
  ONE   OP_30 VAD
  MONE  OP_30 V
  CE1   OP_8  V

RHS_WAD 3
  ONE   OP_30 WAD
  MONE  OP_30 W
  CE1   OP_8  W

RHS_TAD 3
  ONE   OP_30 TAD
  MONE  OP_30 TEMP
  CE1   OP_8  TEMP
```

```
UAD_EQUATION_SET 5:

JAC_UAD_by_UAD 1
  MONE OP_30
JAC_UAD_by_VAD 0
JAC_UAD_by_WAD 0
JAC_UAD_by_TAD 0
.
JAC_VAD_by_VAD 1
.
JAC_TAD_by_TAD 1
  MONE OP_30

NO_NEU_BC_TYPE_UAD 1

NEU_RHS_UAD 1
  CE1 OP_93 DUBDY
.
NO_NEU_BC_TYPE_VAD 1

NEU_RHS_VAD 1
 CE1 OP_93 DVBDY
.
NO_NEU_BC_TYPE_WAD 1

NEU_RHS_WAD 1
  CE1 OP_93 DWBDY
.
NO_NEU_BC_TYPE_TAD 1
.
UBCE_EQUATION_SET 6:
.
EQUATIONS 4

RHS_UBCE UBCE -444
  UAD

RHS_VBCE VBCE -444
  VAD

RHS_WBCE WBCE -444
  WAD

RHS_TBCE TBCE -444
  TAD
END SYSTEM
```

Appendix B

Hypersonic Parabolic Navier–Stokes:

parabolic time averaged compressible NS for hypersonic shock layer aerothermodynamics

B.1 High Speed External Aerodynamics

The content of the main chapters addresses incompressible-thermal Navier–Stokes (NS) conservation principle systems. The *aerodynamics* class spans subsonic through hypersonic universally characterized by a *unidirectional* flowfield. Chapter 4 details weak form algorithm validations for the admitted *parabolic* (PNS) approximations to subsonic incompressible NS and RaNS PDE systems for confined and external aerodynamics. This Appendix extends weak form PNS algorithm theorization/validation to compressible RaNS applicable to external *aerothermodynamics* at hypersonic freestream Mach number.

The CFD literature spans three decades, documenting algorithms for compressible PNS theorizations, laminar and turbulent. The pioneering Brooklyn Polytechnic Institute publications detailing a supersonic laminar cartesian $n=2$, 3 finite difference (FD) algorithm, Rubin and Lin (1971), is followed by Lin and Rubin (1982) and Rubin (1982). Collaboration between NASA Ames and Iowa State University, Tannehill *et al.* (1981), Rakich (1983), led to $n=3$ compressible PNS FD theory generalization to a body-fitted coordinate system. They additionally pioneered theoretical analysis of *departure solution* suppression, the PNS assumption-generated elliptic pressure instability.

Coincidental was the derivation of *implicit approximation factorization* (AF) linear algebra jacobians, Chaussee *et al.* (1991), an adaptation of the original AF algorithm for compressible Euler (inviscid NS) PDE system, Beam and Warming (1978). Thus emerged the AF Wright Aeronautical Laboratory (AFWAL) PNS code, Shanks *et al.* (1982). Parallel

Optimal MODIFIED CONTINUOUS Galerkin CFD, First Edition. A. J. Baker.

Companion Website: www.wiley.com/go/baker/GalerkinCFD

activities generated the AF Ballistic Missile Office (BMO) HYCOM PNS code, Hall (1987), based on the implicit FD theory of Lubard and Helliwell (1974). In this time frame NASA Lewis developed the PEPSIG PNS code implementing the linearized-block-implicit (LBI) FD algorithm, Briley and McDonald (1977), applicable to subsonic/transonic bounded domain combustion aerodynamics.

The AFWAL and HYCOM PNS codes, regularly supporting hypersonic aerodynamics analyses, proved deficient regarding drag and heat transfer prediction accuracy, Stalnaker *et al.* (1986), also numerics possessing excessive artificial diffusion. This Appendix documents the $m\mathrm{GWS}^h + \theta\mathrm{TS}$ steady aerothermodynamics compressible hypersonic time averaged *parabolic* RaNS (PRaNS) algorithm and details available verification/benchmark/validation *a posteriori* data.

Truly unique to steady hypersonic PRaNS theory is the *state variable* definition, established via Favre time averaging that leads to axial momentum appearance throughout. Also unique is the requirement for PRaNS *m*PDE system to be cast in a *body fitted* coordinate system, leading to contravariant vector resolution and an AF jacobian iteration strategy generated via matrix tensor products.

Further specific is a revised theory for *departure solution* suppression, the parabolic approximation pressure instability resident in the shock layer wall-adjacent boundary layer (BL) subsonic reach. Finally, hypersonic aerothermodynamics necessitates a reacting *real gas*, variable gamma equation of state for air species dissociation/generation, detached bow shock trajectory prediction, a turbulence closure model and an algorithm to self generate the shock layer PRaNS domain state variable initial condition (IC).

B.2 Compressible Navier–Stokes PDE System

The topic is weak form algorithm derivation, implementation and assessments for compressible reacting real gas high onset Mach number (Ma_∞) *hypersonic* missile aerothermodynamics with shock capturing. The missile essential geometry with slice and flair, pertinent to trajectory control, is illustrated in Figure B.1.

From Chapter 1, the continuum compressible NS PDE system remains (1.11–1.13)

$$\mathrm{D}M\colon \frac{\partial \rho}{\partial t} + \nabla \cdot \rho \mathbf{u} = 0$$

$$\mathbf{DP}\colon \frac{\partial \rho \mathbf{u}}{\partial t} + \nabla \cdot \rho \mathbf{u}\mathbf{u} = \rho \mathbf{g} + \nabla \mathbf{T}$$

$$\mathrm{D}E\colon \frac{\partial \rho e}{\partial t} + \nabla \cdot (\rho e + p)\mathbf{u} = s - \nabla \cdot \mathbf{q}$$

and requires constitutive closure for traction vector $\mathbf{T}$ and heat flux vector $\mathbf{q}$. For laminar flow $\mathbf{T}$ contains pressure and a viscosity hypothesis involving the Stokes velocity strain-rate tensor. For variable density ρ and multiplying through by ∇ the alteration to (1.14) is

$$\nabla \mathbf{T} = -\nabla p + 2\nabla \bullet \left(\mu \mathbf{S} - \frac{2\mu}{3} \mathrm{S}\delta \right), \quad \mathbf{S} \equiv \frac{1}{2}\left(\nabla \mathbf{u} + \nabla^T \mathbf{u}\right) \tag{B.1}$$

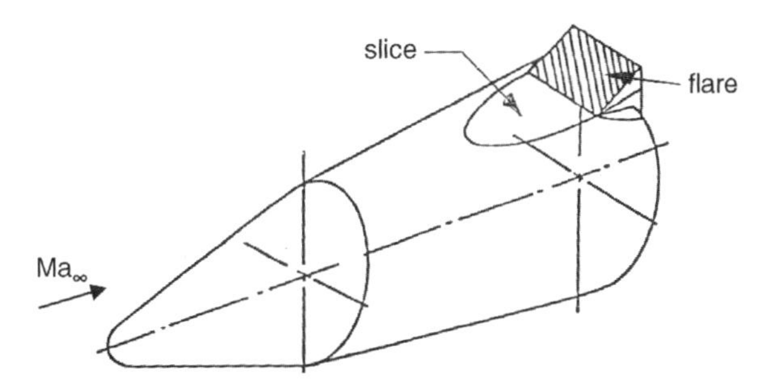

Figure B.1 Missile geometry essence with slice and flair

where p is pressure, μ is fluid absolute viscosity and **S** is Stokes tensor dyadic symbolized. The companion Fourier conduction hypothesis for heat flux vector **q** is unaltered (1.15)

$$\nabla \cdot \mathbf{q} = -\nabla \bullet k\nabla T$$

where k is fluid thermal conductivity and T is temperature.

In preferred PDE *divergence form* the compressible NS state variable is $\{q(\mathbf{x}, t)\} = \{\rho, \mathbf{m}, E, Y\}^T$ for density ρ, momentum vector $\mathbf{m} \equiv \rho\mathbf{u}$, total internal energy $E = \rho e$ and mass fraction Y. As with incompressible NS, the conservation principle system describes a balance of unsteadiness, convective and diffusive processes with delineation identified by non-dimensionalization.

The reference time, length and velocity scales are τ, L and U. Buoyancy usually plays no role in hypersonic aerodynamics and common practice defines $\tau \equiv \mathrm{L/U}$. Since all closure coefficients are temperature dependent, specific heat, viscosity and thermal conductivity reference states are required identified. Once these are stated and inserting (B.1) and (1.15), the non-dimensional (non-D) compressible laminar NS PDE system is

$$\mathrm{D}M\text{:}\ \frac{\partial \rho}{\partial t} + \nabla \cdot \mathbf{m} = 0, \quad \mathbf{m} \equiv \rho\mathbf{u} \tag{B.2}$$

$$\mathrm{D}\mathbf{P}\text{:}\ \frac{\partial \mathbf{m}}{\partial t} + \nabla \cdot \mathbf{mm}/\rho + \mathrm{Eu}\nabla p - \frac{1}{\mathrm{Re}}\nabla \cdot 2\mu^*\left(\mathbf{S} - \frac{2}{3}\mathbf{S}\delta\right) = 0 \tag{B.3}$$

$$\mathrm{D}E\text{:}\ \frac{\partial E}{\partial t} + \nabla \cdot (E + p)\mathbf{m}/\rho - \frac{1}{\mathrm{RePr}}\nabla \cdot k^*\nabla T - s_E^* = 0 \tag{B.4}$$

$$\mathrm{D}Y\text{:}\ \frac{\partial Y}{\partial t} + \nabla \cdot Y\,\mathbf{m}/\rho - \frac{1}{\mathrm{Re}Sc}\nabla \cdot \mathrm{D}^*\nabla Y - s_Y^* = 0 \tag{B.5}$$

In (B.2–B.5) superscript * denotes non-dimensionalized by the reference value. The Euler, Reynolds, Prandtl, Schmidt and Eckert numbers parameterizing the compressible NS PDE system involve the reference state and (typically) farfield onset freestream symbolized by subscript ∞. The definitions are $\mathrm{Eu} = \rho_{\mathrm{ref}}\mathrm{U}^2/p_{\mathrm{ref}}$, $\mathrm{Re} \equiv \rho_{\mathrm{ref}}\mathrm{UL}/\mu_{\mathrm{ref}}$, $\mathrm{Pr} \equiv \mu_{\mathrm{ref}}c_{p\mathrm{ref}}/k_{\mathrm{ref}}$, $\mathrm{Sc} \equiv \mathrm{D}_{\mathrm{ref}}/\mu_{\mathrm{ref}}$ for $\mathrm{D}_{\mathrm{ref}}$ the reference binary diffusion coefficient for species Y and $\mathrm{Ec} \equiv \mathrm{U}^2/c_{p\mathrm{ref}}\mathrm{T}_{\mathrm{ref}}$. For compressible NS the Peclet number Pe does not replace RePr.

The laminar compressible NS PDE system is of little practical use as by definition $\mathrm{U} \equiv \mathrm{U}_\infty$ is *very* large at hypersonic Mach number. As with incompressible NS the time averaged alteration of (B.3) introduces a Reynolds stress tensor. The *density weighted* velocity time average, denoted by superscript tilde (∼), Favre (1965), is the alteration of (4.52)

$$\tilde{u}_i(x_k) \equiv \left(\frac{1}{\bar{\rho}}\right) \lim_{T\to\infty} \frac{1}{T}\int_t^{t+T} \rho(x_k,\tau)u_i(x_k,\tau)\,d\tau \tag{B.6}$$

where superscript bar denotes conventional time average. Via this definition and moving to tensor notation the time average of (B.2) is

$$D\overline{M}: \quad \frac{\partial\bar{\rho}}{\partial t} + \frac{\partial \overline{m}_j}{\partial x_j} = 0, \quad \overline{m}_j \equiv \bar{\rho}\tilde{u}_j \tag{B.7}$$

Definition (B.6) eliminates the explicit appearance of density fluctuations in the compressible RaNS PDE system. Proceeding to **DP** the Favre resolution for time averaged velocity is

$$u_i(x_k,t) \equiv \tilde{u}_i(x_k) + u_i''(x_k,t) \tag{B.8}$$

To form the time average of the convection term in **DP** multiply (B.8) by ρ and proceed through the process (B.6). The end result is

$$\overline{\rho u_i} \equiv \bar{\rho}\tilde{u}_i + \overline{\rho u_i''} \tag{B.9}$$

However, from the definition following the comma in (B.7) the Favre time average of the fluctuating velocity vanishes, hence in (B.9) $\overline{\rho u_i''} = 0$. With this observation the time average of the convection nonlinearity in **DP** is

$$\overline{\rho u_j u_i} = \bar{\rho}\tilde{u}_j\tilde{u}_i + \overline{\rho u_j'' u_i''} \equiv \overline{m}_i\,\overline{m}_j/\bar{\rho} + \tau_{ij} \tag{B.10}$$

The second equality in (B.10) expresses divergence form state variable members and τ_{ij} is the compressible RaNS Reynolds stress tensor. For hypersonics the legacy literature preference for Reynolds stress closure is an eddy viscosity model

$$\overline{\rho u_i'' u_j''} \approx \tau_{ij} \equiv -2\mu^t\left[\overline{S}_{ij} - \frac{2}{3}\left(\overline{S}_{kk} + \bar{\rho}\bar{k}\right)\delta_{ij}\right] \tag{B.11}$$

where $\bar{k}$ denotes the turbulent kinetic energy time average. The Stokes tensor time average definition is

$$\overline{S}_{ij} \equiv \frac{1}{2}\left(\frac{\partial\tilde{u}_i}{\partial x_j} + \frac{\partial\tilde{u}_j}{\partial x_i}\right) \tag{B.12}$$

Combined with the molecular diffusion term in (B.3) and the definition of turbulent Reynolds number

$$\mathrm{Re}^t \equiv \frac{\mu^t}{\mu_{\mathrm{ref}}} \tag{B.13}$$

the non-D compressible RaNS PDE diffusion term is

$$f\left(\mu^*, \overline{\rho u_i'' u_j''}\right) \Rightarrow \frac{2}{\text{Re}}(\mu^* + \text{Re}^t)\left[\overline{S}_{ij} - \frac{2}{3}\left(\overline{S}_{kk} + \bar{\rho}\bar{k}\right)\delta_{ij}\right] \tag{B.14}$$

The compressible RaNS system requires closure for turbulent eddy viscosity μ^t also turbulent thermal conductivity k^t. Boussinesq formulations of BL type, Baldwin–Lomax (1978), Baldwin–Barth (1990), were specifically derived for external transonic/supersonic aerodynamics CFD practice. Adaptation of k-ϵ or k-ω two PDE closures, Section 4.11, is of course pertinent and certainly more appropriate for piecewise continuous (*only*) surfaces, specifically the body–slice–flair regions in Figure B.1.

A *real gas* equation of state $p = p\ (\rho, E, Y)$ is required to close the compressible RaNS PDE system. For a variable γ air model, or a reacting mixture obeying Dalton's Law

$$p(\mathbf{x}, t) = \mathcal{R}\rho T \sum_i \frac{Y_i}{\text{M}_i} \tag{B.15}$$

wherein $\mathcal{R}$ is universal gas constant and M_i the molecular mass of the ith species with mass fraction Y_i. At hypersonic Mach number flight in the upper atmosphere the real gas air species are $Y_i = (\text{O, N, NO, O}_2, \text{N}_2)$, $1 \le i \le 5$.

Thermodynamic properties for this reacting air model are available, Tannehill and Mugge (1974). The compressible NS total internal energy definition is

$$\frac{1}{\rho}(E + p) - \mathbf{u} \bullet \mathbf{u}/2 = \sum_{i=1}^{5} Y_i c_{p\,i} T + \sum_{i=3}^{5} \frac{\mathcal{R}\theta_v^i/\text{M}_i}{\exp\left(\theta_v^i/T\right) - 1} + \sum_{i=1}^{3} Y_i h_i^0 \tag{B.16}$$

wherein $c_{pi} = (5/2)\mathcal{R}/\text{M}_i$ for $i = 1$, 2 and $(7/2)\mathcal{R}/\text{M}_i$ for $i = 3$, 4, 5. Further, h_i^0 is formation enthalpy, θ_v^i is vibrational temperature with $\mathbf{u} \bullet \mathbf{u}/2 = u^2/2$ kinetic energy per unit mass. In thermal equilibrium θ_v^i is defined identical to translational-rotational temperature T.

The *modified* flux vector *m*PDE system for compressible time averaged RaNS is

$$\mathcal{L}^m(\{q\}) = [m]\frac{\partial\{q\}}{\partial t} + \frac{\partial}{\partial x_j}\left[\{f_j\} - \{f_j^d\}\right] - \{s\} = \{0\} \tag{B.17}$$

For time averaged state variable $\{q(\mathbf{x}, t)\} \equiv \{\rho, m_i, E, Y_\alpha\}^T$, eliminating the defining superscript notations, the non-D definitions for (B.17) are

$$[m] \equiv \left[1 - \frac{\gamma \Delta t^2}{6}\frac{\partial}{\partial x_j}\left(u_j u_k \frac{\partial}{\partial x_k}\right)\right] \tag{B.18}$$

$$\{f_j(\{q\})\} = \left\{\begin{matrix} m_j \\ (m_j/\rho)m_i + \text{Eu}\,p\,\delta_{ij} \\ (m_j/\rho)(E + p) \\ (m_j/\rho)Y_\alpha \end{matrix}\right\} \tag{B.19}$$

$$\left\{f_j^d(\{q\})\right\} = \left\{\begin{array}{c} \frac{\beta \Delta t}{2}\left(u_j u_k \frac{\partial \rho}{\partial x_k}\right) \\ \frac{2}{\mathrm{Re}}(\mu^* + \mathrm{Re}^t)\left(S_{ij} - \frac{2}{3}(S_{kk} + \rho k)\delta_{ij}\right) + \frac{\beta \Delta t}{2}\left(u_j u_k \frac{\partial m_i}{\partial x_k}\right) \\ \frac{1}{\mathrm{Re}}\left(\frac{k^*}{\mathrm{Pr}} + \frac{\mathrm{Re}^t}{\mathrm{Pr}^t}\right)\frac{\partial E}{\partial x_j} + \frac{\beta \Delta t}{2}\left(u_j u_k \frac{\partial E}{\partial x_k}\right) \\ \frac{1}{\mathrm{Re}}\left(\frac{\mathrm{D}^*}{\mathrm{Sc}} + \frac{\mathrm{Re}^t}{\mathrm{Sc}^t}\right)\frac{\partial Y_\alpha}{\partial x_j} + \frac{\beta \Delta t}{2}\left(u_j u_k \frac{\partial Y_\alpha}{\partial x_k}\right) \end{array}\right\} \tag{B.20}$$

$$\{s\} = \left\{\begin{array}{c} 0 \\ 0 \\ -\tau_{ij}^{\mathrm{D}} \frac{\partial u_i}{\partial x_j} \\ s(Y_\alpha) \end{array}\right\} \tag{B.21}$$

As for incompressible NS, mPDE $\gamma \equiv -0.5$ is *optimal* for a $k=1$ TP/NC basis mGWS+θTS algorithm (B.17–B.21) implementation. However, since the topic is the steady parabolic approximation, the unsteady term in (B.17), hence also (B.18), become deleted. In (B.20) the mPDE theory dissipation term coefficient is $\beta\Delta t/2$, Baker and Kim (1987). This selection results from parabolic compressible RaNS algorithm completion prior to the optimal NS theory derivation, Kolesnikov and Baker (2001). Recalling Section 5.8, for this parabolic mGWS+θTS algorithm Δt *is* replaced by the local time scale $|u_i|h_i/u^2 = |\hat{u}_i|h_i/|\mathbf{u}|$, a non-negative number, summation on repeated index with h_i the mesh measure parallel to velocity vector u_i.

B.3 Parabolic Compressible RaNS PDE System

Altering compressible RaNS PDE system (B.17–B.21) to the parabolic approximation requires assuming a steady unidirectional (aerodynamic) flowfield. As detailed in the Reynolds ordering analyses leading to incompressible BL – PNS algorithms, Chapter 4, handling pressure in regions of subsonic flow requires specific attention to avoid ill-posedness. The *symbol theory* predictions for iterative stability, Armfield and Fletcher (1986), Platfoot and Fletcher (1991), are chosen for the compressible parabolic RaNS (PRaNS) mPDE system.

The hypersonic aerothermodynamics PRaNS PDE system is required cast in a *body fitted* coordinate system $\eta_k = \eta_k(x_j)$ such that the velocity vector is dominantly parallel to one member, typically η_1. Invoking this global transformation alters (B.17) to, Chaussee *et al.* (1991)

$$\mathcal{L}^m(\{q\}) = [m]\frac{\partial \det\{q\}}{\partial t} + e_{kj}\frac{\partial}{\partial \eta_k}\left[\{f_j\} - \{f_j^d\}\right] - \det\{s\} = \{0\} \tag{B.22}$$

for transformation jacobian determinant (det) time-invariant. The array e_{kj} contains the transformation matrix entries and (B.22) can be recast in divergence form by inserting e_{kj}

inside the flux vector derivative operator. With this preamble compressible RaNS *m*PDE system alteration to PRaNS space marching *m*PDE system requires:

- assuming steady unidirectional flow
- aligning the η_1 coordinate with missile body axis
- replacing NS state variable E with volume specific stagnation enthalpy ρH introducing Eckert number Ec
- augmenting PRaNS *m*PDE system with subsonic reach pressure gradient moderation parameters
- a real gas, variable γ aerothermodynamics algorithm, equilibrium assumption eliminates handling DY, (B.5)
- selecting a turbulence closure model, turbulent Prandtl number Pr^t definition
- generating a bow shock fitting algorithm for sphere–cone geometries at hypersonic angle of attack with yaw
- deriving a PRaNS state variable initial condition (IC) self-generator.

In the PRaNS space marching *m*PDE system, lower case signifies vector resolution in global cartesian system x_j. A capital letter denotes vector *contravariant* resolution in the η_k body-fitted coordinate system. The derivative transformation involves the chain rule

$$\frac{\partial}{\partial x_j} = \frac{\partial \eta_k}{\partial x_j}\frac{\partial}{\partial \eta_k} \tag{B.23}$$

leading to contravariant velocity and momentum resolutions

$$U_i \equiv e_{i\,j} u_j = \eta_{ij} \det u_j, \quad M_i \equiv \rho U_i \tag{B.24}$$

The dimensionless $e_{k\,j}$ array data for (B.24) are calculated using the $k = 1$ TP basis coordinate transformation metric data, Section 3.9.

Eliminating the unsteady and source terms in (B.22) leads to the compressible PRaNS *m*PDE system flux vector statement in body-fitted coordinate system η_k

$$\mathcal{L}^m(\{q\}) = \frac{\partial}{\partial \eta_k}\left[\{f_k\} - \{f_k^d\}\right] = \{0\} \tag{B.25}$$

The state variable for (B.25) is $\{q(\mathbf{x})\} = \{M_1,\ u_j M_1,\ \rho H M_1\}^T$. For tensor index range $1 \le (j,\ k) \le 3$ the kinetic flux vector definition is

$$\{f_k(\{q\})\} \equiv \left\{ \begin{array}{l} M_k \\ u_1 M_k + \omega\, \mathrm{Eu} \dfrac{\partial \eta_k}{\partial x_1} p \\ u_2 M_k + \varepsilon\, \mathrm{Eu} \dfrac{\partial \eta_k}{\partial x_2} p \\ u_3 M_k + \mathrm{Eu} \dfrac{\partial \eta_k}{\partial x_3} p \\ \rho H\, M_k \end{array} \right\} \tag{B.26}$$

In (B.26) axial pressure gradient PNS stabilization parameter ω is due to Vigneron *et al.* (1978). The wall-normal pressure gradient stability parameter ϵ is due to Armfield and Fletcher (1986), Armfield (1987).

The dissipative flux vector in (B.25) involves spatial derivatives *only* in the plane transverse to the η_1 coordinate, the parabolic assumption key attribute. For notational consistency, recall Section 4.10, hereon invoke the tensor index reduced summation convention $2 \leq (l, m) \leq 3$, hence in (B.25) $\partial\{f_k^d\}/\partial\eta_k \Rightarrow \partial\{f_m^d\}/\partial\eta_m$.

The *m*PDE coefficient $\beta\Delta t/2$ replacement is $\beta|\hat{u}_i|h_i/|\mathbf{u}|$. The parabolic assumption renders the unit vector always parallel to η_1 hence the mesh measure h_i therein becomes the space marching integration step size $\Delta\eta_1$. Via PNS ordering, Section 4.10, $|\mathbf{u}| \approx \tilde{u}_1$ which is always positive as axial recirculation is not admissible. This *m*PDE dissipation term coefficient is an adaptation of the original derivation, Baker and Kim (1987). In D**P** the coefficient u_j is moved outside the embedded state variable derivative. A further alteration defines the term coefficient to be state variable member *dependent*, that is $\beta \Rightarrow \beta_I \equiv C_I\Delta\eta_1/\tilde{u}_1$ for $1 \leq I \leq 5$ denoting PRaNS state variable member.

The resultant PRaNS dissipative flux vector definition for (B.25) is

$$\{f_m^d(\{q\})\} \equiv \left\{\begin{matrix} \beta_1\left(U_m u_l \dfrac{\partial M_1}{\partial x_l}\right) \\ \dfrac{2}{\mathrm{Re}}(\mu^* + \mathrm{Re}^t)\left(S_{1\,m} - \dfrac{2}{3}(S_{l\,l} + \rho\, k)\delta_{1\,l}\right)\det + \beta_2\left(U_m u_l \dfrac{\partial M_1}{\partial x_l}\right) \\ \dfrac{2}{\mathrm{Re}}(\mu^* + \mathrm{Re}^t)\left(S_{2\,m} - \dfrac{2}{3}(S_{l\,l} + \rho k)\delta_{2l}\right)\det + \beta_3\left(U_m u_l \dfrac{\partial M_1}{\partial x_l}\right) \\ \dfrac{2}{\mathrm{Re}}(\mu^* + \mathrm{Re}^t)\left(S_{3\,m} - \dfrac{2}{3}(S_{l\,l} + \rho k)\delta_{3\,l}\right)\det + \beta_4\left(U_m u_l \dfrac{\partial M_1}{\partial x_l}\right) \\ \dfrac{\mathrm{Ec}}{\mathrm{Re}}(\mu^* + \mathrm{Re}^t)S_{j\,m}\det u_j + \dfrac{\det}{\mathrm{Re}}\left(\dfrac{k^*}{\mathrm{Pr}} + \dfrac{\mathrm{Re}^t}{\mathrm{Pr}^t}\right)\dfrac{\partial T}{\partial x_m} + \beta_5\left(U_m u_l \dfrac{\partial \rho H M_1}{\partial x_l}\right) \end{matrix}\right\} \tag{B.27}$$

Computing experience confirms that the resultant PRaNS algorithm is stable for *m*PDE theory term definitions in (B.27). The stagnation enthalpy *m*PDE derivative state variable product also results from computing experience. Finally, in the second row of (B.27) the term multiplied by the Kronecker delta δ_{1l} is always zero; it is retained for appearance consistency.

B.4 Compressible PRaNS *m*PDE System Closure

Closure of PRaNS PDE system (B.25–B.27) requires defining the stability parameters, a decision on turbulence closure model and implementing real gas air thermodynamics. The algorithm for sphere–cone bow shock fitting at hypersonic angle of attack, which leads to automated PRaNS IC generation, is detailed in the following section.

Regarding PRaNS stability, symbol theory predicts the requirement to iteratively retard pressure gradient updates in wall adjacent subsonic reaches of the shock layer. The Platfoot and Fletcher (1991) theory improvement on the original Vigneron theory for iterative retardation of the axial (η_1) pressure gradient is selected.

Theory implementation for axial pressure gradient is

$$\frac{\partial p}{\partial \eta_1} \Rightarrow \omega \left[\frac{p_j^n - p_{j-1}^n}{\xi_j - \xi_{j-1}} \right] + (1 - \omega) \left[\frac{p_{j+1}^{n-1} - p_j^n}{\xi_{j+1} - \xi_j} \right] \tag{B.28}$$

wherein subscript j denotes current axial station in the space marching direction, that is, $\xi_j \equiv \eta_1(0) + n\Delta\eta_1$. PRaNS space marching is initiated at a coordinate plane $\eta_1(0)$ intersecting the sphere–cone afterbody juncture. Superscript n denotes PRaNS algorithm current space marching domain sweep with $n - 1$ the *a posteriori* data from the previous PRaNS algorithm domain sweep.

The symbol theory, Armfield and Fletcher (1986), modifies the Vigneron parameter definition for (B.28) to

$$\omega = \left(\frac{\gamma \left(\mathrm{Ma}_\xi - \mathrm{Ma}_\xi^* \right)^2}{1 + (\gamma - 1)\left(\mathrm{Ma}_\xi - \mathrm{Ma}_\xi^* \right)^2} \right) \left[1 - \exp(-n/N) \right] \tag{B.29}$$

wherein Ma_ξ is the u_1 Mach number at station ξ_j. The offset Mach number Ma_ξ^* and N are theory parameters, suggested values 0.4 and 50.

Additional to axial pressure gradient factor ω, the parameter ϵ moderates wall normal pressure gradient $\partial p / \partial \eta_2$ implementation, (B.26). The definition is, Armfield and Fletcher (1986)

$$\varepsilon \geq \frac{2\left(1 - \mathrm{Ma}_\xi^2 \right)^{1/2}}{1 + (\gamma - 1)\mathrm{Ma}_\xi^2} \tag{B.30}$$

The symbol theory leading to derivation of (B.28–B.30) is thoroughly detailed in Baker *et al.* (1994, Appendix B).

The parabolic approximation requires a unidirectional axial velocity and admits secondary vortex structures in planes transverse thereto, recall Section 4.13. The sphere–cone geometry at angle of attack leads to this velocity vector field structure. The PRaNS **DP** closure for (B.11) is an external compressible mixing length eddy viscosity model, Baldwin and Lomax (1978). It is extended to D*E*(ρ*H*) via a turbulent Prandtl number Pr^t assumption.

The Baldwin–Lomax model is of inner/outer BL type with crossover at a distance r_C perpendicular to the cone–cylinder body surface where both models predict the same μ^t. The inner model is a van Driest length scale damped Prandtl mixing length hypothesis. The alteration in (4.60) is

$$\mu_{inner}^t \equiv \rho\, l^2 |\omega| \tag{B.31}$$

where $|\omega|$ is transverse plane vorticity magnitude

$$|\omega| \equiv \left[\left(\frac{\partial u_1}{\partial \eta_2} - \frac{\partial}{\partial \eta_1} \left(u_2^2 + u_3^2 \right)^{1/2} \right)^2 \right]^{1/2} \tag{B.32}$$

The alteration in (4.60) for Prandtl mixing length is

$$l \equiv \kappa r(1 - \exp(-r^+/\mathrm{A}^+), \quad r^+ \equiv \frac{\rho_w u_1(r) r}{\mu_w} = \frac{\sqrt{\rho_w \tau_w r}}{\mu_w}$$
$$\tau_w \equiv \tau(r=0) = \mu_w \frac{\partial u_1}{\partial x_l} \tag{B.33}$$

where $\kappa = 0.435$ remains the Karman constant, $\mathrm{A}^+ \equiv 26$ and subscript w signifies the value at the missile body surface (wall).

The outer eddy viscosity model is a Klebanoff formulation with alteration of the definition in (4.60) to

$$\begin{aligned}
\mu^t_{outer} &\equiv \rho K C_{cp} F_{wake} F_{Kleb}(r), \\
F_{Kleb}(r) &= \left(1 - 5.5(r C_{Kleb}/r_{\max})^6\right)^{-1}, \\
F_{wake} &= \min\left(r_{\max} F_{\max}, C_{wake}\, r_{\max}\, u^2_{diff}/F_{\max}\right), \\
F_{\max} &= \max(F_r), F_r \equiv l(r)|\omega|/\kappa, \\
u^2_{diff} &\equiv (u_i u_i)_{\max} - (u_i u_i)_{min}, \; for\; 1 \le i \le 3
\end{aligned} \tag{B.34}$$

The model constants for (B.34) are $C_{cp} = 0.09$, $C_{wake} = 0.25$ and $C_{Kleb} = 0.3$. Finally, the Baldwin–Lomax model restricts μ^t by equating it to zero whenever the computed minimum value in the axial velocity BL profile is less than $\mathrm{C}_{mu}\, \mu_\infty$ for $\mathrm{C}_{mu} = 14.0$ and μ_∞ the free-stream value of absolute viscosity.

The real gas variable γ equilibrium aerothermodynamic closure involves the air dissociation–recombination reactions

$$\begin{aligned}
O_2 + \mathrm{M} &\leftrightarrow 2O + \mathrm{M} \\
N_2 + \mathrm{M} &\leftrightarrow 2N + \mathrm{M} \\
N_2 + O_2 + \mathrm{M} &\leftrightarrow 2NO + \mathrm{M}
\end{aligned} \tag{B.35}$$

where M denotes a collision factor which is any of the species present. For real gas air model species $Y_i = (\mathrm{O, N, NO, O_2, N_2})$, $1 \le i \le 5$, the *Law of Mass Action* associated with (B.35) is, Anderson (1989)

$$\frac{Y_1^2}{Y_4} = \frac{\mathrm{M}_1^2}{\mathrm{M}_4}\frac{K_1(T)}{\rho}, \quad \frac{Y_2^2}{Y_5} = \frac{\mathrm{M}_2^2}{\mathrm{M}_5}\frac{K_2(T)}{\rho}, \quad \frac{Y_3^2}{Y_4 Y_5} = \frac{\mathrm{M}_3^2}{\mathrm{M}_4 \mathrm{M}_5} K_3(T) \tag{B.36}$$

where $K_i(T)$ are the equilibrium relations, Park (1985), and M_i denotes the molecular mass of species Y_i.

Equation (B.36) provides three constraints on the five mass fractions Y_i. The additional constraints are global conservation of mass

$$\sum_{i=1}^{5} Y_i = 1, \quad \text{since} \quad Y_i \equiv \frac{\rho_i}{\rho} \text{ and } \sum_{i=1}^{5} \rho_i \equiv \rho \tag{B.37}$$

and local preservation of relative proportion of oxygen–nitrogen molecules for air

$$\left(\frac{Y_1}{M_1}+\frac{Y_3}{M_3}+\frac{2Y_4}{M_4}\right)/21=\left(\frac{Y_2}{M_2}+\frac{Y_4}{M_4}+\frac{2Y_5}{M_5}\right)/79 \tag{B.38}$$

The seven aero*t*hermodynamics state variables $\{q_T\}=\{Y_i,\ p,\ T\}^T$ of the real gas air model are coupled via (B.35–B.38), (B.15–B.16). Substitution of (B.15) into (B.16) eliminates pressure while introducing compressible RaNS state variable $\{q_{\text{RaNS}}\}=\{\rho,\ M_i,\ E\}^T$. Additionally, linear (B.38) substituted into (B.36) reduces out Y_3, Y_4 and Y_5 as independent unknowns.

The end result is the *reduced* $\{q_T^r(\{q_{\text{RaNS}}\})\}$ functional with entries Y_1, Y_2 and T

$$F\left(q_T^r(\{q_{\text{RaNS}}\})\right)=F\left(\rho,m_i,E,Y_1(\{q_{\text{RaNS}}\}),Y_2(\{q_{\text{RaNS}}\}),T(\{q_{\text{RaNS}}\})\right)=\{0\} \tag{B.39}$$

Notationally replacing $\{q_{\text{RaNS}}\}$ by the state variable members q_i, $1\leq i\leq 5$, the differential of (B.39)

$$\sum_{i=1}^{5}\left[\left(\frac{\partial F}{\partial q_T^r}\right)_{q_i}\frac{\partial q_T^r}{\partial q_i}+\left(\frac{\partial F}{\partial q_i}\right)_{q_T^r}\right]\mathrm{d}q_i=0 \tag{B.40}$$

expresses the linear combination of linearly independent differentials $\mathrm{d}q_i$, $1\leq i\leq 5$, via the square bracket coefficient array.

Hence (B.40) holds *iff* this bracket vanishes for each *i*. These constraints lead to five algebraic equations

$$\left\{\frac{\partial q_T^r}{\partial q_i}\right\}=-\left[\left(\frac{\partial F}{\partial q_T^r}\right)_{q_i}\right]^{-1}\left\{\frac{\partial F}{\partial q_i}\right\}_{q_T^r},\quad 1\leq i\leq 5 \tag{B.41}$$

The jacobian matrix inverse in (B.41) contains invariant entries, and provides the relationships required to cast solution of (B.39) as a Newton iteration algorithm for the three independent variables in $\{F(\{q_T^r\})\}=\{0\}$. The jacobian $\partial\{F(\{q_T^r\}\}/\partial\{q_T^r\}$ is formed analytically leading to robust quadratic convergence, Baker *et al.* (1994, App. A).

B.5 Bow Shock Fitting, PRaNS State Variable IC

The PRaNS shock layer domain is bounded by the missile surface aft of sphere–cone juncture, the IC plane and the bow shock. The PRaNS PDE system being initial-value in η_1 requires an IC at an axial coordinate just downstream of sphere–cone juncture. A planar slice through an axisymmetric sphere–cone missile, Figure B.2, details the key geometrical features and parameters of this problem statement.

PRaNS state variable IC generation requires an axisymmetric, laminar real gas compressible NS solution algorithm. For hypersonic PRaNS compatibility, viscosity and thermal conductivity temperature dependence is accounted for via Sutherland's interpolation,

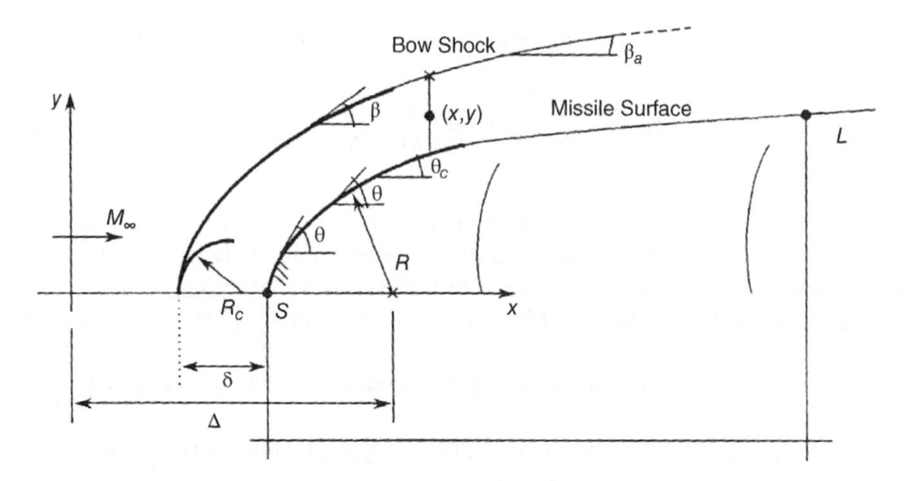

Figure B.2 Axisymmetric sphere–cone bow shock characterization

van Driest (1956), along with the detailed five species real gas air model, Iannelli (1991), Freels (1992).

The compressible NS initiation accurately accounts for the sharp entropy gradient in the thin shock layer, Anderson (1989). Referring to Figure B.2 the bow shock geometry bounding the full NS domain is defined via a spherical forebody geometry correlation, Billig (1967)

$$x = -|R| - |\delta| + \Delta + \frac{R_c}{\tan^2 \beta_\alpha}\left[\left(1 + \frac{y^2 \tan^2 \beta_\alpha}{R_c^2}\right)^{1/2} - 1\right] \tag{B.42}$$

The data for (B.42) include spherical nose radius R, bow shock standoff distance δ, the x coordinate of the spherical nose center is Δ, and the bow shock vertex radius of curvature R_c.

Via conical flow theory, Anderson (1990, Ch. 10), β_α is bow shock asymptotic angle to the freestream for given freestream Mach number, Ma_∞, and sphere to cone transition surface angle θ_c. The geometric parameters for detached bow shock sphere–cone bodies are

$$\frac{\delta}{\mathrm{R}} = 0.142 \exp\left(3.24/\mathrm{Ma}_\infty^2\right), \quad \frac{R_c}{R} = 1.143 \exp\left(0.54/(\mathrm{Ma}_\infty - 1)^{1.2}\right) \tag{B.43}$$

Therefore, for R, Δ, θ_c and onset Mach number Ma_∞ the bow shock location and shock surface trajectory are estimable. The local bow shock angle is

$$\frac{\mathrm{d}y}{\mathrm{d}x} \equiv \frac{1}{\tan \beta} = y/\sqrt{\left(R_c^2 + y^2 \tan \beta^2\right)} \tag{B.44}$$

and as $y \Rightarrow \infty$ $\tan \beta \Rightarrow \tan \beta_\alpha$.

The bow shock pressure rise and post-shock velocity vector resolution are estimated using the Rankine–Hugoniot shock jump relations, Anderson (1989). As a function of β

and freestream state

$$\frac{p_s}{\rho_\infty U_\infty^2} = \frac{p_\infty}{\rho_\infty U_\infty^2}\left[1 + \frac{2\gamma}{\gamma+1}\left(\mathrm{Ma}_\infty^2 \sin^2\beta - 1\right)\right]$$
$$\frac{u_s}{U_\infty} = 1 - \frac{2\left(\mathrm{Ma}_\infty^2 \sin^2\beta - 1\right)}{(\gamma+1)\mathrm{Ma}_\infty^2} \tag{B.45}$$
$$\frac{v_s}{U_\infty} = \frac{1}{\tan\beta}\frac{2\left(\mathrm{Ma}_\infty^2 \sin^2\beta - 1\right)}{(\gamma+1)\mathrm{Ma}_\infty^2}$$

Hence, for any RaNS nodal DOF with abscissa coordinate x, the corresponding bow shock coordinate y_s is determinable from (B.43–B.44). Thereby β is known and the Rankine–Hugoniot relations (B.45) are evaluable.

The IC for unsteady axisymmetric laminar NS state variable DOF in the shock layer is generated via interpolation between the generated bow shock surface (x_s, y_s) data and the body surface (x_b, y_b). Surface flow tangency is assumed, hence Newtonian pressure coefficient $C_p = 2\sin 2\theta$ leads to the surface pressure estimate

$$p_b(x_b, y_b) = \left(\frac{p}{\rho_\infty U_\infty^2}\right)_b \approx \frac{p_\infty}{\rho_\infty U_\infty^2}\sin^2\theta \tag{B.46}$$

The corresponding body surface tangent velocity vector resolution is

$$V_b(x_b, y_b) = \left(\frac{x_b - x_s}{x_L - x_s}\right)U_L \tag{B.47}$$

where U_L is the conical flow theory prediction at $x = L$.

Assuming adiabatic flow in the shock layer the cartesian velocity vector resolution in terms of body angle $\theta(x_b, y_b)$ is

$$\frac{u_b}{U_\infty} = V_b(x_b)\cos\theta\,(x_b, y_b), \quad \frac{v_b}{U_\infty} = V_b(x_b)\sin\theta\,(x_b, y_b) \tag{B.48}$$

For assumed uniform stagnation enthalpy and perfect gas law

$$\frac{\rho_b}{\rho_\infty} = \frac{\dfrac{2\gamma}{\gamma-1}\left(\dfrac{p_b}{\rho_\infty U_\infty^2}\right)}{\dfrac{2\gamma}{\gamma-1}\left(\dfrac{p_\infty}{\rho_\infty U_\infty^2}\right) + 1 - \left(\dfrac{V_b}{U_\infty}\right)^2} \tag{B.49}$$

leading to specific internal energy estimate

$$\frac{e_b}{\rho_\infty U_\infty^2} = \frac{1}{\gamma-1}\left(\frac{p_b}{\rho_\infty U_\infty^2}\right) + \frac{\rho_b}{2\rho_\infty}\left[\left(\frac{u_b}{U_\infty}\right)^2 + \left(\frac{v_b}{U_\infty}\right)^2\right] \tag{B.50}$$

Definitions (B.42–B.50) provide all shock layer data necessary to generate the axisymmetric laminar NS state variable DOF IC via interpolation. The NS solution domain outflow boundary coincides with the missile geometry sphere–cone transition denoted by θ_c in Figure B.2. For reported *a posteriori* data NS solutions were generated using a $k = 1$ TP basis mGWS^h+θTS algorithm compressible laminar NS code, Iannelli (1991).

Once converged, the NS state variable DOF distribution on the shock layer domain outflow plane is converted to PRaNS state variable definition DOF. The transition from laminar NS to turbulent BL profile is accomplished by "turning on" the Baldwin–Lomax model and marching the PRaNS algorithm downstream a sufficient number of small increments $\Delta\eta_1$. This execution omits axial and normal pressure gradient symbol theory modifications. The PRaNS IC coordinate is selected the upstream knurl boundary layer trip of the experimental data supporting validation.

The terminal PRaNS requirement is an algorithm to predict the angular distribution of the bow shock over the cone at angle of attack and yaw. Prior practice employed an iterative strategy, Chaussee (1982), based on estimated pressure downstream of the bow shock at each successive axial station. It accounted for angle of attack but was devoid of yaw. This approach is replaced with a rigorous characteristics explicit formulation for angle of attack and yaw, thoroughly detailed in Baker *et al.* (1994, Appendix C).

B.6 The PRaNS *m*GWS^h+θTS Algorithm

The mGWS^h+θTS algorithm for the PRaNS mPDE system (B.25–B.41) seeks the *approximation* $\{q^N(\mathbf{x}, \eta_1)\}$ to the Favre time averaged state variable $\{q(\mathbf{x}, \eta_1)\} = \{M_1, u_j M_1, \rho H M_1\}^T$. As always, $\{q^N(\mathbf{x}, \eta_1)\}$ is the matrix product of a *trial space* $\{\Psi(\mathbf{x}(\eta_m))\}$, a set of N specified functions lying in H^1 ($k = 1$ basis) with a set of initial value-dependent coefficients. Therefore, in the *continuum*

$$\{q(\mathbf{x}, \eta_1)\} \approx \{q^N(\mathbf{x}, \eta_1)\} \equiv \mathrm{diag}\left[\{\Psi(\mathbf{x})\}^T\right]\{Q(\eta_1)\} \tag{B.51}$$

where diag[•] denotes a *diagonal* matrix with entries $\{\Psi(\mathbf{x})\}$. The column matrix $\{Q(\eta_1)\}$ contains the η_1-dependent coefficients, the sought *degrees-of-freedom* (DOF) of the approximation.

The Galerkin criterion requires the *approximation error* in (B.51), specifically $\{e^N(\mathbf{x}, \eta_1\} \equiv \{q(\mathbf{x}, \eta_1)\} - \{q^N(\mathbf{x}, \eta_1)\}$, be *orthogonal* to the trial space $\{\Psi(\mathbf{x})\}$. Via an interpolation and *extremization*, Section 2.2, the weak form scalar is converted into a matrix statement of order identical to the DOF in $\{Q(\eta_1)\}$.

The resultant *m*odified continuous Galerkin weak statement is

$$\begin{aligned}\{m\mathrm{GWS}^N(\eta_1)\} &\equiv \int_\Omega \mathrm{diag}\left[\{\Psi(\mathbf{x})\}\right]\mathcal{L}^m(\{q^N\})\mathrm{d}\tau \equiv \{0\} \\ &= [\mathrm{m}]\frac{\mathrm{d}\{Q\}}{\mathrm{d}\eta_1} + \left\{\mathrm{RES}\left(\{f_k^N(\bullet)\} - \{f_k^{d\,N}(\bullet)\}\right)\right\}\end{aligned} \tag{B.52}$$

an order $N \times \mathrm{DOF}(\{q^N\})$ nonlinear matrix ordinary differential equation (ODE) system. The nonlinear functional dependencies (•) in (B.52) correlate one-to-one with the mPDE system (B.25–B.27).

Conceptually solving (B.52) for $d\{Q(\eta_1)\}/d\eta_1$ by clearing $[m]^{-1}$ generates the data necessary for completing the Taylor series (TS) underlying ODE integration algorithms. As optimal accuracy is required for *all* parabolic NS algorithms, the $\theta \equiv 0.5$ implicit single step Euler ODE algorithm, subscript n denoting timing, generates the terminal nonlinear algebraic equation system

$$\begin{aligned}&\{m\text{GWS}^N(\eta_1)\} + \theta\text{TS} \equiv \{\text{F}(\{Q\})\} = \{0\}\\&= [\text{m}]\left(\{Q\}_{n+1} - \{Q\}_n\right) + \frac{\Delta\eta_1}{2}\left(\{\text{RES}\}_{n+1} + \{\text{RES}\}_n\right)\end{aligned} \tag{B.53}$$

For iteration index p, the nonlinear matrix iterative solution process for the strongly coupled DOF embedded in matrix statement (B.53) is

$$\begin{aligned}&[\text{JAC}(\{Q\})]\{\delta Q\}^{p+1} = -\{\text{F}(\{Q\})\}^p\\&\{Q\}^{p+1}_{n+1} = \{Q\}^p_{n+1} + \{\delta Q\}^{p+1} = \{Q\}_n + \sum_{\alpha=0}^{p}\{\delta Q\}^{\alpha+1}\\&[\text{JAC}(\{Q\})] \equiv \frac{\partial\{\text{F}(\{Q\})\}}{\partial\{Q\}}\end{aligned} \tag{B.54}$$

To enable the integral evaluations in (B.52), the PRaNS spatial domain of dependence with boundary spanned by η_m is *discretized* as $\Omega \cup \partial\Omega \approx \Omega^h \cup \partial\Omega^h$ by a mesh of measure h for $\cup$ denoting union. This admits replacing the global-span trial space $\{\Psi(\mathbf{x})\}$ with the *much* smaller *trial space basis* $\{N_k(\eta_m)\}$. The PRaNS algorithm choice is the $n=2$, $k=1$ tensor product (TP) basis.

With subscript e denoting the *generic* finite element domain Ω_e and recalling $\Omega^h \equiv \cup\ \partial\Omega_e$, the transition from continuum $\{q^N(\mathbf{x}, \eta_1)\}$ to *discrete* $\{q^h(\mathbf{x}, \eta_1)\}$ solution approximation is

$$\begin{aligned}&\{q^N(\mathbf{x},\eta_1)\} \equiv \text{diag}\left[\{\Psi(\mathbf{x})\}^T\right]\{Q(\eta_1)\} \Rightarrow \{q^h(\mathbf{x},\eta_1)\} \equiv \cup_e\{q(\mathbf{x},\eta_1)\}_e\\&\{q(\mathbf{x},\eta_1)\}_e \equiv \text{diag}\left[\{N_k(\eta_m(\mathbf{x}))\}^T\right]\{Q(\eta_1)\}_e\end{aligned} \tag{B.55}$$

Via definition (B.55) evaluation of the integrals in (B.52) is replaced with integrations on the generic element domain Ω_e. All computed matrices, each of order element DOF, are projected into the global $N\times$DOF-order algebraic matrix statement via *assembly*, denoted S_e, the n-D coupled DOF extension on the Section 2.3 exposition.

Thereby, the strategy for forming (B.52) is completed and the PRaNS algorithm (B.52–B.54) becomes the computable operational sequence

$$\begin{aligned}&\{\text{F}^h(\{Q\})\}^p \Rightarrow \text{S}_e\{\text{F}(\{Q\})\}^p_e\\&[\text{JAC}^h(\{Q\})] \Rightarrow \text{S}_e[\text{JAC}(\{Q\})]_e \equiv \text{S}_e\left(\frac{\partial\{\text{F}(\{Q\})\}_e}{\partial\{Q\}_e}\right)\\&\{\text{F}(\{Q\})\}^p_e \equiv [\text{m}]_e\left(\{Q\}^p_{e,n+1} - \{Q\}_{e,n}\right) + \frac{\Delta\eta_1}{2}\left(\{\text{RES}\}^p_{e,n+1} + \{\text{RES}\}_{e,n}\right)\end{aligned} \tag{B.56}$$

Note all matrices defined in (B.56) lie on the $n=2$ plane spanned by the η_m coordinate system hence will have element matrix prefix B.

The PRaNS mGWS^h+θTS+algorithm discrete state variable for (B.56) is $\{q^h(\mathbf{x},t)\} = \{M_1^h, (u_j M_1)^h, (\rho H M_1)^h\}^T$. For corresponding DOF ordering $\{QI(\eta_1\}$, $1 \le I \le 5$, the element matrix statement for the initial value term in (B.56) is

$$[\mathrm{m}]_e\Big(\{QI\}^p_{e,n+1} - \{QI\}_{e,n}\Big) = \det{}_e[\mathrm{B200}]\Big(\{QI\}^p_{e,n+1} - \{QI\}_{e,n}\Big) \tag{B.57}$$

In a departure from incompressible NS mPDE practice, to avoid differentiating PRaNS state variable products the Green–Gauss divergence theorem is implemented on the transverse plane resolution of both flux vectors in (B.25). The continuum weak form for these terms in (B.52) is

$$\begin{aligned}\{m\mathrm{GWS}^N(\eta_1)\} &\Rightarrow \int_\Omega \mathrm{diag}\left[\{\Psi(\mathbf{x})\}\right]\frac{\partial}{\partial\eta_m}\left(\{f_m^N\} - \{f_m^{d\,N}\}\right)\mathrm{d}\tau \\ &= -\int_\Omega \mathrm{diag}\left[\frac{\partial\{\Psi(\mathbf{x})\}}{\partial\eta_m}\right]\left(\{f_m^N\} - \{f_m^{d\,N}\}\right)\mathrm{d}\tau \\ &\quad + \int_{\partial\Omega}\mathrm{diag}\left[\{\Psi(\mathbf{x})\}\right]\left(\{f_m^N\} - \{f_m^{d\,N}\}\right)\bullet \mathbf{n}_m \mathrm{d}\sigma \\ &\equiv \left\{\mathrm{RES}\left(\{f_k^N(\bullet)\} - \{f_k^{d\,N}(\bullet)\}\right)\right\}\end{aligned} \tag{B.58}$$

where $\mathbf{n}_m$ is the outwards pointing unit normal vector on the transverse plane closure $\partial\Omega$.

The resulting discrete implementation for (B.56) is

$$\begin{aligned}\{\mathrm{RES\_}QI\}_e = &-\int_{\Omega_e}\frac{\partial\{N_1(\eta_l(\mathbf{x}))\}}{\partial\eta_m}\left(\{f_m^h\}_e - \{f_m^{dh}\}_e\right)\mathrm{d}\tau \\ &+\int_{\partial\Omega_e\cup\partial\Omega}\{N_1(\eta_l(\mathbf{x}))\}\left(\{f_m^h\}_e - \{f_m^{dh}\}_e\right)\bullet\mathbf{n}_m\mathrm{d}\sigma\end{aligned} \tag{B.59}$$

The alteration of the kinetic flux vector (B.26) for (B.59) is

$$\{f_m(\{QI\}_e)\}_e \equiv \begin{Bmatrix} M_m \\ u_1 M_m + \omega\,\mathrm{Eu}\dfrac{\partial\eta_m}{\partial x_1}p \\ u_2 M_m + \varepsilon\,\mathrm{Eu}\dfrac{\partial\eta_m}{\partial x_2}p \\ u_3 M_m + \mathrm{Eu}\dfrac{\partial\eta_m}{\partial x_3}p \\ \rho H\, M_m \end{Bmatrix}_e \tag{B.60}$$

The dissipative flux vector (B.27) is unaltered.

For visual clarity matrix entries in $\{\mathrm{RES\_}QI\}_e$ in (B.59–B.60) utilize state variable labels $\{QI(\eta_1)\} \equiv \{\mathrm{M1, u1M1, u2M1, u3M1, \rho HM1}\}^T$ with $2 \le (l, m) \le 3$ changed to $2 \le (L, M) \le 3$. In the DOF order $\{QI(\eta_1\}$, $1 \le I \le 5$, the element matrices for $\{\mathrm{RES\_}QI\}_e$ in (B.59–B.60) are

$$
\begin{aligned}
\{\mathrm{RES\_Q1}\}_e &\equiv \{\mathrm{RES\_M1}\}_e \\
&= -[\mathrm{B2}L0]_e\{\mathrm{M}L\}_e + l_e[\mathrm{A200}]\{\mathrm{M}L \bullet \mathbf{n}L\}_e \\
&\quad + \beta_1\{\mathrm{U}M\mathrm{u}L\}_e^T[\mathrm{B30}ML]_e\{\mathrm{M1}\}_e
\end{aligned}
\tag{B.61}
$$

$$
\begin{aligned}
\{\mathrm{RES\_Q2}\}_e &\equiv \{\mathrm{RES\_u1M1}\}_e \\
&= -[\mathrm{B2}L0]_e\{\mathrm{u1M}L\}_e + l_e[\mathrm{A200}]\{\mathrm{u1M}L \bullet \mathbf{n}L\}_e \\
&\quad + \mathrm{Eu}\left(\frac{\partial \eta_1}{\partial x_1}\right)_e \det{}_e[\mathrm{B200}]\{\Delta \mathrm{P}(\omega)/\Delta \eta_1\}_e \\
&\quad + \mathrm{Eu}\left(\left(\frac{\partial \eta_2}{\partial x_1} - \varepsilon\right)_e [\mathrm{B202}]_e + \frac{\partial \eta_3}{\partial x_1}[\mathrm{B203}]_e\right)\{\mathrm{P}\}_e \\
&\quad + \frac{2\det_e}{\mathrm{Re}}\{\mu^* + \mathrm{RET}\}_e^T[\mathrm{B30}L0]_e\{\mathrm{S1}L\}_e \\
&\quad + \beta_2\{\mathrm{U}M\mathbf{u}L\}_e^T[\mathrm{B30}ML]_e\{\mathrm{M1}\}_e
\end{aligned}
\tag{B.62}
$$

$$
\begin{aligned}
\{\mathrm{RES\_Q3}\}_e &\equiv \{\mathrm{RES\_u2M1}\}_e \\
&= -[\mathrm{B2}L0]_e\{\mathrm{u2M}L\}_e + l_e[\mathrm{A200}]\{\mathrm{u2M}L \bullet \mathbf{n}L\}_e \\
&\quad + \mathrm{Eu}\left(\frac{\partial \eta_1}{\partial x_2}\right)_e \det{}_e[\mathrm{B200}]\{\Delta \mathrm{P}(\omega)/\Delta \eta_1\}_e \\
&\quad + \mathrm{Eu}\left(\frac{\partial \eta_2}{\partial x_2}[\mathrm{B202}]_e + \frac{\partial \eta_3}{\partial x_2}[\mathrm{B203}]_e\right)\{\mathrm{P}\}_e \\
&\quad + \frac{2\det_e}{\mathrm{Re}}\{\mu^* + \mathrm{RET}\}_e^T[\mathrm{B30}L0]_e\{\mathrm{S2}L\}_e \\
&\quad - \frac{4\det_e}{3\mathrm{Re}}\{\mu^* + \mathrm{RET}\}_e^T[\mathrm{B30}L0]_e\{SMM + \rho k\}_e\delta_{2L} \\
&\quad + \beta_3\{\mathrm{U}M\mathrm{u}L\}_e^T[\mathrm{B30}ML]_e\{\mathrm{M1}\}_e
\end{aligned}
\tag{B.63}
$$

$$
\begin{aligned}
\{\mathrm{RES\_Q4}\}_e &\equiv \{\mathrm{RES\_u3M1}\}_e \\
&= -[\mathrm{B2}L0]_e\{\mathrm{u3M}L\}_e + l_e[\mathrm{A200}]\{\mathrm{u3M}L \bullet \mathbf{n}L\}_e \\
&\quad + \mathrm{Eu}\left(\frac{\partial \eta_1}{\partial x_3}\right)_e \det{}_e[\mathrm{B200}]\{\Delta \mathrm{P}(\omega)/\Delta \eta_1\}_e \\
&\quad + \mathrm{Eu}\left(\frac{\partial \eta_2}{\partial x_3}[\mathrm{B202}]_e + \frac{\partial \eta_3}{\partial x_3}[\mathrm{B203}]_e\right)\{\mathrm{P}\}_e \\
&\quad + \frac{2\det_e}{\mathrm{Re}}\{\mu^* + \mathrm{RET}\}_e^T[\mathrm{B30}L0]_e\{\mathrm{S3}L\}_e \\
&\quad - \frac{4\det_e}{3\mathrm{Re}}\{\mu^* + \mathrm{RET}\}_e^T[\mathrm{B30}L0]_e\{SMM + \rho k\}_e\delta_{3L} \\
&\quad + \beta_4\{\mathrm{U}M\mathrm{u}L\}_e^T[\mathrm{B30}ML]_e\{\mathrm{M1}\}_e
\end{aligned}
\tag{B.64}
$$

$$
\begin{aligned}
\{\mathrm{RES\_Q5}\}_e \equiv \{\mathrm{RES\_\rho HM1}\}_e \\
&= -[\mathrm{B2}L0]_e\{\rho\mathrm{HM}L\}_e + l_e[\mathrm{A200}]\{\rho\mathrm{HM}L \bullet \mathbf{n}L\}_e \\
&+ \frac{\mathrm{Ecdet}_e}{\mathrm{Re}}\{\mu^* + \mathrm{RET}\}_e^T[\mathrm{B30}L0]_e\{\mathrm{S}JL{,}u_J\}_e \\
&+ \frac{\mathrm{det}_e}{\mathrm{RePr}}\{k^* + \mathrm{RET}\}_e^T[\mathrm{B30}LL]_e\{\mathrm{TEMP}\}_e \\
&+ \beta_5\{\mathrm{U}M\mathrm{u}L\}_e^T[\mathrm{B30}ML]_e\{\rho\mathrm{HM1}\}_e
\end{aligned}
\tag{B.65}
$$

Comments pertinent to terms in the $\{\mathrm{RES\_Q}I\}_e$ element matrices:

- (B.62), (B.63): axial pressure gradient implementation $\{\Delta\mathrm{P}(\omega)/\Delta\eta_1\}_e$ is defined in (B.28) as a function of ω, (B.29), and $\Delta\eta_1$ responds to subscript indexing on ξ_j
- (B.62): wall normal pressure gradient implementation is moderated by ϵ, (B.30)
- (B.63), (B.64): dissipative flux vector Stokes tensor terms $\{SMM + \rho k\}_e\delta_{2L}$, $\{SMM + \rho k\}_e\delta_{3L}$ contain time averaged density-turbulence kinetic energy product ρk
- (B.65): Pr^t is assumed identical to Pr

The Stokes strain rate tensor constitutes *data* for (B.62–B.65). The GWS^h algorithm on the *differential definition* replacement of (B.12) in PRaNS reduced tensor index form yields (9.92) alteration

$$
\begin{aligned}
\mathrm{GWS}^N\left(\overline{S}_{i\,l}\right) &= \int_\Omega \Psi_\beta(x)\left(\overline{S}_{i\,l} - \frac{1}{2}\left(\frac{\partial\tilde{u}_i}{\partial x_l} + \frac{\partial\tilde{u}_l}{\partial x_i}\right)\right)^N \mathrm{d}\tau = \{0\} \\
&\equiv S_e\int_{\Omega_e}\{N_1\}\left(\overline{S}_{i\,l} - \frac{1}{2}\left(\frac{\partial\tilde{u}_i}{\partial x_l} + \frac{\partial\tilde{u}_l}{\partial x_i}\right)\right)^h \mathrm{d}\tau
\end{aligned}
\tag{B.66}
$$

The resultant element matrix statements are

$$
\begin{aligned}
\{\mathrm{F\_S}IL\}_e &= \mathrm{det}_e[\mathrm{B200}]\{\mathrm{S}IL\}_e - \frac{1}{2}\begin{pmatrix}[\mathrm{B20}L]_e\{\mathrm{u}I\}_e \\ +[\mathrm{B20}I]_e\{\mathrm{u}L\}_e\end{pmatrix}, 2 \le (I, L) \le 3 \\
\{\mathrm{F\_S1}L\}_e &= \mathrm{det}_e[\mathrm{B200}]\{\mathrm{S1}L\}_e - \frac{1}{2}\begin{pmatrix}[\mathrm{B20}L]_e\{\mathrm{u1}\}_e \\ +[\mathrm{B200}]_e\{\mathrm{du}L/\mathrm{d}\eta_1\}_e\end{pmatrix}, 2 \le L \le 3
\end{aligned}
\tag{B.67}
$$

In the second line of (B.67) the element data $\{\mathrm{du}L/\mathrm{d}\eta_1\}_e$ are formed via a suitable FD expression, recall (4.33).

Aside from the [B200] matrix in (B.57), also (B.67), and the normal efflux matrix [A200] with length measure l_e, matrix subscript e in (B.61–B.67) indicates Ω_e dependence. Replacement with element-independent library matrices is via the $k = 1$ TP basis coordinate transformation. Retaining text index notation convention, Section 7.4, the matrix library statement replacements for the PRaNS algorithm are

$$
\begin{aligned}
[\mathrm{B20}J]_e &= \left(\frac{\partial \eta_k}{\partial x_j}\right)_e [\mathrm{B20}k], [\mathrm{B2}J0]_e = \left(\frac{\partial \eta_k}{\partial x_j}\right)_e [\mathrm{B2}k0] \\
[\mathrm{B30}J0]_e &= \left(\frac{\partial \eta_k}{\partial x_j}\right)_e [\mathrm{B30}k0], [\mathrm{B30}JK]_e = \det_e^{-1}\left(\frac{\partial \eta_l}{\partial x_j}\frac{\partial \eta_m}{\partial x_k}\right)_e [\mathrm{B30}lm]
\end{aligned}
\quad (\mathrm{B.68})
$$

B.7 PRaNS $m\mathrm{GWS}^h$+θTS Algorithm Completion

The $m\mathrm{GWS}^h$+θTS algorithm for PRaNS *m*PDE system is complete pending iterative matrix algebra procedure (B.54–B.56) identification. As section B.1 cited, this era of compressible aerodynamics CFD algorithm code methodology transitioned from explicit time-marching to implicit grid-sweeping AF jacobian methods on boundary-fitted coordinate system meshes.

Unique to a Galerkin weak formulation, the initial-value $[\mathrm{B200}]_e$ element (mass) matrix in (B.57) enables *matrix tensor product* definition of an AF block tri-diagonal grid-sweeping jacobian for (B.54), Baker (2012, Chapter 10.10)

$$
[\mathrm{JAC}^h\_\{Q\}] \cong [\mathrm{JAC2}^h\_\{Q\}] \otimes [\mathrm{JAC3}^h\_\{Q\}] \quad (\mathrm{B.69})
$$

In (B.69) ⊗ denotes matrix tensor product, and direct substitution into (B.54) defines the iterative algebraic statement replacement

$$
\begin{aligned}
&[\mathrm{JAC}\_\{Q\}]\{\delta Q\}^{p+1} \Rightarrow \\
&\quad [\mathrm{JAC2}^h\_\{Q\}] \otimes [\mathrm{JAC3}^h\_\{Q\}]\{\delta Q\}^{p+1} = -\{\mathrm{F}\_\{Q\}\}^p
\end{aligned}
\quad (\mathrm{B.70})
$$

The solution sequence defined by (B.70) is

$$
\begin{aligned}
&[\mathrm{JAC2}^h\_\{Q\}]\{\mathrm{P}\} = -\{\mathrm{F}\_\{Q\}\}^p \\
&[\mathrm{JAC3}^h\_\{Q\}]\{\delta Q\}^{p+1} \equiv \{\mathrm{P}_S\}
\end{aligned}
\quad (\mathrm{B.71})
$$

for $\{\mathrm{P}\}$ a generated matrix and subscript S on $\{\mathrm{P}_S\}$ indicating a matrix row–column interchange prior to the second solution step is required.

The key to AF efficiency, hence speed, is identifying matrices of much *lower order* than apparent in definition (B.69). Realizing (B.69) is assembled as defined in (B.56), the PRaNS algorithm $k=1$ TP basis $n=2$ element matrix $[\mathrm{B200}]_e$ is exactly formed as the *tensor matrix product* of two $n=1$ mass matrices $[\mathrm{A200}]_e$

$$
\begin{aligned}
[\mathrm{A200}]_e \otimes [\mathrm{A200}]_e &= \frac{l_x}{6}\begin{bmatrix} 2 & 1 \\ 1 & 2 \end{bmatrix} \otimes \frac{l_y}{6}\begin{bmatrix} 2 & 1 \\ 1 & 2 \end{bmatrix} \\
&= \frac{l_x l_y}{36}\begin{bmatrix} 4 & 2 & 1 & 2 \\ 2 & 4 & 2 & 1 \\ 1 & 2 & 4 & 2 \\ 2 & 1 & 2 & 4 \end{bmatrix} = [\mathrm{B200}]_e
\end{aligned}
\quad (\mathrm{B.72})
$$

The row–column order in $[\mathrm{B200}]_e$, second line of (B.72), reflects the counter-clockwise DOF order convention for the $n=2$, $k=1$ TP basis mass matrix. Matrix tensor products of $[\mathrm{A200}]_e$ and $n=1$ element matrices with derivatives generate $n=2$ directional derivative equivalents, specifically

$$\begin{aligned}[\mathrm{A200}]_e \otimes [\mathrm{A201}]_e &= \frac{l_x}{6}\begin{bmatrix} 2 & 1 \\ 1 & 2 \end{bmatrix} \otimes \frac{1}{2}\begin{bmatrix} -1 & 1 \\ -1 & 1 \end{bmatrix} \\ &= \frac{l_x}{12}\begin{bmatrix} -2 & 2 & 1 & -1 \\ -2 & 2 & 1 & -1 \\ -1 & 1 & 2 & -2 \\ -1 & 1 & 2 & -2 \end{bmatrix} = [\mathrm{B201}]_e \end{aligned} \tag{B.73}$$

Inserting PRaNS reduced summation index convention $L=2$, 3, the AF jacobians $[\mathrm{JAC}L]_\mathrm{e}$, defined in (B.69) are formed on $k=1$ TP basis $n=1$ elements Ω_e. Recalling the calculus jacobian derivation process, the *generic* AF jacobian for the PRaNS algorithm $\{\mathrm{RES\_Q}I\}_e$ definitions (B.61–B.65) plus $[\mathrm{m}]_e$ in (B.53) is

$$[\mathrm{JAC}L]_e \cong l_e[\mathrm{A200}] + \theta\Delta\eta_1\left(-[\mathrm{A2}L0] + \mathrm{Pa}^{-1}[\mathrm{A30}L0] + [\mathrm{BC}L]\right) \tag{B.74}$$

where $[\mathrm{BC}L]$ is the efflux BC placeholder.

While individual element matrices are tensor product generated accurately, the tensor product of (B.74) with itself, written in the other coordinate direction, generates error matrices. Labeling these directions Y and Z and invoking shorthand symbols M, C, D, BC for mass, convection, diffusion and BC matrices

$$\begin{aligned}[\mathrm{JAC2}]_e \otimes [\mathrm{JAC3}]_e &\equiv \left(l_e[\mathrm{MY}] + \theta\Delta\eta_1[\mathrm{CY} + \mathrm{DY} + \mathrm{BCY}]\right) \otimes \\ &\qquad \left(l_e[\mathrm{MZ}] + \theta\Delta\eta_1[\mathrm{CZ} + \mathrm{DZ} + \mathrm{BCZ}]\right) \\ &= l_e[\mathrm{MY}] \otimes l_e[\mathrm{MZ}] + l_e[\mathrm{MY}] \otimes \theta\Delta\eta_1[\mathrm{CZ} + \mathrm{DZ} + \mathrm{BCZ}] \\ &\qquad + l_e[\mathrm{MZ}] \otimes \theta\Delta\eta_1[\mathrm{CY} + \mathrm{DY} + \mathrm{BCY}] \\ &\quad + (\theta\Delta\eta_1)^2[\mathrm{CY} + \mathrm{DY} + \mathrm{BCY}] \otimes [\mathrm{CZ} + \mathrm{DZ} + \mathrm{BCZ}] \\ &= \mathrm{A}_e[\mathrm{B200}] + \theta\Delta\eta_1([\mathrm{B210}] + [\mathrm{B220}] + [\mathrm{B3010}] + [\mathrm{B3020}] + [\mathrm{BC}])_e \\ &\qquad + (\theta\Delta\eta_1)^2\ [\text{Error Matrices}] \end{aligned} \tag{B.75}$$

identifies the matrix tensor product sum error as $O(\theta\Delta\eta_1)^2$. This limits maximum space marching step size, in practice fully compensated by the minimal time required to complete the iteration (B.71) using (B.75).

Forming AF jacobian approximations in (B.56) for the PRaNS algorithm remains via calculus. Replacing each B-indexed matrix with the corresponding A-index entity the self-coupling matrix common to all DOF, $\{\mathrm{Q}I(\eta_1\}$, $1 \le I \le 5$, resident on the block tri-diagonal AF jacobian diagonal is

$$
\begin{aligned}
\frac{\partial\{\mathrm{F\_\{Q}I\})\}_e}{\partial\{\mathrm{Q}I\}_e} &\cong [\mathrm{JAC}L\_II]_e, \quad \text{no sum on } I, L = 2, 3 \\
&= [\mathrm{A200}]\mathrm{diag}[\det_e^{1/2}] \\
&+ \frac{\Delta\eta_1}{2}\begin{pmatrix} -[\mathrm{A2}L0] + l_e[\mathrm{A10}]\delta_{nL} \\ +(\mathrm{Pa}l_e)^{-1}\left(\overline{\mathrm{RET}}_{eff}\right)_e[\mathrm{A211}]\dfrac{\partial\{\mathrm{S}IL\}_e}{\partial\{\mathrm{Q}I\}_e} \\ +\dfrac{\beta_I}{l_e}\left(\overline{\mathrm{ULu}L}\right)_e[\mathrm{A211}]\dfrac{\partial\{\mathrm{M1}\}_e}{\partial\{\mathrm{Q}I\}_e} \end{pmatrix}
\end{aligned}
\tag{B.76}
$$

In (B.76) the l_e multiplier on [A200] in (B.74), which leads to A_e[B200] in (B.75), is replaced by a diagonal matrix with entries the square root of the $n = 2$ Ω_e coordinate transformation determinant. The Kronecker delta $\delta_{\mathbf{n}L}$ engages the divergence theorem-generated efflux BC, Pa is the DOF pertinent non-D parameter and the turbulence closure and *m*PDE term hypermatrices are reduced to the multiplier average times the diffusion matrix [A211]. Signified by the overbar, this is the net action of the $n = 1$, $k = 1$ basis implementation, Table 4.3. Index definition is $L = 2$, 3, for each respective grid sweep and $\overline{\mathrm{RET}}_{eff} = (\mu^*(k^*) + \mathrm{RET})_{avg}$ on Ω_e.

Evaluating the derivative $\partial\{\mathrm{M1}\}_e/\partial\{\mathrm{Q}I\}_e$ leads to the following diagonal matrix multipliers for (B.76)

$$
\begin{aligned}
\frac{\partial\{\mathrm{M1}\}_e}{\partial\{\mathrm{Q}I\}_e} &\cong \mathrm{diag}[1], I = 1 \\
&= \mathrm{diag}[1/\mathrm{u}I]_e, I = 2, 3, 4 \\
&= \mathrm{diag}[1/H]_e, I = 5
\end{aligned}
\tag{B.77}
$$

The derivative by $\{\mathrm{Q}I\}_e$ of the Stokes tensor is evaluated from definition (B.12) substituted into its appearance in (B.27)

$$
2\det_e\left(S_{ij} - \frac{2}{3}(S_{kk} + \rho\, k)\delta_{ij}\right)_{\mathrm{PRaNS}} \Rightarrow \det_e\left(\left(\frac{\partial u_i}{\partial x_l} + \frac{\partial u_l}{\partial x_i}\right)^h - \frac{1}{3}\left(\frac{\partial u_m}{\partial x_m}\right)^h \delta_{il}\right)
\tag{B.78}
$$

This generates the [A211] diffusion matrix in (B.76). The jacobian diagonal block element matrices for (B.76) are

$$
\begin{aligned}
\frac{\partial\{\mathrm{S}IL\}_e}{\partial\{\mathrm{Q}I\}_e} &\Rightarrow \frac{\partial}{\partial\{\mathrm{Q}I\}_e}\left[\left(\frac{\partial u_i}{\partial x_l} + \frac{\partial u_l}{\partial x_i}\right)^h - \frac{1}{3}\left(\frac{\partial u_m}{\partial x_m}\right)^h \delta_{i\,l}\right]\det_e \\
&= \mathrm{diag}[0], I = 1, 5 \\
&\cong \mathrm{diag}\left[\det_e^{1/2}/(\mathrm{M1})\right]_e \det_e, I = 2 \\
&\cong \mathrm{diag}\left[2\det_e^{1/2}(\mathrm{M1})/3\right]_e, I = 3, L = 2;\ \mathrm{diag}\left[\det_e^{1/2}/(\mathrm{M1})\right]_e, I = 3, L = 3 \\
&\cong \mathrm{diag}\left[2\det_e^{1/2}(\mathrm{M1})/3\right]_e, I = 4, L = 3;\ \mathrm{diag}\left[\det_e^{1/2}/(\mathrm{M1})\right]_e, I = 4, L = 2
\end{aligned}
\tag{B.79}
$$

The off-diagonal element matrices for the block tri-diagonal AF jacobian are derived via similar calculus operations. The coupling to $\{\mathrm{Q1}\}_e$ is

$$\left.\frac{\partial\{\mathrm{F\_\{Q}I\})\}_e}{\partial\{\mathrm{Q1}\}_e}\right| \cong [\mathrm{JACL\_}I1]_e$$

$$= \frac{\Delta\eta_1}{2}\left(\frac{\beta_I}{l_e}\left(\overline{\mathrm{U}L\mathrm{u}L}\right)_e[\mathrm{A211}]\mathrm{diag}[\mathrm{u}I]_e\right), I = 2,3,4$$

$$= \frac{\Delta\eta_1}{2}\left(\frac{\beta_5}{l_e}\left(\overline{\mathrm{U}L\mathrm{u}L}\right)_e[\mathrm{A211}]\mathrm{diag}[H]_e\right), I = 5 \tag{B.80}$$

As required for (B.79), cross-coupling for DOF $\{\mathrm{Q}I(\eta_1\}_e$, $2 \le I \le 4$, not equal to J involves Stokes tensor differentiation

$$\left.\frac{\partial\{\mathrm{F\_\{Q}I\})\}_e}{\partial\{\mathrm{Q}J\}_e}\right|_{I \neq J} \cong [\mathrm{JAC}L\mathrm{\_}IJ]_e = \frac{\Delta\eta_1}{2}(\mathrm{Pa}l_e)^{-1}\left(\overline{\mathrm{RET}}_{eff}\right)_e[\mathrm{A211}]\frac{\partial\{\mathrm{S}IL\}_e}{\partial\{\mathrm{Q}J\}_e} \tag{B.81}$$

The diagonal matrix post-multipliers for the Stokes tensor derivative in (B.81) are

$$\frac{\partial\{\mathrm{S}IL\}_e}{\partial\{\mathrm{Q}J\}_e} \Rightarrow \frac{\partial}{\partial\{\mathrm{Q}J\}_e}\left[\left(\frac{\partial u_i}{\partial x_l}+\frac{\partial u_l}{\partial x_i}\right)^h - \frac{1}{3}\left(\frac{\partial u_m}{\partial x_m}\right)^h \delta_{i\,l}\right]\mathrm{det}_e\Big|_{1\le i\le 3,\ J \neq I}$$

$$= \mathrm{diag}\left[\mathrm{det}_e^{1/2}\mathrm{du}L/\mathrm{d}\eta_1/(\mathrm{M1})\right]_e, I = 2$$

$$= \mathrm{diag}\left[2\,\mathrm{det}_e^{1/2}(\mathrm{M1})/3\right]_e, I = 3, L = 2;\ \mathrm{diag}\left[\mathrm{det}_e^{1/2}/(\mathrm{M1})\right]_e, I = 3, L = 3$$

$$= \mathrm{diag}\left[2\,\mathrm{det}_e^{1/2}(\mathrm{M1})/3\right]_e, I = 4, L = 3;\ \mathrm{diag}\left[\mathrm{det}_e^{1/2}/(\mathrm{M1})\right]_e, I = 4, L = 2 \tag{B.82}$$

The similar operation for the $\{\mathrm{Q5}\}_e$ cross-coupling jacobian starts with

$$\frac{\partial\{\mathrm{F\_\{Q5\}})\}_e}{\partial\{\mathrm{Q}J\}_e} \cong [\mathrm{JAC}L\mathrm{\_}5J]_e = \frac{\mathrm{Ec}\ \mathrm{det}_e^{1/2}}{\mathrm{Re}}\left(\overline{\mathrm{RET}}_{eff}\right)_e[\mathrm{A211}]\frac{\partial\{\mathrm{S}IL \times uI\}_e}{\partial\{\mathrm{Q}J\}_e} \tag{B.83}$$

substituting index i (I) for j in the Stokes tensor dot product to avoid confusion. The differentiation process is more transparent noting the state variable with DOF $\{\mathrm{Q}J\}_e$ for $2 \le J \le 4$ is $\{u_i M_1\}_e$. Via the chain rule, one set of Stokes tensor derivative diagonal matrix post-multipliers are (B.82) times the average $(\overline{u_i})_e$ on Ω_e

$$\frac{\partial\{\mathrm{S}IL \times \mathrm{u}I\}_e}{\partial\{\mathrm{Q}J\}_e} \Rightarrow \frac{\partial}{\partial\{u_i M_1\}_e}\left[\left(\frac{\partial u_i}{\partial x_l}+\frac{\partial u_l}{\partial x_i}\right)^h - \frac{1}{3}\left(\frac{\partial u_m}{\partial x_m}\right)^h \delta_{i\,l}\right]u_i\,\mathrm{det}_e\Big|_{1\le(i,I)\le 3}$$

$$= (\overline{u_1})_e\mathrm{diag}\left[\mathrm{det}_e^{1/2}\mathrm{du}L/\mathrm{d}\eta_1/(\mathrm{M1})\right]_e, J = 2$$

$$= (\overline{u_2})_e\mathrm{diag}\left[2\,\mathrm{det}_e^{1/2}(\mathrm{M1})/3\right]_e, J = 3, L = 2;\ (\overline{u_2})_e\mathrm{diag}\left[\mathrm{det}_e^{1/2}/(\mathrm{M1})\right]_e, J = 3, L = 3$$

$$= (\overline{u_3})_e\mathrm{diag}\left[2\,\mathrm{det}_e^{1/2}(\mathrm{M1})/3\right]_e, J = 4, L = 3;\ (\overline{u_3})_e\mathrm{diag}\left[\mathrm{det}_e^{1/2}/(\mathrm{M1})\right]_e, J = 4, L = 2 \tag{B.84}$$

The second set is the average of the Stokes tensor times the diagonal matrix post-multiplier

$$\frac{\partial\{SIL \times uI\}_e}{\partial\{QJ\}_e} \Rightarrow \left[\left(\frac{\partial u_i}{\partial x_l} + \frac{\partial u_l}{\partial x_i}\right)^h - \frac{1}{3}\left(\frac{\partial u_m}{\partial x_m}\right)^h \delta_{il}\right] \frac{\partial u_i}{\partial\{u_i M_1\}_e} \det{}_e \Bigg|_{1\le(i,I)\le 3}$$

$$= \left(\overline{SIL}\right)_e \mathrm{diag}\left[\det{}_e^{1/2}/(\mathrm{M1})\right]_e, J = 2, 3, 4 \qquad \text{(B.85)}$$

The final step to PRaNS algorithm closure is determination of pressure on the missile surface. Elsewhere throughout the PRaNS domain pressure is determined from the equation of state and state variable DOF. However, since the missile surface BC for PRaNS state variable DOF $\{QI\}$, $1 \le I \le 5$, is uniformly homogeneous Dirichlet, an additional relationship is required identified for wall pressure.

A TS manipulation on the wall-reduced normal momentum PDE, analogous to the wall vorticity BC derivation, Section 6.2, generates a wall pressure DOF approximation involving current and previous sweep pressure and state variable DOF at the two nearest-wall nodes, Baker *et al.* (1994, Section 3.5). Admissible surface temperature DOF BCs are heat flux, thermal convection and specified (Dirichlet).

The theoretical (linearized) asymptotic error estimate pertinent to solutions generated by the PRaNS $k=1$ TP basis $m\mathrm{GWS}^h+\theta\mathrm{TS}$ algorithm remains (7.8) for $\gamma \equiv \min(k, r-1)$ with initial value truncation error $O(\Delta\eta_1)^3$ for $\theta = 0.5$. The reference does not report convergence study *a posteriori* data.

B.8 PRaNS Algorithm IC Generation

The PRaNS I-EBV *m*PDE system is initial value in η_1, hence requires an IC for the state variable DOF $\{QI(\eta_1\} \equiv \{\mathrm{M1}, \mathrm{u1M1}, \mathrm{u2M1}, \mathrm{u3M1}, \rho\mathrm{HM1}\}^T$ distribution on the solution initiation transverse plane. This is generated via an $m\mathrm{GWS}^h+\theta\mathrm{TS}$ NS algorithm AF jacobian compressible laminar real gas axisymmetric code on NS state variable $\{q(\mathbf{x}, t)\} = \{\rho, \mathbf{m}, E\}^T$, Iannelli (1991, 2006). Validation geometry is an $R = 0.5$ inch radius sphere/10.5° cone afterbody, Figure B.2. The NS algorithm solution process employs Mach number *continuation* to increment farfield onset through $2 \le \mathrm{Ma}_\infty \le 8$.

The BCs for the NS simulation are: state variable $\{q(\mathbf{x}, t)\}$ fixed at upstream Ma_∞, $(\mathbf{n} \bullet \nabla)\{q(\mathbf{x}, t)\} = \{0\}$ on the symmetry axis, $(\mathbf{n} \bullet \nabla)\{\rho, \mathrm{E}\} = \{0\}$ on the sphere/cone surface, $(\mathbf{n} \bullet \nabla)\{\rho, \mathbf{u}, \mathrm{E}\} = \{0\}$ at the NS domain outflow plane selected the sphere–cone juncture. On this plane pressure DOF are unconstrained at all mesh nodes where $\mathrm{Ma} > 1$ is computed. At nodes with computed $\mathrm{Ma} < 1$, pressure DOF are set equal to the BL sonic point pressure, in agreement with BL Reynolds ordering.

The $\mathrm{Ma}_\infty = 2$ NS state variable IC is generated via interpolation of section B.5 definitions onto the NS domain, terminally adjusted to flow tangency on sphere/cone surface. Using the AF jacobian and Euler TS $\theta = 1$, algorithm convergence occurs in $O(10^3)$ variable size time steps. Algorithm implicitness enables extremum axial Courant number (referenced CFL *number* in the literature) of $O(10^2)$. At convergence, the mesh is solution adapted if necessary, and DOF distribution again converged to generate the IC for the onset Ma_∞ increment, and so on, to produce the requisite $\mathrm{Ma}_\infty = 8$ solution.

The thus generated *ideal gas* NS algorithm steady $\mathrm{Ma}_\infty = 8$ solution normalized bow shock standoff prediction is $\Delta/R = 0.15$. This agrees within 2% of experimental data and constitutes quantitative validation. Commenting out the NS PDE system viscous terms and altering the surface BC to tangency, the resultant Euler (inviscid NS) PDE $Ma_\infty = 8$ solution bow shock standoff prediction is $\Delta/R = 0.135$, a reduction of 10% from that for NS. This quantifies the requirement for NS algorithm generation of the PRaNS state variable IC.

Solution adapted mesh refinement during Mach number continuation is invoked to sharpen bow shock interpolation and boundary layer resolution. The *real gas* laminar NS algorithm $Ma_\infty = 8$ steady velocity solution *unit vector* DOF distribution on the terminal solution adapted, non-cartesian $M = 64 \times 32$ $k = 1$ TP basis mesh is graphed in Figure B.3. Unit vector DOF density distributions fully quantify oscillation free interpolation ("capture") of the bow shock and quality resolution of the laminar boundary layer.

The stagnation capability of the typical wind tunnel precludes experimental generation of real gas effects for blunt body flows at $Ma_\infty = 8$. Conversely, the atmospheric at-altitude hypersonic stagnation state of $p_\infty = 0.03$ atm, $T_\infty = 221$K, is sufficient for detached bow shock temperature rise to generate measureable real gas effects. Directly comparative ideal/real gas Euler algorithm temperature DOF distributions on the stagnation streamline, continued onto the sphere/10.5° cone surface, quantifies the critical importance of real gas aerothermodynamics at $Ma_\infty = 8$.

Altering the Euler state variable onset BC to at-altitude stagnation, Mach number continuation convergence predicts real gas bow shock standoff $\Delta/R = 0.11$, an 18% *decrease* from the ideal gas prediction, Figure B.4 top. Thus is generated a very significant(!) ~700K *decrease* in post-shock temperature extremum, which leads to a uniformly lower temperature DOF distribution on the cone surface. This reduction is due to energy absorption by molecular species dissociation. The stagnation streamline–sphere/10.5 cone surface

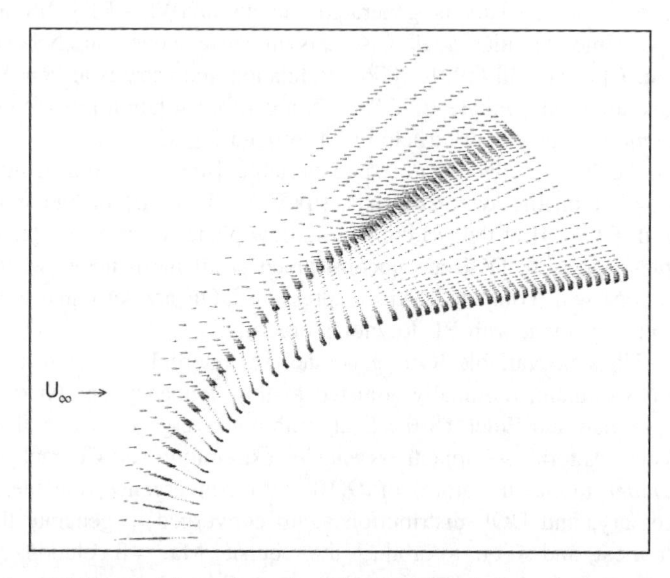

Figure B.3 Sphere/10.5° cone $Ma_\infty = 8$ steady axisymmetric solution velocity unit vector DOF distribution, solution adapted mesh, mGWS^h+θTS laminar compressible real gas NS algorithm

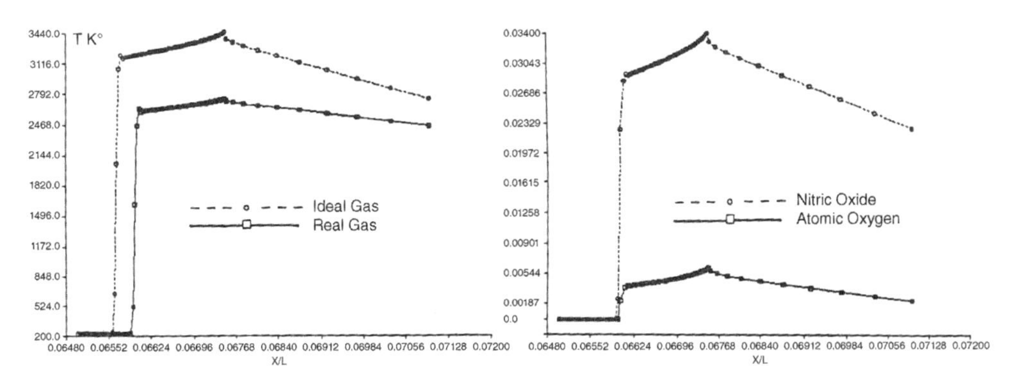

Figure B.4 Stagnation streamline–sphere/10.5° cone surface ideal-real gas aerothermodynamics closure DOF distributions, at altitude $\mathrm{Ma}_\infty = 8$ stagnation state, Euler axisymmetric $m\mathrm{GWS}^h$+θTS algorithm: left, temperature K; right, real gas NO, O mass fraction in percent

Euler real gas predicted distributions of atomic oxygen and nitric oxide are graphed in Figure B.4 right.

B.9 PRaNS $m\mathrm{GWS}^h$+θTS Algorithm Validation

With laminar NS prediction of DOF data adaptable to a PRaNS state variable IC confirmed, the algorithm validation specification is $\mathrm{Ma}_\infty = 8$ zero angle of attack onset to a sharp 7° cone/cylinder/7° aftcone wind tunnel model. Each model section is nominally one foot long, and in American standard units wind tunnel stagnation reference is $p_0 = 850$ psia, $T_0 = 1350$°R, total dynamic pressure $q_0 = 3.9$ psia, $p_\infty = 0.087$ psia, $\mathrm{Re} = 3.7 \times 10^6$/ft and $\mathrm{Re}_R = 1.5 \times 10^5$, Martellucci and Weinberg (1982, p. 72). Validation requirements include comparison with wind tunnel experiment and alternative CFD *a posteriori* data, both restricted to ideal gas, surface pressure distribution data for slice and flare, and real gas parabolic assumption breakdown prediction as appropriate.

A sharp cone at hypersonic zero angle of attack generates an axisymmetric conical shock layer. For $\mathrm{Ma}_\infty = 8$ the attached shock angle is constant at $\beta = 10.2°$, hence bow shock trajectory is not required computed. Via conical flow theory the post-shock reference state is $\mathrm{Ma}_{\mathrm{shock}} = 7.037$, $p_{\mathrm{shock}} = 0.1892$ psia and $T_{\mathrm{shock}} = 333.5$°R. These data, manipulated into PRaNS state variable definition as appropriate, constitute Dirichlet BCs on the post bow shock surface trajectory.

Helical grooves, knurled circumferentially starting at forecone half-span, act as the BL trip in the wind tunnel experiment, Figure B.5 top, Martellucci and Weinberg (1982, Figure 2). The two lower graphics present a cut-away view of the PRaNS domain axial bounding surface distributions, also an illustrative (only!) wall adapted $k = 1$ TP basis transverse plane mesh axial evolution in η_1.

The upstream η_1 coordinate of the knurled section is the PRaNS IC plane. The $m\mathrm{GWS}^h$+θTS algorithm validation requirements include:

- turbulent BL solution adapted mesh resolution adequacy
- NS solution DOF transition to PRaNS state variable IC

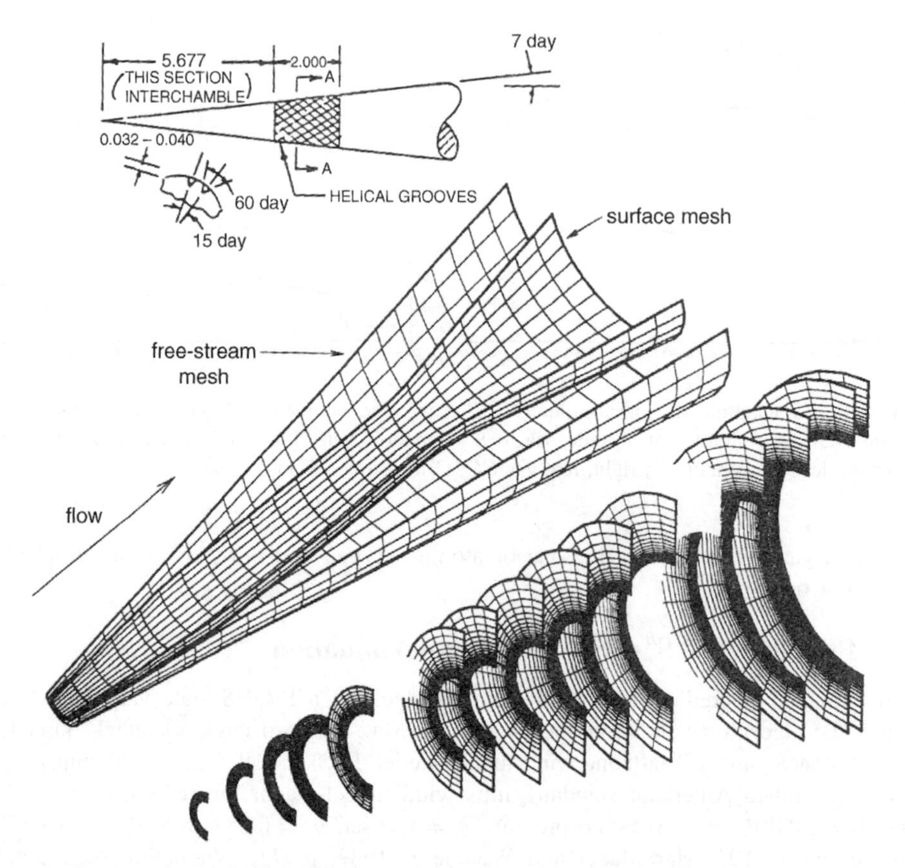

Figure B.5 Sharp 7° cone/cylinder/7° aftcone validation geometry: top, wind tunnel model with BL trip knurl; middle, bow shock and body surface PRaNS domain boundary illustration, $\text{Ma}_\infty = 8$; bottom, shock layer transverse plane $k = 1$ TP basis mesh axial evolution illustration

- AF jacobian iterative convergence stability
- real gas aerothermodynamics closure
- PRaNS wall pressure DOF prediction
- wall temperature Dirichlet BC wall node heat flux prediction
- *m*PDE β term dispersion error elimination efficiency
- multiple sweep algorithm, symbol theory implementation
- parabolic assumption breakdown as slice/flare appropriate.

Acceptable element aspect ratio limits wall-normal solution adapted meshing. At the forecone knurl leading edge the wind tunnel measured shock layer thickness is ~0.008 ft (~0.1 in). Compute experiments confirm a wall element span of $\sim 10^{-5}$ in on this knurl plane results for a wall normal pr = 1.07 adapted, half-azimuthal uniform $M = 64 \times 18$ $k = 1$ TP basis mesh.

The generated NS laminar $\text{Ma}_\infty = 8$ steady solution is altered to PRaNS state variable DOF *approximate* IC on this plane mesh by NS state variable member $\{E\}$ alteration to volume specific stagnation enthalpy $\{\rho H\}$. Final conversion to the turbulent BL PRaNS DOF IC is accomplished by algorithm execution with Baldwin–Lomax model "turned on." Marching the forecone knurled section length $\Delta x = 0.025$ ft (3.0 in) completes this DOF transition. The turbulent BL similarity coordinate of first off-wall node is $\text{Re}^+ \approx 0.01$, verifying fully adequate turbulent BL resolution. Re^+ is the compressible turbulent BL similarity coordinate equivalent of incompressible y^+, (4.80).

To match conical bow shock stand-off evolution during downstream integration the radial span of the $M = 64 \times 18$ mesh is *arbitrary lagrangian-eulerian* (ALE) algorithm stretched, as detailed in Section 7.13. Symbol theory departure solution control is inoperative during PRaNS algorithm first sweep. Pressure DOF at all BL subsonic nodes are equated to the BL sonic point pressure DOF at each η_1 station, as $O(\text{Re}^{-1/2})$ consistent with BL theory, (4.67).

For wind tunnel experiment hot (~stagnation) and cold wall Dirichlet BCs, PRaNS *real gas* aerothermodynamics algorithm non-D {T} DOF distributions through the shock layer at $\eta_1 = 1.9$ ft, just upstream of cylinder transition to aftcone, are compared in Figure B.6, top and bottom. The ordinate log scale ranges five decades to visually quantify the truly significant distinctions. The {T} DOF distribution slope discontinuities are located in the BL subsonic reach where the {u1} DOF Mach number is $\text{Ma}_1 \approx 0.5$.

Data ordinate intersection defines BL *thermal thickness*, with that for cold wall BC ~50% larger than for the ~stagnation wall BC. Wall normal temperature DOF gradients are *totally* distinct, hence also is surface heat flux. Accurate wall heat flux prediction is weak form *analytical* using the GWS^h Dirichlet BC$\Leftrightarrow$flux DOF algorithm, Section 2.3, rather than a non-uniform FD nodal TS.

The turbulent axial velocity {u1} DOF distributions are five-decade semi-log compared in Figure B.7. The near wall DOF profiles are essentially indistinguishable, hence so would be the wall shear stress predictions. The notable difference is hot wall axial velocity BL thickness Δ is ~40% smaller than for cold wall BC, hence the thermal and velocity BL thicknesses are distinct. Both {T} and {u1} DOF mesh resolution is quantified fully adequate.

Order 600 integration steps $\Delta\eta_1$ are required for the PRaNS real gas aerothermodynamics $m\text{GWS}^h\text{+}\theta\text{TS}$ algorithm to march the ~ 3 ft long 7° cone/cylinder/7° aftcone domain. The AF jacobian requires 4–5 iterations/step to converge to $\max|\{\Delta QI\}| \leq \text{E}-04$ during BL transition to turbulent. Thereafter, for this convergence requirement three iterations/step is the norm in reaching the model termination plane at $\eta_1 = 3.3$ ft.

Each PRaNS state variable member contains contravariant axial momentum M1, that is, $\{QI(\eta_1)\} \equiv \{\text{M1, u1M1, u2M1, u3M1, }\rho\text{HM1}\}^T$. The algorithm axial {u1} and {M1} DOF distributions are computed *totally distinct* as five-decade semi-log scale compared at $\eta_1 = 3.3$ ft for cold wall BC $T_{\text{wall}} = 560°\text{R}$, Figure B.8. The entire variation in {M1} occurs within the BL subsonic reach below $\text{Ma}_1 \approx 0.4$. Mesh resolution in the horizontal linear variation plateau of {M1} DOF on $0.2 < \text{Ma}_1 < 0.4$ appears coarse, conversely is fully adequate for {u1} turbulent BL resolution. The {M1} DOF distribution above $\text{Ma}_1 \approx 0.4$ is uniformly constant, equal to the post-shock surface Dirichlet BC. The {u1} BL thickness is essentially unchanged in progression from $\eta_1 = 1.9$ to $\eta_1 = 3.3$, compare to Figure B.7 top.

The axial velocity {u1} Mach number DOF distribution $\{\text{Ma}_1\}$ at $\eta_1 = 3.3$ ft is graphed in Figure B.9 top, with BL sonic point noted. The PRaNS state variable definitions leading to {M1} and {u1} differ by density, (B.24). The computed *totally disparate* DOF

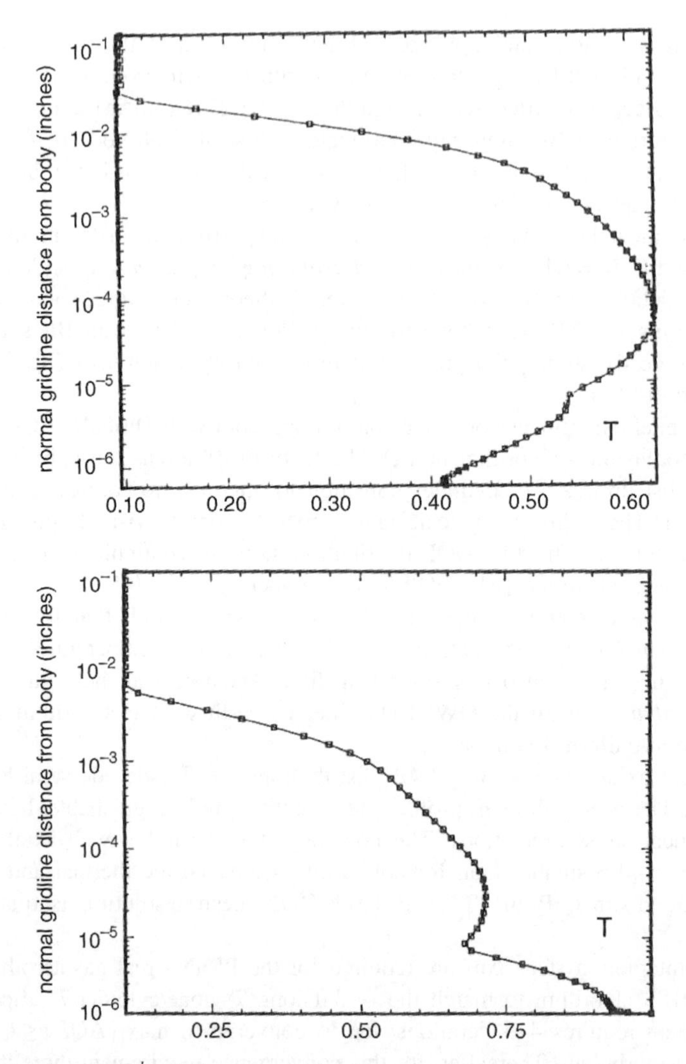

Figure B.6 PRaNS $m\text{GWS}^h + \theta\text{TS}$ real gas algorithm single sweep non-D temperature {T} DOF distributions, $\eta_1 = 0.9$ ft on 7° forecone, $\text{Ma}_\infty = 8$, semi-log scaled for wall temperature Dirichlet BCs: top, $T_{\text{wall}} = 560°\text{R}$; bottom, $T_{\text{wall}} = 1100°\text{R}$ ($T_0 = 1350°\text{R}$)

distributions are the consequence of density temperature-dependence in the *lowest* subsonic reach of the $\text{Ma}_\infty = 8$ shock layer turbulent BL. Density extrema are distributed in sharp *inflection* at $\text{Ma}_1 \approx 0.4$, Figure B.9 bottom, magnitudes at BL sonic points are identical to the subsonic reach minimum and variation through the BL exceeds a factor of three.

These $\text{Ma}_\infty = 8$ PRaNS *a posteriori* data confirm criticality of real gas variable γ aerothermodynamics closure replacing ideal gas, recall shock layer temperature evolution distinctions, Figure B.4. Further, that these thermodynamic variations occur in the BL low subsonic reach confirms that mathematically precise handling of subsonic reach pressure ellipticity, (B.28–B.30), is imperative.

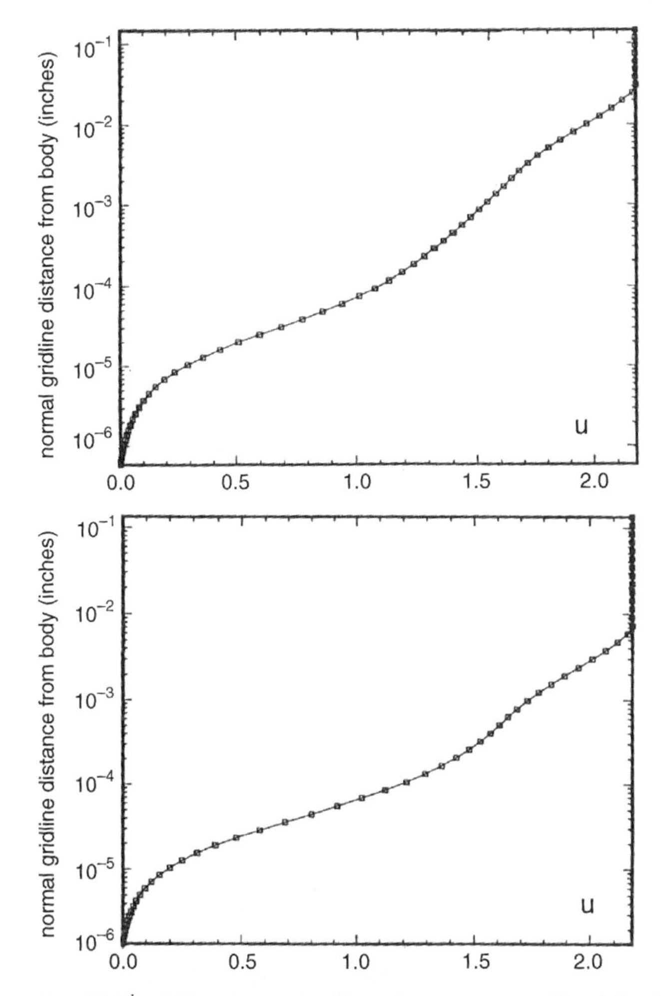

Figure B.7 PRaNS mGWS^h+θTS real gas algorithm, first sweep non-D axial velocity {u1} DOF distributions, semi-log scaled at $\eta_1 = 1.9$ ft on 7° forecone, $Ma_\infty = 8$, for Dirichlet wall temperature BCs: top, $T_{wall} = 560°R$; bottom, $T_{wall} = 1100°R$

A PRaNS algorithm necessary requirement is generation of mirror symmetric transverse plane momentum and velocity vector resolution {ML} and {uL} DOF distributions The exactly symmetric {uL} DOF distributions, Figure B.10, attest to GWS^h natural homogeneous Neumann BC accuracy on domain cutting planes, Figure B.5 middle, also at the post-shock surface domain boundary. Body fitted coordinate node columns confirm BL thickness distribution is uniform and quantify amply adequate mesh resolution with extrema matching BL $O(\mathrm{Re}^{-1/2})$ requirements.

PRaNS real gas aerothermodynamics algorithm fidelity in addressing the hypersonic shock layer problem class is thus confirmed. Rigorous theory validation requires multiple sweep execution comparison with wind tunnel surface pressure measurements. Pertinent data exist for a sharp 7° cone/cylinder model modified with cylinder aft section planar slice

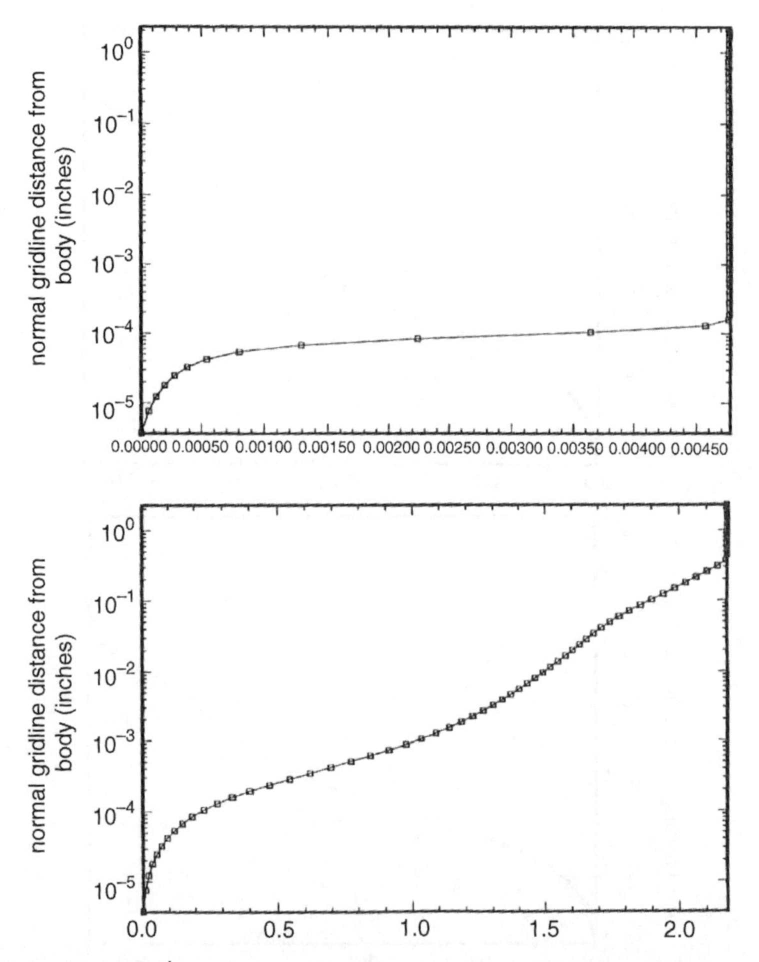

Figure B.8 PRaNS mGWS^h+θTS real gas algorithm, state variable non-D DOF distributions semi-log scaled at 7° aftcone $\eta_1 = 3.3$ ft, $Ma_\infty = 8$, Dirichlet BC $T_{wall} = 560°R$: top, axial momentum {M1}; bottom, axial velocity {u1}

communicating with an adjustable flare (the "flap"), Figure B.11. This model dimensionality is similar to the assessed 7° cone/cylinder/7° cone wind tunnel model.

At the cylinder planar slice-flap juncture, located at $\eta_1 = 28.2$ in, under "DATA" for downstream flap angle $\delta = -7°$ the non-D surface pressure p_{wall}/p_∞ experiences a sharp decrease (circular symbols) with ensuing flap pressure negative slope similar to that on the slice. For flap angle $\delta = +10°$ a sharp pressure increase is recorded *just upstream* of the slice-flap juncture (square symbols) with pressure gradient thereafter positive. For extreme flap angle $\delta = +20°$, the single measured (diamond symbol) pressure increase confirms massive BL separation is likely. The solid lines adjacent to data are ideal gas AFWAL-PNS code single sweep non-D surface pressure p_{wall}/p_∞ *a posteriori* data.

The 7° cone/cylinder with slice-flap wind tunnel model data for flap angles $\delta = -7°$ and $\delta = +10°$ are pertinent to characterizing multiple sweep PRaNS predictions. First sweep

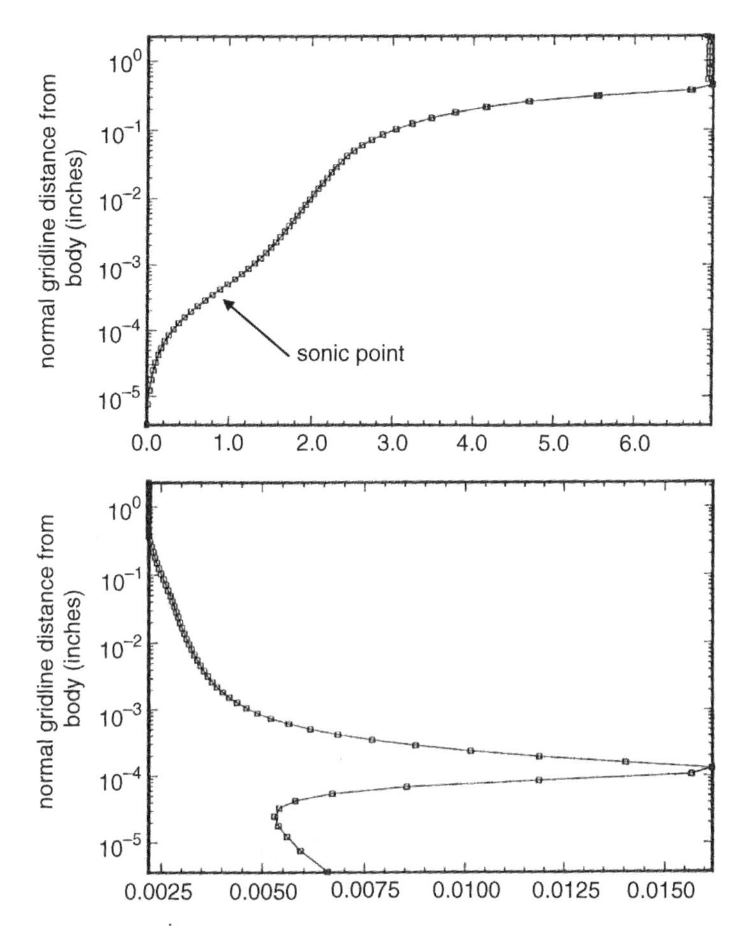

Figure B.9 PRaNS $m\text{GWS}^h$+θTS real gas algorithm, solution non-D DOF distributions semi-log scaled at 7° aftcone $\eta_1 = 3.3$ ft, $\text{Ma}_\infty = 8$, Dirichlet BC $T_{\text{wall}} = 560$°R: top, {u1} velocity Mach number $\{\text{Ma}_1\}$; bottom, density {RHO}

PRaNS 7° cone/cylinder/7° cone with BC $T_{\text{wall}} = 560$°R is obviously without departure solution theory implementation. As detailed, the algorithm solution proceeds to wind tunnel model terminal plane at $\eta_1 = 3.3$ ft. Starting again at the IC plane, the PRaNS algorithm second sweep implements the departure theory algorithms (B.28–B.30). Execution also proceeds to the model terminal plane, $\eta_1 = 3.3$ ft.

Starting anew at the IC plane, the PRaNS third sweep fully implements departure theory control. The solution process evidences instability in approaching the model cylinder/7° aftcone juncture located at $\eta_1 = 2.0$ ft. Execution continues until axial velocity separation is predicted at the first off-wall {u1} DOF, just downstream of this juncture, hence integration stops.

Surface axial pressure $p_{\text{wall}} - p_\infty$ (psia) DOF distributions computed by the PRaNS algorithm 7° cone/cylinder/7° cone model first, second and third integration sweeps are compared in Figure B.12. This model coordinate $\eta_1 = 1.0$ ft correlates with the 7° cone/cylinder

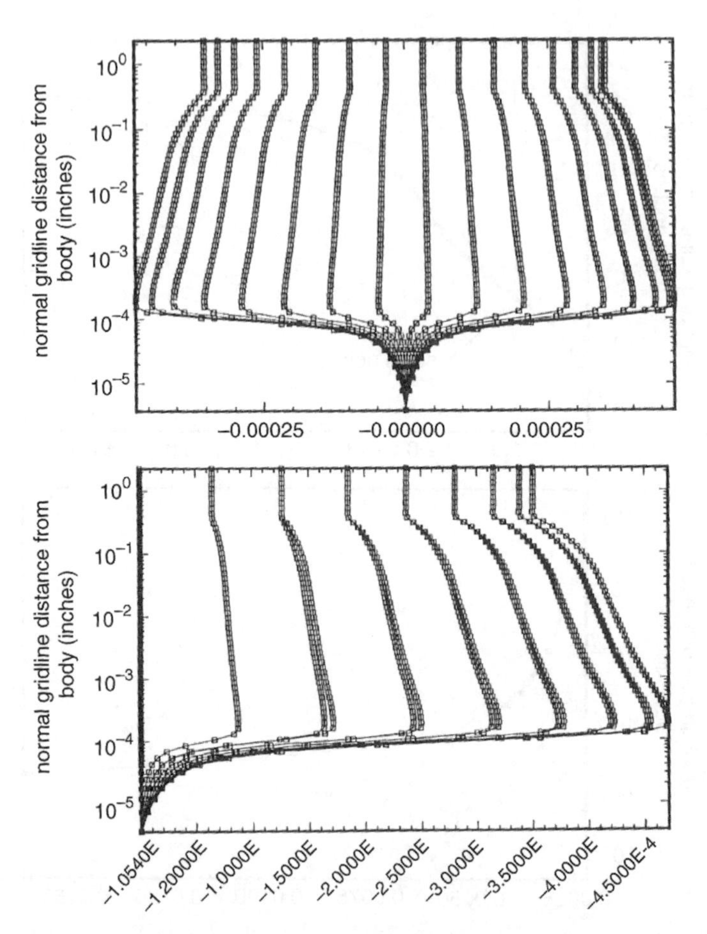

Figure B.10 PRaNS $m\text{GWS}^h$+θTS real gas algorithm solution non-D DOF distributions, semi-log scaled on 7° aftcone $\eta_1 = 3.3$ ft, $\text{Ma}_\infty = 8$, Dirichlet BC $T_{\text{wall}} = 560°\text{R}$: top, wall normal velocity {u2}; bottom, crossflow velocity {u3}

with slice-flap coordinate $\eta_1 = 28.2$ in, Figure B.11, for flap angle $\delta = -7°$. Conversely, this model coordinate $\eta_1 = 2.0$ ft correlates with the 7° cone/cylinder with slice-flap coordinate $\eta_1 = 28.2$ in, Figure B.11, for flap angle $\delta = +10°$.

Essentially no distinction exists between PRaNS first and second sweep surface pressure distributions, while the third sweep solution differs substantially immediately following $\eta_1 = 1.0$ ft, in approaching $\eta_1 = 2.0$ ft and thereafter on the aftcone. The third sweep first distinction is the sharp pressure decrease (circular symbols) predicted just downstream of the 7° cone/cylinder juncture at $\eta_1 = 1.0$ ft. This model geometry change is essentially that of the cylinder–slice model with $\delta = -7°$ flap. The *quantitative* agreement of these data, circular symbols each side of $\eta_1 = 28.2$ in, Figure B.11, with the PRaNS third sweep data constitute an algorithm validation.

The second key PRaNS third sweep validation is prediction of a sharp pressure decrease just upstream of model cylinder/7° aftcone juncture, $\eta_1 = 2.0$ ft, followed immediately by a

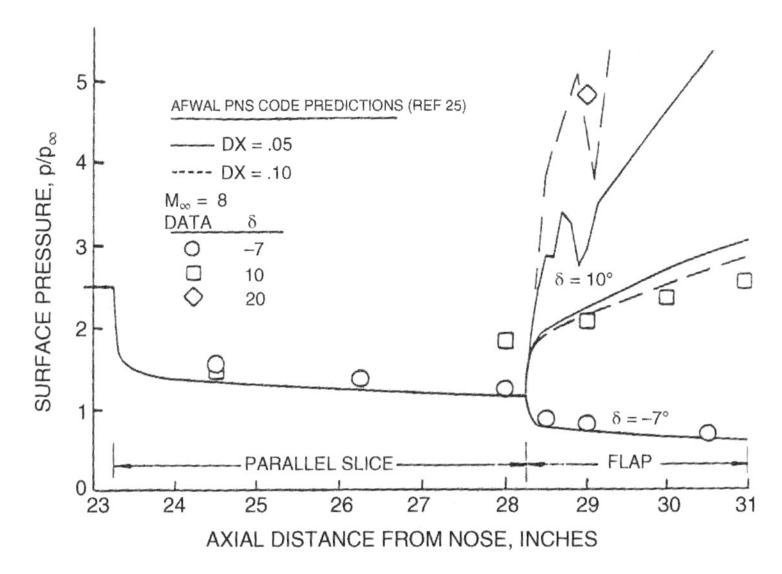

Figure B.11 Wind tunnel experiment, measured non-D surface pressure p_{wall}/p_{∞} distributions, sharp 7° cone/cylinder aft altered with slice and flare (flap), $Ma_{\infty} = 8$, flap angles (DATA) of −7°, +10°, +20°. AFWAL-PNS code single sweep solution p_{wall}/p_{∞} distributions are solid lines, from Martellucci and Weinberg (1982, Figure 53)

sharp pressure increase and large(!) positive pressure gradient thereafter on the 7° aftcone. This surface pressure DOF distribution agrees quantitatively with the $\delta = 10°$ flap data, square symbols in Figure B.11, aside from that first data symbol appearing 0.2 in upstream of the $\eta_1 = 28.2$ in juncture. This constitutes validation that PRaNS departure theory implementation is imperative to prediction fidelity. The summary validation is algorithm separation prediction soon after $\eta_1 = 2.0$ ft.

Planar slice graphs of the second and third sweep PRaNS domain pressure quantify shock layer *a posteriori* data distributions, Figure B.13 top/bottom. These grayscales of color floods denote BL edge, shock layer sonic line and $Re^+ = 1$ trajectory. The third sweep domain graph, shorter due to separation, firmly quantifies wall adjacent elevated pressure intrusion forward of the cylinder/7° aftcone juncture at $\eta_1 = 2.0$ ft. Original color graphics including the boxed region separation close-up are in Baker *et al.* (1994, Figure 4.12–4.13).

B.10 Hypersonic Blunt Body Shock Trajectory

A spherical nose missile in hypersonic freestream generates a detached bow shock enclosure of the shock layer. PRaNS requires a prediction algorithm for post bow shock surface downstream evolution admitting freestream angle of attack and yaw. Legacy algorithms relying on classic normal and oblique shock relationships, cf. Chaussee (1982), hence require estimation of post-shock static pressure evolution and a shock propagation direction.

The requisite ideal/real gas characteristics theory algorithm for hypersonic detached shock trajectory prediction without pre-selections is derived, Baker *et al.* (1994,

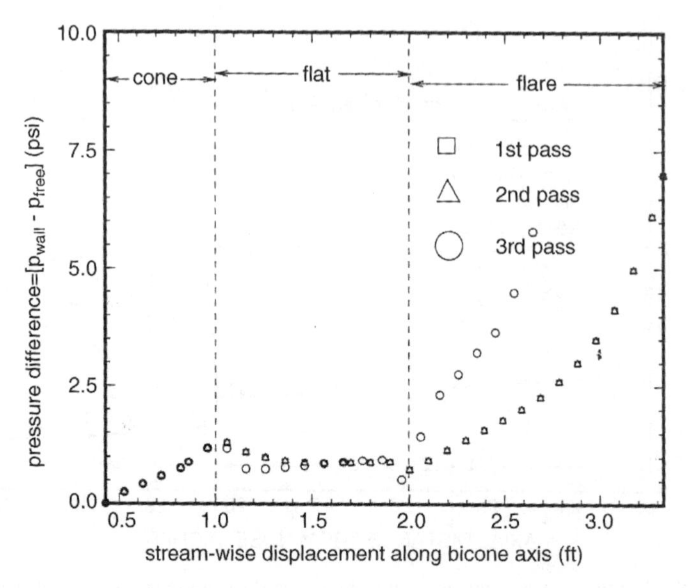

Figure B.12 PRaNS mGWS^h+θTS real gas algorithm, 7° cone/cylinder/7° aftcone, $\text{Ma}_\infty = 8$, computed surface axial pressure difference $p_{\text{wall}} - p_\infty$ (psia) distributions, domain integration sweeps 1,2,3

Appendix C). The multi-dimensional jump conditions generate a first-order *partial differential* for shock surface expressed in pressure ratio only for an ideal gas, and pressure and enthalpy ratios for reacting real gas. Consequently, upon prediction of only these ratios the resultant characteristics solution directly generates the shock propagation direction *and* shock intersection locus of coordinates with planar surfaces orthogonal to the missile body axis.

The classic weak formulation for arbitrary bow shock surface multi-dimensional jump conditions for steady hyperbolic conservation law conservation statement is the starting point. The steady compressible Euler PDE flux vector form

$$\mathcal{L}(\{q\}) = \frac{\partial\{f_j\}}{\partial x_j} = \{0\}, \{f_j(\{q\})\} = \begin{Bmatrix} \rho u_j \\ \rho u_j u_i + p\,\delta_{ij} \\ u_j(E+p) \end{Bmatrix} \tag{B.86}$$

is written on state variable $\{q(\mathbf{x}, t)\} = \{\rho, u_i, E\}^T$. The weak form (WF) for (B.86), recall Section 2.2, following Green–Gauss divergence theorem imposition is

$$\begin{aligned} \text{WF} &\equiv \int_\Omega w(\mathbf{x})\mathcal{L}(\{q\})\text{d}\tau \equiv 0, \forall w(x) \\ &= -\int_\Omega \frac{\partial w(\mathbf{x})}{\partial x_j} f_j(\{q\})\text{d}\tau + \int_{\partial\Omega} w(\mathbf{x}) f_j(\{q\})\hat{n}_j \text{d}\sigma \end{aligned} \tag{B.87}$$

for $\hat{n}_j$ the outwards pointing unit vector normal to the surface $\partial\Omega$ enclosing the domain Ω.

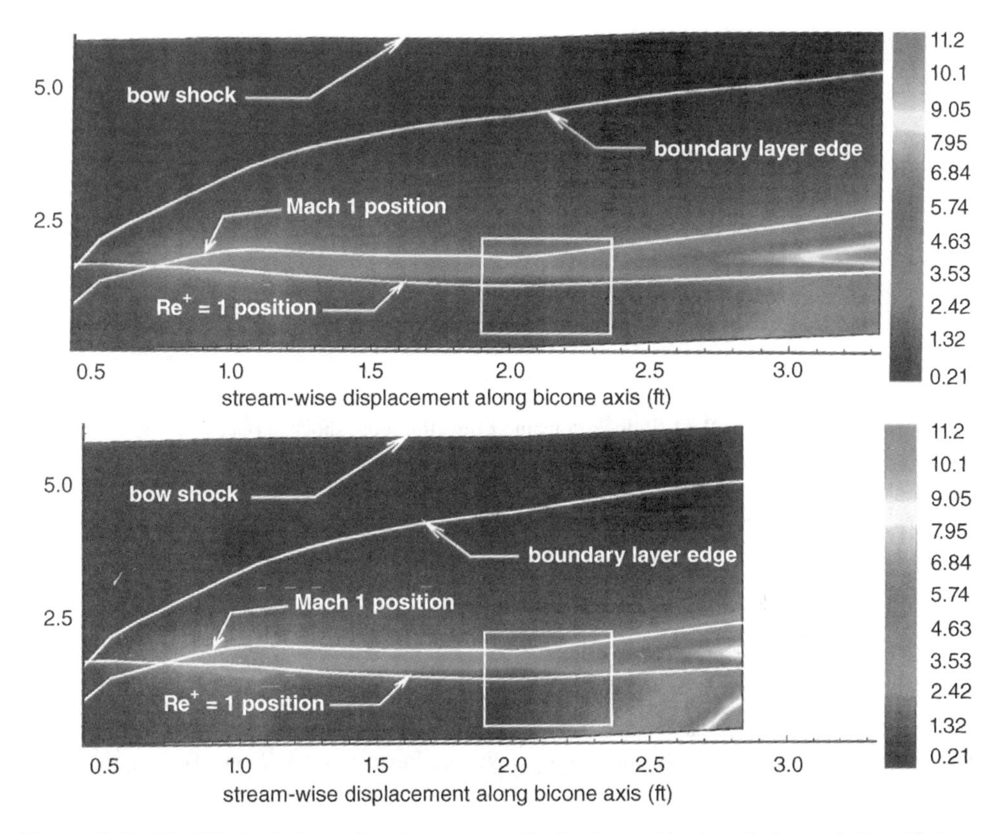

Figure B.13 PRaNS shock layer domain pressure distribution unitized scale interpolation, 7° forecone/cylinder/7° aftcone, $Ma_\infty = 8$: top, second sweep; bottom, third sweep, with bow shock, BL edge, sonic line and $Re^+ = 1$ trajectories

For coordinate x parallel to the body axis generating the bow shock, select the pillbox shape domain Ω, Figure B.14, that:

- spans the bow shock
- pillbox exterior face unit normal parallels the local shock surface normal, and
- shrinks to zero measure in the direction of unit vector $\hat{n}_j$.

For these restrictions (B.87) becomes

$$\sum_{j=1}^{3} w(\mathbf{x}) f_j(\{q\}) \hat{n}_j = 0 \tag{B.88}$$

and since w($\mathbf{x}$) is absolutely arbitrary (B.88) generates the generalized shock jump statement

$$\sum_{j=1}^{3} f_j(\{q\}) \hat{n}_j = 0 \equiv \sum_{j=1}^{3} \Delta f_j(\{q\}) \tag{B.89}$$

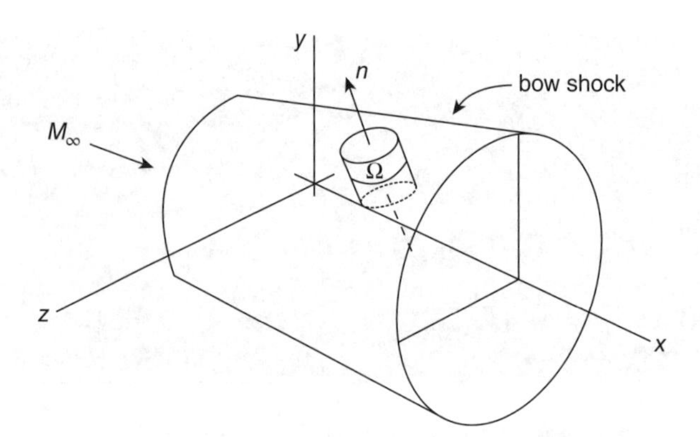

Figure B.14 Pillbox domain Ω bisecting bow shock surface

For bow shock surface parameterization $F(\mathbf{x})=0$, via calculus the unit vector $\hat{n}_j$ normal thereto is

$$\hat{n}_j = \frac{\nabla F}{|\nabla F|} = \frac{\dfrac{\partial F}{\partial x}\hat{i} + \dfrac{\partial F}{\partial y}\hat{j} + \dfrac{\partial F}{\partial z}\hat{k}}{|\nabla F|} \tag{B.90}$$

For Figure B.14 bow shock orientation the local directional derivatives for any point on the surface are $\tan\theta_{xy} = \partial x/\partial y$ and $\tan\theta_{xz} = \partial x/\partial z$. Further, since $F(\mathbf{x})=0$ is equivalently $x=x(y, z)$ then

$$\frac{\partial x}{\partial y} = -\frac{\partial F}{\partial y}\Big/\frac{\partial F}{\partial x}, \quad \frac{\partial x}{\partial z} = -\frac{\partial F}{\partial z}\Big/\frac{\partial F}{\partial x} \tag{B.91}$$

which leads to the sought expression for (B.90)

$$\hat{n}_j = \left(\hat{i} - \frac{\partial x}{\partial y}\hat{j} - \frac{\partial x}{\partial z}\hat{k}\right)/|\nabla F|\frac{\partial F}{\partial x} \tag{B.92}$$

Substituting (B.92) into the shock jump WF (B.89) and expanding generates the general n-D shock jump relationship

$$\sum_{j=1}^{3}\Delta f_j(\{q\}) = 0 \Rightarrow \Delta f_1(\{q\}) = \Delta f_2(\{q\})\frac{\partial x}{\partial y} + \Delta f_3(\{q\})\frac{\partial x}{\partial z} \tag{B.93}$$

Expanding (B.93) leads to five algebraic equations parameterized by the directional derivatives for Euler flux vector definition (B.86). Restricting to zero angle of attack for the moment, denote the freestream state with subscript 1, hence $v_1=0=w_1$. The post-shock state label is subscript 2, and for $f_j(\{q\}) = \rho u_j$ the expansion of (B.93) is

$$\rho_1 u_1 - \rho_2 u_2 = -\rho_2 v_2\frac{\partial x}{\partial y} - \rho_2 w_2\frac{\partial x}{\partial z} \tag{B.94}$$

Useful relationships result upon forming ratios of equations generated by (B.93). For example, dividing (B.94) through by $\rho_1 u_1$ leads to the density ratio equation

$$\frac{\rho_2}{\rho_1} = \left[\frac{u_2}{u_1} - \frac{v_2}{v_1}\frac{\partial x}{\partial y} - \frac{w_2}{w_1}\frac{\partial x}{\partial z}\right]^{-1} \tag{B.95}$$

Multiplying (B.94) by u_1 and subtracting the result from the $i = 1$ momentum equation in (B.93) leads to

$$\frac{u_2}{u_1} = 1 - \frac{(p_2/p_1) - 1}{\gamma \mathrm{Ma}_1^2} \tag{B.96}$$

where Ma_1 is Mach number associated with u_1.

Similar multiplications by v_2 and w_2 and subtracting the results from the $i = 2, 3$ momentum equations in (B.93) generates

$$\frac{v_2}{u_1} = 1 - \frac{(p_2/p_1) - 1}{\gamma \mathrm{Ma}_1^2}\left(\frac{\partial x}{\partial y}\right) \tag{B.97}$$

$$\frac{w_2}{u_1} = 1 - \frac{(p_2/p_1) - 1}{\gamma \mathrm{Ma}_1^2}\left(\frac{\partial x}{\partial z}\right) \tag{B.98}$$

Altering state variable member E to stagnation enthalpy H in (B.93) and proceeding through similar operations leads to

$$\left(\frac{H_2}{H_1}\right)^{-1} = \frac{u_2}{u_1} - \left[\frac{v_2}{u_1}\left(\frac{\partial x}{\partial y}\right) + \frac{w_2}{u_1}\left(\frac{\partial x}{\partial z}\right)\right] \tag{B.99}$$

The n-D shock jump relations (B.95–B.99) are valid for any thermodynamic equation of state, ideal or equilibrium real gas. For an ideal gas

$$p = (\gamma - 1)\left(E - \frac{\rho}{2}\left(u^2 + v^2 + w^2\right)\right) \tag{B.100}$$

and following detailed manipulations, Baker *et al.* (1994, Appendix C) insertion of (B.100) leads to elimination of the pressure ratios in (B.95–B.99). Defining the *geometric factor*

$$\text{factor} \equiv 1 + \left(\frac{\partial x}{\partial y}\right)^2 + \left(\frac{\partial x}{\partial z}\right)^2 \tag{B.101}$$

leads to ideal gas shock jump compact expressions

$$\frac{\rho_2}{\rho_1} = \frac{(\gamma + 1)\mathrm{Ma}_1^2}{(\gamma - 1)\mathrm{Ma}_1^2 + 2\,\text{factor}} \tag{B.102}$$

$$\frac{u_2}{u_1} = 1 - \frac{2\left(\left(\mathrm{Ma}_1^2/\text{factor}\right) - 1\right)}{(\gamma + 1)\mathrm{Ma}_1^2} \tag{B.103}$$

$$\frac{v_2}{u_1} = \frac{(2\,\mathrm{Ma}_1^2/\mathrm{factor}) - 1}{(\gamma + 1)\mathrm{Ma}_1^2}\left(\frac{\partial x}{\partial y}\right) \tag{B.104}$$

$$\frac{w_2}{u_1} = \frac{(2\,\mathrm{Ma}_1^2/\mathrm{factor}) - 1}{(\gamma + 1)\mathrm{Ma}_1^2}\left(\frac{\partial x}{\partial z}\right) \tag{B.105}$$

$$\frac{p_2}{p_1} = 1 + \frac{2\gamma}{(\gamma + 1)}\left((\mathrm{Ma}_1^2/\mathrm{factor}) - 1\right) \tag{B.106}$$

the n-D generalization of the classic oblique shock relations, Anderson (1990).

They also revert to the plane shock jump conditions for restriction $\partial x/\partial z = 0$ and $\partial x/\partial y = \tan\theta$, whereupon $\mathrm{factor} = 1/\sin^2\beta$ for β the shock angle for wedge angle θ. This reduction generates the wind tunnel 7° sharp cone model $\mathrm{Ma}_\infty = 8$ shock angle β, Section B.9.

The n-D generalization for angle of attack α and yaw adds manipulation algebraic complexity as freestream onset u_1 now possesses non-vanishing components v_1 and w_1. The replacement of (B.94) for the first of equations (B.93) is

$$\rho_1 u_1 - \rho_2 u_2 = (\rho_1 v_1 - \rho_2 v_2)\frac{\partial x}{\partial y} + (\rho_1 w_1 - \rho_2 w_2)\frac{\partial x}{\partial z} \tag{B.107}$$

Replacing definition (B.101) with factor α (for angle of attack α)

$$\mathrm{factor}\,\alpha \equiv 1 - \left(\frac{v_1}{u_1}\right)\left(\frac{\partial x}{\partial y}\right) - \left(\frac{w_1}{u_1}\right)\left(\frac{\partial x}{\partial z}\right) \tag{B.108}$$

the ideal and/or equilibrium real gas shock jump relations for angle of attack α and yaw replace (B.102–B.106) with

$$\frac{\rho_2}{\rho_1} = \mathrm{factor}\,\alpha\left[\frac{u_2}{u_1} - \frac{v_2}{v_1}\frac{\partial x}{\partial y} - \frac{w_2}{w_1}\frac{\partial x}{\partial z}\right]^{-1} \tag{B.109}$$

$$\frac{u_2}{u_1} = 1 - \frac{(p_2/p_1) - 1}{\gamma\mathrm{Ma}_1^2(\mathrm{factor}\,\alpha)} \tag{B.110}$$

$$\frac{v_2}{u_1} = \frac{v_1}{u_1} + \frac{(p_2/p_1) - 1}{\gamma\mathrm{Ma}_1^2(\mathrm{factor}\,\alpha)}\left(\frac{\partial x}{\partial y}\right) \tag{B.111}$$

$$\frac{w_2}{u_1} = \frac{w_1}{u_1} + \frac{(p_2/p_1) - 1}{\gamma\mathrm{Ma}_1^2(\mathrm{factor}\,\alpha)}\left(\frac{\partial x}{\partial z}\right) \tag{B.112}$$

$$\left(\frac{\partial x}{\partial y}\right)^2 + \left(\frac{\partial x}{\partial z}\right)^2 = \frac{\gamma\mathrm{Ma}_1^2((H_2/H_1) - 1)}{\dfrac{H_2}{H_1}\left(\dfrac{p_2}{p_1} - 1\right)}(\mathrm{factor}\,\alpha)^2 - 1 \tag{B.113}$$

B.11 Shock Trajectory Characteristics Algorithm

The characteristics analysis of (B.86) predicts that only one eigenvalue is positive at a shock point. Hence, one state variable member is required determined from solution of (B.86) while the remaining members are computed from (B.109–B.113). System inspection confirms knowledge of the pressure and enthalpy ratios p_2/p_1 and H_2/H_1 admits explicit determination of u_2/u_1, v_2/u_1, w_2/u_1 at downstream integration station $\eta_1 + \Delta\eta_1$.

The characteristics algorithm formulation involves manipulation of (B.113) to a non-linear first-order PDE written on the directional derivatives. Subsequent organization into a coupled ODE system is directly solvable via classic characteristics theory, Courant and Hilbert (1989), Garabedian (1986).

Via coefficient coalescence the manipulation to PDE generates

$$\begin{aligned}\left(\frac{\partial x}{\partial y}\right)^2\left(1-\left(\frac{v_1}{u_1}\right)^2 g\right)+\left(\frac{\partial x}{\partial z}\right)^2\left(1-\left(\frac{w_1}{u_1}\right)^2 g\right)-2\left(\frac{\partial x}{\partial y}\right)\left(\frac{\partial x}{\partial z}\right)\left(\frac{v_1}{u_1}\right)\left(\frac{w_1}{u_1}\right)g \\ +2\left(\frac{\partial x}{\partial y}\right)\left(\frac{v_1}{u_1}\right)g+2\left(\frac{\partial x}{\partial z}\right)\left(\frac{w_1}{u_1}\right)g=g-1\end{aligned} \tag{B.114}$$

for the definition

$$g \equiv \frac{\gamma \mathrm{Ma}_1^2((H_2/H_1)-1)}{\dfrac{H_2}{H_1}\left(\dfrac{p_2}{p_1}-1\right)} \tag{B.115}$$

Letting l_y and l_z denote $\partial x/\partial y$ and $\partial x/\partial z$ for convenience, the resultant classically solvable ODE system is

$$\frac{\mathrm{d}l_y}{\mathrm{d}x}=a\left(\frac{\partial g}{\partial y}+l_y\frac{\partial g}{\partial x}\right)/d \tag{B.116}$$

$$\frac{\mathrm{d}l_z}{\mathrm{d}x}=a\left(\frac{\partial g}{\partial z}+l_z\frac{\partial g}{\partial x}\right)/d \tag{B.117}$$

$$\frac{\mathrm{d}y}{\mathrm{d}x}=\left(-2l_y\left(1-\left(\frac{v_1}{u_1}\right)^2 g\right)+2l_z\left(\frac{v_1}{u_1}\right)\left(\frac{w_1}{u_1}\right)g-2\left(\frac{v_1}{u_1}\right)g\right)/d \tag{B.118}$$

$$\frac{\mathrm{d}z}{\mathrm{d}x}=\left(-2l_z\left(1-\left(\frac{w_1}{u_1}\right)^2 g\right)+2l_y\left(\frac{v_1}{u_1}\right)\left(\frac{w_1}{u_1}\right)g-2\left(\frac{v_1}{u_1}\right)g\right)/d \tag{B.119}$$

for definitions

$$\begin{aligned}a &\equiv 2l_y\left(\frac{v_1}{u_1}\right)+2l_z\left(\frac{w_1}{u_1}\right)-1-\left(l_y\left(\frac{v_1}{u_1}\right)+l_z\left(\frac{w_1}{u_1}\right)\right)^2 \\ d &\equiv 2(g-1)+2l_y\left(\frac{v_1}{u_1}\right)+2l_z\left(\frac{w_1}{u_1}\right)g\end{aligned} \tag{B.120}$$

For g, (B.115), sufficiently smooth in the integration interval $x+\Delta x$, equivalently $\eta_1+\Delta\eta_1$

$$g(x,y,z) = g(x_0,y_0,z_0) + \sum_{i=1}^{3} \frac{\partial g}{\partial x_i}(x_i - x_{i0}) + O\left(\Delta x_i^2\right) \tag{B.121}$$

where $g(x_0, y_0, z_0)$ is known from the Euler NS blunt body solution at the PRaNS IC plane $x=x_0$. Then assuming Δx $(\Delta\eta_1)$ sufficiently small such that $\partial g/\partial x_i \cong \partial g/\partial x_i|_0$ is adequately accurate, hence *a priori* known *data*, (B.121) enables direct prediction of state variable *ratio* jumps without the requirement to determine the individual state variable jumps.

The ICs for ODE system variables l_y and l_z at plane $x=x_0$ are determined from (B.108), (B.111–B.112) as

$$l_{y0} \equiv \left(\frac{\partial x}{\partial y}\right)_0 = \gamma \mathrm{Ma}_1^2\left(\frac{v_2}{u_1} - \frac{v_1}{u_1}\right)/\mathrm{factor}s \tag{B.122}$$

$$l_{z0} \equiv \left(\frac{\partial x}{\partial z}\right)_0 = \gamma \mathrm{Ma}_1^2\left(\frac{w_2}{u_1} - \frac{w_1}{u_1}\right)/\mathrm{factor}s \tag{B.123}$$

with factor*s* (for *s*hock) definition

$$\mathrm{factor}s \equiv (p_2/p_1) + \gamma \mathrm{Ma}_1^2\left(\frac{v_2}{u_1} - \frac{v_1}{u_1}\right)\left(\frac{v_1}{u_1}\right) + \gamma \mathrm{Ma}_1^2\left(\frac{w_2}{u_1} - \frac{w_1}{u_1}\right)\left(\frac{w_1}{u_1}\right) \tag{B.124}$$

With these ICs (B.116–B.120) is closed and well posed for solution at PRaNS solution plane $x=x_0$, thereby all subsequent solution planes. Importantly, this shock coordinate solution strategy intrinsically satisfies the *domain-of-dependence* requirements of characteristic theory. The solution sequence is:

- insert solutions (B.122–B.124), the jump conditions dependent only on known flow state, into (B.109–B.112)
- these solutions replace those potentially generated for the continuity and momentum components of (B.86)
- in agreement with characteristics theory requirement that one state variable be determined from (B.86), simultaneously solve D*E* state variable component therein with (B.109–B.112).

The $\theta=0.5$ second-order ODE algorithm is preferred in keeping with the PRaNS mGWS^h+θTS algorithm decision, (B.53). The characteristics theory matrix algorithm jacobian is analytically derivable and linear which renders the defined ODE solution process efficient. In closing, the reference details the characteristics algorithm exact solution for simplification to zero angle of attack and no yaw. For further simplification to oblique shock the formulation recovers the classic solution, Anderson (1989).

B.12 Blunt Body PRaNS Algorithm Validation

The spherical nose missile geometry model for angle of attack validation, Figure B.2, is an $R_c = 0.04167$ ft (0.5 in) sphere/10.5°/7° bicone with planar slice pair oppositely disposed on the 7° aftcone, Martellucci and Weinberg (1982). PRaNS domain perspective with mesh overlay defining missile and post bow shock bounding surfaces for $Ma_\infty = 8.0$, $\alpha = 10°$ and zero yaw is Figure B.15 right.

Angle of attack requires a circumferentially dense body-fitted solution adapted transverse plane discretization. The selected $M = 65 \times 65$ $k = 1$ TP basis mesh coupled ALE-characteristics bow shock trajectory algorithm prediction of η_1 evolution is illustrated for ten PRaNS planes separated by 50 uniform integration steps at $\Delta\eta_1 = 0.001$ ft, Figure B.15 left. The symmetrically disposed slice leading edges start between planes 8 and 9, planes 10 and 2 do not show well as the color original is yellow and the terminal plane graph is for 25 $\Delta\eta_1$ steps.

The wind tunnel experiment 660°R fixed surface temperature matches $Ma_\infty = 8.0$ post-shock static. The ALE-characteristics bow shock trajectory algorithm computation of post-shock PRaNS state variables are implemented as Dirichlet BCs, a theory consistency check replacement for homogeneous Neumann BCs. The PRaNS IC is generated via an axisymmetric compressible NS algorithm, Section B.8, with DOF distribution stretched to fit the slightly non-symmetric PRaNS IC plane domain, the tiny first plane graphic in Figure B.15 left. Following DOF conversion to volume specific stagnation enthalpy, IC DOF alteration to turbulent BL profile is accomplished as described, Section B.9.

The sphere/10.5°/7° bicone-slice model, devoid of flare, enables successive PRaNS integration sweeps leading to a departure theory implemented converged solution at the model terminal plane. Departure theory algorithms are implemented by saving computed transverse plane pressure distributions at 50 $\Delta\eta_1$ step intervals for each subsequent sweep.

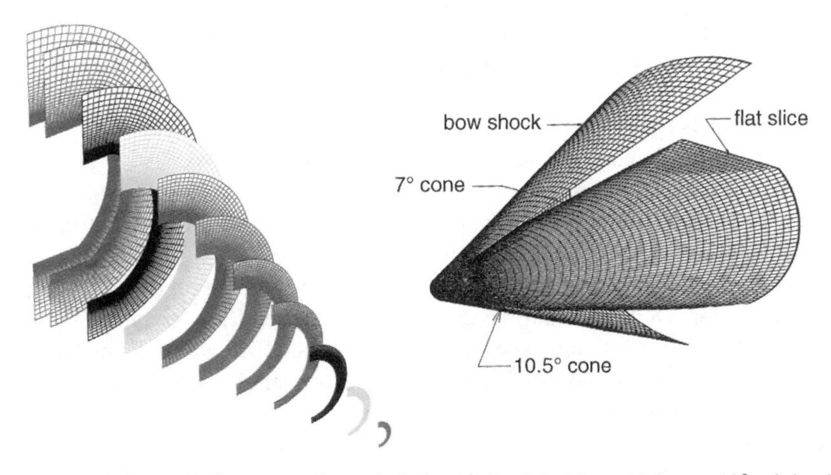

Figure B.15 Sphere/10.5°/7° bicone-slice wind tunnel model, $Ma_\infty = 8.0$, $\alpha = 10°$: right, PRaNS domain bounding surfaces; left, coupled ALE-characteristic shock trajectory algorithm transverse plane body-fitted $k = 1$ TP basis mesh η_1 evolution, 10 planes from IC separated by 50 $\Delta\eta_1 = 0.001$ ft integration steps, last separation is 25 $\Delta\eta_1$ steps

PRaNS generated missile model surface pressure DOF *a posteriori* data interpolations for integration sweeps 1–4 are graphed in aft perspective grayscale floods with contours (originals are in color), Figures B.16, B.17 top, bottom respectively. Each graphic terminal plane is "transparent" to the viewer to enable confirmation of mirror symmetric pressure distributions predicted on the aftcone reach following aftcone slice initiation.

These data confirm that the PRaNS algorithm multiple sweep process generates negligible alterations to surface pressure distributions from bicone model IC plane to aftcone slice initiation. Downstream thereof, after plane 8, Figure B.15 bottom, the first two PRaNS sweep predictions of aftcone slice influence on missile side surface pressure distributions are significantly incorrect, compare sweep 1, 2, 3 DOF data interpolations.

Conversely, PRaNS sweep 3 and 4 surface pressure distributions are visually indistinguishable confirming generation of the subsonic BL reach ellipticity-enforced solution. In concert these multi-sweep data solidly *validate* the PRaNS algorithm requirement for departure theory implementation to account for BL subsonic reach pressure. Finally, that algorithm sweep 3 and 4 surface pressure *a posteriori* DOF data distributions are indistinguishable confirms algorithm convergence.

Additional validation data for coupled IC generation-ALE-characteristics shock trajectory prediction PRaNS reacting real gas algorithm at $\alpha = 10°$ are reported. Figures B.16–B.17 verify that the sphere/10.5°/7° bicone PRaNS solutions upstream of slice initiation are independent of algorithm sweep. Hence selected DOF distributions are just downstream of slice initiation, plane 9 in Figure B.15.

Radial node column $\{u1\}$ velocity Mach number $\{Ma_1\}$ DOF distributions are four-decade semi-log scaled for 45° meridional rays in Figure B.18. The turbulent BL thickness distribution ranges $0.03 < \delta < 0.05$ in, windward to leeward, with associated post-shock layer Mach number range $5 \leq Ma_1 \leq 12$, a ~50% variation on onset $Ma_\infty = 8$. The windward BL edge transition $\{Ma_1\}$ DOF modest oscillations, also the last few node DOF transitions to characteristics post bow shock Mach number, reflect real gas aerothermodynamics speed of sound sensitivity as $\{u1\}$ DOF distributions are strictly monotone, Figure B.19 bottom. This constitutes *quantitative* validation of theory implementation consistency pervading the close coupled hypersonic real gas NS-ALE-characteristics PRaNS algorithm suite.

The distinctions between non-D axial momentum $\{M1\}$ and velocity $\{u1\}$ DOF distributions, observed for 7° cone/cylinder/7° cone *a posteriori* data, persist, Figure B.19 top/bottom. The $\{u1\}$ DOF BL edge levels asymptote monotonically to post-shock surface Dirichlet BCs. The horizontal plateau in $\{M1\}$ distribution has moved to the turbulent BL $Ma_1 \sim 2.0+$ supersonic reach, reference Figure B.18.

The PRaNS solution plane 9 non-D transverse plane momentum vector $\{ML\}$ radial-crossflow resolution DOF distributions are four-decade semi-log graphed, Figure B.20. The radial resolution exhibits requisite exact mirror symmetries with 90° meridional plane DOFs a computed zero. Resolution data extrema are of identical magnitudes and are consistent with $O(\mathrm{Re}^{-1/2})$ BL theory Reynolds ordering.

The characteristics bow shock trajectory algorithm post-shock surface predictions are altered to PRaNS state variable DOF Dirichlet BCs. The last few DOF in each non-zero $\{ML\}$ resolution adjust to these BCs with magnitudes $O(\mathrm{Re}^{-1/2})$ insignificant in comparison with $\{M1\}$. These near shock DOF alterations are the measure of error accumulation in bow shock axial trajectory prediction due to the slope linearization in

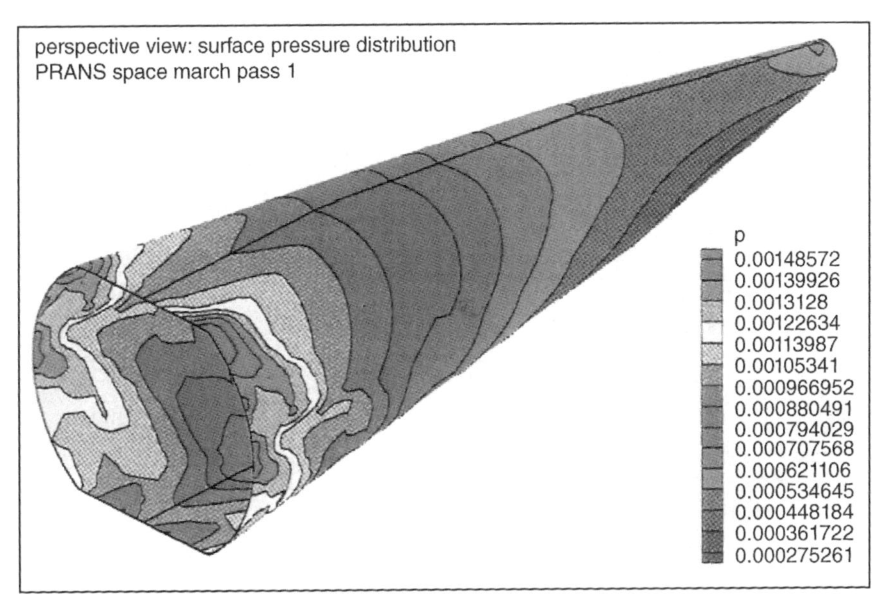

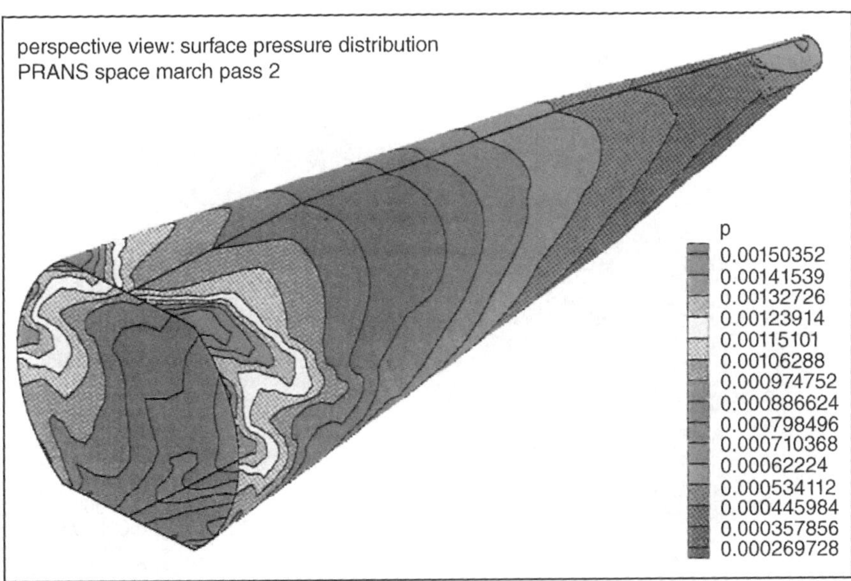

Figure B.16 Sphere/10.5°/7° bicone-slice wind tunnel model, $Ma_\infty = 8.0$, $\alpha = 10°$, $T_{wall} = 660°R$, PRaNS solution surface pressure *a posteriori* DOF data interpolation perspective with contours, aft viewpoint, transparent end plane: top, first sweep; bottom, second sweep

implementing (B.121). The angle of attack requirement of sufficiently refined circumferential mesh is clearly met.

Plane 9 PRaNS domain DOF distributions for state variable density $\{\rho\}$ and Dirichlet BC $T_{wall} = 660°R$ aerothermodynamics closure parameter $\{T\}$, extracted from volume

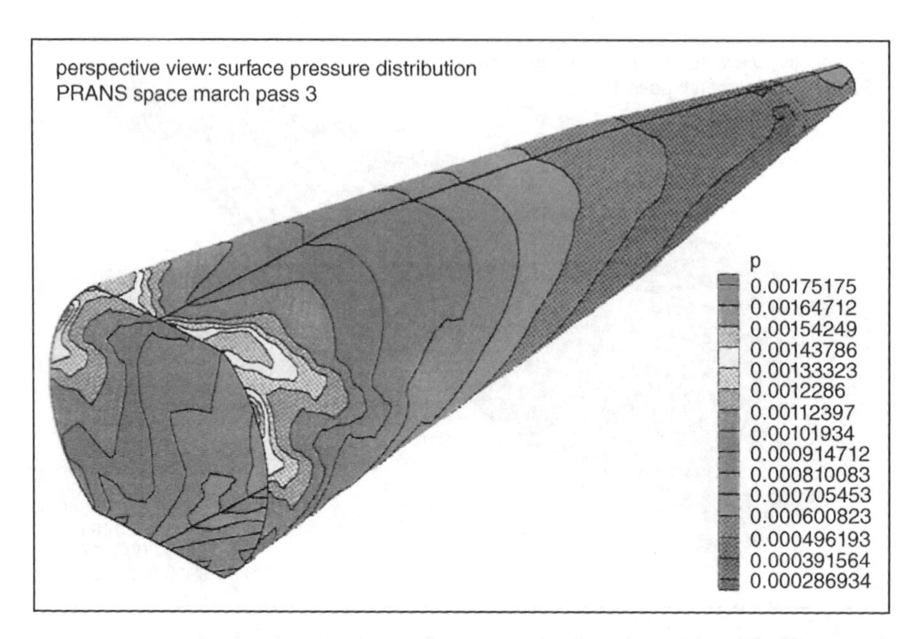

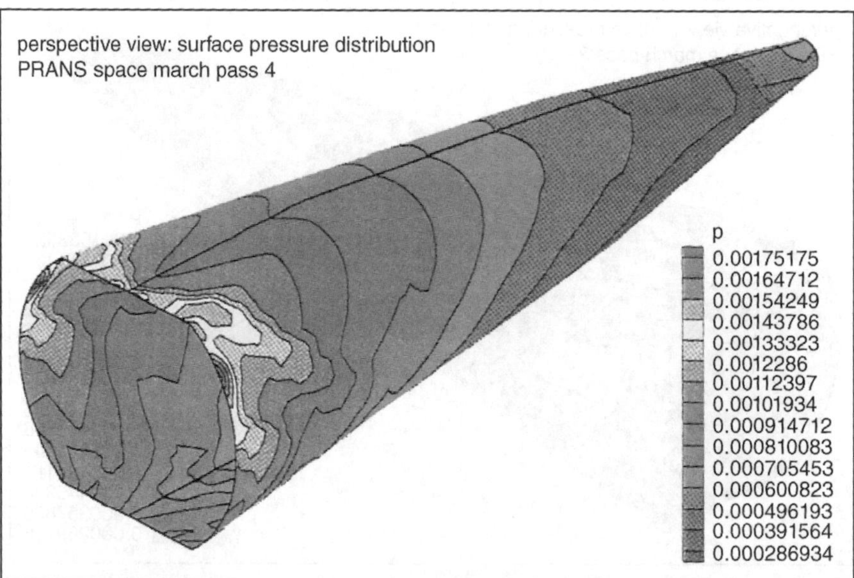

Figure B.17 Sphere/10.5°/7° bicone-slice wind tunnel model, $Ma_\infty = 8.0$, $\alpha = 10°$, $T_{wall} = 660°R$, PRaNS solution surface pressure *a posteriori* DOF data interpolation perspective with contours, aft viewpoint, transparent end plane: top, third sweep; bottom, fourth sweep

specific stagnation enthalpy, are graphed in Figure B.21 top, bottom. Density $\{\rho\}$ DOF magnitude distributions range one order with all extrema located in the turbulent BL subsonic reach, as occurred for the 7° cone/cylinder/7° cone prediction, Figure B.9. In BL supersonic range $Ma_1 \sim 2.0+$ the distinction between $\{M1\}$ and $\{u1\}$ is clearly due

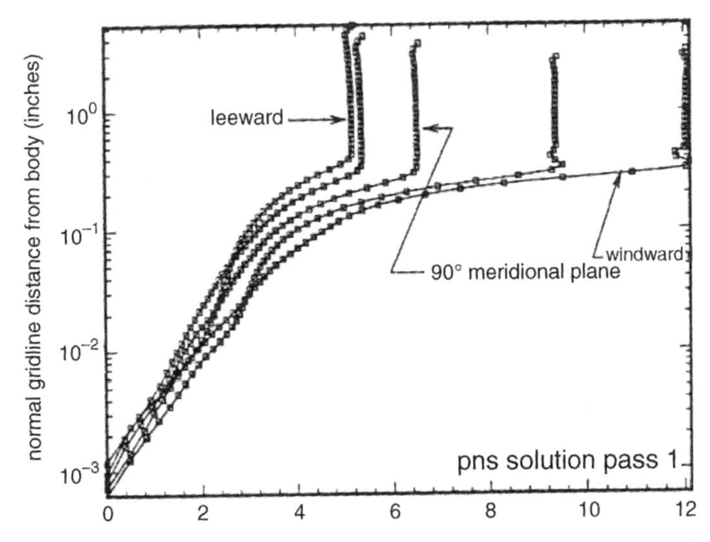

Figure B.18 Sphere/10.5°/7° bicone-slice wind tunnel model, $\mathrm{Ma}_\infty = 8.0$, $\alpha = 10°$, $T_{wall} = 660°\mathrm{R}$, shock layer PRaNS domain plane 9 {u1} velocity Mach number {Ma_1} DOF distributions, 45° meridional increments windward to leeward

to {ρ} DOF variations which are rather insensitive to meridional ray. The {T} DOF distributions reach their extrema in this reach, with leeward to windward ray maxima differing by a factor of three. As with {M*L*} the {ρ} last DOF node pair display modest adjustments to post-shock characteristics BC data. The {T} DOF distributions asymptote monotonically to post-shock static temperature, the wind tunnel experiment $T_{wall} = 660°\mathrm{R}$ design point.

Figure B.22 top graphs PRaNS solution plane 9 domain pressure non-D DOF {p} distributions. The associated local extrema reside in the BL from near surface to supersonic {M1} plateau at $\mathrm{Ma}_1 \sim 2.0+$, similar to those for {ρ} DOF. The plane 9 axial pressure gradient DOF {$dp/d\eta_1$} distributions are nil compared to that in the windward ray vicinity. The extremum occurs at the BL edge δ, with magnitude sufficient to accelerate the near windward ray velocity to local axial Mach number $\mathrm{Ma}_1 = 12$, Figure B.18.

B.13 Summary

A suite of algorithms addressing CFD prediction of hypersonic shock layer *aerothermodynamics* is detailed. The consequential theoretical fluid mechanics scope expansion on text content is in thermodynamics, specifically a real gas variable γ closure for air. The assumption of equilibrium eliminates handling mass transport PDE system (B.5). The derived modified continuous Galerkin $m\mathrm{GWS}^h$+θTS algorithm, a direct extension on Chapter 4 content, addresses the steady, three-dimensional, body-fitted, Favre time averaged compressible turbulent *parabolic* RaNS (PRaNS) *m*PDE system for external hypersonic aerothermodynamics.

Supersonic/hypersonic shock layer aerodynamics was a cold war engendered major CFD research focus in the 1980s into the 1990s. Various steady compressible NS PDE system approximations were then mandated, for example *parabolic* RaNS, *thin layer* NS/RaNS, to

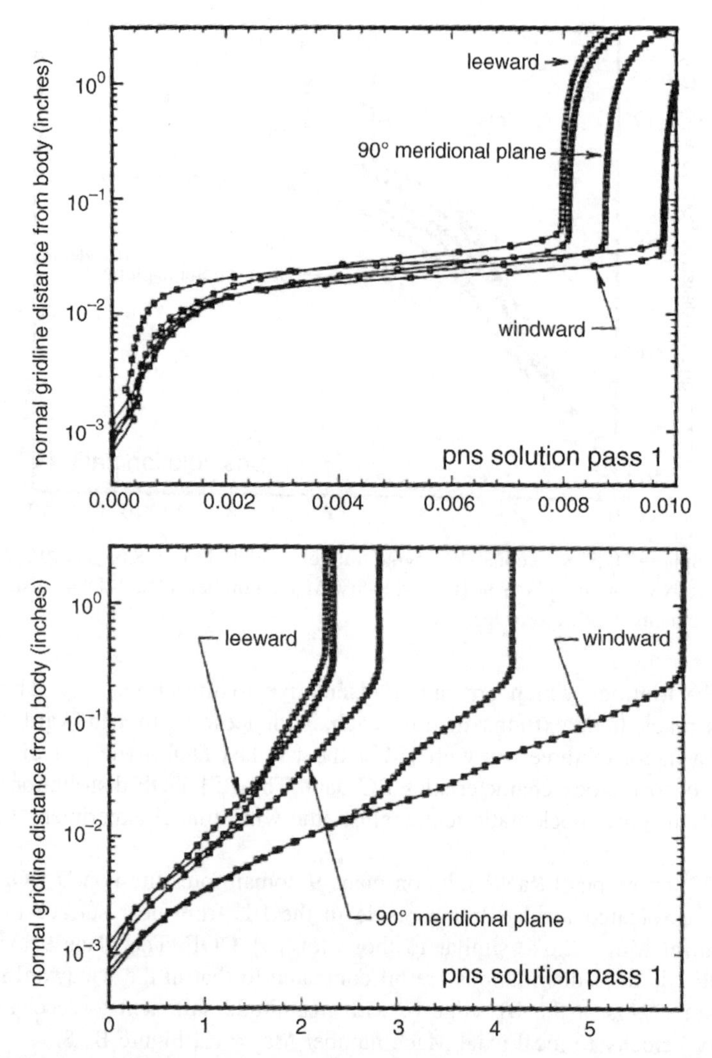

Figure B.19 Sphere/10.5°/7° bicone-slice wind tunnel model, $Ma_\infty = 8.0$, $\alpha = 10°$, $T_{wall} = 660°R$, shock layer PRaNS domain plane 9 non-D DOF distributions: top, axial momentum {M1}; bottom, axial velocity {u1}

enable generation of sufficiently dense mesh 3-D aerodynamics solutions using time-frame compute hardware capability. This historical constraint is now totally vanished.

This appendix archives in context and in one place the numerous fluid dynamics topic expansions required for turbulent compressible external supersonic/hypersonic aerothermodynamics CFD prediction including:

- Favre time averaged compressible NS state variable
- Favre NS/RaNS/PRaNS PDEs body-fitted, contravariant vector resolution, pervasive coordinate transformation embedding
- real air dissociation-recombination variable γ thermodynamics

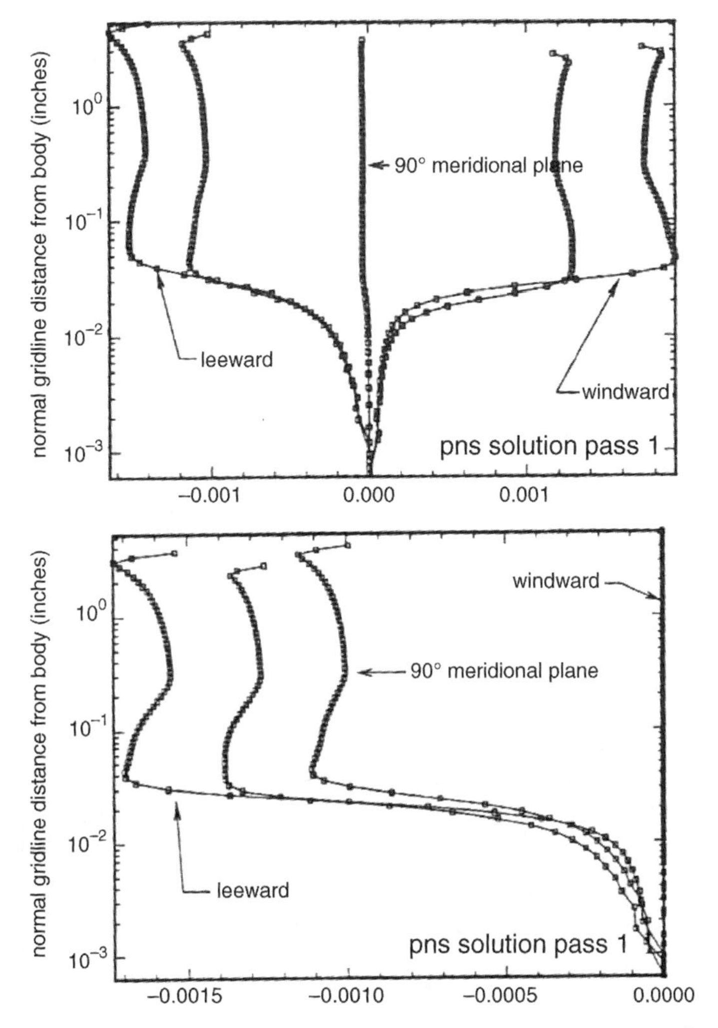

Figure B.20 Sphere/10.5°/7° bicone-slice wind tunnel model, $Ma_\infty = 8.0$, $\alpha = 10°$, $T_{wall} = 660°R$, shock layer PRaNS domain plane 9 non-D momentum DOF distributions: top, wall normal momentum {M2}; bottom, circumferential momentum {M3}

- tensor matrix product AF jacobians, body-fitted coordinate linear algebra grid-sweeping algorithm, extensible to $n = 3$
- PRaNS state variable initial condition (IC) self-generation:
  1. detached bow shock geometry correlation
  2. weak form theory generalized Rankine–Hugoniot shock jump relations for angle of attack with yaw
  3. rigorous characteristics theory algorithm to predict detached bow shock surface geometry evolution
  4. laminar compressible NS algorithm Mach number continuation generation of PRaNS initiation plane IC

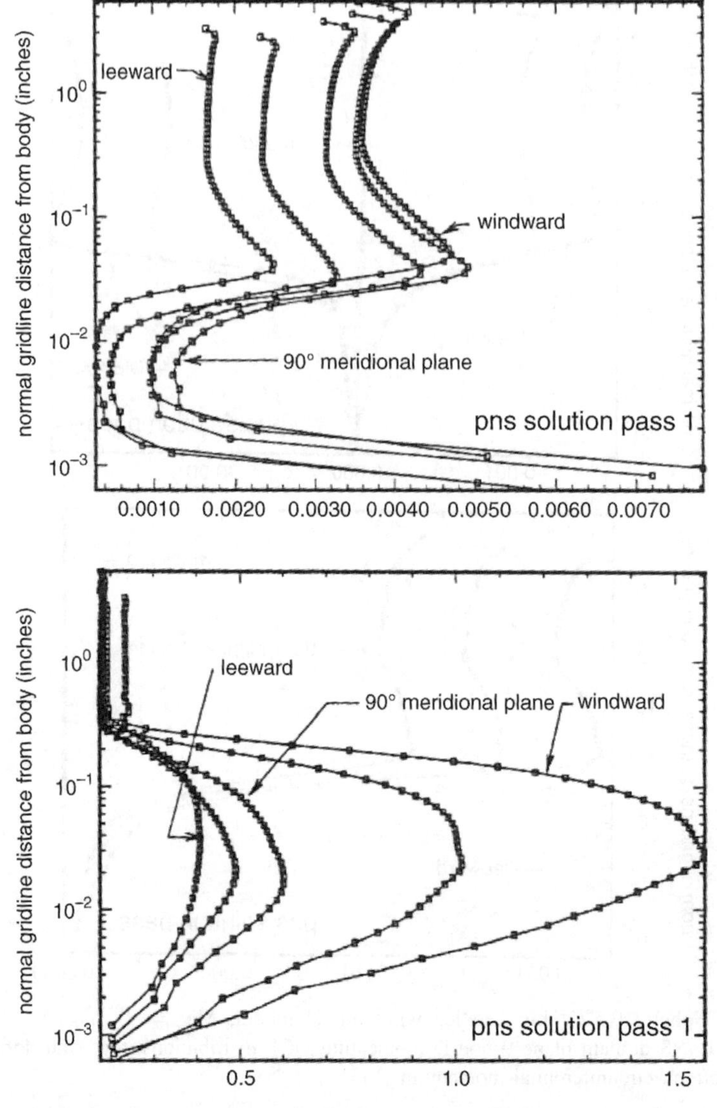

Figure B.21 Sphere/10.5°/7° bicone-slice wind tunnel model, $Ma_\infty = 8.0$, $\alpha = 10°$, $T_{wall} = 660°R$, shock layer PRaNS domain plane 9 non-D DOF distributions: top, density $\{\rho\}$; bottom, temperature $\{T\}$

5. NS laminar transition to PRaNS turbulent IC mimics wind tunnel experimental practice
6. TS algorithm for wall pressure DOF determination

- compressible RaNS state variable $\{q(\mathbf{x}, t)\} = \{\rho, \mathbf{m}, E, Y\}^T$ transition to steady PRaNS state variable $\{q(\mathbf{x})\} = \{M_1, u_j M_1, \rho H M_1\}^T$, Green–Gauss theorem kinetic flux vector imposition, alteration to volume specific stagnation enthalpy
- departure theory addressing of hypersonic turbulent BL subsonic reach pressure ellipticity.

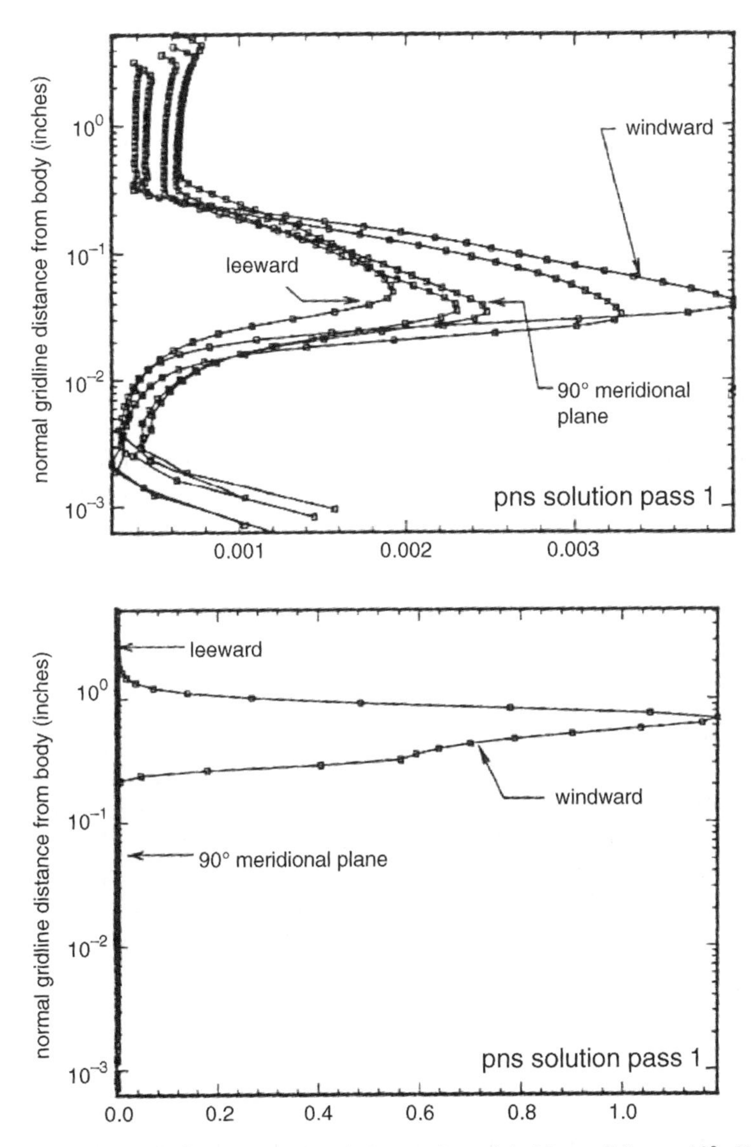

Figure B.22 Sphere/10.5°/7° bicone-slice wind tunnel model, $\mathrm{Ma}_\infty = 8.0$, $\alpha = 10°$, $T_{wall} = 660°\mathrm{R}$, shock layer PRaNS domain plane 9 non-D DOF distributions: top, pressure $\{p\}$; bottom, axial pressure gradient $\{dp/d\eta_1\}$

The resultant closely coupled support suite enabled generation of detailed *a posteriori* data for the $m\mathrm{GWS}^h + \theta(=0.5)\mathrm{TS}$, $k = 1$ TP basis PRaNS algorithm for identified mPDE system for hypersonic shock layer aerothermodynamics $\mathrm{Ma}_\infty = 8.0$ onset to sharp and spherical nose wind tunnel models. Direct comparison with available experimental data yields algorithm validation including departure theory formulation. Support suite *a posteriori* data comparisons with classic hypersonic fluid dynamics theory yields additional validations.

Hypersonic aerodynamics with non-equilibrium combustion aerothermodynamics continues a forefront topic, Ehrlich *et al.* (2012), Musielak (2012). The USAF/Boeing/NASA/University jointly evolved NPARC Alliance Wind-US flow solver, Nelson (2010), represents current compressible aerodynamics CFD capability. Solicitations for Wind-US code theory/closure improvements appear regularly, generated in response to user-observed inadequacies, Georgiadis *et al.* (2009, 2011). Cited topics include transitional laminar-turbulent BL closure model improvement, shock-turbulent BL interaction modeling, kinetics-turbulent scalar transport closure model sensitivity, combustor-exhaust system modeling, compressible mass transport mixing models, closure replacements for constant turbulent Prandtl–Schmidt number assumptions, addition of radiation requiring rapid prediction of dense view factor distributions.

Specific to viable supersonic *commercial aviation*, NASA-led staged research programs in sonic boom moderation, N + 1 (near term), and N + 2, N + 3 (technology ready in 2020–2025 and 2030–2035 timeframes), require forefront CFD advances enabling *very* precisely detailed flowfield prediction, Wilson (2013). Of direct pertinence, a weak form *focused domain* mGWS^h+RK-TS (Runge–Kutta) CFD algorithm for laminar NS PDE system (B.2–B.4) generates state variable *a posteriori* DOF distributions *within* the shock itself for $1 < \mathrm{Ma}_\infty \leq 10$, Iannelli (2011).

Clearly, CFD algorithm advances for compressible, non-equilibrium reacting supersonic/hypersonic aerothermodynamics require highly pertinent theoretical innovations for cited forefront challenges. The documented *optimal* phase accuracy/dispersion error *annihilation* of modified *continuous* mGWS theory is undoubtedly extensible to compressible NS/RaNS/LES/*ar*LES mPDE *smooth* solution prediction on $n = 3$. Coupling with *discontinuous* Galerkin weak form theory, cf. Cockburn (1999), will add *non-smooth* solution shock capturing/BL interaction capability to encompass the full fluid dynamics spectrum requirement.

Equation-driven, integrated CFD theory implementation in multi-parallel clustered CPU-GPU computing environments is the ultimate requirement. The incompressible space filtered NS well-posed *ar*LES theory conversion to high performance computing practice, Sekachev (2013), is exemplary. Theory advances coalesced into this type platform practice will ultimately lead to aerothermodynamics CFD topical completion, a stimulating prospect.

Exercises

B.2.1 Via (B.12) rewrite (B.1) in tensor index notation, hence confirm its validity.

B.2.2 For the definitions (B.6–B.9) confirm that $\overline{\rho u_i''} = 0$.

B.2.3 Verify the correctness of (B.14).

B.6.1 Verify the Green–Gauss divergence theorem implementation on the kinetic flux vector (B.58–B.60).

B.6.2 Verify $\{\mathrm{RES\_Q1}\}_e$, (B.61).

B.6.3 Verify the Reynolds stress tensor contributions in (B.63–B.65).

B.6.4 Verify the Galerkin weak statement algorithm for Reynolds strain rate tensor (B.67).

In Section B.7 rename the DOF in $\{QI(\eta_1\}$ using the PRaNS state variable labels $\{q^h(\mathbf{x},t)\} = \{M_1^h, (u_j M_1)^h, (\rho H M_1)^h\}^T$. Form the [B . . .]$_e$ jacobian element matrices then confirm the following AF jacobian replacements:

B.7.1 (B.76).

B.7.2 (B.77).

B.7.3 (B.79).

B.7.4 (B.81–B.82).

References

Anderson, J.D. (1989). *Hypersonic and High Temperature Gas Dynamics*, McGraw-Hill, New York.

Anderson, J.D. (1990). *Modern Compressible Flow, with Historical Perspective*, McGraw-Hill, New York.

Armfield, S.W. and Fletcher, C.A.J. (1986). "Pressure Related Instabilities of Reduced Navier–Stokes Equations for Internal Flows," *Comm. Numerical Methods Engineering*, V. 2, pp. 377–383.

Armfield, S.W. (1987). "Numerical Simulation of Incompressible Turbulent Swirling Flow in Conical Diffusers," PhD dissertation, University of Sydney, Australia.

Baker, A.J. and Kim, J.W., (1987). "A Taylor Weak Statement Algorithm for Hyperbolic Conservation Laws," *J. Numerical Methods Fluids*, V. 7, pp. 489–520.

Baker, A.J., Freels, J.D., Iannelli, G.S. and Manhardt, P.D. (1994). "A Finite Element Code for Real Gas Aerodynamics Simulation (PRaNS)," Vol. I. Algorithm and Theory, Technical Report BMO:TR-91-39.

Baker, A.J. (2012). *Finite Elements ⇔ Computational Engineering Sciences*, John Wiley, London.

Baldwin, B.S. and Lomax, H. (1978). "Thin-layer approximation and algebraic model for separated turbulent flows," Technical Paper AIAA:78-257.

Baldwin, B.S. and Barth, T.J. (1990). "A one-equation turbulence transport model for high Reynolds number wall-bounded flows," Technical Report NASA:TM-102847, Technical Paper AIAA:91-610.

Beam, R.M. and Warming, R.F. (1978). "An Implicit Finite Difference Algorithm for Hyperbolic Systems of Conservation Laws," *AIAA Journal*, V. 16, pp. 393–401.

Billig, F.S. (1967). "Shock Wave Shapes Around Spherical and Cylindrical-Nosed Bodies," *J. Spacecraft and Rockets*, V. 4, pp. 822–823.

Briley, W.R. and McDonald, H. (1977). "Solution of the Multi-Dimensional Navier–Stokes Equations by a Generalized Implicit Method," *J. Computational Physics*, V. 24, pp. 372–397.

Chaussee, D.S., (1982). "Euler/Navier–Stokes Code Development with Shock Capturing vs Shock Fitting," course notes, *Annual Short Course on Advances in Computational Fluid Dynamics*, University of Tennessee Space Institute/Tullahoma.

Chaussee, D.S., Patterson, J.L., Kutler, P., Pulliam, T.H. and Steger, J.L. (1991). "A Numerical Simulation of Hypersonic Viscous Flows over Arbitrary Geometries at High Angle of Attack," Technical Paper AIAA:91-0050.

Cockburn, B. (1999). "Discontinuous Galerkin methods for convection-dominated problems," in Barth, T. and Deconink, H. (eds), *High Order Methods for Computational Physics*, Lecture Notes in Computational Science and Engineering, V.9, pp. 69–224, Springer Verlag.

Courant, R. and Hilbert, D. (1989). *Methods of Mathematical Physics*, Vol. II, Wiley, New York.

Ehrlich, C. *et al.* (2012). "Hypersonics technologies and aerospace planes," *Aerospace America*, December, p. 73.

Favre, A. (1965). "Equations des Gaz Turbulents Compressibles: 1. Formes Generales," *J. Mechanique*, V. 4, pp. 361–390.

Freels, J.D. (1992). "A Taylor Weak Statement Finite Element Algorithm for Real-Gas Compressible Navier–Stokes Simulation," PhD Dissertation, University of Tennessee/Knoxville.

Garabedian, P.R. (1986). *Partial Differential Equations*, Chelsea Publishing Co., New York.

Georgiadis, N.J., Yoder, D.A., Towne, C.S., Engblom, W.A., Bhagwandin, V.S., Power, G.D., Lankford, D.W. and Nelson, C.C. (2009), "Wind-US Code Physical Modeling Improvements to Complement Hypersonic Testing and Evaluation," Technical Paper AIAA:2009-193, also NASA/TM-2009-215615.

Georgiadis, N.J., Yoder, D.A., Vyas, M.A. and Engblom, W.A., (2011). "Status of Turbulence Modeling for Hypersonic Propulsion Flowpaths," Technical presentation at AIAA/ASME/SAE/ASEE Joint Propulsion Conference, San Diego CA.

Hall, D.W. (1987). Technical Report BMO:TR-87-99.

Iannelli, J. (1991). A Globally Well-posed Accurate and Efficient Finite Element CFD Algorithm for Compressible Aerodynamics," PhD dissertation, University of Tennessee/Knoxville.

Iannelli, J. (2006). *Characteristics Finite Element Methods in Computational Fluid Dynamics*, Springer, Heidelberg.

Iannelli, J. (2011). " An Implicit Galerkin Finite Element Runge-Kutta Algorithm for Shock Structure Investigations," *J. Computational Physics*, V. 230, pp. 260–286.

Kolesnikov, A. and Baker, A.J. (2001). "An Efficient High Order Taylor Weak Statement Formulation for the Navier–Stokes Equations," *J. Computational Physics*, V. 173, pp. 549–574.

Lin, A. and Rubin, S.B. (1982). "Three-Dimensional Supersonic Viscous Flow over a Cone at Incidence," *AIAA Journal*, V. 20, pp. 1500–1507.

Lubard, S.G., and Helliwell, W.S. (1974). "Calculation of the Flow on a Cone at High Angle of Attack," *AIAA Journal*, V. 12, pp. 965–974.

Martellucci, A. and Weinberg, S. (1982). "MAT Program Test Summary Report – Biconic Body with Slice/Flap," Technical Report BMO:TR-82-28.

Musielak, D. (2012). "National Hypersonic Centers: fast track to truly fast flight," *Aerospace America*, June, pp. 40–45.

Nelson, C.C. (2010). "An Overview of the NPARC Alliance's Wind-US Flow Solver," Technical Paper AIAA:2010-27.

Park, C. (1985). "On Convergence of Computation of Chemically Reacting Flows," Technical Paper AIAA:85-0247.

Platfoot, R.A. and Fletcher, C.A.J., (1991). "Gas Flows within Turning Sections," *J. Numerical Methods Heat and Fluid Flow*, V. 1, pp. 19–29.

Rakich, J.V. (1983). "Iterative PNS Method for Attached Flows with Upstream Influence," Technical Paper AIAA:83-1955.

Rubin, S.G. (1982). *Numerical and Physical Aspects of Aerodynamic Flow*, Springer-Verlag, Heidelberg.

Rubin, S.G. and Lin, A. (1971). "Numerical Methods for Two- and Three-dimensional Viscous Flow Problems," Technical Report PIBAL:71-8, Polytechnic Institute of Brooklyn.

Sekachev, M. (2013). "Essentially analytical theory closure for thermal-incompressible space filtered Navier–Stokes PDE system on bounded domains," PhD dissertation, University of Tennessee/Knoxville.

Shanks, S.P., Srinivasan, G.R. and Nicolet, W.E. (1982). "Parabolized Navier–Stokes Code: Formulation and User's Guide," Technical Report AFWAL: TR-82-3034.

Stalnaker, J.G., Nicholson, L.A., Hanline, D.S. and McGraw, E.H. (1986). "Improvements to the AFWAL PNS Code Formulation," Technical Report AFWAL:TR-86-3076.

Tannehill, J.C. and Mugge, P.H. (1974). "Improved Curve Fits for the Thermodynamic Properties of Equilibrium Air Suitable for Numerical Computation using Time-dependent or Shock Capturing Methods," Technical Report NASA:CR-2470.

Tannehill, J.C., Venkatapathy, E. and Rakich, J.V. (1981). "Numerical Solution of Two-dimensional Viscous Blunt Body Flows with an Impinging Shock," *AIAA Journal*, V. 20, pp. 203–213.

vanDriest, E.R. (1956). "The problem of aerodynamic heating," *Aero. Engr. Review*, V. 15, pp. 26–41.

Vigneron, Y.C., Rakich, J.V. and Tannehill, J.C. (1978). "Calculation of Supersonic Viscous Flows over Delta Wings with Sharp Subsonic Leading Edges," Technical Paper AIAA:78-1137.

Wilson, J.B. (2013). "SST Research, Breaking New Barriers," *Aerospace America*, January, pp. 26–31.

Author Index

Optimal MODIFIED CONTINUOUS Galerkin CFD, First Edition. A. J. Baker.

Companion Website: www.wiley.com/go/baker/GalerkinCFD

Subject Index

Optimal MODIFIED CONTINUOUS Galerkin CFD, First Edition. A. J. Baker.

Companion Website: www.wiley.com/go/baker/GalerkinCFD